IUTAM SYMPOSIUM ON NEW APPLICATIONS OF NONLINEAR AND CHAOTIC DYNAMICS IN MECHANICS

SOLID MECHANICS AND ITS APPLICATIONS
Volume 63

Aims and Scope of the Series

The fundamental questions arising in mechanics are: *Why?*, *How?*, and *How much?* The aim of this series is to provide lucid accounts written by authoritative researchers giving vision and insight in answering these questions on the subject of mechanics as it relates to solids.

The scope of the series covers the entire spectrum of solid mechanics. Thus it includes the foundation of mechanics; variational formulations; computational mechanics; statics, kinematics and dynamics of rigid and elastic bodies; vibrations of solids and structures; dynamical systems and chaos; the theories of elasticity, plasticity and viscoelasticity; composite materials; rods, beams, shells and membranes; structural control and stability; soils, rocks and geomechanics; fracture; tribology; experimental mechanics; biomechanics and machine design.

The median level of presentation is the first year graduate student. Some texts are monographs defining the current state of the field; others are accessible to final year undergraduates; but essentially the emphasis is on readability and clarity.

IUTAM Symposium on

New Applications of Nonlinear and Chaotic Dynamics in Mechanics

Proceedings of the IUTAM Symposium
held in Ithaca, NY, U.S.A.,
27 July–1 August 1997

Edited by

FRANCIS C. MOON

College of Engineering,
Cornell University,
Ithaca, NY, U.S.A.

KLUWER ACADEMIC PUBLISHERS
DORDRECHT / BOSTON / LONDON

A C.I.P. Catalogue record for this book is available from the Library of Congress.

ISBN 0-7923-5276-9

Published by Kluwer Academic Publishers,
P.O. Box 17, 3300 AA Dordrecht, The Netherlands.

Sold and distributed in the North, Central and South America
by Kluwer Academic Publishers,
101 Philip Drive, Norwell, MA 02061, U.S.A.

In all other countries, sold and distributed
by Kluwer Academic Publishers,
P.O. Box 322, 3300 AH Dordrecht, The Netherlands.

Printed on acid-free paper

Printed in the Netherlands.

PREFACE

It is two decades since Mitchell Feigenbaum's landmark papers on period doubling and the modern beginnings of what is now called "Chaos Theory". From the very beginning, mechanics has been a central focus for modern nonlinear dynamical systems. Fluid, structural, machine and rigid body dynamics has been a fertile field for nonlinear phenomena and chaos in particular. Early experimental evidence for chaotic phenomena in mechanics gave the new chaos theory a mark of credibility, importance, and relevance that its earlier sister field, catastrophe theory, did not achieve. The fact that mechanics straddles both physics and engineering also meant that mechanics became a pathway for direct application of chaos theory to applied problems.

So what is new in nonlinear dynamics and mechanics today? First the scope of applications in solid mechanics has broadened to cover material processing, inelasticity and fracture mechanics. In rigid body dynamics, more complex systems such as vehicles, robotics and controlled machines have come into the purview of nonlinear dynamicists. On the mathematics side of nonlinear dynamics, it is now recognized that spatio-temporal problems, hysteretic and time delay problems are the new frontiers in this field. Also the term "complexity" has been added to the lexicon of chaos theory to describe the dynamics of many interacting sub systems which can exhibit self organization and evolution. Complexity analysis has gained a foothold in biological and some social sciences as well in fluid and chemical physics. It remains to be seen what impact it will have in applied mechanics and engineering.

Linear dynamics is a mathematical theory that reduces all motion to a combination of eigenmodes and associated frequencies. It is such a powerful methodology that it has dominated the analysis, design and educational courses in dynamical systems for over a century. However, in the last 20 years this reductionist view of dynamics has been challenged by new nonlinear theory and accompanying experimental evidence. Namely, there exists motions of physical and biological systems which cannot be explained by the synthesis of modes. Nonlinear systems exhibit collective phenomena such as solitons, chaotic transients, strange attractors, quasi-periodic motions and more. The period of 1975-90 was one of analyzing and cataloging these new types of motion. However, in the last decade attention has been focused on application of this new knowledge.

In this book we review a selection of applications of nonlinear dynamics and chaos theory in mechanics and mechanical systems. These applications span the machining of metals to vehicle dynamics, from elastic structures to plastic deformation and impact. Unlike linear dynamics, the use of nonlinear tools by industry has been slow, due perhaps to a lack of trained personnel in industry who have a working knowledge of nonlinear dynamics. But as evidenced by some of the authors in this book, this too is changing with increasing numbers of nonlinear scientists finding positions in industrial and government laboratories.

The IUTAM Symposium in Ithaca, New York was the third of a series devoted to new discoveries in nonlinear dynamics and chaos related to mechanics, with Stuttgart 1990, and London, 1993 the prior venues for this meeting. These three symposia were focused on non-fluid phenomena especially solid and structural mechanics and multibody machine systems. However, a fair number of papers were devoted to fluid-elastic and aero-elastic nonlinear dynamics. Nonlinear fluid mechanics is usually dominated by the quadratic nonlinear acceleration in the Navier Stokes equation. However, in solid, structural and machine mechanics there exists a greater variety of nonlinearities such as hysteresis, delay, nonsmooth discontinuities such as impact, and higher order nonlinearities in elastic structures as in shells and buckled structures.

Mechanics applications have been in the forefront of modern nonlinear dynamics. For example, the chaotic dynamics of buckled structures first studied by Phillip Holmes at Southampton and Cornell (1974-1979) was one of the first practical applications of homoclinic bifurcation theory and Melnikov theory was used to obtain the first criterion for chaotic behavior in a physical engineering system. Also, in this time period, Michael Thompson applied the new chaos theory to the capsize of ships and Earl Dowell applied it to aeroelasticity. Later in the early 1980's, Friedrich Pfeiffer sought to explain machine noise in gears using chaos theory. These early mechanics applications of modern nonlinear dynamical theory expanded the catalog of chaotic phenomena and paradigms beyond period doubling and the Lorenz equations, and opened up new tools of analysis and experimentation in nonlinear dynamics. As a result of this earlier research, undergraduate mechanical engineers at Cornell University and many other engineering schools are taught how to take and use a Poincaré map, and have basic awareness of nonlinear phenomena.

The present volume highlights both new mathematical ideas as well as new applications of nonlinear dynamics. The authors include three academic generations including the early pioneers, the first generation of students, now quite distinguished themselves as well as novitiates to the field.

Rather than present the papers in the order they were delivered at the Symposium, I have chosen to group them according to topic and application. This ordering is not unique as some papers may fit in several categories. The order of these sub-sections of this book will be as follows:

I. Analysis Tools, Bifurcation Theory
II. Elastica, Cables and Shell Dynamics
III. Inelasticity Materials Processing
IV. Machine System Dynamics
V. Impact and Non-smooth Systems
VI. Circuits, Control, Cardiac Modelling
VII. Vehicle and Ship Dynamics
VIII. Fluid-elastic and Fluid Systems
IX. Spatio-temporal Chaos

The four keynote lectures, lecture and poster presentations are indicated in the Table of Contents.

The editor wishes to acknowledge the excellent administrative help of Debbie De Camillo, as well as Cora Jackson for her help in organizing the manuscripts for publication.

F.C. Moon

Welcome Address by the President of IUTAM

Dear Colleagues from all over the world,
Ladies and Gentlemen,

True science does not recognize economical or political systems. Cooperation between scientists from different countries and parts of the world has a long tradition.

Organized meetings between scientists in the field of mechanics were initiated 75 years ago, namely in 1922, when Prof. Theodore von Kármán and Prof. Tullio Levi-Civita organized the world's first conference in hydro- and aero-mechanics. Two years later, in 1924, the First International Congress encompassing all fields of mechanics, that means analytical, solid and fluid mechanics, including their applications, was held in Delft, The Netherlands. From then on, with exception of the year 1942, International Congresses in Mechanics have been held every four years.

The disruption of international scientific cooperation caused by the Second World War was deeper than that caused by the First World War, and the need for reknotting ties seemed stronger than ever before when the mechanics community reassembled in Paris for the Sixth Congress in 1946. Under these circumstances, at the Sixth Congress in Paris, it seemed an obvious step to strengthen bonds by forming an international union, and as a result IUTAM was created and statutes were adopted. Then, the next year, in 1947, the Union was admitted to ICSU, the International Council of Scientific Unions. This council coordinates activities among various other scientific unions to form a tie between them and the United Nations Educational, Scientific and Cultural Organization, well known as UNESCO.

Today, IUTAM forms the international umbrella organization of nearly 50 national organizations of mechanics from nations all over the world. Furthermore, a large number of international scientific organizations of general or more specialized branches of mechanics are connected with IUTAM as Affiliated Organizations. As a few examples, let me mention: the European Mechanics Society (EUROMECH), the International Association of Computational Mechanics (IACM), the International Association for Vehicle System Dynamics (IAVSD). Within IUTAM the only division used so far is related to solid and fluid mechanics as indicated by our two Symposia Panels. But recently six Working Groups have been established by the General Assembly devoted to specific areas of mechanics. These areas are:

- Mechanics of Non-Newtonian Fluids,
- Dynamical Systems,
- Fracture Mechanics and Damage,
- Mechanics of Materials,
- Electromagnetic Processing,
- Computational Mechanics.

It is expected that in the long run the Working Groups will be developed into Standing Committees. In addition, other specific areas of mechanics may be identified to support the international cooperation in more branches of mechanics.

IUTAM carries out an exceptionally important task of scientific cooperation on mechanics on the international scene. Each national organization of IUTAM, like the U.S. National Academy of Sciences, is represented by a number of scientists in IUTAM's General Assembly. In particular, the U.S. delegates with IUTAM are

Professor Achenbach from Northwestern University,
Professor Aref from the University of Illinois at Urbana-Champaign,
Professor Dowell from Duke University, who is with us this week,
Professor Hodge Jr. from Stanford University and
Professor Leal from the University of California, Santa Barbara.

Mechanics is a very well developed science in the U.S. represented at most universities and national laboratories. Since 1949 there has been held 229 IUTAM symposia world-wide. Out of them 28 IUTAM symposia where organized in the U.S., and one IUTAM Symposium was hosted before here in Ithaca. As early as 1959 the IUTAM Symposium on Fluid Mechanics in the Ionosphere took place at Cornell in cooperation with the Unions IUGG and IAU.

As I mentioned before, IUTAM organizes international congresses and symposia all over the world. Thus, the Twentieth International Congress of Theoretical and Applied Mechanics will be held in Chicago, USA, from 27 August - 2 September 2000, what means in three years from now. Announcements of this forthcoming congress will be widely distributed and published in many scientific journals, and in due time, a final announcement of the congress will be available to all who have responded to prior announcements.

The present Symposium is exceptionally interesting because it is within the heart of mechanics. The Symposium covers important approaches: Theory, Experiments and Application of Nonlinear Dynamics is not only very timely, but is also very well hosted given the outstanding research carried out in this field at Cornell University. The Cornell proposal for the Symposium was readily accepted and granted by the General Assembly of IUTAM. This is the third IUTAM Symposium devoted to Nonlinear Dynamics after the Symposium held 1989 in Stuttgart, Germany and the Symposium organized in London, U.K., four years ago as already mentioned by Professor Francis Moon. Thus, there is no doubt that IUTAM considers Nonlinear Dynamics as an essential field of mechanics.

On behalf of IUTAM, I wish to express my sincere thanks to Cornell University, in particular to the Dean of the Engineering College, Professor Hopcroft, for the invitation to host this significant scientific event, and I wish to welcome all the invited participants for their readiness to come and to contribute to the success of the Symposium by very active participation in the lectures and the scientific discussions, as well as in the social program.

Finally, I would like to mention that to sponsor a scientific meeting is one thing, to organize one is another. A heavy burden is placed on the shoulders of the Chairman who is in charge of the scientific and the practical local arrangements, and of the associates who are assisting him. All who have tried this before know perfectly well how much work has to be done in organizing a meeting like this one.

Thus, we should be thankful, not only to the International Scientific Committee, but also very much to the Local Organizing Committee, and in particular to the Chairman of both Committees, Professor Francis Moon, who carried the heaviest load and responsibility.

It is up to us now, Ladies and Gentlemen, to harvest the fruits of the Organizer's work. Let us contribute our share to make this IUTAM Symposium a meeting that will be long remembered as a very successful one!

On behalf of IUTAM, I greet you all and wish you great success!

Werner Schiehlen
Professor of Mechanics
University of Stuttgart, Germany

CONTENTS

KL - Key Lecture, P - Poster Presentation

I. ANALYSIS TOOLS, BIFURCATION THEORY

These papers explore a variety of new analytical tools in nonlinear dynamical systems. These topics include

- •bifurcation theory for stochastic systems
- •wavelets
- •cell mapping
- •bursts and chaotic transients
- •global bifurcation theory
- •reduction of order techniques
- •internal resonances
- •quasi-periodic systems

Certainly in the future we should expect to see further research on the relation between stochastic nonlinear systems and deterministic chaos. Also, techniques using wavelets and reduction of order are likely to see further applications.

BIFURCATION IN STOCHASTICALLY PERTURBED DYNAMICAL SYSTEMS

S.T. Ariaratnam
Department of Civil Engineering, University of Waterloo
Waterloo, Ontario, Canada

1 Introduction

We consider first a deterministic dynamical system governed by the ordinary differential equation

$$\dot{x} = f(x, \alpha), \quad x(0) = x_0 \in \mathrm{R}^d, \tag{1}$$

where $f(x, \alpha)$ is a continuously differentiable function with respect to its arguments. Suppose the equation possesses a steady-state solution $x_s(t, \alpha)$. As the parameter α is varied it is possible that this solution may become unstable at some value α_c and a further bifurcating solution emerges for α in the vicinity of α_c. The stability of the original solution and of the bifurcating solution is usually examined by investigating the stability of the trivial solution of the linearisation of equation (1) around each of these solutions. The standard procedure for this purpose is the evaluation of the maximal Lyapunov exponent which determines the exponential rate of growth or decay of some norm of the solution; the sign of the maximal Lyapunov exponent determines the stability of the trivial solution.

Now suppose that the parameter α, instead of being constant, is subjected to a random fluctuation in time. The resulting dynamical system is then described by the *stochastic* differential equation

$$\dot{x} = f(x, \alpha + \sigma\xi_t), \quad x(0) = x_0 \in \mathrm{R}^d, \tag{2}$$

where ξ_t is a stochastic process and σ is an intensity parameter. The stationary solution $x_s(t, \alpha)$ of equation (2), if it exists, is the analogue of the steady-state solution of the deterministic equation (1). Just as in the deterministic case, it is natural to linearise equation (2) around the steady-state solution $x_s(t, \alpha)$ and investigate the stability, in some sense, of the trivial solution of the resulting linear stochastic differential equation

$$\dot{v} = A(x_s(t, \alpha), \xi_t)v, \quad v(0) = v_0 \in \mathrm{R}^d, \tag{3}$$

where $A(x_s, \xi_t)$ is the $d \times d$ Jacobian matrix $(\partial f_i / \partial x_j)$ evaluated at $x = x_s(t, \alpha)$. In equation (3), the random processes $x_s(t, \alpha)$ and ξ_t appear as coefficients. The

F. C. Moon (ed.),
IUTAM Symposium on New Applications of Nonlinear and Chaotic Dynamics in Mechanics, 3–11.

average exponential rate of separation of neighboring solutions of equation (3) as time t becomes large is given, with probability one, by the Lyapunov exponent λ defined by

$$\lambda(\alpha, v_0) = \overline{\lim_{t\to\infty}} \frac{1}{t} \log |v(t, \alpha, v_0)|. \quad (4)$$

In general, there will be d such Lyapunov exponents depending on the initial random vector v_0 which, under certain mild ergodicity conditions on the random matrix A, are deterministic real numbers (Oseledec [8]). The sign of the maximal Lyapunov exponent determines the stability with probability one (w.p.1) of the trivial solution of equation (3) and hence of the stationary solution $x_s(t, \alpha)$ of equation (2). A dynamical (or D-) bifurcation from $x_s(t, \alpha)$ is said to occur at the value α_c of α at which the maximal Lyapunov exponent changes sign from negative to positive. A new stationary solution may now emerge for α near α_c and the stability of this new solution must be ascertained by a similar procedure. This is the concept of stochastic bifurcation that has been adopted by dynamical system theorists (Arnold [1]) and is in accord with the concept of bifurcation for the corresponding deterministic system when stochastic fluctuation of the parameter is set to zero.

On the other hand, many scientists (Horsthemke and Lefever [5]) have for some time employed a different criterion which is based on a change in the form of the stationary probability density function of the system response. Under this criterion, a stochastic phenomenological (or P-) bifurcation is said to occur when this probability density undergoes a qualitative change from a unimodal to a bimodal or multi-modal density function. The value of the bifurcation parameter at which such a transition in the density function occurs has been shown (Baxendale [3], Arnold and Crauel [2]) to be related to the nontrivial zero of the pth moment Lyapunov exponent $\Lambda(p, \alpha)$ of the solution of equation (3) defined as

$$\Lambda(p, \alpha) = \lim_{t\to\infty} \frac{1}{t} \log E|v(t, \alpha, v_0)|^p, \quad p \in \mathrm{R} \quad (5)$$

where E signifies the mathematical expectation of the quantity that follows. The function $\Lambda(p, \alpha)$ is a convex function in p and has the property $\Lambda'(0, \alpha) = \lambda_{max}(\alpha)$ where λ_{max} is the maximal Lyapunov exponent. By definition $\Lambda(0, \alpha) = 0$ and if $p^*(\alpha)$ denotes the second zero of $\Lambda(p, \alpha)$ then the value α_c^* of the bifurcation parameter at which a P-bifurcation occurs is given by the relation $p^*(\alpha_c^*) = -d$, where d is the dimension of the linear dynamical system described by equation (3). The number p^* is referred to as the stability index. This remarkable result enables one to determine the point of P-bifurcation without the need to consider the nonlinear terms in the governing differential equation.

In this paper, the aforementioned concepts are illustrated through the example of the Rayleigh-Van der Pol oscillator which, in the absence of stochastic perturbation, exhibits a Hopf bifurcation. A noteworthy feature of the system considered is that the stationary probability density of the response can be obtained exactly by solving the associated Fokker-Planck equation. From this density function any qualitative change in its form can be readily acertained as the parameter α is varied.

2 The Rayleigh-Van der Pol Oscillator

The deterministic Rayleigh-Van der Pol oscillator is described by the equation

$$\ddot{x} - [\alpha - \beta(x^2 + \dot{x}^2)]\dot{x} + x = 0 \tag{6}$$

where α, β are constants and the stiffness parameter is normalized to unity by a suitable scaling of the time t. When β is kept fixed at a positive value and the damping parameter α is increased, the system exhibits a limit cycle of amplitude $(\alpha/\beta)^{1/2}$ for small values of α, β. If the system is perturbed by noise, the resulting system will be governed by the stochastic differential equation

$$\ddot{x} - [\alpha - \beta(x^2 + \dot{x}^2) + \sigma_1\xi_t]\dot{x} + (1 + \sigma_2\eta_t)x = \sigma_0\zeta_t \tag{7}$$

where $\sigma_1, \sigma_2, \sigma_0$ are intensity parameters and ξ_t, η_t, ζ_t are taken to be independent 'physical' white noise processes of unit intensity and zero mean value. Equation (7) is to be interpreted in the sense of Stratonovich. In terms of state variables $x_1 = x, x_2 = \dot{x}$, equation (7) can be written

$$\begin{aligned} dx_1 &= x_2 dt \\ dx_2 &= [\{\alpha - \beta(x_1^2 + x_2^2)\}x_2 - x_1]dt \\ &\quad + \sigma_1 x_2 \circ dw_{1t} - \sigma_2 x_1 \circ dw_{2t} + \sigma_0 \circ dw_{3t} \end{aligned} \tag{8}$$

where w_{1t}, w_{2t}, w_{3t} are independent 'unit' Wiener processes. We shall first consider the case when $\sigma_0 = 0$, i.e. in the absence of additive perturbation.

When $\sigma_1 = \sigma_2 = \sigma$, the Fokker-Planck equation for the probability density $f(x_1, x_2)$ of the stationary solution of equation (8) has been obtained by Dimentberg [4] as

$$f_s(x_1, x_2) = C(x_1^2 + x_2^2)^{-\gamma} \exp\left[\frac{-\beta(x_1^2 + x_2^2)}{\sigma^2}\right] \tag{9}$$

provided $\alpha > -\sigma^2/2$, the normalisation constant C being given by

$$C^{-1} = \pi\sigma^{2-4\gamma}\beta^{\gamma-1}\Gamma(1-\gamma) \tag{10}$$

where $2\gamma = 1 - 2\alpha/\sigma^2$. In terms of polar coordinates $x_1 = a\cos\Phi$, $x_2 = -a\sin\Phi$, $\Phi = t + \phi$, this density transforms to

$$f_s(a, \phi) = Ca^{1-2\gamma} \exp\left(\frac{-\beta a^2}{\sigma^2}\right) \tag{11}$$

where $a = (x_1^2 + x_2^2)^{1/2}$ is the amplitude and also the Euclidean norm $|x|$ of the response. The evolution of the density function (9) as α is increased from negative to positive values is shown in Fig. 1 for $\sigma = 0.8$. For $\alpha < -\sigma^2/2$, f_s is the Dirac delta function centered at the origin; for $-\sigma^2/2 < \alpha < \sigma^2/2$, it is unimodal with an integrable singularity at the origin; for $\alpha > \sigma^2/2$, it is zero at the origin with a ring of peaks given by $x_1^2 + x_2^2 = (2\alpha - \sigma^2)/(2\beta)$. Thus one may infer that a P-bifurcation occurs for $\alpha = \sigma^2/2$. The mean square of the amplitude of the stationary response is calculated to be

$$E|x_s|^2 = \frac{2\alpha + \sigma^2}{2\beta}, \quad \alpha > -\sigma^2/2 \tag{12}$$

where x denotes the vector $(x_1, x_2)^T$. It may be noted that $Ex_1^2 = Ex_2^2 = (E|x_s|^2)/2$.

In order to determine the parameter value corresponding to a D-bifurcation it is necessary to evaluate the Lyapunov exponents of the linearized equation around the stationary reference solution $x_s(t, \alpha)$. From the linearisation of equation (8) and the use of the so-called trace or Liouville formula (Oseledec [8]), it may be inferred that the sum of the two Lyapunov exponents is

$$\lambda_1 + \lambda_2 = \alpha - 2\beta E|x_s|^2. \tag{13}$$

For the trivial solution $x_s = 0$, we get

$$\lambda_1^0 + \lambda_2^0 = \alpha, \tag{14}$$

and for the non-trivial solution, substituting for $E|x_s|^2$ from equation (12), we have

$$\lambda_1^s + \lambda_2^s = -(\alpha + \sigma^2). \tag{15}$$

It is now necessary to find one of the two Lyapunov exponents so that the other may be found from equations (14) and (15). For this purpose, the method of stochastic averaging will be employed. The polar coordinate transformation from $(x, \dot{x})$ to (a, ϕ) of equation (7) leads to

$$\begin{aligned} \dot{a} &= (\alpha - \beta a^2 + \sigma\xi_t)a\sin^2\Phi + \sigma\eta_t a \sin\Phi\cos\Phi \\ \dot{\phi} &= (\alpha - \beta a^2 + \sigma\xi_t)\sin\Phi\cos\Phi + \sigma\eta_t\cos^2\Phi \end{aligned} \tag{16}$$

which are exactly equivalent to equation (7). If it is assumed that α, β and σ^2 are of the same order of smallness compared to unity, say $O(\varepsilon)$, stochastic averaging (Stratonovich [9], Kas'minskii [6]) may be performed on the right-hand side terms of equations (16) to obtain the following Itô equations for the averaged amplitude and phase:

$$da = \frac{1}{2}(\alpha + \sigma^2 - \beta a^2)a\,dt + 2^{-1/2}\sigma a\,db_{1t} \tag{17}$$

$$d\phi = 2^{-1/2}\sigma\,db_{2t}, \tag{18}$$

where db_{1t}, db_{2t} are independent Wiener processes with unit intensity parameter. The solutions of the averaged equation converge in the weak sense to the solutions of the original equations over a time interval of length $O(\varepsilon^{-1})$ as $\varepsilon \to 0$. If $a_s(t)$ denotes a stationary solution of equation (17), letting $a(t) = a_s(t) + r(t)$, the linearised equations in r and $\rho = \log r$ in Itô form are

$$dr = \frac{1}{2}(\alpha + \sigma^2 - 3\beta a_s^2)r\,dt + 2^{-1/2}r\sigma\,db_{1t} \tag{19}$$

$$d\rho = \frac{1}{2}(\alpha + \frac{1}{2}\sigma^2 - 3\beta a_s^2)dt + 2^{-1/2}\sigma\,db_{1t}. \tag{20}$$

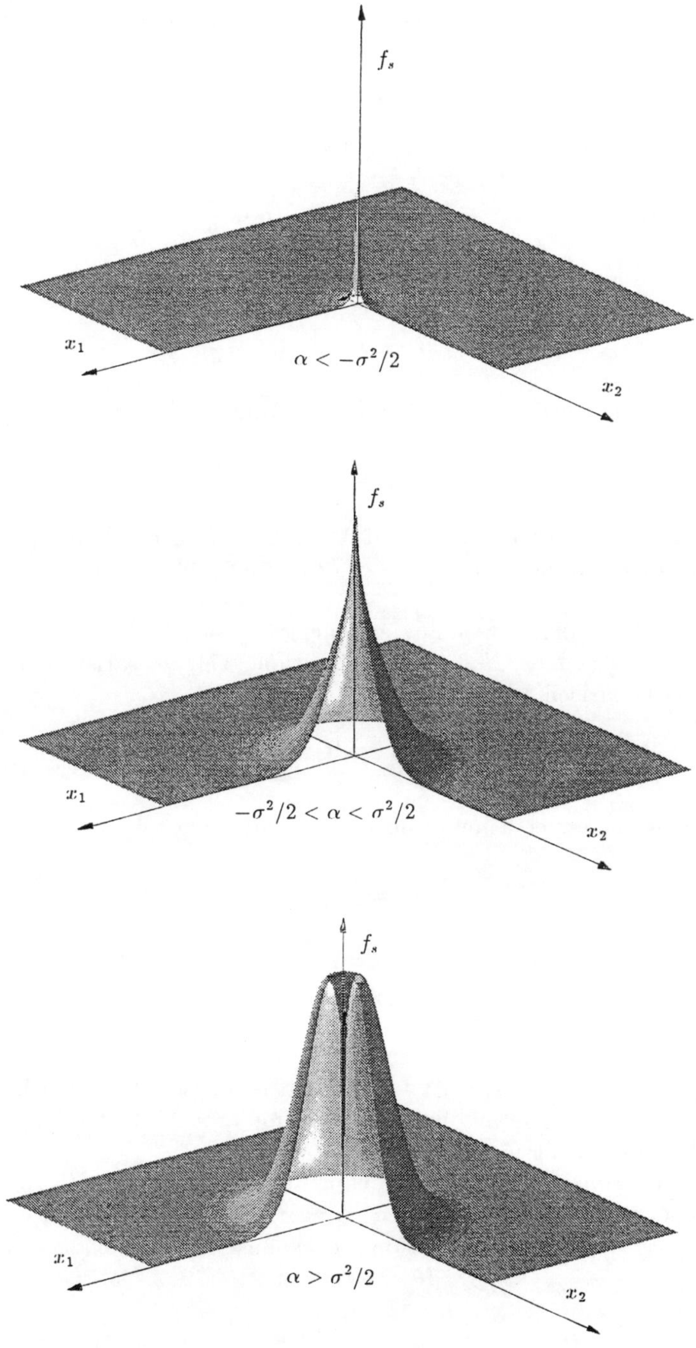

Figure 1: Stationary probability density ($\sigma_0 = 0$).

From equation (20) the Lyapunov exponent λ is obtained w.p.1 as

$$\lambda_1 = \frac{1}{2}(\alpha + \frac{1}{2}\sigma^2 - 3\beta E a_s^2) \tag{21}$$

which for the trivial solution $a_s = 0$ gives

$$\lambda_1 := \lambda_1^0 = \frac{1}{2}(\alpha + \frac{1}{2}\sigma^2) \tag{22}$$

and for the non-trivial solution $Ea_s^2 = (2\alpha + \sigma^2)/(2\beta)$ gives

$$\lambda_1 := \lambda_1^s = -(\alpha + \frac{1}{2}\sigma^2). \tag{23}$$

From equations (14) and (15), the remaining Lyapunov exponents are found to be

$$\lambda_2^0 = \frac{1}{2}(\alpha - \frac{1}{2}\sigma^2), \quad \lambda_2^s = -\frac{1}{2}\sigma^2. \tag{24}$$

Thus $\lambda_{max}^0 = \frac{1}{2}(\alpha + \frac{1}{2}\sigma^2)$ and $\lambda_{max}^s < 0$. Hence a D-bifurcation occurs at the parameter value $\alpha := \alpha_c = -\sigma^2/2$ and a stable nontrivial solution emerges for $\alpha \geq \alpha_c$.

The pth moment Lyapunov exponent can be obtained from the differential equation satisfied by Ea_s^p. From equation (17), a^p satisfies the following Itô equation linearised around $a = 0$:

$$da^p = \frac{1}{2}p(\alpha + \frac{1}{2}\sigma^2 + \frac{1}{2}\sigma^2 p)a^p dt + 2^{-1/2}\sigma\, p\, a^p\, db_{1t}. \tag{25}$$

Taking the expectation of both sides and using the definition (5), we obtain

$$\Lambda(p, \alpha) = \frac{1}{2}p(\alpha + \frac{1}{2}\sigma^2 + \frac{1}{2}\sigma^2 p) \tag{26}$$

whose zero other than $p = 0$ is $p := p^* = -(1 + 2\alpha/\sigma^2)$.

Since the dimension d of the system considered is 2, according to Baxendale's result, a P-bifurcation should occur when $p^* = -2$, i.e. when $\alpha = \sigma^2/2$. This is exactly in accord with the earlier finding from the exact expression for the stationary probability density function. It is worth noting that here the result is obtained simply from the linearised equation around the trivial solution without consideration of the nonlinear terms.

We now consider the system defined by equation (7) with the additive noise term $\sigma_0\zeta_t$ included. When $\sigma_1 = \sigma_2 = \sigma$, the Fokker-Planck equation for the probability density of the stationary response has the exact solution (Lin and Cai [7])

$$f_s(x_1, x_2) = C[\sigma^2(x_1^2 + x_2^2) + \sigma_0^2]^{-\gamma} \exp[-\beta(x_1^2 + x_2^2)/\sigma^2] \tag{27}$$

where $2\gamma = 1 - 2(\alpha\sigma^2 + \beta\sigma_0^2)/\sigma^4$ and the constant C is given by

$$C^{-1} = \pi\sigma^{2-4\gamma}\beta^{1-\gamma}\Gamma(1 - \gamma, \beta\sigma_0^2/\sigma^4)\exp(-\beta\sigma_0^2/\sigma^4) \tag{28}$$

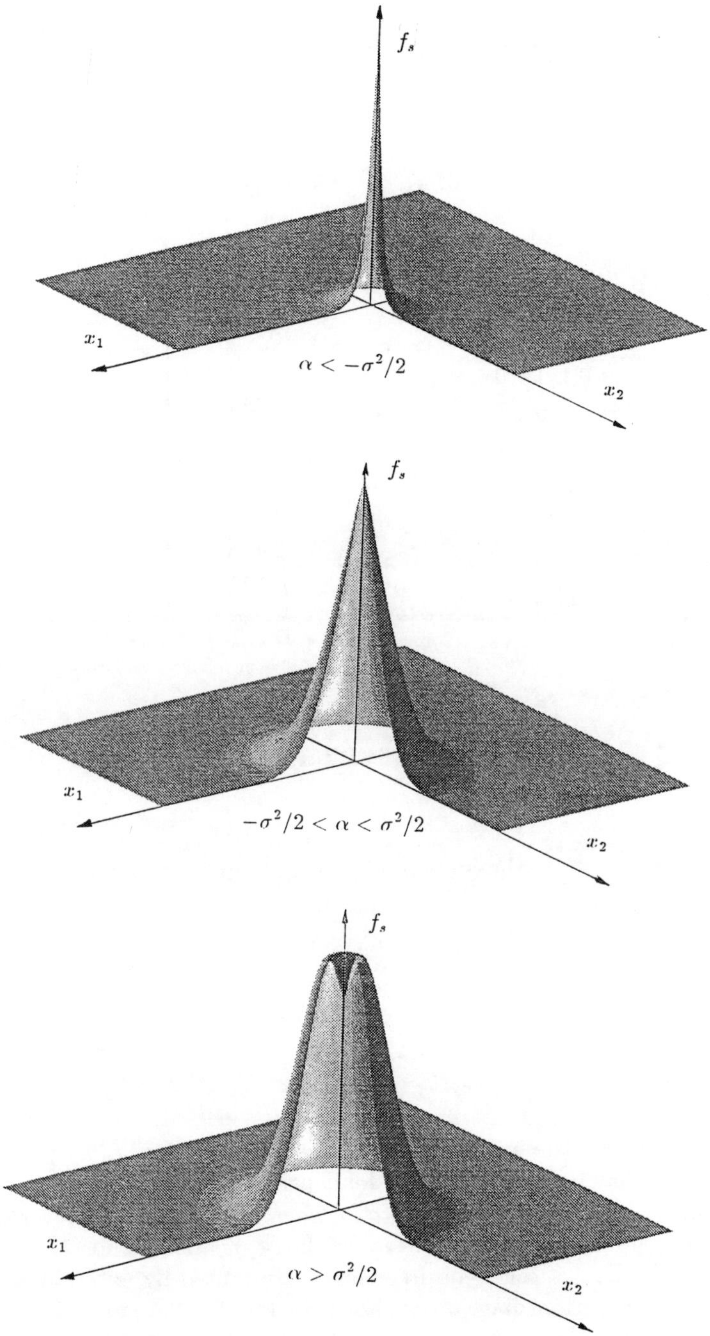

Figure 2: Stationary probability density ($\sigma_0 \neq 0$).

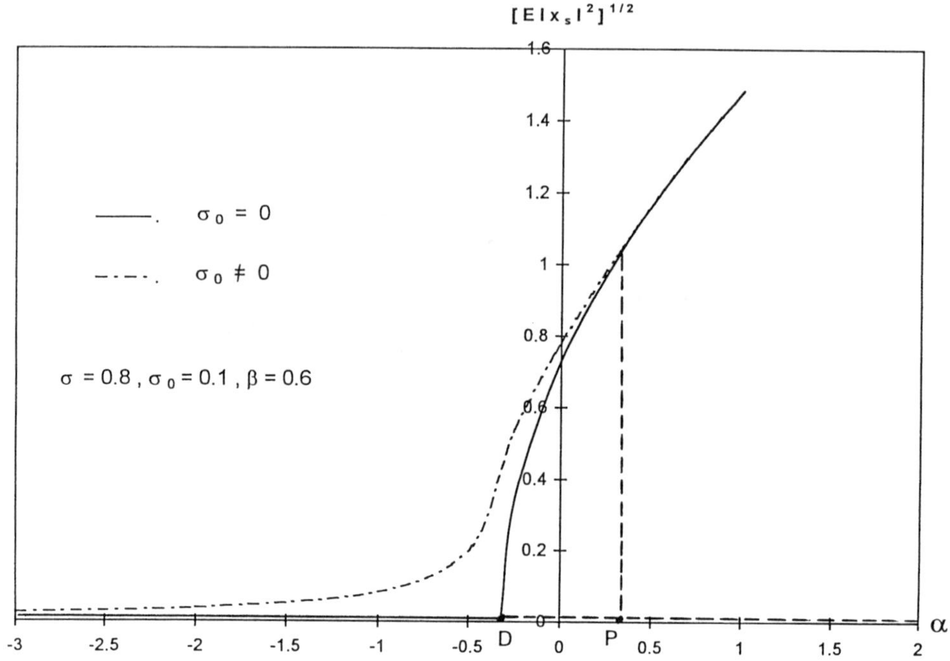

Figure 3: Bifurcation diagram.

where $\Gamma(a, x)$ is the incomplete gamma function defined as

$$\Gamma(a, x) = \int_x^\infty t^{a-1} e^{-t} dt. \tag{29}$$

The expected value of the square of the norm of x_s is, from equation (27), found to be

$$E|x_s|^2 = \frac{\sigma^2 \Gamma(2-\gamma, \beta\sigma_0^2/\sigma^4)}{\beta\Gamma(1-\gamma, \beta\sigma_0^2/\sigma^4)} - \frac{\sigma_0^2}{\sigma^2}. \tag{30}$$

Fig.2 shows the evolution of the density function f_s as the parameter α is varied. It may be observed that the density has a single finite peak at the origin for values of $\alpha < \alpha_c^*$; for $\alpha > \alpha_c^*$ it has a crater near the origin and a ring of peaks along the circle $x_1^2 + x_2^2 = (2\alpha - \sigma^2)/2\beta$. Thus the P-bifurcation is preserved whereas the D-bifurcation is eliminated by external additive noise.

The bifurcation diagram is shown in Fig.3. The heavy lines correspond to the homogeneous system with $\sigma_0 = 0$ while the dotted lines are for the system with additional additive noise ($\sigma_0 \neq 0$).

3 Conclusions

Some concepts of stochastic bifurcation are illustrated through an example of a two-dimensional system subject to parametric and external noise fluctuations. The example chosen is that of the Rayleigh-Van der Pol oscillator which exhibits a Hopf bifurcation in the absence of noise. As the mean linear viscous damping parameter is increased it is shown that first a dynamical (or D-) bifurcation occurs followed by a phenomenological (or P-) bifurcation. The addition of noise in the form of an external forcing function destroys the D-bifurcation but preserves the P-bifurcation. The roles of the Lyapunov exponent and the moment Lyapunov exponent of the linearised system in determining the bifurcation parameter values are illustrated.

References

[1] L Arnold. Lyapunov exponents of nonlinear stochastic systems. In F. Ziegler and G.I. Schueller, editors, *Nonlinear Stochastic Dynamic Engineering Systems*, pages 181–201. Springer Verlag, Berlin Heidelberg New York, 1988.

[2] L. Arnold and H. Crauel. Random dynamical systems. In L. Arnold, H. Crauel, and J.-P. Eckmann, editors, *Lyapunov Exponents, Oberwolfach 1990*, volume 1486 of *Lecture Notes in Mathematics*, pages 1–22. Springer Verlag, Berlin Heidelberg New York, 1991.

[3] P.H. Baxendale. Invariant measures for nonlinear stochastic differential equations. *ibid*, pages 123–140, 1991.

[4] M.F. Dimentberg. An exact solution to a certain nonlinear random vibration problem. *Int. J. Nonlinear Mech.*, 17(4):231–236, 1982.

[5] W. Horsthemke and R. Lefever. *Noise-Induced Transitions.* Springer Verlag, Berlin Heidelberg New York, 1984.

[6] R.Z. Khas'minskii. A limit theorem for the solutions of differential equations with random right-hand sides. *Theory of Prob. and Applications*, 11(3):390–406, 1966. (English translation).

[7] Y.K. Lin and G.Q. Cai. *Probabilistic Structural Dynamics. Advanced Theory and Applications.* McGraw-Hill, New York, 1995.

[8] V.I. Oseledec. A multiplicative ergodic theorem. Lyapunov characteristic numbers for dynamical systems. *Trans. Moscow Math. Soc.*, 19:197–231, 1968. (English translation).

[9] R.L. Stratonovich. *Topics in the Theory of Random Noise, Volume I.* Gordon and Breach, New York-London, 1963.

NONSTATIONARY RESPONSES IN TWO DEGREE-OF-FREEDOM NONLINEAR SYSTEMS WITH 1-TO-2 INTERNAL RESONANCE

ANIL. K. BAJAJ
School of Mechanical Engineering
Purdue University, West Lafayette, IN 47907

BAPPADITYA BANERJEE
Automated Analysis Corporation
423 S. Washington St.
Peoria, IL 61615

PATRICIA DAVIES
School of Mechanical Engineering
Purdue University, West Lafayette, IN 47907

Abstract The dynamics of two degree-of-freedom dynamical systems with 1:2 subharmonic internal resonance and weak quadratic nonlinearities is analyzed in the neighbourhood of bifurcation points, when the excitation frequency varies slowly through the region of primary resonance. The slowly evolving averaged equations are numerically studied for motions initiated in the vicinity of the stationary responses. An analytical technique, based on the *Dynamic Bifurcation Theory* is then developed to explain the numerical observations for slow frequency sweep through the bifurcations, and the effects of system parameters on the dynamics near a Pitchfork Point are determined.

1 Introduction

Much work has been done on the nonstationary response of single degree-of-freedom nonlinear systems with slowly varying parameters. Interesting effects have been observed as the excitation parameters are swept slowly through regions of resonant response (Lewis, 1932; Mitropolskii, 1965). Most studies in engineering literature have used the asymptotic method of averaging, coupled with direct numerical integration of the nonautonomous averaged equations, to determine the nonstationary responses. The focus of these investigations has been to characterize the deviations of system responses, due to a slow evolution of system or excitation parameters, from the responses in the stationary case. Nonstationary

F. C. Moon (ed.),
IUTAM Symposium on New Applications of Nonlinear and Chaotic Dynamics in Mechanics, 13–22.

vibrations in externally excited multi degree-of-freedom systems have received much less attention in the literature, with Evan-Iwanowski (1976) and Shyu *et al*(1993) being notable exceptions. In the mathematics community, much effort has been made to understand the dynamics of Hamiltonian systems, as some parameters evolve through a resonance region (Kevorkian and Cole, 1996).

Many investigators have studied the effects of slow evolution of parameters through regions where bifurcations arise in the stationary case (Haberman, 1979; Davies and Krishna, 1996). A concept of slowly varying equilibrium solutions was introduced, and in the neighborhood where the expressions for the slowly varying equilibrium solutions break down, inner layers and matched asymptotic expansions were utilized to construct approximations to the slowly varying solutions. Raman *et al* (1996) used the techniques of *Dynamic Bifurcation Theory* to study analytically the effects of linear frequency sweep through resonance regions in harmonically excited Duffing and Mathieu-Duffing oscillators. Their analysis uses the results for first-order time varying systems (Haberman, 1979) together with Center Manifold Theory (Carr, 1982; Wiggins, 1991) extended to systems with slowly varying parameters. Using these, they were able to predict the delay in jumps and penetration into regions of instability observed in numerical simulations.

Here, we apply these ideas from *Dynamic Bifurcation Theory* to study the nonstationary responses in weakly nonlinear multi degree-of-freedom systems that exhibit internal resonances. Specifically, we study the nonstationary responses for systems with quadratic nonlinearities and harmonic forcing when the excitation frequency slowly evolves through the resonance region.

2 Equations of Motion and Averaging

Many examples of mechanical and structural systems that can be modeled by two degree-of-freedom systems with quadratic nonlinearities, have been discussed by Nayfeh and Balachandran (1989). The equations of motion for a general two degree-of-freedom system subjected to a slowly evolving external excitation, can be written in generalized modal coordinates ζ_j, j=1,2, as

$$\begin{aligned} \ddot{\zeta}_1 + \omega_1^2\zeta_1 + \epsilon C_1(\dot{\zeta}_1, \dot{\zeta}_2) + \epsilon Q_1(\zeta_1, \dot{\zeta}_1, \zeta_2, \dot{\zeta}_2) &= \epsilon F_1(t, \epsilon t), \\ \ddot{\zeta}_2 + \omega_2^2\zeta_2 + \epsilon C_2(\dot{\zeta}_1, \dot{\zeta}_2) + \epsilon Q_2(\zeta_1, \dot{\zeta}_1, \zeta_2, \dot{\zeta}_2) &= \epsilon F_2(t, \epsilon t), \end{aligned} \tag{1}$$

where ϵ, $0 < \epsilon \ll 1$, is an arbitrary small parameter; C_j's are the linear velocity proportional damping terms; ω_j's are the linear natural frequencies of the two oscillators; F_j's are the external excitation terms; and Q_j's are some general quadratic nonlinear terms dependent on the generalized coordinates and velocities. Also, a dot over ζ_j denotes the derivative of ζ_j with respect to the fast time "t". The forms of the various system dynamic terms are

$$\begin{aligned} C_1(\dot{\zeta}_1, \dot{\zeta}_2) &= c_{12}\dot{\zeta}_1 + c_{14}\dot{\zeta}_2, \quad C_2(\dot{\zeta}_1, \dot{\zeta}_2) = c_{22}\dot{\zeta}_1 + c_{24}\dot{\zeta}_2, \\ Q_1(\zeta_1, \dot{\zeta}_1, \zeta_2, \dot{\zeta}_2) &= a_{11}\zeta_1^2 + a_{13}\zeta_1\zeta_2 + a_{22}\dot{\zeta}_1^2 + a_{24}\dot{\zeta}_1\dot{\zeta}_2 + a_{33}\zeta_2^2 + a_{44}\dot{\zeta}_2^2, \end{aligned}$$

$$Q_2(\zeta_1,\dot{\zeta}_1,\zeta_2,\dot{\zeta}_2) \quad = \quad b_{11}\zeta_1^2 + b_{13}\zeta_1\zeta_2 + b_{22}\dot{\zeta}_1^2 + b_{24}\dot{\zeta}_1\dot{\zeta}_2 + b_{33}\zeta_2^2 + b_{44}\dot{\zeta}_2^2, \quad (2)$$

where c_{jk} (j,k =1,2) and a_{jk}, b_{jk} (j,k = 1, 2, 3, 4) are constants. The external excitations are chosen to be harmonic with frequencies and amplitudes that slowly evolve over time:

$$\begin{aligned} F_1 &= F_{11}\cos(r\theta(t)) + F_{12}\sin(r\theta(t)), \\ F_2 &= F_{21}\cos(r\theta(t)+\psi) + F_{22}\sin(r\theta(t)+\psi), \end{aligned} \quad (3)$$

where, the amplitudes F_{jk}, (j,k = 1, 2) and the instantaneous frequency $\Omega = \Omega(\epsilon t) = \frac{d\theta}{dt}$ are a function of the slow time ϵt, and r is a positive integer.

The method of averaging was applied by Mitropolskii (1965) to multi degree-of-freedom systems with slowly varying parameters and mono frequency oscillations, where he considered these parameters as constants during the averaging process, and as functions of slow time ϵt in the derivatives. Essentially, a near identity transformation was used to separate the motions on the fast time t from the motions on the slow time ϵt. Then, the equations were averaged along an integral manifold of the equations which have been put in the standard form (Wiggins, 1990) for averaging. The asymptotic expansion approach of Mitropolskii (1965) was extended by Evan-Iwanowski (1976) to multi degree-of-freedom systems exhibiting some combination resonance involving more than one degree-of-freedom. Essentially, the same approach is followed here in deriving the averaged equations for the system under consideration. Details of the derivations are fully described in Banerjee (1996). This process ultimately results in the following scaled equations for the system:

$$\begin{aligned} a_1' &= -a_1\xi_1 - \tfrac{a_1a_2}{2}\sin(\beta_2 - 2\beta_1), \\ a_1\beta_1' &= a_1\tfrac{\sigma_2-\mu}{2} + \tfrac{a_1a_2}{2}\cos(\beta_2 - 2\beta_1), \\ a_2' &= -a_2\xi_2 + a_1^2\sin(\beta_2 - 2\beta_1) - E_4\sin\beta_2, \\ a_2\beta_2' &= a_2\sigma_2 + a_1^2\cos(\beta_2 - 2\beta_1) - E_4\cos\beta_2, \end{aligned} \quad (4)$$

and

$$\sigma_2' = \epsilon\sigma_{2r}, \quad (5)$$

where "$\prime$" denotes derivative with respect to the slow time "ϵt"; a_i, β_i (i=1,2) are the amplitudes and phases of responses of the two oscillators; σ_1 and σ_2 are the scaled mistunings (or detuning) of the excitation frequency Ω from the linear natural frequencies of the first and the second modes, respectively; $\mu = \sigma_2 - 2\sigma_1$ is the internal mistuning between the two oscillators; ξ_i (i=1,2) are the dampings in the two modes; σ_{2r} is the rate of sweep of the mistuning σ_2; and E_4 is a scaled amplitude of the external excitation. The initial conditions for the averaged system are denoted by some $(a_{10}, \beta_{10}, a_{20}, \beta_{20})$ and σ_{20}. It is interesting to note here that the parameter σ_2 has been assumed to vary linearly in slow time. This averaging process is, however, valid and will result in the same amplitude equations, if the frequency sweep is assumed to be a periodic variable of slow time, or if the amplitude of the external excitation E_4 is also assumed to be a slowly evolving function of time.

Our interest now is in the study of solutions of equations (4) and (5) as a function of the system parameters, when the frequency mistuning rate, σ_{2r}, is nonzero. The solutions for the stationary case, $\sigma_{2r} = 0.0$, have been well studied in the literature (Nayfeh and Balachandran, 1989; Banerjee, 1996). In the non-stationary case there are no standard techniques available and most investigations have resorted to numerical simulations of these averaged equations.

3 Numerical Studies of the Averaged System

We first consider a typical example with stationary responses, as shown in Figure 1. For $\sigma_{2r} = 0.0$, the averaged equations have two types of steady state solutions: the *single mode solution* where a_1, the amplitude of the lower frequency mode, is zero, and the *coupled mode solution* in which both a_1 and a_2 are nonzero. The single mode solution is unique resembling the response of a linear oscillator, and it exists for all excitation frequencies σ_2. At some excitation force level, the single

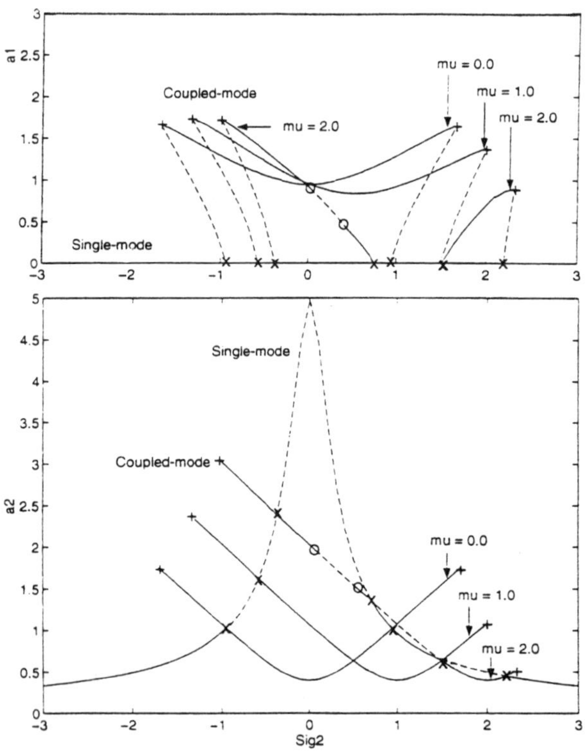

Figure 1. Stationary responses for the coupled oscillators; $E_4 = 1.0$, $\xi_1 = \xi_2 = 0.2$.

mode solution undergoes a Pitchfork bifurcation (Wiggins, 1990) to a coupled mode solution. There exist two coupled mode solutions over some excitation frequency σ_2, and the two solutions join at a turning point. Thus, the coupled mode solutions arise via a Pitchfork bifurcation from the single mode solution and get annihilated by each other at a turning point. The steady state solutions as a function of the excitation frequency σ_2 also depend on the dampings, ξ_i (i=1,2), the internal mistuning, μ, and the forcing amplitude, E_4.

In Figure 1 are shown the stationary amplitude response curves for the two modes of vibration for $\xi_1 = \xi_2 = 0.2$, and $E_4 = 1.0$. The Pitchfork points in the single mode branch are denoted by "x", the turning points in the coupled mode solutions are denoted by "+", and the Hopf bifurcation points in the coupled mode branch are denoted by "o". Between the two Pitchfork points, the system response where the first oscillator is not responding is unstable. Very interestingly, when the two given oscillators are not tuned exactly ($\mu \neq 0$), the response curves lose their symmetry about $\sigma_2 = 0$. Also, the frequency interval over which the single mode solution is unstable shifts to higher values. For sufficiently large μ, there arise two aditional Pitchfork bifurcation points on the single mode solution. For negative μ, the responses are mirror images of those for positive μ. The dynamics of the averaged equations for parameter values in the Hopf bifurcation interval have been explored extensively in the literature (Banerjee, 1996), and it is known that limit cycles, periodic doubling bifurcations, chaotic solutions, as well as Silnikov type phenomena can arise. Thus, in the frequency interval between the two Hopf points, complex dynamics exists for the stationary system.

We now consider solutions of equations (4) and (5) when the sweep rate σ_{2r} is nonzero. The response depends on the initial conditions $(a_{10}, \beta_{10}, a_{20}, \beta_{20})$ for the amplitude variables, and on the frequency mistuning μ. Since $a_{10} = 0.0$ is always a solution of these equations, the (a_2, β_2) plane is an invariant set, and

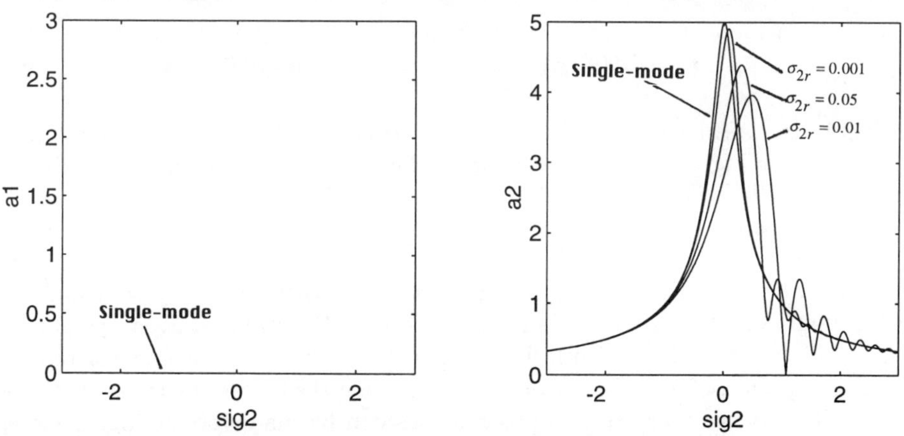

Figure 2. Nonstationary responses for the coupled oscillators for different sweep rates; $a_{10} = 0.0$, $E_4 = 1.0$, $\xi_1 = \xi_2 = 0.2$, $\mu = 0.0$.

any solution started with initial conditions in that set remains within the set for all times. Thus, the nonstationary response with initial condition $a_{10} = 0.0$ is the nonstationary response of a linear oscillator with linear frequency sweep through the resonance region (Mitropolskii, 1965; Kevorkian and Cole, 1996); this is illustrated in Figure 2 for different sweep rates and for $\mu = 0.0$. In subsequent figures, some stationary responses are shown for comparison. Also, the nonstationary solutions are initiated either on or close to the stationary solution away from the resonance region (at $\sigma_2 = -3.0$).

Nonstationary responses for initial conditions out of the (a_2, β_2) plane are shown in Figures 3 and 4. For a very slow positive sweep rate (Figure 3), $\sigma_{2r} = 0.001$, the solution essentially follows along the single mode stationary solution even across the Pitchfork bifurcation point near $\sigma_2 \approx -1.0$. There is a slow deviation of the a_2 component away from the stationary solution until $\sigma_2 \approx 0$, at which point the solution begins to deviate from the single mode solution exponentially and approaches the stable coupled mode solution. It then

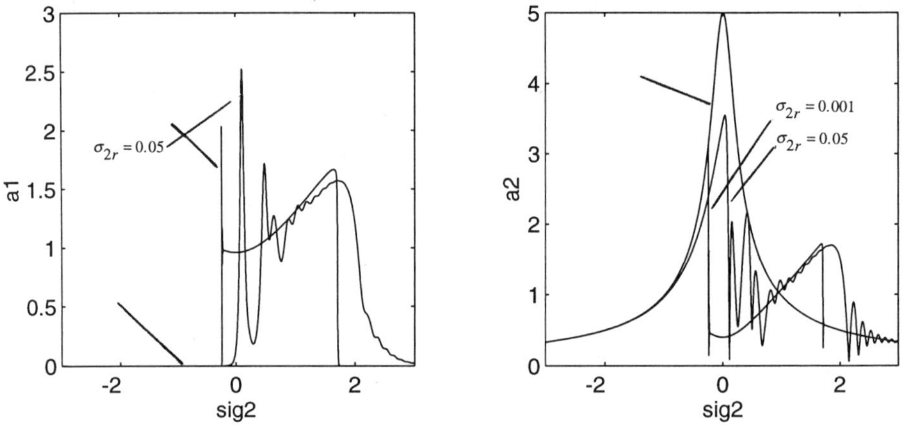

Figure 3. Nonstationary responses for the coupled oscillators for different sweep rates; $a_{10} = 0.001$, $E_4 = 1.0$, $\xi_1 = \xi_2 = 0.2$, $\mu = 0.0$.

overshoots the stationary coupled mode solution, and finally settles down near its vicinity, staying in the neighborhood of the stationary coupled mode solution even beyond the saddle-node or the turning point in the coupled mode branch, where it jumps down to the stable single mode solution. For a slow frequency evolution through the Pitchfork point, there is penetration into the so called 'unstable region'. After the transition away from the stationary solution, the nonstationary solution settles to a neighborhood of the stable stationary solution in an oscillatory manner. Similarly, the nonstationary solution persists beyond the turning point before it jumps down to approach the stable stationary solution in an oscillatory manner. Similar observations can be made for the higher sweep rate, $\sigma_{2r} = 0.05$.

The faster the sweep rate, the greater is the penetration into the region of instability, and longer is the time taken for the response to approach the neighborhood of the stationary and stable coupled mode equilibrium solution. The transition from the neighborhood of one stationary solution to that of another is less abrupt for a higher sweep rate. The nonstationary solution may not even approach the vicinity of a stationary solution, if the sweep rate is high and a change of stability occurs. The delay in a jump from one stable stationary branch to another stable stationary branch in the vicinity of a turning point is strongly affected by the sweep rate.

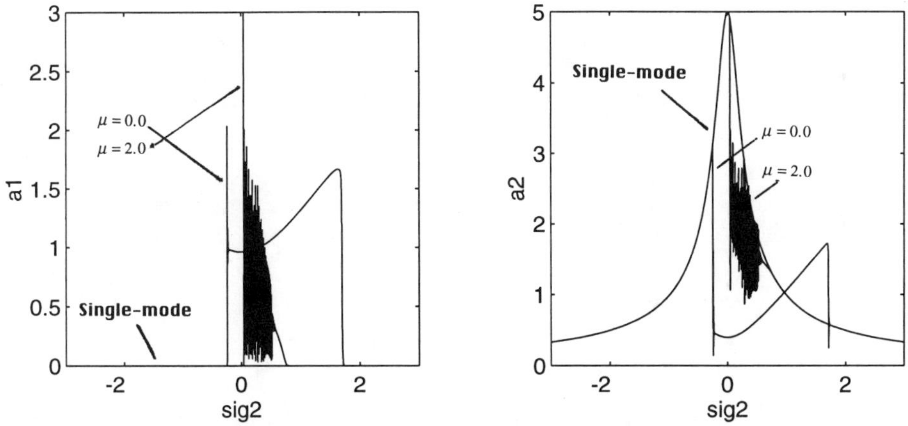

Figure 4. Nonstationary responses for different internal mistunings; $a_{10} = 0.001$, $E_4 = 1.0$, $\xi_1 = \xi_2 = 0.2$, $\sigma_{2r} = 0.001$.

The effects of the internal mistuning μ on the nonstationary responses are shown in Figure 4 for $\mu = 0.0$ and $\mu = 2.0$. As suggested by the shift in Pitchfork points for the stationary responses (Figure 1), the presence of internal mistuning results in considerable delay in transition from the vicinity of the unstable single mode solution. Consequently, the maximum response attained in the stationary case can also be achieved in the nonstationary case. When the solution does leave the neighborhood of the single mode solution, the coupled mode stationary solutions are Hopf unstable. This explains the complex oscillatory response of the system as the excitation frequency is slowly evolving. By the time the Hopf frequency interval is traversed, the stationary coupled mode solution joins the single mode solution, and the nonstationary solution settles in the neighborhood of the stable single mode solution.

When the two oscillators have a large internal mistuning, $\mu = 2.0$, the effects of different sweep rates are shown in Figure 5. At the higher sweep rate, $\sigma_{2r} = 0.05$, the solution contains none of the characteristics of the Hopf unstable stationary solutions, and resembles the response for initial conditions with $a_{10} = 0.0$. Thus, at high sweep rates, one can pass through a complex dynamics region without initiating the internal coupling between the oscillators, at the expense of increase in the amplitude of the directly excited oscillator.

Numerical simulations of the averaged equations for different initial conditions, but for the same sweep rate, show that closer the initial conditions are to the single mode equilibrium solution, the longer the response of the system stays near this equilibrium after the bifurcation. Thus, there is more penetration into the instability region when initial conditions are closer to the stationary solution. The effect near the turning point in the nonstationary response is much less significant (Banerjee, 1996).

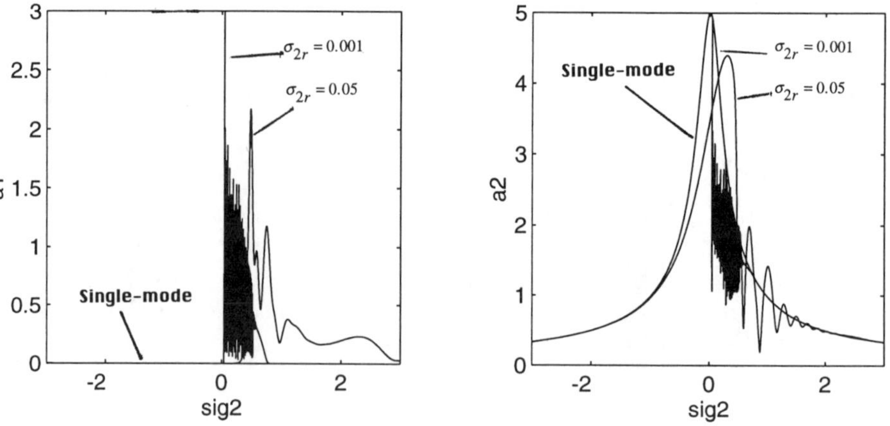

Figure 5. Nonstationary responses for different sweep rates at $\mu \neq 0$; $a_{10} = 0.001$, $E_4 = 1.0$, $\xi_1 = \xi_2 = 0.2$, $\mu = 2.0$.

4 Slowly Varying Equilibrium and Center Manifold Analysis

We would like to explain analytically, some of the observations made above. We are specially interested in the solutions with slow sweep rates and those that remain near the stationary solutions. For this, we closely follow the developments in Raman *et al* (1996). The amplitude equations (4) and (5) for the system with slowly varying parameters can be written, in terms of the slow time scale $T = \epsilon t$, as:

$$\epsilon \frac{d\underline{x}}{dT} = \underline{f}(\underline{x}, \sigma_2(T)). \tag{6}$$

When $\epsilon = 0$, this reduces to the system for the stationary solutions. Let a stationary solution of equations (6) be denoted by $\underline{x}_0 \equiv \underline{x}_0(0,T)$. It then satisfies the equation $\underline{f}(\underline{x}_0, \sigma_2(T)) = 0$. For small and nonzero ϵ, some solutions of equations (6) can be constructed in terms of a power series in ϵ, with the $O(1)$ term being given by one of the branches of the stationary solution. This solution is called the *slowly varying equilibrium solution.*

One asymptotic expansion for the slowly varying equilibrium solution is

$$\underline{x}(\epsilon, T) = \underline{x}_0(0,T) + \epsilon \underline{x}_1(0,T) + \ldots \tag{7}$$

Substituting equation (7) into equation (6) and comparing coefficients of different powers of ϵ results in a sequence of nonlinear and linear equations. As shown by

Raman *et al* (1996), the second term in the expansion

$$\underline{x}_1 = \left(\frac{d\underline{f}}{d\underline{x}}|_{\underline{x}=\underline{x}_0}\right)^{-1} \frac{d\underline{x}_0}{dT}. \tag{8}$$

Clearly, around bifurcation points, $\underline{x}_1$ becomes singular and hence the expansion in equation (7) can no longer be used to approximate the nonstationary response. If the singularity of the Jacobian arises due to some eigenvalues going through zero, a center manifold of dimension equal to the number of zero eigenvalues can be constructed around the bifurcation point. The essential singular behavior is then captured by the equations on the center manifold, and the solutions for these nonstationary equations on the center manifold can be used to obtain an asymptotic approximation to the nonstationary response for the averaged system. This theory for systems with slowly parameters was developed by Raman *et al* (1996).

For the averaged equations (4) and (5), the breakdown in the slowly varying equilibrium solutions occurs near the Pitchfork and saddle-node bifurcation points. The subcritical Pitchfork bifurcation points in the single mode branch, as well as the turning points in the coupled mode branches, arise with one zero eigenvalue of the Jacobian, the essential local behaviour for slow parameter variation is captured by equations on a 1-dimensional center manifold. The form of this equation depends on the nature of the instability (a Pitchfork or a turning point). For details of this analysis, we refer the reader to Banerjee (1996), and present here only the final results in the case of the Pitchfork points.

The lowest order approximation to the equation on the center manifold in the neighborhood of a Pitchfork point is of the form

$$z_1' = \sigma_{2r}\gamma_1 T z_1 + \gamma_3 z_1^3, \tag{9}$$

where γ_1 and γ_3 are coefficients dependent on the system parameters at the Pitchfork bifurcation point. This Bernoulli equation captures the essential features described for slow transitions across Pitchfork bifurcations (Haberman, 1979). Its solution for some nonzero initial condition, $z_1(T_0) = \Delta$, is

$$z_1(T) = \left(-2\gamma_3 e^{-\gamma_1\sigma_{2r}T^2} \int_{T_0}^{T} e^{\gamma_1\sigma_{2r}\bar{T}^2} d\bar{T} + \frac{e^{\gamma_1\sigma_{2r}({T_0}^2 - T^2)}}{\Delta^2}\right)^{-1/2}. \tag{10}$$

For a subcritical pitchfork bifurcation the solution in equation (10) becomes unbounded at a finite value of $T = T_u$, given by

$$\int_{T_0}^{T_u} e^{\gamma_1\sigma_{2r}\bar{T}^2} d\bar{T} = \frac{e^{\gamma_1\sigma_{2r}T_0^2}}{2\gamma_3\Delta^2}. \tag{11}$$

The integral $\int_{T_0}^{T_u} e^{\gamma_1\sigma_{2r}\bar{T}^2} d\bar{T}$, increases monotonically with T_u. Hence, for a fixed initial time, T_0, and sweep rate, σ_{2r}, the smaller the initial displacement Δ, the larger the time for the solution to become unbounded. For a fixed initial time, T_0, and initial displacement Δ, the larger the sweep rate, σ_{2r}, the larger is T_u, see Figure 3. The time for the solution to become unbounded can be related to the system parameters through the expressions for the coefficients γ_1 and γ_2 and equation (11), and these details can be found in Banerjee (1996).

5 Summary and Conclusions

The response of two degree-of-freedom nonlinear systems with quadratic nonlinearities and 1-to-2 internal resonance was studied when the external excitation frequency slowly evolves through a resonance. The asymptotic method of averaging was used to derive the nonstationary amplitude equations. Numerical solutions of the averaged equations show that the deviations of the nonstationary responses from the corresponding stationary solutions are very much influenced by the frequency sweep rate, the initial conditions, and the internal frequency mistuning. The effects of these parameters on the delayed instability near the Pitchfork points are explained using *Dynamic Bifurcation Theory.*

6 References

1. F. M. Lewis, "Vibration during acceleration through a critical speed". *Transaction ASME* **54**, 253-261 (1932).
2. V. A. Mitropolskii, *Problems of the Asymptotic Theory of Nonstationary Vibrations*, Israel Program for Scientific Translations, Jerusalem (1965).
3. R. M. Evan-Iwanowski, *Resonance Oscillations in Mechanical Systems*, Elsevier Scientific Publishing Co., New York (1976).
4. I.-M.Shyu, D. Mook and R. H. Plaut, "Whirling of a forced cantilever beam with static deflection-iii: passage through resonance". *Nonlinear Dynamics* **4**, 461-481 (1993).
5. J. Kevorkian and J. D. Cole, *Multiple Scale and Singular Perturbation Methods*, Springer-Verlag, New York (1996).
6. R. Haberman, "Slowly varying jump and transition phenomena associated with algebraic bifurcation problems" *SIAM Journal on Applied Mathematics* **37**,69-106 (1979).
7. Huw G. Davies and Rangavajhula Krishna, "Nonstationary response near generic bifurcations", *Nonlinear Dynamics* **10**, 235-250 (1996).
8. A. Raman, A. K. Bajaj and P. Davies, "On the slow transition across instabilities in nonlinear dissipative systems". *Journal of Sound and Vibration* **192**(4), 835-865 (1996).
9. J. Carr, *Applications of Centre Manifold Theory*, Springer-Verlag, New York (1981).
10. S. Wiggins, *Introduction to Applied Nonlinear Dynamical Systems and Chaos*, Springer-Verlag, New York (1990).
11. A. H. Nayfeh and B. Balachandran, "Modal interactions in dynamical and structural systems". *Applied Mechanics Reviews* **42 part 2** (11), S175-S201 (1989).
12. Bappaditya Banerjee, *Dynamics of Multi Degree-of-Freedom Stationary and Nonstationary Nonlinear Systems*, Ph. D. Dissertation, Purdue University, West Lafayette, IN (1996).

SINUSOIDALLY VARYING PERIOD-DOUBLING BIFURCATIONS

Huw G. Davies
University of New Brunswick
Canada

1. Introduction

Bifurcations of dynamic systems may be affected in a variety of ways if the parameter that controls the bifurcation is varied slowly. The variation may alter, delay, or eliminate the bifurcation, introduce new types of bifurcation, or change stability regions. The variation may be imposed deliberately to control the response of the system. Typical applications include passage through critical speeds of machines, through laser threshold values, chemical reactions with a degrading catalyst, bursting oscillations and control. Here we consider sinusoidal variation through a period-doubling bifurcation.

Raman, Bajaj and Davies (1996) have treated slow linear transition across bifurcations in some classical nonlinear systems by reducing the systems to generic forms on centre manifolds. Periodic variation of similar subcritical and supercritical normal forms were studied by Davies and Rangavajhula (1996). These forms allow exact solutions (Haberman (1979)), and both local and global stability can be examined. The centre manifold reduction lends itself to bifurcations associated with fixed points of systems. Limit cycles, including in particular period-doubling cascades of limit cycles, are less easily handled. For these period-doubling bifurcations we choose to study associated one dimensional discrete maps. Numerical simulations show that a periodically varying p-1 response may be maintained even when penetrating an autonomous period-doubling cascade, provided the variation is sufficiently rapid. We concentrate here on an analytic solution for such a sinusoidally varying response just in the neighbourhood of the first p-1 to p-2 bifurcation. We obtain uniformly valid series solutions, and the conditions which control whether or not the bifurcation takes place.

2. A Nonautonomous Unimodal Map

We consider a general one dimensional map, with sinusoidal variation of the bifurcation parameter:

F. C. Moon (ed.),
IUTAM Symposium on New Applications of Nonlinear and Chaotic Dynamics in Mechanics, 23–29.

$$y_{n+1} = f(y_n, x_n)$$
$$\text{with} \quad x_n = x_0 - \alpha \cos \epsilon n \qquad \text{where} \quad \epsilon = 2\pi/N \ll 1 \tag{1}$$

Baesens (1991) has studied the case of linear variation rather thoroughly. By following her general approach, but by using matched asymptotic expansions, Davies and Rangavajhula (1997) obtained simple explicit uniformly valid expressions for the p-1 and p-2 manifolds. They further used a triple-deck MAE to describe the transition of trajectories from the vicinity of the repelling p-1 manifold to the attracting p-2 manifold. We now use this approach for sinusoidal variation of the parameter x_n.

3. Period-1 Response

A Taylor series expansion of the map yields the purely p-1 outer expansion with sinusoidal variation of x_n.

$$y^o(x) = y^* \pm \epsilon Q\, y^{*\prime} / \left(\frac{\partial f}{\partial y}(y^*, x) - 1 \right) + 0(\epsilon^2) \tag{2}$$

where $Q = (\alpha^2 - (x - x_0)^2)^{1/2}$. Note that for most of our results it is easier to work with x rather than n as the independent variable. Away from critical points, y^o remains $0(\epsilon)$ close to the corresponding autonomous value $y^* = f(y^*, x)$. The expansion is singular when $\partial f/\partial y = 1$. If the x_n variation reaches this point, then the behaviour near the singularity (y_s, x_s) is examined by stretching the variables in the form $x = x_s + \epsilon^m \eta$, $y = y_s + \epsilon^n z(\eta)$.

We find the following forms of bifurcations:

(a) saddle node, with m = 1, n = 1/2

$$(2\alpha\eta)^{1/2} z' = \eta f_x + \frac{1}{2} z^2 f_{yy} \tag{3}$$

(Throughout this paper, subscripts indicate partial derivatives evaluated at the bifurcation point.)

(b) transcritical, with m = n = 2/3

$$(2\alpha\eta)^{1/2} z' = \eta z f_{xy} + \frac{1}{2} z^2 f_{yy} \tag{4}$$

(c) pitchfork, with m = 2/3, n = 1/3

$$(2\alpha\eta)^{1/2}\, z' = \eta z f_{xy} + \frac{1}{6} z^3 f_{yyy} \tag{5}$$

The $\eta^{1/2}$ term in each equation results from the sinusoidal variation of x. In each case, the inner solutions $y^i(\eta)$ can be matched to the outer solution (2) in the usual van Dyke (1964) manner. For example, the solution of equation (3) can be written in the form

$$z(\eta) = b(U' + kV') \;/\; (U + kV) \tag{6}$$

where $U(0, a\eta^{1/2})$ and $V(0, a\eta^{1/2})$ are parabolic cylinder functions (using the notation of Abramowitz and Stegun) where $a = (-4f_x f_{yy}/\alpha)^{1/4}$, $b = a(2\alpha)^{1/2}/f_{yy}$ and k is a constant to be determined by the matching process. In the usual way we expand (2) in terms of the inner variable η and expand (6) in terms of the outer variable x using two-term asymptotic expansions of U and V, to show $(y^o)^i = (y^i)^o$. The composite expansion $y^c = y^o + y^i - (y^o)^i$ is uniformly valid for $x \geq x_s$.

Composite expansions for the transcritical and pitchfork bifurcations are found in a similar way. The inner expansions can be expressed in terms of the integrals

$$I^{\pm}(\eta) = \exp(\mp\beta\eta^{3/2}) \int_0^{\eta} \exp(\pm\beta\zeta^{3/2})\zeta^{-1/2}\, d\zeta \tag{7}$$

where $\beta = \frac{2}{3} f_{xy}\, (2\alpha)^{-1/2}$.

4. Period-2 Response

A period-1 to period-2 bifurcation occurs when $\partial f/\partial y = -1$. Within the p-2 region, we work with the iterate of the map, F(y, x) = f(f(y, x), x). The p-2 expansion is similar to equation (2). Again stretched variables are used in the neighbourhood of the period-doubling bifurcation point. In this case we find, for purely p-2 motion, a pitchfork bifurcation, as expected, with m = 2/3 and n = 1/3. The inner solution satisfies

$$2(2\alpha\eta)^{1/2}\, z' = \eta z F_{xy} + \frac{1}{6} z^3 F_{yyy} \tag{8}$$

Solutions are obtained in terms of the integrals (7), and matched as before to the outer p-2 solution. Note that with this approach we obtain just one of the p-2 branches. The other is obtained in a similar way or by using equation (1).

5. Combined Period-1 to Period-2 Response

Sinusoidally varying trajectories as they sweep through the p-1 to p-2 bifurcation may, or may not, exhibit transitions to p-2. Whether, and where, a sometimes rapid transition occurs depends on the range of variation parameters x_0 and α, the sweep frequency ϵ, and also rather critically on the inherent noise level of the system. There are several cases to consider.

5.1. TRANSITION FROM WELL BELOW TO WELL ABOVE

The response in the neighbourhood of the bifurcation point is described in a boundary layer of $0(\epsilon^{1/2})$. There are two distinguished limits:

(a) an inner solution $y^{i}(\eta) = y_s + \epsilon^{1/4} z$ where z satisfies equation (8) but with the term $(2\alpha\eta)^{1/2}$ replaced by the constant $B = (\alpha^2 - (x_s - x_o)^2)^{1/2}$.

(b) an inner-inner solution $y^{ii} = y_s + \epsilon^{1/2} z_0 + \epsilon z_1 + 0(\epsilon^{3/2})$ where

$$2Bz_0^{\prime} = z_0 \eta F_{xy} + \frac{1}{2}\eta^2 F_{xx} + Bf_x \tag{9}$$

with similar equations for z_1, etc.

Apart from the addition of the constant B, this is the same as the linear case (Davies and Rangavajhula, 1997). The inner solution is written in terms of Dawson integrals (sweeping up) or error functions (sweeping down), and matches to the outer solution as before with $(y^0)^i = (y^i)^0$. Equation (9) has solution

$$z_0(\eta) = z_0(0)\exp(F_{xy}\eta^2/4B) + \frac{1}{2}\eta f_x \tag{10}$$

Now, the term $\frac{1}{2}\eta f_x$ is the $0(\epsilon^{1/2})$ description of the period-1 manifold near the p-1 to p-2 bifurcation. If we start (at $\eta = 0$) on the manifold, then y^{ii} generates the p-1 manifold. But noise or a perturbation of $0(\epsilon^n)$ at $\eta = 0$ grows away from the p-1 manifold in the form

$$y^{ii}(\eta) = y^{p1} + \epsilon^n \exp(F_{xy}\eta^2/4B) \tag{11}$$

y^{ii} then matches to y^i in an overlap region (Bender and Orzag) where the rapid transition p-1 to p-2 occurs.

Figure 1 shows numerical simulations of several cycles for the particular case of the logistic map $f(x, y) = x(y - y^2)$. The rapid transition occurs in our simulation near x =

3.25; the scatter indicates that random noise triggers the transition. The curves marked ϵ^2, ϵ^4, ϵ^6 are from equation (11). An imposed perturbation $\epsilon^2 \cong 1.5 * 10^5$ at x = 3 during one cycle shows excellent agreement.

To describe the sweeping down part of the cycle, only the outer and inner solutions are required. The matching imposes an "initial" condition at $x = x_s$. The value is marked on Figure 1.

5.2. TRANSITION TO JUST ABOVE X_S

The boundary layer for $x > x_s$ is $0(\epsilon^{2/3})$, with the upper limit $x = x_0 + \alpha$ within this boundary layer. Inner $0(\epsilon^{1/3})$, and inner-inner $0(\epsilon^{4/3})$ solutions are used to describe the response.

5.3. SINUSOIDAL VARIATION VERY CLOSE TO X_S

For trajectories within an $0(\epsilon)$ boundary layer about $x = x_s$ we write $x = x_s + \epsilon d_0$ and $\alpha = \epsilon d_1$. It is now more convenient to work with the original independent variable $\zeta = \epsilon\eta$. We find an inner-inner solution $y^{ii} = y_s + \epsilon z$ where

$$z(\zeta) = z(0)\exp(F_{xy}(d_0\zeta - d_1\sin\zeta)) + \frac{1}{2}f_x(d_0 - d_1\cos\zeta) \qquad \textbf{(12)}$$

The second term is the p-1 manifold. Now, given an arbitrary perturbation z(0), we have that y^{ii} either reduces to, or grows away from, the p-1 manifold depending only on the sign of d_0. An inner solution $y^i = y_s + \epsilon^{1/2}z$ satisfies a modified pitchfork equation.

$$z = z(d_0 - d_1\cos\zeta)F_{xy} + \frac{1}{2}z^3F_{yyy} \qquad \textbf{(13)}$$

Similar conclusions apply. To this order the autonomous bifurcation is maintained. Stable purely p-1 or purely p-2 response occurs depending only on whether $x_0 < x_s$ or $x_0 > x_s$ respectively. An example from the logistic map is shown in Figure 2.

6. Summary

Period-1 and period-2 trajectories described by a one-dimensional unimodal map have been analyzed as a control variable is varied sinusoidally. Noise is seen to be critical in determining whether, and where, sometimes rapid transition occurs from p-1 to p-2. The autonomous bifurcation is maintained for small $0(\epsilon)$ motion about the bifurcation point.

Work supported by the Natural Science and Engineering Research Council of Canada.

7. References

Abramowitz, M. and Stegun, I.A. (1968) *Handbook of Mathematical Functions*, National Bureau of Standards.

Baesens, C. (1991) Slow sweep through a period doubling cascade, *Physica* D 53, 319-375.

Bender, C.M. and Orzag, S.A. (1978) *Advanced mathematical methods for scientists and engineers*. New York: McGraw Hill.

Davies, H.G. and Rangavajhula, K. (1996) Nonstationary response near generic bifurcations. *Nonlinear Dynamics* 10, 235-250.

Davies, H.G. and Rangavajhula, K. (1997) Dynamic period-doubling bifurcations of a unimodal map. *Proc. Roy. Soc A*. (accepted for publication).

Haberman, R. (1979) Slowly varying jump and transition phenomena. *SIAM J. App. Matt.* 53, 1045-1058.

Raman, A., Bajaj, A.K. and Davies, P. (1996) On the slow transition across stabilities in nonlinar dissipative systems. *J. Sound Vib.* 192, 835-865.

van Dyke, M. (1964) *Perturbation methods in Fluid Dynamics*, Academic Press.

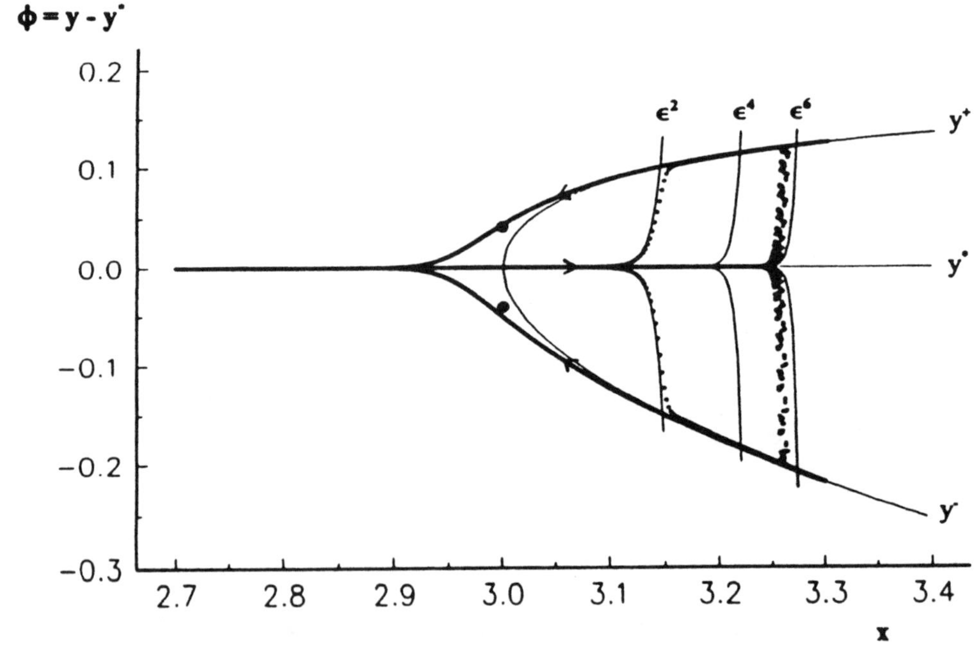

Figure 1. Numerical simulation of several cycles of the logistic map with $x_n = 3 - 0.3\cos(\pi n/800)$. The curves marked ϵ^2, ϵ^4, ϵ^6 are from equation (11). The rapid transition near $x = 3.15$ during one of the cyles is triggered by an imposed perturbation of magnitude ϵ^2 at $x = 3$.

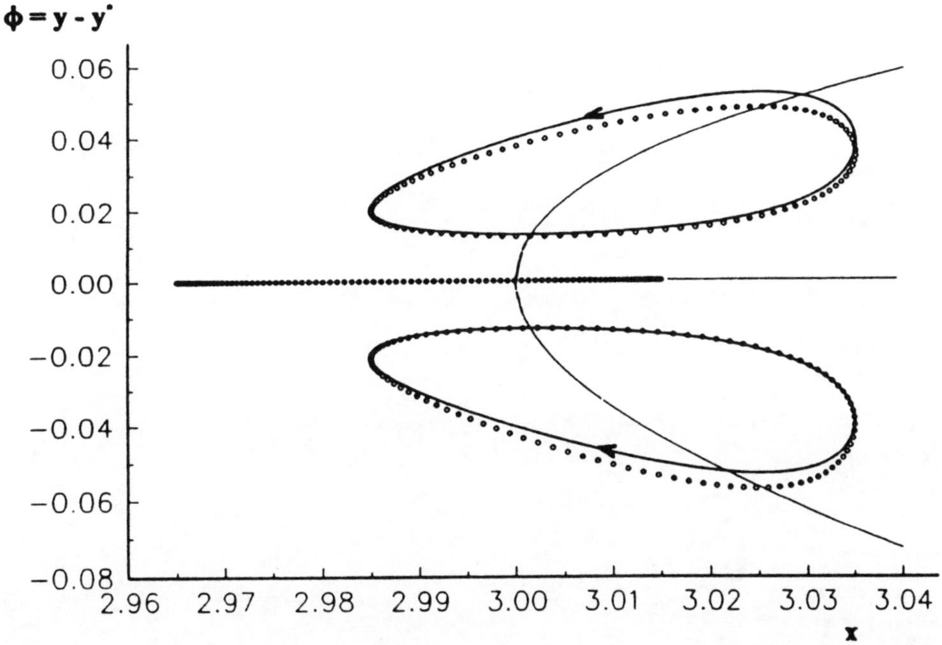

Figure 2. Numerical simulation of the logistic map for the two cases $x_n = 2.99 - 0.025 \cos(\pi n/100)$ and $x_n = 3.01 - 0.025 \cos(\pi n/100)$. Dots are numerical values. The lines are from the solution of equation (13).

WAVELET APPROACH TO MECHANICAL PROBLEMS. SYMPLECTIC GROUP, SYMPLECTIC TOPOLOGY AND SYMPLECTIC SCALES

A.N. FEDOROVA AND M.G. ZEITLIN
Institute of Problems of Mechanical Engineering,
Russian Academy of Sciences, Russia, 199178, St. Petersburg,
V.O., Bolshoj pr., 61, e-mail: zeitlin@math.ipme.ru

Abstract. We present the applications of methods from wavelet analysis to polynomial approximations for a number of nonlinear problems. According to the orbit method and by using approach from the geometric quantization theory we construct the symplectic and Poisson structures associated with generalized wavelets by using metaplectic structure. We consider wavelet approach to the calculations of Melnikov functions in the theory of homoclinic chaos in perturbed Hamiltonian systems, for parametrization of Arnold–Weinstein curves in Floer variational approach and characterization of symplectic Hilbert scales of spaces.

1. Introduction. In this paper we consider the application of powerful methods of wavelet analysis to polynomial approximations of nonlinear physical and mechanical problems. In the related paper in this volume we considered the general approach for constructing wavelet representation for solutions of nonlinear polynomial dynamical problems. But now we are interested in underlying Hamiltonian structures. Therefore, we need to consider generalized wavelets, which allow us to take into account the corresponding hidden symplectic, Poissonian, quasicomplex structures, instead of simple compactly supported wavelet representation. By using the orbit method and constructions from the geometric quantization theory we consider the symplectic and Poisson structures associated with Weyl–Heisenberg wavelets by using metaplectic structure and the corresponding polarization in §2–§6. In §7 we consider applications to construction of Mel-

F. C. Moon (ed.),
IUTAM Symposium on New Applications of Nonlinear and Chaotic Dynamics in Mechanics, 31–40.

nikov functions in the theory of homoclinic chaos in perturbed Hamiltonian systems. In §8 we consider the generalization of our variational wavelet approach [1]–[9] to the symplectic invariant calculation of Arnold-Weinstein curves (closed loops) in Hamiltonian systems. In §9 we consider wavelet characterization of symplectic Hilbert scales of spaces, which is important in problems related with KAM perturbations.

2. Metaplectic Group and Representations. In wavelet analysis the following three concepts are used now: 1). a square integrable representation U of a group G, 2). coherent states over G, 3). the wavelet transform associated to U.

We have three important particular cases: a) the affine $(ax + b)$ group, which yields the usual wavelet analysis: $[\pi(b, a)f](x) = 1/\sqrt{a}f\left((x - b)/a\right)$, b). the Weyl-Heisenberg group which leads to the Gabor functions, i.e. coherent states associated with windowed Fourier transform: $[\pi(q, p, \varphi)f](x) = \exp(i\mu(\varphi - p(x - q))f(x - q)$, in both cases time-frequency plane corresponds to the phase space of group representation. c). also, we have the case of bigger group, containing both affine and Weyl-Heisenberg group, which interpolate between affine wavelet analysis and windowed Fourier analysis: affine Weyl–Heisenberg group [10]. But usual representation of it is not square–integrable and must be modified: restriction of the representation to a suitable quotient space of the group (the associated phase space in that case) restores square – integrability: $G_{aWH\longrightarrow}$ homogeneous space. Also, we have more general approach which allows to consider wavelets corresponding to more general groups and representations [11], [12]. Our goal is applications of these results to problems of Hamiltonian dynamics and as consequence we need to take into account symplectic nature of our dynamical problem. Also, the symplectic and wavelet structures must be consistent (this must be resemble the symplectic or Lie-Poisson integrator theory). We use the point of view of geometric quantization theory (orbit method) instead of harmonic analysis. Because of this we can consider (a) – (c) analogously.

Let $Sp(n)$ be symplectic group, $Mp(n)$ be its unique two- fold covering – metaplectic group. Let V be a symplectic vector space with symplectic form (,), then $R \oplus V$ is nilpotent Lie algebra - Heisenberg algebra:

$$[R, V] = 0, \quad [v, w] = (v, w) \in R, \quad [V, V] = R.$$

$Sp(V)$ is a group of automorphisms of Heisenberg algebra.

Let N be a group with Lie algebra $R \oplus V$, i.e. Heisenberg group. By Stone– von Neumann theorem Heisenberg group has unique irreducible unitary representation in which $1 \mapsto i$. This representation is projective: $U_{g_1}U_{g_2} = c(g_1, g_2) \cdot U_{g_1 g_2}$, where c is a map: $Sp(V) \times Sp(V) \to S^1$, i.e. c is S^1-cocycle.

But this representation is unitary representation of universal covering, i.e. metaplectic group $Mp(V)$. We give this representation without Stone-von Neumann theorem. Consider a new group $F = N' \bowtie Mp(V)$, $\bowtie$ is semidirect product (we consider instead of $N = R \oplus V$ the $N' = S^1 \times V$, $S^1 = (R/2\pi Z)$). Let V^* be dual to V, $G(V^*)$ be automorphism group of V^*.Then F is subgroup of $G(V^*)$, which consists of elements, which acts on V^* by affine transformations.
This is the key point!

Let $q_1, ..., q_n; p_1, ..., p_n$ be symplectic basis in V, $\alpha = pdq = \sum p_i dq_i$ and $d\alpha$ be symplectic form on V^*. Let M be fixed affine polarization, then for $a \in F$ the map $a \mapsto \Theta_a$ gives unitary representation of G: $\Theta_a : H(M) \to H(M)$

Explicitly we have for representation of N on H(M):

$$(\Theta_q f)^*(x) = e^{-iqx} f(x), \quad \Theta_p f(x) = f(x - p)$$

The representation of N on H(M) is irreducible. Let A_q, A_p be infinitesimal operators of this representation

$$A_q = \lim_{t \to 0} \frac{1}{t}[\Theta_{-tq} - I], \quad A_p = \lim_{t \to 0} \frac{1}{t}[\Theta_{-tp} - I],$$

$$\text{then} \qquad A_q f(x) = i(qx) f(x), \quad A_p f(x) = \sum p_j \frac{\partial f}{\partial x_j}(x)$$

Now we give the representation of infinitesimal basic elements. Lie algebra of the group F is the algebra of all (nonhomogeneous) quadratic polynomials of (p,q) relatively Poisson bracket (PB). The basis of this algebra consists of elements $1, q_1, ..., q_n, p_1, ..., p_n, q_i q_j, q_i p_j, p_i p_j, \quad i, j = 1, ..., n, \quad i \le j,$

$$PB \; is \quad \{f, g\} = \sum \frac{\partial f}{\partial p_j}\frac{\partial g}{\partial q_i} - \frac{\partial f}{\partial q_i}\frac{\partial g}{\partial p_i} \quad \text{and} \quad \{1, g\} = 0 \quad for \text{ all } g,$$

$$\{p_i, q_j\} = \delta_{ij}, \quad \{p_i q_j, q_k\} = \delta_{ik} q_j, \quad \{p_i q_j, p_k\} = -\delta_{jk} p_i,$$

$$\{p_i p_j, p_k\} = 0, \quad \{p_i p_j, q_k\} = \delta_{ik} p_j + \delta_{jk} p_i, \quad \{q_i q_j, q_k\} = 0,$$

$$\{q_i q_j, p_k\} = -\delta_{ik} q_j - \delta_{jk} q_i$$

so, we have the representation of basic elements $f \mapsto A_f : 1 \mapsto i, q_k \mapsto ix_k$,

$$p_l \mapsto \frac{\delta}{\delta x^l}, p_i q_j \mapsto x^i \frac{\partial}{\partial x^j} + \frac{1}{2}\delta_{ij}, \qquad p_k p_l \mapsto \frac{1}{i}\frac{\partial^k}{\partial x^k \partial x^l}, q_k q_l \mapsto i x^k x^l$$

This gives the structure of the Poisson manifolds to representation of any (nilpotent) algebra or in other words to continuous wavelet transform.

3. The Segal-Bargman Representation. Let $z = 1/\sqrt{2} \cdot (p - iq)$, $\bar{z} = 1/\sqrt{2} \cdot (p + iq)$, $p = (p_1, ..., p_n)$, F_n is the space of holomorphic functions of n complex variables with $(f, f) < \infty$, where

$$(f, g) = (2\pi)^{-n} \int f(z)\overline{g(z)} e^{-|z|^2} dp dq$$

Consider a map $U : H \to F_n$, where H is with real polarization, F_n is with complex polarization, then we have

$$(U\Psi)(a) = \int A(a, q)\Psi(q) dq, \qquad \text{where} \quad A(a, q) = \pi^{-n/4} e^{-1/2(a^2+q^2)+\sqrt{2}aq}$$

i.e. the Bargmann formula produce wavelets.We also have the representation of Heisenberg algebra on F_n :

$$U \frac{\partial}{\partial q_j} U^{-1} = \frac{1}{\sqrt{2}} \left(z_j - \frac{\partial}{\partial z_j} \right), \qquad U q_j U^{-1} = -\frac{i}{\sqrt{2}} \left(z_j + \frac{\partial}{\partial z_j} \right)$$

and also : $\omega = d\beta = dp \wedge dq$, where $\beta = i\bar{z}dz$.

4. Orbital Theory for Wavelets. Let coadjoint action be $< g \cdot f, Y > = < f, Ad(g)^{-1} Y >$, where $<,>$ is pairing $g \in G$, $f \in g^*$, $Y \in \mathcal{G}$. The orbit is $\mathcal{O}_f = G \cdot f \equiv G/G(f)$. Also, let A=A(M) be algebra of functions, V(M) is A-module of vector fields, A^p is A-module of p-forms. Vector fields on orbit is

$$\sigma(\mathcal{O}, X)_f(\phi) = \frac{d}{dt}(\phi(\exp tXf))\Big|_{t=0}$$

where $\phi \in A(\mathcal{O})$, $f \in \mathcal{O}$. Then $\mathcal{O}_f$ are homogeneous symplectic manifolds with 2-form $\Omega(\sigma(\mathcal{O}, X)_f, \sigma(\mathcal{O}, Y)_f) = < f, [X, Y] >$, and $d\Omega = 0$. PB on $\mathcal{O}$ have the next form $\{\Psi_1, \Psi_2\} = p(\Psi_1)\Psi_2$ where p is $A^1(\mathcal{O}) \to V(\mathcal{O})$ with definition $\Omega(p(\alpha), X) = i(X)\alpha$. Here $\Psi_1, \Psi_2 \in A(\mathcal{O})$ and $A(\mathcal{O})$ is Lie algebra with bracket $\{,\}$. Now let N be a Heisenberg group. Consider adjoint and coadjoint representations in some particular case. $N = (z, t) \in C \times R$, $z = p + iq$; compositions in N are $(z, t) \cdot (z', t') = (z + z', t + t' + B(z, z'))$, where $B(z, z') = pq' - qp'$. Inverse element is $(-t, -z)$. Lie algebra n of N is $(\zeta, \tau) \in C \times R$ with bracket $[(\zeta, \tau), (\zeta', \tau')] = (0, B(\zeta, \zeta'))$. Centre is $\tilde{z} \in n$ and generated by (0,1); Z is a subgroup $\exp \tilde{z}$. Adjoint representation N on n is given by formula $Ad(z, t)(\zeta, \tau) = (\zeta, \tau + B(z, \zeta))$ Coadjoint: for $f \in n^*$, $g = (z, t)$, $(g \cdot f)(\zeta, \zeta) = f(\zeta, \tau) - B(z, \zeta) f(0, 1)$ then orbits for which $f|_{\bar{z}} \neq 0$ are plane in n^* given by equation $f(0,1) = \mu$. If $X = (\zeta, 0)$, $Y = (\zeta', 0)$, $X, Y \in n$ then symplectic structure is

$$\Omega(\sigma(\mathcal{O}, X)_f, \sigma(\mathcal{O}, Y)_f) =< f, [X, Y] >= f(0, B(\zeta, \zeta'))\mu B(\zeta, \zeta')$$

Also we have for orbit $\mathcal{O}_\mu = N/Z$ and $\mathcal{O}_\mu$ is Hamiltonian G-space.

5. Kirillov Character Formula or Analogy of Gabor Wavelets.

Let U denote irreducible unitary representation of N with condition $U(0,t) = \exp(it\ell) \cdot 1$, where $\ell \neq 0$,then U is equivalent to representation T_ℓ which acts in $L^2(R)$ according to

$$T_\ell(z,t)\phi(x) = \exp\left(i\ell(t + px)\right)\phi(x - q)$$

If instead of N we consider E(2)/R we have S^1 case and we have Gabor functions on S^1.

6. Oscillator Group. Let O be an oscillator group, i.e. semidirect product of R and Heisenberg group N. Let H,P,Q,I be standard basis in Lie algebra o of the group O and H^*, P^*, Q^*, I^* be dual basis in o^*. Let functional f=(a,b,c,d) be $aI^* + bP^* + cQ^* + dH^*$. Let us consider complex polarizations $h = (H, I, P + iQ)$, $\bar{h} = (I, H, P - iQ)$ Induced from h representation, corresponding to functional f (for $a > 0$), unitary equivalent to the representation

$$W(t,n)f(y) = \exp(it(h - 1/2)) \cdot U_a(n)V(t),$$

$$\text{where} \qquad V(t) = \exp[-it(P^2 + Q^2)/2a], \quad P = -d/dx, \quad Q = iax,$$

and $U_a(n)$ is irreducible representation of N, which have the form $U_a(z) = exp(iaz)$ on the center of N. Here we have: U(n=(x,y,z)) is Schrödinger representation, $U_t(n) = U(t(n))$ is the representation obtained from previous by automorphism (time translation) $n \longrightarrow t(n)$; $U_t(n) = U(t(n))$ is also unitary irreducible representation of N. $V(t) = \exp(it(P^2 + Q^2 + h - 1/2))$ is an operator, which according to Stone–von Neumann theorem has the property $U_t(n) = V(t)U(n)V(t)^{-1}$.

This is our last private case, but according to our approach we can construct by using methods of geometric quantization theory many "symplectic wavelet constructions" with corresponding symplectic or Poisson structure on it. Very useful particular spline–wavelet basis with uniform exponential control on stratified and nilpotent Lie groups was considered in [12]. In particular case of Heisenberg group $N \ni (p, q, t)$ we may use as bases in our space of representations $L^2(N)$ the following orthonomal bases [12] $\psi_{i,j,k}$

generated from a finite set of 15 regular, localized and oscillating functions $\psi_i : \psi_{i,j,k} = 4^j\psi_i(2^j p - k_1, 2^j q - k_2, 2^j t - k_3 + 2^j k_2 p - 2^j k_1 q)$.

7. Melnikov Functions Approach. We give now some points of applications of wavelet methods from the preceding consideration to Melnikov approach in the theory of homoclinic chaos in perturbed Hamiltonian systems. We consider some points of our program of understanding routes to chaos in Hamiltonian systems in the wavelet approach [1]-[9]. All points are: 1. A model. 2. A computer zoo. The understanding of the computer zoo. 3. A naive Melnikov function approach. 4. A naive wavelet description of (hetero) homoclinic orbits (separatrix) and quasiperiodic oscillations. 5. Symplectic Melnikov function approach. 6. Splitting of separatrix... ⟶stochastic web with magic symmetry, Arnold diffusion and all that.

1. As a model we have two frequencies perturbations of two-mode Galerkin approximations to beam oscillations in liquid [7]:

$$
\begin{aligned}
\dot{x}_1 &= x_2, \qquad \dot{x}_3 = x_4, \qquad \dot{x}_5 = 1, \qquad \dot{x}_6 = 1,\\
\dot{x}_2 &= -ax_1 - b[\cos(rx_5) + \cos(sx_6)]x_1 - dx_1^3 - mdx_1x_3^2 - px_2 - \varphi(x_5)\\
\dot{x}_4 &= ex_3 - f[\cos(rx_5) + \cos(sx_6)]x_3 - gx_3^3 - kx_1^2x_3 - gx_4 - \psi(x_5)
\end{aligned}
$$

or in Hamiltonian form

$$
\dot{x} = J\cdot\nabla H(x) + \varepsilon g(x,\Theta),\ \dot{\Theta} \ = \ \omega,\ (x,\Theta)\in R^4\times T^2,\ T^2 = S^1\times S^1,
$$

for $\varepsilon = 0$ we have:

$$
\dot{x} = J\cdot\nabla H(x), \qquad \dot{\Theta} = \omega \tag{1}
$$

2. For pictures and details one can see [4], [7]. The key point is the splitting of separatrix (homoclinic orbit) and transition to fractal sets on the Poincare sections.

3. For $\varepsilon = 0$ we have homoclinic orbit $\bar{x}_0(t)$ to the hyperbolic fixed point x_0. For $\varepsilon \neq 0$ we have normally hyperbolic invariant torus T_ε and condition on transversally intersection of stable and unstable manifolds $W^s(T_\varepsilon)$ and $W^u(T_\varepsilon)$ in terms of Melnikov functions $M(\Theta)$ for $\bar{x}_0(t)$. $M(\Theta) = \int_{-\infty}^{\infty} \nabla H(\bar{x}_0(t))\wedge g(\bar{x}_0(t),\omega t+\Theta)dt$. This condition has the next form:

$M(\Theta_0) = 0,\quad \sum_{j=1}^{2}\omega_j\frac{\partial}{\partial\Theta_j}M(\Theta_0)\neq 0$. According to the approach of Birkhoff-Smale-Wiggins we determined the region in parameter space in which we observe the chaotic behaviour [7].

4. If we cannot solve equations (1) explicitly in time, then we use the

wavelet approach from our other paper for the computations of homoclinic (heteroclinic) loops as the wavelet solutions of system (1). For computations of quasiperiodic Melnikov functions: $M^{m/n}(t_0) = \int_0^{mT} DH(x_\alpha(t)) \wedge g(x_\alpha(t), t+t_0)dt$, we used periodization of wavelet solution [3], [4].

5. We also used symplectic Melnikov function approach:

$M_i(z) = \lim_{j\to\infty} \int\limits_{-T_j^*}^{T_j} \{h_i, \hat{h}\}_{\Psi(t,z)}dt, d_i(z,\varepsilon) = h_i(z_\varepsilon^u) - h_i(z_\varepsilon^s) = \varepsilon M_i(z) + O(\varepsilon^2)$, where $\{,\}$ is the Poisson bracket, $d_i(z,\varepsilon)$ is the Melnikov distance. So, we need symplectic invariant wavelet expressions for Poisson brackets. The computations are produced according to §2–§6.

6. Some hypothesis about strange symmetry of stochastic web in multi-degree-of freedom Hamiltonian systems [8].

8. Symplectic Topology and Wavelets. Now we consider the generalization of our wavelet variational approach to symplectic invariant calculation of closed loops in Hamiltonian systems [13].

We also have the parametrization of our solution by some reduced algebraical problem but in contrast to the general case where the solution is parametrized by construction based on scalar refinement equation, in symplectic case we have parametrization of the solution by matrix problems – Quadratic Mirror Filters equations [14].

The action functional for loops in the phase space is [13]

$$F(\gamma) = \int_\gamma pdq - \int_0^1 H(t,\gamma(t))dt$$

The critical points of F are those loops γ, which solve the Hamiltonian equations associated with the Hamiltonian H and hence are periodic orbits. By the way, all critical points of F are the saddle points of infinite Morse index, but surprisingly this approach is very effective. This will be demonstrated using several variational techniques starting from minimax due to Rabinowitz and ending with Floer homology. So, (M,ω) is symplectic manifolds, $H: M \to R$, H is Hamiltonian, X_H is unique Hamiltonian vector field defined by $\omega(X_H(x), v) = -dH(x)(v), \quad v \in T_xM, \quad x \in M$, where ω is the symplectic structure. A T-periodic solution $x(t)$ of the Hamiltonian equations: $\dot{x} = X_H(x)$ on M is a solution, satisfying the boundary conditions $x(T) = x(0), T > 0$. Let us consider the loop space $\Omega = C^\infty(S^1, R^{2n})$, where $S^1 = R/\mathbf{Z}$, of smooth loops in R^{2n}. Let us define a function $\Phi : \Omega \to R$ by setting

$$\Phi(x) = \int_0^1 \frac{1}{2} < -J\dot{x}, x > dt - \int_0^1 H(x(t))dt, \quad x \in \Omega$$

The critical points of Φ are the periodic solutions of $\dot{x} = X_H(x)$. Computing the derivative at $x \in \Omega$ in the direction of $y \in \Omega$, we find

$$\Phi'(x)(y) = \frac{d}{d\epsilon}\Phi(x + \epsilon y)|_{\epsilon=0} = \int_0^1 < -J\dot{x} - \nabla H(x), y > dt$$

Consequently, $\Phi'(x)(y) = 0$ for all $y \in \Omega$ iff the loop x satisfies the equation

$$-J\dot{x}(t) - \nabla H(x(t)) = 0,$$

i.e. $x(t)$ is a solution of the Hamiltonian equations, which also satisfies $x(0) = x(1)$, i.e. periodic of period 1. Periodic loops may be represented by their Fourier series: $x(t) = \sum_{k \in \mathbf{Z}} e^{k2\pi Jt} x_k, \quad x_k \in R^{2k}$, where J is quasicomplex structure. We give relations between quasicomplex structure and wavelets in §9. But now we use the construction [14] for loop parametrization. It is based on the theorem about explicit bijection between the Quadratic Mirror Filters (QMF) and the whole loop group: $LG : S^1 \to G$. In particular case we have relation between **QMF**-systems and measurable functions $\chi : S^1 \to U(2)$ satisfying

$$\chi(\omega + \pi) = \chi(\omega) \begin{bmatrix} 0 & 1 \\ 1 & 0 \end{bmatrix},$$

in the next explicit form

$$\begin{bmatrix} \hat{\Phi}_0(\omega) & \hat{\Phi}_0(\omega + \pi) \\ \hat{\Phi}_1(\omega) & \hat{\Phi}_1(\omega + \pi) \end{bmatrix} = \chi(\omega) \begin{bmatrix} 0 & 1 \\ 1 & 0 \end{bmatrix} + \chi(\omega + \pi) \begin{bmatrix} 0 & 0 \\ 0 & 1 \end{bmatrix},$$

where $\left|\hat{\Phi}_i(\omega)\right|^2 + \left|\hat{\Phi}_i(\omega + \pi)\right|^2 = 2, \quad i = 0, 1.$

Also, we have symplectic structure on LG

$$\omega(\xi, \eta) = \frac{1}{2\pi} \int_0^{2\pi} < \xi(\theta), \eta'(\theta) > d\theta$$

So, we have the parametrization of periodic orbits (Arnold–Weinstein curves) by reduced QMF equations.

9. Symplectic Hilbert Scales via Wavelets. We can solve many important dynamical problems such that KAM perturbations, spread of energy to higher modes, week turbulence, growths of solutions of Hamiltonian equations only if we consider scales of spaces instead of one functional space. For Hamiltonian system and their perturbations for which we need take into account underlying symplectic structure we need to consider symplectic scales of spaces. So, if $\dot{u}(t) = J\nabla K(u(t))$ is Hamiltonian equation we need wavelet description of symplectic or quasicomplex structure on the level of functional spaces. It is very important that according to [16] Hilbert basis is in the same time a Darboux basis to corresponding symplectic structure. We need to provide Hilbert scale $\{Z_s\}$ with symplectic structure [15], [17]. All what we need is the following. J is a linear operator, $J: Z_\infty \to Z_\infty$, $J(Z_\infty) = Z_\infty$, where $Z_\infty = \cap Z_s$. J determines an isomorphism of scale $\{Z_s\}$ of order $d_J \geq 0$. The operator J with domain of definition Z_∞ is antisymmetric in Z: $< Jz_1, z_2 >_Z = - < z_1, Jz_2 >_Z, z_1, z_2 \in Z_\infty$. Then the triple $\{Z, \{Z_s | s \in R\}, \quad \alpha =< \bar{J}dz, dz >\}$ is symplectic Hilbert scale. So, we may consider any dynamical Hamiltonian problem on functional level. As an example, for KdV equation we have $Z_s = \{u(x) \in H^s(T^1) | \int_0^{2\pi} u(x)dx = 0\}$, $s \in R$, $\quad J = \partial/\partial x$, J is isomorphism of the scale of order one, $\bar{J} = -(J)^{-1}$ is isomorphism of order -1. According to [18] general functional spaces and scales of spaces such as Holder–Zygmund, Triebel–Lizorkin and Sobolev can be characterized through wavelet coefficients or wavelet transforms. As a rule, the faster the wavelet coefficients decay, the more the analyzed function is regular [18]. Most important for us example is the scale of Sobolev spaces. Let $H_k(R^n)$ is the Hilbert space of all distributions with finite norm $\|s\|^2_{H_k(R^n)} = \int d\xi (1 + |\xi|^2)^{k/2} |\hat{s}(\xi)|^2$. Let us consider wavelet transform

$$W_g f(b,a) = \int_{R^n} dx \frac{1}{a^n} \bar{g}(\frac{x-b}{a}) f(x), \quad b \in R^n, \quad a > 0$$

w.r.t. analyzing wavelet g, which is strictly admissible, i.e.

$$C_{g,g} = \int_0^\infty \frac{da}{a} |\hat{g}(\bar{ak})|^2 < \infty.$$

Then there is a $c \geq 1$ such that

$$c^{-1} \|s\|^2_{H_k(R^n)} \leq \int_{H^n} \frac{dbda}{a} (1 + a^{-2\gamma}) |W_g s(b,a)|^2 \leq c \|s\|^2_{H_k(R^n)}$$

This shows that localization of the wavelet coefficients at small scale is linked to local regularity.

Extended version and related results may be found in [1]-[9].

We would like to thank Zohreh Parsa (BNL) for many discussions and continued encouragement during and after workshop "New Ideas for Particle Accelerators" and Institute for Theoretical Physics, University of California, Santa Barbara for hospitality.

This research was supported in part under "New Ideas for Particle Accelerators Program" NSF- Grant No. PHY94-07194.

References

1. Fedorova A.N., Zeitlin M.G. (1997) Nonlinear dynamics of accelerator via wavelet approach, in New Ideas in Accelerator Physics, AIP, in press.
2. Fedorova A.N., Zeitlin M.G., Parsa Z. (1997) Wavelet approach to accelerator problems, parts 1-3, Proc. Particle Accelerator Conf., Vancouver, AIP, in press.
3. Fedorova, A.N., and Zeitlin, M.G. (1996) Wavelet approach to polynomial dynamics, Proc. of 23 Summer School 'Nonlinear Oscillations in Mechanical Systems' St. Petersburg, pp. 322–333.
4. Fedorova, A.N., and Zeitlin, M.G. (1996) Wavelets in dynamics, optimal control and Galerkin approximations, Proc. 7th IEEE DSP Workshop, Norway, pp. 409–412.
5. Fedorova, A.N., and Zeitlin, M.G. (1997) Wavelets in optimization and approximations, Proc. 2nd IMACS Symp. on Math. Modelling, ARGESIM Report **11**, Austria, pp. 1083–1088; Math. Comp. in Simulation, in press.
6. Fedorova, A.N., and Zeitlin, M.G. (1996) Routes to chaos in perturbed Hamiltonian systems, Proc. EUROMECH-2nd European Nonlinear Oscillations Conf., pp. 79–82.
7. Fedorova, A.N., and Zeitlin, M.G. (1996) Nonlinear oscillation of a beam, Proc. EUROMECH-2nd European Nonlinear Oscillations Conf. pp. 153–156.
8. Fedorova A.N., Zeitlin M.G. (1997) Wavelet approach to nonlinear problems, parts 1-4, ZAMM, in press.
9. Fedorova A.N., Zeitlin M.G. (1997) Nonlinear optimal control problem via wavelet approach, Proc. of European Control Conf., Brussels, vol. 5, pp. 1–6.
10. Kalisa, C., and Torresani, B. (1992) *N-dimensional Affine Weyl–Heisenberg Wavelets*: preprint CPT-92 P.2811 Marseille.
11. Kawazoe, T. (1995) Wavelet transform associated to principal series representation, Proc. Japan Acad. **71** Ser. A, pp. 154–160.
12. Lemarie P.G. (1991) Wavelets, spline interpolation and Lie groups, Proc. Int. Math. Congr., Satellite Symp., pp. 154–164.
13. Hofer E., Zehnder E. (1994) *Symplectic Topology*, Birkhauser.
14. Holschneider M., Pinkall U. (1994) Quadrature mirror filters and loop groups, CPT-94 P.3017, Marseille.
15. Kuksin, S. (1993)*Nearly integrable Hamiltonian systems*, Springer.
16. Kuksin, S. (1995) Infinite-dimensional symplectic capacities, Comm.Math.Phys. **167**, pp. 531–552.
17. Bourgain, J. (1996) On the growth in time of Sobolev norms, IMRN, **6** pp. 277–304.
18. Holschneider, M. (1996) Wavelet analysis of PDO, CPT-96/P3344, Marseille.

ON THE GENERALIZED CELL MAPPING

R. GUDER, E. KREUZER
Ocean Engineering II - Mechanics
Technical University of Hamburg-Harburg
Eissendorfer Str. 42
D-21073 Hamburg
Germany

1. Introduction

Nonlinear dynamical systems are characterized by a variety of behavior ranging from periodic, quasiperiodic to chaotic motions. Different kinds of motion may even coexist for the same values of system's parameters. The coexistence of different long-term behavior of dissipative engineering systems can cause practical problems because it depends on the initial conditions which attractor will eventually determine the motion. Moreover, state disturbances may result in a transition from a prescribed motion to a different neighboring solution if the domain of attraction is small. A technical system should, however, be robust against such disturbances. Therefore, it is of practical importance to know both the different types of long-term behavior and their domains of attraction. In recent years a lot of work has been devoted in order to understand and characterize nonlinear dynamical systems in this respect.

For determining the attractors and their domains of attraction numerically within a practically relevant range of state variables as well as to analyze the behavior on the attractor and the transient motion, the generalized cell mapping theory (Hsu, 1980), (Hsu, 1987) proved to be very helpful, see (Hsu, 1992), (Kreuzer, 1987).

Based upon a Poincaré mapping, the generalized cell mapping has been developed for the global analysis of nonlinear dynamical systems. The mathematical description of an autonomous generalized cell mapping leads to

Dedicated to Professor Dr. techn. Dr. h.c. Franz Ziegler on the occasion of his 60th birthday.

F. C. Moon (ed.),
IUTAM Symposium on New Applications of Nonlinear and Chaotic Dynamics in Mechanics, 41–50.

a finite, discrete, stationary Markov chain where the transition probability matrix completely determines the behavior of the system. In the next section we will briefly introduce the generalized cell mapping. For more details on this method we refer to (Hsu, 1987) and (Kreuzer, 1987). A problem of the generalized cell mapping is that in general it can only be approximated. In section 3 we present an adaptive method for the approximation of the generalized cell mapping where the adaption is controlled by an adjustable error. Specially for higher dimensional systems the computational effort for the generalized cell mapping is enormous. Therefore, we introduce in section 4 an adaptive refinement algorithm which is controlled by the major dynamics of the system. This algorithm starts on a very rough cell space and refines only those cells where the long-term dynamics of the system take place.

2. The generalized cell mapping

This section supplies some motivation and a brief survey on basic concepts of the generalized cell mapping. Since cell mapping theory is not a widespread tool in applied mathematics or is used under other names, the following introduction is intended to summarize the essential ingredients of the generalized cell mapping.

2.1. BASIC CONCEPTS AND THE THEORY OF PROBABILITIES

The evolution equations of dynamical systems may be described by the Poincaré map

$$\mathbf{x}(n+1) = \boldsymbol{\tau}(\mathbf{x}(n)), \ \boldsymbol{\tau} : \mathbb{R}^p \to \mathbb{R}^p, \ \mathbf{x} \in \mathbb{R}^p, \ n \in \mathbb{N}, \tag{1}$$

where one should think of n as the discrete time. For dissipative systems, the flow of the system contracts the volume. We shall deal exclusively with dissipative systems.

Due to limitations on the representation of numbers in the computer, however, it is not really possible in computer simulations to treat the state space as a continuum of points. Considering this effect, Hsu (Hsu, 1980) developed the concept of the cell mapping based on a coarse discretization of the state space.

After choosing a domain of interest $\Omega \subset \mathbb{R}^p$ in the state space, we partition this domain in N sets $J_i \subset \Omega$. The sets J_i are the so-called regular cells. The region $J_0 = \mathbb{R}^p \backslash \Omega$ outside the domain of interest constitutes one single cell, the so-called sink cell or 0th cell. Regular cells and the sink cell form the cell space $S = \{0, 1, ..., N\}$. With this procedure the continuous state space is replaced by a discrete cell space.

By defining a discrete state or random variable ξ

$$\xi(n) = i \in S \Leftrightarrow \mathbf{x}(n) \in J_i, \tag{2}$$

the state of the dynamical system at time $t = n$ is no longer described by a point $\mathbf{x}(n)$ in the state space, but by the probability $\zeta_i(n)$ that the point $\mathbf{x}(n)$ lives in cell J_i at time n:

$$\zeta_i(n) = W[\xi(n) = i], \; i \in S. \tag{3}$$

Writing the probabilities $\zeta_i(n)$ as the cell probability vector $\boldsymbol{\zeta}(n) = [\zeta_1(n), \zeta_2(n), ..., \zeta_N(n)]^t$, the evolution of the system is described by:

$$\boldsymbol{\zeta}(n+1) = \mathbf{P}_N \boldsymbol{\zeta}(n), \; n \in \mathbb{N}, \tag{4}$$

where $\mathbf{P}_N$ is the matrix of the conditional transition probabilities between the cells, that is

$$p_{ij} = [\mathbf{P}_N]_{ij} = W[\xi(n+1) = i | \; \xi(n) = j], \; i, j \in S. \tag{5}$$

The matrix $\mathbf{P}_N$ is called the transition probability matrix of the generalized cell mapping. Clearly, $\mathbf{P}_N$ is a stochastic matrix and (4) defines a finite, discrete, stationary Markov chain on the cell space S.

2.2. THE GENERALIZED CELL MAPPING AND MEASURE THEORY

In the next section we will present an algorithm to approximate the stochastic matrix $\mathbf{P}_N$ for a given cell-partition of Ω. For this it is useful to recall that the transition probabilities p_{ij} can be described in terms of the mapping $\boldsymbol{\tau}$ by a discretization of the *Frobenius-Perron operator* $\mathcal{P}$ (Li, 1976). This operator $\mathcal{P} : L^1(\Omega) \to L^1(\Omega)$ is implicitly defined by

$$\int_B (\mathcal{P}h) \, dm = \int_{\boldsymbol{\tau}^{-1}(B)} h \, dm \quad \text{for all measurable } B \subset \Omega$$

where m denotes the Lebesgue measure, that is, $m(A) = \int_A 1 \, dx$. Using piecewise constant functions (based on the cell-structure) for the discretization of $\mathcal{P}$ it is not difficult to see that the p_{ij}'s are given by

$$p_{ij} = \frac{m(\boldsymbol{\tau}^{-1}(J_i) \cap J_j)}{m(J_j)}. \tag{6}$$

For more details the reader is referred to (Lasota and Mackey, 1994).

3. An Adaptive Method for the Approximation of the Generalized Cell Mapping

In general the computation of the corresponding transition probabilities p_{ij} in (6) cannot be done analytically. In fact, as soon as $\boldsymbol{\tau}$ is a mapping on $\mathbb{R}^p$ with $p \geq 2$ the set $\boldsymbol{\tau}^{-1}(J_i) \cap J_j$ could in principle have a very complicated geometric shape. Therefore, we have to compute the p_{ij}'s numerically. A new developed algorithm allows to approximate these numbers in an adaptive way.

We will compare the efficiency of our method with Monte-Carlo approaches. These approaches have the major advantage that they are very easy to implement, e.g. (Hunt, 1993). All Monte-Carlo approaches use single points to approximate sets, and correspondingly they have the disadvantage that no *deterministic* a priori error estimate is available. In (Guder et al., 1996) we overcome this problem in two steps by using an exhaustion technique: first we determine a criterion which guarantees that the image of a subset I of J_j is completely contained in some J_i, that is $\boldsymbol{\tau}(I) \in J_i$. Then we proceed recursively to construct an adaptive algorithm. For details of the adaptive algorithm we refer to (Guder et al., 1996).

By means of an example the efficiency of our algorithm is demonstrated next. Moreover, we compare its performance with the Monte-Carlo method described in (Hunt, 1993).

We consider the two-dimensional mapping $\boldsymbol{\tau} : [0,1]^2 \to [0,1]^2$,

$$\boldsymbol{\tau}(x,y) = (\tau_s(x), \tau_s(y)),$$

where $\tau_s : [0,1] \to [0,1]$ is the so-called *spike mapping* (Hunt, 1993)

$$\tau_s(z) = \begin{cases} \frac{1}{\sqrt{2}} - \sqrt{2}|\frac{1}{2} - z| & \forall z \in [0, \frac{1}{\sqrt{8}}] \cup [1 - \frac{1}{\sqrt{8}}, 1], \\ 1 - \frac{1}{\sqrt{2}}\sqrt{1 - (1 - |1 - 2z|)^2} & \forall z \in [\frac{1}{\sqrt{8}}, 1 - \frac{1}{\sqrt{8}}]. \end{cases} \tag{7}$$

The two-dimensional mapping $\boldsymbol{\tau}$ has an absolutely continuous invariant measure with the density $f : [0,1]^2 \to \mathbb{R}_+$,

$$f(x,y) = (2 - |2 - 4x|)(2 - |2 - 4y|). \tag{8}$$

The exhaustion technique is used to compute an approximation of the transition probabilities p_{ij}. Hence, a first choice for a comparison of the different numerical methods would be to compare the approximated transition probabilities. However, we directly compare the L^1-error made in the computation of the stationary density of the generalized cell mapping, since this is a significant quantity of the dynamics of the system. Moreover, typically these errors are much bigger than the errors in the entries of the matrices.

We denote the approximation of the densities obtained via the exhaustion technique and the Monte-Carlo method by f_{adap} and f_{MC} respectively. In Table 1 the following characterizing quantities for the description of the numerical results are used:

♯cells	The total number of cells in each coordinate direction. All computations are based on uniform cells.
δ_{adap} [%]	The relative error $\frac{\|f-f_{adap}\|_{L^1}}{\|f\|_{L^1}}$ in percent.
$\sharp points_{adap}$	The total number of $\boldsymbol{\tau}$-evaluations needed by the exhaustion algorithm.
δ_{MC} [%]	The relative error $\frac{\|f-f_{MC}\|_{L^1}}{\|f\|_{L^1}}$ in percent.
$\sharp points_{MC}$	The total number of $\boldsymbol{\tau}$-evaluations needed by the Monte-Carlo method.

TABLE 1. Comparison of numerical results obtained by the adaptive algorithm and a Monte-Carlo approach.

♯cells	δ_{adap} [%]	$\sharp points_{adap}$	δ_{MC} [%]	$\sharp points_{MC}$
16 × 16	10.8	51	19.33	110
16 × 16	0.52	127	9.46	5,000
16 × 16	0.03	298	7.73	10,000
32 × 32	16.3	51	21.0	80
32 × 32	1.58	124	9.14	4,000
32 × 32	0.05	298	6.95	10,000

The results illustrate that for this example our method is much more efficient than the numerical technique based on the Monte-Carlo approach. Taking into account that in applications the number of $\boldsymbol{\tau}$-evaluations is the most time-consuming part in the computations, the numbers in Table 1 are particularly convincing.

4. Adaptive refinement algorithm

For many dynamical systems, especially for systems with a state space dimension bigger than two, it is not useful to work with a collection of uniform cells, because the computation of the transition propability matrix becomes very time consuming. Therefore, it is more appropriate to work with cells of different sizes. Starting with a coarse discretization of the state space, the problem is to make an appropriate choice which cells should be refined. The main distinctive mark between cells are their transient behavior. There are the transient cells and the persistent cells. With respect to the cell mapping approach persistent cells are considered to be the attractors. On this attractor the system has a given propability approximated by the stationary density of the generalized cell mapping. In order to approximate parts of the attractor which are highly frequented by a typical trajectory of the dynamical system, it is useful to subdivide just cells where the stationary density of the generalized cell mapping is not zero.

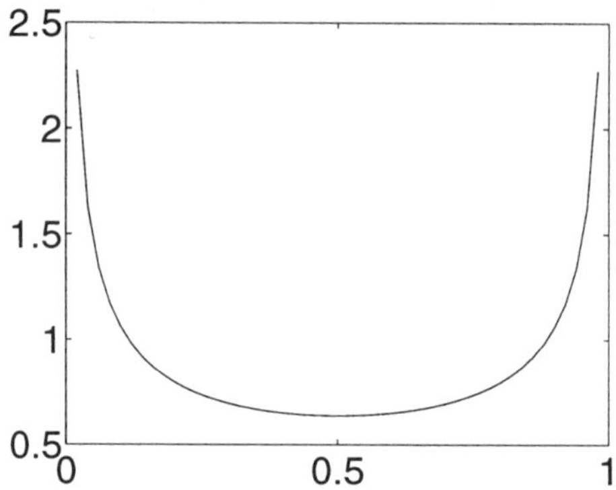

Figure 1. The exact density function f of the logistic mapping with parameter 4.

Now the question is, should all of the cells that belong to the support of the stationary density function of the generalized cell mapping be subdivided? In order to illustrate the situation, we take a look at the density function of the logistic mapping $\tau : \mathbb{R}^1 \to \mathbb{R}^1$ $\tau : x \to 4x(1-x)$. The graph of the associated density function of the invariant measure of the logistic mapping is drawn in Fig. 1. The density has the property that it has a very small derivative in the middle part and high derivative at both ends of the intervall $[0, 1]$. Since the approximation of this density by the stationary density of the generalized cell mapping means that the function is approximated by a piecewise constant function, it might be appropriate to subdivide those cells where the derivative of the density function is high.

We assume that the intervall $[0, 1]$ is subdivided into N uniform cells $G = \{J_1, J_2, \ldots, J_N\}$ and that the cells are so small that the exact density function f can be approximated by a linear function f_{lin} on each cell with the property $f'_{lin}(x)|_{x \in J_i} = f'(m_i)$ where m_i is the midpoint of J_i. Furthermore, on each cell there has to be at least one point which belongs to f_{lin} as well as to the stationary density function. The stationary density function

f_N of the generalized cell mapping is described by

$$f_N = \sum_{i=1}^{N} a_i \chi_i,$$

where $a_i \in \mathbb{R}_+^0$ and χ_i is the characteristic function of the cell J_i. The upper bound for the local L^1 error $\hat{\Delta}_i$ of $\|f_N - f_{lin}\|_{L^1(J_i)}$ on J_i is $\hat{\Delta}_i = \frac{1}{2} diam^2(J_i)|f'(m_i)|$. Accordingly, the upper bound for the total error $\hat{\Delta}$ is then computed to

$$\hat{\Delta} = \sum_{i=1}^{N} \frac{1}{2} diam^2(J_i)|f'(m_i)|. \tag{9}$$

In practice the density functions f or f_{lin} are not given, so we need to approximate the terms $|f'_{lin}(m_i)|$. In order to do that, we define the index set of all adjacent cells of J_i with $N(J_i) = \{j \in \{1, \ldots, N\} : \partial J_i \cap \partial J_j \neq \varnothing, j \neq i\}$. Since the stationary density of the generalized cell mapping is an approximation of the exact density function, we approximate the terms $|f'_{lin}(m_i)|$ by

$$|f'_{lin}(m_i)| \approx \max_{j \in N(J_i)} \frac{|f_N(m_i) - f_N(m_j)|}{|m_i - m_j|}. \tag{10}$$

We take the maximum in order to get an upper bound of the error (9). We use (10) to approximate the local error $\hat{\Delta}_i$ and the total error $\hat{\Delta}$ and denote the associated approximations by Δ_i and Δ.

In order to determine which cells should be subdivided, we compute the mean error $\overline{\Delta}$ of the approximation, that is $\overline{\Delta} = \frac{1}{N}\Delta$ and define the index set of all cells which will be subdivided $Sub(S)$ by:

$$Sub(S) = \{i \in \{1, \ldots, N\} : \Delta_i \geq \overline{\Delta}\}. \tag{11}$$

The adaptive refinement algorithm works as follows. First, we choose an initial cell space S_1. Then we compute with the algorithm from section 3 the generalized cell mapping and the associated stationary density function. With eq. (10) we obtain Δ_i, Δ, and $\overline{\Delta}$ and determine the set $Sub(S)$. Finally, all cells $J_i, i \in Sub(S)$ are subdivided to obtain the new cell space S_2, and by recursion S_3, S_4, ...

It is convenient to illustrate the efficiency of the adaptive refinement algorithm by two examples. First, we consider a one-dimensional mapping where the associated density function is known analytically and also has been solved in (Dellnitz and Junge, 1996). Furthermore, we compare our method with the one obtained by the adaptive refinement described in (Dellnitz and Junge, 1996). Additionally, we consider the Duffing equation as a two-dimensional system and show the cell refinement process at certain steps and the stationary density function obtained by the generalized cell mapping.

The following steps are used in the computation:

- For the computation of the transition probability matrix the exhaustion technique described in chapter three is employed.
- The refinement is always done just in one coordinate direction in order to keep the storage requirement small.
- The geometry of the cells is not stored. Instead of that we store the cells in a binary tree. The only geometric information considered is the initial hypercube Ω.

Let us first consider the map $\tau_1 : [0,1] \to [0,1]$,

$$\tau_1(x) = \left(\frac{1}{8} - 2 \left| x - \frac{1}{2} \right|^3 \right)^{\frac{1}{3}} + \frac{1}{2},$$

with the invariant density $f_1(x) = 12(x - \frac{1}{2})^2$. In (Dellnitz and Junge, 1996) this map is denoted with f_3 and the associated density with h_3. In Fig. 2 the graph of the map τ_1 and the density function f_1 are shown.

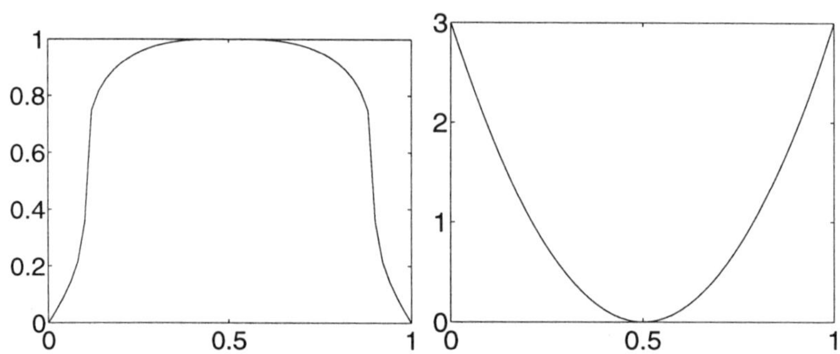

(a) The map τ_1 (b) The density f_1

Figure 2. The Graphs of τ_1 (a) and its density function f_1 (b).

In Table 2 we present the results obtained by the two algorithms. In the first two columns are the number of cells and in the last two columns the L^1-error of the two algorithms. We use *Box* as a synonym for the algorithm described in (Dellnitz and Junge, 1996) and *Cell* for the algorithm described in this paper.

TABLE 2. Comparison of the results for the map τ_1.

number of cells		L^1-error	
Box	Cell	Box	Cell
30	32	0.2015	0.0489
174	179	0.0571	0.0079
816	728	0.0064	0.0019

Finally, we apply the adaptive refinement algorithm to a Poincaré map of the Duffing equation

$$\left.\begin{array}{rc} \dot{x}_1 = & x_2, \\ \dot{x}_2 = & x_1 - x_1^3 - dx_2 + a\cos\omega\theta, \\ \dot{\theta} = & 1, \end{array}\right\} \; (x_1, x_2, \theta) \in \mathbb{R}^2 \times S^1,$$

with the paramaters $a = 0.3, d = 0.15$, and $\omega = 1.0$. For these values the Duffing equation has a period one solution and a strange attractor depending on the initial condition. Figure 3 (b) — (c) show discretizations of the square $[-2, 2]^2$ obtained by the adaptive refinement algorithm for the initial cell space (a). In addition each cell is shaded with its associated value of the stationary density function of the generalized cell mapping. Black means the stationary density function on this cell is zero and white means the stationary density function on this cell is the maximum. The stationary density function is computed with the algorithm described in section 3. Furthermore, we present in Fig. 4 a perspective view of the stationary density function of the generalized cell mapping. The stationary density function is visualized with gray shaded cubes where the height of a cube is 10% of the value of the stationary density function.

Literatur

Dellnitz, M.; Junge, O.: Adaptive Box Refinement in Subdivision Techniques for the Approximation of Dynamical Behavior, submitted (1996).

Guder, R.; Kreuzer, E.: Using the generalized cell mapping to approximate invariant measures on compact manifolds, to appear in *International J. of Bif. and Chaos* **7** 11 (1997).

Guder, R.; Dellnitz, M.; Kreuzer, E.: An Adaptive Method for the Approximation of the Generalized Cell Mapping. In: *Chaos, Solitons and Fractals* **8**,4 (1997), p. 525 — 534.

Hsu, C. S.: A theory of cell-to-cell mapping dynamical systems. In: *J. Appl. Mech.* **47** (1980) p. 931–939.

Hsu, C. S.: Cell-to-Cell Mapping — A Method of Global Analysis for Nonlinear Systems. New York/...: Springer-Verlag 1987.

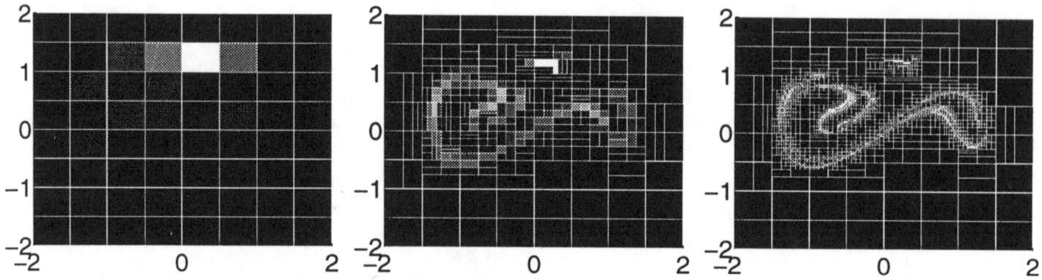

(a) 64 cells (b) 324 cells (c) 1639 cells

Figure 3. Three different cell meshes (a) — (c).

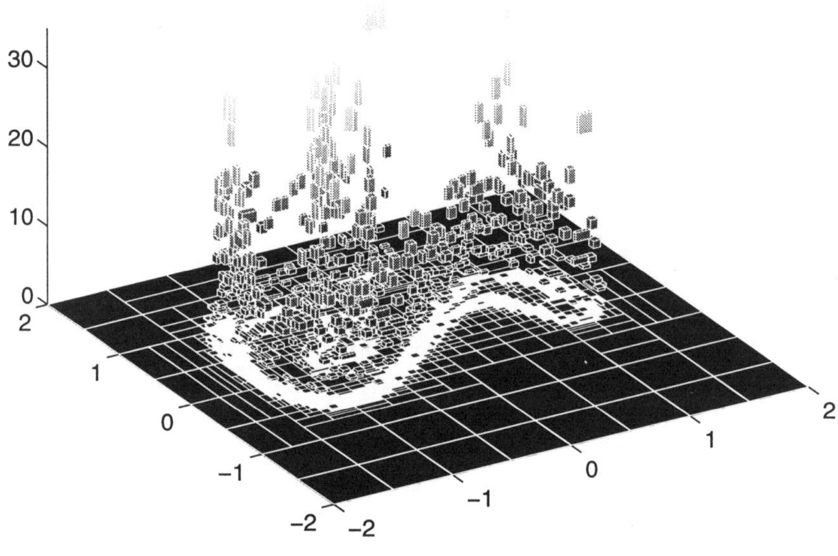

Figure 4. Stationary density of the approximation with 1093 cells.

Hsu, C. S.: Global analysis by cell mapping. In: *Int. J. Bif. Chaos* **2** (1992) p. 727–771.

Hunt, F.Y.: A Monte-Carlo approach to the approximation of invariant measures, National Institute of Standards and Technology, *NISTIR* **4980**, (1993).

Kreuzer, E.: Numerische Untersuchung nichtlinearer dynamischer Systeme. Berlin/...: Springer-Verlag 1987.

Lasota, A.; Mackey C.: Chaos, Fractals and Noise. New York/...: Springer-Verlag, 1994.

Li, T. Y.: Finite approximation of the Frobenius-Perron operator, a solution to Ulam's conjecture. In: *J. Approximation Theory* **17** (1976), p. 177–186.

BURSTS

EDGAR KNOBLOCH AND JEFF MOEHLIS
Department of Physics
University of California, Berkeley, CA 94720

1. Introduction

A thorough understanding of imperfections is critical for many engineering applications. At the linear level, local imperfections can trap modes and be responsible for a new class of potential instabilities. At the weakly nonlinear level, imperfections are known to play an important role near degeneracies in parameter space. Nominally symmetric systems are always degenerate in this sense since the presence of symmetries typically eliminates certain terms from the amplitude equations describing the evolution of instabilities in such systems. Consequently, such systems are almost always sensitive to small symmetry-breaking imperfections. Particularly dangerous are imperfections that destroy continuous symmetries, such as translation or rotation invariance. Such imperfections are typically responsible for the introduction of global bifurcations into the dynamics, and these are likely to be responsible for the appearance of chaos in the imperfect system (Knobloch 1996). However, in certain cases the loss of discrete symmetry can have a similar effect, at least if the symmetry group is large enough. This is the case in the class of systems discussed below.

We consider here systems with approximate D_4 symmetry undergoing a Hopf bifurcation. This symmetry arises frequently in applications. As an example, consider a system of partial differential equations defined on a square domain, and suppose that the equations (and boundary conditions) are invariant under reflections and rotations of the square by 90°. A Hopf bifurcation in such a system will be described by the amplitude equations for the Hopf bifurcation with D_4 symmetry. In normal form these equations have an additional S^1 phase shift symmetry which allows an essentially complete discussion of their dynamics. In particular, it is possible to show that the third order truncation of this normal form cannot exhibit chaotic dynamics. The proof proceeds by showing that the resulting equations can be written as a dynamical system defined on the surface of a three-dimensional

F. C. Moon (ed.),
IUTAM Symposium on New Applications of Nonlinear and Chaotic Dynamics in Mechanics, 51–60.

sphere in phase space, together with two other decoupled equations (Swift 1988). When the D_4 symmetry is broken by turning the domain into a rectangular one, albeit with a nearly square cross-section, this decoupling no longer occurs, and the dynamics becomes fundamentally three-dimensional. A closely related problem is provided by the Faraday system in a nearly square container. In this system, gravity-capillary waves are excited on the surface of a viscous fluid by vertical vibration of the container, usually as a result of a subharmonic resonance. Because of the parametric forcing the S^1 symmetry is now absent, but the amplitude equations describing the interaction of roll-like states oriented parallel to the sides continue to have approximate D_4 symmetry. In a square container, careful experiments by Simonelli and Gollub (1989) uncovered no chaotic dynamics in this system. On the other hand, in a rectangular but nearly square container the situation is quite different. Here Simonelli and Gollub uncovered the presence of a new class of oscillations, hereafter called "bursts". These are oscillations in the *amplitudes* of the two competing modes in which the energy builds up to a high value before undergoing an abrupt collapse to a small amplitude state. Such bursts can occur either periodically or irregularly, depending on parameters. It is this behavior that is of interest in the present paper. Similar behavior was also noted in experiments on convection in He^3/He^4 mixtures at cryogenic temperatures (Sullivan and Ahlers 1988). In these experiments, performed in a $34 : 6.9 : 1$ cell, the heat transport through the system exhibited irregular large amplitude bursting only 0.03% above threshold for the convective instability. Here, however, the origin of the approximate D_4 symmetry is quite different. Because of the slender shape of the container we may idealize the system as effectively two-dimensional. Such a system will undergo an oscillatory instability to either a mode of even or odd parity with respect to reflection in the extended direction. In the (formal) limit in which the aspect ratio of the system is allowed to become large, the distinction between even and odd modes is lost, and the amplitude equations describing the interaction of these two modes acquire an additional "interchange" symmetry (Landsberg and Knobloch 1996). As discussed further below, this interchange symmetry together with the reflection generates the group D_4. In a finite container this symmetry is never exact, however, since one or other of the two competing modes sets in first. Consequently the mode interaction in a system of large but finite extent is described by the normal form equations for the Hopf bifurcation with broken D_4 symmetry. This picture is supported by numerical simulations of the partial differential equations describing two-dimensional binary fluid convection in systems with aspect ratio $L = 16.25$ (Jacqmin and Heminger 1994). These simulations also reveal the presence of bursts, typically irregular and sometimes of very large amplitude, but in addition provide diagnostics that demonstrate convincingly that they involve primarily two adjacent modes of opposite parity. These results provide the primary motivation for studying the Hopf bifurcation with broken D_4 symmetry.

2. Derivation of the Equations

In this section we sketch the derivation of the amplitude equations describing the interaction of adjacent Hopf modes in a one-dimensional container of length L with identical boundary conditions at $x = \pm L/2$. Such a system has a reflection symmetry about $x = 0$; the primary Hopf modes are either even or odd under under this reflection (Dangelmayr and Knobloch 1991). We consider the interaction of an even mode with an adjacent odd mode in the formal limit $L \to \infty$. Let (z_+, z_-) be the complex amplitudes of the two modes. The requirement that a reflected state also be a state of the system translates into the requirement that the amplitude equations be equivariant with respect to the group action

$$\kappa_1 : (z_+, z_-) \to (z_+, -z_-). \tag{1}$$

Moreover, as argued by Landsberg and Knobloch (1996), the equations for the formally infinite system cannot distinguish between the two modes, i.e. in this limit the amplitude equations must also be equivariant with respect to the group action

$$\kappa_2 : (z_+, z_-) \to (z_-, z_+). \tag{2}$$

These two operations generate the group $D_4 = \langle \kappa_1, \kappa_2 \rangle$. For a container with large but finite length, this symmetry will be weakly broken; in particular, the even and odd modes may become unstable at slightly different Rayleigh numbers and with slightly different frequencies. The resulting equations are thus close to those for a 1:1 resonance, but with a special structure dictated by the proximity to D_4 symmetry. Because of the normal form symmetry

$$\hat{\sigma} : (z_+, z_-) \to e^{i\sigma}(z_+, z_-), \qquad \sigma \in [0, 2\pi), \tag{3}$$

the resulting equations have an additional S^1 symmetry. If these equations are truncated at third order we obtain (Landsberg and Knobloch 1996)

$$\dot{z}_+ = [\lambda + \Delta\lambda + i(\omega + \Delta\omega)]z_+ + A(|z_+|^2 + |z_-|^2)z_+ + B|z_+|^2 z_+ + C\bar{z}_+ z_-^2 \tag{4}$$

$$\dot{z}_- = [\lambda - \Delta\lambda + i(\omega - \Delta\omega)]z_- + A(|z_+|^2 + |z_-|^2)z_- + B|z_-|^2 z_- + C\bar{z}_- z_+^2 \tag{5}$$

Here $\Delta\omega$ measures the difference in frequency between the two modes at onset, and $\Delta\lambda$ measures the difference in their linear growth rates. Under appropriate nondegeneracy conditions (which we assume here) we may neglect all interchange symmetry-breaking contributions to the nonlinear terms. In terms of new variables defined by

$$z_+ = \rho^{-1/2}\cos(\theta/2)e^{i(\phi+\psi)/2}, \qquad z_- = \rho^{-1/2}\sin(\theta/2)e^{i(-\phi+\psi)/2}, \tag{6}$$

and a new time τ defined by $d\tau/dt = \rho^{-1}$, these equations take the more convenient form

$$\frac{d\rho}{d\tau} = -\rho[2A_R + B_R(1+\cos^2\theta) + C_R\sin^2\theta\cos 2\phi] - 2(\lambda + \Delta\lambda\cos\theta)\rho^2 \tag{7}$$

$$\frac{d\theta}{d\tau} = \sin\theta[\cos\theta(-B_R + C_R\cos 2\phi) - C_I\sin 2\phi] - 2\Delta\lambda\sin\theta\rho \tag{8}$$

$$\frac{d\phi}{d\tau} = \cos\theta(B_I - C_I\cos 2\phi) - C_R\sin 2\phi + 2\Delta\omega\rho, \tag{9}$$

where $A = A_R + iA_I$, etc. Equations (7-9) are invariant under the operations $\theta \to \theta + 2\pi$ and $\phi \to \phi + \pi$, symmetries which are related to the (broken) $D_4 \times S^1$ symmetry of the full system. A consequence of these symmetries is that if $(\rho_0, \theta_0, \phi_0)$ is a fixed point of equations (7-9), then so are the points $(\rho_0, \theta_0 + m\pi, \phi_0 + 2n\pi)$, where m and n are integers. This is so also for periodic solutions. We say that such fixed points and periodic solutions are *symmetry-related.*

The behavior of the decoupled variable ψ may be found by solving the appropriate differential equation for ψ evaluated at the solutions to the three-dimensional system. In particular, ψ (modulo 4π) is periodic for fixed points and periodic solutions to the three-dimensional system; thus, fixed points and periodic solutions in the three-dimensional system correspond to periodic solutions and tori in the full four-dimensional system (4,5), respectively. In the following we use the variable $r \equiv \rho^{-1} = |z_+|^2 + |z_-|^2$ as a useful measure of the energy in a burst.

3. Fixed points and periodic solutions of the three-dimensional system

3.1. THE PERFECT SYSTEM $\Delta\lambda = \Delta\omega = 0$

The D_4-symmetric equations $\Delta\lambda = \Delta\omega = 0$ have been analyzed by Swift (1988). In this case the $(\theta(\tau), \phi(\tau))$ equations decouple from the rest, and describe dynamics on the surface of a sphere of variable radius. For fixed points (θ_0, ϕ_0) of this two-dimensional system this radius becomes constant as $\tau \to \infty$, and depends on the quantity

$$F(\theta, \phi) = 2A_R + B_R(1+\cos^2\theta) + C_R\sin^2\theta\cos 2\phi. \tag{10}$$

Specifically,

(a) if $\lambda < 0, F(\theta_0, \phi_0) < 0$ then $r \to 0$,
(b) if $\lambda F(\theta_0, \phi_0) < 0$ then $r \to r_0 > 0$,
(c) if $\lambda > 0, F(\theta_0, \phi_0) > 0$ then $r \to \infty$.

Provided case (b) holds, there are three types of fixed points (θ_0, ϕ_0) with nontrivial symmetry which exist for all coefficient values (Swift, 1988). In addition there are open regions of coefficient space with nonsymmetric fixed points (that is, fixed points with trivial isotropy) and others with a unique limit cycle $(\theta^*(\tau), \phi^*(\tau))$. For the limit cycles we define

$$\bar{F} = \frac{1}{T_\tau} \int_0^{T_\tau} F(\theta^*(\tau'), \phi^*(\tau'))d\tau', \qquad (11)$$

where T_τ is the period, and conclude (cf. van Gils and Silber 1995) that
(a) if $\lambda < 0, \bar{F} < 0$ then $r \to 0$,
(b) if $\lambda \bar{F} < 0$, there exists a nonzero, finite $r(\tau)$ with $r(\tau + T_\tau) = r(\tau)$,
(c) if $\lambda > 0, \bar{F} > 0$ then $r \to \infty$.

Because the associated spherical system (8,9) is two-dimensional, no complex dynamics is possible unless the S^1 normal form symmetry is broken, as in the Faraday system. We do not pursue this possibility here, and instead focus on the effects of breaking the D_4 symmetry. We find that the possibilities (c) are responsible for the bursts present in this system.

3.2. THE IMPERFECT SYSTEM $\Delta\lambda \neq 0$ AND/OR $\Delta\omega \neq 0$

For the equations with broken D_4 symmetry ($\Delta\lambda \neq 0$ and/or $\Delta\omega \neq 0$) only the fixed points with even and odd parity remain as primary branches; the analogs of the remaining primary branches may bifurcate in secondary bifurcations from these, and are most easily found numerically (Landsberg and Knobloch 1996). However, there is another class of fixed points as well; these are crucial for understanding the bursting behavior. Their existence follows from the restriction of equations (7-9) to the invariant subspace $\rho = 0$. The resulting equations are identical to equations (8,9) with $\Delta\lambda = \Delta\omega = 0$; thus the fixed points of the associated spherical system that governs the dynamics of the perfect system continue to have significance for the imperfect system but now correspond to *infinite* amplitude fixed points. Infinite amplitude nonsymmetric fixed points and infinite amplitude limit cycles may also exist depending on the values of the parameters, exactly as in the perfect problem. Indeed, for $\Delta\lambda = \Delta\omega = 0$, infinite amplitude fixed points are amplitude-stable for $\lambda > 0, F(\theta_0, \phi_0) > 0$, while infinite amplitude limit cycles are amplitude-stable for $\lambda > 0, \bar{F} > 0$.

It is a simple matter to obtain the eigenvalues s_1, $s_\pm$ of the fixed points (ρ, θ, ϕ) of equations (7-9) and the corresponding eigenvectors. For the infinite amplitude fixed point $(\rho, \theta, \phi) = (0, \frac{\pi}{2}, 0)$ we obtain

$$s_1 = -2A_R - B_R - C_R, \quad s_\pm = \frac{1}{2}\left\{B_R - 3C_R \pm \sqrt{B_R^2 + 8B_I C_I - 8C_I^2 + 2B_R C_R + C_R^2}\right\},$$

with eigenvectors

$$\left(\frac{-2A_R^2 - 3A_R B_R - B_R^2 + B_I C_I - C_I^2 + A_R C_R + B_R C_R}{(C_I - B_I)\Delta\lambda + 2(A_R + B_R)\Delta\omega}, \frac{(C_R - 2A_R - B_R)\Delta\lambda + 2C_I \Delta\omega}{(C_I - B_I)\Delta\lambda + 2(A_R + B_R)\Delta\omega}, 1\right),$$

$$\left(0, \frac{1}{2(C_I - B_I)}\left\{B_R + C_R \pm \sqrt{B_R^2 + 8B_I C_I - 8C_I^2 + 2B_R C_R + C_R^2}\right\}, 1\right).$$

The corresponding results for $(0, \frac{\pi}{2}, \frac{\pi}{2})$ follow from the parameter symmetry $C \to -C$. Since s_1 can be negative and the corresponding eigenvector has a nonzero component in the ρ direction, a trajectory starting near the stable manifold will approach infinite amplitude. Moreover, since the eigenvectors of the remaining eigenvalues lie in the invariant plane $\rho = 0$ such a trajectory will evolve, after reaching infinite amplitude, in this plane until it encounters a fixed point (limit cycle) possessing an unstable eigenvalue (Floquet multiplier) that ejects it from it. This is our picture of the bursting mechanism, and we now substantiate it with explicit computations.

4. Genesis of a burst

As an illustrative example, consider the following parameter values:

$$A_R = 1.0, \quad B_R = -2.8, \quad B_I = 5.0, \quad C_R = 1.0, \quad C_I = 1.0,$$

with λ treated as the bifurcation parameter. We do not need to specify A_I which enters into the equation for ψ only. When $\Delta\lambda = \Delta\omega = 0$, the $(\rho_0, \frac{\pi}{2}, \frac{\pi}{2})$ and $\theta = 0$ fixed points are supercritical, while the $(\rho_0, \frac{\pi}{2}, 0)$ is subcritical; all are unstable. For these parameter values nonsymmetric fixed points do not exist. For $\lambda > 0$ there is, in the associated spherical system, a stable periodic solution surrounding the $(\rho_0, \frac{\pi}{2}, \frac{\pi}{2})$ fixed point.

As symmetry breaking parameters we choose $\Delta\lambda = 0.03$ and $\Delta\omega = 0.02$. The results of a detailed numerical study of these parameter values are summarized in the remarkable bifurcation diagram shown in Fig. 1. This figure shows

$$< r > \equiv \frac{1}{T}\int_0^T r dt = \frac{T_\tau}{\int_0^{T_\tau} \rho d\tau} \tag{12}$$

as a function of the bifurcation parameter λ for finite amplitude fixed points and finite amplitude periodic solutions only. Here T is the appropriate period of the solution in the original time t, and T_τ is the period of the solution to equations (7-9) found in terms of the rescaled time τ. In the

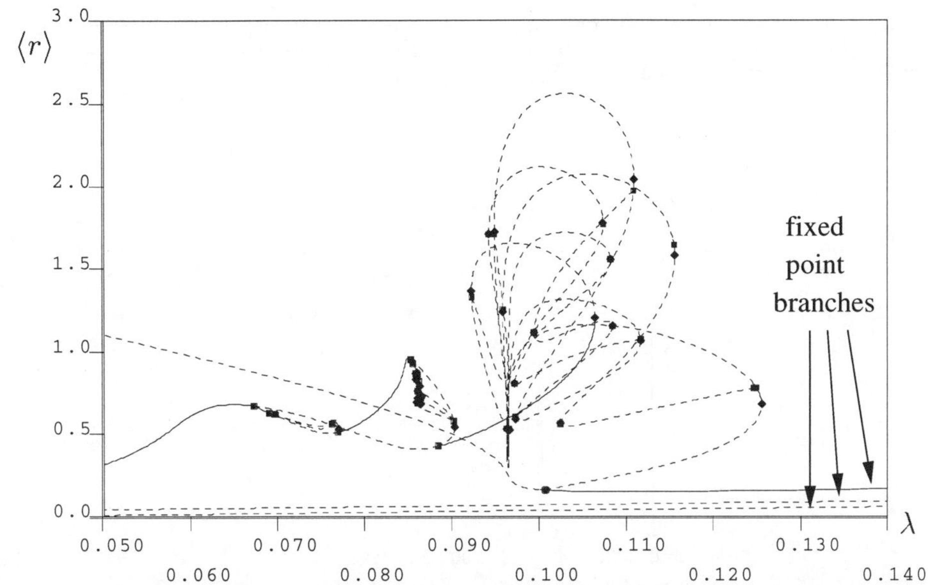

Figure 1. Bifurcation diagram for $A_R = 1.0$, $B_R = -2.8$, $B_I = 5.0$, $C_R = 1.0$, $C_I = 1.0$ and $\Delta\lambda = 0.03$, $\Delta\omega = 0.02$.

figure solid (broken) lines indicate stable (unstable) solutions; circles, diamonds, and squares indicate Hopf, saddle-node, and period-doubling bifurcations, respectively. Many period-doubled branches are omitted. The periodic branches all correspond to periodic bursts since the corresponding trajectories make periodic excursions towards the invariant plane $\rho = 0$, as shown in Figs. 2 and 3. Note that because the bursts are fast events in the original time t the average $\langle r \rangle$ for a sequence of large amplitude bursts may in fact be quite small. At some values of λ, such as $\lambda = 0.072$, no stable periodic branches are present, and irregular bursts are found as shown in Fig. 4. This figure includes a plot of successive maxima against one another; the map appears one-dimensional, and shows unambiguously that the bursts are chaotic. A number of global bifurcations in which a periodic solution approaches a finite amplitude fixed point are found to occur near $\lambda = 0.0965$ and manifest themselves as cusps in Fig. 1.

Fig. 1 omits a very important class of global bifurcations involving *infinite* amplitude solutions, i.e. solutions with $\rho = 0$. For these coefficients, $s_1(0, \frac{\pi}{2}, 0) = -0.2$ with eigenvector $(-3.96, -0.49, 1)$. Moreover, $s_+(0, \frac{\pi}{2}, 0) = 0.068$ and $s_-(0, \frac{\pi}{2}, 0) = -5.87$; consequently the infinite amplitude fixed point is a saddle point whose unstable manifold forms a structurally stable connection with an infinite amplitude limit cycle around $(0, \frac{\pi}{2}, \frac{\pi}{2})$. This limit cycle is attracting in the $\rho = 0$ invariant plane but

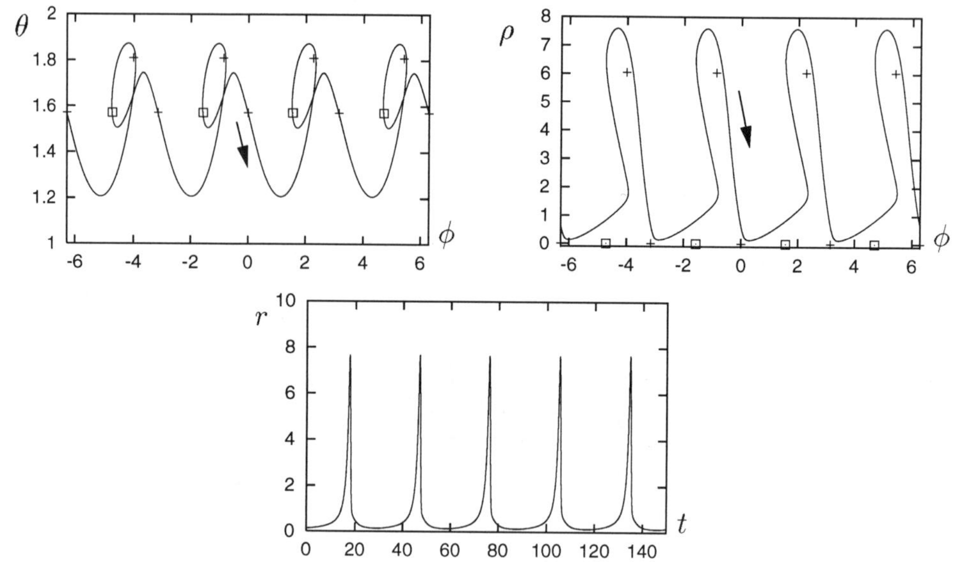

Figure 2. Periodic bursts from successive excursions towards symmetry-related infinite amplitude fixed points. For this burst sequence $\lambda = 0.1$, $\langle r \rangle \approx 0.7$.

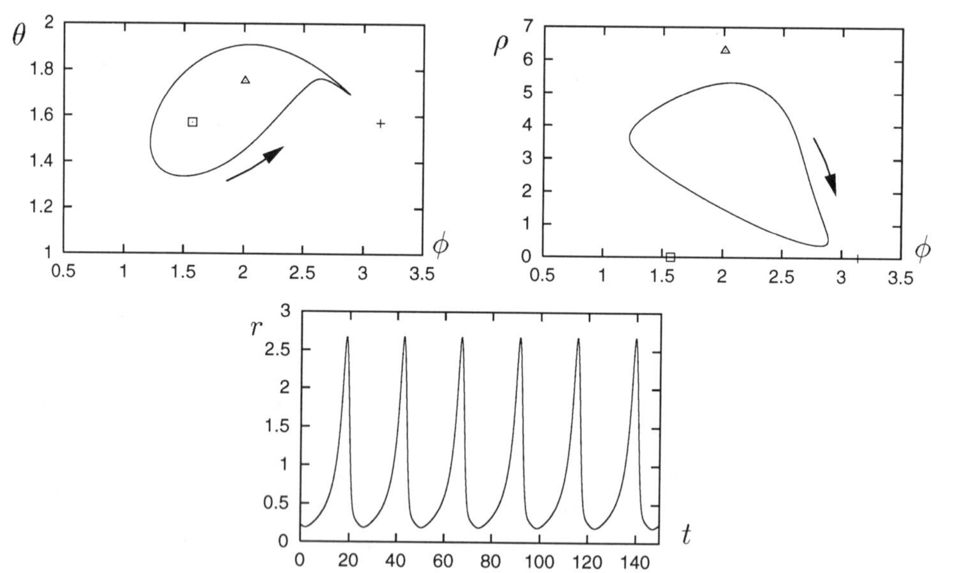

Figure 3. Periodic bursts from successive excusrions towards the same infinite amplitude fixed point. For this burst sequence $\lambda = 0.1253$, $\langle r \rangle \approx 0.75$.

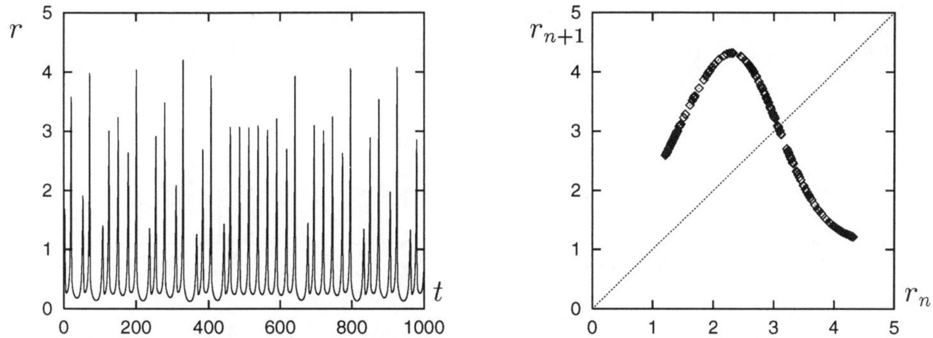

Figure 4. Chaotic bursts for $\lambda = 0.072$.

for these coefficient values possesses an unstable Floquet multiplier in the ρ direction and hence has a two-dimensional unstable manifold. These manifolds together describe a single infinite amplitude burst (see Fig. 5). Recurrent bursts occur if this unstable manifolds intersects the stable manifold of the fixed point or its translates (cf. Figs. 2,3). This is a codimension-one phenomenon, and can be studied using a Shil'nikov-like analysis; this analysis bears substantial similarity to that carried out by Hirschberg and Knobloch (1993) for the Shil'nikov-Hopf bifurcation, and is particularly relevant to very large bursts such as those present at $\Delta\lambda = 0.06$, $\Delta\omega = -0.01$ (Fig. 6). A similar geometrical scenario is responsible for the generation of chaotic traveling waves in systems with O(2) symmetry (Knobloch and Moore 1991). In the present case, however, the trajectories typically also approach close to a λ-dependent finite amplitude fixed point and consequently the analysis of the global bifurcations at infinity provides only a partial description of the bifurcation diagram of Fig. 1. Further details can be found in Moehlis and Knobloch (1997).

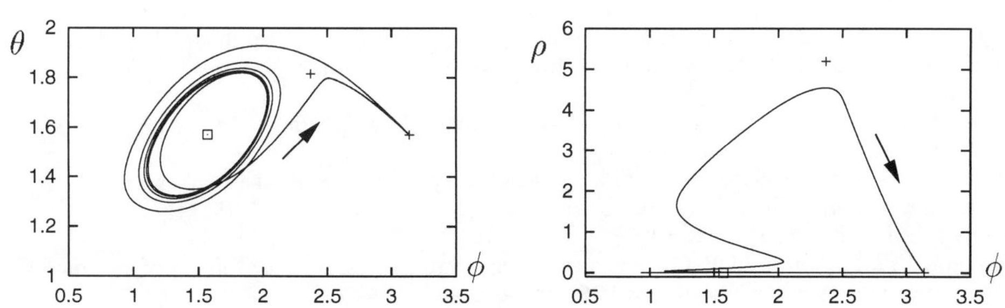

Figure 5. Heteroclinic cycle involving an infinite amplitude fixed point and an infinite amplitude limit cycle for $\lambda = 0.0974$.

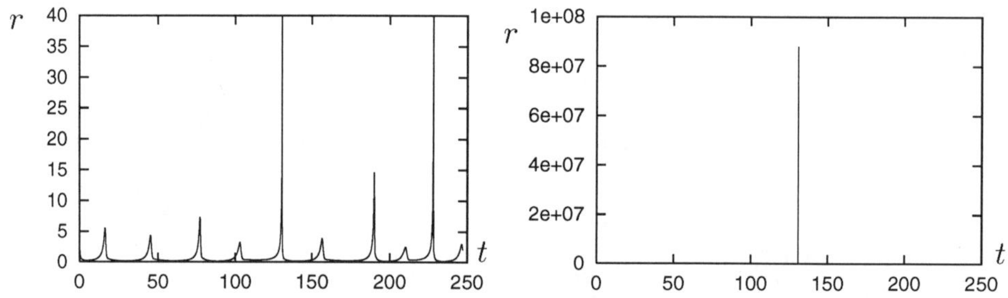

Figure 6. Very large amplitude burst for $\Delta\lambda = 0.06$, $\Delta\omega = -0.01$, $\lambda = 0.1$.

5. Conclusion

The bursting mechanism outlined above arises in a natural way in slender systems supporting oscillations of even and odd parity. The resulting dramatic response can lead to material fatigue or, in the fluid context, to the breakdown of laminar motion into intermittent turbulent bursts. Although neither development is captured by the "frozen" spatial structure responsible for the burst mechanism, we have invoked such potential applications in naming this interesting phenomenon.

Acknowledgement: This work was supported by the National Science Foundation under grant DMS-9703684.

References

Dangelmayr, G. and Knobloch, E. (1991) Hopf bifurcation with broken circular symmetry, *Nonlinearity* **4**, pp. 399–427

van Gils, S.A. and Silber, M. (1995) On the uniqueness of invariant tori in $D_4 \times S^1$ symmetric systems, *Nonlinearity* **8**, pp. 615–628

Hirschberg, P. and Knobloch, E. (1993) Šil'nikov-Hopf bifurcation, *Physica D* **62**, pp. 202–216

Jacqmin, D. and Heminger, J. (1994) Double-diffusion with Soret effect in a rectangular geometry: Linear and nonlinear traveling wave instabilities. Unpublished

Knobloch, E. (1996) System symmetry breaking and Shil'nikov dynamics, *Fields Inst. Comm.* **5**, pp. 271–279

Knobloch, E. and Moore, D.R. (1991) Chaotic travelling wave convection, *Eur. J. Mech., B/Fluids* **10**, no. 2-Suppl., pp. 37–42

Landsberg, A.S. and Knobloch, E. (1996) Oscillatory bifurcation with broken translation symmetry, *Phys. Rev. E* **53**, pp. 3579–3600

Moehlis, J. and Knobloch, E. (1997) In preparation

Simonelli, F. and Gollub, J.P. (1989) Surface wave mode interactions: effects of symmetry and degeneracy, *J. Fluid Mech.* **199**, pp. 471–494

Sullivan, T.S. and Ahlers, G. (1988) Nonperiodic time dependence at the onset of convection in a binary liquid mixture, *Phys. Rev. A* **38**, pp. 3143–3146

Swift, J.W. (1988) Hopf bifurcation with the symmetry of the square, *Nonlinearity* **1**, pp. 333–377

DYNAMICS OF A QUASIPERIODICALLY-FORCED MATHIEU OSCILLATOR

RICHARD RAND
Department of Theoretical and Applied Mechanics
Cornell University
Ithaca, NY 14853

AND

RANDOLPH ZOUNES AND RACHEL HASTINGS
Center for Applied Mathematics
Cornell University
Ithaca, NY 14853

1. Introduction

Mathieu's equation [6],

$$\ddot{x} + (\delta + \epsilon \cos t)\, x = 0, \tag{1}$$

is the paradigm for problems in parametric excitation in which an autonomous linear structure is driven by a periodic forcer, usually in a direction perpendicular to the direction of motion (e.g., the vertically forced pendulum, dynamic buckling of an elastic column, water waves in a vertically driven channel).

This work concerns a natural extension of such problems to cases in which the forcer is quasiperiodic. We investigate the quasiperiodic (QP) Mathieu equation,

$$\ddot{x} + (\delta + \epsilon \cos t + \epsilon \cos \omega t)\, x = 0. \tag{2}$$

For a given set of parameters $(\delta, \omega, \epsilon)$, eq.(2) is said to be **stable** if all solutions are bounded, and **unstable** otherwise. We use numerical integration and Lyapunov exponents to determine regions of stability in the δ-ω plane for fixed ϵ. In addition, we obtain approximate analytic expressions for the transition curves bounding regions of stability using two distinct methods:

F. C. Moon (ed.),
IUTAM Symposium on New Applications of Nonlinear and Chaotic Dynamics in Mechanics, 61–70.

regular perturbations and harmonic balance [8],[9], [10]. Finally, we use singular perturbations to study the stability of eq.(2) for small ω and ϵ when δ is close to $\frac{1}{4}$ [4].

Eq.(2) has been investigated by perturbation methods in [1], [3], [7]. These authors have noted the failure of perturbations for small ϵ when ω takes on resonant values due to small-divisors. We shall show how the method of harmonic balance can be used to avoid these difficulties.

2. Numerical Integration

In order to obtain an approximate stability chart, we numerically integrated eq.(2) forward in time from arbitrarily chosen initial conditions at $t = 0$ up to $t = 20,000$. At each step we computed the amplitude $\sqrt{x(t)^2 + \dot{x}(t)^2}$ and judged a motion to be unstable if its amplitude became greater than a million times its initial value for any t between 0 and 20,000, and stable otherwise. Our results are displayed in Fig.1.

We note that the structure of the stability regions for the QP Mathieu equation is much more complicated than for the Mathieu equation [6].

3. Lyapunov Exponents

Lyapunov exponents [5] provide a second approach based on numerical integration with which we can obtain an approximate stability chart for eq.(2). The *Lyapunov exponent* of the solution $x(t)$ is defined as the following limit:

$$\lambda = \limsup_{t\to\infty} \frac{1}{t} \ln |x(t)|. \tag{3}$$

A positive Lyapunov exponent, $\lambda > 0$, corresponds to an unstable solution, and since there is no dissipation in eq.(2), stable solutions correspond to $\lambda = 0$. After numerically integrating eq.(2), we found it difficult to distinguish between small positive Lyapunov exponents and those which were truly zero. We resolved this by noting that Lyapunov exponents which are genuinely nonzero should maintain their values as the time of integration is increased. On the other hand, Lyapunov exponents that are genuinely zero will have computed approximations that tend to zero as the time of integration is increased. Our results are displayed as a contour plot in Fig.2.

4. Regular Perturbations

In the case of Mathieu's equation, regular perturbations have been used to provide approximate analytic expressions for the transition curves in the δ-ϵ plane which separate regions of stability from regions of instability [6]. The procedure, which is valid for small values of ϵ, is based on the result from Floquet theory that along a transition curve, there exist solutions with period 2π or 4π. This leads to the conclusion that when $\epsilon = 0$, regions of instability ("Arnold tongues") emanate from the δ axis at points given by

$$\delta = \delta_0 = \frac{n^2}{4}, \qquad n = 1, 2, 3, \ldots \tag{4}$$

Thus in the case of Mathieu's equation, we set

$$\delta = \delta_0 + \epsilon\delta_1 + \epsilon^2\delta_2 + \cdots \tag{5}$$

$$x(t) = x_0(t) + \epsilon x_1(t) + \epsilon^2 x_2(t) + \cdots \tag{6}$$

where $x_0(t)$ is separately equated to $\sin nt/2$ and $\cos nt/2$, and together with eq.(4), each yields a transition curve. The pair of curves associated with a given value of n yields an approximation to the corresponding Arnold tongue. (In the case of $n = 0$, however, a single transition curve occurs corresponding to $x_0(t) = 1$.) After substituting these power series expansions into Mathieu's equation, the coefficient δ_i is obtained by eliminating secular terms from the differential equation for $x_i(t)$.

We shall apply a similar scheme to the QP Mathieu equation. The key ansatz is the generalization of the transition points of eq.(4) to

$$\delta_0 = \frac{(n + m\omega)^2}{4}, \qquad n = 0, 1, 2, ..., \; m = 0, \pm 1, \pm 2, ... \tag{7}$$

The expansions (5), (6) are substituted into the QP Mathieu equation where $x_0(t)$ is separately equated to $\sin(\frac{n+m\omega}{2}t)$ and $\cos(\frac{n+m\omega}{2}t)$, and together with eq.(7), each yields a transition curve. Once again, the coefficient δ_i is obtained by eliminating secular terms from the differential equation for $x_i(t)$.

As an example of the kind of results obtained under this method (using computer algebra), take the case $(n, m) = (1, -1)$:

$$\delta = \frac{1}{4}(1 - \omega)^2 + \frac{1}{2}\frac{\epsilon^2\,(\omega^2 - \omega + 1)}{\omega\,(2\omega - 1)(\omega - 2)} + O(\epsilon^4) \tag{8}$$

$$\delta = \frac{1}{4}(1 - \omega)^2 - \frac{3}{2}\frac{\epsilon^2\,(\omega^2 - 3\omega + 1)}{\omega\,(2\omega - 1)(\omega - 2)} + O(\epsilon^4) \tag{9}$$

Plots of transition curves generated under this method for $n = 0, 1, 2$ and $m = 0, \pm 1, \pm 2$, all valid to $O(\epsilon^4)$, are displayed in Fig.3. Note that the expressions for the transition curves are not valid in neighborhoods of $\omega = 0, \frac{1}{3}, \frac{1}{2}, \frac{2}{3}, 1, \frac{3}{2}, 2$, and 3 since there are terms that have vanishing denominators at these resonant values, cf., eqs.(8), (9). For this reason, these portions of the transition curves are omitted in Fig.3. Additional resonances will show up at higher order truncations.

5. Harmonic Balance

Another approach which has been applied successfully to Mathieu's equation is harmonic balance [6]. This method is again based on the result from Floquet theory that along a transition curve there exist solutions with period 2π or 4π, leading to the following Fourier expansion for solutions $x(t)$:

$$x(t) = A_0 + \sum_{k=1}^{\infty} A_k \cos \frac{k}{2}t + B_k \sin \frac{k}{2}t. \tag{10}$$

Substitution of eq.(10) into eq.(1) and collecting terms (i.e., balancing harmonics) leads to an infinite set of linear, homogeneous equations for the coefficients $\{A_k, B_k\}$. For a nontrivial solution, the infinite (or Hill's) determinant of the associated coefficient matrix must vanish. By truncating this infinite system, we obtain an approximate implicit equation for the transition curves in the δ-ϵ parameter plane.

We shall apply a similar scheme to the QP Mathieu equation. The key ansatz is the generalization of the Fourier series eq.(10) to the quasiperiodic form:

$$x(t) = \sum_{n=0}^{\infty} \sum_{m=-\infty}^{\infty} A_{nm} \cos(\frac{n+m\omega}{2}t) + B_{nm} \sin(\frac{n+m\omega}{2}t). \tag{11}$$

In practice, approximate results are obtained when the infinite sums in eq.(11) are replaced by sums from 0 to N and from $-N$ to N, respectively. Since the forcing term in eq.(2) is an even function of t, the solution space can be spanned by an even solution and an odd solution. This permits us to take first B_{nm} and then A_{nm} as zero in eq.(11), thereby reducing the size of the (truncated) determinant by half. In the former case, we substitute eq.(11) with $B_{nm} = 0$ into eq.(2). Using computer algebra, we perform a trigonometric reduction and collect terms to give the following system of equations for the A_{nm}:

$$A_{n,m}\left(\delta - \frac{1}{4}(n+m\omega)^2\right) + \frac{\epsilon}{2}\left(A_{n+2,m} + A_{n-2,m} + A_{n,m+2} + A_{n,m-2}\right) = 0. \tag{12}$$

Eqs.(12), when truncated at the Nth harmonic, represent $2N^2 + 2N + 1$ simultaneous equations. For example, in the case of $N = 4$, the matrix of coefficients has dimension 41, and the evaluation of the corresponding determinant and the generation of graphical plots took an hour using Maple on a Sun SPARCstation 10. As an example of the kind of results obtained by this method, we present the determinant corresponding to the cosine-series solution, displayed in factored form for the truncation $N = 2$:

$$
\begin{aligned}
\det\ = 0 = \ & 512\,(-16\,\epsilon^2 - 2\,\omega^2 + 1 - 8\,\delta - 8\,\omega^2\,\delta + 16\,\delta^2 + \omega^4) \\
& \Big(8\,\epsilon^3 + (32\,\omega^2 - 48\,\delta + 12)\ \epsilon^2 \\
& + (-16\,\omega^2 + 32\,\delta^2 - 64\,\omega^2\,\delta - 16\,\delta + 2 + 32\,\omega^4)\ \epsilon \\
& + 64\,\omega^4\,\delta - 48\,\delta^2 - 16\,\omega^4 - 128\,\omega^2\,\delta^2 - 1 + 64\,\delta^3 + 8\,\omega^2 + 12\,\delta\Big) \\
& \Big(-8\,\epsilon^3 + (-12\,\omega^2 + 48\,\delta - 32)\ \epsilon^2 \\
& + (16\,\omega^2 - 32 + 16\,\omega^2\,\delta + 64\,\delta - 2\,\omega^4 - 32\,\delta^2)\ \epsilon \\
& - 12\,\omega^4\,\delta + 128\,\delta^2 - 8\,\omega^4 + 48\,\omega^2\,\delta^2 - 64\,\delta - 64\,\delta^3 + 16\,\omega^2 + \omega^6\Big) \\
& \Big((5\,\omega^4\,\delta - 1 + 4\,\delta^3 - 8\,\omega^2\,\delta^2 + \omega^2 + \omega^4 + 5\,\delta - \omega^6 + 2\,\omega^2\,\delta - 8\,\delta^2)\ \epsilon^2 \\
& + 2\,\omega^6\,\delta^2 + 2\,\delta^2 - 2\,\omega^2\,\delta + 4\,\omega^4\,\delta + 6\,\delta^4\,\omega^2 - 6\,\delta^3 - 2\,\omega^6\,\delta \\
& + 2\,\omega^2\,\delta^2 - 6\,\delta^3\,\omega^2 + 6\,\delta^4 - 2\,\delta^5 - 6\,\omega^4\,\delta^3 + 2\,\omega^4\,\delta^2\Big)
\end{aligned}
$$

Fig.4 shows both the sine and cosine solutions for $N = 4$. Note the absence of the small-divisor problem, in contrast to the perturbation method results of Fig.3.

6. Singular Perturbations

In order to better understand the nature of the dynamics of the QP Mathieu equation, we present the following singular perturbation method, valid for small values of the parameter ω and for values of δ which are close to $\frac{1}{4}$. We set

$$\omega = k\epsilon \quad \text{and} \quad \delta = \frac{1}{4} + \epsilon\,\delta_1 + \cdots \tag{13}$$

and we use the two-variable expansion method [2] in which $\xi = t$ and $\eta = \epsilon t$, whereupon eq.(2) becomes,

$$x_{\xi\xi} + 2\epsilon x_{\xi\eta} + \epsilon^2 x_{\eta\eta} + (\delta + \epsilon\cos\xi + \epsilon\cos k\eta)\,x = 0. \tag{14}$$

Expanding $x = x_0 + \epsilon\,x_1 + \cdots$ and collecting terms, we obtain

$$x_{0\xi\xi} + \frac{1}{4}\,x_0 = 0 \tag{15}$$

$$x_{1\xi\xi} + \frac{1}{4}\,x_1 = -2\,x_{0\xi\eta} - \delta_1\,x_0 - (\cos\xi + \cos k\eta)\,x_0. \tag{16}$$

Eq.(15) has the solution

$$x_0 = R\,\cos(\frac{\xi}{2} + \theta), \tag{17}$$

where R and θ are functions of the slow time, η. Substituting eq.(17) into eq.(16), we obtain

$$x_{1\xi\xi} + \frac{1}{4}x_1 = \frac{dR}{d\eta}\sin(\frac{\xi}{2}+\theta) + R\frac{d\theta}{d\eta}\cos(\frac{\xi}{2}+\theta)$$

$$-\delta_1 R\cos(\frac{\xi}{2}+\theta) - R\cos k\eta\cos(\frac{\xi}{2}+\theta) \quad (18)$$

$$-\frac{R}{2}[\cos((\frac{3\xi}{2}+\theta)+\cos((\frac{\xi}{2}-\theta)].$$

Removal of secular terms dictates that we set to zero the coefficients of $\sin(\frac{\xi}{2}+\theta)$ and $\cos(\frac{\xi}{2}+\theta)$, giving

$$\frac{dR}{d\eta} - \frac{R}{2}\sin 2\theta = 0, \qquad R\frac{d\theta}{d\eta} - \delta_1 R - R\cos k\eta - \frac{R}{2}\cos 2\theta = 0, \quad (19)$$

where we have used the identity

$$\cos(\frac{\xi}{2}-\theta) = \cos(\frac{\xi}{2}+\theta - 2\theta) = \cos(\frac{\xi}{2}+\theta)\cos 2\theta + \sin(\frac{\xi}{2}+\theta)\sin 2\theta. \quad (20)$$

The first of eqs.(19) can be solved in closed form,

$$R = C\, e^{\frac{1}{2}\int \sin 2\theta\, d\eta}, \quad (21)$$

where C is an arbitrary constant. Eq.(21) gives R as a function of θ, and θ itself is determined by the second of eqs.(19), i.e., by

$$\frac{d\theta}{d\eta} = \delta_1 + \frac{1}{2}\cos 2\theta + \cos k\eta\,. \quad (22)$$

Although the R and θ variables are therefore uncoupled, the boundedness of x_0 is governed by the behavior of $R(\eta)$, cf., eq.(17). We are therefore interested in how the behavior of θ, determined by eq.(22), influences the boundedness of R via eq.(21).

The answer is this: If eq.(22) exhibits a **limit cycle** on the $\theta - \eta$ phase torus, then R is an exponential function of η and the x_0 motion is unstable. This follows because if $\theta(\eta)$ is a periodic function mod $n\pi$, then so is $\sin 2\theta$, but with non-zero average value taken over one orbit of the limit cycle (in general). Thus the integral in eq.(21), $\int \sin 2\theta\, d\eta$, taken over one cycle, will not in general be zero, and R will grow or decay exponentially in η. Since the Wronskian of eq.(2) is constant in t, an exponentially decaying solution must be accompanied by a second linearly independent solution which is exponentially growing. Thus a limit cycle in eq.(22) corresponds to the occurrence of an unbounded solution in eq.(2). If, on the other hand, the torus flow eq.(22) is equivalent to an irrational flow, then the integral $\int \sin 2\theta \, d\eta$ will tend to zero on the average, R will remain bounded as $t \to \infty$, and the x_0 motion will be stable.

So the question of the stability of the QP Mathieu equation (2) is reduced to the question of whether eq.(22) has a limit cycle. Unfortunately, a closed form solution of eq.(22) is unavailable. Nevertheless, we may numerically integrate eq.(22) for given values of the parameters δ_1 and k, and determine whether the phase flow exhibits a limit cycle by inspection. Before doing so, we change variables so that the phase torus is $2\pi \times 2\pi$: we set $\phi = 2\theta$ and $\tau = k\eta$, giving

$$\frac{d\phi}{d\tau} = \frac{2\delta_1 + \cos\phi + 2\cos\tau}{k}. \tag{23}$$

The process of deciding whether a given phase portrait exhibits a limit cycle can be automated by numerically generating a Poincare map corresponding to the surface of section $\tau = \pi$. A stable limit cycle then corresponds to a stable fixed point of the associated one-dimensional circle map. Fig.5 shows the result of such a procedure, in which thousands of points were chosen at random in the (δ_1, k) parameter space, and the presence of a dot represents stability, i.e., the absence of a stable limit cycle.

7. Conclusions

A comparison of Figs.1–4 reveals general agreement between the numerical and analytical approximations to the stability charts of the quasiperiodic Mathieu equation. The method of Lyapunov exponents in Fig.2 shows greater detail than the simple numerical integration of Fig.1, although Fig.1 involved four times more computation than Fig.2! Fig.3, based on the perturbation method, displays gaps where portions of the transition curves (which lie near the resonant values of ω) have been omitted, while Fig.4 generated by the method of harmonic balance does not suffer from this singular behavior. All four methods show that for small ϵ the two largest instability regions lie in the neighborhood of the curves $\delta = \frac{1}{4}$ and $\delta = \frac{\omega^2}{4}$, each of which represents a 2:1 resonance between the respective driving frequency (1 and ω) and the unforced frequency ($\sqrt{\delta}$). The analytical methods show that the thickness of these two instability regions is approximately ϵ. However it is evident from both the numerical plots as well as from the analytical results that there are many additional smaller instability regions. For example, the point of intersection between these two largest instability regions, namely $\delta = \frac{1}{4}, \omega = 1$, is the birthplace of a number of smaller instability regions; (see Figs.1 and 2). For this choice of parameters, both driving frequencies are in 2:1 resonance with the unforced frequency and we may expect the unfolding of such a degeneracy to be accompanied by a diversity of bifurcations.

We also presented a singular perturbation method which is valid for small ω, in the neighborhood of $\delta = \frac{1}{4}$. The criterion for stability was the presence of an irrational torus flow in eq.(23), (cf., Fig.5). These results may be compared with those obtained by numerical integration by enlarging the region around $\delta = \frac{1}{4}$ in Fig.1, as shown in Fig.6. We note that there is excellent agreement between the perturbation results of Fig.5 and the direct numerical treatment of Fig.6.

8. References

1. Abel, J.M., 1970, "On an Almost Periodic Mathieu Equation", *Quart. Appl. Math.* vol.28, pp.205-217.
2. Bender, C.M. and Orszag, S.A., 1978, *Advanced Mathematical Methods for Scientists and Engineers*, McGraw-Hill, New York.
3. Davis, S.H. and Rosenblat, S., 1980, "A quasiperiodic Mathieu-Hill equation", *SIAM J.Appl.Math.* vol.38, pp.139-155.
4. Rand, R. and Hastings, R., 1995, "A Quasiperiodic Mathieu Equation", Proceedings of the 1995 Design Engineering Technical Conferences, Boston, Massachusetts, Sept.17-20, 1995, Vol. 3, Part A: *Vibration of Nonlinear, Random and Time-Varying Systems*, vol.DE-Vol.84-1, pp.747–758, American Soc. Mech. Engineers.
5. Rand, R.H., 1994, *Topics in Nonlinear Dynamics with Computer Algebra*, Gordon and Breach, Langhorne, PA.
6. Stoker, J.J., 1950, *Nonlinear Vibrations in Mechanical and Electrical Systems*, Wiley, New York.
7. Vrscay, E.A., 1991, "Irregular behavior arising from quasiperiodic forcing of simple quantum systems: insight from perturbation theory", *J.Phys.A:Math.Gen.* vol.24, pp.L463-L468.
8. Zounes, R.S., 1997, "An Analysis of the Nonlinear Quasiperiodic Mathieu Equation", Ph.D. thesis, Center for Applied Mathematics, Cornell University.
9. Zounes, R.S. and Rand, R.H., 1996, "Transition Curves in the Quasiperiodic Mathieu Equation", in *Nonlinear Dynamics and Controls*, editors:A.K.Bajaj, N.Sri Namachchivaya, and M.A.Franchek, ASME Book No.G01025, pp.1–6, American Soc. Mech. Engineers.
10. Zounes, R.S. and Rand, R.H., 1998, "Transition Curves for the Quasiperiodic Mathieu Equation", SIAM J.Applied Math. (to appear).

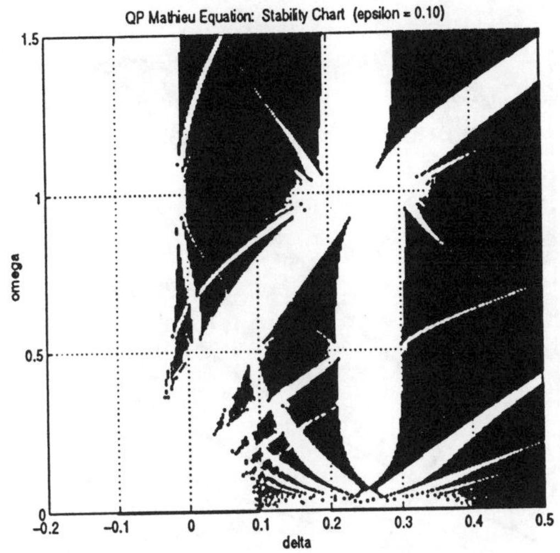

Fig.1. Stability of eq.(2) with $\epsilon = 0.1$ as determined directly from numerical integration. Points (δ, ω) in the blackened regions of the δ-ω parameter plane correspond to stable (bounded) solutions.

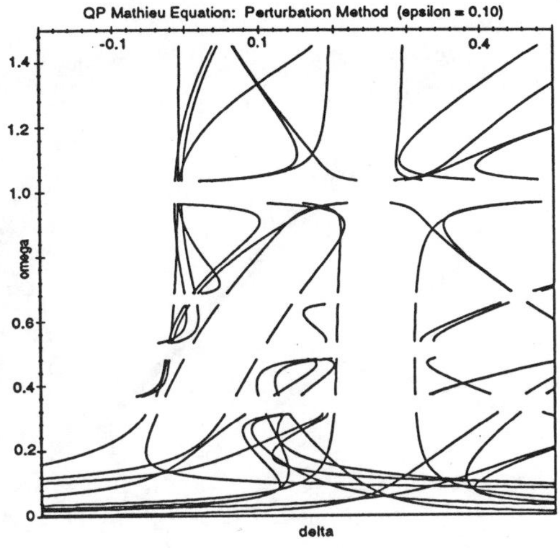

Fig.3 Transition curves of eq.(2) as determined by the perturbation method with $\epsilon = 0.1$. The regions around resonant values of ω have been omitted.

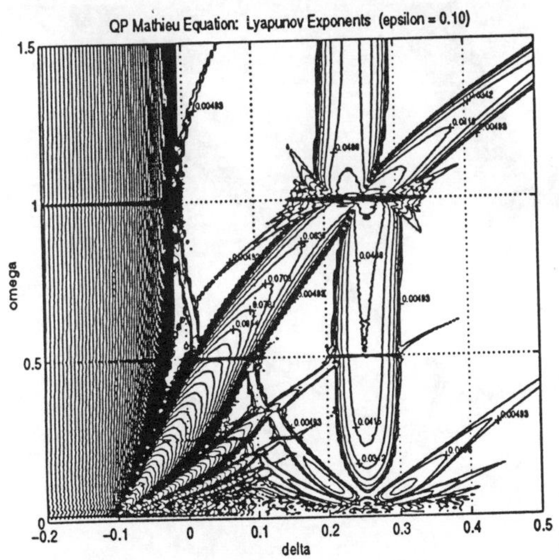

Fig.2. Contour plot of Lyapunov exponents for eq.(2) with $\epsilon = 0.1$. The level curves correspond to constant values of $\lambda > 0$ for which eq.(2) is unstable.

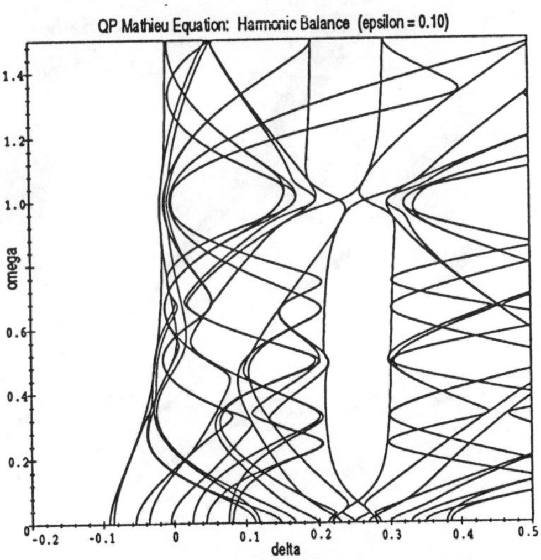

Fig.4. Transition curves of eq.(2) with $\epsilon = 0.1$ as determined by the method of harmonic balance for truncation order $N = 4$.

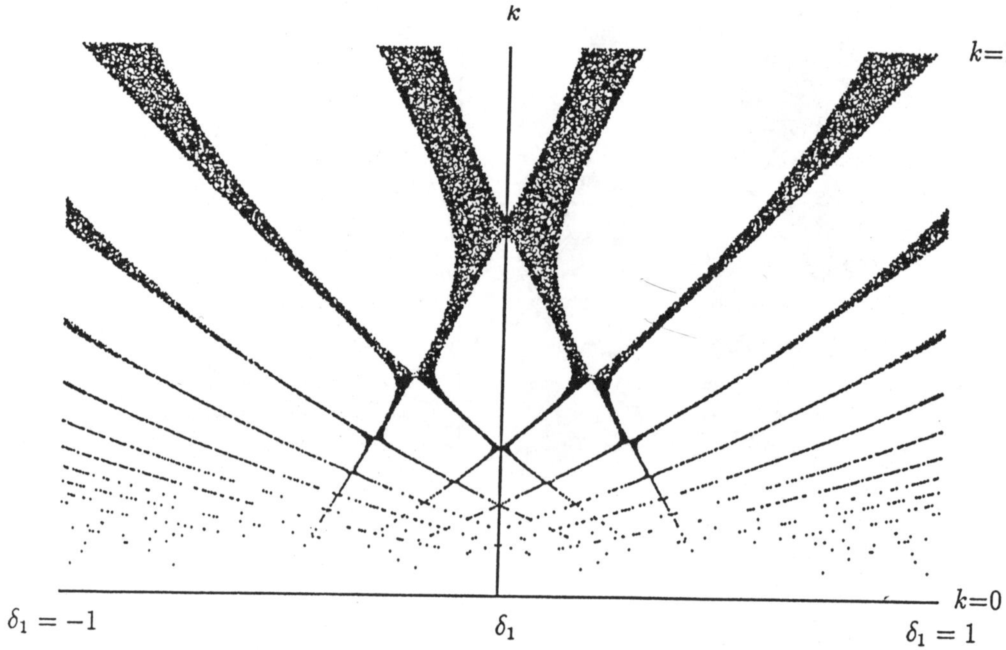

Fig.5. Stability of Eq.(2) as obtained by singular perturbation method.

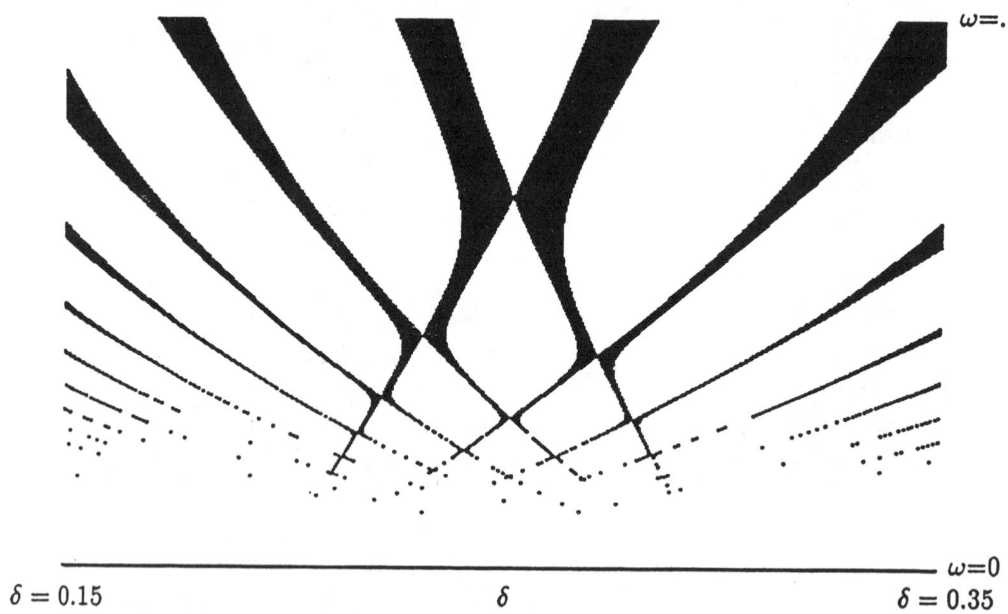

Fig.6. Stability of Eq.(2) for $\epsilon = 0.1$ from numerical integration.

IMPROVED GALERKIN METHOD IN THE DIMENSION REDUCTION OF NONLINEAR DYNAMICAL SYSTEMS

Alois Steindl, Hans Troger and Josef V. Zemann
Vienna University of Technology
A-1040 Vienna, Austria

Abstract. We apply the Karhunen-Loeve modes to improve the usual Galerkin reduction and compare this result with various other choices of ansatz functions. Moreover we also compare results obtained from a flat Galerkin reduction with those from a nonlinear Galerkin reduction making use of the Approximate Inertial Manifold method.

1. Introduction

An introduction into the problem of dimension reduction of high dimensional or even infinite dimensional dissipative nonlinear dynamical systems can be found in Coullet and Spiegel (1983), Aceves et al. (1986) and specifically related to this paper in Steindl et al. (1998). There it is explained that in cases where Center Manifold theory is not applicable basically either a nonlinear Galerkin reduction may be performed via the Approximate Inertial Manifold (AIM) method or a flat Galerkin reduction may be performed making use of Karhunen-Loeve (KL)-modes (Holmes et al. (1996), Rodriguez and Sirovich (1990)). The presentation of this latter method in its engineering application is the main aim of this paper.

2. Galerkin approximation

Some general comments referring to Galerkin methods may be found in Holmes et al. (1996) and specifically related to this paper in Steindl et al. (1998).

In general, applying Galerkin methods the field variable $\boldsymbol{u}(\boldsymbol{x},t)$ is projected onto a set of basis vectors (functions) $\boldsymbol{\psi}_j(\boldsymbol{x})$ called the Galerkin basis in the form

$$\boldsymbol{u}(\boldsymbol{x},t) = \sum_{j=1}^{m} q_j(t)\boldsymbol{\psi}_j(\boldsymbol{x}). \quad (1)$$

The m basis vectors $\boldsymbol{\psi}_j$, the *ansatz-functions*, belong to a complete set of functions and are usually chosen to satisfy the boundary conditions. Their proper choice is the first question to be answered. A second important question concerns the choice of m.

F. C. Moon (ed.),
IUTAM Symposium on New Applications of Nonlinear and Chaotic Dynamics in Mechanics, 71–80.

The *Karhunen-Loeve* method supplies answers to these two questions insofar that if, first, KL-functions are chosen as ψ_j, a better approximation is achieved than with any other set of ansatz functions of the same dimension m and, second, information on how large m should be taken is supplied in form of an energy estimate.

2.1. KARHUNEN-LOEVE METHOD

The Karhunen-Loeve method, which in the mathematical literature is also called *Proper Orthogonal Decomposition* method has been quite successfully applied for the study of turbulence and coherent structures in fluid flow problems (Sirovich (1987), Holmes et al. (1996)). From an ensemble of p data functions u_i, obtained from experiments or by simulation, it generates a set of deterministic basis functions ϕ_j based on second-order statistics. These basis functions can be used in a Galerkin approximation. In Sirovich (1987), Rodriguez and Sirovich (1990) it is pointed out that this method is especially important if the experiment or the simulation is done at a fixed set of parameter values but then with the reduced low dimensional system the dynamic investigation can be extended to a wide range of parameter values, for example, to perform a bifurcation analysis. However, our results indicate that this might not be generally true.

It is not necessary to present the basic ideas of the KL-method here because this is excellently done in section 3.1 in Holmes et al. (1996). We only remark that an eigenvalue problem of the form

$$(R(x,x'),\phi(x')) = \int_D R(x,x')\phi(x')dx' = \lambda\phi(x). \tag{2}$$

has to be solved. The generalized function $R(x,x') = E\{u(x)u(x')\}$ is the *covariance* or the *autocorrelation* of $u(x)$. It is assumed that $u(x)$ is a regular function absolutely integrable on any finite region D.

We will show in the following section that the solution of (2) supplies, first, the set of optimal modes ϕ_i and, second, the corresponding eigenvalues λ_i which can be interpreted as a measure of the energy content carried by the corresponding mode. By optimal modes or vectors it is understood that they approximate the data in such a way that they are parallel to the axis of the inertia ellipsoid of the cloud of data points (see, for example, Fig. 1). We note that mathematically, besides the statistical aspect, a problem is given analogous to the calculation of the principal moments of inertia of a body in $\mathbb{R}^3$ (Parkus (1966), section 2.5).

2.1.1. *Karhunen-Loeve method applied to Galerkin approximation*

In this section, following Kirby (1993), we explain in detail how one has to treat a specific dynamic problem for which the equations of motion are given.

We assume that an ensemble $\{\boldsymbol{u}_i\}_{i=1}^{p}$ of p pattern vectors is obtained from numerical simulation, where each $\boldsymbol{u}_i \in V = I\!R^n$. Let B be an arbitrary basis $B = \{\boldsymbol{v}_1, \ldots, \boldsymbol{v}_n\}$ such that any $\boldsymbol{u}_i$ can be represented by

$$\boldsymbol{u}_i = \sum_{j=1}^{n} b_{ij} \boldsymbol{v}_j \; . \tag{3}$$

We define now the ensemble average of a set of p pattern vectors by

$$E\{\boldsymbol{u}\} = \langle \boldsymbol{u} \rangle = \frac{1}{p} \sum_{i=1}^{p} \boldsymbol{u}_i \; . \tag{4}$$

Often in practical problems a time continuous quantity $\boldsymbol{u}(t)$ is either measured or computed as we have done it for the examples to follow. Then, typically, values of this quantity are stored at discrete time steps t_i. If the time steps are equidistant, one can write $\boldsymbol{u}(t_i) = \sum_{j=1}^{n} b_j(t_i)\boldsymbol{v}_j$. The averaging process $\langle \boldsymbol{u} \rangle = \lim_{t\to\infty} \frac{1}{T} \int_0^T \boldsymbol{u}(t)dt$ yields now, replacing the integral by a finite sum, $\langle \boldsymbol{u} \rangle = \langle \boldsymbol{u}(t_i) \rangle = \frac{1}{p\Delta t} \sum_{i=1}^{p} \left(\sum_{j=1}^{n} b_j(t_i)\boldsymbol{v}_j \right) \Delta t$. Since Δt cancels out we again end up with the notation of (4) if we use $\boldsymbol{u}_i = \boldsymbol{u}(t_i)$. The aim of our analysis is to choose a new basis $\widetilde{B} = (\boldsymbol{\phi}_1, \ldots, \boldsymbol{\phi}_n)$ in which we represent $\boldsymbol{u}_i$ in the form

$$\boldsymbol{u}_i = \sum_{j=1}^{n} a_{ij} \boldsymbol{\phi}_j . \tag{5}$$

This new basis $\{\boldsymbol{\phi}_j\}$ should be optimal in the sense that the ensemble average error e_{ms} of the truncated expansion $\boldsymbol{u}_i \approx \boldsymbol{u}_{im} = \sum_{j=1}^{m} a_{ij}\boldsymbol{\phi}_j$ will have minimal error. In the general theory (Holmes et al. (1996), Lumley (1970)) it is explained that in order to formulate a meaningful minimization problem the quadratic expression $e_{ms} = \langle \|\boldsymbol{u} - \boldsymbol{u}_m\|^2 \rangle$ for the mean square error e_{ms} of an ensemble of pattern vectors must be introduced. The task to be performed can now be formulated such that we want to perform an optimal decomposition of the vectors $\boldsymbol{u}_i = \boldsymbol{u}_{im} + \boldsymbol{e}_i$, where $\boldsymbol{u}_{im} \in W_m$, $\boldsymbol{e}_i \in W_m^{\perp}$ with $V = W_m \oplus W_m^{\perp}$, such that the mean square error of the truncation is minimal. With (5) we obtain $\boldsymbol{e}_i = \boldsymbol{u}_i - \boldsymbol{u}_{im} = \sum_{j=m+1}^{n} a_{ij}\boldsymbol{\phi}_j$. Inserting into the expression for e_{ms} we have

$$\begin{aligned} e_{ms} &= \left\langle \left(\sum_{j=m+1}^{n} a_{ij}\boldsymbol{\phi}_j, \sum_{k=m+1}^{n} a_{ik}\boldsymbol{\phi}_k \right) \right\rangle = \left\langle \sum_{j,k=m+1}^{n} a_{ij}a_{ik}(\boldsymbol{\phi}_j, \boldsymbol{\phi}_k) \right\rangle \\ &= \left\langle \sum_{j=m+1}^{n} (a_{ij})^2 \right\rangle = \left\langle \sum_{j=m+1}^{n} (\boldsymbol{u}, \boldsymbol{\phi}_j)^2 \right\rangle . \end{aligned} \tag{6}$$

Here we have assumed that $(\boldsymbol{\phi}_j, \boldsymbol{\phi}_k) = \delta_{jk}$. We further note that we may write $(\boldsymbol{u}, \boldsymbol{\phi})^2 = (\boldsymbol{u}^T\boldsymbol{\phi})^T(\boldsymbol{u}^T\boldsymbol{\phi}) = \boldsymbol{\phi}^T\boldsymbol{u}\boldsymbol{u}^T\boldsymbol{\phi} = (\boldsymbol{\phi}, \boldsymbol{u}\boldsymbol{u}^T\boldsymbol{\phi})$. This expression allows to introduce the *ensemble averaged covariance matrix*

$$\mathbf{C} = \langle \boldsymbol{u}\boldsymbol{u}^T \rangle, \tag{7}$$

where $\boldsymbol{u}$ is given by (3). Provided the $\boldsymbol{v}_j$ in B are orthonormal the entries of the matrix $\mathbf{C}$ are given by the quadratic terms $b_{ij}b_{ik}$.

Example 1: Let $\boldsymbol{u}_i = b_{i1}\boldsymbol{v}_1 + b_{i2}\boldsymbol{v}_2$, where $\boldsymbol{v}_1 = (1,0)^T$ and $\boldsymbol{v}_2 = (0,1)^T$. Then

$$\boldsymbol{u}_i\boldsymbol{u}_i^T = (b_{i1})^2\boldsymbol{v}_1\boldsymbol{v}_1^T + b_{i1}b_{i2}\boldsymbol{v}_1\boldsymbol{v}_2^T + b_{i2}b_{i1}\boldsymbol{v}_2\boldsymbol{v}_1^T + (b_{i2})^2\boldsymbol{v}_2\boldsymbol{v}_2^T .$$

Under consideration of

$$\boldsymbol{v}_1\boldsymbol{v}_1^T = \begin{pmatrix} 1 & 0 \\ 0 & 0 \end{pmatrix}, \qquad \boldsymbol{v}_1\boldsymbol{v}_2^T = \begin{pmatrix} 0 & 1 \\ 0 & 0 \end{pmatrix},$$
$$\boldsymbol{v}_2\boldsymbol{v}_1^T = \begin{pmatrix} 0 & 0 \\ 1 & 0 \end{pmatrix}, \qquad \boldsymbol{v}_2\boldsymbol{v}_2^T = \begin{pmatrix} 0 & 0 \\ 0 & 1 \end{pmatrix}$$

one obtains

$$\boldsymbol{u}_i\boldsymbol{u}_i^T = \begin{pmatrix} (b_{i1})^2 & b_{i1}b_{i2} \\ b_{i2}b_{i1} & (b_{i2})^2 \end{pmatrix}.$$

Averaging over i according to (4) results in the matrix $\mathbf{C}$ given by (7).

We continue now with the formulation of the minimization problem of e_{ms} by rewriting e_{ms} in the form

$$e_{ms} = \left\langle \sum_{j=m+1}^{n} (\boldsymbol{\phi}_j, \boldsymbol{u}\boldsymbol{u}^T\boldsymbol{\phi}_j) \right\rangle = \sum_{j=m+1}^{n} (\boldsymbol{\phi}_j, \langle \boldsymbol{u}\boldsymbol{u}^T \rangle \boldsymbol{\phi}_j) = \sum_{j=m+1}^{n} (\boldsymbol{\phi}_j, \mathbf{C}\boldsymbol{\phi}_j).$$

For the minimization of e_{ms} in addition the orthonormality condition for the $\boldsymbol{\phi}_j$ must hold. Hence, we form the functional of a constrained minimization problem by means of the Lagrange multiplier method

$$\begin{aligned} L(\boldsymbol{\phi}) &= e_{ms} - \sum_{j=m+1}^{n} \lambda_j((\boldsymbol{\phi}_j, \boldsymbol{\phi}_j) - 1) = \\ &= \sum_{j=m+1}^{n} \left[(\boldsymbol{\phi}_j, \mathbf{C}\boldsymbol{\phi}_j) - \lambda_j((\boldsymbol{\phi}_j, \boldsymbol{\phi}_j) - 1) \right] . \end{aligned}$$

The necessary condition for an extremal value is given by setting the (Fréchet) derivative equal to zero. This yields

$$(\mathbf{C} - \lambda_j \mathbf{I})\boldsymbol{\phi}_j = \mathbf{0}\,, \qquad j = m+1, \ldots, n\,.$$

The solution of the problem of the search for the optimal eigenvectors yields for all vectors $\boldsymbol{\phi}_j$ the same eigenvalue problem $\mathbf{C}\boldsymbol{\phi} = \lambda\boldsymbol{\phi}$, where the $n \times n$ matrix $\mathbf{C}$ is defined by (7).

We mention two useful properties of the KL-method:

Property 1: Since $a_k = (\boldsymbol{u}, \boldsymbol{\phi}_k)$ we obtain

$$\begin{aligned} \langle a_k a_j \rangle &= \langle (\boldsymbol{u}, \boldsymbol{\phi}_k)(\boldsymbol{u}, \boldsymbol{\phi}_j) \rangle = \langle (\boldsymbol{\phi}_k, \boldsymbol{u}\boldsymbol{u}^T \boldsymbol{\phi}_j) \rangle = (\boldsymbol{\phi}_k, \langle \boldsymbol{u}\boldsymbol{u}^T \rangle \boldsymbol{\phi}_j) \\ &= (\boldsymbol{\phi}_k, \mathbf{C}\boldsymbol{\phi}_j) = (\boldsymbol{\phi}_k, \lambda_j \boldsymbol{\phi}_j) = \lambda_j \delta_{kj} \,. \end{aligned}$$

Hence the eigenvalue $\lambda_j = \langle a_j a_j \rangle$ corresponds to the statistical variance or can be considered to be a measure of the "kinetic energy" in the j-th KL-mode. If the eigenvalues are normalized as probabilities the percentual energy content of an approximation can be estimated.

Property 2: The eigenvalues of $\mathbf{C}$ are non-negative. From Property 1 we have

$$\lambda_j = \langle a_j^2 \rangle \geq 0 \,, \qquad j = 1, \ldots, m \,.$$

The basis $\{\boldsymbol{\phi}_j\}$ generated by the eigenvectors of $\mathbf{C}$ is ordered in accordance to the spectrum of $\mathbf{C}$ such that $\lambda_1 \geq \lambda_2 \geq \ldots \geq \lambda_m \geq 0$, where $\boldsymbol{\phi}_i$ corresponds to λ_i. Obviously e_{ms} will be a minimum due to the ordering since

$$e_{ms} = \sum_{j=m+1}^{n} (\boldsymbol{\phi}_j, \mathbf{C}\boldsymbol{\phi}_j) = \sum_{i=m+1}^{n} (\boldsymbol{\phi}_j, \lambda_j \boldsymbol{\phi}_j) = \sum_{j=m+1}^{n} \lambda_j .$$

3. Applications

3.1. DOUBLE PENDULUM WITH FOLLOWER FORCE

We study the planar large amplitude motion of a double pendulum (Fig. 1) consisting of two rigid rods of same length ℓ, connected by linearly viscoelastic hinges and loaded by a follower force F, which always has the direction of the second rod (Troger and Steindl (1991), section 4.1). As degrees of freedom the angles ψ_1 and ψ_2 measuring the deflection of the two rods against the unstrained vertical position are introduced. The equations of motion are given in (Troger and Steindl (1991), section 4.1).

To study the stability problem of the equilibrium $\psi_1 = \psi_2 = 0$ the equations of motion are linearized about the equilibrium position. Fixing the stiffness and damping parameters to $\gamma_1 = \gamma_2 = k_1 = 1.0$ and $k_2 = 0.6$ the eigenvalue problem yields the critical value of the follower force $F = F_c =$

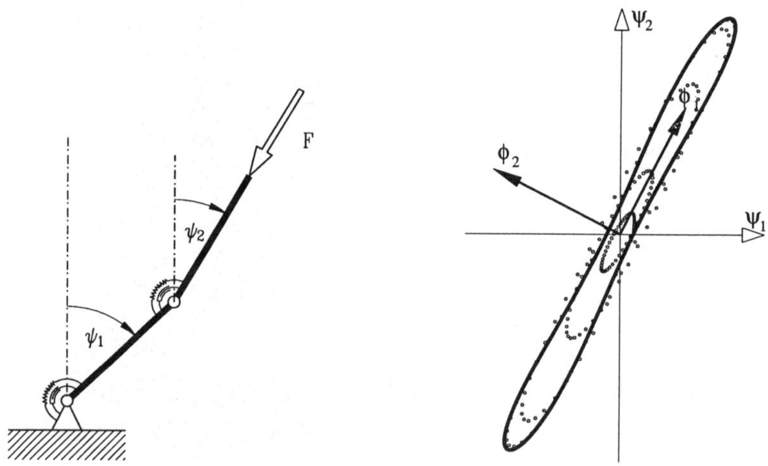

Figure 1. Left: Double pendulum with two degrees of freedom ψ_1 and ψ_2, lodaded with a follower force F; right: Cloud of simulation results given by the limit cycle and the dots marking the the initial transient motion and the corresponding KL-eigenvectors $\boldsymbol{\phi}_1$ and $\boldsymbol{\phi}_2$

2.196 for loss of stability of the streched state. It is well known (Troger and Steindl (1991)) that for such a double pendulum due to a Hopf bifurcation at $F = F_c$ a stable (supercritical) limit cycle bifurcates from the equilibrium. Now we simulate the nonlinear equations of motion for $F = 3.0 > F_c$, that is in the domain where the stable limit cycle exists. Using the representation $\boldsymbol{u} = (\psi_1, \psi_2)^T$ we take $p = 2000$ samples with the stepsize $\Delta t = 0.25$ sec from the simulation data. According to $\boldsymbol{u}_i = \boldsymbol{u}(t_i)$ and (4) we calculate the covariance matrix (7) for minimizing the error in position (p) and velocity (v). These covariance matrices are

$$\mathbf{C}_p = \begin{pmatrix} 0.2952 & 0.5587 \\ 0.5587 & 1.1000 \end{pmatrix}, \qquad \mathbf{C}_v = \begin{pmatrix} 0.1050 & 0.1965 \\ 0.1965 & 0.3946 \end{pmatrix}.$$

Solving the respective eigenvalue problem yields

$$\begin{aligned} \lambda_1^p &= 1.386, & \boldsymbol{\phi}_1^p &= \begin{pmatrix} .4559 \\ .8900 \end{pmatrix}, \\ \lambda_2^p &= 0.8974 \cdot 10^{-2}, & \boldsymbol{\phi}_2^p &= \begin{pmatrix} -.8900 \\ .4559 \end{pmatrix}, \\ \lambda_1^v &= .4939, & \boldsymbol{\phi}_1^v &= \begin{pmatrix} .4510 \\ .8925 \end{pmatrix}, \\ \lambda_2^v &= .5693 \cdot 10^{-2}, & \boldsymbol{\phi}_2^v &= \begin{pmatrix} -.8925 \\ .4510 \end{pmatrix}. \end{aligned}$$

The vector $\boldsymbol{\phi}_2$ is rotated by 90° against $\boldsymbol{\phi}_1$. We see that the flutter motion is strongly coupled and that the first mode is more significant than the second

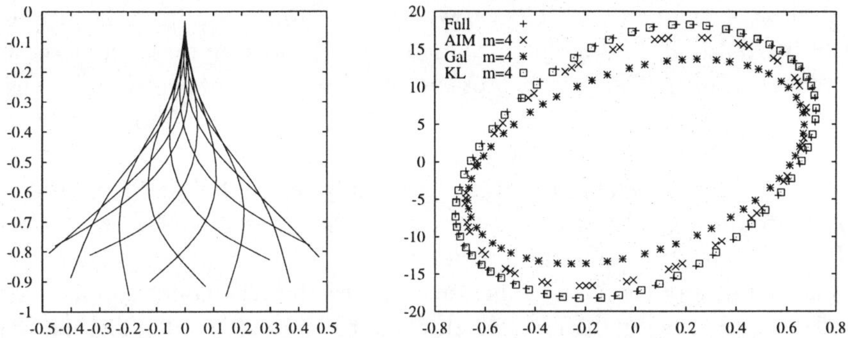

Figure 2. Left: Simulation of the fluid conveying tube for $\varrho = 10.0$ for a discretization with $n = 32$ elements; right: Comparison of the simulation results with 4-dimensional approximations with flat Galerkin (KL-modes and eigenmodes) and AIM (eigenmodes)

one. This is best seen by normalizing the eigenvalues as probabilities:

$$(p) \quad \tilde{\lambda}_1^p = 0.9778, \quad \tilde{\lambda}_2^p = 0.0222, \qquad (v) \quad \tilde{\lambda}_1^v = .9886, \quad \tilde{\lambda}_2^v = .0113.$$

Moreover from Fig. 1 it can be nicely seen how the KL-vectors (the optimal vectors) point into the direction of the axes of the inertia ellipsoid of the simulation data (the cloud of data points). The scattered points in Fig. 1 represent the transient motion, after loss of stability of the straight position, which finally runs into the limit cycle which is given by the full line. The slight differences between the results following from the position and velocity data can be explained by the fact that higher modes may be of different significance in one or the other of these.

3.2. Oscillation of a Fluid Conveying Tube

We study the self excited planar large amplitude oscillations of the fluid conveying tube treated also in Steindl et al. (1998) where a short description of the mechanical model and the derivation of the equations of motion is presented.

Analogously to the behaviour of the double pendulum for the dimensionless flow rate $\varrho = \varrho_c = 8.027$, a loss of stability of the downhanging tube by a supercritical Hopf bifurcation takes place (Troger and Steindl, 1991).

In the left frame in Fig. 2 we present simulation results for the tube oscillations at the flow rate $\varrho = 10.0 > \varrho_c = 8.027$. These results were obtained from a Finite Difference dicretization of the continuous tube with 16 elements resulting in a 32 dimensional system. From these data the KL-basis is calculated from 1000 data points taken at equidistant intervals Δt.

In the right frame of Fig. 2 results of the full 32-dimensional model designated by $FULL$ are compared with a flat Galerkin (Gal) approximation with $m = 4$, a nonlinear Galerkin with $m = 4$ following from the Approximate Inertial Manifold method (Steindl et al. (1998)). In these cases the first eigenmodes of the linearized problem at $\varrho = \varrho_c$ have been used as ansatz functions. Further a flat Galerkin using the first four KL-modes is shown. The reason that we had to use four KL-modes, instead of two, is that with two modes the trajectory converged to a steady state solution. We see that the KL approximation gives the best match to the simulation data but also the AIM approximation gives a very good result. Worst is the flat Galerkin. As a third possibility for ansatz functions, besides the KL-modes and the eigenmodes we also use in the results below the beam modes of the clamped tube (without flowing fluid) for comparison.

In order to see which modes make relevant contributions to the motion of the tube we consider the limit cycle of Fig. 3 and calculate the average contribution $\overline{a}_k$ of the k-th mode by

$$\overline{a}_k^2 = \frac{1}{J}\sum_{i=1}^{J} a_k^2(t_i).$$

Here $J = 1000$ is used. The results are shown for three different choices of modes in Fig. 3. Namely for the eigenmode basis in the left upper frame, for the beam modes, which are sorted according to their node number, in the right upper frame and for the Karhunen Loeve modes in the left lower frame. Obviously there are strong qualitative differences. Whereas for the beam modes (right upper frame) and for the Karhunen Loeve modes (left lower frame) the contribution of the higher modes decays monotonically, this is not the case for the eigenmode basis, obtained from the solution of the linearized fluid conveying tube problem and which is sorted according to the magnitude of the realparts of the eigenvalues at ϱ_c (left upper frame). It is clearly visible that there are some eigenvectors (e. g. modes 19 and 20) located further left in the complex plane which in average contribute more than several ones located closer to the imaginary axis. To try to explain this remarkable phenomenon in this frame also the contribution of the various modes at larger values of the flow rate are depicted. It is clearly visible that for a Center Manifold reduction at $\varrho = \varrho_c$ the relevant contribution comes from the first two modes. However for increasing values of the flow rate ϱ the contribution from modes corresponding to eigenvalues located further left in the complex plane at ϱ_c increases. This, certainly, is a nonlinear effect. Our explanation is that these eigenvectors are a small perturbation of low order beam modes and are strongly excited by the nonlinearities. Finally we present the normalized eigenvalues of the KL-approximation in the table in Fig. 3. The energy distribution of the approximation on the various modes is clearly visible.

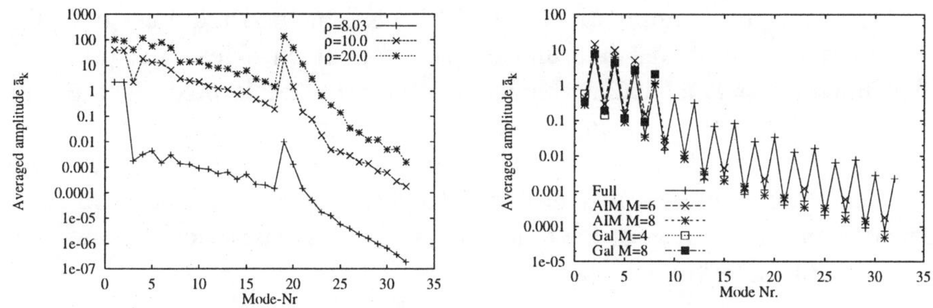

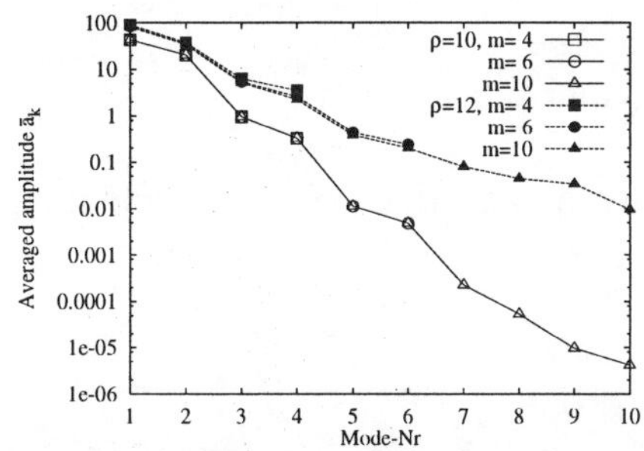

$m = 4$	6	10
0.8147	0.8155	0.8156
0.1849	0.1840	0.1840
0.407e-3	0.409e-3	0.408e-3
0.466e-6	0.464e-6	0.464e-6
	0.559e-9	0.558e-9
	0.103e-9	0.103e-9

Figure 3. Averaged contribution of each of the first 32 eigenvectors (modes) to the limit cycle oscillation. Left upper frame: eigenmodes at various values for the flow rate ϱ; right upper frame: beam modes at $\varrho = 10$; left lower frame: KL modes at $\varrho = 10$ and 12; right lower frame: normalized KL eigenvalues at $\varrho = 10$

4. Conclusions

The numerical results show that for the large amplitude oscillations of a fluid conveying tube modelled by an infinite dimensional mechanical system and described by a set of nonlinear partial differential equations a reduction to a finite dimensional system of amplitude equations is possible using different sets of ansatz functions and different methods of reduction. Among the different sets of ansatz functions the Karhunen-Loeve modes in our case are by far the best choice for a flat Galerkin approximation, because the contribution of the higher modes decays most rapidly. Selecting the method of reduction certainly the nonlinear Galerkin method will be superior to the flat Galerkin method for an arbitrary choice of the ansatz functions. This is especially important if the dimension of the reduced system is chosen to be very small, which always will be the goal of an reduction process, because only for a low dimensional system a reasonable qualitative analysis can be performed. That the results of the KL approximation for the tube oscillations are so good could be a consequence that the limit cycle, from which we have taken the data, can be very well approximated by a planar structure. Therefore only two modes make an essential contribution to the energy content.

Concerning the computational effort the KL-method requires, first, the generation of simulation data and the solution of an eigenvalue problem. Then, however, only a flat Galerkin reduction must be performed. For the nonlinear Galerkin method the ansatz functions are usually given. In the case that the eigenfunctions of the linearized problem are used they are obtained from the calculation of the stability problem. However for the nonlinear Galerkin reduction the calculation of the reduced system is at least twice as expensive as for the flat Galerkin method.

Acknowledgement

This research project has been supported by the Austrian Science Foundation (FWF), under the project P 10705–MAT.

References

Aceves, A., Adachihara, H., Jones, C., Lerman, J. C., McLaughlin, D. W., Moloney, J. V., Newell, A. C., 1986. Chaos and Coherent Structures in Partial Differential Equations. *Physica* D 18, 85–112.

Coullet, P. H., Spiegel, A., 1983. Amplitude Equations for Systems with Competing Instabilities. *SIAM J. Appl. Math.* 43, 776–821.

Holmes, P.J., Lumley J.L. and Berkooz G. 1996. *Turbulence, Coherent Structures, Dynamical Systems and Symmetry Cambridge Monographs on Mechanics.* Cambridge University Press, Cambridge UK.

Kirby, M., 1993. *Lecture Notes on Pattern Analysis,* TU-Wien.

Lumley, J. L., 1970. *Stochastic Tools in Turbulence,* Academic Press, New York – London.

Parkus, H., 1966. *Mechanik fester Körper.* Springer Verlag, Wien – New York.

Rodriguez, J. D., Sirovich, L., 1990. Low-Dimensional Dynamics for the Complex Ginzburg-Landau Equation. *Physica* D 43, 77–86.

Sirovich, L., 1987. Turbulence and the Dynamics of Coherent Structures. Part I: Coherent Structures; Part II: Symmetries and transformations; Part III: Dynamics and Scaling. *Quaterly of Appl. Math.,* vol. XLV, 561–590.

Steindl, A., Troger, H., 1996. Equations of Motion of a Fluid Conveying Tube. In K. Kirchgässner, O. Mahrenholtz, and R. Mennicken, editors, *ICIAM 95*, volume 87 of *Mathem. Research*, pages 533–536. Akademie-Verlag, 1996.

Steindl, A., Dimension Reduction Methods for a Fluid Conveying Tube. *ZAMM* 77, S321-S322.

Steindl, A., Troger, H., Zemann V. 1998. Nonlinear Galerkin Methods Applied in the Dimension Reduction of Vibrating Fluid Conveying Tubes, Proceedings of the 4th Int. FSE, AE & FIV + N Symposium at the 1997 ASME Congress in Dallas.

Troger, H., Steindl, A., 1991. *Nonlinear Stability and Bifurcation Theory, An Introduction for Engineers and Applied Scientists.* Springer Verlag, Wien – New York.

SEQUENCES OF GLOBAL BIFURCATIONS AND MULTIPLE CHAOTIC TRANSIENTS IN A MECHANICAL DRIVEN OSCILLATOR

W. Szemplińska-Stupnicka, E. Tyrkiel
Institute of Fundamental Technological Research
Polish Academy of Sciences
Świętokrzyska 21, 00-049 Warsaw, POLAND

ABSTRACT: In the paper we study the phenomena of multiple chaotic transient motions in a driven, dissipative nonlinear oscillator. The numerical analysis shows how the occurrence of multiple chaotic transients is related to development of various patterns of the fractal structure of basins of attraction of coexisting attractors, and how the metamorphoses of the basin boundaries are dictated by a sequence of various types of global bifurcations.

1. Introduction

The phenomenon of chaotic transient motion can be dangerous for the dynamic engineering systems which are exposed to sudden perturbations, such as blows of wind, water waves or other sources. In linear or weakly nonlinear vibrating system the transient motion is predictable: one can estimate the rate of decay and the time the transient motion lasts if the coefficient of damping and the magnitude of perturbation are given. In contrast, the transient motion that is chaotic is unpredictable: one is unable to foresee how long it will last and what is the final outcome. This type of „transients" can look like steady-state chaos for a long time and suddenly can settle onto one or another coexisting attractor. Moreover, the phenomenon can be generated at the level of amplitude of forcing term which is significantly lower than that at which the system exhibits permanent chaos. This makes the question of necessary conditions for the chaotic transient to occur to be an essential point in the analysis of dynamic responses in engineering systems.

It is known that the chaotic transients are strictly related to the existence of a fractal structure of basins of attraction of the coexisting attractors and that the fractal basin boundaries are generated by global (homoclinic and heteroclinic) bifurcations. It follows that the two basic concepts of the theory of nonlinear dynamics: the global bifurcation and the metamorphoses of basins of attraction, play an essential role in the investigations of the critical thresholds in the system control parameter values, for which chaotic transients can occur.

In this paper the global bifurcations and the related new types of chaotic transients are exemplified by using an archetypal twin-well potential Duffing oscillator governed by the equation:

F. C. Moon (ed.),
IUTAM Symposium on New Applications of Nonlinear and Chaotic Dynamics in Mechanics, 81–91.

$$\ddot{u} + k\dot{u} - \alpha u + \beta u^3 = A\cos\varpi\tau, \qquad k, \alpha, \beta > 0, \qquad T = 2\pi/\omega. \tag{1}$$

At $F = 0$ the system has the two stable rest positions $u^0_{1,2} = \pm\sqrt{\alpha/\beta}$ and the unstable one $u^0_3 = 0$.

An attention is focused on the control parameters region at which the single-well T-periodic attractors do exist, that is, at the driving frequency lower than that of the saddle-node bifurcation of the nonresonant attractor (to the left of the *fold A* in Fig. 1b). In particular, two domains of the control parameters are studied in detail [1]-[4]:

- the domain where the hysteresis of the primary resonance occurs;
- the domain where the T-periodic cross-well attractor exists and, due to a codimension-two bifurcation, coexists with one of the two different types of attractors (cross-well chaotic attractor or the single-well T-periodic nonresonant attractor).

2. Multiple chaotic transient motions in the region of the primary resonance hysteresis

To reduce the number of parameters we transform the equation (1) to the standard form by introducing new time scale and a new independent space variable. We do this by applying the nondimensional time $t = \tau\,\Omega_0$, where $\Omega_0 = \sqrt{2\alpha}$ stands for the natural frequency of small oscillations around the stable equilibrium position; the nondimensional displacement is assumed as $x = u\sqrt{\beta/\alpha}$.

The twin-well Duffing equation reduced to the standard form is:

$$\ddot{x} + h\dot{x} - \tfrac{1}{2}u + \tfrac{1}{2}u^3 = F\cos\omega t, \tag{2}$$

where: $$h = \frac{k}{\sqrt{2\alpha}}, \qquad F = \frac{A}{2\alpha}\sqrt{\frac{\beta}{\alpha}}, \qquad \omega = \frac{\varpi}{\sqrt{2\alpha}}.$$

The two stable rest positions (at $F = 0$) are now situated at the values $x^0_{1,2} = \pm 1$, and the natural frequency of small oscillations within a single potential well takes the value $\Omega_0 = 1$. We fix throughout on the value of damping coefficient $h = 0.1$.

First we consider the region of the control parameters F, ω where the primary resonance hysteresis occurs. Due to the softening nature of the stiffness the nonlinear amplitude-frequency response curves skew towards lower frequencies than that of the linear natural frequency and thus, in a certain zone of the driving frequency, there coexist two attractors within each potential well: resonant attractor S_r and nonresonant attractor S_n (Fig. 1a). In the system control parameter plane F - ω the two T-periodic attractors coexist within the triangle-like domain confined by the arcs denoted as *fold B*, *fold A* and *PD* (Fig. 1b). Within the narrow strip - between the *PD* and *crisis* lines - the resonant attractor S_r undergoes cascade of period-doubling bifurcations, turns into a single-well chaotic attractor, and finally is destroyed in a boundary crisis scenario. Therefore, the smooth response curve with the *fold B* at the top of the resonance occurs only within the region of the forcing parameter $F_0 < F < F_1$, and the curve where the period-doubling and vanishing of the resonant attractor is observed occurs at $F_1 < F < F_2$. The

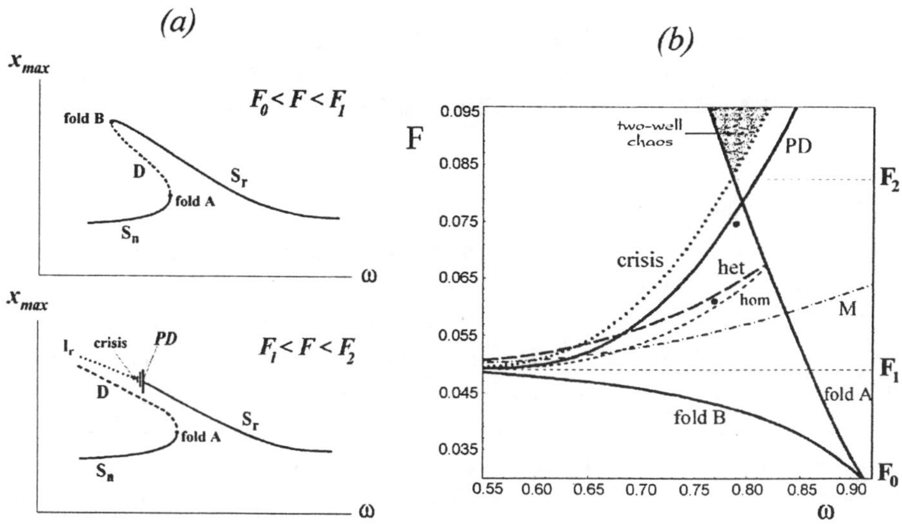

Fig.1. (a) Response-frequency diagram at two levels of the forcing parameter F; (b) bifurcation diagram in the $F-\omega$ plane.

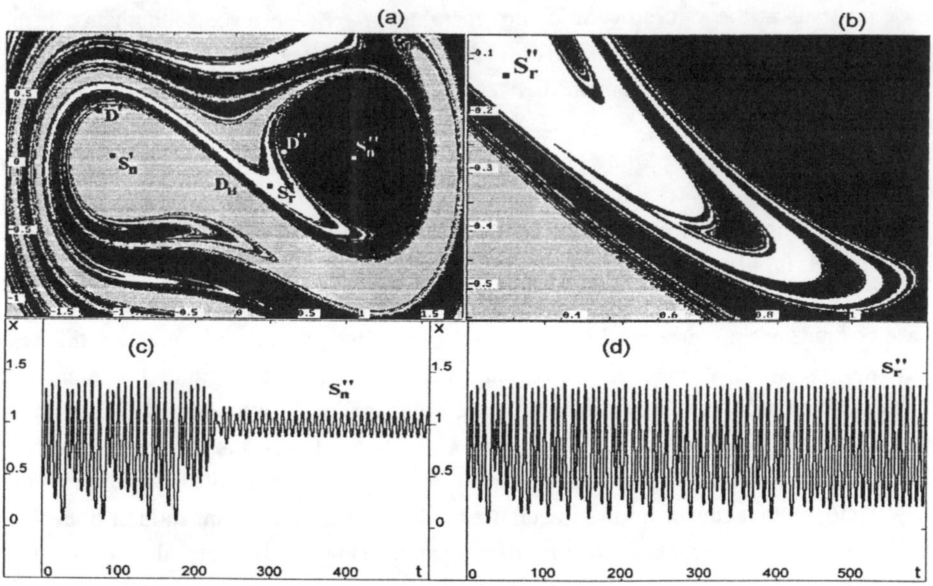

Fig.2. $F = 0.058$, $\omega = 0.77$; (a) basins of attractor at $F_{hom} < F < F_{het}$; (b) a blow-up of the basin boundary in the neighborhood of the attractor S_r^-; (c) chaotic transients for initial conditions $x(0) = 0.7941$, $\dot{x}(0) = -0.377$; (d) chaotic transients for initial conditions $x(0) = 0.7941$, $\dot{x}(0) = -0.3768$.

two response resonance curves which refer to the two regions of the forcing parameter are depicted in Fig. 1a. Note, that at $F > F_2$ the boundary crisis of the single-well chaotic attractor generates the cross-well chaotic attractor, and both single-well periodic orbits cease to exist.

To determine the necessary conditions for an occurrence of chaotic transient motion, we find numerically the sequence of the global bifurcations that occurs due to a coexistence of the hilltop saddle D_H and the single-well saddles D' and D'' (in the left and right potential well, respectively), and then explore effects of the bifurcations on the development of a fractal structure of the basins of attraction. The sequence is demonstrated in the F - ω plane (Fig. 1b). The curve M denotes the homoclinic bifurcation of the hilltop saddle D_H (known as the Melnikov criterion), and this is the only global bifurcation which occurs in the domain of the resonance hysteresis as well as in the domain on the right of the *fold A*, where only single resonant attractor exists in each potential well. With an increase of the forcing parameter F the homoclinic bifurcation of each of the single-well saddles (D', D'') is observed and denoted as *hom*. It is followed by the heteroclinic bifurcation which involves the insets of the hilltop saddle D_H and one of the outsets of each single-well saddle, namely that one which tends towards the resonant attractor S_r. The latter bifurcation is denoted as *het* in Fig. 1b.

To explore effects of the sequence of global bifurcations on the generation of the fractal structure of basins of attraction we demonstrate and discuss separately the basin-phase portraits at the values of the control parameters F, ω in three subdomains: I. the subdomain between the Melnikov (M) and in-well (*hom*) homoclinic bifurcation, $F_M < F < F_{hom}$; II. the subdomain between the in-well homoclinic (*hom*) and the heteroclinic (*het*) bifurcation, $F_{hom} < F < F_{het}$; III. the subdomain above the heteroclinic bifurcation (*het*), but prior to the loss of stability of both single-well attractors (*PD, fold A*), $F_{het} < F < F_{PD} \cap F_{het} < F < F_{foldA}$.

It turns out that the effects of the homoclinic Melnikov bifurcation within the subdomain I on the fractal structure of the basins' boundaries, although emerging, remain negligible on a macroscopic level of observation, and therefore, the related basin-phase portraits are not presented in the paper. Significant effects are brought by the *hom* bifurcation in the subdomain II: the regular basin boundary that separated in each well the basins of attraction of the resonant S_r and nonresonant S_n attractors is broken and a layer of the fractal basin boundary is formed (Fig. 2a,b). This leads to sensitivity of the system response to initial conditions and so, to the unpredictability of the final outcome: the trajectories that start from the fractal area exhibit chaotic transient motion until they restabilize onto the resonant or nonresonant attractor (Fig. 2c,d). Note, that this type of chaotic transient is confined to the motion within a single potential well and, therefore, the phenomenon is not restricted to the twin-well potential oscillator.

One may expect that due to the Melnikov homoclinic bifurcation the system has also possibility of generating the cross-well transient chaos. But the fractal layer of the basin

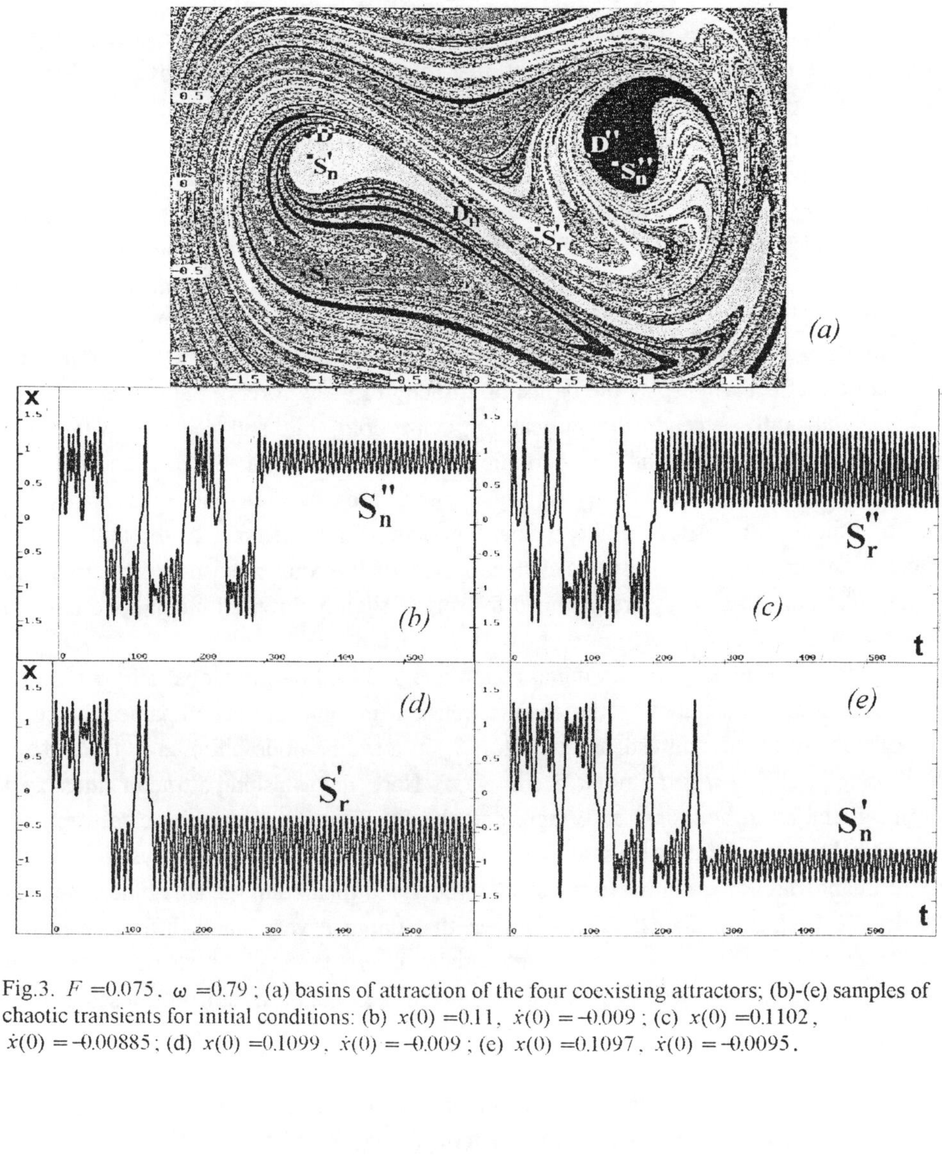

Fig.3. $F = 0.075$, $\omega = 0.79$; (a) basins of attraction of the four coexisting attractors; (b)-(e) samples of chaotic transients for initial conditions: (b) $x(0) = 0.11$, $\dot{x}(0) = -0.009$; (c) $x(0) = 0.1102$, $\dot{x}(0) = -0.00885$; (d) $x(0) = 0.1099$, $\dot{x}(0) = -0.009$; (e) $x(0) = 0.1097$, $\dot{x}(0) = -0.0095$.

boundary which involves the basins of attractors from both potential wells is so thin (and situated close to the edges of the basins), that this type of transients is neglected in the subdomain considered.

The fractal structure of the basin-phase portrait that form a tangled mixture of the basins of all four coexisting attractors (two attractors in each well) develops dramatically after crossing the threshold values F, ω which correspond to the heteroclinic bifurcation (line *het* in Fig. 1b) - subdomain III. As the parameters F, ω become more distant from the critical *het* values, the effects of the heteroclinic bifurcation are growing, that is, the subset of initial conditions corresponding to the fractal domain of the basins is increasing, and gradually begins to cover a significant area of the phase space $x(0) - \dot{x}(0)$. To highlight the effect we display the basins of attraction for the forcing parameter value F, which significantly exceeds the critical F_{het} threshold, but still remains within the triangle-like domain where all four periodic attractors coexist (Fig. 3a). In this case, the highly intertwined structure of the basins gives rise to cross-well chaotic transients, which can result in as many as four different final outcomes. Four examples of the time history of chaotic transients which settle finally onto each of the four coexisting attractors, are displayed in Fig. 3b-e. They are obtained for sets of slightly different initial conditions in the close neighborhood of $x(0) = 0.11$, $\dot{x}(0) = -0.009$.

The sensitive dependence on initial conditions, induced by the global bifurcations of the basins, is also illustrated by the chaotic transient motion that accompanies the strict loss of stability of the nonresonant attractor S_n in a saddle-node bifurcation (*fold A*) at the forcing level $F > F_{het}$ ($F = 0.075$, Fig. 4a,b). Here, the vanishing attractor finds itself on the fractal basin boundary and, therefore, the final outcome after the catastrophe is unpredictable. The bifurcation was realized numerically by setting the system on the nonresonant solution just prior to the saddle-node, and then adding a small increment of the driving frequency, $\Delta\omega$. It is clearly seen that under a very small difference in the increment value (0.001) the outcome is essentially different: after long chaotic cross-well transient, the ensuing trajectory restabilizes onto the resonant attractor in the same or in the opposite well (Fig. 4c,d).

3. Effects of the homoclinic bifurcation on chaotic transients in the region of existence of the cross-well T-periodic attractor (Large Orbit)

In this section we focus on the phenomena of transient chaos in the region of system control parameters, where the cross-well T-periodic attractor (called „Large Orbit" in the subsequent text) exists, and can coexist with the cross-well chaotic attractor or with the nonresonant single-well attractors (see Fig. 2 in [2]). In this region the homoclinic bifurcation of the saddle D_L, that is, the saddle associated with the Large Orbit S_L, gives rise to new types of cross-well chaotic transients. We consider the domain of F - ω plane where a codimension-two bifurcation occurs, that is, where there is an intersection of the *fold A* and the critical line of the homoclinic tangency of the saddle D_L (Fig. 5). In this

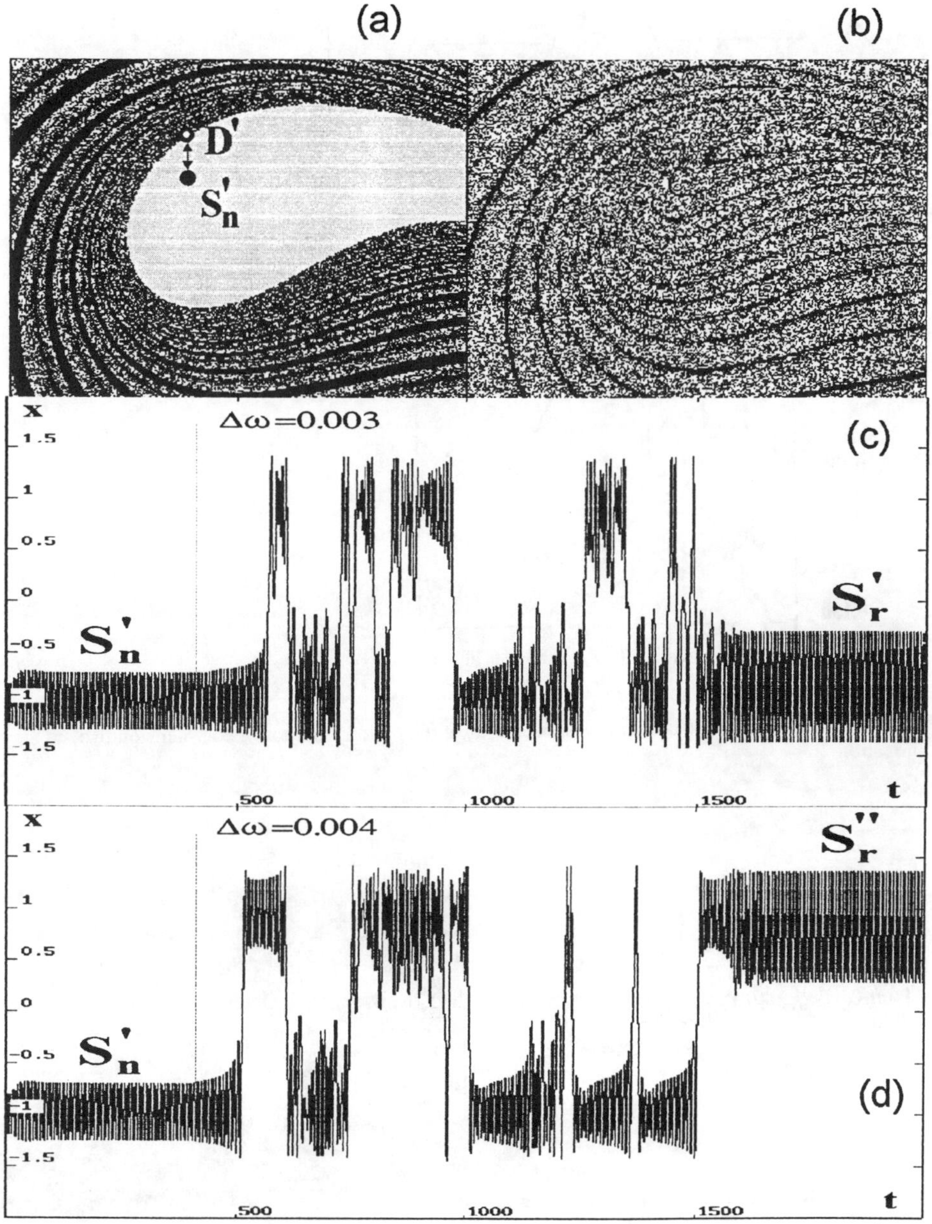

Fig.4. Tangled saddle-node bifurcation of the nonresonant attractor $S_n^{'}$ with unpredictable outcomes. $F = 0.075$ (a) basin phase-portrait prior to the bifurcation, $\omega = 0.8$; (b) basin phase-portrait just after the bifurcation, $\omega = 0.804$; (c), (d) two time histories of the chaotic transients.

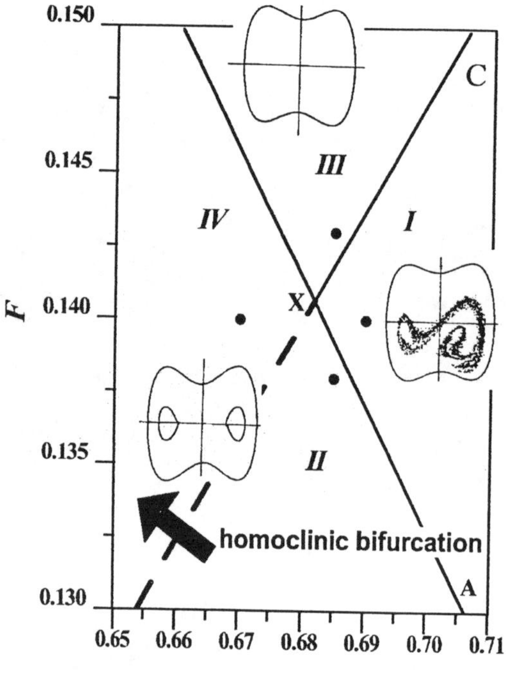

Fig.5. Representation of the bifurcation subdomains in the region close to the codimension two bifurcation

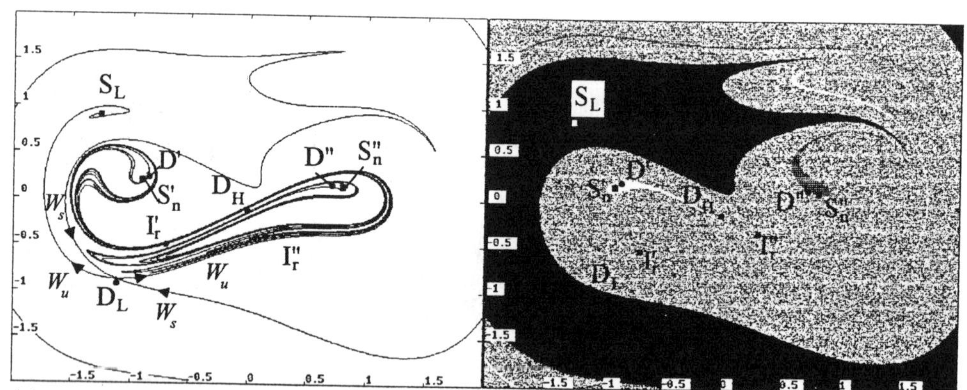

Fig.6. Subdomain II; (a) structure of the invariant manifolds of the saddle D_L; (b) basins of attraction of the three periodic attractors.

figure we denote four bifurcation subdomains: in the subdomain I the Large Orbit coexists with the cross-well chaotic attractor, in the subdomain III the Large Orbit is a unique attractor, and in the subdomains II and IV the coexistence of the Large Orbit and the single-well nonresonant attractors $S_n^{'}, S_n^{''}$ is observed. The homoclinic bifurcation of the saddle D_L is indicated by a solid line and denoted C in the upper part of the figure, while it is drawn by a dashed line in the bottom part (below the intersection point X corresponding to a codimension-two bifurcation). The solid line C defines also the boundary of existence of the cross-well chaotic attractor, which vanishes in the crisis scenario. In contrast, the portion of the homoclinic bifurcation curve which is denoted by the dashed line does not affect the stability and so the existence of any of the coexisting attractors.

The homoclinic bifurcation of the saddle D_L in the two subdomains (II and IV in Fig. 5) manifests itself by a dramatic change of the basins of attraction associated with the Large Orbit S_L and the single-well attractors $S_n^{'}, S_n^{''}$. The effect is illustrated in Fig. 6a,b and Fig. 7a-c. First we show the invariant manifolds of the saddle D_L prior to the homoclinic bifurcation (Fig. 6a): it is seen that the insets and outsets of the saddle do not intersect. It follows that the insets define a regular boundary between the basins of attraction of S_L and the pair of single-well attractors $S_n^{'}$ and $S_n^{''}$ (Fig. 6b). In contrast, the boundary between the basins of attraction of the left and right single-well attractors is fractal. This is, of course, due to the homoclinic bifurcation of the hilltop saddle D_H which occurs at significantly lower value of the forcing parameter F than the value considered.

It is useful to interpret the basins of attraction also in a different way, by treating the sum of the single-well attractors $S_n^{'}$ and $S_n^{''}$ as a union. This splits the whole phase-space into only two basins: the basin of the Large Orbit S_L and the basin of the union $S_n^{'}, S_n^{''}$ (Fig. 7a)

Finally, we move from the subdomain II to the subdomain IV crossing the critical F, ω values for which the homoclinic bifurcation occurs, and observe both the structure of the invariant manifolds of the saddle D_L and the bifurcation of the basins of attraction of S_L and of the union $S_n^{'}, S_n^{''}$. Now the insets and outsets of D_L form a highly intertwined structure, and therefore the regular boundary between the two types of periodic attractors is broken (Fig. 7b,c). The effect of the homoclinic bifurcation on the basin-phase portrait manifests itself by an explosion of the basin of S_L and its invasion into the area formerly occupied by the basin of the union $S_n^{'}$ and $S_n^{''}$. The interior of the latter basin becomes now a tangled mixture of the basins of attraction of the three coexisting attractors - S_L, $S_n^{'}$ and $S_n^{''}$ (except the small tadpole-shape regions, which remain as „regular portions" of the basin of attraction of $S_n^{'}$ and $S_n^{''}$).

Analogously as in the region of resonant hysteresis, the trajectories that start with initial conditions situated in the highly fractal domain of the phase-space $x(0) - \dot{x}(0)$ can be very complex and look like long lasting, cross-well chaotic transients with

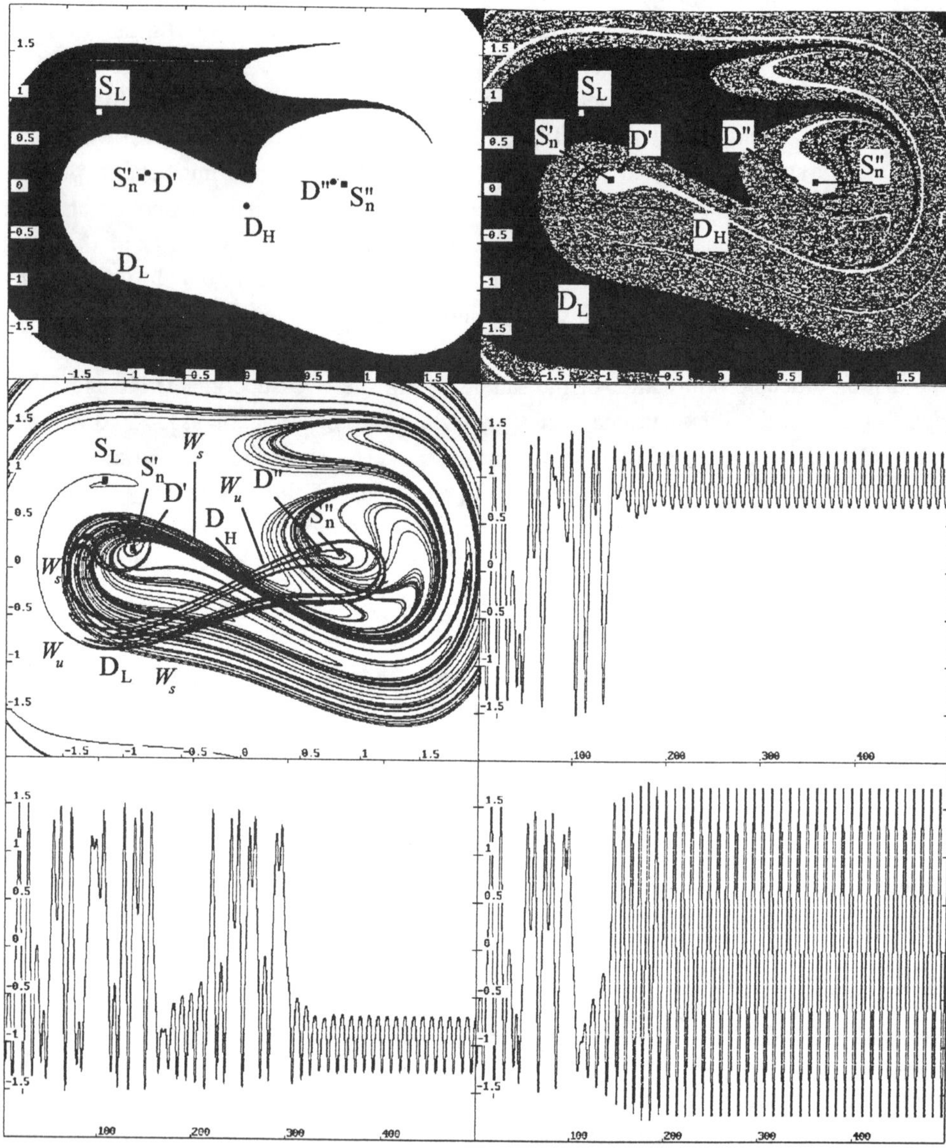

Fig.7. (a), (b) Basins of attraction of the attractor S_L and the union of $S_n^{'}$ and $S_n^{''}$: (a) in the subdomain II, (b) in the subdomain IV; (c) structure of the invariant manifolds of the saddle D_L after the homoclinic bifurcation, subdomain IV; (d)-(f) examples of chaotic transients with different final outcomes for initial conditions close to $x(0) = -1$, $\dot{x}(0) = -0.28$.

unpredictable final outcomes. A sample of the set of chaotic transients that start with initial conditions close to the bottom of the left potential well, each transient settling onto one of the three available coexisting attractors, is presented in Fig. 7d-f.

4. Conclusions

The study of the phenomena of multiple chaotic transient motions in the twin-well potential Duffing oscillator was concentrated on the region of driving frequencies which are close to the primary resonance, but are lower than the critical frequency of the nonresonant saddle-node bifurcation. Numerical exploration reveals that in this region of control parameters the system behavior is especially rich in the phenomena of various types of chaotic transients. This is explained in terms of the multiple global bifurcations that occur in the domain considered, and which guarantee the generation of various patterns of fractal structure of basins of attraction of the coexisting attractors.

At the region of the nonlinear resonance hysteresis (low values of the forcing parameter F), the additional types of global bifurcations and, hence, the new types of chaotic transients are related to the coexistence of the two periodic (resonant and nonresonant) attractors and the accompanying cyclic saddle inside each of the potential well.

At higher values of the forcing parameter F, where the T-periodic cross-well attractor (Large Orbit) appears, the new types of chaotic transients are due to the homoclinic bifurcation of the associated cross-well T-periodic saddle. The bifurcation brings a dramatic change in the structure of the basins of attraction; after this global bifurcation the basin of the cross-well T-periodic attractor begins to cover a large domain of the phase-space. It follows, that now even the trajectories that start close to the bottom of the potential well can settle, after a long cross-well chaotic transient, onto the Large Orbit.

References

1. W. Szemplińska-Stupnicka and J. Rudowski (1993) „Steady-states in the twin-well potential oscillator: computer simulations and approximate analytical studies", *CHAOS, Int. J. Nonlinear Science,* **3** (3), 375 - 385.

2. W. Szemplińska-Stupnicka and K. Janicki (1997) „Basin boundary bifurcations and boundary crisis in the twin-well Duffing oscillator: scenarios related to the saddle of the large resonant orbit", *Int. J. Bifurcation and Chaos*, 7 (1), 129-146.

3. W. Szemplińska-Stupnicka and E. Tyrkiel (1997) „Sequences of global bifurcations and the related outcomes after crisis of the resonant attractor in a nonlinear oscillator", to appear in *Int. J. Bifurcation and Chaos*, 7 (10).

4. References in the above listed papers.

BIFURCATION OF HOMOCLINIC ORBITS IN AUTONOMOUS SYSTEMS AND IN CHAOTIC BLUE SKY CATASTROPHE

M. BELHAQ & F. LAKRAD
Laboratory of Mechanics, Faculty of Sciences Aïn Chock, BP 5366, Mâarif, Casablanca, Morocco

1. Introduction

Homoclinic and heteroclinic orbits are of great interest from applied point of view. In reaction-diffusion problems they form the profiles of traveling wave solutions. Their existence can also cause complicated dynamics in the three-dimensional systems, as in The Shilnikov example. The classical approach to predict the bifurcations of these homoclinic orbits, in driven or planar autonomous systems, deals with the Andronov-Melnikov method. This approach is mainly based on the distance between manifolds. Recently, another analytical method to predict the homoclinic bifurcations for autonomous systems was presented in [1]. This approach consists of attacking directly the period of periodic solution. More precisely, the condition we considered at such bifurcations is the limit of the period goes to infinity. The purpose of this paper is to present a new criterion to approximate such bifurcations. This criterion involves principally the periodic orbit and the hyperbolic point (saddle). The period is not needed in this approach. We will apply this new criterion to predict a chaotic blue sky catastrophe. Comparisons to numerical simulations and previous methods ([1], [2] and [3]) are reported.

2. Criterion

Consider the planar differential system in the form

$$\dot{x} = f_1(x, y, \mu); \dot{y} = f_2(x, y, \mu) \tag{1}$$

where f_1 and f_2 are supposed to be sufficiently smooth with respect to their arguments and $\mu = (\mu_1, \mu_2)$. This equation (1) can model the dynamic behavior of many systems in mechanics. We assume that the system (1) possesses a homoclinic or a heteroclinic orbit Γ to a saddle $S(x_s, y_s)$ at $\mu = \mu_c$, where μ_c is the critical parameter value corresponding to the homoclinic bifurcation. Define $(x_A(t, a(\mu)), y_A(t, a(\mu)))$ as an approximation of the periodic solution with the

F. C. Moon (ed.),
IUTAM Symposium on New Applications of Nonlinear and Chaotic Dynamics in Mechanics, 93–100.

amplitude $a(\mu)$ and frequency $\Omega(a(\mu)) = \frac{2\Pi}{T(a(\mu))} = \Omega_\mu$. The criterion to predict the homoclinic bifurcations given in [1], consists of computing explicitly, using a perturbation method [4], an approximation of the frequency Ω_μ. In the limit $\mu \rightarrow \mu_c$, the periodic orbit approaches S and so, the period T_μ tends to infinity at the bifurcation. Therefore, the natural condition we considered was merely given by $T_\mu \rightarrow \infty$ or $\Omega_\mu = 0$. We will now present the new technique. Instead of using the period or the frequency, we attack directly the periodic solution and the saddle S. Denote by $\bar{X}_A(\bar{x}_A(a(\mu)), \bar{y}_A(a(\mu)))$ the value of the periodic solution located, for $t = 0$, on an axis (Δ) connecting S to the focus F, see figure 1. In the limit $\mu \rightarrow \mu_c$, the periodic orbit passes closer and closer to the saddle ($\bar{X}_A \rightarrow S$). At the bifurcation, alias $\mu = \mu_c$, the condition to be satisfied is

$$\bar{X}_A(\bar{x}_A, \bar{y}_A) = S(x_s, y_s) \tag{2}$$

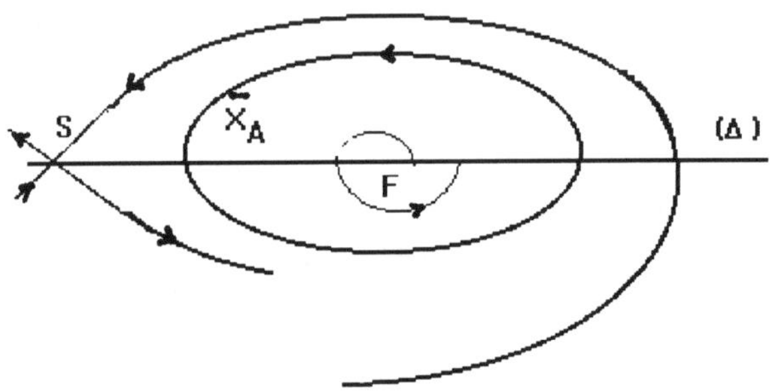

Figure 1: *Homoclinic bifurcation according to the criterion (2)*

3. Application and comparison

In this section, we apply the new approach (2) to some cases. As a first example, we consider the oscillator with quadratic nonlinearities

$$\begin{aligned} \dot{x} &= y, \\ \dot{y} &= -\omega^2 x + \epsilon(\tilde{\alpha} y + \tilde{c} x^2 - \tilde{\beta} xy + \tilde{\gamma} y^2), \end{aligned} \tag{3}$$

where $\alpha = \epsilon\tilde{\alpha}$, $c = \epsilon\tilde{c}$, $\beta = \epsilon\tilde{\beta}$ and $\gamma = \epsilon\tilde{\gamma}$ (α, c, β, and γ are the original parameters and ϵ is supposed to be small); the parameter ω denotes the natural frequency of the system. Applying a perturbation method [5], the approximation to second order of the general solution is written as

$$x_A(t) = a cos(\phi) + \epsilon U_1(a, \phi) \tag{4}$$

$$\frac{da}{dt} = \epsilon\frac{\tilde{\alpha}a}{2} - \epsilon^2[\frac{\tilde{\beta}a^3}{8\omega^2}(\tilde{c}+\tilde{\gamma}\omega^2)]$$
$$\frac{d\phi}{dt} = \omega - \epsilon^2[(\frac{5\tilde{c}^2}{12\omega^3}+\frac{5\tilde{\gamma}\tilde{c}}{12\omega}+\frac{\tilde{\gamma}^2\omega}{6}+\frac{\tilde{\beta}^2}{24\omega})a^2] \quad (5)$$

Using the criterion (2) to the first order approximation, alias $x_A(t) = a\cos(\phi)$ and $y_A(t) = \dot{x}(t) = -a\dot{\phi}\sin(\phi)$, we obtain

$$a\cos(\phi) = \frac{\omega^2}{c}; -a\Omega\sin(\phi) = 0 \quad (6)$$

Resolving these equations leads to the relation $a = \frac{\omega^2}{c}$. Taking into account the expression of the amplitude $a = \sqrt{\frac{4\alpha\omega^2}{\beta(c+\gamma\omega^2)}}$ given in [1], the homoclinic bifurcation curve is approximated by

$$\alpha = \frac{\beta(c+\gamma\omega^2)\omega^2}{4c^2} \quad (7)$$

At the second order, for $\phi = 0$, the criterion (2) gives $a+\frac{a^2}{2}(\gamma+\frac{c}{\omega^2})+\frac{a^2}{6}(\gamma-\frac{c}{\omega^2}) = \frac{\omega^2}{c}$. This system leads to the second order approximation of the homoclinic curve (for $\gamma = 0$)

$$\alpha = \frac{\beta\omega^2}{4c}(\frac{\sqrt{21}-3}{2})^2 \quad (8)$$

Using the approach given in [1], we obtain, for $\gamma = 0$, $\alpha = \frac{6\beta c\omega^2}{10c^2+\beta^2\omega}$. Applying the Melnikov technique, we have $\alpha = \frac{\beta\omega^2}{7c}$ [6]. In figure 2, we compare these last results for the values $\gamma = 0$, $c = 0.4$ and $b = 0.2$.

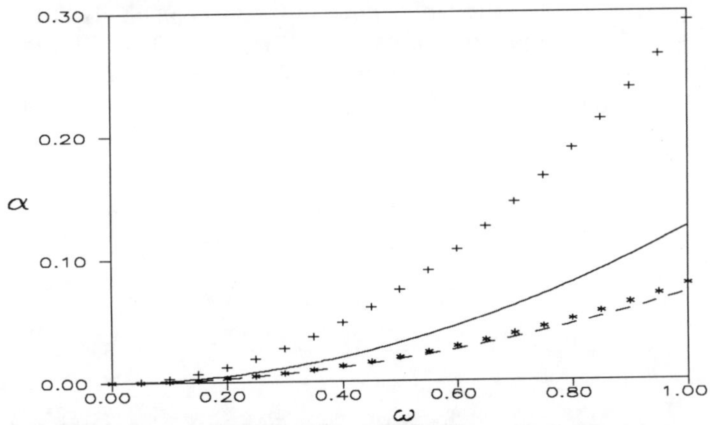

Figure 2: *Comparisons of homoclinic bifurcation curves of (3), - - - Melnikov method, +++ The criterion [1], ____ criterion(2) to first order, *** criterion(2) to second order*

Note that to improve the precision of the criterion (2), it is quite easy to perform higher order approximations of the periodic solution (4)-(5) using symbolic calculations.

The second example is the Arnold-Bogdanov-Takens system

$$\dot{x} = y; \dot{y} = \mu_1 + \epsilon\tilde{\mu_2}y + \epsilon x^2 + \epsilon xy \tag{9}$$

with $\mu_1 \leq 0$ and $\mu_2 = \epsilon\tilde{\mu_2} \geq 0$; ϵ is small parameter. The fixed points of (9) are $(x, y) = (\pm\sqrt{-\mu_1}, 0)$; the positive one is a saddle point. For illustrating the criterion (2) to predict the homoclinic connection in this example, it is more convenient to apply the multiple scales method. Performing this method to second order, the periodic orbit may be sought in the form

$$\begin{aligned} x_A(t,\epsilon) &= x_0(T_0,T_1) + \epsilon x_1(T_0,T_1) + \epsilon^2 x_2(T_0,T_1) + ..., \\ y_A(t,\epsilon) &= y_0(T_0,T_1) + \epsilon y_1(T_0,T_1) + \epsilon^2 y_2(T_0,T_1) + ... \end{aligned} \tag{10}$$

The approximation to second order of the periodic orbit is given by

$$x_A(t) = a\cos(\phi) - \frac{\mu_2}{2}; y_A(t) = -a\sqrt{2\mu_2}\sin(\phi) \tag{11}$$

where $a^2 = -4(\mu_2^2 + \mu_1)$ and $\phi = (\sqrt{2\mu_2}t + \theta)$. We immediately see that the criterion (2), written as $a\cos(\phi) - \frac{\mu_2}{2} = -\mu_1$; $-a\sqrt{2\mu_2}\sin(\phi) = 0$, leads to the homoclinic bifurcation curve

$$\mu_1 = -(\frac{1+\sqrt{52}}{6})^2\mu_2^2 \tag{12}$$

The Melnikov technique [3] gives $\mu_1 = -(\frac{7}{5})^2\mu_2^2$. For comparisons, see figure 3. On the other hand, to predict the saddle node bifurcation, one only has to equal the coordinates of the fixed points to obtain the curve $\mu_1 = 0$. In addition, since at the Hopf bifurcation the amplitude of the orbit vanishes, we obtain from the expression of the amplitude given above, the Hopf curve $\mu_1 = -\mu_2^2$, as in [3].

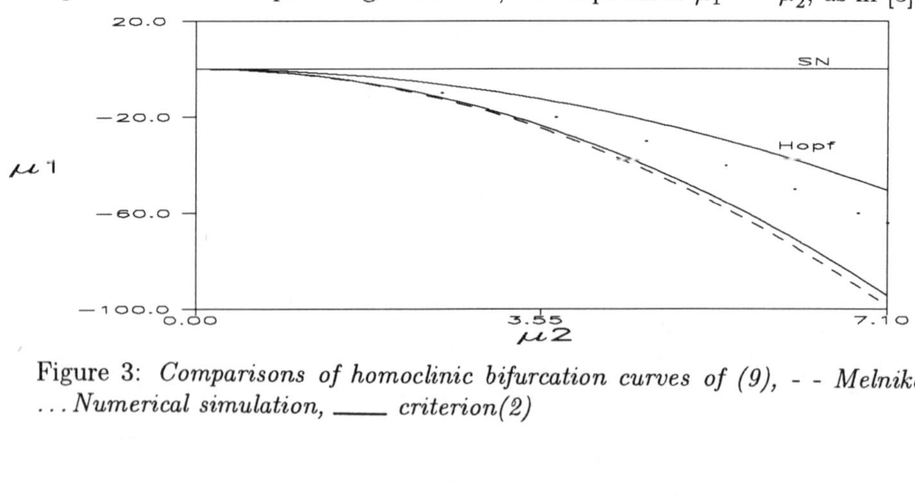

Figure 3: *Comparisons of homoclinic bifurcation curves of (9), - - Melnikov, ...Numerical simulation, ____ criterion(2)*

The next example is the acceleration Van der Pol system with DC bias [8]

$$\dot{x} = ky + \epsilon\tilde{\mu}bx - \epsilon\tilde{\mu}xy^2; \dot{y} = -x + c \tag{13}$$

For $c \neq 0$, the phase portrait of (13) has two fixed points. The coordinates of the saddle are $(c, \frac{k}{2\mu c} + \sqrt{(\frac{k}{2\mu c})^2 + b})$. As c increases, the saddle point moves toward the limit cycle, and at some value of c, it collides with the limit cycle causing its disappearance from the phase portrait.
Using the multiple scale method, the approximation to second order of the periodic solution is

$$\begin{aligned} x_A(t) &= a\cos(\phi) + c + \epsilon U_1(a, c, \phi) \\ y_A(t) &= -\frac{a}{\sqrt{k}}\sin(\phi) + \epsilon V_1(a, c, \phi) \end{aligned} \tag{14}$$

where, $a = \sqrt{4kb}$ and $\phi = \sqrt{k}t + \theta$. The criterion (2) to the first order gives

$$\begin{aligned} a\cos(\phi) + c &= c \\ -\frac{a}{\sqrt{k}}\sin(\phi) &= (\frac{k}{2\mu c}) + \sqrt{(\frac{k}{2\mu c})^2 + b} \end{aligned} \tag{15}$$

Resolving these equations leads to the first order approximation of the homoclinic bifurcation curve

$$b = (\frac{2k}{3\mu c})^2 \tag{16}$$

The second order approximation of the homoclinic bifurcation is given by resolving the following equations $x_A(t) = c$; $y_A(t) = (\frac{k}{2\mu c}) + \sqrt{(\frac{k}{2\mu c})^2 + b}$.
Figure 4 illustrates the comparisons between the analytical prediction (16) and numerical simulations for the values $k = 0.7$ and $\mu = 10$.

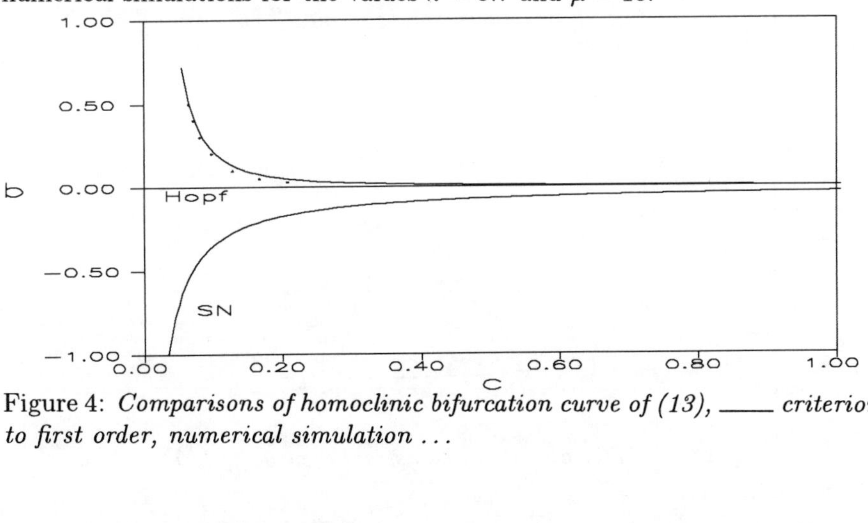

Figure 4: *Comparisons of homoclinic bifurcation curve of (13),* ——— *criterion(2) to first order, numerical simulation ...*

In the last example, we show that the criterion (2) may also be applied to predict the disappearance of a Birkoff-Shaw chaotic attractor (blue sky catastrophe) by collision with a saddle type object. This bifurcation can be considered as a generalization of the blue sky of a limit cycle discussed above. It can occur in the following Van der Pol asymmetric forced system with DC bias ([8])

$$\dot{x} = ky + \epsilon\tilde{\mu}bx - \epsilon\tilde{\mu}xy^2; \dot{y} = -x - f\sin(\Omega t) + c \tag{17}$$

Using the multiple scale method, the approximation of the quasi-periodic attractor of (17) is given by

$$\begin{aligned} x_A(t) &= a\cos(\phi) + \frac{fk}{(k-\Omega^2)}\sin(\Omega t) + c \\ y_A(t) &= -\frac{a}{\sqrt{k}}\sin(\phi) + \frac{f\Omega}{(k-\Omega^2)}\cos(\Omega t) \end{aligned} \tag{18}$$

where $a = \sqrt{4k(b - \frac{f^2\Omega^2}{2(k-\Omega^2)^2})}$ and $\phi = \sqrt{k}t + \theta$. Assume that for small periodic driving, the saddle type object of (18) still qualitatively near the saddle of the autonomous case (i.e.$f = 0$) in the Poincaré-section. Set $t = p\frac{\Pi}{\Omega}$, $p = 1, 3, 5, \ldots.$ Applying this transformation to (18), and using the criterion (2) by equating the resulting equations to the coordinates of the saddle, we obtain

$$\begin{aligned} a\cos(\phi_1) + c &= c \\ -\frac{a}{\sqrt{k}}\sin(\phi_1) - \frac{f\Omega}{(k-\Omega^2)} &= \frac{k}{2c\mu} + \sqrt{(\frac{k}{2c\mu})^2 + b} \end{aligned} \tag{19}$$

where ϕ_1 is the discritized phase. The resolution of these equations gives the following approximation of the chaotic blue sky catastrophe curve (see figure 5)

$$c = \frac{ks}{\mu(s^2 - b)}; s = -(\frac{f\Omega}{k - \Omega^2}) + \frac{a}{\sqrt{k}} \tag{20}$$

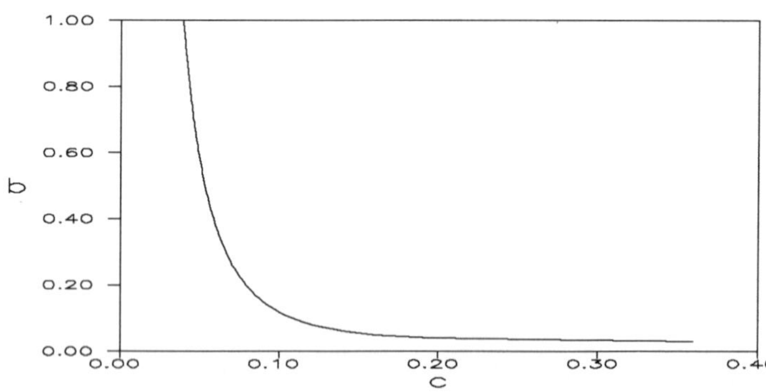

Figure 5: *Chaotic blue sky catastrophe curve of (17)*

To compare with the numerical simulation following [8], we set $k = 0.7$, $\mu = 10$, $b = 0.1$, $f = 0.25$ and $\Omega = 1.5$. Numerically $c = 0.09$ and analytically by (20) we find $c = 0.10$. To be more pricise, the position of the saddle point approximated in (19) by the saddle of the autonomous case, can also be calculated numerically.

4. Remarks

Recently, a technique to predict homoclinic bifurcations was established in [1]. This technique, mainly considers the limit of the periodic orbit tends to infinity at the homoclinic connection. In this paper we have focused attention on the periodic orbit itself and the saddle as well. After all, the principal object involved in this bifurcation is precisely the periodic orbit. We have presented a new analytical criterion based on the analysis of the periodic orbit moving toward the saddle point when approaching the homoclinic bifurcation. In the limit, the periodic orbit and the saddle collide.
The comparisons shown here, reveal the good prediction given by the new technique of this work. A generalization of these techniques to predict homoclinic bifurcations in periodically driven oscillators was proposed in [7]. This idea of attacking the periodic orbit can be applied to investigate the chaotic blue sky catastrophe. The saddle node bifurcation for fixed points or for periodic orbits can also be studied using the approximation of the periodic orbit, as seen in the Arnold-Bogdanov-Takens example. The technique of this work may be applied to study homoclinic bifurcations in three-dimensional systems. The challenge here is to construct a good approximation of the periodic orbit.

5. References

1. Belhaq, M. and Fahsi, A., Homoclinic bifurcations in self-excited oscillators, Mech. Res. Comm. 23 (1996) 4, 381-386.
2. Doelman, A. and Verhulst, F., Bifurcations of strongly nonlinear self-excited oscillations, Preprint, (1992).
3. Guckenheimer, J. and Holmes, P.J., Nonlinear oscillations, dynamical systems and bifurcation of vector fields. Springer in Appl. Math. Sc. 42, New York (1983).
4. Bogolioubov, N. and Mitropolsky, I., Les méthodes asymptotiques en théorie des oscillations non linéaires. Gauthier-Villars, Paris, (1962).
5. Belhaq, M. and Fahsi, A., Higher-order approximation of subharmonics close to strong resonances in forced oscillators. Computers Math. Applic., vol 33 (1997) 8.
6. Belhaq, M., Clerc, R. L. and Hartmann, C., Bifurcations homocliniques pour des équations de Liénard forcées priodiquement. Journal de Mécanique

Théorique et Appliquée, 6 (1987) 6, 865-877.
7. Belhaq, M. and Fahsi, A., Repeated resonances to chaos in forced oscillators. Mech. Res. Comm. 21 (1994) 5, 415-422.
8. Thomson, J.M.T. and Stewart, H.B., Nonlinear dynamics and chaos. Geometrical methods for engineers and scientists. John Wiley, New York (1986).

WAVELET APPROACH TO POLYNOMIAL MECHANICAL PROBLEMS

A.N. FEDOROVA AND M.G. ZEITLIN
Institute of Problems of Mechanical Engineering,
Russian Academy of Sciences, Russia, 199178, St. Petersburg,
V.O., Bolshoj pr. 61, e-mail: zeitlin@math.ipme.ru

Abstract. We give wavelet description for nonlinear optimal dynamics (energy minimization in high power electromechanical system). We consider two cases of our general construction. In a particular case we have the solution as a series on shifted Legendre polynomials, which is parametrized by the solution of reduced algebraical system of equations. In the general case we have the solution as a multiresolution expansion from wavelet analysis. In this case the solution is parametrized by solutions of two reduced algebraic problems, one as in the first case and the second is some linear problem, which is obtained from one of the next wavelet constructions: Fast Wavelet Transform, Stationary Subdivision Schemes, the method of Connection Coefficients.

1. Introduction. Many important physical and mechanical problems are reduced to the solving of the systems of nonlinear differential equations with polynomial type of nonlinearities. In this paper and related paper in this volume we consider applications of methods of wavelet analysis to such problems. Wavelet analysis is a relatively novel set of mathematical methods, which gives us possibility to work with well-localized bases in functional spaces and with general type of operators (including pseudodifferential) in such bases. Many examples may be found in papers [9]–[18]. Now we apply our approach to a case of constrained variational problem: the problem of energy minimization in electromechanical systems. We consider a synchronous electrical machine and a mill as load (in this approach

F. C. Moon (ed.),
IUTAM Symposium on New Applications of Nonlinear and Chaotic Dynamics in Mechanics, 101–108.

we can consider instead of a mill any mechanical load with polynomial approximation for the mechanical moment).

We consider the problem of "electrical economizer" as an optimal control problem. As result of the first stage we give the explicit time description of the optimal dynamics for that electromechanical system. As result of the second stage we give the time dynamics of our system via construction based on the set of switched type functions (Walsh functions), which can be realized on the modern thyristor technique. In this paper using the method of analysis of dynamical process in the Park system [1], which we developed in ref. [9], [10], we consider the optimal control problem in that system. As in [9], [10], our goal is to construct explicit time solutions, which can be used directly in microprocessor control systems. Our consideration is based on the Integral Variational Method, which was developed in [21]. As we shall see later, we can obtain explicit time dependence for all dynamical variables in our optimal control problem. It is based on the fact that optimal control dynamic in our case is given by some nonlinear system of equations which is the extension of initial Park system. Moreover, the equations of optimal dynamics also is the system of Riccati type (we use the quadratic dependence of the mechanical moment). It should be noted that this system of equations is not the pure differential system but it is the mixed differential-algebraic or functional system of equations [19].

In §2 we consider the description of our variational approach, which can be generalized in such a way that allows us to consider it in Hamiltonian (symplectic) approach [15].

In §3 we consider the form of explicit time solutions.

Our initial dynamical problem (without control) is described by the system of 6 nonlinear differential equations, which has the next Cauchy form (for definitions see [9], [10])

$$\frac{di_k}{dt} = \sum_{\ell} A_\ell i_\ell + \sum_{r,s} A_{rs} i_r i_s + A_k(t)$$

where $A_\ell, A_{rs} (\ell, r, s = \overline{1,6})$ are constants, $A_k(t), (k = \overline{1,5})$ are explicit functions of time, $A_6(i_6, t) = a + di_6 + bi_6^2$ is analytical approximation for the mechanical moment of the mill. At initial stage of the solution of optimal control problem in both methods we need to select from initial set of dynamical variables $i_1, \dots, i_6$ the controlling and the controllable variables. In our case we consider i_1, i_2 as the controlling variables. Because we consider the energy optimization, we use the next general form of energy functional in our electromechanical system $Q = \int_{t_0}^{t} [K_1(i_1, i_2) + K_2(\dot{i}_1, \dot{i}_2)]dt$, where K_1, K_2 are quadratic forms. Thus, our functional is the quadratic functional on the variables i_1, i_2 and its derivatives. Moreover, we may consider the optimization problem with some constraints which are motivated by technical reasons [9], [13]. Then after standard manipulations from the theory of optimal control, we reduce the problem of energy minimization to some extended nonlinear system of equations. As a result of the solution of equations of optimal dynamics we have: 1). the explicit time dependence of the controlling variables $u(t) = \{i_1(t), i_2(t)\}$ which give 2). the optimum of corresponding functional of the energy and 3). explicit time dynamics of the controllable variables $\{i_3, i_4, i_5, i_6\}(t)$. The obtained solutions are given in the next form: $i_k(t) = i_k(0) + \sum_{i=1}^{N} \lambda_k^i X_i(t)$, where in our first case (§3) we have $X_i(t) = Q_i(t)$, where $Q_i(t)$ are shifted Legendre polynomials [21] and λ_k^i are the roots of reduced algebraic system of equations. In our second case (§4)–the general wavelet case $X_i(t)$ correspond to multiresolution expansion in the basis of compactly supported wavelets and λ_k^i are the roots of corresponding algebraic Riccati systems with coefficients, which are given by Fast Wavelet Transform (FWT) [2], Stationary Subdivision Schemes(SSS) [6], the method of Connection Coefficients (CC) [23].

Giving the controlling variables in the explicit form, we have optimal, according to energy, dynamics in our electromechanical systems. Evidently the technical realization of controlling variables as an arbitrary continuous functions of time is impossible, but we can replace them by their re-expansions on the base of switching type functions, which can be realized now on the modern thyristor technique. We considered this re-expansion in [9], [13], where we used Walsh and Haar functions [3] as a base set of switching type functions. This is a special case of general sequency analysis [20]. About practical realization of expansions from §4 on the base of general wavelet packet basis [4] see [18].

2. Polynomial Dynamics. Our problems may be formulated as the systems of ordinary differential equations $dx_i/dt = f_i(x_j, t)$, $(i, j = 1, ..., n)$ with fixed initial conditions $x_i(0)$, where f_i are not more than polynomial functions of dynamical variables x_j and have arbitrary dependence of time. Because of time dilation we can consider only next time interval: $0 \leq t \leq 1$. Let us consider a set of functions $\Phi_i(t) = x_i dy_i/dt + f_i y_i$ and a set of functionals $F_i(x) = \int_0^1 \Phi_i(t)dt - x_i y_i \mid_0^1$, where $y_i(t)(y_i(0) = 0)$ are dual variables. It is obvious that the initial system and the system $F_i(x) = 0$ are equivalent. In the second paper in this volume we consider the symplectization of this approach. Now we consider formal expansions for x_i, y_i:

$$x_i(t) = x_i(0) + \sum_k \lambda_i^k \varphi_k(t) \quad y_j(t) = \sum_r \eta_j^r \varphi_r(t), \tag{1}$$

where because of initial conditions we need only $\varphi_k(0) = 0$. Then we have the following reduced algebraical system of equations on the set of unknown coefficients λ_i^k of expansions (1):

$$\sum_k \mu_{kr} \lambda_i^k - \gamma_i^r(\lambda_j) = 0 \tag{2}$$

Its coefficients are $\mu_{kr} = \int_0^1 \varphi_k'(t)\varphi_r(t)dt$, $\gamma_i^r = \int_0^1 f_i(x_j, t)\varphi_r(t)dt$. Now, when we solve system (2) and determine unknown coefficients from formal expansion (1) we therefore obtain the solution of our initial problem. It should be noted if we consider only truncated expansion (1) with N terms then we have from (2) the system of $N \times n$ algebraical equations and the degree of this algebraical system coincides with degree of initial differential system. So, we have the solution of the initial nonlinear (polynomial) problem in the form

$$x_i(t) = x_i(0) + \sum_{k=1}^{N} \lambda_i^k X_k(t), \tag{3}$$

where coefficients λ_i^k are roots of the corresponding reduced algebraical problem (2). Consequently, we have a parametrization of solution of initial problem by solution of reduced algebraical problem (2). But in general case, when the problem of computations of coefficients of reduced algebraical system (2) is not solved explicitly as in the quadratic case, which we shall consider below, we have also parametrization of solution (1) by solution of corresponding problems, which appear when we need to calculate coefficients of (2). As we shall see, these problems may be explicitly solved in wavelet approach.

3. The solutions. Next we consider the construction of explicit time solution for our problem. The obtained solutions are given in the form (3), where in our first case we have $X_k(t) = Q_k(t)$, where $Q_k(t)$ are shifted Legendre polynomials and λ_k^i are roots of reduced quadratic system of equations. In wavelet case $X_k(t)$ correspond to multiresolution expansions in the base of compactly supported wavelets and λ_k^i are the roots of corresponding general polynomial system (2) with coefficients, which are given by FWT, SSS or CC constructions. According to the variational method to give the reduction from differential to algebraical system of equations we need compute the objects γ_a^j and μ_{ji}, which are constructed from objects:

$$\begin{aligned}
\sigma_i &\equiv \int_0^1 X_i(\tau)d\tau = (-1)^{i+1}, \qquad (4)\\
\nu_{ij} &\equiv \int_0^1 X_i(\tau)X_j(\tau)d\tau = \sigma_i\sigma_j + \frac{\delta_{ij}}{(2j+1)},\\
\mu_{ji} &\equiv \int X_i'(\tau)X_j(\tau)d\tau = \sigma_j F_1(i,0) + F_1(i,j),\\
& \qquad F_1(r,s) = [1-(-1)^{r+s}]\hat{s}(r-s-1), \quad \hat{s}(p) = \begin{cases} 1, & p \geq 0 \\ 0, & p < 0 \end{cases}\\
\beta_{klj} &\equiv \int_0^1 X_k(\tau)X_l(\tau)X_j(\tau)d\tau = \sigma_k\sigma_l\sigma_j + \\
& \qquad \alpha_{klj} + \frac{\sigma_k\delta_{jl}}{2j+1} + \frac{\sigma_l\delta_{kj}}{2k+1} + \frac{\sigma_j\delta_{kl}}{2l+1},\\
\alpha_{klj} &\equiv \int_0^1 X_k^* X_l^* X_j^* d\tau = \frac{1}{(j+k+l+1)R(1/2(i+j+k))} \times \\
& \qquad R(1/2(j+k-l))R(1/2(j-k+l))R(1/2(-j+k+l)),
\end{aligned}$$

if $j+k+l = 2m, m \in Z$, and $\alpha_{klj} = 0$ if $j+k+l = 2m+1$; $R(i) = (2i)!/(2^i i!)^2$, $Q_i = \sigma_i + P_i^*$, where the second equality in the formulae for $\sigma, \nu, \mu, \beta, \alpha$ hold for the first case.

4. Wavelet computations. Now we give construction for computations of objects (4) in the wavelet case. We use some constructions from multiresolution analysis [8]: a sequence of successive approximation closed subspaces V_j: $\dots V_2 \subset V_1 \subset V_0 \subset V_{-1} \subset V_{-2} \subset \dots$ satisfying the following properties: $\bigcap_{j\in\mathbf{Z}} V_j = 0$, $\overline{\bigcup_{j\in\mathbf{Z}} V_j} = L^2(\mathbf{R})$, $f(x) \in V_j \Longleftrightarrow f(2x) \in V_{j+1}$ There is a function $\varphi \in V_0$ such that $\{\varphi_{0,k}(x) = \varphi(x-k)_{k\in\mathbf{Z}}\}$ forms a Riesz basis for V_0. We use compactly supported wavelet basis: orthonormal basis for functions in $L^2(\mathbf{R})$. As usually $\varphi(x)$ is a scaling function, $\psi(x)$ is a wavelet function, where $\varphi_i(x) = \varphi(x-i)$. Scaling relation that defines φ, ψ are

$$\begin{aligned}
\varphi(x) &= \sum_{k=0}^{N-1} a_k\varphi(2x-k) = \sum_{k=0}^{N-1} a_k\varphi_k(2x),\\
\psi(x) &= \sum_{k=-1}^{N-2} (-1)^k a_{k+1}\varphi(2x+k)
\end{aligned}$$

Let be $f : \mathbf{R} \longrightarrow \mathbf{C}$ and the wavelet expansion is

$$f(x) = \sum_{\ell \in \mathbf{Z}} c_\ell \varphi_\ell(x) + \sum_{j=0}^{\infty} \sum_{k \in \mathbf{Z}} c_{jk} \psi_{jk}(x) \tag{5}$$

The indices k, ℓ and j represent translation and scaling, respectively: $\varphi_{jl}(x) = 2^{j/2}\varphi(2^j x - \ell)$, $\psi_{jk}(x) = 2^{j/2}\psi(2^j x - k)$. The set $\{\varphi_{j,k}\}_{k\in\mathbf{Z}}$ forms a Riesz basis for V_j. Let W_j be the orthonormal complement of V_j with respect to V_{j+1}. Just as V_j is spanned by dilation and translations of the scaling function, so are W_j spanned by translations and dilation of the mother wavelet $\psi_{jk}(x)$. If in formulae (5) $c_{jk} = 0$ for $j \geq J$, then $f(x)$ has an alternative expansion in terms of dilated scaling functions only $f(x) = \sum_{\ell \in \mathbf{Z}} c_{J\ell}\varphi_{J\ell}(x)$. This is a finite wavelet expansion, it can be written solely in terms of translated scaling functions. We use wavelet $\psi(x)$, which has k vanishing moments $\int x^k \psi(x)dx = 0$, or equivalently $x^k = \sum c_\ell \varphi_\ell(x)$ for each k, $0 \leq k \leq K$. Also we have the shortest possible support: scaling function DN (where N is even integer) will have support $[0, N-1]$ and $N/2$ vanishing moments. There exists $\lambda > 0$ such that DN has λN continuous derivatives; for small $N, \lambda \geq 0.55$. To solve our second associated linear problem we need to evaluate derivatives of $f(x)$ in terms of $\varphi(x)$. Let be $\varphi_\ell^n = d^n\varphi_\ell(x)/dx^n$. We derive the wavelet - Galerkin approximation of a differentiated $f(x)$ as $f^d(x) = \sum_\ell c_l \varphi_\ell^d(x)$ and values $\varphi_\ell^d(x)$ can be expanded in terms of $\varphi(x)$:

$\phi_\ell^d(x) = \sum_m \lambda_m \varphi_m(x)$, $\lambda_m = \int_{-\infty}^{\infty} \varphi_\ell^d(x)\varphi_m(x)dx$ The coefficients λ_m are 2-term connection coefficients. In general we need to find $(d_i \geq 0)$

$$\Lambda_{\ell_1 \ell_2 \ldots \ell_n}^{d_1 d_2 \ldots d_n} = \int_{-\infty}^{\infty} \prod \varphi_{\ell_i}^{d_i}(x)dx \tag{6}$$

For Riccati case we need to evaluate two and three connection coefficients $\Lambda_\ell^{d_1 d_2} = \int_{-\infty}^{\infty} \varphi^{d_1}(x)\varphi_\ell^{d_2}(x)dx$, $\Lambda^{d_1 d_2 d_3} = \int_{-\infty}^{\infty} \varphi^{d_1}(x)\varphi_\ell^{d_2}(x)\varphi_m^{d_3}(x)dx$.

According to CC method [23] we use the next construction. When N in scaling equation is a finite even positive integer the function $\varphi(x)$ has compact support contained in $[0, N-1]$. For a fixed triple (d_1, d_2, d_3) only some $\Lambda_{\ell m}^{d_1 d_2 d_3}$ are nonzero : $2 - N \leq \ell \leq N - 2$, $2 - N \leq m \leq N - 2$, $|\ell - m| \leq N - 2$. There are $M = 3N^2 - 9N + 7$ such pairs (ℓ, m). Let $\Lambda^{d_1 d_2 d_3}$ be an M-vector, whose components are numbers $\Lambda_{\ell m}^{d_1 d_2 d_3}$. Then we have the first key result: Λ satisfy the system of equations ($d = d_1 + d_2 + d_3$): $A\Lambda^{d_1d_2d_3} = 2^{1-d}\Lambda^{d_1d_2d_3}$, $A_{\ell,m;q,r} = \sum_p a_p a_{q-2\ell+p} a_{r-2m+p}$. By moment equations we have created a system of $M + d + 1$ equations in M unknowns. It has rank M and we can obtain unique solution by combination of LU decomposition and QR algorithm. The second key result gives us the 2-term connection coefficients: $A\Lambda^{d_1d_2} = 2^{1-d}\Lambda^{d_1d_2}$, $d = d_1 + d_2$, $A_{\ell,q} = \sum_p a_p a_{q-2\ell+p}$. For nonquadratic case we have analogously additional linear problems for objects (6). Also, we use FWT [2] and SSS [6] for computing coefficients of reduced algebraic systems. We use for modelling D6, D8, D10 functions and programs RADAU and DOPRI for testing [19].

As a result we obtained the explicit time solution (3) of our problem. In comparison with wavelet expansion on the real line which we use now and in calculation of Galerkin approximation, Melnikov function approach, etc also we need to use periodized wavelet expansion, i.e. wavelet expansion on finite interval. Also in the solution of perturbed system we have some problem with variable coefficients. For solving last problem we need to consider one more refinement equation for scaling function $\phi_2(x)$: $\phi_2(x) = \sum_{k=0}^{N-1} a_k^2 \phi_2(2x - k)$ and corresponding wavelet expansion for variable coefficients $b(t)$: $\sum_k B_k^j(b)\phi_2(2^j x - k)$, where $B_k^j(b)$ are functionals supported in a small neighborhood of $2^{-j}k$ [7].

The solution of the first problem consists in periodizing. In this case we use expansion into periodized wavelets defined by $\phi^{per}_{-j,k}(x) = 2^{j/2}\sum_Z \phi(2^j x + 2^j\ell - k)$ [5]. All these modifications lead only to transformations of coefficients of reduced algebraic system, but general scheme remains the same.

Extended version and related results may be found in [9]–[18].

We would like to thank Zohreh Parsa (BNL) for many discussions and continued encouragement during and after workshop "New Ideas for Particle Accelerators" and Institute for Theoretical Physics, University of California, Santa Barbara for hospitality.

This research was supported in part under "New Ideas for Particle Accelerators Program" NSF- Grant No. PHY94-07194.

References

1. Anderson P.M., Fouad A.A. (1977) *Power system control and stability*, Iowa State Univ. Press, USA.

2. Beylkin G., Coifman R. and Rokhlin V. (1991) Fast wavelet transform, Comm. Pure Appl.Math., **44**, pp. 141–183.
3. Blachman N.M. (1974) Sequentional analysis, Proc. IEEE, **62**, pp. 72–82.
4. Coifman R.R., Wickerhauser M.V. (1993) Wavelets analysis and signal processing, in *Wavelets*, SIAM, pp. 153–178.
5. Cohen A., Daubechies I., Vial P. (1993) Wavelets on the interval and fast wavelet transforms, preprint.
6. Dahlke S., Weinreich I. (1993) Wavelet–Galerkin methods, Constructive approximation, **9**, pp 237–262.
7. Dahmen, W., Micchelli C.A. (1993) Using the Refinement Equation for Evaluating Integrals of Wavelets, SIAM J. Numer.Anal., **30**, pp. 507–537.
8. Daubechies I. (1988) Orthonomal bases of compactly supported wavelets, Comm. Pure Appl. Math., **41**, pp. 906–1003.
9. Fedorova A.N., Zeitlin M.G. (1993) Variational Analysis in Optimal Control of Synchronous Drive of Mill, Preprint IPME, **no. 96**.
10. Fedorova A.N., Rubashev G.M., Zeitlin M.G. (1990) An algorithm of Solving of Equations of Synchronous Drive of Mill, Electrichestvo (Electricity), in Russian, **no. 6**, pp. 40–45.
11. Fedorova A.N, Zeitlin M.G. (1996) Wavelets in Dynamics, Optimal Control and Galerkin Approximations, IEEE Digital Signal Processing Workshop, Norway, pp. 409–412.
12. Fedorova A.N., Zeitlin M.G. (1997) Wavelets in optimization and approximations, Proc. 2nd IMACS Symp. on Math.Modelling, ARGESIM Report 11, Austria, pp. 1083–1088; Math.Comp. in Simulation, in press.
13. Fedorova A.N., Zeitlin M.G. (1995) Variational Analysis in the Optimal Control Problem, Proc. of 22 Summer School'Nonlinear Oscillations in Mechanical Systems', St. Petersburg, pp. 89–96.
14. Fedorova A.N., Zeitlin M.G. (1996) Wavelet Approach to Polynomial Dynamics, Proc. of 23 Summer School 'Nonlinear Oscillations in Mechanical Systems', St. Petersburg, pp. 322-333.
15. Fedorova A.N., Zeitlin M.G. (1997) Wavelet approach to nonlinear problems, parts 1-4, ZAMM, in press.
16. Fedorova A.N., Zeitlin M.G. (1995) Wavelets and Variational Analysis in Optimal Control Problem, 2-nd Russian– Swedish Control Conference, St. Petersburg, pp. 204-208.
17. Fedorova A.N., Zeitlin M.G. (1996) Routes to chaos in perturbed Hamiltonian systems. In direction to symplectic wavelets, EUROMECH-2nd European Nonlinear Oscillations Conf., pp. 79–82.
18. Fedorova A.N., Zeitlin M.G. (1997) *Proc. of 25 Summer School 'Nonlinear Oscillations in Mechanics'*, St. Petersburg, in press.
19. Hairer E., Lubich C., Roche M. (1989) *Lecture Notes in Mathematics*, **Vol. 1409**.
20. Harmuth H.F. (1977) *Sequency theory*, Academic press.
21. Hitzl D.L., Huynh T.V, Zele F. (1984) Integral Variational Method, Physics Letters, **104**, pp. 447–451.
22. Hofer H., Zehnder E. (1994) *Symplectic Invariants and Hamiltonian Dynamics*, Birkhauser.
23. Latto A., Resnikoff H.L. and Tenenbaum E. (1991) The Evaluation of Connection Coefficients, Aware Technical Report AD910708.

JUMPS TO RESONANCE PHENOMENA IN NONLINEAR MECHANICS: FRACTAL BASINS, CHAOTIC TRANSIENTS AND UNPREDICTABILITY

MOHAMED S. SOLIMAN

Department of Engineering
Queen Mary College, University of London,
London, E1 4NS, United Kingdom

Introduction

One of the classical features exhibited in nonlinear engineering systems is the *jump phenomenon*, in which under the variation of a control parameter, such as the frequency, in the direction of instability, there is jump to a disconnected attractor, resulting not only in a qualitative, but often a substantial quantitative change in the response. Often such jumps to resonance, are purely deterministic in that there is always restabilization onto the large amplitude attractor which lies on the resonant branch of the response curve. However we show that in situations in which there is more than one remote co-existing attractor at the local bifurcation, jumps may be indeterminate in the sense that we cannot predict to which solution the system will settle upon. This is often a result fractal structure in the vicinity of the critical point resulting in chaotic transients until the settles onto one of the co-existing attractors

In this article we show that indeterminate bifurcations are a common phenomenon in a wide range of nonlinear systems: they occur at saddle-node folds in *softening* or *hardening* systems where the restoring force is not proportional to the deformation; at Hopf bifurcations in systems with *varying damping characteristics*; at trans-critical bifurcations in *parametrically excited systems*.

F. C. Moon (ed.),
IUTAM Symposium on New Applications of Nonlinear and Chaotic Dynamics in Mechanics, 109–115.

Indeterminate jumps to and from resonace in softening and hardening systems

One of the most typical forced nonlinear systems is where the restoring force is not proportional to the deformation often modelling by equations of the type:

$$\ddot{x} + \mu\dot{x} + \omega^2 x + f(x) = F \sin \omega t$$

When *f(x)* is a softening type nolinearity the resonace response curve skews towards lower frequencies than that of the linearised natrutral frequncy; on the other hand when *f(x)* is a hardening type nolinearity the resonace response curve skews towards higher frequencies than that of the linearised natural frequency (Figure 1).

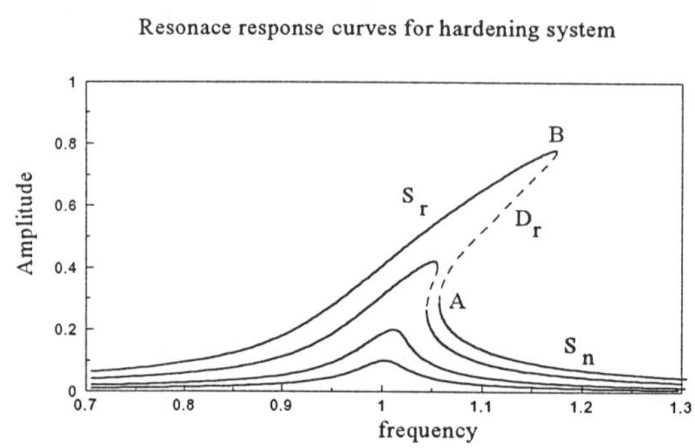

Figure 1 : Typical resonace response curves in nonlinear systems

Classical studies have shown that jumps to and from resonace are typical and that there is restablization onto the upper or lower brach of the reponse curve. However since multiple solutions may co-exist. at to point of instability we cannot always say this is the case. As shown in Soliman [1993] for softening systems and Soliman [1997b] for hardening system unpredictable jumps *to* and *from* resonace can also occur in these type of systems. Figure 2 shows typical examples of basin organization prior to indeterminate jumps occuring in hardening and softening systems:

Co-existing basins just before indeterminate jumps in softening systems

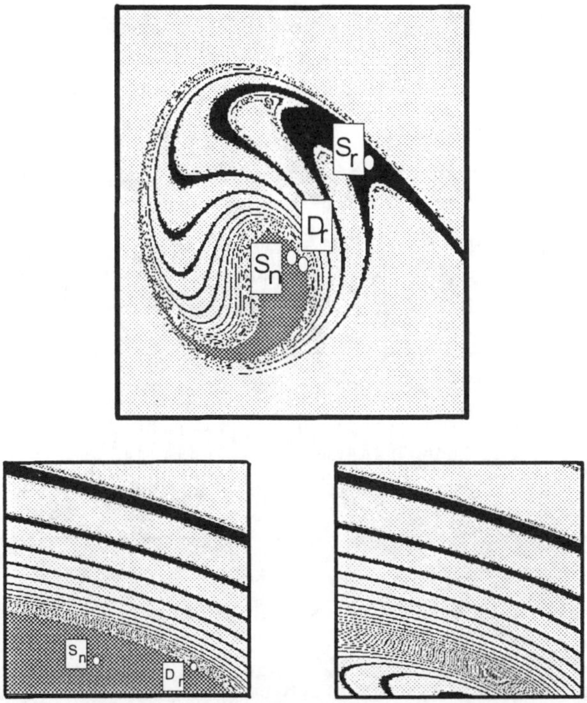

Co-existing basins just before indeterminate jumps in hardening systems

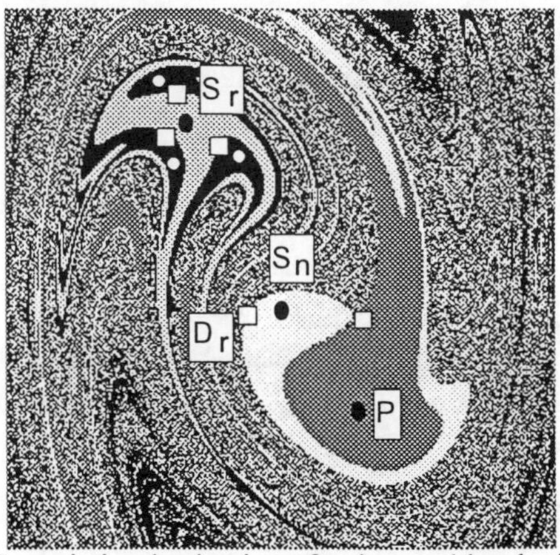

Figure 2 :Co-existing basins in softening and hardening systems

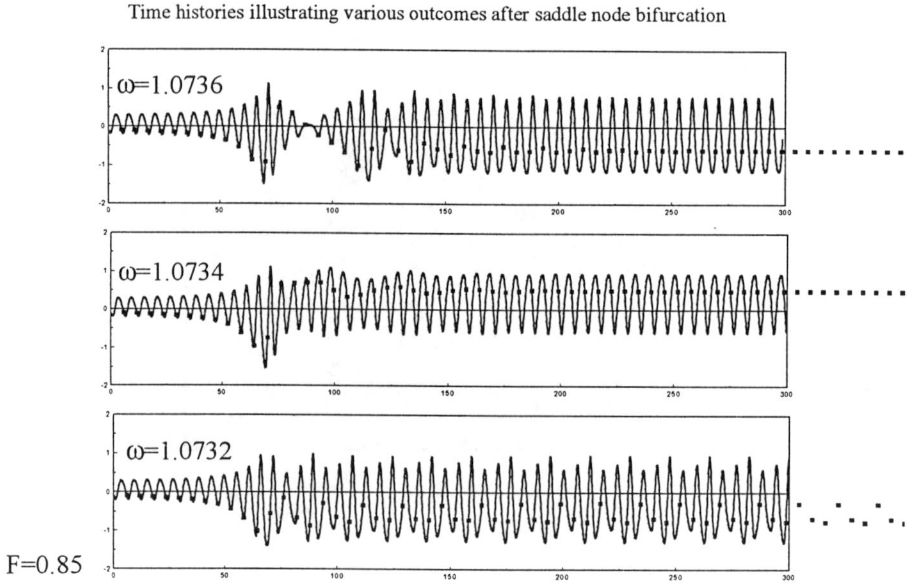

Figure 3: Various possible outcomes in hardening system

Unpredictable jumps systems in self-excited systems with nonlinear damping

A typical example that exhibits self-sustained oscillations was introduced by Van der Pol who modelled an electrical circuit with a triode valve which possessed nonlinear resistive (damping) properties; at low current the system possesses negative resistance which gradually becomes positive as the current increases. A forced Van-der-Pol type oscillator is for example modelled by

$$\ddot{x} + \mu\dot{x} + \omega^2 x + f(x,\dot{x}) = F\,sin\,\omega t$$

For $F=0$ the equilibrium can lose it stability at Hopf bifurcation. For $F \neq 0$ a periodic cycle can lose it stability at a secondary Hopf bifurcation. As shown in more detail in Thompson [1992], *indeterminate Hopf bifurcations* may occur and Soliman [1997b] *secondary Hopf bifurcations* may occur where the outcome is extremely sensitive to any finite perturbation given to the system.

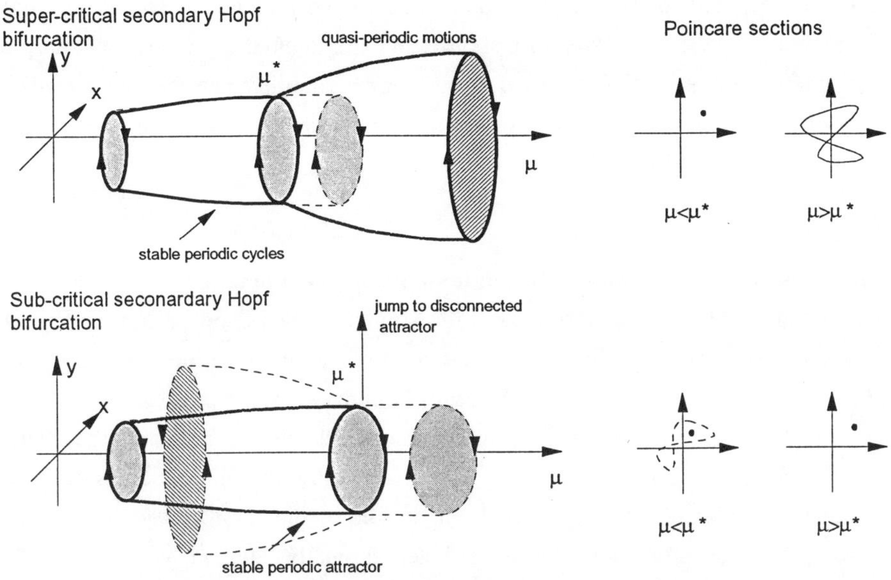

Figure 4: Secondary Hopf Bifurcations

Unpredictable jumps in parametrically excited systems

Parametrically excited systems are often modelled by Mathieu type equations.

$$\ddot{x} + \mu\dot{x} + (1 + F \sin \omega t) f(x) = 0$$

When *f(x)* is nonlinear, equilibria lying on fractal basin boundaries have also been shown to be a common ingredient. As shown in Soliman & Thompson [1992] here too they jumps to resonance with an unpredictable outcome may occur.

Indeterminate dynamic bifurcations in nonstationary-systems: transitions with an unpredictable outcome

For the experimentalist when dealing real physical systems where it is difficult to control the imperfections, such as the material inhomogenities, the initial conditions or parameters, 'steady-states' do not always correspond to the norm; transitions with an indeterminate outcome imply that repetition under 'almost identical' conditions may result in qualitatively different types of response. In experimental situations for example, where parameters are often slowly varied, a loss of repeatability may be interpreted in context of the inherent indeterminacy properties of the system, rather than be attributed as experimental error [Soliman, 1995].

Simplified method to determine indeterminate jumps to resonance

In order to analyse if such indeterminate bifurcational phenomenon occurs one has to study the local and *global dynamics* of the systems. If at the bifurcation, the non-resonant solution finds itself on a fractal basin boundary, in the presence of even an infinitesimal perturbation, we cannot predict to which of the remote attractors the system will jump since the response is infinitely sensitive to how the bifurcation is realized. Such a situation can arise when the unstable manifold of the resonant saddle, which is associated with the bifurcation, becomes heteroclinically tangled with the stable manifold of a remote saddle which itself is homoclinically tangled such that a fractal boundary exists between two remote attractors. *Chaotic transients* result until the system settles onto one of the co-existing attractors.

Such an analysis is often quite complex since it involves determining both the main local and global bifurcations (the homoclinic and heteroclinic connections that induce a fractal basin structure at the bifurcation) that can lead to an indeterminate jump to resonance. Soliman [1994] outline an method which allows us to identify approximate regimes of indeterminate jumps to resonance, may either be used as an initial study for further detailed analysis, or as a simplified engineering approach in the analysis of systems liable to this type of resonant behaviour.

References

M.S. SOLIMAN [1993] "Jumps to resonance: long chaotic transient, unpredictable outcome and the probability of restabilization *ASME Journal of Applied Mechanics* **60**, 669-676.

M.S. SOLIMAN [1994] "Predicting regimes of indeterminate jumps by assessing fractal boundaries in control space" *International. Journal of Bifurcation and Chaos* **4** 1645-1653.

M.S. SOLIMAN [1995] "Dynamic bifurcations in nonstationary-systems: transitions with an unpredictable outcome" *Proceedings of the Royal Society Series A.* **451,** 471-485.

M.S. SOLIMAN [1996] "Jump phenomena resulting in unpredictable dynamics in the driven damped pendulum" *International Journal of Nonlinear Mechanics* **31**, 167-174.

M.S. SOLIMAN [1997a] "Nonlinear resonance of hardening systems; chaotic dynamics and unpredictable jumps to and. from resonance" *Journal of Sound and Vibration* (to appear)

M.S. SOLIMAN [1997b] "Indeterminate secondary Hopf bifucations in nonlinear oscillators", *Physical Review E (*to appear)

M.S. SOLIMAN & J.M.T. THOMPSON [1992] "Indeterminate trans-critical bifurcations in parametrically excited systems resonance" Proceedings of the Royal Society Series **A439**, 601-610.

J.M.T. THOMPSON [1992] "Global unpredictability in nonlinear dynamics: capture , dispersal , indeterminate bifurcations" *Physica* **58D**, 260-272.

GLOBAL STABLE OSCILLATIONS NEAR UNSTABLE EQUILIBRIUM POSITIONS: THE HILLTOP EFFECT

M. ZAKRZHEVSKY
Institute of Mechanics, Riga Technical University
1 Kalku St., LV-1658, Riga, Latvia

Abstract. This paper describes new results in the dynamics of driven oscillators, focusing on the stability problem of the systems with several equilibrium positions. Early it was found that in such systems as the twin-well potential Duffing system, systems with piecewise linear restoring forces, the forced pendulum system and others, there are such parameter regions when oscillations between two bottoms of the potential well are stable. In this paper we shall describe a new nonlinear phenomenon when these oscillations are global stable ones. It is common situation that chaotic global stable hilltop attractors also exist in the systems under consideration.

1. Introduction

Two new nonlinear phenomena in the archetypal nonlinear dissipative periodically excited dynamical systems with several equilibrium positions are discussing: the global stable periodic oscillations near unstable equilibrium positions and chaotic attractors near unstable equilibrium positions.

Let us give some definitions. In the case when oscillations take place between two bottoms of the potential well (see Fig. 1), we say that such oscillations are hilltop oscillations and they have hilltop orbits. E.g. the hilltop oscillations take place in the twin-well potential periodically excited Duffing system with $f(x) = -x + x^3$ if the stationary solution for all time is $-1 < x(t) < 1$. In a case when the hilltop oscillations are the unique stable oscillations in the system we say that such oscillations are the global stable hilltop oscillations and this new nonlinear phenomenon we name as the hilltop effect.

F. C. Moon (ed.),
IUTAM Symposium on New Applications of Nonlinear and Chaotic Dynamics in Mechanics, 117–124.

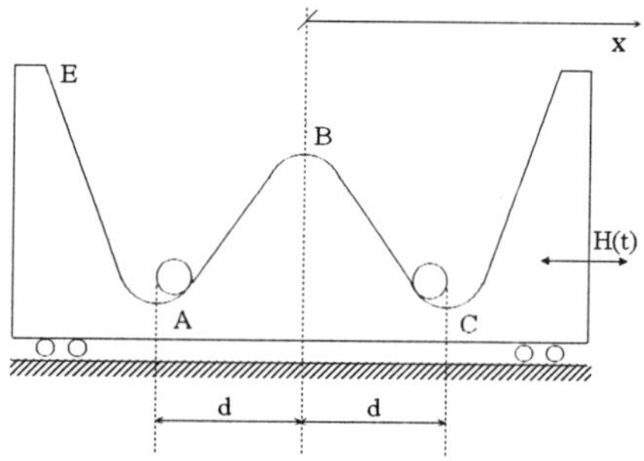

Figure 1. A mechanical model of the twin-well oscillatory system.

Early it was found (Zevin, 1986; Zakrzhevsky & Zevin, 1989), that period-1T stable hilltop oscillations may be in different twin-well potential forced or parametric periodically excited systems. Stable hilltop solutions occur in dissipative and nondissipative, symmetric and asymmetric systems, with piecewise linear or analytical nonlinear force $f(x)$, including Duffing's equation, pendulum systems and so on. These results were obtained by qualitative methods, numerical investigations, and exact analytical solutions for piecewise linear systems, such as trilinear, bilinear and more complex ones. Using bifurcation analysis it was found (Zakrzhevsky & Frolov, 1996) that in some parameter regions, there was multiformity of the different hilltop orbits. Such bifurcations of the hilltop oscillations as symmetry breaking, period doubling, and fold bifurcations also were found.

But in all previous investigations we obtained that the domains of attraction of all these stable hilltop oscillations were small in comparison with domains of attraction of the large orbit LO stable regimes. It was seemed that such situation is natural and common. In this paper our new results are discussing which extend previous conclusions. It will be shown that the stable hilltop oscillations may have large domains of attraction, or even be the global stable oscillations. It will be also shown that chaotic hilltop attractors take place in the systems with several equilibrium positions.

2. Mathematical models

The hilltop oscillations that we shall describe arise quite universally in driven oscillators with two or more potential wells, of the form (Moon, 1987; Thompson & McRobie, 1993)

$$\ddot{x} + g(x, \dot{x}) + p_1(t)f(x) = p_2(t) \qquad (1)$$

where $g(x, \dot{x})$ is some damping function, $f(x)$ is a nonlinear restoring force whose integral gives two or more potential wells, and $p_1(t), p_2(t)$ are parametric and direct forcing functions of period T, which may or may not be sinusoidal. To make the discussion concrete, we shall focus our attention in this paper on several archetypal oscillators within this general class. The first model is so-called twin-well Duffing oscillator (Moon, 1987; Lansbury et al., 1992; Szemplinska-Stupnicka et al., 1992)

$$\ddot{x} + b\dot{x} - x + x^3 = h_1 \cos(\omega t + \varphi) \tag{2}$$

The second model is a trilinear system with negative stiffness c_1 in the middle part of the restoring force $f(x)$

$$\begin{aligned} &\ddot{x} + b\dot{x} + c_2(x + d) = h_1 \cos \omega t, && x < -\Delta \\ &\ddot{x} + b\dot{x} + c_1 x = h_1 \cos \omega t, && -\Delta \leq x \leq \Delta \\ &\ddot{x} + b\dot{x} + c_2(x - d) = h_1 \cos \omega t, && x > \Delta \end{aligned} \tag{3}$$

It is common for equations (2) and (3) that they have an unstable equilibrium point at $x = 0$, and two stable points. The hilltop stable oscillations in the system (3) will exist if any stable stationary solution $-d < x(t) < d$. Then two simple forced pendulum systems also will be shortly discussed:

$$\ddot{x} + b\dot{x} + A\sin(\pi x) = h_1 \cos \omega t \tag{4}$$

$$\ddot{x} + b\dot{x} + A\sin(\pi x) = h_1 \cos \omega t \cos(\pi x) \tag{5}$$

where A is a negative coefficient. The pendulum systems (4) and (5) have additional stable and unstable periodic equilibrium points.

3. Global stable hilltop oscillations in the twin-well Duffing system

We start to discuss our new results for the twin-well potential Duffing system, governed by equation (2). Only oscillations with period-1T in the (b, h_1) control space were investigated. The aim was to understand scenarios of the birth of the global stable hilltop oscillations. Four bifurcation diagrams are shown in Fig. 2, where x is the coordinate of fixed points, h_1 is an amplitude of harmonic forcing, and b is a damping constant. Right columns in Fig. 2 are shown the same bifurcation diagrams, but for amplitudes Am of period-1T. Special attention was paid for receiving all stable and unstable solutions to see main topological changes of bifurcation curves.

The hilltop oscillations we indicate as TO. It occurs near symmetry breaking bifurcation point when amplitude Am is less than 1. The global stable hilltop oscillations take place when hysteresis disappear after damping

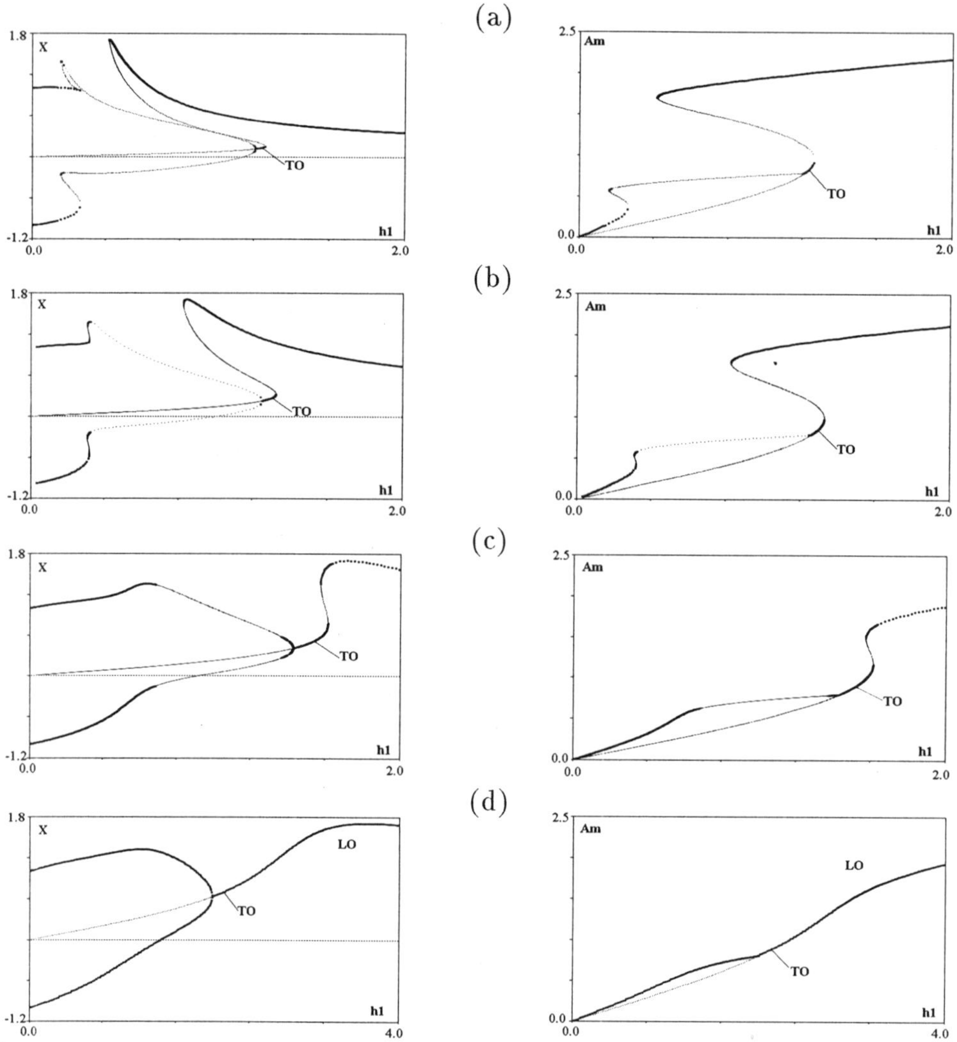

Figure 2. Bifurcation diagrams (x, h_1) for the twin-well Duffing equation (2), showing the birth of the global hilltop oscillations and the beginning of the doubling cascade of hilltop orbits; parameters: $\omega = 1.0, \varphi = \pi/2$; damping constant b : $(a)0.25, (b)0.50, (c)1.0, (d)2.0$.

coefficient b has increased enough; see Fig. 2,c,d. For the latest case ($b = 2.0$) several phase portraits for different amplitude of forcing h_1 are given in Fig. 3. For $b = 2.0$ the global hilltop period-1T symmetric oscillations exist approximately for $2.00 < h_1 < 2.333$.

It is clear from Fig. 2 and Fig. 3 that the first scenario of the birth of the global stable hilltop oscillations is very simple. For large damping the stable SO oscillations move from the wells forwards to each other, when h_1 is increasing, and near the symmetry breaking bifurcation point only the

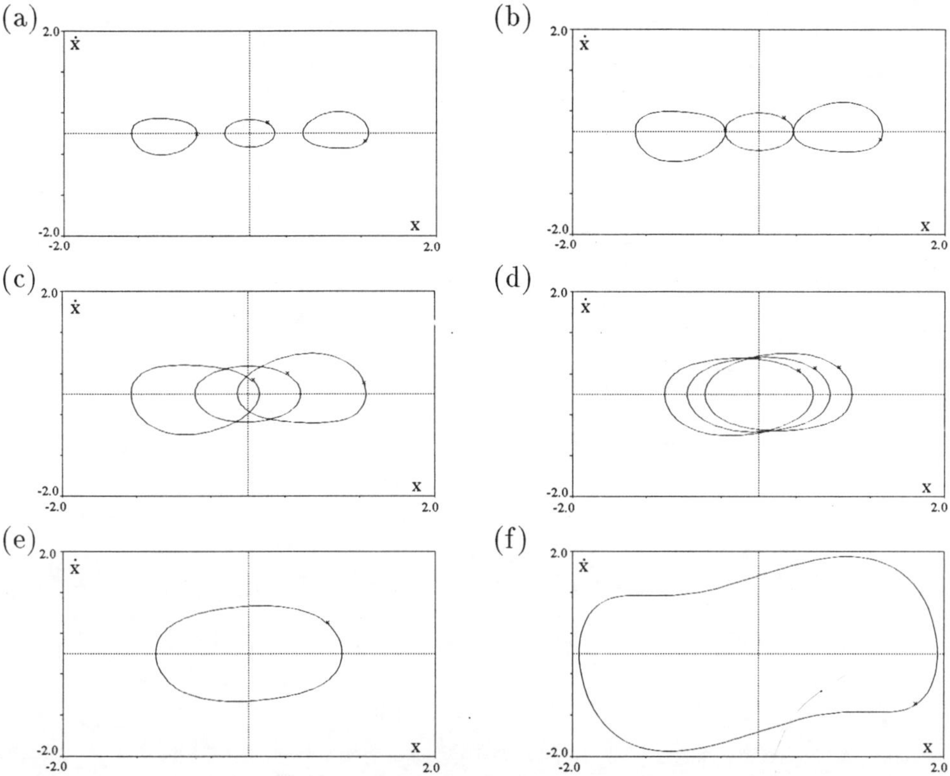

Figure 3. Phase portraits, showing the birth of the global hilltop oscillations (see Fig. 2,d); amplitudes of forcing h1 : (a) 0.75, (b) 1.0, (c) 1.5, (d) 1.919, (e) 2.333, (f) 4.0.

TO stable solution stays in the system. Increasing h_1 furthermore these oscillations became LO.

4. Global stable hilltop oscillations in the piecewise twin-well trilinear system

The global stable hilltop oscillations in the piecewise trilinear system (3) was found rather unexpectedly. For the system with trilinear force f(x) it was found such a situation when there were only two stable solutions in the system: the period-3T large orbit subharmonical solution and period-1T hilltop solution (Zakrzhevsky & Frolov, 1996). Trying to increase the domain of attraction of the hilltop solution, parameter continuation technique was used. It was found that increasing damping in the system it was possible to destroy the 3T-solution and stay only one stable hilltop 1T-oscillations. So the global stable hilltop oscillations has been found in the system. Fig. 4 illustrates the birth of global stable period-1T hilltop oscillations in this case. One can see how domains of attraction of the

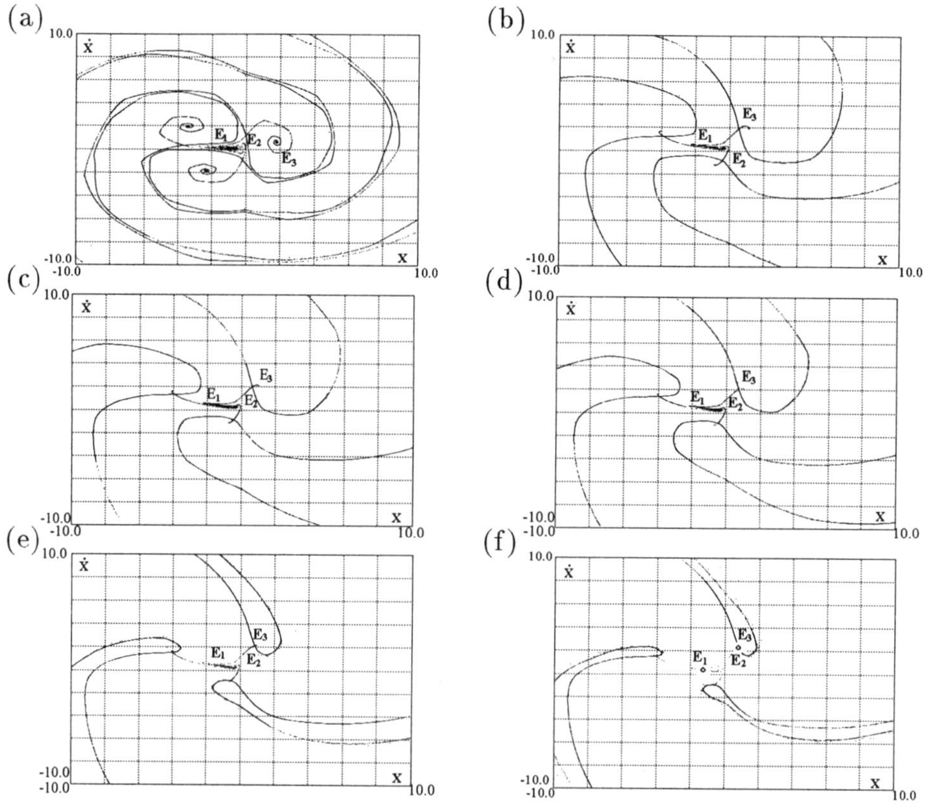

Figure 4. Domains of attraction of the hilltop period-1T (fixed point E_1) and period-3T (fixed pointE_3) large orbit subharmonic solutions. The birth of the global stable hilltop oscillations in equation (3) with trilinear force $f(x)$, when damping constant b is increasing.Parameters: $h_1 = 4.5, \omega = 2.0, c_1 = -2.5, c_2 = 1.0, \Delta = 0.4, d = 1.4$; damping constant b: (a) 0.10000, (b) 0.34585, (c) 0.36505, (d) 0.37625, (e) 0.38815, (f) 0.38888.

period-3T solution disappear when damping coefficient increases. This scenario is alike to the first, but here it is necessary to destroy not the period-1T LO stable solution, but the period-3T LO subharmonic solution.

5. Chaotic hilltop oscillations

It has found that chaotic hilltop oscillations are usual for periodically excited systems with several equilibrium positions as the well-known chaotic small or large orbit oscillations. The chaotic hilltop oscillations were found in Duffing's equation as well as in the systems with piecewise linear force f(x), and in pendulum systems. It was found two different but similar Feigenbaum's scenarios for creating chaotic hilltop oscillations: the first for period-1T oscillations, and the second for the period-3T hilltop oscillations with (3,3)-orbit. Fig 5 illustrate the first scenario for piecewise system (3).

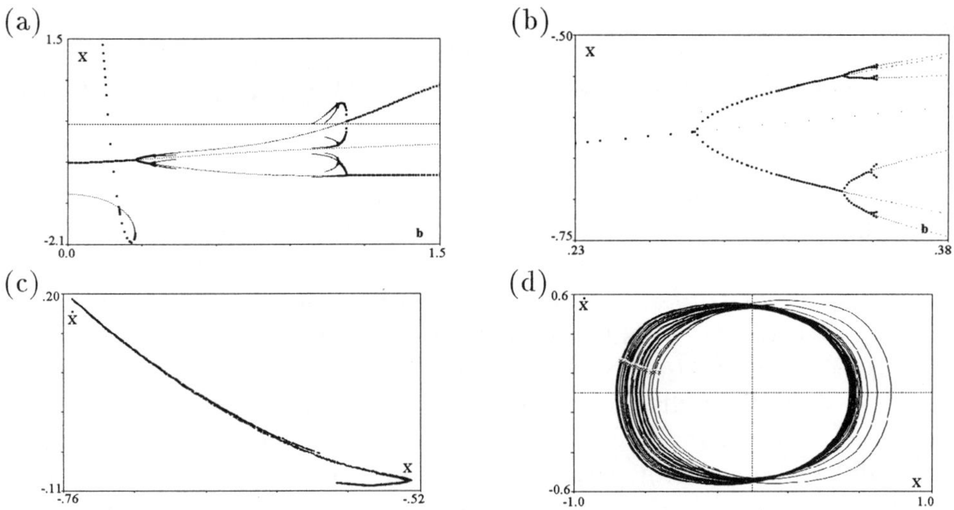

Figure 5. The global stable chaotic hilltop oscillations for trilinear system (3); parameters: $h_1 = 1.0, \omega = 0.9, c_1 = -1.0, c_2 = 1.0, \Delta = 0.5, d = 1.0$; (a) bifurcation diagram (x, b); (b) cascade to chaos; (c) Poincaré mapping of the chaotic hilltop attractor, damping constant $b = 0.36$; (d) phase portrait of the chaotic hilltop attractor, $b = 0.36$.

6. Global stable hilltop oscillations in pendulum systems

Two above discussed systems always have limited oscillations. In pendulum systems (4) or (5) it is possible not only oscillations but also rotations. But the global stable hilltop oscillations exist in such system as well. We show in Fig. 6 one example of such period-1T oscillations in the system (4). It was shown, using contour mapping from 3000 initial conditions, that these identical stable hilltop oscillations were unique in the system. In Fig.6 five transient processes from different initial conditions are plotted. The similar global stable oscillations are possible in the system (5).

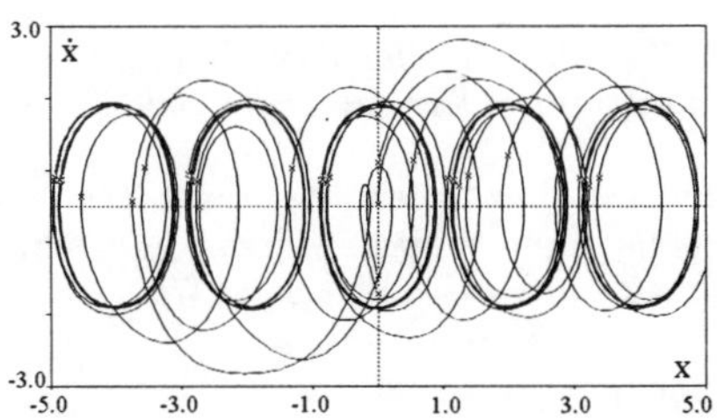

Figure 6. The global stable hilltop solution in the pendulum system (4) and transient processes; parameters: $b = 0.5, A = -0.5, h_1 = 4.0, \omega = 2.0$.

7. Conclusions

The global stable hilltop oscillations (the hilltop effect) is a typical nonlinear phenomenon for forced and/or parametric excited systems with two, several, or many potential wells. The phenomenon takes place in the limited parameter region where dissipation is rather large. The hilltop effect is a structurally stable nonlinear phenomenon. It occurs for different types of periodic excitations (forced, parametric, or complex; harmonic, impulse, etc.), and different dissipative forces. The phenomenon takes place in the systems with one, two and, we suppose, with many degrees of freedom, as well as in the systems governed by partial differential equations. We are sure that different useful applications of the hilltop oscillations in mechanics, engineering and other disciplines will be found. The possibility of the global stable hilltop periodic and chaotic oscillations in the archetypal nonlinear systems such as the twin-well potential Duffing system, forced pendulum system, piecewise linear system, etc. allows to predict that they may occur in mechanics of ship capsize, in the problems of power generators synchronization, in dynamics of switches, slender and thin-walled structures, and other nonlinear oscillatory systems.

References

Lansbury, A.N., Thompson, J.M.T., Stewart, H.B. (1992) Basin erosion in the twin-well Duffing oscillator: two distinct bifurcation scenarios, *Int. J. Bifurcation & Chaos*, 2 505-532.

Moon, F.C. (1987) *Chaotic Vibrations: An Introduction for Applied Scientists and Engineers*, John Willey & Sons.

Szemplinska-Stupnicka, W. and Rudowski, J. (1992) Local Methods in Predicting an Occurrence of Chaos in Two-Well Potential System: Superharmonic Frequency Region. *J Sound Vib.*, 152, 57-72.

Thompson, J.M.T. and McRobie, F.A. (1993) Indeterminate Bifurcations and the Global Dynamics of Driven Oscillators. *Proceeding of the 1st European Nonlinear Oscillations Conference*, Berlin: Akad.Verl., 107-128.

Zakrzhevsky, M. and Frolov, V. (1996) Bifurcation Analysis of the Stable Forced Oscillations Near Unstable Equilibrium Position, *Proceeding of the 2nd European Nonlinear Oscillations Conference*, Prague, Vol.2, p.265-267.

Zakrzhevsky, M. and Zevin, A. (1989) The Forced and Parametric Oscillations Near Unstable Equilibrium Position, *XIX Scientific School on Nonlinear Oscillations*, Kaunas.

Zevin, A.A. (1986) Non-local Criteria for the Existence and Stability of Periodic Oscillations in Autonomous Hamiltonian systems, AN SSSR, *Prikl. Mat. Mekh.* 50, 1, 45-51.

II. ELASTICA, CABLES AND SHELL DYNAMICS

The papers in this section explore the rich tradition of nonlinear elastic and shell theory. Both P. Holmes and J.M.T. Thompson and co-workers have published earlier works on the analogy between nonlinear dynamics of a spinning top and spatial complexity in the twisted elastica. In this book they show how homoclinic orbit theory can be applied to the study of solutions of the elastica.

The dynamics of hanging cables are extremely important in cable car transportation. The paper by F. Benedettini and G. Rega extend some of their earlier results on the nonlinear bifurcations of forced motion of cables.

Elastic shell theory and nonlinear dynamics is relatively unexplored. The papers of G.J. Lord et al. and P.B. Gonçalves provide an introduction to this subject.

F. C. Moon (ed.),
IUTAM Symposium on New Applications of Nonlinear and Chaotic Dynamics in Mechanics, 125.

HOMOCLINIC ORBITS, SPATIAL CHAOS AND LOCALIZED BUCKLING

J.M.T. THOMPSON and G.H.M. VAN DER HEIJDEN
Centre for Nonlinear Dynamics,
University College London,
Gower Street, London WC1E 6BT, UK.

1. Introduction

New insights and phenomena are emerging from an application of phase space concepts to the localization of structural post-buckling deformations [1]. Such localization is an important feature in the buckling of struts on nonlinear foundations (modelling, for example, railway tracks and sea-bed pipelines [2] under mechanical or thermal loading) and compressed cylindrical shells. It is also observed in the flanges of thin-walled sections, and a number of recent shipping disasters seem to be associated with the localized buckling of corrugated bulk-heads.

In this paper, we review progress and present some new results, looking especially at the spatial deformations of thin rods: relevant to marine pipe-lines, drill-strings and communication cables [3]; and to the writhing and super-coiling of DNA which is an important biological topic [4, 5]. Here a fundamental problem concerns the static deformations of a stretched and twisted elastic rod, equivalent to the motions of a spinning top. In this static-dynamic analogy of Kirchhoff, a homoclinic orbit of the top corresponds to a spatially localized post-buckling state. An *isotropic* rod (with a cross-section bounded by a circle, an equilateral triangle, a square, or any regular polygon) has equal principal bending stiffnesses and is an archetypal model on which simple experiments can be performed [6]. Equivalent to the *symmetric* top, it is integrable, and the closed-form solution offers unique insights into localization: modulation is governed by a nonlinear oscillator, like that obtained by multiple-scales perturbation of a strut on a foundation [7]. An *anisotropic* rod with unequal bending stiffnesses is non-integrable: it exhibits spatial chaos with an infinite number of localized homoclinic post-buckling paths. This complexity also arises [8] in the response of isotropic rods with an initial curvature: this may be important for cables and pipelines that can acquire initial curvature while they are stored in a coil.

A significant new application of phase space techniques to localization in axially compressed cylindrical shells has recently emerged from the pioneering work of Hunt and his co-workers [9, 10]: see [11], and the paper by Lord, Champneys & Hunt in these proceedings.

F. C. Moon (ed.),
IUTAM Symposium on New Applications of Nonlinear and Chaotic Dynamics in Mechanics, 127–138.

2. Beam on a Nonlinear Foundation

The long axially-compressed beam on a nonlinear foundation serves as a useful introduction to structural localization, as nicely seen in a series of papers by Hunt and his colleagues: see [12], and references therein. Engineers are familiar with bending solutions decaying from laterally loaded ends, as depicted in Fig 1. Meanwhile, on a softening foundation, we can have a central localization at a sub-critical value of the compression. In the dynamic analogy this represents an orbit homoclinic to the trivial state which is a saddle-focus with a 2D spiralling outset and a 2D spiralling inset.

Equation: $y'''' + Py'' + y - y^2 = 0$ Linear buckling: $P^C = 2$, $y = \cos x$

Homoclinic post-buckling: $y \approx s\,(6/\sqrt{19})\,\text{sech}\,(sx/2)\cos x$, where $s \equiv \sqrt{(P^C - P)}$

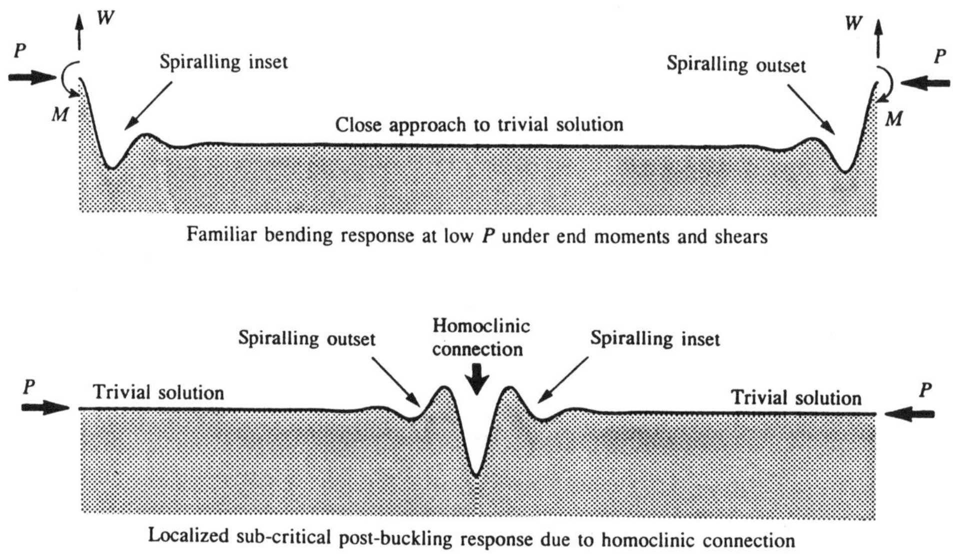

Figure 1. Beam on a nonlinear elastic foundation.

One way of analyzing this localization is to make a multiple-scales perturbation analysis [7]. This generates the equation of an undamped, undriven, Duffing oscillator governing the slowly-varying modulation amplitude. The nonlinear stiffness of this oscillator is cubic, even when the foundation is quadratic. Its phase portraits represent periodic, quasi-periodic and localized paths emerging from the critical buckling point. An identical picture (Fig 4) is obtained by closed-form analysis of the isotropic rod.

While useful, perturbation analysis misses the global complexity of the nonlinear problem. Buffoni, et al, [13] show that the beam has an infinite number of localized post-buckling paths, as in Fig 2. In this diagram, the beam has four complex eigenvalues, giving a saddle-focus in the trivial state, for $-2<P<2$, where $P^C = 2$ is the critical buckling load. An infinite number of homoclinic paths approach arbitrarily close to P^C. It is this spatial chaos that we address here by the static-dynamic analogy. For problems where there is no *real* physical analogy, as with a top, governing ODEs can always be viewed as those of a *virtual* dynamical system.

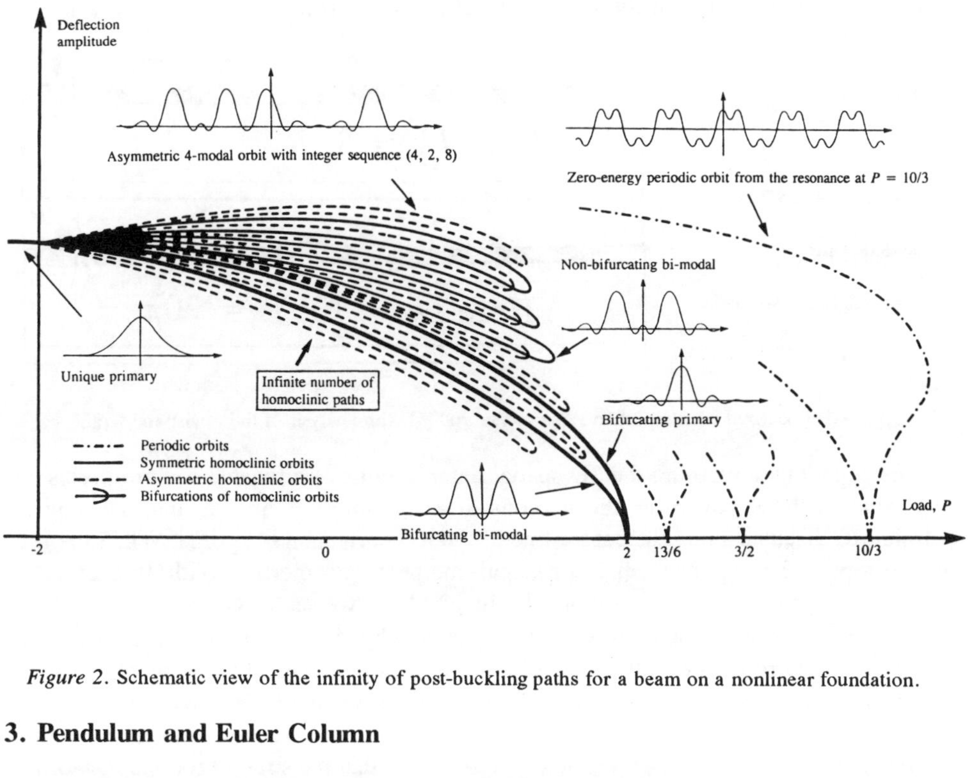

Figure 2. Schematic view of the infinity of post-buckling paths for a beam on a nonlinear foundation.

3. Pendulum and Euler Column

The static-dynamic analogy is seen in its simplest *real* form relating the in-plane deformations of an elastic strut (the elastica) to the motions of a pendulum [14]. Here a post-buckling path (for pinned ends, a fixed load, and variable length, say) would involve a family of orbits, with different values of the pendulum's Hamiltonian. The saddle connection of the pendulum corresponds to a single loop of the strut under tension. If the strut were long, and we ignored the boundary conditions, all the orbits of the phase portrait would represent possible equilibrium states at a fixed load. There would be a family of solutions at a given load, as in Fig 4 for the rod.

4. Perfect Isotropic Rods under Torsion

A long elastic rod loaded at its end by a wrench, namely an applied tension, T, and a twisting moment, M, is a useful archetypal model for localized buckling.

Initial experiments [6] show that a circular rod exhibits a continuous process of localization, leading to self-contact and subsequent writhing. A small initial helical deflection with 1 twist per wave is explained later in terms of an initial curvature of the rod. The final buckling mode (Fig 3) has approximately 3 twists per localized helical wave, agreeing with the theoretical prediction of $2(1+\nu)$, where ν is Poisson's ratio. More recent experiments, on a square silicone rubber rod with no initial curvature, show no sign of the initial 1 twist per wave [8].

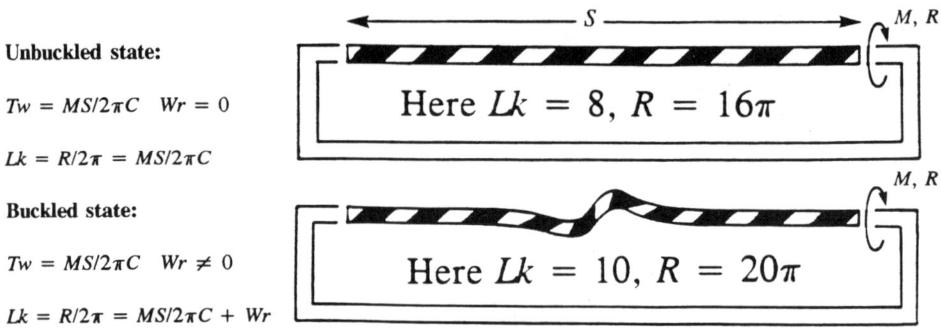

Figure 3. Localized buckling of an experimental rod [6] with analysis of link, twist and writhe.

The equations governing the 3D spatial deformations of a twisted rod are precisely the same as those governing the motions of a spinning top. In this analogy, due to Kirchhoff, distance along the centre-line of the rod corresponds to time. When the rod is isotropic, the top has equal principal moments of inertia. With this circular symmetry, the equations are completely integrable as we see in §6.

With equal principal bending stiffnesses denoted by $A = B$, and torsional stiffness, C, an isotropic rod buckles theoretically into a helix at a critical value of M given by $m_c = M/\sqrt{(BT)} = 2$. At this critical load (independent of C) a pair of complex conjugate eigenvalues meet on the imaginary axis and become imaginary. This corresponds to the onset of precession in a top. Note that the straight rod loses stability as m increases through m_c, while the vertical sleeping top loses stability as m decreases through m_c: in static-dynamic analogies, stability regions are usually reversed.

With $A = B$, the full equations can be reduced, without approximation, to those of a nonlinear oscillator, like that generated by perturbation analysis of the beam. In the reduced space of this oscillator, the Hamiltonian Hopf bifurcation at m_c appears as a pitchfork, subcritical for the rod, supercritical for the top. For the rod, post-buckling paths emerge from the bifurcation, as in Fig 4. There is one path of helical states, a continuum of periodically modulated states, and one path of localizing homoclinics.

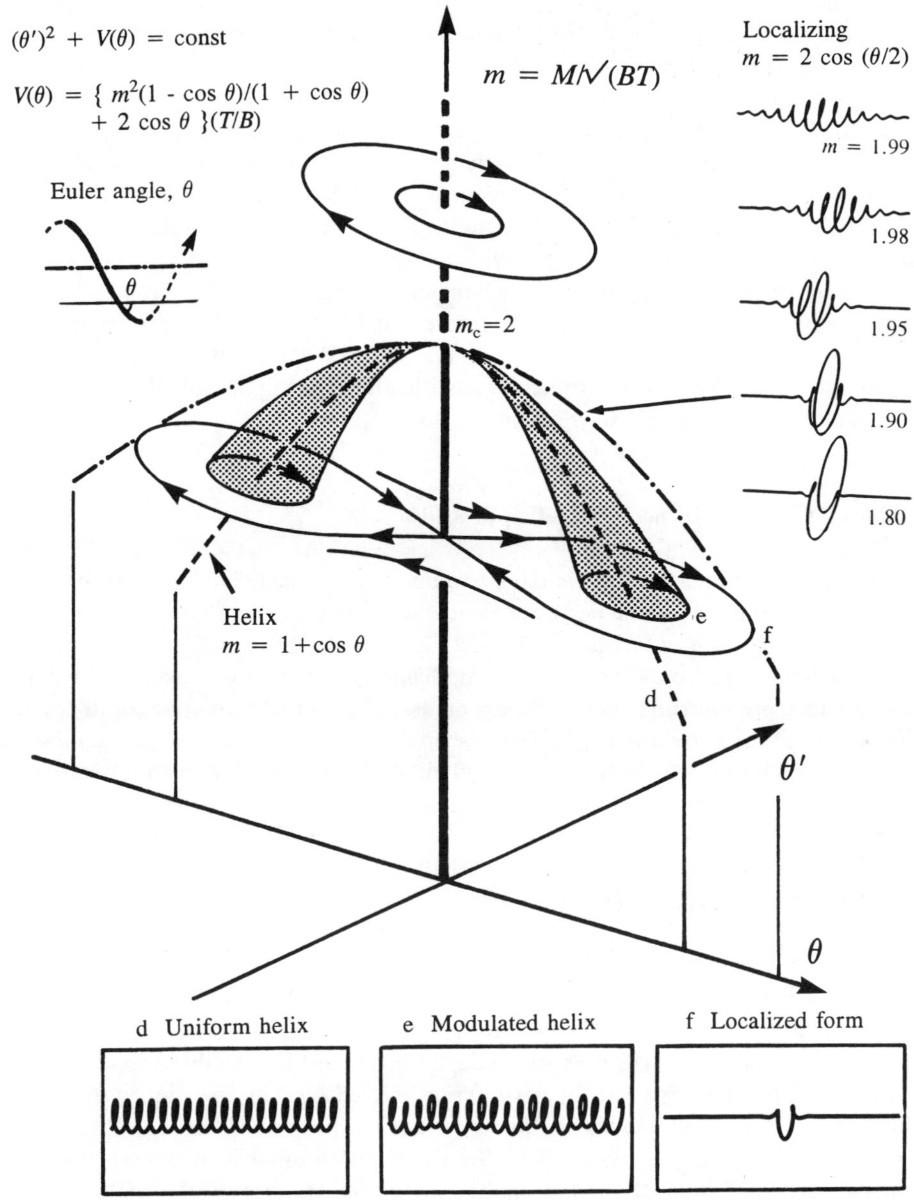

Figure 4. Equivalent nonlinear oscillator [17] in exact solution of a twisted isotropic rod.

The spatial forms in Fig 4 show that the continuous process of localization on the homoclinic path develops rapidly as we move away from the bifurcation [15]. The helical and localized post-buckling paths are predicted to be unstable under both dead and rigid loading: with a backwards-sloping post-buckling path, as in the buckling of shells. Energy analysis and experiments [6] show that the localized solution is the preferred mode of buckling, and offers the minimum energy barrier against premature buckling. For this reason, we shall now focus mainly on the homoclinic paths.

A useful topological result [16], developed to explain super-coiling of DNA, is $Lk = Tw + Wr$. Here link, Lk, corresponds to the end rotation, R, through which M does work; twist, Tw, is the integral of the twist rate, $\tau/2\pi$; writhe, Wr, is a property of the space curve of the centre-line. For the homoclinic solution of an isotropic rod, τ is constant along the rod (see §6) and is therefore equal to M/C: Tw is then just $MS/2\pi C$ where S is the length of the rod. So during homoclinic buckling there is no change in Tw, and M simply does work through Wr, as illustrated in Fig 3. Note that if $C = \infty$, there is no twist before or after buckling at finite M.

5. Spatial Chaos in Imperfect Isotropic Rods

An isotropic rod with a constant initial curvature lies in a plane circle in its unstressed state. Such a rod has a 1 twist per wave helical solution under an applied wrench. This is a pre-buckling state onto which a 3 twist per wave buckling mode is subsequently super-imposed. A test on a long circular rod hanging in a shallow catenary [6] shows these distinct pre-buckling and buckling modes. The initial curvature destroys the integrability of the perfect rod, generating spatially complex forms, similar to those of the beam shown in Fig 2. More will be said about these forms for anisotropic rods in §7. This recent discovery [8] may be significant for circular rods stored in a coil.

6. Phase-space Theory of Twisted Rods

Consider a long elastic rod with principal bending stiffnesses, A, B, and torsional stiffness, C, loaded at its ends by equal and opposite applied wrenches, (T, M). We use as dependent variables components of the forces and moments acting across a cut of the rod at arc-length s, taking as our independent time-like variable the scaled arc-length $t \equiv s(M/B)$. Specifically, we write the change in the forces from the fundamental (loaded but unbuckled) state, divided by the applied tension, T, as: $x_1(t)$, $x_2(t)$, shear forces; and $x_3(t)$, tension. Similarly, the change in moments from the fundamental state, divided by the applied moment, M, are: $x_4(t)$, $x_5(t)$, bending moments; and $x_6(t)$, twisting moment.

With linear elasticity, and ignoring shear and axial deformation, we can follow the formulations of Love and Antman [17]. Equilibrium of the rod is then governed by the six first-order equations [15]

$$\dot{x}_1 = (1 + \nu)\, x_2\, (1 + x_6) - x_5\, (1 + x_3) \tag{1}$$

$$\dot{x}_2 = (1 + \rho)\, x_4\, (1 + x_3) - (1 + \nu)\, x_1\, (1 + x_6) \tag{2}$$

$$\dot{x}_3 = x_1\, x_5 - (1 + \rho)\, x_2\, x_4 \tag{3}$$

$$\dot{x}_4 = \nu\, x_5\, (1 + x_6) + x_2 \,/\, m^2 \tag{4}$$

$$\dot{x}_5 = (\rho - \nu)\, x_4\, (1 + x_6) - x_1 \,/\, m^2 \tag{5}$$

$$\dot{x}_6 = - \rho\, x_4\, x_5\ , \tag{6}$$

where $m = M/\sqrt{(BT)}$, $\rho = (B/A) - 1$, $\nu = (B/C) - 1$, and a dot denotes differentiation with respect to t. For a solid circular rod (only), ν can be identified as Poisson's ratio. A solution of these equations can be post-processed using the *Frenet-Serret* equations of differential geometry to give the spatial form of the rod.

These equations define a dynamical system, **D**, with parameters (m, ρ, ν), which also governs the motions of a spinning top, non-symmetric if $A \neq B$. Mielke & Holmes [18] show that **D** is Hamiltonian even for more general constitutive laws, and including shear and axial deformation.

System **D** has 2 independent integrals of motion, contained implicitly within (1)-(6). These can be derived directly by overall statics by considering the resultant force and moment vectors, $\boldsymbol{F}$, $\boldsymbol{G}$, at any cut through the rod. Ignoring signs, vector $\boldsymbol{F}$ must equal vector $\boldsymbol{T}$, so $|\boldsymbol{F}| = T$, giving

$$x_1^2 + x_2^2 + (1 + x_3)^2 = 1\ . \tag{7}$$

Meanwhile the component of vector $\boldsymbol{G}$ along the wrench axis must equal $\boldsymbol{M}$. Hence $\boldsymbol{F} \cdot \boldsymbol{G} \,/\, T = M$, giving

$$x_1\, x_4 + x_2\, x_5 + (1 + x_3)\, (1 + x_6) = 1\ . \tag{8}$$

These two constraints imply that all motions lie in a *4D sub-space* which is independent of (m, ρ, ν).

Next, using (1)-(6), it is easily verified that $\mathrm{d}H/\mathrm{d}t = 0$ where H is the Hamiltonian,

$$H = m^2[\, (1 + \rho)x_4^2 + x_5^2 + (1 + \nu)(1 + x_6)^2] + 2x_3 = \text{const.} \tag{9}$$

A given motion evolves at constant $H = H(m, \rho, \nu)$. The corresponding Lagrangian equals the energy density of the rod, as in the pendulum-strut analogy [7].

When $A \neq B$, all motions lie in the fixed 4D sub-space, and a given motion is constrained to evolve at constant H. However these conditions do not represent three independent *isolating integrals*, and the system is not *completely integrable* in the classical sense. Chaos is possible, as we shall see.

However, when $A = B$ we have $\rho = 0$ and equation (6) gives $x_6 =$ const. This states that the twisting moment, and hence the twist rate, τ, are constant along any equilibrium state of the rod. The dynamical system now has three independent isolating integrals, and is completely integrable. In particular, chaos is impossible.

Linearizing (1)-(6) about the trivial solution, $x_i = 0$, for $A \neq B$ gives a characteristic equation of the form $\lambda^6 + c_4\lambda^4 + c_2\lambda^2 = 0$. Two degenerate eigenvalues, $\lambda=0$, have eigenvectors which point out of the 4D sub-space. The remaining 4 eigenvalues give bifurcation arcs, $m_c(\rho, \nu)$, shown in Fig 5 for the case of $B/C = 4/3$, $\nu = 1/3$. Recall that ν has no relation to Poisson's ratio once $\rho \neq 0$.

This diagram shows that the buckling locus changes from a Hopf bifurcation to a pitchfork bifurcation at a co-dimension-two point, C. The movement of the eigenvalues in the (m, ρ) plane is indicated. Note in particular that we have complex eigenvalues in the dotted regime [15]. Here the rod has an infinity of homoclinics, as we shall see.

Looking first at the homoclinic paths bifurcating from the trivial solution, we recall that the isotropic rod has just one localizing post-buckling path bifurcating at $m_c = 2$. This path has approximately $2(1+\nu)$ twists per wave. For mildly anisotropic rods, with $0<\rho<\rho_c$, the bifurcating manifold of primary homoclinics at $A = B$ is split, giving two bifurcating localized paths. This agrees with normal form analysis [19] using results from [20]. Notice that this analysis uses two reversibilities, corresponding to the D_2-symmetry of the problem. Strongly anisotropic rods, with $\rho_c<\rho$, buckle into a tape-like form with one twist per wave. A normal form analysis (confirmed by numerics) reveals that at C there is a change in multiplicity of the localized buckling solutions. Of the two paths at moderate anisotropy, only the energetically favourable 'flat' deformation remains at $\rho_c<\rho$, there being no initial 'standing' equilibrium mode.

7. Spatial Complexity of Anisotropic Rods

For a wide class of rods, Mielke and Holmes [18] made a rigorous Melnikov analysis for a perturbation from $\rho = 0$ to prove the existence of transversal intersections, and hence horseshoes and chaos. A numerical study of spatial chaos was made in [21].

Normal form analysis yields the existence of homoclinic orbits near (and below) the buckling locus m_c. If we assume the persistence of these homoclinic orbits, Devaney's theorem allows us to establish the existence of horseshoes and chaos in the entire region of complex eigenvalues. Wiggins quotes this theorem in the form: *For a 4D Hamiltonian system with a transverse homoclinic orbit to a saddle-focus, a suitable Poincaré map contains a Smale horseshoe.* The homoclinic connection, from a 2D spiralling outset to a 2D spiralling inset, is illustrated schematically in Fig 6(a).

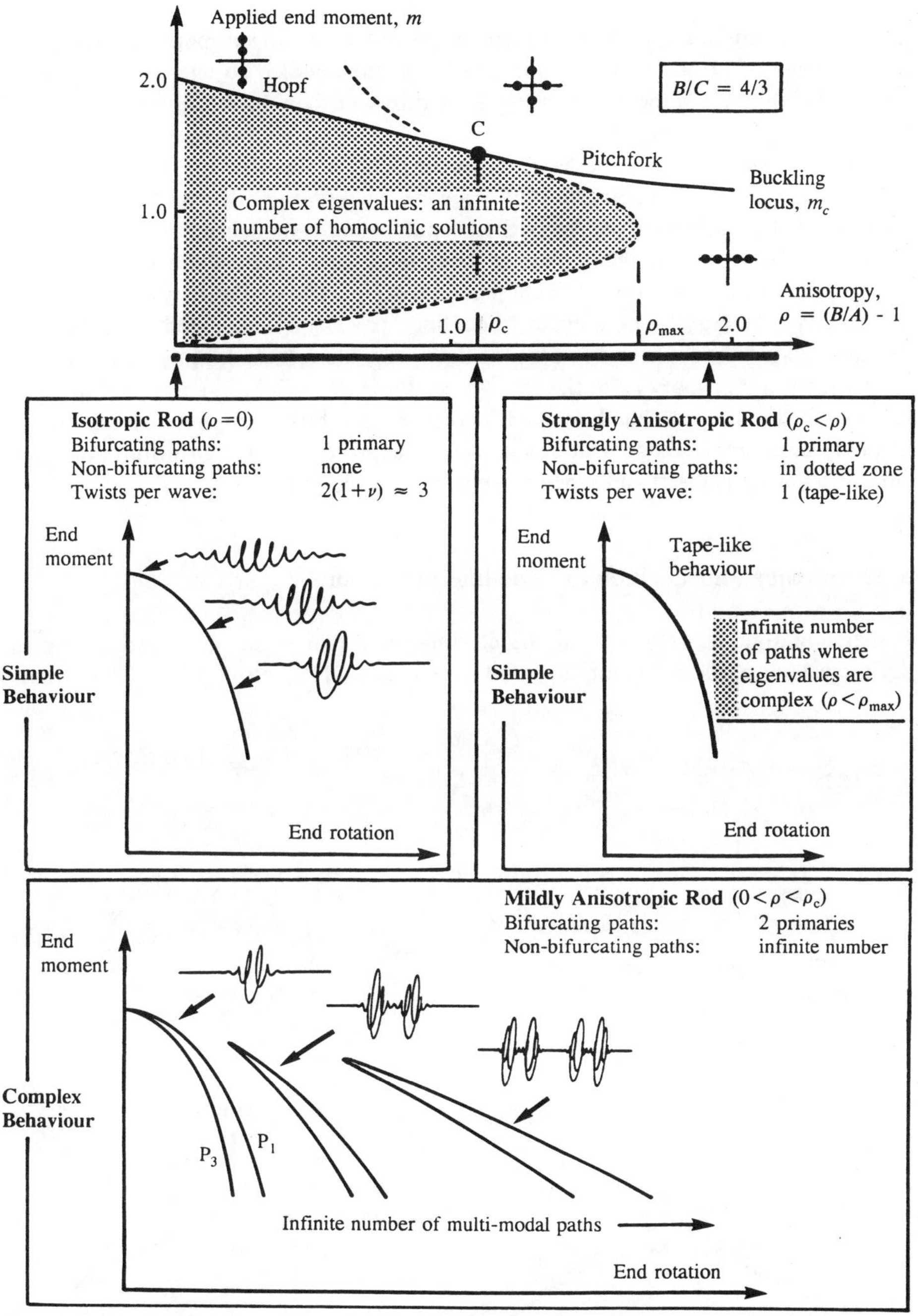

Figure 5. Bifurcation arcs, and three response graphs for homoclinics of anisotropic rods.

The theorem implies spatial chaos with an infinity of localized modes looking like multiple copies of primary modes as in Fig 5, but the detailed structure of the multi-modal paths, shown for the beam in Fig 2, requires further results along the lines of Belyakov and Silnikov [22] and Buffoni, et al, [13]. A numerical study of the global bifurcations of the localized modes is given in [23]. A shooting method, exploiting reversibility, was used to locate the primary buckling modes and families of multi-modals. It was shown that, as in the beam, the multi-modals do not persist right up to the buckling load, but die in pairs at saddle-node folds. Load-deflection diagrams were computed by continuation of the homoclinic orbits. The multiplicity of paths gives a multiplicity of energy barriers against buckling for a stable unbuckled rod subjected to dynamic disturbances, as in Fig 6(b) for $0<\rho<\rho_c$. Here the solid line, P_3, is a low-energy bifurcating primary path; the short-dashed line, P_1, is a high-energy bifurcating primary path; the long-dashed line is a high-energy non-bifurcating multi-modal path.

Strongly anisotropic rods, with $\rho_c<\rho$, can be expected to have an infinite number of non-bifurcating paths if and when the eigenvalues go complex.

8. Intermittency and Lock-on to Tape-like Behaviour

Consider, finally, the transition at the co-dimension-two point, C. Approaching C under increasing ρ, the rod exhibits spatial intermittency [19] during which it becomes

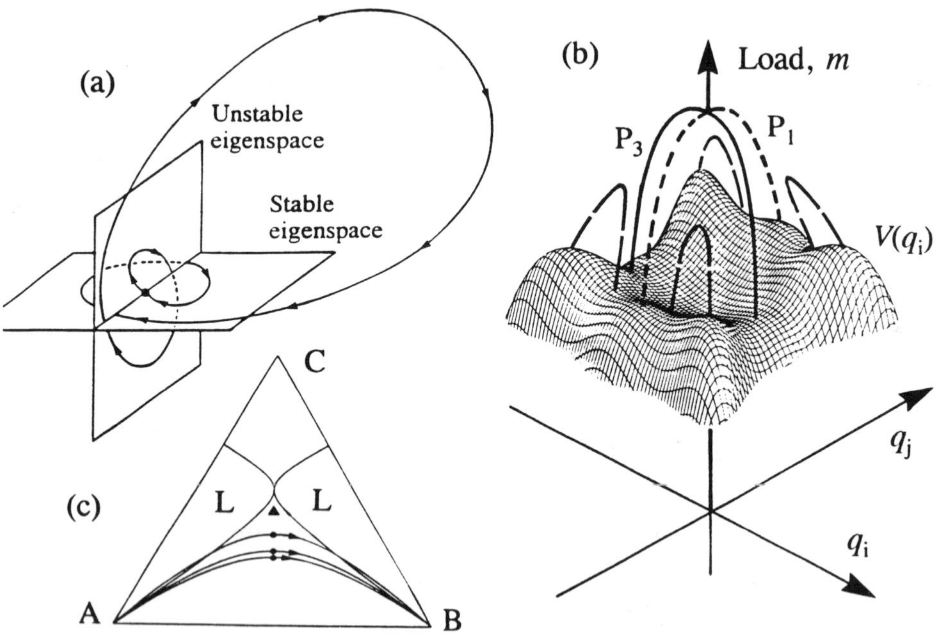

Figure 6. (a) Homoclinic to a saddle-focus (b) Potential energy surface (c) Ratios $A{:}B{:}C$ for lock-on

progressively locked onto one twist per helical wave, as in Fig 7. Just before lock-on, the centre-line is a modulated helix, over most of which the internal twist, f, is approximately a multiple of π, implying a nearly-locked state. As the amplitude of the helix drops to zero, synchronization is temporarily lost, and f changes rapidly by π to become again almost tape-like as the amplitude increases again. Beyond C, buckling at the pitchfork is tape-like, the rod bending only about its weak axis with $f = 0$.

Dropping constraint $B/C = 4/3$ of Fig 5, the locus of bifurcation C is displayed, [19], in the triangular space of Fig 6(c), representing the ratios $A{:}B{:}C$. Here lock-on to tape-like behaviour is in regions L. The 3 continuous arcs from A to B represent 3 families of solid elliptical rods, parametrized by aspect ratio. Top arc is for Poisson's ratio 0, middle arc for ratio 1/3, and lower arc for ratio 1/2. Elliptical rods only reach

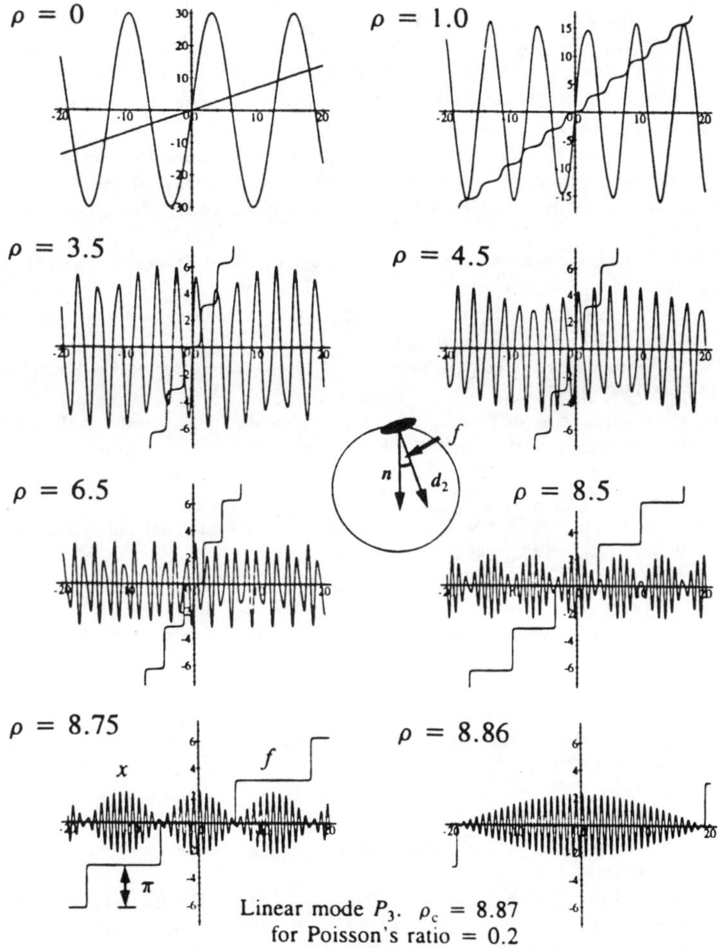

Figure 7. Approach to lock-on: side views of linear buckling mode, and internal twist angle.

lock-on at finite aspect ratio if Poisson's ratio is less than 1/3. The triangle at $A = B = 2C/3$ is an averaged value for DNA (158-base-pair) used by Manning, et al [24].

References

[1] Champneys, A.R., Hunt, G.W. & Thompson, J.M.T. (editors), *Localization and solitary waves in solid mechanics*, Theme Issue of *Phil. Trans. R. Soc. Lond.*, in press, 1997.

[2] Maltby, T.C. & Calladine, C.R., An investigation into upheaval buckling of buried pipelines, I & II, *Int. J. Mech. Sci.* **37**, 943-963 and 965-983 (1995).

[3] Coyne, J., Analysis of the formation and elimination of loops in twisted cable, *IEEE Journal of Oceanic Engineering* **15**, 72-83 (1990).

[4] Bauer, W.R., Crick, F.H.C. & White, J.H., Supercoiled DNA, *Sci. American* **243**, 118-133 (1980).

[5] Calladine, C.R. & Drew, H.R., *Understanding DNA*, Academic Press, London (1992).

[6] Thompson, J.M.T. & Champneys, A.R., From helix to localized writhing in the torsional post-buckling of elastic rods, *Proc. R. Soc. Lond.* **A452**, 117-138 (1996).

[7] Hunt, G.W., Bolt, H.M. & Thompson, J.M.T., Structural localization phenomena and the dynamical phase-space analogy, *Proc. R. Soc. Lond.* **A425**, 245-267 (1989).

[8] Champneys, A.R., van der Heijden, G.H.M. & Thompson, J.M.T., Spatially complex localization after one-twist-per-wave equilibria in twisted circular rods with initial curvature, in *Theme Issue* [1] above.

[9] Hunt, G.W. & Neto, E.L., Localized buckling in long axially loaded cylindrical shells, *J. Mech. Phys. Solids* **39**, 881-894 (1991).

[10] Hunt, G.W. & Neto, E.L., Maxwell critical loads for axially loaded cylindrical shells, *J. Appl. Mech.* **60**, 702-706 (1993).

[11] Lord, G.J., Champneys, A.R. & Hunt, G.W., Computation of localized post-buckling in long axially-compressed cylindrical shells, in *Theme Issue* [1] above.

[12] Hunt, G.W. & Wadee, M.K., Comparative lagrangian formulations for localized buckling, *Proc. R. Soc. Lond.* **A434**, 485-502 (1991).

[13] Buffoni, B., Champneys, A.R. & Toland, J.F., Bifurcation and coalescence of a plethora of homoclinic orbits for a Hamiltonian system, *J. Dyn. Diff. Eq.* **8**, 221-281 (1996).

[14] Thompson, J.M.T. & Virgin, L.N., Spatial chaos and localization phenomena in nonlinear elasticity, *Phys. Lett.* **A126**, 491-496 (1988).

[15] Champneys, A.R. & Thompson, J.M.T., A multiplicity of localized buckling modes for twisted rod equations, *Proc. R. Soc. Lond.* **A452**, 2467-2491 (1996).

[16] Fuller, F.B., The writhing number of a space curve, *Proc. Nat. Acad. Sci. USA* **68**, 815-819 (1971).

[17] Antman, S.S., *Nonlinear problems of elasticity*, Springer-Verlag, Berlin (1995).

[18] Mielke, A. & Holmes, P., Spatially complex equilibria of buckled rods, *Arch. Rational Mech. Anal.* **101**, 319-348 (1988).

[19] van der Heijden, G.H.M. & Thompson, J.M.T., Lock-on to tape-like behaviour in the torsional buckling of anisotropic rods, *Physica D*, in press, 1997.

[20] Iooss, G. & Pérouème, M.C., Perturbed homoclinic solutions in reversible 1:1 resonance vector fields, *J. Diff. Eqns.* **102**, 62-88 (1993).

[21] Davies, M.A. & Moon, F.C., Spatial chaos in the elastica and the spinning top: Kirchhoff analogy, *Chaos* **3**, 93-99 (1993).

[22] Belyakov, L.A. & Silnikov, L.P., Homoclinic curves and complex solitary waves, *Selecta Mathematica Sovietica* **9**, 219-228 (1991).

[23] van der Heijden, G.H.M., Champneys, A.R. & Thompson, J.M.T., The spatial complexity of localized buckling in rods with non-circular cross-section, *SIAM J. Appl. Math.*, in press, 1997.

[24] Manning, R.S., Maddocks, J.H. & Kahn, J.D., A continuum rod model of sequence-dependent DNA structure, *J. Chem. Phys.* **105**, 5626-5646 (1996).

NONREGULAR REGIMES OF MONODIMENSIONAL MECHANICAL SYSTEMS WITH INITIAL CURVATURE: EXPERIMENTS AND TIME SERIES ANALYSIS

F. BENEDETTINI* AND G. REGA**
** Dipartimento di Ingegneria delle Strutture, Acque e Terreno, Universita' dell'Aquila, Monteluco Roio, 67040 L'Aquila, Italy*
*** Dipartimento di Ingegneria Strutturale e Geotecnica, Universita' di Roma "La Sapienza" Via A. Gramsci 53, 00197 Roma, Italy*

1. Introduction

The 3D finite dynamics of monodimensional elastic structures with initial curvature is particularly rich and varied because of the presence of both even and odd nonlinearities, the former being directly linked to the initial curvature [1]. This work is concerned with the analysis of experimental models of two different systems belonging to that class. The first is a discrete model of an elastic suspended cable excited by vertical, sinusoidally varying, motion of the hanging points. The second is a steel model of a double hinged circular arch excited by a vertical, sinusoidally varying, concentrated force at its tip. In strongly developed nonlinear regimes, interesting phenomena linked to the nonlinear modal interaction appear in such systems over a wide range of excitation frequencies, owing to the very close sequence of primary and secondary resonance conditions. Moreover, changing the sag to span ratio of the cable or adding a vertical dead load on the tip of the arch it is possible to obtain various internal resonance conditions which further exhalt those nonlinear interaction phenomena.

Generally speaking, the first step of a nonlinear dynamics analysis is devoted to the detection of the more robust classes of regular motion around the external resonance conditions. For the experimental cable, overall behavior charts and bifurcations accounting for transitions between periodic contiguous classes of motion were drawn in previous works [2,3]. For the arch, the main interest was in obtaining the external conditions (amplitude and frequency of the vertical, symmetric driving force) under which antisymmetric modes are excited, furnishing in such a way the boundary of the instability region of the simple unimodal symmetric response [4,5].

In this work the aim is to detect some bifurcations between simple periodic and complex quasi-periodic and chaotic classes of motion, and to characterize some of the latters. Different levels of complexity inside regions of nonregular evolutions are analyzed first qualitatively by means of phase portraits, Poincaré maps and power spectra of a given point of the system, and then quantitatively measuring strangeness and complexity. More than one scalar measure on the system are used as well.

When working with experimental models all the quantitative measures concerning strangeness and chaoticity are obtained in a reconstructed pseudo-phase space being,

F. C. Moon (ed.),
IUTAM Symposium on New Applications of Nonlinear and Chaotic Dynamics in Mechanics, 139–148.

under certain reconstruction conditions, in a one-to-one correspondence with the real, unaccessible phase space [6]. Several implementations of the delay map technique and choices of the reconstruction parameters are considered [7].

When flexible experimental models show a spatio-temporal complexity, a set of simultaneous measures on different points of the model should be acquired and processed, which is actually unpractictable because of the high cost of no-contact displacement transducers. Thus, for the cable, only pairs of time laws are simultaneously acquired in different positions, and are then processed with the usual time averaging to obtain the elements of the spatial correlation matrix of the samples. In any case, even with this restriction, a significant improvement of the reconstruction finalized to the attractor dimension estimation is obtained, and the use of the proper orthogonal decomposition procedure allows to verify which are the spatial shapes much more visited by the system during the chaotic evolution.

Fig. 1 Suspended cable/mass model (a); circular arch model (b).

2. Tools and techniques for nonlinear experimental analysis

2.1 STRANGENESS AND CHAOTICITY

When dealing with experimental tests on flexible multi-d.o.f. or continuous systems, the problem concerned with the inaccessibility of the whole state space immediately arises.

The possibility to measure only one or few target positions makes necessary the use of the state space reconstruction techniques mainly in the case of complex time evolutions in time and space. Under certain reconstruction conditions and according to the Takens theorem [6], the delay embedding procedure allows to obtain a pseudo-phase space having the same topological properties and invariant measures as the real, inaccessible phase space. The basic steps to be performed to obtain a meaningful reconstruction are summarized as follows:

- Using the measure of the *mutual average information,* an estimate of the right delay τ to be used in the reconstruction is first made.
- Using the *false nearest neighbors detector,* the right value of the embedding dimension m in which the underlying attractor is unfolded is then obtained.

- Using a suitable *representation basis* obtained by the covariance matrix of the samples, a trajectory resolution showing the best signal-to-noise ratio is obtained.
- Using the *Grassberger and Procaccia algorithm* applied to the reconstructed pseudo-vectors the correlation dimension (the slope of the lg[C(r)] vs lg(r) curves) is found for different embedding dimensions in the proximity of the right *m* estimated as in the second point. The saturation of the correlation dimension value gives eventually an indirect confirmation of the procedure consistency.
- An improvement of the width of the scaling region obtained during the last step can be obtained when using two or more time series simultaneously acquired from different points on the system in comparison with the standard procedure working with only one point of measure.

The delay embedding procedure is of special interest for allowing reliable estimations of the system dimensionality: indeed, it furnishes indications on the smaller number of d.o.f. taking part in the system nonlinear dynamics. The reconstruction conditions were already described and applied to the experimental model of suspended cable in [7] for analyzing some sample responses. Herein, they are applied to both the suspended cable and the circular arch in different regions of chaotic response.

Besides obtaining an estimation of the possible fractal nature of the underlying motion attractors with the aformentioned procedure, similar techniques are used for the two systems to estimate the whole spectrum of Lyapunov exponents, or alternatively, the maximum one. While the classical Wolf et al. algorithm used in estimating the maximum exponent from a time series is rather stable with respect to variations of algorithmic parameters, in the case of time series acquired from the experimental models shown in Fig.1, not so good results are found when dealing with more complex algorithms furnishing an estimation of the local jacobian and the whole Lyapunov spectrum. A good qualitative and quantitative validation of the results given by the Wolf algorithm can be obtained by using the very simple, fast and parameter-insensitive algorithm proposed by Rosenstein in a recent paper [8].

2.2 SPATIAL COHERENCE

When a multi-d.o.f. or a continuous system shows a complex dynamic evolution, beyond the time complexity evidenced by the aformentioned invariant measures, it also exhibits a spatial complexity. With the aim of analyzing it in terms of the spatial configurations more visited on average during a complex (space and time) evolution, the proper orthogonal decomposition procedure applied to spatially contemporaneous measures on the system is used. With respect to the delay embedding procedure, it allows to have indications not only on the number of d.o.f. taking meaningful part in the nonlinear response, but also on their mechanical meaning (spatial shape). Combined use of both procedures can thus result of major interest for a proper selection of minimal theoretical models able to reproduce the complex dynamics of experimental systems. The basic steps to be performed to obtain a meaningful spatial analysis are summarized as follows:

- *Evaluation of the coherence matrix* whose elements are the averaged values of the cross-correlations between all possible couples of different and synchronous time series when targeting the system with a finite (small) number of points of measure.

- *Calculation of eigenvalues and eigenvectors* giving respectively the mean square amplitudes (or the power) associated to every proper orthogonal mode, and the modes themselves.
- All the eigenvectors (*proper orthogonal modes*) representing the spatial shapes visited on average proportionally to the corresponding eigenvalues form the spatial basis capturing more power per mode than any other basis.
- Being by construction the coherence matrix an Hermitian tensor having only positive eigenvalues, possible negative results for the eigenvalues are evidence of experimental errors.

The basic ideas and one of the first applications of the procedure in solid mechanics are reported in [9]. In the following, the procedure will be applied to the cable in Fig.1a for obtaining proper orthogonal modes in different regions of steady chaotic regime. Because of the high cost of several synchronous experimental measures, using only two acquisition devices and averaging several times on long time series (2500 forcing period) it is possible to have a quite good estimation of the coherence matrix with a reasonable experimental effort.

3. Nonregular 3D dynamics of the elastic suspended cable

A discrete model of an elastic suspended cable (a nylon wire carrying a finite number of concentrated masses) excited by vertical, harmonic, in-phase or out-of-phase motion of the hanging points, is first considered (Fig. 1a). A basic dynamical feature of such system in the neighbourhood of the first crossover point [3] is the occurrence of a nearly 2:2:1:2 simultaneous resonance condition involving the first in-plane and out-of-plane symmetric and antisymmetric modes. This entails very rich local and global finite dynamics, which have been widely analyzed through both overall behavior charts [2] and specific response diagrams [3] obtained in various resonance zones of the excitation parameter plane. Therein, though mainly focusing the attention on the features of competing classes of multimode regular motions, several regions of occurrence of QP and chaotic responses were also detected mostly on the basis of qualitative measures of the inherent dynamics. In the sequel, we first present some reconstructions of global properties of nonregular attractors, and then quantitatively analyze the evolution of the kind of chaotic response occurring within a region previously identified as chaotic without further specifications. Finally, we report sample results obtained by means of the proper orthogonal decomposition technique, aimed at detecting some major and different contributions from the underlying modes of the system in the chaotic responses obtained in different regions of the excitation parameter plane.

3.1 RECONSTRUCTION OF GLOBAL PROPERTIES IN SAMPLE CHAOTIC REGIONS

In Fig.2a-g we report some qualitative (upper) and quantitative (lower) properties of two reconstructed chaotic attractors occurring in the zone of the order ½-subharmonic resonance of the crossover frequency, under out-of-phase (left) and in-phase (right) support motion of comparable amplitude. Different degrees of complexity are shown by the two attractors. The former exhibits features of a fuzzy torus, as revealed by its

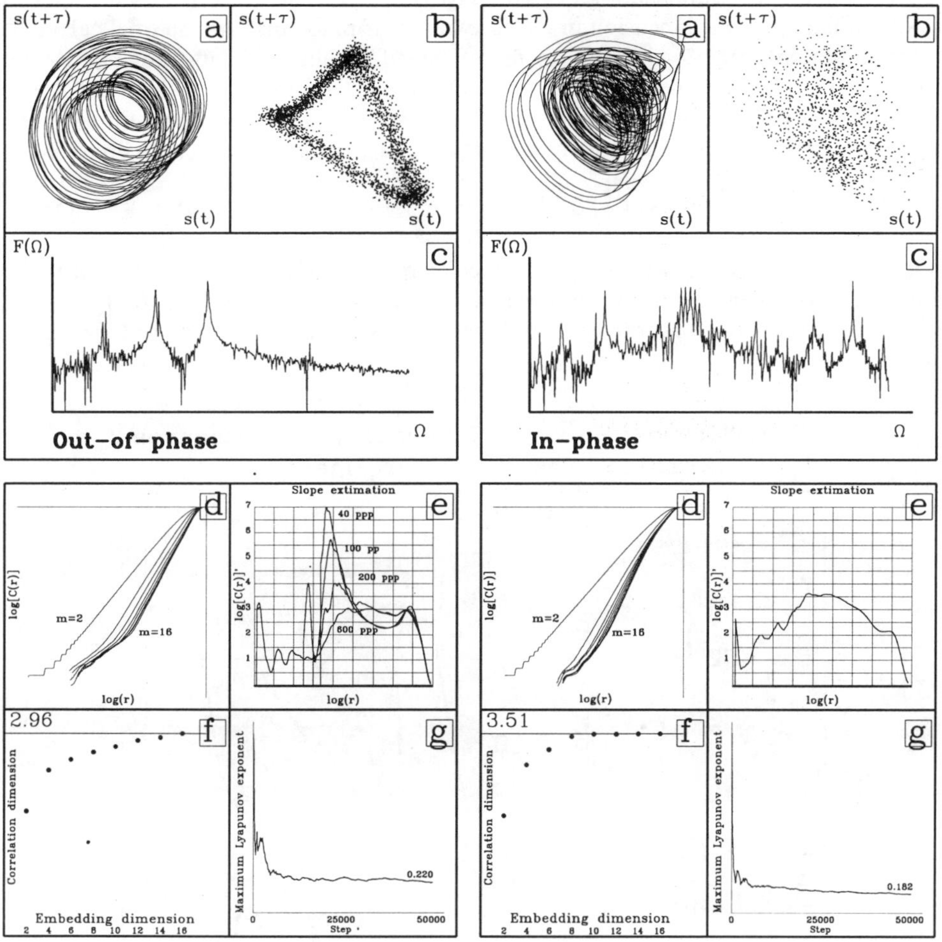

Fig. 2 Suspended cable/mass. Nonregular reconstructed attractors under in-phase (left) and out-of-phase (right) support motion: delay phase portrait (a) and Poincaré map (b), frequency response spectrum (c), correlation dimension (d) and relevant slope (e), embedding dimension for saturation (f), and Lyapunov exponent (g).

scattered thickening around a closed curve in the Poincaré map of delayed coordinates: it corresponds to a chaotically modulated motion of the system. The latter, instead, looks like a fully developed chaotic attractor. These different strangeness features of the two attractors are confirmed by the corresponding saturated values of the correlation dimension, calculated for varying values of the embedding dimension (f): they are identified by the plateau levels of the two slope diagrams (e) of the basic log [C(r)] vs log(r) plots (d). Though improved resolutions of the fuzzy torus surface - obtained by considering higher acquisition rates of the experimental data - allow to obtain increased values of the attractor dimension in the case of out-of-phase motion, the final calculated value (2.96) still remains lower than that obtained for the other attractor (3.51), thus highlight higher strangeness of the latter. This circumstance, however, does not directly reflect into the measure of chaoticity: indeed, the values of the maximum (positive) Lyapunov exponent (g) calculated in the two cases are rather close to each other.

3.2 EVOLUTION OF RESPONSE SCENARIO IN A GIVEN CHAOTIC REGION

It is worth quantitatively analyzing the global properties of the chaotic attractor within a given chaotic region. We focus the attention on one zone (Fig. 3a) of the overall response chart obtained near the 1/2-subharmonic resonance under out-of-phase support motion [2].

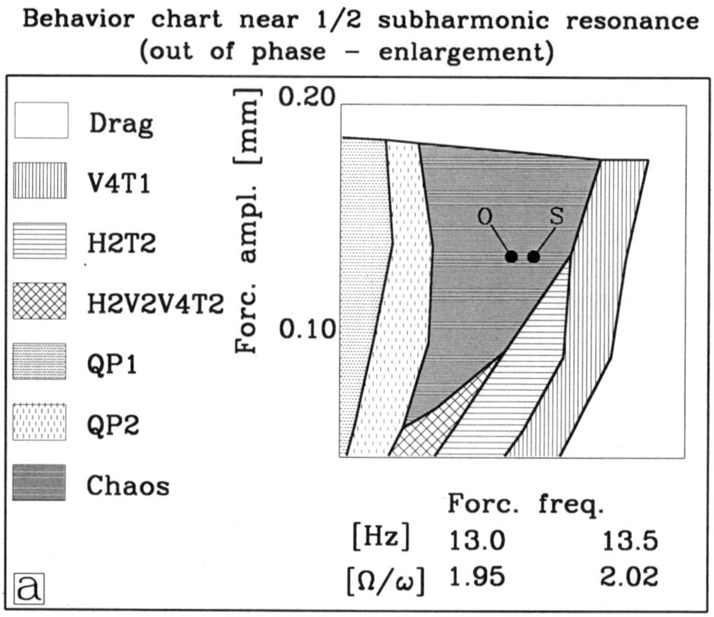

Fig. 3a. 1/2-subharmonic resonance, out-of-phase support motion.

When varying one control parameter, meaningful variations of those properties can be observed. For instance, by calculating the correlation dimension D_C at different values of the excitation frequency in a rather wide range of chaotic response occurring at a fixed excitation amplitude, we have detected two fairly well distinct zones. In the former, the D_C value oscillates around 2.55 (see the plateau level of the slope diagram in Figure 3b), in the latter - corresponding to slightly larger excitation frequencies - a D_C value around 4.50 is deduced from the rather differently shaped slope diagram (Figure 3c). This almost sudden increase of strangeness also reflects in an increase of chaoticity, as testified by the two values (0.077 vs 0.19) of the maximum Lyapunov exponent λ highlighting notably major speed of diverging trajectories in the second zone. This increase is likely to be connected with some enrichments and widenings of pre-existing broadbands in the relevant frequency response spectra (Figures 3d,e), as well as with the appearance of new broadbanded peaks in the second spectrum evidencing contributions to the system response from further harmonics.

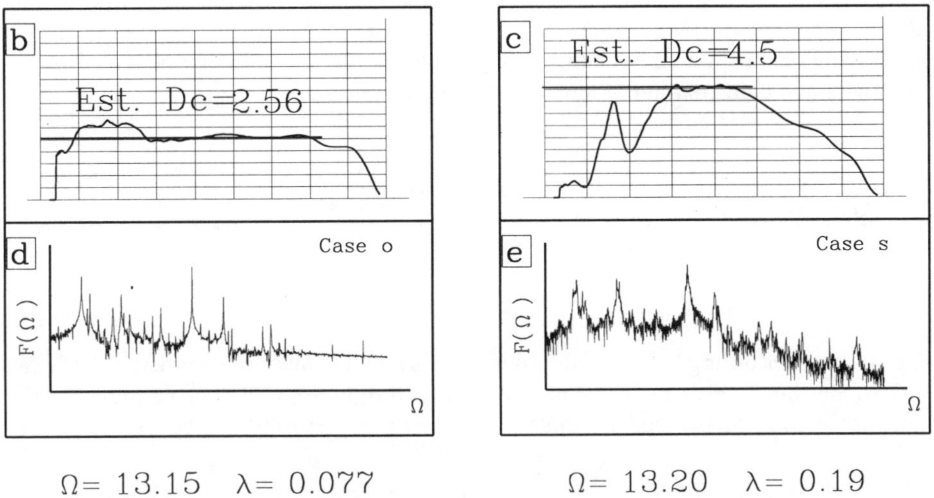

Fig. 3b-c. Two attractors (left and right) of different strangeness (b, c) and chaoticity (d, e) in the chaotic region of Fig.3a.

3.3 PROPER ORTHOGONAL MODES

A clearer understanding of the meaning of variations in the global properties of the reconstructed attractors within a given region of nonregular response can be pursued by investigating about the number and kind (spatial shape) of degrees-of-freedom giving significative contributions to the system response. This first requires estimating the system dimensionality based on a multimeasure time series analysis, and second measuring the spatial coherence between different points of the system to obtain informations about the configurations more visited on average by the system during a complex time evolution. If more than one single experimental measure are available, the proper orthogonal decomposition technique can be used to the latter purpose.

Based on averaging over several pairs of two-point measures of the vertical displacement made in two masses of the system with two follower optical cameras, proper orthogonal modes have been calculated in a number of highlighted chaotic zones. Figure 4a shows just those taking part in one system chaotic response under out-of-phase support motion at 1/2-subharmonic resonance, previously discussed (see Figure 3e). The strong prevalence of a proper orthogonal mode looking very similar to the second planar antisymmetric linear mode of the cable is worthily noticed. This confirms a conjecture about meaningful contributions from second planar (V4) and nonplanar (H4) (herein non-observable) antisymmetric modes already made in [3], based on mostly qualitative observations of the dynamics. Other calculated proper orthogonal modes (Figs. 4b,c) refer to responses pertaining to two chaotic regions identified for the system excited few right of the primary resonance condition (with respect to the first crossover frequency) under either in-phase or out-of-phase support motion [2]. In the two cases, proper orthogonal modes very similar to the eigenvectors of the regular V1 response (first planar symmetric mode evolving from one to three half-waves) and of the regular V2 response (first planar antisymmetric mode), respectively occurring in neighbouring regions to the chaotic ones, are observed. Again, this results in a confirmation of already made conjectures.

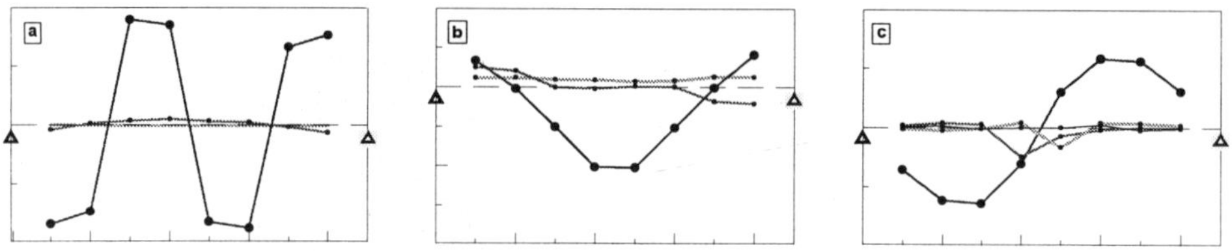

Fig. 4. Proper orthogonal modes looking like second planar antisymmetric mode (a) (1/2-subharmonic resonance, out-of-phase support motion), first planar symmetric (b) and antisymmetric (c) modes (primary resonance, in-phase (b) and out-of-phase (c) support motion).

4. Nonregular 2D dynamics of the elastic circular arch

An experimental model of an elastic double hinged circular arch shown in Fig.1b has been constructed with the aim of finding experimental evidence of the already studied problem [4] of the occurrence, in the planar dynamics, of antisymmetric strong vibrations under a vertical (symmetric) force applied at the tip of the arch. The system can exhibit an autoparametric resonance condition for particular parameters choice: indeed, there is natural possible occurrence of a 2:1 internal resonance between the first symmetric and the first antisysmmetric planar modes depending on internal elasto-geometric parameters. In this case an interesting series of nonlinear dynamical phenomena are present in the system, mostly the occurrence of a sub-zone of instability of any periodic solution with onset of quasi-periodic and chaotic states, inside the principal instability zone of the unimodal symmetric solution. The study has been conducted just with the aim of finding evidence of all those interesting nonlinear phenomena.

It starts with the construction of a behavior chart in the control plane of the excitation parameters. In Fig.5 the occurrence of zones of different dynamical behavior are presented: in the a) zone characterized by low excitation amplitudes only the symmetric planar solution is stable, in the b) zone the system exhibits coexistence of the purely symmetric and coupled (symm. antisymm.) solutions depending on i.c., in the c) zone only the coupled solution is stable, while in the d) region every periodic solution becomes unstable after an Hopf bifurcation and the time evolution is quasi-periodic before eventually evolving in a fully developed chaotic state. A local analysis for a QP (upper) and a chaotic (lower) state is reported in Fig.6. Parts a) and d) represent the relevant orbits in the physical plane of horizontal and vertical components of the motion of the arch tip, parts b) and e) are concerned with the estimation procedure for the embedding dimensions, and the parts c) and f) show the values of the correlation dimension for the two cases, corresponding to the occurrence of a two-torus attractor (upper, Dc=2) and a chaotic attractor (lower, Dc=2.9).

It is worth noticing how, as partially reported in Fig.5, the QP cases are found around the boundary line between b) and c) region in the chart, while the chaotic states belong to the core of the c) region.

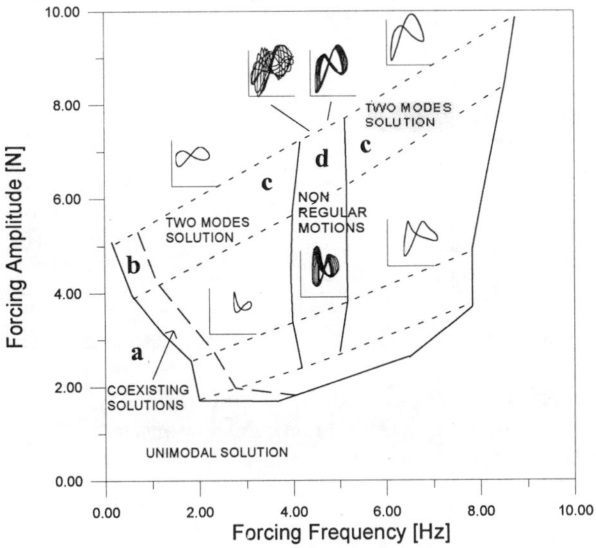

Fig.5. The arch: experimental zone of instability of simple unimodal planar solution

5. Summary and conclusions

The nonregular experimental finite dynamics of two different monodimensional elastic systems with initial curvature have been investigated. They appear particularly rich and varied over a wide range of control parameter values due to the onset of complex phenomena of nonlinear modal interaction linked with the occurrence of meaningful external and internal resonance conditions.

The attention has been focused on both temporal and spatial characterization of nonlinear regimes. The delay embedding tecnique furnishing global indicators of the dynamics has been used for disclosing the topological properties of the underlying attractors, their dimensionality and their time complexity.

The proper orthogonal decomposition procedure highlighting the shape more visited on average during the chaotic evolution has been used for characterizing the spatial complexity of the observed responses. Hints are obtained both on the number and kind

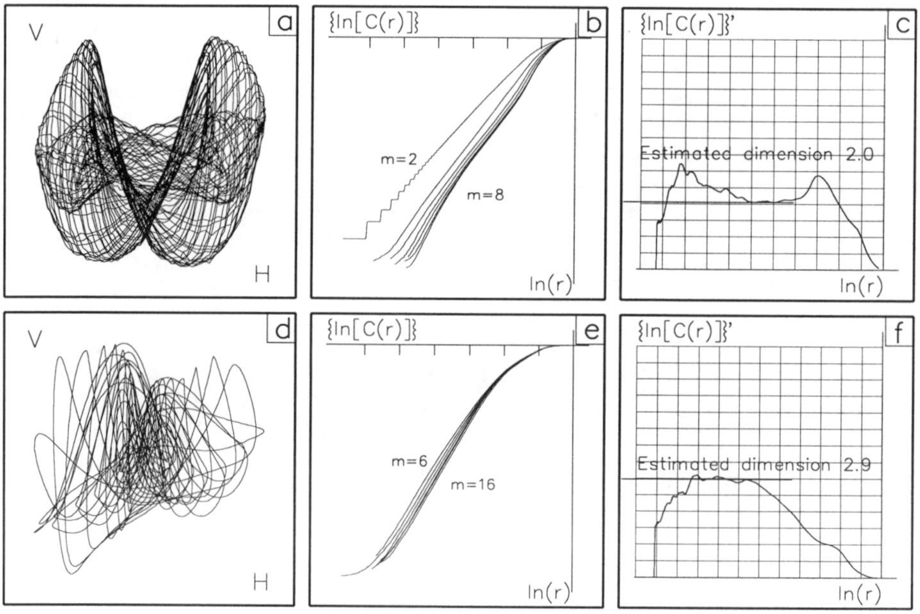

Fig.6. The arch: experimental evidence of QP (upper) and chaotic (lower) responses

of degrees-of-freedom necessary to correctly model the system dynamics, and on possible classification criteria of different complex responses.

6. References

1. Rega, G. (1997) Nonlinearity, Bifurcation and Chaos in the Finite Dynamics of an Elastic Structural System, *Proceedings of 2nd European Nonlinear Oscillations Conference,* Prague, 1, 45-46.
2. Benedettini, F., and Rega, G. (1997) Experimental Investigation of the Nonlinear Response of a Hanging Cable. Part II: Global analysis, *Nonlinear Dynamics* 14, 119-138.
3. Rega, G., Alaggio, R. and Benedettini, F. (1997) Experimental Investigation of the Nonlinear Response of a Hanging Cable. Part I: Local Analysis, *Nonlinear Dynamics* 14, 89-117.
4. Thomsen, J.J. (1992) Chaotic Vibrations of Non-Shallow Arches, *Journal of Sound and Vibration* 153(2), 239-258.
5. Benedettini, F. (1997) Planar Finite Forced Dynamics of a Double Hinged Circular Arch: Theory and Experiments, *Proceedings of ASME Design Engineering Technical Conferences,* Sacramento.
6. Takens, F. (1981) Detecting Strange Attractors in Turbulence, *Dynamical Systems and Turbulence,* D.A. Rand and L.S. Yang, eds., Springer Lecture Notes in Mathematics, Springer Verlag, New York, 898, 266-281.
7. Benedettini, F., (1996) An experimental time series analysis approach in the classification of non periodic motions in nonlinear structural dynamics, *Proceedings of the Third European Conference on Structural Dynamics: EURODYN 96,* Augusti, Borri, Spinelli eds., A.A. Baklema, Rotterdam, 415-421.
8. Rosenstein, M.T., Collins, J.J. and De Luca, C.J. (1993) A Practical Method for Calculating Largest Lyapunov Exponents from Small Data Sets, *Physica D* 65, 117-134.
9. Cusumano, J.P. and Bai, B.Y. (1993) Period-infinity Periodic Motions, Chaos, and Spatial Coherence in a 10 Degree of Freedom Impact Oscillator, *Chaos Solitons & Fractals* 3 (5), 515-535.

CONSTRAINED EULER BUCKLING: LINE CONTACT SOLUTIONS

PHILIP HOLMES AND JOHN SCHMITT
Program in Applied and Computational Mathematics,
Department of Mechanical and Aerospace Engineering,
Princeton University, Princeton, NJ 08544, U.S.A.

AND

GABOR DOMOKOS
Department of Strength of Materials,
Technical University of Budapest, H-1521 Hungary.

Abstract.
We consider elastic buckling of an inextensible beam with hinged ends and fixed end displacements, confined to the plane, and in the presence of rigid, frictionless sidewalls which constrain overall lateral displacements. We formulate the geometrically nonlinear (Euler) problem, and derive complete analytical results for the case of line contacts with the sidewalls. In contrast to the unconstrained problem, we find a rich bifurcation structure, with multiple branches.

1. Introduction

This work, which continues that of [3], arose from a desire to understand buckling of polypropylene fibers in a "stuffer box" manufacturing environment for non-woven fabrics. The wavelengths and buckling modes selected are important in determining fabric properties. Since buckling onset and modal selection appear to be determined by the initial elastic behavior, we consider elastic (Euler) planar buckling of a beam subject to hard loading (displacement boundary conditions) and constrained to lie within rigid, parallel, frictionless sidewalls set at distances $\pm h$ from the centerline. Assuming a linear constitutive law, after nondimensionalisation the equilibrium equation for each "free" element of the beam is

$$\theta'' + \lambda \sin\theta + \mu \cos\theta = 0, \tag{1}$$

F. C. Moon (ed.),
IUTAM Symposium on New Applications of Nonlinear and Chaotic Dynamics in Mechanics, 149–158.

where $\theta(s)$ is the tangent angle, $\theta'' = \frac{d^2\theta}{ds^2}$, λ and μ are axial and lateral forces, and $s \in [0,1]$ is the nondimensional arclength. Equation (1) is subject to overall zero-moment boundary conditions

$$\theta'(0) = 0 = \theta'(1), \tag{2}$$

(in the hinged-end case), and the axial displacement constraint:

$$x(1) - x(0) = \int_0^1 \cos\theta(s)ds = d \in (-1,1]. \tag{3}$$

At each contact point s_j, μ undergoes a jump $\mu_{j+1} - \mu_j$ equal to the (normal, lateral) constraining load at that point, so within the beam a sequence of BVPs must be solved and matched at contact points, which are themselves determined by constraint equations such as:

$$y(s_1) - y(s_0 = 0) = \int_0^{s_1} \sin\theta(s)ds = \pm h, \tag{4}$$

$$y(s_j) - y(s_{j-1}) = \int_{s_j}^{s_{j-1}} \sin\theta(s)ds = \pm 2h, \text{ etc.} \tag{5}$$

where h is the sidewall distance from the centerline. Here the unknowns include the contact points s_j and the loads λ and μ_j. Since (1) has a first integral, exact solutions may be found in terms of elliptic functions, but they are awkward except in certain cases with symmetries or single contacts, and in the class of line contact solutions considered here. Consequently, we developed a numerical method [2] to predict load-displacement curves (λ vs. $1-d$) in the general case. In [3], to which the reader should refer for further details, we discuss this model extensively, comparing analytical, experimental and numerical results.

Viewed as a sequence of phase-plane problems, (1)–(5) is an example of a hybrid dynamical system [4], and relatively little is known about the bifurcations such systems can exhibit. We approach the problem by considering the superposition of "elementary" bifurcation (= solution branch) diagrams in $(\lambda, \mu_j, D = 1-d)$-space for the unconstrained and a sequence of less-constrained problems. Bifurcation points for the full problem correspond to intersections of solution branches of these less-constrained problems. In this paper we consider the class of problems constrained to have line contact segments at one or both sidewalls. To appreciate the richness of the general problem, the reader should turn to Figure 6, which shows a partial bifurcation diagram computed by the methods of [2].

2. Unconstrained buckling: elemental solutions

We start by recalling the solution for the unconstrained Euler problem, and then use this to assemble line contact solutions of arbitrary complexity.

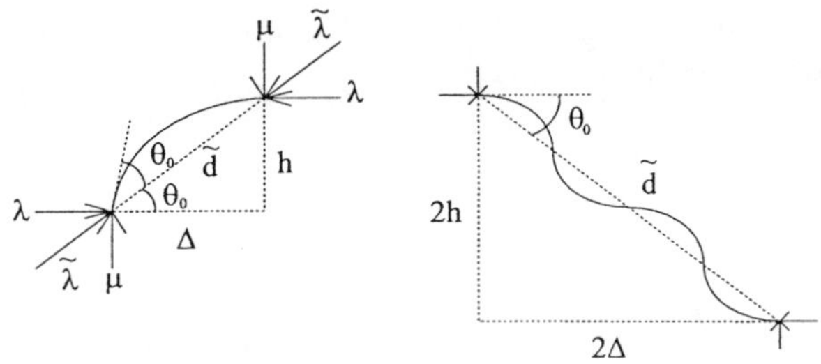

Figure 1. End and interior elements between line contact segments.

Buckling of an axially end-loaded strut of length l corresponds to solving (1) with $\mu = 0$, $\lambda = \tilde{\lambda}$, and boundary conditions $\theta'(0) = 0 = \theta'(l)$. Using the first integral

$$\frac{\theta'^2}{2} - \tilde{\lambda}\cos\theta = -\tilde{\lambda}\cos\theta_0 \ ,$$

the resulting quadrature may be evaluated to yield

$$\tilde{\lambda} = \left[\frac{2(m+1)K(k)}{l}\right]^2 \ , \tag{6}$$

where K denotes the complete elliptic integral of the first kind, $k = \sin(\theta_0/2)$ is the elliptic modulus, and $m+1$ is the modal number; $m \geq 0$ denoting the number of internal inflection points. Moreover, from (3), we have for the distance between the end points:

$$\tilde{d} = \left[\frac{2E(k)}{K(k)} - 1\right] l \ , \tag{7}$$

where E denotes the complete elliptic integral of the second kind. For future use, we denote the element's (arbitrary) length as l, the end point separation $\tilde{d}$, and the load $\tilde{\lambda}$. See the appendix of [3] for more details.

From (6-7) we may now construct elemental solutions from which general line contact equilibria can be built. The key observation is that, at each end of a line contact segment in which $\theta(s) \equiv 0$, necessarily $\theta(s) = \theta'(s) = 0$. Thus, referring to Figure 1, any element spanning end point to contact point or contact point to contact point, obeys (6-7) and additional geometrical constraints that follow from the fact that the resultant of axial and lateral forces must pass through both end points:

$$\tan\theta_0 = \frac{h}{\Delta} = \frac{\mu}{\lambda} \ . \tag{8}$$

The resultant load is $\tilde{\lambda} = \sqrt{\mu^2 + \lambda^2}$, and using $\tilde{d}^2 = \Delta^2 + h^2$, and trignometrical identities, we have

$$\mu = \frac{2k\sqrt{1-k^2}}{1-2k^2}\lambda\,, \quad \Delta = \frac{1-2k^2}{2k\sqrt{1-k^2}}h\,. \tag{9}$$

Substitution into (6-7) yields:

$$\lambda = \frac{4(m+1)^2(1-2k^2)K^2(k)}{l^2}\,, \quad l = \frac{hK(k)}{2k\sqrt{1-k^2}(2E(k)-K(k))}\,, \tag{10}$$

which may be reduced to a single equation soluble for k, given λ:

$$\lambda = \frac{16(m+1)^2}{h^2}k^2(1-k^2)(1-2k^2)(2E(k)-K(k))^2 \stackrel{\text{def}}{=} (m+1)^2\Lambda_h(k)\,, \tag{11}$$

after which l, Δ, μ may be found from (9-10). Observe that, while λ must be the same thoughout all elements of the beam, including those in contact with the wall, μ is generally different at each free element end point.

Figure 2 shows the function $\Lambda_h(k)$. Since $2E(k) < K(k)$ for $k > k^* \approx 0.908$, solutions in this range correspond to "everted" shapes with $\tilde{d} < 0$ and one or more interior loops and self-intersections (cf. [5], Sect. 263, Figs 54-55). We shall be only concerned with solutions which develop as the displacement $D = 1 - d$ increases from 0, and since the distance $\tilde{d}$ between end point and contact point (or contact point and contact point), is necessarily positive ($\geq h$ or $\geq 2h$), we shall ignore such solutions here. Thus there are 1 or 2 solutions for each λ in the admissible range $(m+1)^2[\Lambda_{h\min}, \Lambda_{h\max}] \approx ((m+1)^2/h^2)[-0.329, 2.75]$.

3. Construction of line contact branches

Before assembling these elemental solutions with intervening line segments to form a complete solution to (1)–(5), we develop a symbolic classification.

We define an *element of order m and length l* as a solution of the unconstrained Euler problem (9-11).

We define the *n'th mode complex* as all those (line contact) solutions with $n-1$ interior zeroes $y(s) = 0$ and n line contact regions. To each branch we assign a sequence $\{m_0, *, m_1, *, \ldots, *, m_n\}$, where m_j is the order of each element and $*$ denotes a line contact segment. Note that $m_0, m_n = 0, 1, 2, \ldots$ can be any integer, but for the interior elements, $m_j = 1, 3, \ldots, 2m-1, \ldots$ are necessarily odd for the sequence to be *admissible*.

We define *regular modes (of order m)* as those with sequences of the form $\{m, *, 2m+1, *, \ldots, *, m\}$. Modes which are not regular are called *irregular*.

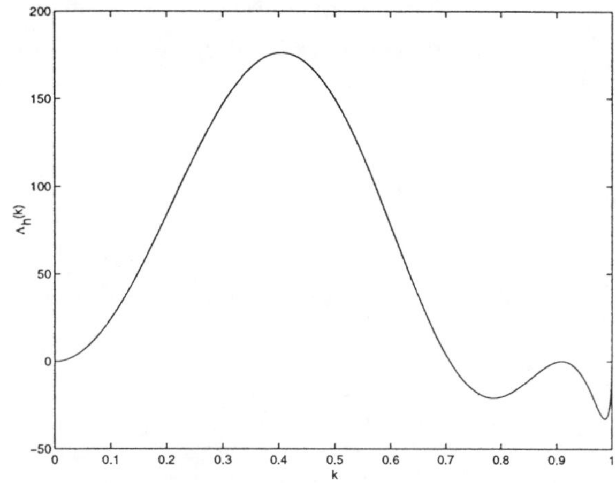

Figure 2. The function $\Lambda_h(k)$, plotted for $h = 0.125$.

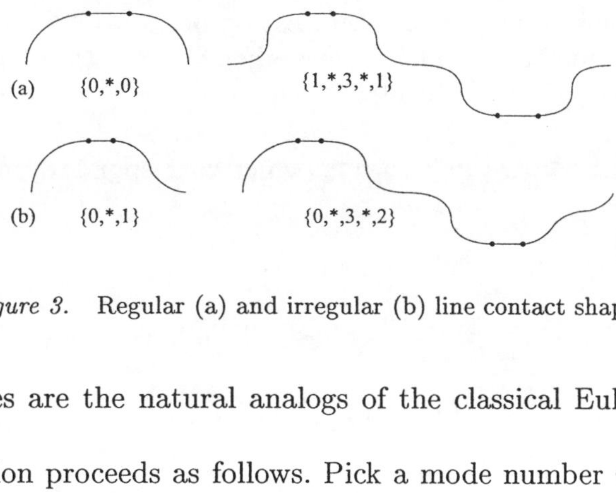

Figure 3. Regular (a) and irregular (b) line contact shapes.

Regular modes are the natural analogs of the classical Euler modes. See Figure 3.

Construction proceeds as follows. Pick a mode number $n \geq 1$, and an admissible sequence of orders $m_0, \ldots, m_n$. Define

$$m^- = \min\{m_0, m_n, (m_j - 1)/2; j = 1, \ldots, n-1\} \ ,$$

pick an admissible axial load $\lambda \in (m^- + 1)^2[\Lambda_{h\min}, \Lambda_{h\max}]$ and solve (11) repeatedly with $m = m_0, \ldots, m_n$ for $k_0, \ldots, k_n$. From (9-10), find $\mu_0, \ldots, \mu_n$, the arclengths of the elements $l_0, \ldots, l_n$, and the axial lengths $\Delta_0, \ldots, \Delta_n$ subtended by them. The total displacement is then

$$D = 1 - d = l_0 - \Delta_0 + 2\sum_{j=1}^{n-1} l_j - \Delta_j + l_n - \Delta_n \ . \qquad (12)$$

Since in general there are 2 solutions of (11) for each λ, we have up to 2^{n+1} branches of solutions in the (λ, D) plane for each $\{m_0, *, \ldots, *, m_n\}$ sequence, some of which may coincide (eg., for regular modes). Observe that sequences can change only at bifurcation points (see below).

4. Physical solutions and branch structures

Thus far we have not used the fact that the total arclength of elements and line contact regions must equal 1, the beam's overall length. This gives the condition for physical solutions:

$$l_0 + 2\sum_{j=1}^{n-1} l_j + l_n \stackrel{\text{def}}{=} l_{\text{tot}} \leq 1\ ; \tag{13}$$

if (13) is satisfied with strict inequality, the balance of arclength can be filled with *any* distribution of line segment elements summing to $1 - l_{\text{tot}}$. Equality in (13) identifies bifurcation points at which the line contact branches intersect point contact and secondary bifurcation branches [3]. Observe that, as h decreases, Λ_h increases and l decreases for given k; thus, as $h \to 0$ (sidewalls approach) a greater "central" region of any given branch becomes physical.

From (9-10) we note the following, which correspond to solutions of (11) with $\lambda = 0$:

$$l, \Delta l \to \infty,\ l - \Delta \to 0 \text{ as } k \to 0\ ; \tag{14}$$

$$l \to \infty,\ \Delta \to \frac{1 - 2k^2}{2k\sqrt{1-k^2}} h \approx -0.853h \text{ as } k \to k^*\ ; \tag{15}$$

$$l = \frac{hK(\frac{1}{\sqrt{2}})}{2E(\frac{1}{\sqrt{2}}) - K(\frac{1}{\sqrt{2}})} \stackrel{\text{def}}{=} \bar{l}h \approx 2.188h,\ \Delta = 0 \text{ at } k = \frac{1}{\sqrt{2}}\ . \tag{16}$$

This implies that branches reach $\lambda = 0$ at integer multiples of $D = \bar{l}h$ or that they tend to $D = \infty$ as $\lambda \to 0^-$. In view of (14-15), the only branches that *cross* $\lambda = 0$ are those for which all k_j are selected to pass through $1/\sqrt{2}$ at this load: thus, one branch alone in each modal complex is continuous at $\lambda = 0$. Also, branches achieve their maxima and minima respectively on the lines $\lambda = (m^- + 1)^2 \Lambda_{h\max}, (m^- + 1)^2 \Lambda_{h\min}$. See the examples below.

Generally modes enjoy no symmetries, but *regular modes* are symmetric if the same root $k_j = k$ of (11) is picked for each element. For $(m + 1)^2 \Lambda_{h\min} < \lambda < (m + 1)^2 \Lambda_h \max$ there are thus precisely two symmetric regular mode solutions of each order m; they form a (λ, D) branch similar to the graph of $\Lambda_h(k)$, with critical points at $(m + 1)^2 \Lambda_{h\max}$ and $(m + 1)^2 \Lambda_{h\min}$. $2^{n+1} - 2$ asymmetric regular modes can be constructed by

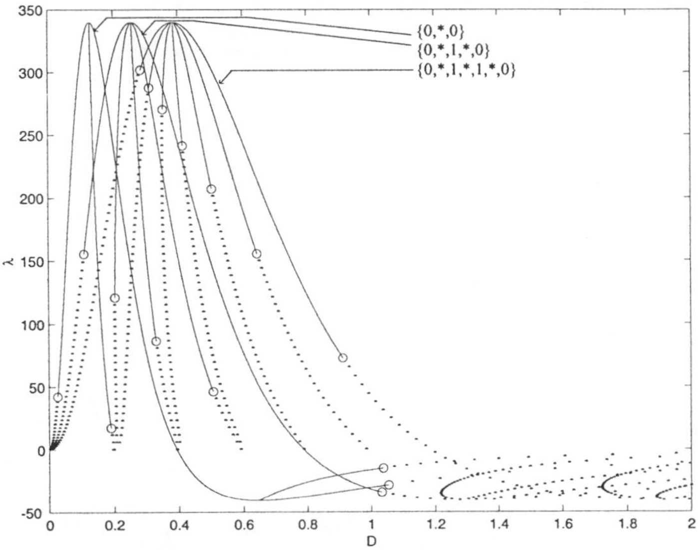

Figure 4. Bifurcation diagrams for the regular $n = 1, 2, 3$; $m = 0$ modal complexes: h = 0.09, physical regions shown solid.

picking all possible remaining choices for the $n + 1$ k_j's, but since there at most two distinct k's (and hence $(l_j - \Delta_j)$'s) for each λ, these form only $2n - 1$ distinct curves on the (λ, D) diagram. From (9-10) and (12), these "mixed" modes lie equidistant between the "outer" symmetric branches, with which they coalesce at $(m+1)^2\Lambda_{h\max}$ and $(m+1)^2\Lambda_{h\min}$. Figure 4 shows such diagrams for the first three regular modal complexes. These and subsequent branches were all computed using the elliptic function routines in *MATLAB*.

Finally, let

$$m^+ = \max\{m_0, m_n, (m_j - 1)/2, j = 1, \ldots, n-1\} , \tag{17}$$

correspond to the highest order element, and consider the sequence of modal complexes generated as $m^+ \to \infty$ with all other m_j fixed. Since $\lambda/(m^+ + 1)^2 \to 0$, we conclude that these complexes approach a limit.

Figures 5-6 show several examples of irregular modal complexes, illustrating the points made above.

5. Conclusions and comments

We have developed explicit solutions for line contact equilibrium states of an inextensible elastic beam subject to end loads and normal constraint forces

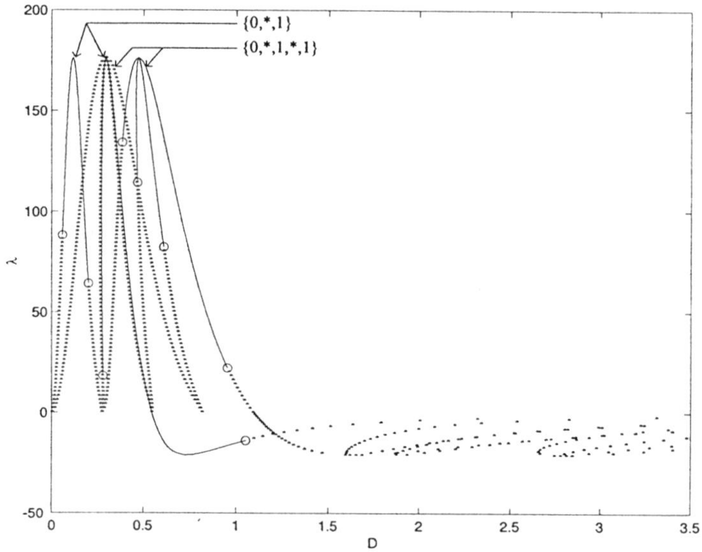

Figure 5. Bifurcation diagrams for some irregular modal complexes: h = 0.125, physical regions shown solid. Note that the right hand branch of {0,*,1} almost coincides with a branch of {0,*,1,*,1}.

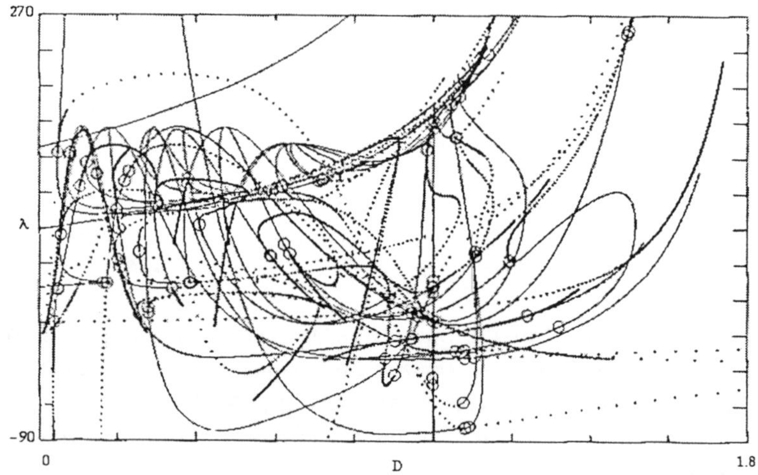

Figure 6. Numerically computed bifurcation diagrams for some unconstrained, point contact and line contact branches: h = 0.125, physical regions shown solid.

at frictionless rigid side walls, equidistant from and parallel to the centerline connecting the (hinged) end points. The solutions involve complete elliptic integrals and yield load displacement diagrams. Although we have plotted only axial load *vs.* displacement here, lateral loads can also be calculated.

It is well-known that for the unconstrained Euler beam with $\lambda > \pi^2$, only the first ($n = 1$) mode is stable to arbitrary small perturbations; all other mode shapes correspond to saddle points in the energy [6, 7, 1]. However, constraints can stabilise certain regular modes of order $m = 0$, provided no line contact region exceeds the critical length $2\pi/\sqrt{\lambda}$ for buckling of a clamped-clamped element (see [3]). Indeed, in [8] stability for the $n = 1, m = 0$ case is proven, and it follows that solutions for all other $n > 1$ and $m = 0$ are neutrally stable (one can freely adjust individual line contact lengths without changing the energy). Thus at least some of the solutions found here are physically observable; in fact in the experiments reported in [3] we have seen regular $n = 1, 2, 3, m = 0$ branches.

A second important class of solutions are those having point contact at the sidewalls, where the moment $\theta'(s) \neq 0$. As shown in [3], solutions of these involve *incomplete* elliptic functions and they are much more difficult to work with, since direct superpositions of individual elements alone do not yield global solutions; one has to simultaneously adjust all the lateral loads μ_j to match moments at each contact point. We plan to discuss this case in a future article.

Acknowledgements

This work was supported by OTKA F021307, MKM 815 (GD), US-Hungarian Joint Fund 656/96 (GD and PH), and DoE DE-FG02-95ER25238 (PH). John Schmitt was partially supported by a DoD Graduate Fellowship and a Wu Fellowship of the School of Engineering and Applied Science, Princeton University.

References

1. G. Domokos. Global description of elastic bars. *Zeitschr. Angew. Math. und Mech.*, 74 (4):T289–T291, 1994.
2. G. Domokos and Zs. Gáspár. A global, direct algorithm for path-following and active static control of elastic bar structures. *Int'l J. of Structures and Machines*, 23 (4):549–571, 1995.
3. G. Domokos, P. Holmes, and B.S.H. Royce. Constrained Euler buckling. *J. Nonlin. Sci.*, 7:1–34, 1997.
4. J. Guckenheimer and S. Johnson. Planar hybrid systems. In *Lecture Notes in Computer Science No. 999*. Springer Verlag, 1995.
5. A.E.H. Love. *A Treatise on the Mathematical Theory of Elasticity*. Cambridge University Press, Cambridge, UK, 1927. Reprinted by Dover Publications, Inc. New York.

6. J.H. Maddocks. Stability of nonlinearly elastic rods. *Arch. Rat. Mech. Anal.*, 85:311–354, 1984.
7. J.H. Maddocks. Stability and folds. *Arch. Rat. Mech. Anal.*, 99:301–328, 1987.
8. E. Miersemann and H.D. Mittelmann. A free boundary problem and stability for the nonlinear beam. *Math. Meth. in the Appl. Sci.*, 8:516–532, 1986.

THE ROLE OF HOMOCLINIC BIFURCATIONS IN UNDERSTANDING THE BUCKLING OF LONG THIN CYLINDRICAL SHELLS

G. J. LORD AND A. R. CHAMPNEYS
Department of Engineering Mathematics
University Walk, University of Bristol
Bristol, BS8 1TR, U.K.
G.J.Lord@bristol.ac.uk, A.R.Champneys@bristol.ac.uk

AND

G. W. HUNT
School of Mechanical Engineering
University of Bath, Bath, BA2 7AY, UK.
G.W.Hunt@bath.ac.uk

Abstract.

Experiments have shown that long cylinders buckle into localized patterns axially. It is argued that traditional linear or nonlinear analysis is unlikely to capture such modes, nor the effective buckling load at which such responses stabilise. However, the inherent translational indeterminacy of localised buckling is well captured by considering infinitely long cylinders and seeking homoclinic solutions of the von Kármán–Donnell equations. This exploits the dynamical analogy of such structural problems, so that symmetry arguments and numerical techniques developed for dynamical systems may be used. The method is illustrated by successful application to a cylinder which has well documented experimental results.

1. Introduction

A fruitful approach to certain statics problems posed over long domains has been to treat them as dynamical systems (see, for example, the contribution by J.M.T. Thompson in these proceedings). Localized responses of such systems, which energy arguments often reveal as being important

F. C. Moon (ed.),
IUTAM Symposium on New Applications of Nonlinear and Chaotic Dynamics in Mechanics, 159–168.

physically, may be modelled by homoclinic solutions of ordinary differential equations differential equations (ODEs) posed on the real line.

We aim to show succinctly that the buckling of long thin, cylindrical shells is a classically hard problem that is well explained using this approach. The equilibrium of the shell is governed by the von Kármán–Donnell partial differential equations (PDEs), which may be viewed as a dynamical system in the axial length of the cylinder. The difficulty for traditional analysis (using modal decomposition or finite element techniques) is that buckling is violently sub-critical, so that linear theory can over-predict the true experimentally observed load by as much as 400% (see, for example, Figure 1(a) where the true buckling load is near the turning point of the post-buckled curve). Therefore, the nature of post buckling behaviour is fundamental in predicting the true failure loads and displacements.

Figure 1 reproduces some experimental work from the 1970s depicting *elastic* buckling deformations of moderately long cylinders under end loading. Several features are worthy of note. First, buckling is typically localized to some portion of the axial length of the cylinder. Secondly, there is a translational indeterminacy axially in the location of the buckled portion (compare (c) and (d)). Thirdly, for each buckle there is a well defined circumferential 'wave number', which is *not* fixed by the geometry of the cylinder. Finally, there are two forms of buckle pattern (cf. Figure 1(b),(d)) which in the following we refer to as *symmetric* and *cross-symmetric* respectively.

In this paper we show how our numerical methods, based on the dynamical systems analogy, captures all the features of Von Eßlinger and Geier's experiments both qualitatively and quantitatively. We consider an infinitely long cylinder and discretize the von Kármán–Donnell equations circumferentially by a Galerkin method. This yields a large system of ODEs in the axial variable x, for which we seek homoclinic, i.e. axially localized, solutions. We seek forms of solutions which are either symmetric or cross-symmetric, implying reversibilities of the ODEs.

In earlier work (Lord, Champneys & Hunt 1997*a*, Lord, Champneys & Hunt 1997*b*), we confirmed numerically the existence of homoclinic orbit solutions to these equations and obtained good agreement with experimental data on a cylinder due to Yamaki (1984). We also described in detail the numerical techniques used to compute those solutions, which are based on extension to the current PDE setting of existing methods for homoclinic solutions or ODEs (e.g. (Beyn 1990, Friedman & Doedel 1994)). We are interested here in "primary" or "uni–modal" homoclinic orbits. However recent work by Peterhof, Sandstede & Scheel (1997) strongly suggests the existence of "multi–modal" or n-pulse homoclinic orbits such as computed in Lord et al. (1997*a*).

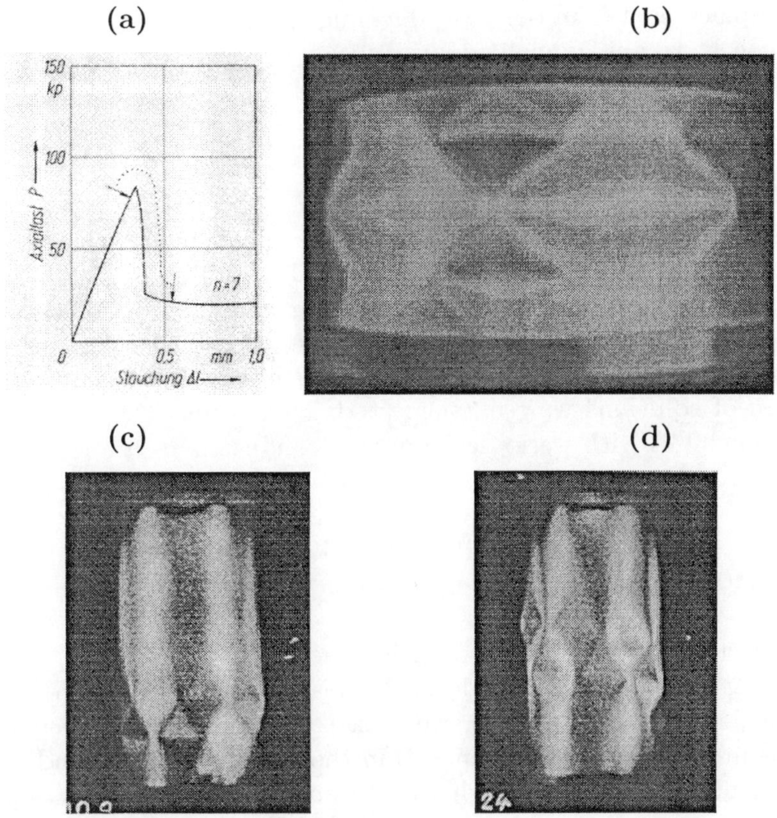

Figure 1. Experimental results after Von Eßlinger 1970 and Von Eßlinger & Geier 1972. (a) Load vs end-shortening bifurcation diagram taken from experiment (1970). (b) Symmetric solution found on a short cylinder (1972). (c) Cross-symmetric solution localized close to one end of cylinder (1970), and (d) cross-symmetric solution localized just off centre of cylinder(1970). In (b) and (d) we see an axial translation of the same buckle pattern.

In Section 2, we describe the von Kármán–Donnell equations, their Galerkin approximation and how to exploit symmetries in the problem. Section 3 then presents numerical results and compares them to the experiments. All our computations were performed using the numerical continuation code AUTO (Doedel, Keller & Kernévez 1991). Finally, Section 4 draws conclusions.

2. The von Kármán–Donnell equations and their approximation as a dynamical system

Consider an infinitely long, thin cylindrical shell of radius R and shell thickness t. The classical equilibrium equations for the in–plane stress function

ϕ and displacement w in the post–buckling regime of the cylinder are given by the von Kármán–Donnell equations:

$$\kappa^2\nabla^4 w + \lambda w_{xx} - \rho\phi_{xx} = w_{xx}\phi_{yy} + w_{yy}\phi_{xx} - 2w_{xy}\phi_{xy} \quad (1)$$
$$\nabla^4\phi + \rho w_{xx} = (w_{xy})^2 - w_{xx}w_{yy}, \quad (2)$$

where ∇^4 denotes the two dimensional bi–harmonic operator; $x \in \mathbb{R}$ is the axial and $y \in [0, 2\pi R)$ is the circumferential co-ordinate. The parameters appearing in (1) and (2) are the curvature $\rho := 1/R$, $\kappa^2 := t^2/12(1-\nu^2)$, where ν is Poisson's ratio, and the load parameter $\lambda := P/Et$, where P is the compressive axial load (force per unit length) and E is Young's modulus. The form of solutions we seek suggest that equations (1) and (2) should be supplemented with periodic boundary conditions in y and asymptotic boundary conditions in the axial direction x:

$$(w,\phi)(x,0) = (w,\phi)(x,2\pi R),$$
$$(w,\phi)(x,y), (w,\phi)_x(x,y), (w,\phi)_{xx}(x,y), (w,\phi)_{xxx}(x,y) \to 0 \text{ as } x \to \pm\infty. \quad (3)$$

The system (1) and (2) has a rich structure of symmetries, see (Hunt, Williams & Cowell 1986, Wohlever & Healey 1995). In accordance with observed deformation patterns, we seek solutions that are even periodic solutions in y and which remain within the subspace corresponding to invariance under rotation through $2\pi/s$. Hence we use the following cosine functions as the basis functions in the Galerkin approximation

$$w(y) = \sum_{m=0}^{\infty} a_m \cos(ms\rho y); \qquad \phi(y) = \sum_{m=0}^{\infty} b_m \cos(ms\rho y), \qquad s \in \mathbb{N}.$$

We refer to $\cos(s\rho y)$ as the *seed mode.*

Substituting into the von Kármán–Donnell equations, taking the L^2 inner product and expanding the nonlinear terms we find a system of ODEs for the Fourier modes a_m and b_m for $m = 0, \ldots, \infty$ which we may formally write as

$$a_m^{iv} = La_m + F_m(a_m), \qquad b_m^{iv} = Lb_m + F_m(b_m), \quad (4)$$

where superscripts denote differentiation with respect to x (see (Lord et al. 1997*a*) for the details). The Galerkin approximation is formed by taking equations (4) for $m = 0, \cdots, M-1$, for some finite M.

We think of equations (4) as a dynamical system, with the axial variable x taking a time-like role. Systems of the form (4), arising from elliptic PDEs, were considered in detail by Mielke (1991), and shown to have a Hamiltonian structure.

There is a further symmetry of von Kármán–Donnell equations that plays an important role in the localised buckling solutions observed physically. The buckling modes observed experimentally, tend to be either *symmetric* (as in Figure 1(b)) or *cross-symmetric* (Figure 2(d)) about a horizontal cross-section of the cylinder. A solution to (1), (2) that is *symmetric* about the cross-section $x = T$ satisfies

$$w(x,y) = w(2T - x, y) \qquad \& \qquad \phi(x,y) = \phi(2T - x, y). \tag{5}$$

Equations (5) impose the natural symmetric conditions on the Fourier modes a_m and b_m for $m = 0, \cdots, M-1$ at $x = T$,

$$a'_m(T) = a''_m(T) = b'_m(T) = b''_m(T) = 0. \tag{6}$$

In contrast, a *cross-symmetric* solution satisfies, for some seed mode s,

$$w\,(x,y) = w\,(2T - x, y + \pi R/s) \;\&\; \phi\,(x,y) = \phi\,(2T - x, y + \pi R/s)\,. \tag{7}$$

Thus, in terms of the Fourier modes, we have that

$$\begin{array}{ll} a'_m(T) = b'_m(T) = a'''_m(T) = b'''_m(T) = 0, & m = 0, 2, 4, \cdots; \\ a_m(T) = b_m(T) = a''_m(T) = b''_m(T) = 0, & m = 1, 3, 5, \cdots. \end{array} \tag{8}$$

It is not difficult to see that the symmetries of (1) and (2) defined by (5) and (7) define a *reversibility* of the ODEs (4), as in (Devaney 1976), with fixed point sets, S, forming $4M$–dimensional sub–manifolds of phase space $\mathbb{R}^{8M}$. Hence we can use as a boundary condition that the solution $a_m(T)$, $b_m(T)$, lie in the the $4M$–dimensional space S.

Note, finally, that there is a "degeneracy" in equations (4) for the zero mode ($m = 0$) such that these could be solved with initial conditions for $a''_0, a'''_0, b''_0, b'''_0$ independently of the initial conditions for a_0, b_0 and a'_0, and b'_0. This corresponds to a trivial translational symmetry in the problem (sometimes termed a rigid body mode). This translation invariance plays an important role. In our formulation the localization may occur at any point along the length of the cylinder. Indeed this translation invariance is observed experimentally (see for example Figures 3 and 4 in Von Eßlinger (1970), reproduced partially in Figure 1.

A standard periodic analysis of the von Kármán–Donnell equations seeks the minimum load $\lambda = \lambda_d$ and corresponding axial and circumferential wavelengths such that a bifurcation occurs. One may easily show in this way that $\lambda_d = 2\rho\kappa$. Wavelengths for which $\lambda = \lambda_d$ lie on a circle in axial/circumferential wave space as first elucidated by Koiter (1945). The weakly nonlinear analysis of Hunt & Lucena Neto (1991), interpreted in the present context, suggests that once a circumferential wave number has been chosen, then there is a small amplitude bifurcation of (a pair of) homoclinic solutions at $\lambda = \lambda_d$ as in the normal form of a Hamiltonian-Hopf bifurcation (Iooss & Peroueme 1993).

In order to compute the true localised buckling load, we numerically extend the weakly nonlinear analysis, by computing homoclinic solutions to (4) truncated for some M. Note that the single mode approximation found by taking $M = 1$ ($m = 0$ only) in (4) yields a linear system for which there are no homoclinic solutions. Thus the simplest approximation we can take that may admit a homoclinic solution is the two mode approximation found by taking $M = 2$ ($m = 0, 1$) in (4).

The numerical method, described in detail in Lord et al. (1997*a*), makes full use of the reversibility and symmetry properties of (4) regarded as a dynamical system. Specifically, we solve for symmetric or cross-symmetric homoclinic solutions as a boundary-value problem (BVP) on a truncated domain, with left-hand projection boundary conditions that place the solution in the linearised unstable manifold of the origin (see, e.g. (Beyn 1990)). At the right-hand boundary conditions, we exploit the symmetric section conditions (6), (8). This BVP can be solved by a regular continuation code to compute load-deflection bifurcation diagrams, but may require careful steps to be taken in order to compute initial approximations.

3. Numerical results

To compare with the experiments of Von Eßlinger & Geier (1972) calculations were performed for a shell with

$$\rho = 0.01\text{mm}^{-1}, \qquad t = 0.190\text{mm}, \qquad \nu = 0.3, \qquad E = 4.11 \text{ GPa}. \qquad (9)$$

In the figures below x is plotted on [0,2T] unless otherwise indicated and x,y,$w(x, y)$ and Fourier coefficients a_k are measured in mm. All computations were performed using the numerical continuation code AUTO (Doedel et al. 1991) and, unless otherwise stated, the number of collocation intervals NTST=20.

Note that the shell we compare to here is of length $L = 100$ mm and so hardly qualifies as long as its aspect ratio (length to diameter) is only $L/2R = 0.5$.

We compare with experiments the ratio of loads λ_m/λ_d where $\lambda_d = 2\rho\kappa \approx 1.1499 \times 10^{-3}$ denotes the smallest value of λ at which the fundamental solution bifurcates, and λ_m is the minimum post buckling load (the first local minimum in the bifurcation diagram, see Figures 2 & 3 (a)). For the numerical simulations, λ_m was taken to be the first limit point on the branch of homoclinics as the loading parameter was decreased from λ_d.

In the computations the half length of the cylinder T was taken to be either $T = 100$ or $T = 250$. We chose a 6 mode approximation (ie

$m = 0, 1, 2, 3, 4, 5$) since it was found in (Lord et al. 1997*a*) that gave a good compromise between spatial convergence and tractability of the problem.

In the bifurcation diagrams, Figure 2 (a) and Figure 3(a), we plot the load λ against a measure of the end shortening defined by arc–length for a symmetric case and a cross-symmetric case. Note that the trivial branch (corresponding to the diagonal line in Figure 1 (a) is equivalent to the λ-axis here - since the von Kármán–Donnell equations factor out the overall squash of the cylinder. The curve of homoclinic orbits originates, in both cases, from the bifurcation point $\lambda_d \approx 1.1499 \times 10^{-3}$ on the λ axis. The limit point λ_m is at the first minimum value of the loading parameter λ along the curve originating at λ_d.

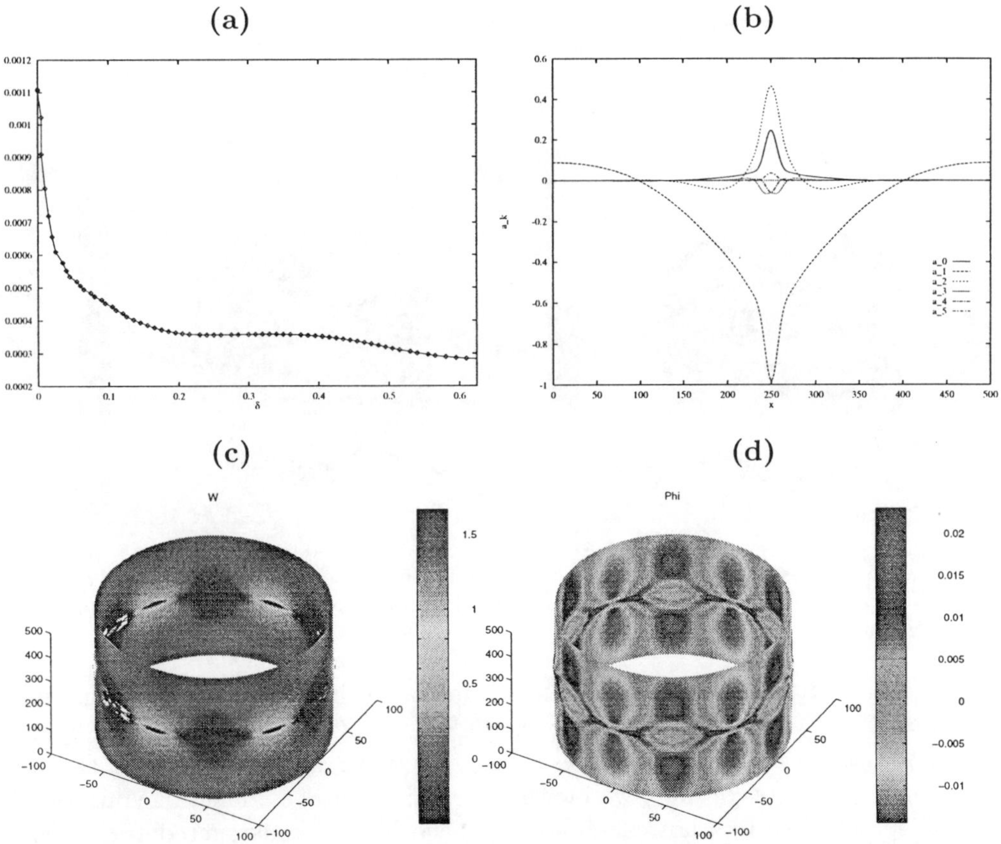

Figure 2. Symmetric form of solution computed for seed $s = 7$ with $M = 6$ modes. (a) Bifurcation diagram, (b) Fourier modes a_k (mm) for displacement, (c) reconstructed displacement w (mm), (d) reconstructed stress function ϕ.

In Figures 2 and 3 (b) we have plotted the Fourier modes a_k for the displacement w for the symmetric and cross symmetric cases respectively.

Note that the Fourier modes a_4 and a_5 are very small - indicating convergence as the spatial resolution is increased.

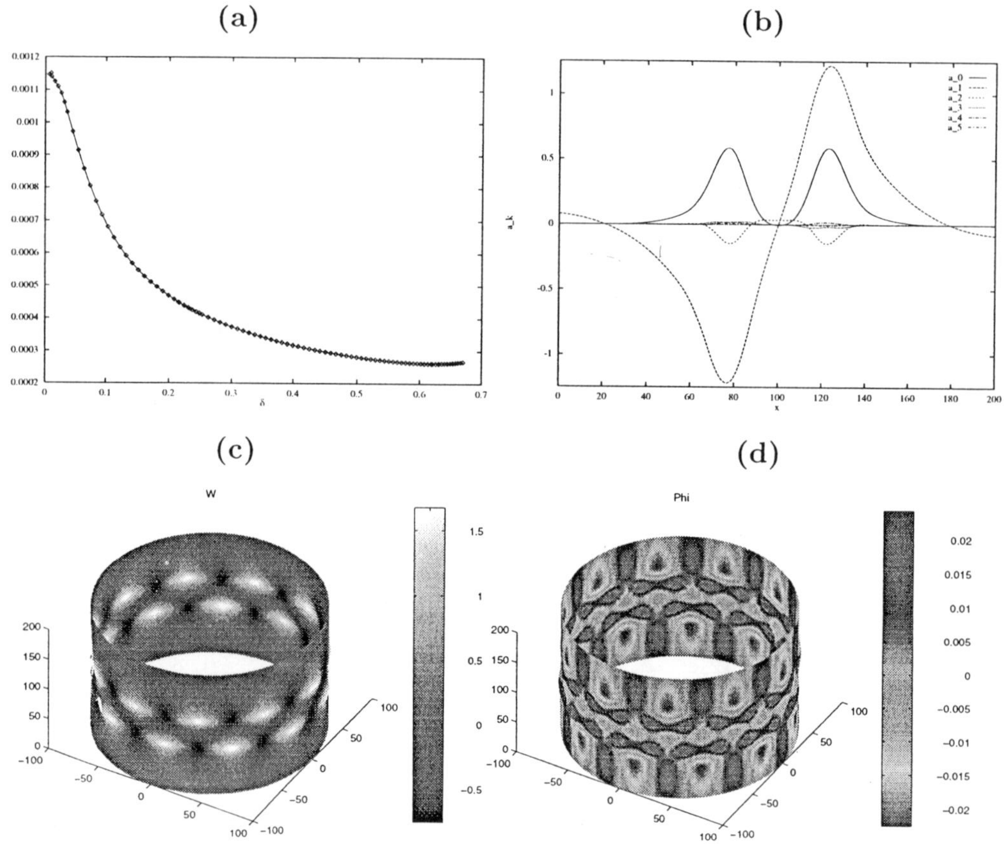

Figure 3. Cross-symmetric form of solution computed for seed $s = 7$ with $M = 6$ modes. (a) Bifurcation diagram, (b) Fourier modes a_m (mm) for displacement, (c) reconstructed displacement w (mm), (d) reconstructed stress function ϕ.

The 3-dimensional plot in Figures 2 and 3 (c) show the full reconstructed displacement $w(x, y)$ plotted over the deformed cylinder. The (cross-) symmetric nature of the solutions is clearly evident. Qualitatively, all the 3-dimensional plots compare well with the experimental evidence of Von Eßlinger & Geier (1972). In Figures 2 and 3 (d) we have reconstructed the stress function ϕ and plotted that over the deformed cylinder.

We now consider a quantative comparison between our numerics and experimental results. For the symmetric case we found a difference of $\approx$ 20% in the experimental to computed ratio λ_m/λ_d. This discrepancy is discussed below. In the cross-symmetric case excellent quantative agreement was obtained, as presented in Table 1.

s	10	11	12	13
Present	0.1843	0.2027	0.2277	0.2538
Experimental	0.197	0.208	0.228	0.257

TABLE 1. Comparison of the ratio λ_m/λ_d of our results with the experiments of Von Eßlinger & Geier (1972) for cross-symmetric forms of the solution.

4. Conclusion

We presented numerical results for the buckling cylinder problem and compared our results with experimental data. Qualitatively, we find good agreement for both the symmetric and cross-symmetric forms of solutions.

In the symmetric case there are some discrepancies between the numerical and experimental results. However, the experiments of Von Eßlinger & Geier (1972) indicate that a symmetric buckle pattern is not observed in longer cylinders. In addition, for the relatively short cylinder presented here, they find that the symmetric modes occur at lower-post buckling loads. This evidence is corroborated in (Yamaki 1984) who finds mostly cross-symmetric solutions. Furthermore symmetric modes appear less localized than cross-symmetric, and we would therefore not expect our analysis on this relatively short cylinder to be as quantatively accurate. For longer cylinders, for which there is little experimental evidence of the symmetric pattern, we would expect better agreement.

For the cross-symmetric case, more commonly observed experimentally, we have found excellent agreement quantatively even for the relatively short cylinder. For longer cylinders we expect a yet closer match to the experiments.

Our results suggest that the buckling of a long thin axially compressed cylinder is well described by a localization theory based on homoclinic solutions, independently of any imperfections in the cylinder. Significantly, this allows for the translational indeterminacy inherent in observations (cf. Figure 1). As such, we claim, it is provides a useful complement to finite element approaches with realistic boundary conditions, which may suffer from multiplicities of (near) solutions. This demonstrates that the corresponding asymptotic boundary conditions are the natural boundary conditions for the computation of buckling solutions of long cylinders.

References

Beyn, W.-J. (1990), 'The numerical computation of connecting orbits in dynamical systems', *IMA Journal of Numerical Analysis* **9**, 379–405.

Devaney, R. (1976), 'Reversible diffeomorphisms and flows', *Trans. Amer. Math. Soc.* (218), 89–113.

Doedel, E., Keller, H. & Kernévez, J. (1991), 'Numerical analysis and control of bifurcation problems', *Int. J. Bifurcation and Chaos* **1**, 493–520, 745–772.

Friedman, M. & Doedel, E. (1994), 'Numerical computational of invariant manifolds connecting fixed points.', *SIAM J. Num. Anal.* **28**, 789—808.

Hunt, G. W. & Lucena Neto, E. (1991), 'Localized buckling in long axially–loaded cylindrical shells', *J. Mech. Phys. Solids* **39**(7), 881–894.

Hunt, G. W., Williams, K. A. J. & Cowell, R. G. (1986), 'Hidden symmetry concepts in the elastic buckling of axially loaded cylinders', *Int. J. Solids Structures* **22**(12), 1501–1515.

Iooss, G. & Perouème, M. C. (1993), 'Perturbed homoclinic solutions in reversible 1:1 resonance vector fields', *J. Diff. Eq.* **102**, 62–88.

Koiter, W. T. (1945), On the Stability of Elastic Equilibrium, PhD thesis, University of Delft. English Translation: Tech. Rep. AFFDL-TR-70-25 Air Force Flight Dyn. Lab. 1970.

Lord, G. J., Champneys, A. R. & Hunt, G. W. (1997*a*), Computation of homoclinic orbits in partial differential equations: An application to cylindrical shell buckling, Technical Report ANM Report 6.97, University of Bristol, Department of Engineering Mathematics, University of Bristol, BS8 1TR, UK. Submitted to SIAM J. Sci. Comp.

Lord, G. J., Champneys, A. R. & Hunt, G. W. (1997*b*), Computation of localized post buckling in long axially–compressed cylindrical shells, *in* A. R. Champneys, G. W. Hunt & M. Thompson, eds, 'Phil. Trans. Localization and Solitary Waves in Solid Mechanics', Royal Society.

Mielke, A. (1991), *Hamiltonian and Lagrangian flows on center manifolds: with applications to elliptic variational problems*, Lecture notes in mathematics; 1489, Springer-Verlag, Berlin.

Peterhof, D., Sandstede, B. & Scheel, A. (1997), 'Exponential dichotomies for solitary–wave solutions of semilinear elliptic equations on infinite cylinders', *J. Diff. Equ.* To Appear.

Von Eßlinger, M. (1970), 'Hochgeschwindigkeitsaufnahmen vom Beulvorgang dünnwandiger, axialbelasteter Zylinder', *Der Stahlbau* **3**, 73–76.

Von Eßlinger, M. & Geier, B. (1972), 'Gerechnete Nachbeullasten als untere Grenze der experimentallen axialen Beullasten von Kreiszylindern', *Der Stahlbau* **12**, 353–359.

Wohlever, J. & Healey, T. (1995), 'A group theoretic approach to the global- bifurcation analysis of an axially compressed cylindrical-shell', *Computer Methods in Applied Mechanics and Engineering* **122**, 315–349.

Yamaki, N. (1984), *Elastic Stability of Circular Cylindrical Shells*, Vol. 27 of *Applied Mathematics and Mechanics*, Elsevier.

DYNAMICS OF A BUCKLED DRILLSTRING ROTATING IN A CURVED OIL WELLBORE

A.K. BELYAEV
Institute of Technical Mechanics, Johannes Kepler University of Linz
Altenbergerstr. 69, A-4040, Linz, Austria

1. Introduction

Drillstring failures caused by unstable rotation of the drilling assembly are continually ranked as one of the most frequent and costly problems in the oil industry. Although many other methods of oilwell drilling have been tried historically, rotary drilling technique dominates the industry. The drillstring components, i.e. drillpipes, are typically 5'' in outside diameter and 4.275'' in inside diameter. Depthwise the assemblies can extend up to 8 km or more downhole, i.e. drillstring is a very flexible slender rod. Directional drilling toward the desired target, which is very popular now, complicates the problem as wells are drilled with substantial curvatures and three-dimensional characteristics. In order to provide pressure for forcing the drillbit into formation, especially in case of horizontal drilling, the drillstring is axially compressed, i.e. it is buckled in its static configuration and rotates in a curved well. The objective of the presentation is to analyse the stability of the drillstring rotation.

2. The governing equations of the problem and their solution

The Kirchhoff-Clebsch equations for a spatial curved bar are as follows

$$\frac{d\boldsymbol{Q}}{ds}+\boldsymbol{q}=\boldsymbol{0}\,, \quad \frac{d\boldsymbol{M}}{ds}+\boldsymbol{t}\times\boldsymbol{Q}+\boldsymbol{m}=\boldsymbol{0}\,, \quad \boldsymbol{M}=\boldsymbol{I}\cdot\left(\boldsymbol{\kappa}-\boldsymbol{\kappa}_0\right) \tag{1}$$

where $\boldsymbol{M}$ and $\boldsymbol{Q}$ are the moment vector and the force vector, respectively, $\boldsymbol{q}$ and $\boldsymbol{m}$ denote the vectors of the distributed moments and forces, respectively, $\boldsymbol{t}$ is the unit vector of the tangent to the curvilinear s-axis of the bar. The components of the vectors $\boldsymbol{\kappa}$ and $\boldsymbol{\kappa}_0$ are the curvatures of the deflected axes and the unit angle of twist in actual and reference configuration, respectively, and $\boldsymbol{I}$ is the tensor of bar's rigidity. Application of the central manifold theory yields the following bifurcation equation

$$\frac{d^2\Theta}{ds^2}-\frac{1+\nu}{\rho_d\,\rho_w}\sin\Theta=\Gamma(s) \tag{2}$$

Here Θ is referred to as the rotation angle of the cross-section, ρ_d and ρ_w are the

F. C. Moon (ed.),
IUTAM Symposium on New Applications of Nonlinear and Chaotic Dynamics in Mechanics, 169–171.

curvature radius of the drillstring and the wellbore, respectively, and ν is Poisson's ratio. A slowly varying function $\Gamma(s)$ depends upon the bending and torsional rigidities of the drillstring, the twists of the drillstring and the wellbore, the friction between the drill string and wellbore, and drillpipe's radius.

Some cases allow for the solution in closed form in terms of the elliptic functions and integrals. The case where the drillstring's buckling form is a helix and the well is a plane circular arc $(0<\psi<\Psi)$ is analysed below. The following figures show the rotation angle after the crooked part of the well $\Theta_1=\Theta(\Psi)$ versus $\Theta_0=\Theta(0)$ in case of zero torque on bit.

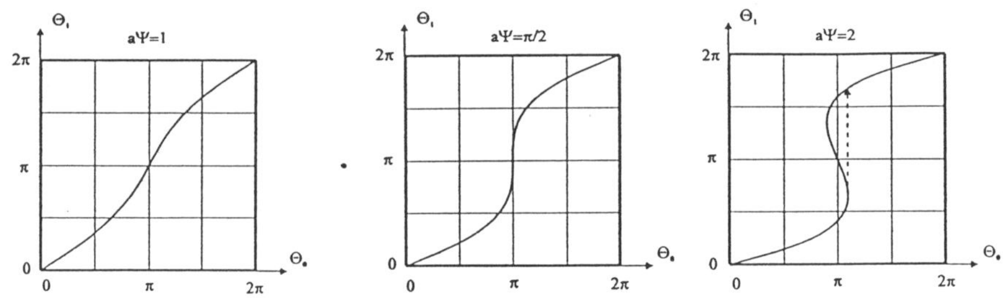

The bifurcation parameter is $a\Psi$ where $a=\sqrt{(1+\nu)\rho_w/\rho_w}$. The rotation is seen to be unstable and accompanied by a dynamic transition to a new configuration. The drillstring rotation angle versus ψ and the snap-through buckling are displayed below

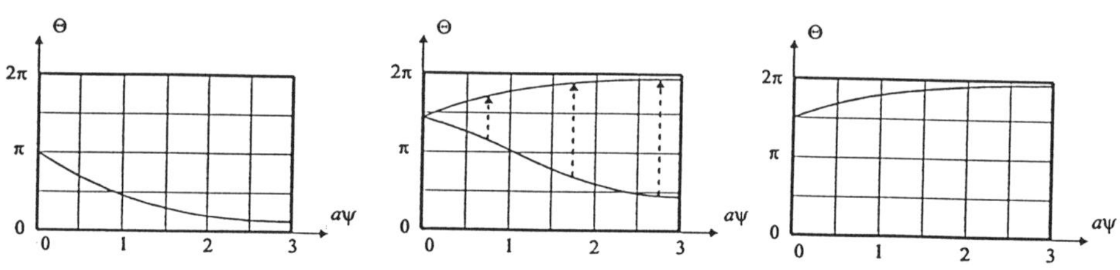

The increasing torque on bit (λ) improves the rotational stability as the following figures show

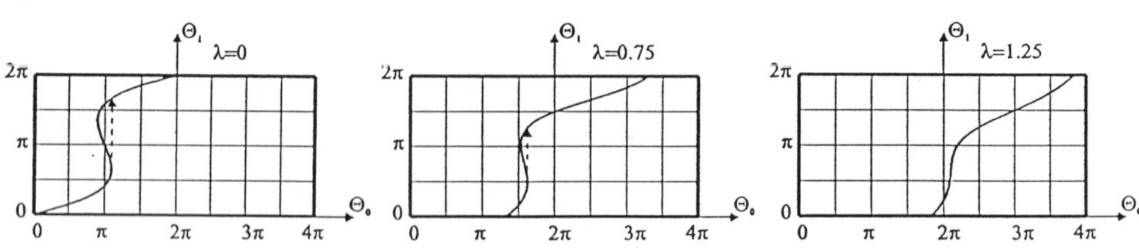

The torque on the top of the drillstring ($\psi = 0$) is shown for some torques on bit λ

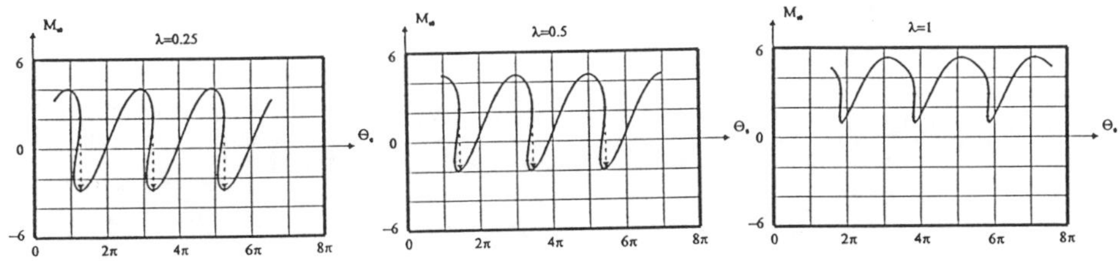

The torque exhibits a behaviour which allows one to interpret the unstable drillstring rotation as stick-slip oscillations or chatter. However, neither wellbore friction nor material damping is present in the model. The instability is caused by the energy exchange between the torsional and bending modes. From the latter figures one finds the critical value of the torque on the top. The drillstring rotation becomes uniform when this critical value is exceeded.

Another interesting effect is that the resistance against the rotation of a free buckled beam in a crooked wellbore is larger than that of a straight drillstring which is clamped at the bit.

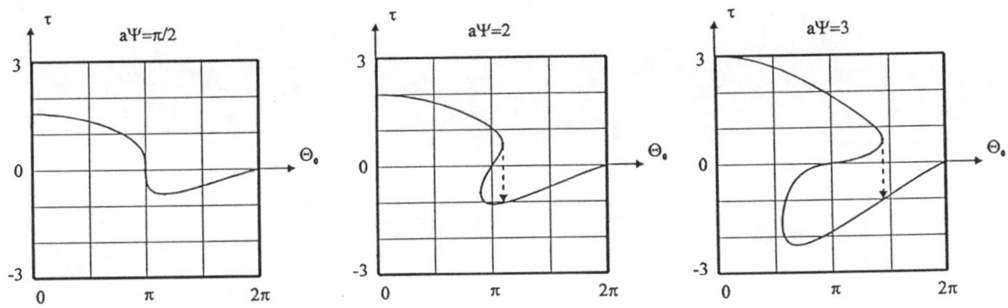

The general case has been studied by means of numerical analysis of the Kirchhoff-Clebsch equations (1) which confirms the previous results.

Reference:

Belyaev, A.K. (1988) Zur Instabilität des rotierenden Ölbohrstranges in einem gekrümmten Bohrloch. *ZAMM*, **77**, S1, 31-32.

THE COMPLEX NON-LINEAR DYNAMICS OF IMPERFECTION SENSITIVE SHELLS

P. B. GONÇALVES
Civil Engineering Department
Catholic University - PUC-Rio
22453-900 Rio de Janeiro, RJ - Brazil
e-mail: paulo@civ.puc-rio.br

1. Introduction

Shell structures have been used for centuries in all engineering fields. This is largely due to the important role played by these shells as efficient loading carrying members leading to very light and slender structures. However, when subjected to a compressive stress state in one or more directions they often exhibit a highly non-linear response and may loose their stability under quasi-static or dynamic conditions at load levels well below those predicted by linearized eigenvalue-problems or by non-linear analyses of a perfect system [1-3]. This is mainly due to the deleterious effects of imperfections (e. g., load, geometrical and boundary imperfections) and, in certain circumstances, to the small degree of safety of these systems on account of the fast decrease or erosion of the basin of attraction associated with the stable solution whose stability must be preserved. This can be explained by the fact that many shells under quasi-static loads display a symmetric unstable or an asymmetric bifurcation along the fundamental equilibrium path [2,3]. In these circumstances there are at least two equilibrium points for load levels lower than the critical load and, induced by the topology of the associated potential function, these shells may display subharmonic and superharmonic bifurcations, period-multiplying bifurcations, multiple solutions, chaotic motions and dangerous jumps caused by the presence of competing potential wells and non-linear resonance curves within each well [4]. Shells under compressive loads display not only cubic geometric non-linearities, like any other structural element, but also strong quadratic non-linearities due to the simultaneous effects of the initial curvature, pre-stresses and geometric imperfections. This strong non-linearity explains the complex dynamics of these systems and controls the various modes of elastic instability.

In order to exploit the efficiency of shells structures, their static and dynamic buckling behavior need to be elucidated. Contrary to static buckling behavior, which is fairly well-understood [1-3], dynamic buckling needs further clarification in view of some lack of pertinent literature.

The aim of the present work is to stress some aspects related to the non-linear dynamics of thin shells under static pre-loading with emphasis on those phenomena connected with the dynamic stability of pre-buckling configurations of the structure under vibration. For small forcing amplitudes, the motions remain in the neighborhood

F. C. Moon (ed.),
IUTAM Symposium on New Applications of Nonlinear and Chaotic Dynamics in Mechanics, 173–180.

of the static reference state, confined into the pre-buckling well. But, for certain values of the control parameters, the motions can no longer remain within this well and the shell jumps into another well, displays large cross-well motions or escape to infinity. These motions may cause undesirable stresses and displacements, leading as a rule to a catastrophic failure of the structural system.

2. Basic Equations

To illustrate the non-linear dynamic behavior of imperfection sensitive shells, let us investigate, as an example, but without loss of generality, the non-linear axisymmetric vibrations of a shallow spherical shell under uniform pressure loading. This represents one of the basic problems in the field of structural stability due to its high imperfection sensitivity. Using the well-known approximations of Marguerre's shallow-shell theory, the system of equations for the transverse vibrations of a shallow imperfect pre-stressed spherical shell can be written in the following non-dimensional form [1,4]

$$\nabla^4 w + w,_{tt} + \bar{c} w,_t = \lambda^2 \alpha^{1/4} \nabla^2 f + \frac{\alpha}{x}\{f,_x (\tilde{w} + w_s + w),_x + f_s,_x w,_x\},_x + Q(t,x)$$

$$\nabla^4 f = -\lambda^2 \alpha^{-1/2} \nabla^2 w - (1/x)\{w,_x w,_{xx} + [(\tilde{w},_x + w_s,_x) w,_x],_x\} \tag{1}$$

where x is the non-dimensional radial coordinate, t is time, w and f are, respectively, the vertical displacement and stress function due to vibrations, w_s and f_s are the displacement field and stress function due to static pre-loading, $\tilde{w}$ is the initial geometric imperfection, Q is a load parameter, $(\ ),_t = \partial(\ )/\partial t$, $\alpha = 12(1-\nu^2)$, $(\ ),_x = \partial(\ )/\partial x$ and λ is a geometrical parameter described by

$$\lambda = \alpha^{1/4} a / \sqrt{Rh} \tag{2}$$

The non-dimensional variables are related to the physical quantities by

$$x = r/a; \qquad w = W/h; \qquad f = F/Eh^3; \qquad Q = q/q_{cl} \tag{3}$$

where $E =$ Young's modulus, $R =$ principal radius of curvature of the sphere, $h =$ shell thickness, $a =$ base radius of the spherical cap, $r =$ radial coordinate and q_{cl} = classical buckling load of a complete spherical shell [1].

An approximate response of the shell to any external dynamic loading $Q(x,t)$ is sought in the separable form

$$w(x,t) = \sum_{i=1}^{\infty} W_i(t)\phi_i(x); \qquad f(x,t) = \sum_{i=1}^{\infty} F_i(t)\psi_i(x) \tag{4}$$

where $\phi_i(x)$ and $\psi_i(x)$ are any suitable functions of the surface coordinate x satisfying the boundary conditions of the shell, and may well be taken as the true free vibration mode shapes of the structure [4]. The functions $W_i(t)$ and $F_i(t)$ are the generalized coordinates to be determined. To simplify the numerical procedure, the pre-stress state and the initial geometric imperfections are assumed in the form

$$w_s(x) = \sum_{i=1}^{\infty} W_{si}\,\phi_i(x); \qquad f(x) = \sum_{i=1}^{\infty} F_{si}\,\psi_i(x) \tag{5}$$

$$\tilde{w}(x) = \sum_{i=1}^{\infty} \tilde{W}_i\,\phi_i(x) \tag{6}$$

Substituting expressions (4-6) into equations (1) and applying a Galerkin minimization procedure, one obtains a set of 2N non-linear ordinary differential equations characterizing the dynamic behavior of the cap. These equations are unfortunately too long to be presented in this work. In some situations the first mode is dominant and the qualitative behavior of the shell can be analyzed with a certain degree of accuracy by a SDOF model. Using in (5) and (6) the number of terms necessary to achieve convergence of the static state and using i=1 in (4), one obtains the following equation of motion and compatibility equation

$$\begin{gathered} W_{1,tt} + \bar{c}W_{1,t} + W_1(\ldots) = F_1(\ldots) + F_1[W_1(\ldots) + \\ \sum(W_{si}(\ldots) + \tilde{W}_i(\ldots))] + W_1\sum F_{si}(\ldots) + Q(t)(\ldots) \end{gathered} \tag{7}$$

$$F_1(t) = W_1(t)\,(\ldots) + W_1^2(t)\,(\ldots) + W_1(t)[W_{si}(\ldots) + \tilde{W}_i(\ldots)] \tag{8}$$

where (...) are coefficients dependent on the definite integrals over the shell surface.

Substituting the value of $F_1(t)$ obtained from equation (8) into equation (7), one obtains the equation of motion in terms of $W_1(t)$ only

$$\begin{gathered} W_{1,tt} + \bar{c}W_{1,t} + W_1\{(\ldots) + \lambda^4(\ldots) + \sum(W_{si} + \tilde{W}_i)(\ldots) + \\ \lambda^2\sum(W_{si} + \tilde{W}_i)(\ldots) + \sum\sum(W_{si} + \tilde{W}_i)(W_{sj} + \tilde{W}_j)(\ldots) + \\ \sum F_{si}(\ldots)\} + W_1^2\{\lambda^2(\ldots) + \sum(W_{si} + \tilde{W}_i)(\ldots)\} + \\ W_1^3(\ldots) = Q(t)(\ldots) \end{gathered} \tag{9}$$

This shell equation displays quadratic and cubic non-linearities. The coefficient of the linear term and, as a consequence, the natural frequency, and the coefficient of the quadratic term are highly dependent on the geometrical parameter λ which is associated with the curvature and slenderness of the shell, the pre-stress state and the geometric imperfections. The cubic term, on the other hand, doesn't depend directly on these parameters. The quadratic term is usually negative and can be particularly large in imperfection-sensitive systems. It is responsible for the unstable post-buckling path and for the softening character of the frequency-amplitude relation. This is illustrated in Table 1 where the stiffness coefficients of the non-linear equation of motion (9) for a clamped spherical cap with $\lambda = 4$ and $\nu = 1/3$ are shown as a function of the level of static pre-loading Q_s.

For a perfect unstressed circular plate $R = \infty$ and, consequently, λ=0 and the non-linear equation (9) is reduced to

$$W_1,_{tt} + \bar{c} W_1,_t + W_1(\ldots) + W_1^3(\ldots) = Q(t)(\ldots) \tag{10}$$

In this case the quadratic term is absent and the cubic term is positive. Thus plates exhibit non-linearity of the hardening type. This explains why plates are imperfection-insensitive and exhibit symmetric bifurcation. The presence of initial imperfections and/or a pre-stress state are, as one can observed, sufficient to induce the appearance of a quadratic term in equation (10).

TABLE 1. Coefficients of the stiffness terms in the equation of motion.

Static Load Q_s	Linear	Quadratic	Cubic
0.0	612	-581	154
0.1	552	-559	154
0.2	486	-534	154
0.3	413	-504	154
0.4	328	-466	154
0.5	219	-410	154

3. Results

The numerical results presented here are for a perfect clamped spherical shell with $\lambda = 4$ and $\nu = 1/3$ and subjected two a uniform static pressure. The dynamic pressure parameter $Q(t)$ is taken as

$$Q(t) = F_e sin(\Omega t) \tag{11}$$

Two approximations are used in the present analysis. A MDOF model where four modes were used in (4) and (5). This was enough to achieve convergence of , respectively, the static and dynamic responses. The second model employed in this analysis is a SDOF model were four modes are used to describe the pre-stress state and one mode (the dominant mode) is used to describe the dynamic response. The interpolating functions are the vibration modes of a perfect unloaded clamped cap [4].

Figure 1 shows the variation of the frequency ratio Ω/Ω_0, where Ω_0 is the natural frequency of the shell, with the vibration amplitude at the apex of the shell for selected static configurations along the non-linear equilibrium path of the cap under uniform pressure-loading, which is shown inset in this Figure. The natural frequency Ω_0 decreases steadily with load as observed in the second column of Table 1 where the listed coefficients are equal to Q_0^2. The shell exhibits a strong softening behavior. The degree of non-linearity of the frequency-response curve varies greatly with the pre-stress state and is particularly large for load levels near the critical load. The load deflection curve shown in Figure 1 where the shell looses its stability at a limit point is typical of many shell structures. For these shells, between the upper and lower limit points (here $Q_U = 0.578$ and $Q_L = 0.458$), there are three equilibrium point, two stable and one unstable. So, at these load levels, one has a non-symmetric potential function. As the load increases the depth of the well associated with the buckled configurations increases steadily while the depth of the well associated with pre-buckling configurations decreases. Similar trend is observed in the associated basin areas [7,8]. So, as the load increases, the probability of dynamic buckling under the unavoidable disturbances which will appear during the service life of the structure increases and may cause, under usually small disturbance levels, a catastrophic failure of the shell. For low load levels there is only one potential well. But, even is this situation, there is a sudden broadening of the well due to the quadratic non-linearity, leading for certain values of the control parameters to large amplitude vibrations and most likely to damages in the system.

A typical bifurcation diagram obtained from the MDOF model is shown in Figure 2 for $Q_S = 0.5$, $\overline{c} = 3$, $\Omega = 12$ and $F_e \in [0.062;0.067]$. As observed, the shell may exhibit period doubling bifurcations, regions of chaotic motions and escape from the potential well associated with the pre-buckling configuration. It is interesting to notice that the first chaotic region appears after a period doubling cascade, while the second region appears suddenly.

The capacity to predict the behavior and load carrying capacity of a given system from the knowledge of its physical and geometrical parameters is the basis of design. Recent results have shown that the capacity to predict the behavior of a non-linear structure is hampered not only by the dependence and sensitivity of the response on the initial conditions and the system parameters but also by the existence of fractal basin and escape boundaries [5-8].

In Figure 3 it is shown a typical stability boundary for an imperfection sensitive shell under harmonic load. Here the results for a shell in a slowly evolving environment are compared with the results where the system after each load increment starts from rest. The escape curve was obtained numerically by increasing slowly the forcing amplitude for each frequency value. In many regions a high sensitivity to the initial conditions is observed. In fact in some of these regions the escape boundary has a fractal-like structure, as shown in Figure 4, and the escape load and the time spent before the shell escapes from the pre-buckling well become unpredictable due to the uncertainties in the initial conditions, imperfection levels and system parameters. The colors in Figure 4 indicate a measure of time to escape out of the safe well. In the first region from left the shell buckles within one forcing cycle, in the second region, within two forcing cycles and so on. The stability boundary in Figure 3 is composed of various "curves", each one associated with a particular bifurcation event. It is observed that for low values of the forcing frequency Ω the dynamic and static escape load practically coincide, while in the region near the natural frequency (here 14.8) the escape load is much lower than the static critical load. In the high frequency region the dynamic escape load can be much higher than the static one. The actual escape load for a real imperfect shell is expected to be much lower than the ones depicted in Figure 1. Firstly

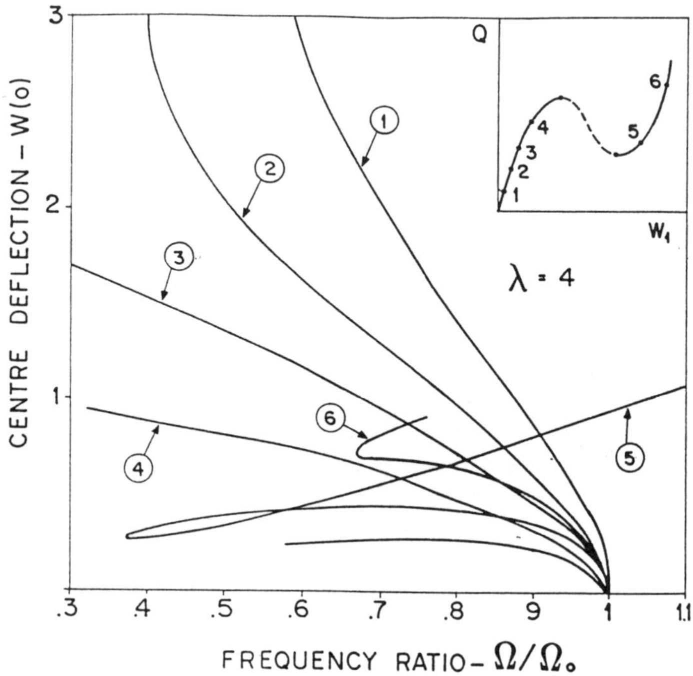

Figure 1 - Influence of the static pre-stress state on the non-linear vibrations of the cap. Inset the non-linear equilibrium path and the corresponding equilibrium configurations

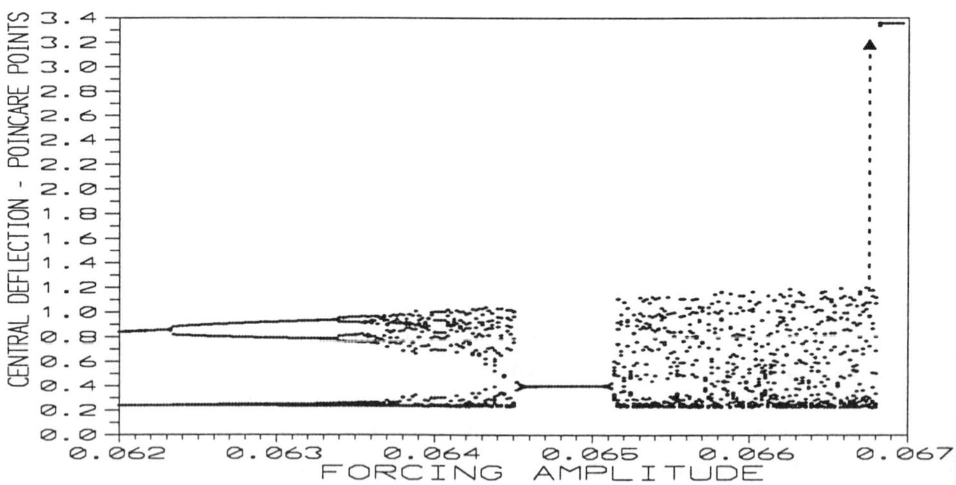

Figure 2 - Bifurcation diagram showing period doubling bifurcations, chaos and escape. MDOF model.

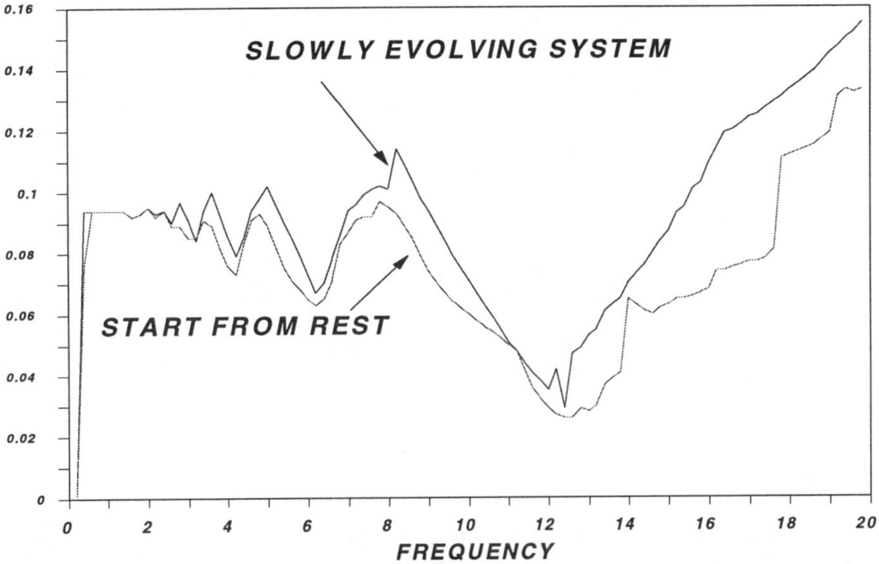

Figure 3 - Stability boundary in control space (frequency vs. forcing amplitude). Influence of initial conditions on the escape load. MDOF model.

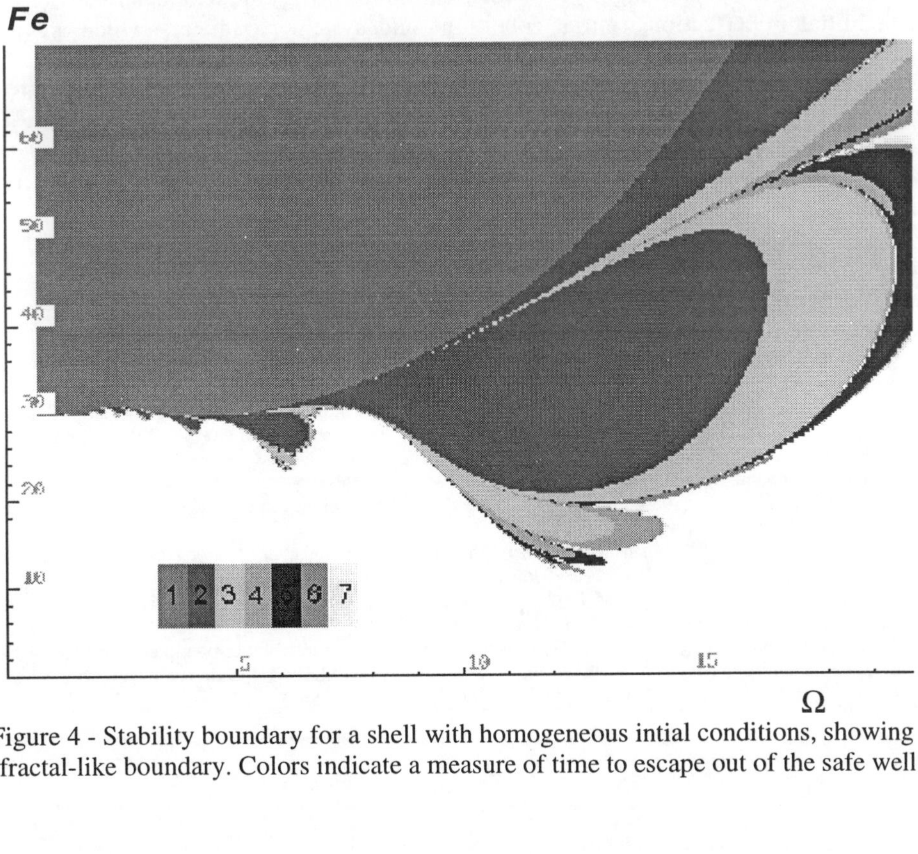

Figure 4 - Stability boundary for a shell with homogeneous intial conditions, showing a fractal-like boundary. Colors indicate a measure of time to escape out of the safe well.

in many regions of the control space for load levels lower than the escape load for the slowly evolving system the basin of attraction becomes fractal and escape unpredictable, secondly the potential well associated with pre-buckling configurations will be in fact much shallower due to the effect of unavoidable imperfections, and, finally, the initial conditions, or even the system parameters may be unknown or exhibit a certain degree of uncertainty. All these facts indicate the difficulties in estimating the load carrying capacity of an imperfection sensitive shell under dynamic load and to the necessity of theoretically well-founded and safe lower bounds for design. Since non-linearities (physical, geometrical, etc.) are becoming more and more important in all engineering structures, these complex non-linear phenomena are more likely to occur and therefore one must study their implications in the safety and integrity of these systems and how to take them into account at the stage of design. This is the major concern of this research project.

4. Conclusions:

The non-linear dynamic behavior of imperfection sensitive shells was analyzed. Special attention was devoted to the determination of critical conditions. It was shown that these systems may exhibit dynamic buckling loads much lower that the static critical load. This load is highly dependent on the post-buckling characteristics of the shell, initial imperfections, initial conditions and system parameters which are not usually known with certainty in a real situation. Also the capacity to predict the behavior of the shell is hampered by the existence of fractal escape and basin boundaries. This illustrates the difficulties in assessing the load carrying capacity of these shells under dynamic loads. So research on the behavior of thin shells under dynamic loads, detailed parametric analyses and experimental investigations are necessary to assess more fully the design implications of the present analysis.

5. References:

1. Gonçalves, P.B. and Croll, J.G.A. (1992) Axisymmetric buckling of pressure-loaded spherical caps. *ASCE J. Struc. Eng.* **118**, 970-985
2. Julien J. F. (Ed) (1991) *Buckling of Shell Structures on Land, in the Sea and in the Air*, Elsevier, NY, USA
3. Fung Y. C. and Sechler E.E. (Eds.) (1974) *Thin Shell Structures*, Prentice-Hall Inc, New Jersey, USA
4. Gonçalves, P.B. (1993) Jump phenomena, bifurcations and chaos in a pressure loaded spherical cap under harmonic excitation. *Appl. Mech. Rev.* **46**, S279-S288
5. Moon, F.C. and Li, X.G. (1985) Fractal basin boundaries and homoclinic orbits for periodic motions in a two well potential, *Physical Rev.* **55**, 1439-1442
6. Thompson, J.M.T. and Soliman, M.S. (1990) Fractal control boundaries of driven oscillators and their relevance to safe engineering design. *Proc. Roy. Soc, Lon.* **A428**, 1-13
7. Soliman, M.S. and Thompson, J.M.T. (1989) Integrity measures quantifying the erosion of smooth and fractal basins of attraction. *J. Sound and Vibration* **135,** 453-475
8. Santee, D. M. and Gonçalves, P. B. (1997) Non-linear vibrations and instabilities of structural systems liable to asymmetric bifurcation, in N. S. Ferguson, H. F. Wolfe and C. Mei (eds.), *Structural Dynamics Recent Advances*, Inst. of Sound and Vibration, Southampton, UK, pp. 743-756

III. INELASTIC MATERIALS, MATERIAL PROCESSING

The four papers in the section provide a good introduction to the application of nonlinear dynamics to machining and chatter control (Davies et al., Nayfeh and Pratt). It is now recognized that machining processes can exhibit subcritical Hopf bifurcations and that practical chatter control must involve an understanding of the nonlinear contributions to the dynamics.

The other two papers (Vestroni and Capecchi, Pratap and Judge) explore the important consequences of hysteresis in dynamic models. In particular the last paper examines elasto-plastic nonlinearities in a nonlinear oscillator.

F. C. Moon (ed.),
IUTAM Symposium on New Applications of Nonlinear and Chaotic Dynamics in Mechanics, 181.

THE DYNAMICS OF CHIP FORMATION IN MACHINING

M. A. DAVIES, T. J. BURNS AND C. J. EVANS
National Institute of Standards and Technology
Gaithersburg MD 20899, USA

1. Introduction

Machining operations such as turning, milling and drilling involve the modification of a *workpiece* through the removal of material. This removal occurs via a concentrated shear flow generated by a wedge-shaped tool as illustrated in Fig. 1(a). The material that is removed is called the *chip*. Because the thermoplastic shear flow that generates the chip is dynamic and highly nonlinear, optimization of machining processes by any means other than trial and error (not uncommon in industry today) requires a fundamental understanding of the underlying physics and new models to predict behavior.

In this paper, we restrict our attention to the *orthogonal cutting* of metals (see Fig. 1(b)). According to Shaw (Shaw, 1984), the defining characteristics of orthogonal cutting are: (1) the tool is a wedge that is much wider in the direction perpendicular to the cutting direction than the material being removed; and (2) the ratio of chip width to uncut chip thickness, b, is at least five. In orthogonal cutting, the material flow can be regarded as essentially two-dimensional.

Observations in many laboratories over the past century have led to the following conventional model of the orthogonal cutting. As illustrated in Fig. 1(a), steady shear flow of the material takes place in a narrow region of thickness h extending from the tool nose to the free surface of the workpiece. This region of shear flow is known as the *primary shear zone.* Material is assumed to remain unloaded as it approaches this zone at the cutting velocity V. It then undergoes a rapid loading which is initially elastic, but the yield stress is exceeded quickly and plastic flow sets in. In many cases, h is quite small (e.g. $b/10$), and the shear zone can be approximately represented by a plane inclined at an angle ϕ (Merchant, 1945) relative to the direction of motion. Contact between the tool and

F. C. Moon (ed.),
***IUTAM Symposium on New Applications of Nonlinear and Chaotic Dynamics in Mechanics*, 183–192.**

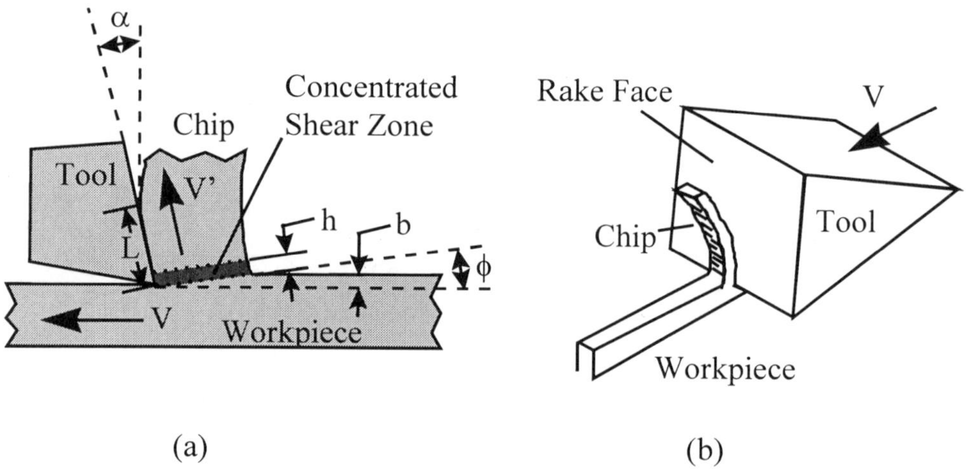

Figure 1. (a) Schematic of a machining operation; and (b) Orthogonal cutting operation.

chip occurs in a region of length L on the *rake face* of the tool. A secondary shear flow develops along this contact zone (Trent, 1991); however, in most models of orthogonal cutting this *secondary shear zone* is ignored, and a chip with thickness $b/\sin\phi$ is assumed to move up the rake face as a coulomb friction slider with constant coefficient of friction μ (Merchant, 1945).

The model described above is based upon the assumption that the chip forms through a process of steady homogeneous shear. However, in many metals the flow will undergo a transformation from steady shear to oscillatory behavior. This oscillatory behavior is manifested in the formation of a *segmented chip* with regions of intense shear separated by regions with relatively little deformation. As pointed out by Recht (Recht, 1964), the formation of segmented chips occurs when the local rate of heat buildup in the material is large enough that thermal softening overwhelms the effects of strain hardening, and the local flow stress drops catastrophically, leading to the formation of an *adiabatic shear band.* Such segmented chips have been observed to form in titanium (Komanduri et al., 1981), nickel-iron alloys (Komanduri and Schroeder, 1986), hardened steels (Komanduri et al., 1982; Shaw, 1993; Konig et al., 1990; Davies et al., 1996; Davies et al., 1997(a)) and amorphous nickel-phosphorus (Davies et al., 1997(a)). However, neither the conventional model nor more recent models of segmented chip formation provide a mechanism for the onset of oscillatory material flow. Understanding the onset and development of this oscillatory behavior is important because segmented chip formation has been correlated with increased tool wear, degradation of workpiece surface finish and less accuracy in the machined part (Komanduri et al., 1982).

In this paper, we provide experimental, numerical and analytical evidence suggesting that the onset of segmented chip formation is the result of a *Hopf bifurcation* in material flow. We modify the conventional model for orthogonal cutting by introducing the concept of a *local plastic deformation zone* that accounts for indentation of the material near the tool tip. A one-dimensional, partial differential equations (PDE) model, similar to those used to describe the formation of a single adiabatic shear band (Clifton et al., 1984; Wright and Walter, 1987), is developed to describe chip formation. Numerical simulations of this model are compared with experimental results; both are suggestive of a Hopf bifurcation and the subsequent development of relaxation oscillations. Based upon these observations, a simplified ordinary differential equations model that treats the shear zone as a control volume moving with the tool is developed. These results lead to an interpretation of metal cutting as a thermomechanical feedback process, which is similar in many ways to an open chemical reactor. Recent work by Molinari and Dudzinski (Molinari and Dudzinski, 1992) is related to our approach; however, they do not consider the connection between local stresses induced by the tool on the workpiece and the global shear stress in the primary shear zone, and they treat the deformation in the shear zone as adiabatic. Also, Marusich and Ortiz (Marusich and Ortiz, 1995) have recently reported results of some large scale Lagrangian finite-element simulations of orthogonal cutting; just as with a laboratory experiment, we view our simpler modeling approach as an attempt to gain insight into the results of such studies.

2. Experimental Results

Orthogonal cutting experiments were conducted on hardened AISI 52100 bearing steel and amorphous nickel-phosphorous using polycrystalline cubic boron nitride (pCBN) tools. A two-axis diamond turning machine with an air-bearing spindle was used in the experiments, which were conducted for a variety of rake angles, uncut chip thicknesses and cutting speeds. Chips were collected and examined in a scanning electron microscope (SEM). By examining the back surface of the chip (side opposite the tool rake face), we were able to make quantitative measurements of the onset of segmented chip formation, as well as the dependence of average segment spacing (spatial frequency) on cutting speed. In addition, by polishing and etching the steel chip cross-sections with a solution of 2% nitric acid in ethyl alcohol the qualitative characteristics of the material flow could be examined. Fig. 2 shows typical results for orthogonal machining of 52100 bearing steel with a rake angle (designated α in Fig. 1) of -10 deg and an uncut chip thickness b of approximately 33 μm. Fig. 2(a) shows polished and etched chip cross-

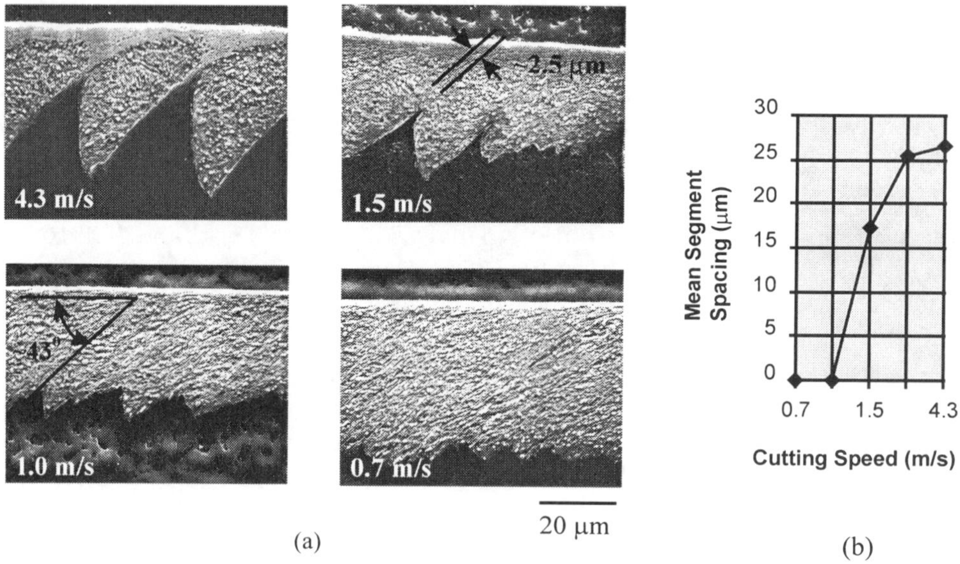

Figure 2. (a) Chip Photos and (b) plot of mean segment sacing as a function of cutitng speed.

sections at cutting speeds ranging from 0.7 m/s to 4.3 m/s. At low speeds, steady, homogeneous shear is evident throughout the chip. As the speed is increased, the shear begins to localize, with segments becoming clearly discernible at 1.5 m/s. For cutting speeds just above this critical speed, the segments form at irregular intervals. However, as the speed is increased, the segments become larger and their spacing becomes very regular. This behavior is summarized in Fig. 2(b), which shows average segment spacing as a function of cutting speed. The shape of this curve is suggestive of a Hopf bifurcation occurring at a critical cutting speed between 1 m/s and 1.5 m/s.

3. One Dimensional PDE Model

The experimental results can be explained with a relatively simple one-dimensional thermoplastic shear model. The proposed model is shown in Fig. 3, where α is the rake angle of the tool, ϕ is the shear angle, V is the cutting speed and b is the uncut chip thickness. The equations of motion for the material flow are derived based upon the following assumptions.

1. A shear plane extending from the tool nose to the free surface and making an angle ϕ with the cutting direction is defined. Material lying ahead of the shear plane ($\tilde{y} < 0$) is unstressed but can conduct heat.

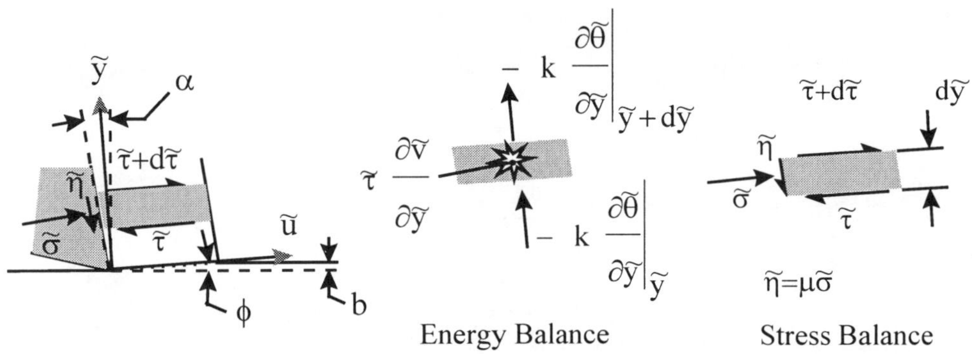

Figure 3. One-dimensional model of chip formation. Tildes represent dimensional quantities.

2. The shearing process is one-dimensional with shear occurring on lines parallel to the shear plane.
3. Heat conduction is one-dimensional and occurs in the direction perpendicular to the shear zone.
4. An Arrhenius-type constitutive law is used to describe the thermal softening of the material. No strain hardening is included.
5. Compressive stresses between the tool and the workpiece are assumed to be elastic-perfectly plastic.
6. Shear stresses on the rake face of the tool are assumed to be proportional to the compressive stress through a simple coulomb friction coefficient. The material slides up the rake face and the frictional stresses act to oppose this motion.
7. No heat is lost to the environment.
8. The momentum of the material is negligible.
9. The elastic component of the shear strain is ignored.

Given these assumptions and utilizing the differential elements shown in Fig. 3, the nondimensional equations describing the thermoplastic deformation of the material can be written as(see (Davies et al., 1997(a); Davies et al., 1997(b))),

$$\begin{aligned}
\frac{\partial \tau}{\partial y} &= -\xi\sigma \qquad (y > 0) && \text{(a)}\\
\frac{\partial \theta}{\partial t} &= \lambda \frac{\partial^2 \theta}{\partial^2 y} + q\tau\Phi(\tau,\theta) && \text{(b)}\\
\frac{\partial v}{\partial y} &= \Phi(\tau,\theta) && \text{(c)}\\
\frac{\partial \sigma}{\partial t} &= \Psi(1 - \exp(\frac{\sigma - 1}{\varepsilon}) - v) && \text{(d)}
\end{aligned} \tag{1}$$

$$\frac{\partial u}{\partial t} = v \qquad \text{(e)}$$

where

$$\begin{aligned}
\xi &= \sqrt{3}\sin\phi(1+\mu\tan(\alpha-\phi)) \\
\lambda &= \frac{k}{\rho c \Phi_o b^2} \\
q &= \frac{\tau_y}{\rho c \theta_o} \\
\Psi &= \frac{\Lambda b}{\sqrt{3}\tau_y \delta}\cos(\alpha-\phi) \\
\Phi_o &= \frac{V_s}{b} \\
V_s &= V\frac{\cos(\alpha)}{\cos(\alpha-\phi)} \\
\Phi &= \exp(\frac{\tau-(1-\nu\theta)}{\epsilon(1+\theta)}).
\end{aligned} \qquad (2)$$

Eq. 1(a) is a statement of static equilibrium between the nondimensional compressive (σ) and shear stresses (η) applied by the tool on the chip and the nondimensional shear stresses (τ) developed parallel to the shear zone. μ is the coefficient of friction between the tool and chip. Eq. 1(b) is the energy balance equation for one-dimensional shear, where θ is the nondimensional temperature, and λ and q are nondimensional groups containing the thermal conductivity, k, heat capacity, c, density, ρ, and yield strength (in shear), τ_y, of the material. Φ_o is the nominal strain rate defined by the ratio of the nominal *shear speed*, V_s, over the uncut chip thickness, b. Eq. 1(c) states that the total nondimensional strain rate is equal to the plastic strain rate Φ. The exponential relationship defining Φ in Eq. 2 is a typical phenomenological constitutive law that specifies the irreversible plastic flow response of the material at high strain rates. It is based upon the assumption that plastic flow in metals is thermally activated and can be modeled by Arrhenius kinetics (Kocks et al., 1975). The coefficients ϵ and ν are the strain-rate sensitivity and thermal softening parameters of the material, respectively. Eq. 1(d) defines the rate of growth of normal contact stresses between the tool and chip material. Initially, for small σ, the rate of growth is proportional (through some appropriate elastic constant Λ) to the quotient of the local velocity difference between the tool and chip ($1-v$ in nondimensional variables) and some appropriate length scale, δ. The local compressive stresses saturate at the yield stress (which is 1 in nondimensional form) due to the exponential term. This equation provides a somewhat crude approximation of "local" elastic-plastic indentation of the workpiece material by the tool. These local stresses are connected to "global" shear stresses in the shear zone by Eq. 1(a). It is the interaction between this indentation and global shear, and the highly nonlinear thermoplastic behavior that can lead to relaxation-like oscillations in the system.

Eq. 1 were discretized using a simple finite-difference method and integrated using a commercial stiff equation solver. Based upon the best available information in the literature, the following parameter values were chosen for hardened steel: $k = 40W/(\mathrm{m}\cdot\deg\mathrm{C})$, $\tau_y = 1600$ MPa, $c = 0.46\mathrm{kJ}/(\mathrm{kg}\cdot\deg\mathrm{C})$, $\rho = 7800\mathrm{kg/m}^3$, $\theta_o = 300\mathrm{K}$, $\nu = 0.73$, $\epsilon = 0.0018$. The rake angle, shear angle and uncut chip thickness b were chosen to approximately match the experiment described above: $\alpha = -10\deg$, $b = 33\ \mu\mathrm{m}$ and $\phi \approx 37\deg$. The values of ν and ϵ were obtained from Burns (Burns, 1990) and the friction coefficient was estimated to be approximately 0.2. The value of $1/\Psi$ was chosen to be order of 10^2. This is based upon the reasonable assumptions that $\Lambda/\tau_y \approx 10^2$ and $b/\delta \approx 1$. Inserting these values into the Eq. 2 yields $\xi = 0.82$ and $q = 1.63$.

The results of the simulations are shown in Fig. 4. Fig. 4(a) corresponds to a wedge speed of 1.5 m/s ($\lambda = 0.16$). The top diagram shows the final chip and tool position. The lower part of Fig. 4(a) shows the temperature and shear stress at the tool tip as a function of time. For this case the shear is homogeneous and the stress and temperature approach a nonoscillatory steady state. Fig. 4(b) corresponds to a tool speed of 1.8 m/s ($\lambda = 0.13$). In this case the material deforms inhomogeneously with regions of localized shear. The dynamics of the temperature and shear stress at the tool tip have undergone a bifurcation to periodic oscillatory behavior. This is in close agreement with the experiments which indicate that a bifurcation occurs between 1 m/s and 1.5 m/s. Further increases in tool speed lead to a more rapid collapse of each shear zone. This causes higher amplitude oscillations in the shear stress and temperature and also leads to the formation of larger segments. The segment spacing was found to saturate with increased cutting speed, as observed in experiments (Davies et al., 1997(b)).

4. Simplified ODE Model: Evidence for a Hopf Bifurcation

Further evidence for a Hopf bifurcation can be obtained from an approximate ODE model for chip formation that is derived by treating the shear zone as a control volume. Fig. 5(a) shows the final velocity and thermal profiles obtained from the simulation shown in Fig. 4(a). The tool speed of $V = 1.5$ m/s is on the borderline of the bifurcation that occurs between $V = 1.5$ m/s and $V = 1.6$ m/s in the simulation. The approximate ODE

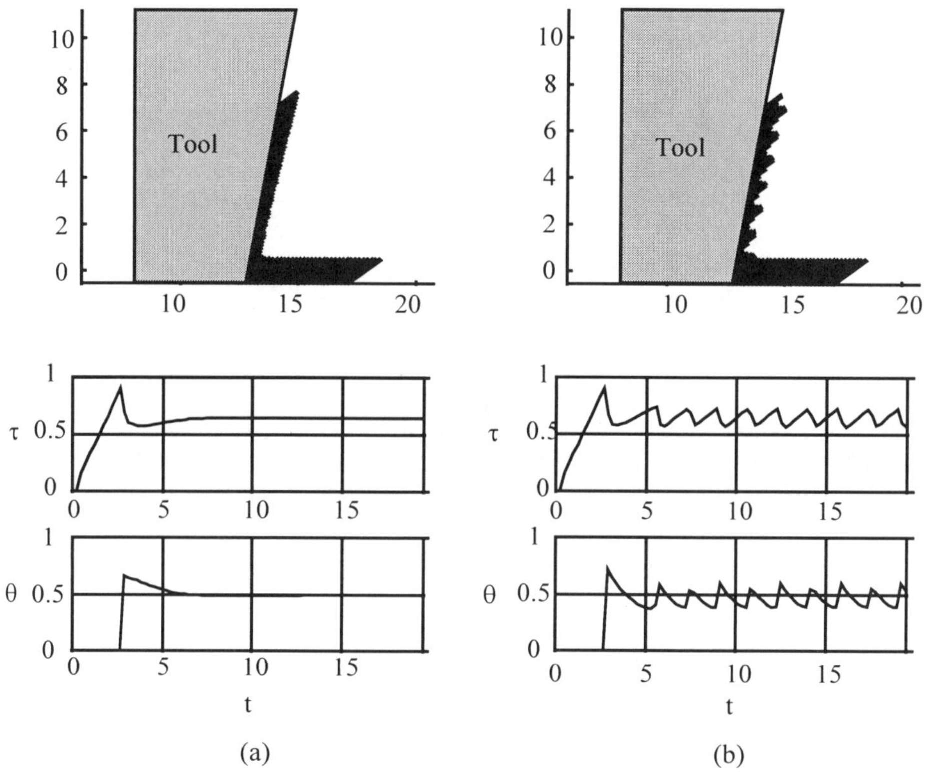

Figure 4. Example simulation results for (a)$\lambda = 0.156$ ($V = 1.5$ m/s) and (b) $\lambda = 0.13$ ($V = 1.8$ m/s).

model is derived by assuming: (1) the thermal profile in the shear zone is approximately piecewise linear, rising from ambient temperature $\tilde{\theta}_0$ to a temperature $\tilde{\theta}$ on exit; (2) the slope of the thermal profile at the entry approximated by $\frac{\tilde{\theta}-\tilde{\theta}_0}{h/2}$ and the slope at the exit is equal to zero; (3) the inactive portion of the chip is a rigid block that is loaded by the tool through linear springs, and this load produces a shear stress on the shear zone; and (4) the heat generation in the shear zone is given by $\tilde{\tau} V_{chip}$ where $V_{chip} = V_s * \Phi$ with Φ defined in Eq. 2. Using these assumptions it can be shown that the the nondimensional equations for the exit temperature and the shear stress on the shear zone have the form (Burns and Davies, 1997),

$$\begin{aligned} \frac{d\tau}{dt} &= 1 - \Phi \\ \frac{d\theta}{dt} &= -\kappa\theta + \gamma\tau\Phi \end{aligned} \tag{3}$$

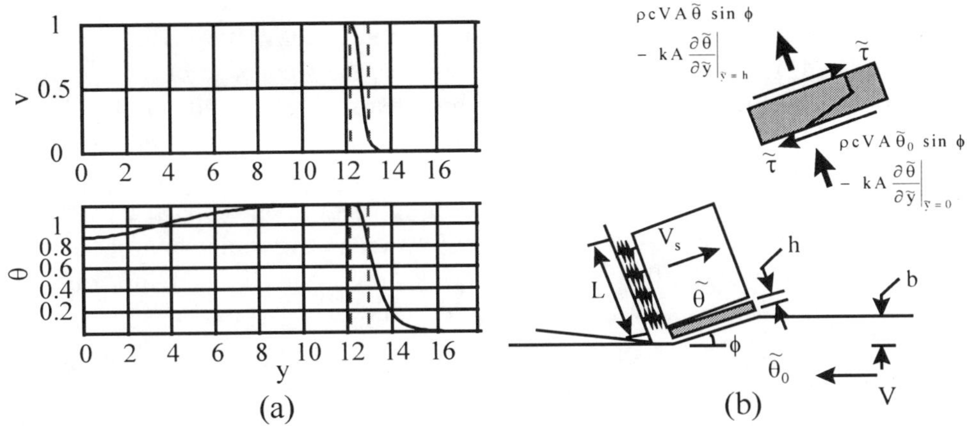

Figure 5. Approximation of the shear zone as a control volume with piecewise linear thermal profile as suggested by simulations. Tildes again represent dimensional quantities.

where θ and τ now represent the nondimensional exit temperature and shear stress on the shear zone respectively, κ and γ are nondimensional groups that can be related to those in Eq. 2 and Φ is defined as above. Eq. 3 are similar in form to those that describe the behavior of an open chemical reactor. The authors have shown that for parameter values corresponding to hardened steel, these equations will undergo a Hopf bifurcation as the cutting speed is increased (Burns and Davies, 1997). It is likely that in the future, this heuristic derivation of Eq. 3 can be made more mathematically rigorous.

5. Conclusions

We have presented two new models of chip formation in machining that strongly suggest that the transition from continuous to segmented chip formation is due to a Hopf bifurcation in the material flow. These models represent a significant improvement over previous analytical models that treat chip formation as a shear localization problem and thus contain no mechanism for oscillation. Our new models lead to an interpretation of metal cutting as a process of thermomechanical feedback with dynamics that are very similar to those of an open chemical reactor. Like the chemical reactor problem, continuous, non-oscillatory behavior occurs when the convective, conductive and heat generation terms are balanced so that the rate of change of temperature in the shear zone (or reactor) is zero. This is the case for low cutting speeds. At higher cutting speeds, the convective term grows relative to the conduction term, and this leads to oscillations. Each cycle of oscillation is characterized by two distinct phases. The first phase consists of a slow buildup of stress and a convective and conductive cooling of the shear zone, while the second phase consists of a sudden, explosive, thermoplastic deformation which leads to a relaxation of the shear stresses. The apparent nonperiodic behavior in the chip segments observed at speeds just above the critical speed for segmentation is still unexplained. However, the addition of a strain hardening term in the constitutive behavior of the approximate ODE model will increase the dimension of the state space to three and allow more complex (possibly chaotic) motions that could explain the observed nonperiodic behavior.

References

Burns, T. J. and Davies, M. A., A nonlinear dynamics model for chip segmentation in machining, *Phys. Rev. Lett.* **79** (1997),447–450.

Burns, T. J., A mechanism for shear band formation in the high strain-rate torsion test, *J. App. Mech.* **57** (1990), 836–844.

Clifton, R. J., Duffy, J., and Hartley, K. A., On critical conditions for shear band formation at high strain rates, *Scripta Metall.* **18** (1984), 443-448.

Davies, M. A., Chou, Y. H., and Evans, C. J. On chip morphology, tool wear and cutting mechanics in finish hard turning, *Ann. CIRP* **45** (1996), 77–82.

Davies, M. A., Burns, T. J., and Evans, C. J. On the dynamics of chip formation in machining hardened materials, to appear, *Ann. CIRP* **46** (1997).

Davies, M. A., Burns, T. J., and Evans, C. J. The dynamics of chip formation, in preparation.

Kocks, U. F., Argon , A. S. and Ashby, M. F., *Thermodynamics and Kinetics of Slip, Progress in Materials Science* Vol. 19, B. Chalmers, et al. (Eds.), Peragamon, Oxford, 1975.

Komanduri, R. and Brown, R. H, On the mechanics of chip segmentation in machining, *J. Eng. Ind.* **103** (1981), 33–51.

Komanduri, R., Schroeder J., von Turkovich, B. F., and Flom, D. G., On the catastrophic shear instability in high-speed machining of an AISI 4340 steel *J. Eng. Ind.* **104** (1982), 121–131.

Komanduri, R., and Schroeder, T., On shear instability in a nickel-iron base superalloy, *J. Eng. Ind.* **108** (1986), 93–100.

Konig, W., Klinger, M. and Link, R., Machining of hard materials with geometrically defined cutting edges, *Ann. CIRP* **39** (1990), 51–64.

Marusich, T. D. and Ortiz, M., A parametric finite element study of orthogonal high-speed machining, in *Proc. 2nd ECCOMASS 96 Conf. on Numerical Methods in Engineering, Paris, France, September 1996.*

Merchant, M. E., Mechanics of the metal cutting process. I. Orthogonal cutting and a type 2 chip *J. App. Phys.* **16** (1945), 267–275.

Molinari, A. and Dudzinski, C. R., Stationary shear band in high-speed machining, *C. R. Acad. Sci. Paris Ser. II* **315** (1992), 399–405.

Recht, R. F. Catastrophic thermoplastic shear *J. App. Mech.* **30** (1964), 384–390.

Shaw, M. E., *Metal Cutting Principles*,Oxford University Press,Oxford, 1984.

Shaw, M. C., Chip formation in the machining of hardened steel *Ann. CIRP* **42** (1993), 29–31.

Trent, E. M. *Metal Cutting, 3rd ed.*, Butterworth Heinemann Ltd, Oxford, 1991.

Wright, T. W., and Walter, J. W. On stress collapse in adiabatic shear bands *Mech. Phys. Sol.* **35** (1987), 701–720.

CHATTER IDENTIFICATION AND CONTROL FOR A BORING PROCESS

ALI H. NAYFEH AND JON R. PRATT
Department of Engineering Science and Mechanics
Virginia Polytechnic Institute and State University
Blacksburg, Virginia 24061-0219

Abstract. Previous work on machine-tool dynamics has identified three mechanisms that lead to the dynamic instability known as chatter, namely the regenerative effect, velocity-dependent friction or stick-slip, and mode-coupling. It has long been recognized that the underlying physics of these mechanisms is nonlinear, yet traditional treatments of the problem focus on linearized reduced-order approaches.

In this paper, we report our efforts to examine and control the complex dynamic response of a boring tool. We have designed and fabricated a boring bar system with sensors and actuators to aid our investigations. Experiments with this so-called "smart" tool reveal evidence of a subcritical transition from stable to unstable cutting. We examine a nonlinear machine-tool dynamics model from the literature that possesses a similar instability. Local analysis of this system suggests that the bifurcation is supercritical. However, the created small-amplitude limit cycle quickly loses stability via a cyclic-fold bifurcation. We find theoretically and experimentally that such a behavior can be controlled by enhancing the structural damping.

1. Introduction

The dynamic response of boring bars has long been recognized for its complexity. Salje [1] examined experimentally the problem of tools with two orthogonally oriented degrees of freedom and revealed rich dynamics. Kuchma [2] considered boring bars with non-circular cross-sections and found that the stability is strongly dependent on the orientation of the plane of symmetry with respect to the normal to the machined surface. Tlusty [3] developed the first simple model for mode-coupled chatter, whereas Tobias [4]

F. C. Moon (ed.),
IUTAM Symposium on New Applications of Nonlinear and Chaotic Dynamics in Mechanics, 193–202.

suggested that regeneration must also be considered in these systems. A variety of other models have been suggested and studied [5-7] that incorporate both effects.

In our experiments [8-10], we have observed jumps from static to dynamic cutting conditions like those reported by Hooke and Tobias [11], for single-point turning, and those observed and modeled by Hanna and Tobias [12], for face milling. Hanna and Tobias [12] used a single-degree-of-freedom model with both cutting-force and structural nonlinearities to account for the jump behavior. Shi and Tobias [13] reproduced the theoretical results of Hanna and Tobias [12] by dropping the structural nonlinearity and considering the effect of the tool leaving the cut, a condition known as "multiple regenerative chatter".

We consider a chatter control method for boring tools based on a Hanna and Tobias-like [12] model. We assume that the boring tool can be approximated by a single-degree-of-freedom system in a direction normal to the machined surface and that it is this mode that is largely responsible for the chip thickness modulations that give rise to chatter. We derive the linear stability boundary and show how it can be extended by increasing the structural damping. We discuss the tangential boundary to the lobed stability curve and demonstrate how it provides a conservative estimate of the tool stability regardless of the nonlinear effects. An active vibration-absorber approach for enhanced tool damping, similar to methods employed by previous investigators [14], is then analyzed.

We present experimental evidence, in the form of various chatter signatures, that regeneration and mode coupling are present in the uncontrolled tool. We then apply active vibration control in the tangential direction and show that, for very light chatter cuts, the normal direction appears to have a subcritical instability that can be quenched using the active vibration-absorber controller.

2. Control Strategy

We develop a control strategy for a simple orthogonal cutting condition. This condition is approximated in boring when the width of cut is approximately five times the feed and the tool nose radius is small [17]. Typically, one assumes that only tool motions directed normal to the machined surface modulate the chip [18] and then models the system using a single-degree-of-freedom approximation for the first lateral bending mode of the tool in this direction.

2.1. MACHINE-TOOL STABILITY

As a test case for our control, we investigate the stability and complicated dynamics of boring by using the following single-degree-of-freedom model [12]:

$$\ddot{x}+2\xi\dot{x}+p^2(x+\beta_2x^2+\beta_3x^3) = -p^2w\left[x - x_\tau + \alpha_2(x-x_\tau)^2 + \alpha_3(x-x_\tau)^3\right] \tag{1a}$$

where

$$x_\tau = x(t-\tau) \tag{1b}$$

Here, w is an effective width of cut, and the time delay τ is the period of one revolution of the workpiece. Our numerical analysis is based on the experimentally determined parameter values [12]: $p = 1088.56$ rad/sec, $\xi = 24792/\omega$ rad^2/sec, $\beta_2 = 479.3$ 1/in, $\beta_3 = 264500$ 1/in^2, $\alpha_2 = 5.668$ 1/in, and $\alpha_3 = -3715.2$ 1/in^2. The equivalent system is shown in Fig. 1.

The linearized form of Eq. (1a) can be written as

$$\ddot{x} + 2\xi\dot{x} + p^2x + p^2w(x - x_\tau) = 0 \tag{2}$$

To study the neutral stability, we substitute

$$x = a\cos\omega t \tag{3}$$

into Eq. (2), set each of the coefficients of $\cos\omega t$ and $\sin\omega t$ equal to zero, and obtain

$$p^2 - \omega^2 + p^2w(1-\cos\omega\tau) = 0 \tag{4}$$

$$2\xi\omega + p^2w\sin\omega\tau = 0 \tag{5}$$

We establish the linear stability boundary in Fig. 2 by solving Eqs. (4) and (5) for w, τ, and ω. It follows from Eqs. (4) and (5) that when $w < 2\xi\omega_c/p^2$, which is 0.0418, the cutting process remains stable.

The line $w = 0.0418$ on the stability plot is known as the tangential boundary to the lobed stability curve. This boundary takes on different shapes if we include the tool penetration rate and the effect of friction along the rake face. Tobias [4] showed that, as the cutting speed is increased, the tangent boundary asymptotically approaches this straight line boundary, even if one considers penetration-rate effects. Wu and Liu [19] explored both penetration-rate and friction effects and found that the penetration effect tends to have a stabilizing influence at low cutting speeds (i.e., an increased process damping effect), whereas friction along the rake face tends to have a destabilizing effect (i.e., a negative damping effect due to friction). Both effects are shown to diminish at high cutting speeds.

It appears from the lobed stability curve that by operating at certain speeds, one can achieve a substantial improvement in the limit width of cut beyond the tangential boundary. However, if one considers the nonlinear stability, this improvement seems questionable.

Consider the case where $\tau = 1/75$. Nayfeh, Chin, and Pratt [15] generated the bifurcation diagram shown in Fig. 3 by solving Eqs. (1) using a seven-term harmonic-balance solution while monitoring the stability using a combination of Floquet theory and Hill's determinant. We see that locally the transition from stable to unstable cutting occurs through a supercritical Hopf bifurcation. However, the created small limit cycle quickly looses stability via a cyclic fold. Thus, the global stability has a subcritical structure, as we confirmed using analog simulation [16]. As a consequence, the nonlinearity reduces the instability limit $w = w_c \approx 0.117072$ predicted by the linear theory to $w \approx 0.0431$, a reduction of about 63%. This limit width of cut is approximately that defined by the tangential boundary. Using the same techniques, Nayfeh, Chin, and Pratt [20] investigated the influence of the time delay on the subcritical instability. In Fig. 4 we repeat their results for the nonlinear response of the tool as a function of width of cut for the four time delays $\tau = 1/60$, $1/64.4$, $1/66$, and $1/75$. We see that large-amplitude solutions always coexist with the trivial solution for widths of cut between the tangential boundary and the limit width of cut predicted by linear stability analysis.

These results indicate that the simple straight-line tangent stability boundary provides a conservative target for designing chatter control because any control that increases this level is an improvement, regardless of the type of cutting forces present. Because this limit is determined wholly by the damping in the structure, our control task is simply that of enhancing the structural damping.

2.2. ACTIVE DAMPING FOR INCREASED STABILITY

If an actuator/sensor pair is available to provide collocated acceleration feedback, second-order compensation is completely analogous to a traditional vibration absorber [21], and the equations of motion for the feedback control system are

$$\ddot{x} + 2\xi\dot{x} + p^2 x = -p^2 w(x - x_\tau) + d(c_a\dot{y} + \omega_a^2 y) \tag{6}$$

$$\ddot{y} + c_a\dot{y} + \omega_a^2 y = -\ddot{x} \tag{7}$$

where y is a second-order compensator with damping c_a and frequency ω_a and d is a coupling gain.

We now consider the linear stability of the controlled system. We take the Laplace transform of Eqs. (6) and (7), assume zero initial conditions, and obtain

$$[s^2 + 2\xi s + p^2 + p^2 w(1 - e^{-s\tau})]X(s) = d(c_a s + \omega_a^2)Y(s) \tag{8}$$

$$(s^2 + 2c_a s + \omega_a^2)Y(s) = -s^2 X(s) \tag{9}$$

Solving for $Y(s)$ in terms of $X(s)$ yields

$$Y(s) = \frac{-s^2 X(s)}{s^2 + 2c_a s + \omega_a^2} \tag{10}$$

Substituting Eq. (10) into Eq. (8), we obtain the characteristic equation of the system as

$$\begin{aligned} s^4 + [2\xi + c_a(1+d)]s^3 + p^2[1 + w(1 - e^{-s\tau})] + 2\xi c_a + \omega_a^2(1+d)s^2 \\ + c_a p^2[1 + w(1 - e^{-s\tau})] + 2\xi\omega_a^2 s + \omega_a^2 p^2[1 + w(1 - e^{-s\tau})] = 0 \end{aligned} \tag{11}$$

We establish the linear stability boundary by substituting $s = i\omega$ into Eq. (11), separating real and imaginary parts, and solving the resulting pair of equations for w, τ, and ω. Choosing compensator values to give $d = 0.35$, $\omega_a/p = 0.7$, and $\zeta = 0.57$, we plot the limit width of cut with and without the absorber as a function of the time delay in Fig. 5. Clearly, the absorber has substantially increased the limit width of cut.

Next, we consider the nonlinear stability. In Fig. 6 we compare the bifurcation diagram of the controlled system for the case $1/\tau = 75$ obtained by the method of harmonic balance with that obtained via analog simulations. The agreement is excellent and suggests that the global subcritical instability is suppressed in the controlled system.

3. Cutting Tests

A series of experiments are performed on a two-axes engine lathe to obtain chatter signatures while using a boring tool with and without vibration control. Time series, autospectra, cross spectra, and coherence are employed to examine the chatter signatures.

3.1. SETUP

The tool is a 12" long and a 1" diameter boring bar with an ISO TNMG 160304 insert. The side rake angle is $-5°$, the back rake angle is $-12°$, and the lead angle is $0°$. The bar is tapped with a series of 10/32 holes to accommodate the attachment of actuators through flexural elements known as "stingers". Magnetostrictive actuators oriented in the normal and tangential direction to the machined surface sting the boring bar at a point $1.75''$ from the tool holder while the motion is sensed in each direction using accelerometers mounted at a point 3″ from the tool holder. The tangentially oriented actuator stud mounts directly to the bar. The normally oriented actuator is attached using a tensioned wire stinger. The arrangement is much stiffer in the tangential than the normal direction. Workpieces are aluminum tubes of nominal inner diameter 4.50″ – 5.25″, outer diameter 6.1″ , and length 5″. Further details of the hardware and compensator design and implementation can be found in [8-10].

3.2. CHATTER SIGNATURES

We investigate boring bar chatter for light, finish style cuts at the minimum feedrate of the lathe, which is 0.0024″ per revolution (ipr). Chatter signatures are obtained for a variety of spindle speeds at nominal widths of cut of $w_0 = 0.002''$ and $w_0 = 0.03''$.

In Fig. 7 the amplitude of the fundamental harmonic of the tool's acceleration tangent to the machined surface is plotted as a function of the workpiece rotational speed in revolutions per minute (rpm) for the case when $w_0 = 0.03''$. We find, as did Salje [1], that the modal characteristics of the tool system vary greatly depending on the clamping conditions. For the data in this plot, the first lateral vibration modes have frequencies and dampings of $f_t = 524$ Hz and $\zeta_t \approx 0.03$ and $f_n = 365$ Hz and $\zeta_n \approx 0.02$, respectively, as determined by impact testing and circle fits of the resulting frequency-response functions.

Two observations can be made from the results in Fig. 7. First, the presence of stable cutting for all workpiece rotational speeds below 165 rpm indicates that the cutting forces are dependent on the mean cutting velocity and contribute a positive damping effect at low speeds, in agreement with Tlusty's results [18] for aluminum. Second, a subcritical-type instability occurs between 165 and 195 rpm, where both chatter and stable cutting conditions coexist.

Next, we consider the time histories and autospectra of tangential and normal accelerations as well as the cross spectra and coherence between these signals for some typical chatter signatures obtained when the tool frequencies and dampings are $f_t = 491$ Hz and $\zeta_t \approx 0.025$ and $f_n = 364$ Hz and $\zeta_n \approx 0.02$. In this series of tests, the tool tends to chatter more easily, and the effect of positive damping due to the cutting process seems to be much less.

Chatter is observed at spindle rotational speeds as low as 90 rpm and for widths of cut as small as 0.002″, as seen in Fig. 8. We note that for the shallow cut of 0.002″, chatter tends to be localized in a direction normal to the machined surface, whereas for a deeper cut of 0.03″, chatter is oriented more along a line tangent to the machined surface, as is evident in Fig. 9. In both cases, the fundamental harmonic of the motion is seen to be linearly coherent, suggesting that mode coupling is present.

3.3. CONTROLLED RESPONSES

In Fig. 10, we show an example of a jump-type phenomenon in the time record of accelerations normal to the machined surface when only the tangential control loop is active. Initially, the control is only active in the tangential direction. The cut appears to be stable. A disturbance is introduced by hitting the bar in the normal direction with a hammer. Clearly, the disturbance grows due to regeneration. After the chatter reaches steady state, control is added in the normal direction, thereby stabilizing the system. Another disturbance is applied but the control maintains stability. In Fig. 11, we show the time traces, autospectra, cross-spectrum, and coherence of the normal and tangential accelerations during steady-state chatter of this type. We see that though chatter is predominantly oriented in the normal direction, a substantial tangential component exists.

Finally, we demonstrate biaxial control in Fig. 12. Large-amplitude chatter is allowed to develop before applying feedback control in both directions that stabilizes the system. We observe a dramatic improvement in the surface finish.

4. Conclusions

A nonlinear machine-tool dynamics model has been applied to boring-bar chatter analysis and its implications for vibration control has been examined. We find that global subcritical-type instabilities due to cutting-force and structural nonlinearities reduce the limit width of cut expected based on local stability analysis to that of the tangential stability boundary. By increasing the structural damping, we show that it is possible to both raise the tangential stability boundary and suppress the subcritical behavior. Experimental evidence of a jump phenomenon in boring bars is presented and controlled via a relatively simple feedback strategy. Chatter signatures, in the forms of time traces, auto- and cross-spectra, and coherence, indicate coupled motions that are essentially periodic. We speculate that a model for boring-bar chatter may be fabricated by using a pair of Hanna and Tobias-like [12] systems that are coupled through the cutting force. Preliminary investigations of such a system appear to be promising and will be reported on in subsequent work.

Acknowledgment

This work was supported by the Army Research Office under Grant No. DAAH04-96-1-0021.

References

1. Salje, E. (1956) Self-Excited Vibrations of Systems with Two Degrees of Freedom, *Trans. ASME*, Vol. 78, pp. 737-748.
2. Kuchma, L. K. (1957) Boring Bars with Improved Resistance to Vibration, *Eng. Digest*, Vol. 18, No. 2, pp. 68-70.
3. Koeingsberger, F., and Tlusty, J. (1970) *Machine Tool Structures*, Vol. 1, Pergamon Press, New Jersey.
4. Tobias, S. A. (1965) *Machine Tool Vibration*, Wiley, New York.
5. Grabec, I. (1988) Chaotic Dynamics of the Cutting Process, *Int. J. Mach. Tools Manuf.*, Vol. 28, pp. 19-32.
6. Wu, D. W., and Liu, C. R. (1985) An Analytical Model of Cutting Dynamics. Part 1: Model Building, *ASME J. Eng. Indust.*, Vol. 107, pp. 107-111.
7. Lin, J. S., and Weng, C. I. (1990) A Nonlinear Dynamic Model of Cutting, *Int. J. Mach. Tools Manuf.*, Vol. 30, pp. 53-64.
8. Pratt, J. R., and Nayfeh, A. H. (1997) Active Vibration Control for Chatter Suppression, in *Proc. of the 38th AIAA Structures, Structural Dynamics, and Materials Conf.*, Kissimmee, FL, AIAA-97-1210, p. 613.
9. Pratt, J. R., and Nayfeh, A. H., (1997) Boring Bar Chatter Control Using A Two-Axes Active Vibration Absorber Scheme, in *Proc. of NoiseCon97*, Book 2, p. 313.
10. Pratt, J. R., and Nayfeh, A. H., (1997) Experimental System Identification and Active Vibration Control of a Smart Machine Tool, to appear in *Proc. of the Eleventh Symposium on Structural Dynamics and Controls*, Blacksburg, VA.
11. Hooke, C. J., and Tobias, S. A. (1964) Finite Amplitude Instability-A New Type of Chatter, in *Proc. of the 4th International M.T.D.R. Conf.*, Manchester, pp. 97-109.
12. Hanna, N. H., and Tobias, S. A., (1974) A Theory of Nonlinear Regenerative Chatter, ASME J. Eng. Indust., Vol. 96, pp. 247-255.
13. Shi, M. and Tobias, S. A. (1984) Theory of Finite Amplitude Machine Tool Instability, *Int. J. Mach. Tool Design Res.*, Vol. 24, pp. 45-69.
14. Tewani, S. G., Rouch, K. E., and Walcott, B. L. (1995) A Study of Cutting Process Stability of a Boring Bar with Active Dynamic Absorber, *Int. J. Mach. Tools Manfu.*, Vol. 35, No. 1, pp. 91-108.
15. Nayfeh, A. H., Chin, C.-M., and Pratt, J. (1997) Perturbation Methods in Nonlinear Dynamics-Applications to Machining Dynamics, to appear *ASME J. Manuf. Sci. Eng.*.
16. Pratt, J., and Nayfeh, A. H. (1996) Experimental Stability of a Nonlinear Time-Delay System,"in *Proc. of the 37th AIAA Structures, Structural Dynamics, and Materials Conf.*, Salt Lake City, UT.
17. Shaw, M. C. (1984) *Metal Cutting Principles*, Oxford University Press, New York.
18. Tlusty, J. (1985) Machine Dynamics, in R. I. King (ed.), *Handbook of High Speed Machining Technology King*, Chapman and Hall, New York, pp. 48-153.
19. Wu, D. W., and Liu, C. R. (1985) An Analytical Model of Cutting Dynamics. Part 2: Verification, *ASME J. Eng. Indust.*, Vol. 107, pp. 107-118.
20. Nayfeh, A. H., Chin, C.-M., and Pratt, J. (in press) Applications of Perturbation Methods to Tool Chatter Dynamics, F. C. Moon (ed.), *Dynamics and Chaos in Manufacturing Processes*, Wiley, New York.
21. Juang, J.-N., and Phan, M. (1995) Robust Controller Designs for Second-Order Dynamic Systems: A Virtual Passive Approach, *J. Guid., Cont., Dyn.*, Vol. 15, No. 5, pp. 1192-1198.

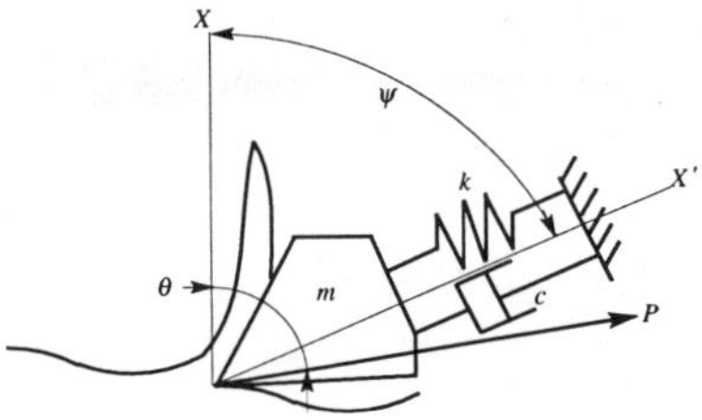

Figure 1. Single-degree-of-freedom tool.

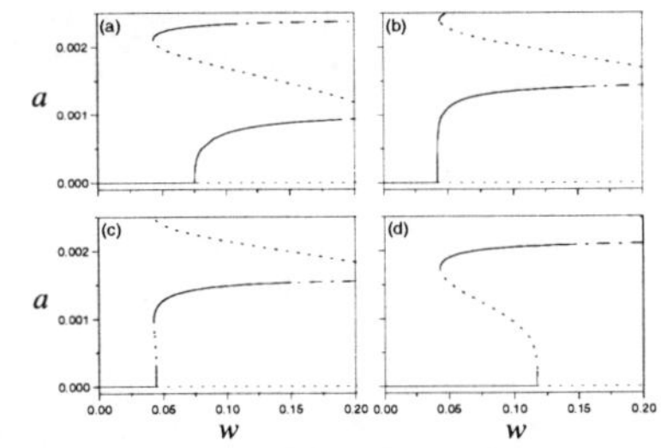

Figure 4. Nonlinear response curves when (a) τ=1/60, (b) τ=1/64.4, (c) τ=1/66, and (d) τ=1/75. Solid lines denote stable and trivial solutions, dotted lines denote unstable trivial or periodic solutions, and dotted-dashed lines denote the region of quasiperiodic and chaotic solutions.

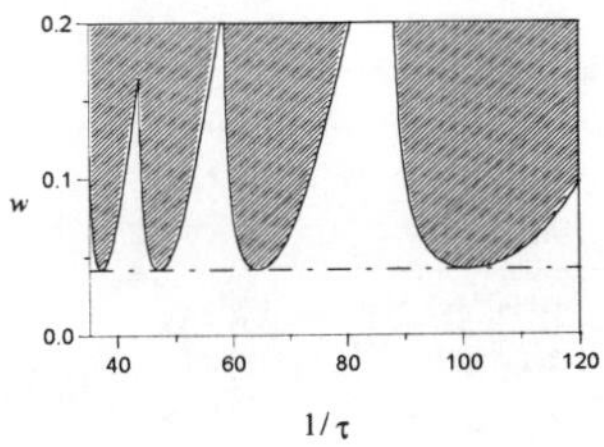

Figure 2. Stability lobes for the model of Hanna and Tobias.

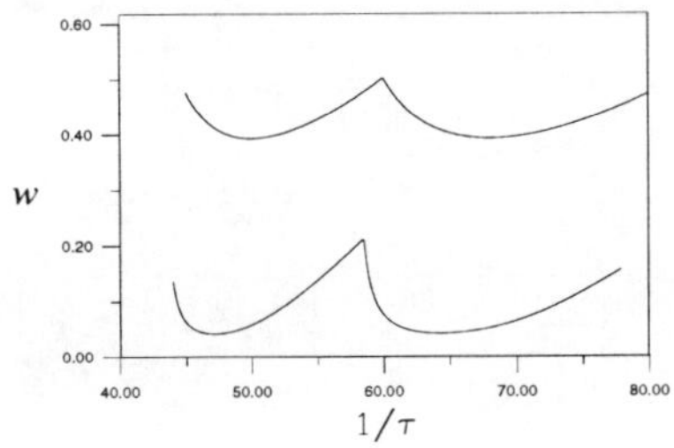

Figure 5. Stability diagram showing the critical width of cut as a function of the rotational frequency. Upper curve corresponds to the controlled system and the lower curve corresponds to the uncontrolled system.

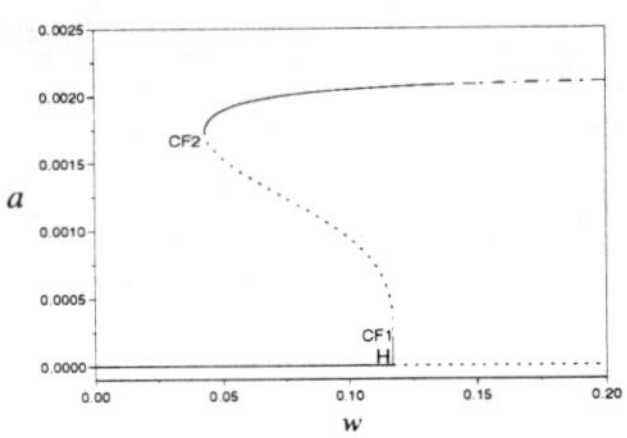

Figure 3. Bifurcation diagram showing the stability as a function of width of cut for a fixed time delay τ=1/75.

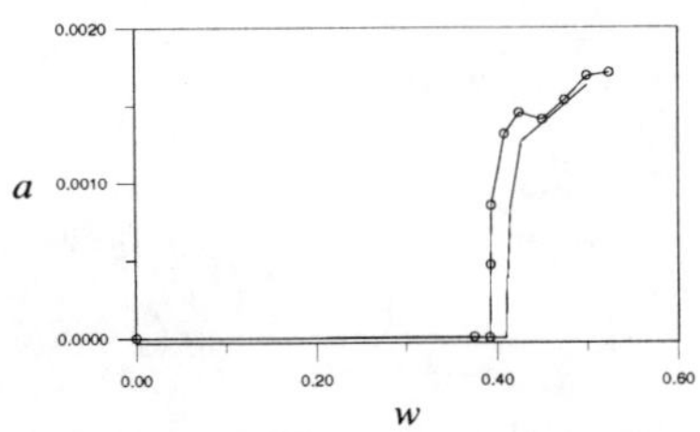

Figure 6. Bifurcation diagram showing the stability as a function of width of cut for τ = 1/75 after addition of an absorber.

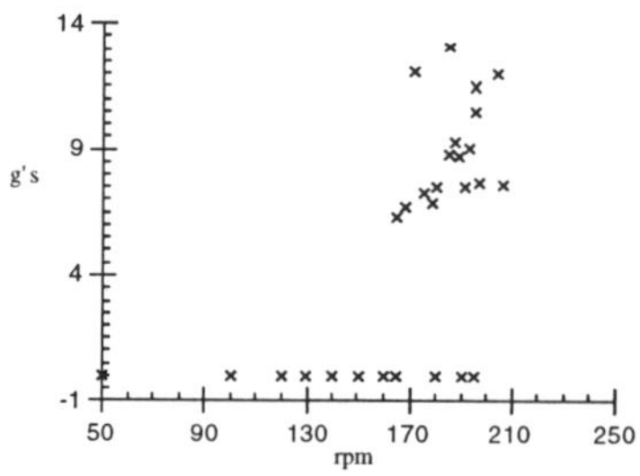

Figure 7. Bifurcation diagram showing the amplitude of the fundamental harmonic of chatter for various spindle speeds when $w = 0.03''$ and feedrate $= 0.0024$ ipr.

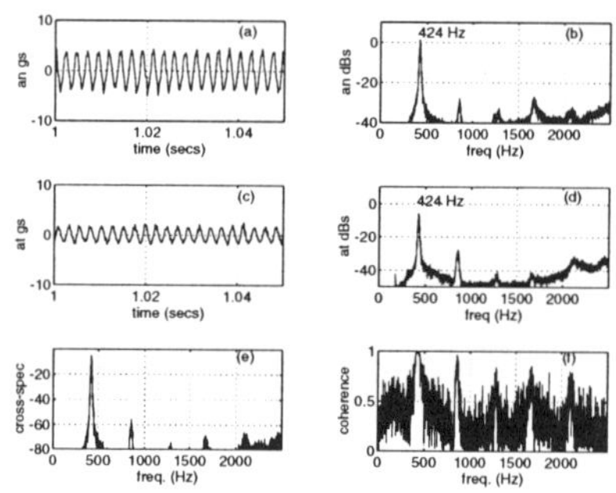

Figure 8. Time traces, autospectra, cross-spectrum, and coherence when $w = 0.002''$, speed $= 90$ rpm, and feedrate $= 0.0024$ ipr.

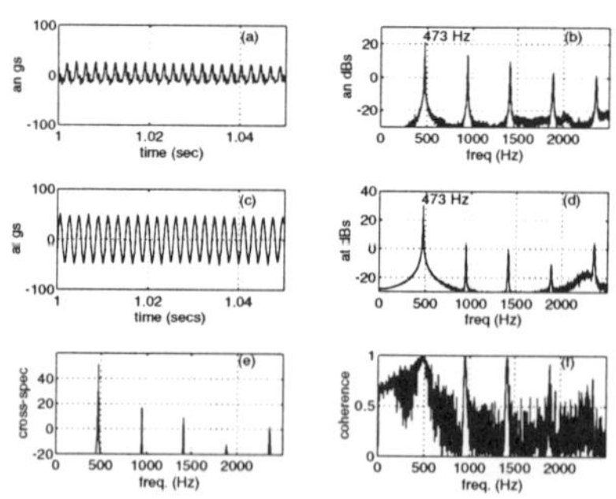

Figure 9. Time traces, autospectra, cross-spectra, and coherence when $w = 0.03''$, speed $= 210$ rpm, and feedrate 0.0024 ipr.

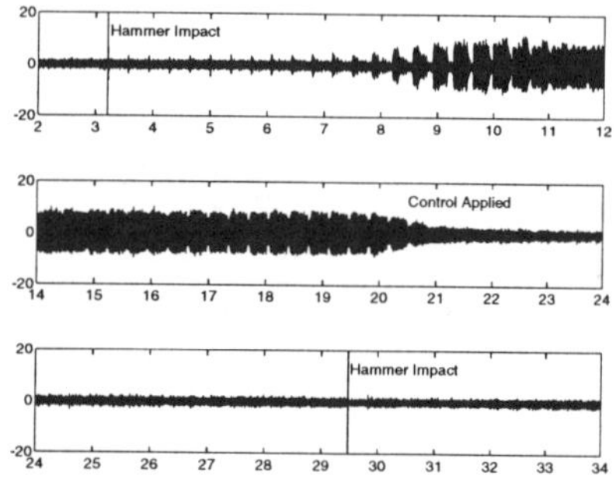

Figure 10. Time trace of tool accelerations normal to the machined surface; initially, only tangential control is active; then, in addition, control is applied in the normal direction to quench the chatter and eliminate the jump-type instability; $w = 0.002''$, speed $= 170$ rpm, and feedrate $= 0.0024$ ipr.

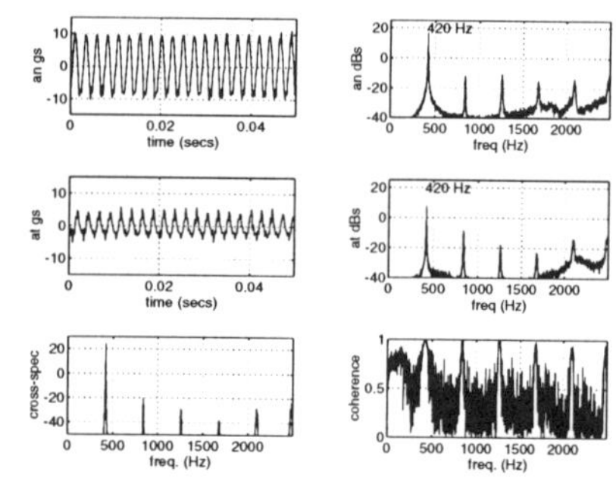

Figure 11. Time traces, autospectra, cross-spectrum, and coherence when $w = 0.002''$, speed $= 180$ rpm, and feedrate $= 0.0024$ ipr.

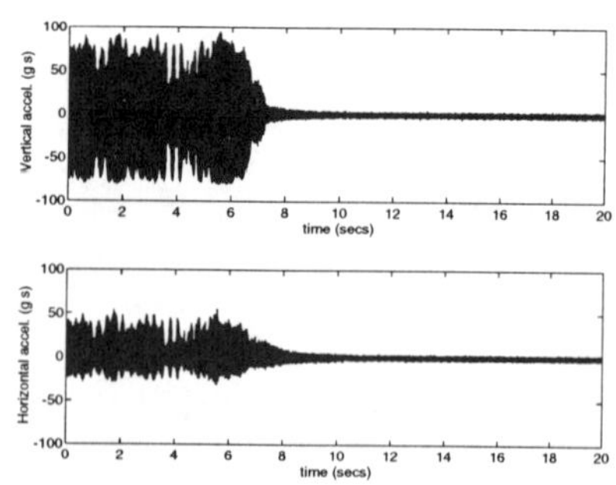

Figure 12. Time trace of normal and tangential tool accelerations; $w = 0.03''$, speed $= 210$ rpm, and feedrate $= 0.0024$ ipr.

COUPLING AND RESONANCE PHENOMENA IN DYNAMIC SYSTEMS WITH HYSTERESIS

F. VESTRONI
Dipartimento di Ingegneria Strutturale e Geotecnica.
University of Rome La Sapienza. Via Eudossiana 18,
00184 Roma; Italy

AND

D. CAPECCHI
Dipartimento di Scienza delle Costruzioni.
University of Naples Federico II. Piazzale V. Tecchio 80,
80125 Napoli; Italy

Abstract. The equations governing the response of hysteretic systems to sinusoidal forces are given as a vector field over an enlarged phase space. Hysteretic Masing and Bouc-Wen models joined to nonlinear elastic models are considered. Depending on the model characteristics it is possible to reduce the dimension of the phase space and to implement efficient algorithms. The paper presents results for one dof systems and for two dof systems. The capacity of the model energy dissipation makes the dynamics more or less complex. The coupling produces a strong modification of the response over a large frequency range with new resonance peaks and bifurcation of quasi–periodic motions.

1. Introduction

In recent years many papers have been published on hysteretic systems [1-17]. However, only one dof oscillators are relatively well studied; because of the non-monodromy of relations, the vector space of velocity and displacement is no longer a proper phase space and tools of nonlinear dynamics cannot be used. It has been shown however that, by introducing a certain number of internal variables, a rigorous analysis of the oscillator dynamics

F. C. Moon (ed.),
IUTAM Symposium on New Applications of Nonlinear and Chaotic Dynamics in Mechanics, 203–212.

lators, labelled as Ramberg-Osgood type, which includes the elastoplastic case, the conjecture is validated that the asymptotic response is only made up of $1T$ periodic symmetric response, stable for any values of the external excitation parameters. For more general hysteretic oscillators a richer behaviour has been found. In [4, 9], with the use of the harmonic balancing, oscillators which exhibit coexisting resonant and nonresonant periodic solutions are analyzed; in [1] a hysteretic oscillator is studied that shows quasi–periodic and perhaps chaotic motions. In [14] an oscillator based on a plastic hinge and nonlinear geometric effects exhibits chaotic motions. Subharmonic orbits of various periodicity are found in [17] for the Bouc-Wen oscillator and in [3] for an oscillator based on Masing rules.

The response of multi dof systems is much less known. Only a few old papers cover hysteretic behaviour and all of them are based on the approximate harmonic balancing [10, 11]. Recent papers [5, 6] study the behaviour of a two dof hysteretic systems system, based on elastoplastic and Masing models which exhibit coupling phenomena not observed for nonlinear elastic systems.

In this paper the response is reported of one and two dof hysteretic systems. The oscillators studied are mainly those based on Masing rules, but reference is also made to oscillators of Bouc-Wen type, joined to an elastic nonlinear model. After the introduction of the vector field that governs the dynamics of the oscillator, which operates over a high dimension manifold, some arguments are introduced that show how the asymptotic motion can be studied over a reduced dimension submanifold where a cheap systematic numerical study is developed, by varying all the control parameters, using concepts derived from the Poincaré map.

2. General Aspects

2.1. HYSTERETIC SYSTEMS

Pure hysteretic systems have the forms:

$$\begin{array}{rl} \text{Canonical} & \text{Noncanonical} \\ z = h(x,\overline{\kappa}) & \dot{z} = h(z,\kappa,\mathrm{sign}(v))v \\ \dot{\overline{\kappa}} = \chi(\overline{\kappa},\mathrm{sign}(v))v & \dot{\kappa} = \chi(\kappa,\mathrm{sign}(v))v \end{array} \tag{1}$$

where z is the hysteretic force, x the displacement, $v = \dot{x}$, $\overline{\kappa}$ and κ are vectors of internal state variables. The formulation in the left hand side, called canonical, is common in the field of continuum mechanics where a thermodynamically consistent law is sought; the formulation in the right side, called non-canonical, is common in the theory of dynamic systems.

Among the hysteretic systems those based on Masing rules are very important in structural mechanics; they are characterized by a skeleton curve around which the hysteretic cycles develop. Since their correct definition requires the introduction of an infinite dimensional phase space, very often approximate implementations are used. One possibility is to use a multilinear discretization of the skeleton curve; there are as many internal variables as linear segments on the skeleton. Another possibility, suggested by Jennings [12] and followed in the paper, consists to assume as internal variables (κ_1, κ_2) the maximum value of z and the value of z at the last unloading point respectively. When the skeleton curve becomes bilinear the Masing model furnishes the elastoplastic kinematic hardening model and the perfectly elastoplastic model as particular cases.

The perfectly plastic model can also be seen as particular case of another important class of hysteretic systems, very used in random vibration analysis. It is the class of Bouc-Wen endocronic models, defined in nondimentional form by the noncanonical expression:

$$\dot{z} = [1 - \beta|z|^n \text{sign}(z)\text{sign}(v)]v \tag{2}$$

For $n = \infty$ the elastoplastic model is obtained. Though the Bouc-Wen model is thermodynamically consistent, it presents very low dissipation about $z = 0$ and for this reason it can violate the Druker stability criterion.

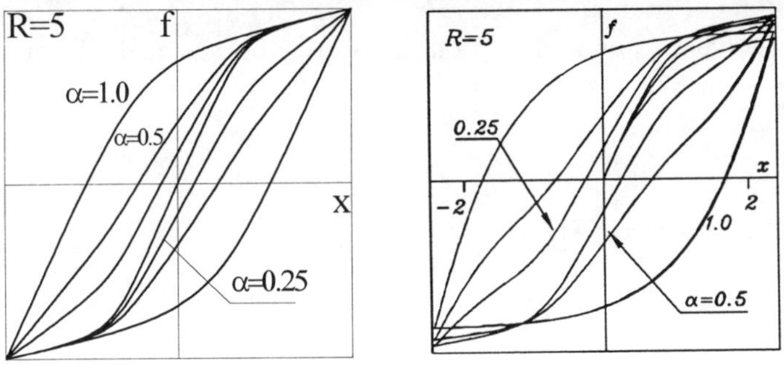

Figure 1. Cycles for different degree of hysteresis

Decomposable models have the form:

$$f = \alpha z + (1 - \alpha)g(x) \tag{3}$$

where α is the degree of hysteresis. A satisfactory class of models is obtained when z is a Masing model and $g(x)$ is its skeleton curve. The choice presents an internal coherence because the elastic component is not arbitrary, but it is directly connected to the pure hysteretic component and allows to obtain models with fixed yielding limit and stiffness while the dissipation of energy in a closed loop can be varied continuously with α. This is no longer true for example for the union between a Bouc-Wen model and a nonlinear elastic model [17]. Figure 1 shows cycles of Masing and Bouc–Wen models for different hysteresis degree α. In the following decomposable models are analyzed, they will be denoted by the name of the corresponding pure hysteretic model.

2.2. VECTOR FIELDS

Reference is here only made to one-dof oscillators subjected to a sinusoidal external excitation of the form $F = \gamma \cos \omega t$; developed concepts can straightforwardly be extended to many-dof systems. With reference to the non-canonical representations, the vector fields associated with the oscillators based on Masing models and Bouc-Wen models are respectively:

MASING

$$\begin{aligned} \dot{x} &= v \\ \dot{v} &= -[\alpha z + (1-\alpha) g(x)] + \gamma \cos \theta \\ \dot{z} &= h(z, \operatorname{sign}(v), \kappa_1, \kappa_2) v \\ \dot{\kappa}_i &= \chi(z, x, \operatorname{sign}(v), \kappa_1, \kappa_2) v \quad i = 1,2 \\ \dot{\theta} &= \omega \end{aligned}$$

BOUC-WEN

$$\begin{aligned} \dot{x} &= v \\ \dot{v} &= -[\alpha z + (1-\alpha) g(x)] + \gamma \cos \theta \\ \dot{z} &= h(z, \operatorname{sign}(v)) v \\ \dot{\theta} &= \omega \end{aligned}$$

The phase space dimension is six, $R^5 \times S$, in the first case and four, $R^3 \times S$, for the second case. For a n-dof chain oscillator the full dimension of the phase space is $5n + 1$ for the first case and $3n + 1$ for the second.

3. Methods

3.1. PERIODIC RESPONSE

Common methods are based on the harmonic balancing and the fixed points of the stroboscopic Poincaré map.

3.1.1. *One dof systems*

With the harmonic balancing the leading unknowns are the Fourier series coefficients of displacement histories. For the Masing oscillators where the force can directly be expressed as function of displacement, an easier algebraic problem comes out.

With the fixed point approach, for the Bouc-Wen oscillators, periodic solutions are found as fixed points of the stroboscopic Poicaré map in the three-dimensional phase space $M \equiv \{p = (x, v, z)\}$. For the Masing oscillators, which have a larger full phase space, things go better; it is shown that all possible asymptotic motions can be obtained assuming initial conditions on the submanifold M_0 of $M \equiv \{p = (x, v, z, \kappa_1, \kappa_2)\}$:

$$M_0 \equiv \{p \in M : p = (x_0, 0, g(x_0), g(x_0), g(x_0), \theta_0), \qquad x_0 \in R, \theta_0 \in S\}$$

In principle a map Q_0 can be defined over M_0 as the map which carries the point (θ_0, x_0), corresponding to an unloading from the skeleton at θ_0, to the point (θ_1, x_1) at θ_1. Though the fixed points of such a map furnish periodic solutions, the most convenient numerical approach consists in the evaluation of the fixed points for the phase flux ϕ_T over a period on the submanifold M_0 [3, 6], accordingly to the equation:

$$\phi_T(x_0, 0, g(x_0), g(x_0), g(x_0), \theta_0) = (x_0, 0, g(x_0), g(x_0), g(x_0), \theta_0) \qquad (4)$$

where T is the period of the sought orbit. In this equation the number of equations is redundant so that it is solved with a least square approach. Stability is determined by the eigenvalues of the map tangent to ϕ_T.

3.1.2. *Two dof systems*
With the harmonic balancing, the unknowns are the Fourier series coefficients of the displacement histories of the two masses, then the number of unknowns is twice as previously.

Also with the fixed point approach the number of unknown is twice that of one dof oscillator. For the Bouc-Wen oscillators, fixed points of the stroboscopic Poicaré map are searched in the six dimensional phase space $M \equiv \{p = (x_1, v_1, z_1, x_2, v_2, z_2)\}$. For the Masing oscillators, the periodic motion can in principle be studied over a submanifold M_1 with dimension four, twice that of M_0. To introduce this manifold it is useful the definition of the vectors: $q = (x, v, z, k_1, k_2)$, $q^0 = (x_0, 0, x_0, x_0, x_0)$. The state variable associated to the i-*th* dof will be consistently denoted by q_i, q_i^0 and the generic point of the phase space is $p = (q_1, q_2, \theta)$. The whole phase space is defined as $M \equiv \{p = (q_1, q_2, \theta), \quad q_1, q_2 \in R^5, \theta \in S\}$ and the four dimensional submanifold M_1 is:

$$M_1 \equiv \{p \in M : p = (q_1^0, q_2, \theta_1^0) \text{ such that } \phi_{\theta_2 - \theta_1}(p) = (q_1, q_2^0, \theta_2), \; \theta_2 \in R\}$$

On the submanifold M_1 it is possible to define a map Q_1 corresponding to Q_0; this map is however very complex to build, because for given q_1^0, q_2^0, θ_1 and θ_2 an inverse problem must be solved to find q_2. From an algorithmic point of view it was felt more convenient to search for the fixed points of the phase flux over a submanifold M_2, of larger dimension, seven instead of four, defined by:

$$M_2 \equiv \{p \in M : p = (q_1^0, q_2, \theta); \quad q_1^0 \in R, q_2 \in R^5, \theta \in S\}$$

Periodic motions are solutions of the algebraic fixed point problem:

$$\phi_T(q_1^0, q_2, \theta) = (q_1^0, q_2, \theta) \tag{5}$$

which is still redundant and has been solved with a least square approach. Stability is determined by the eigenvalues of the map tangent to ϕ_T.

3.2. NONPERIODIC RESPONSE

For one dof Masing oscillators all possible asymptotic motions can be obtained with the aid of attracting basins on the cylinder (x_0, θ_0) isomorphic to M_0. The approach is still practicable though the computation time for an exhaustive study becomes very large. For the Bouc-Wen one-dof oscillator and for the two dof oscillators in general, the evaluation of attracting basins is not practicable because of the large dimension of the phase space. A possible procedure to search for nonperiodic motions, consists in evaluating the frequency-response curves, or alternatively the characteristic curves, for the $1T$ periodic solution and, during continuation, to look for the bifurcation points around which a step-by-step numerical analysis is carried out to investigate the nature of the bifurcated solutions.

4. Phenomena in the asymptotic response

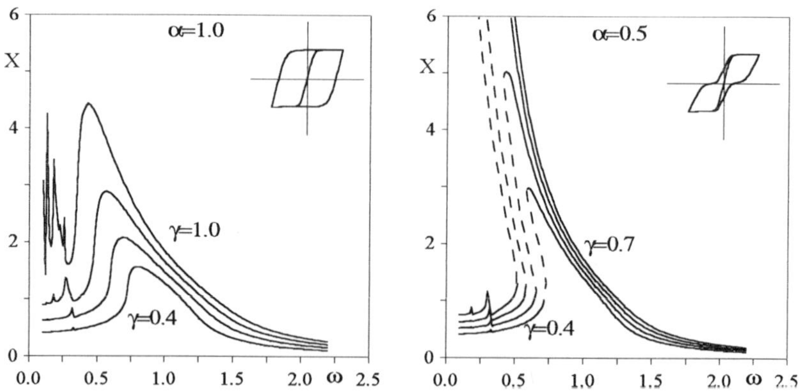

Figure 2. Frequency response curves of 1dof hysteretic oscillator

Nonlinear dynamic phenomena are strongly dependent on the characteristics of the shape of the oscillator restoring force. Hysteretic models with different hysteresis capacity, governed by the parameter α, without or with positive small hardening, are considered in the following. It will be shown that most of the various phenomena exhibited by the elastic models are not found in many cases.

For the 1dof systems, the always singlevalued frequency response curves, obtained in [12] using bilinear and Ramberg–Osgood models, diffused the idea that hysteresis makes jump phenomenon to disappear. This result is however only connected with those specific models; actually multivalued f.r.c.'s characterize a large class of hysteretic oscillators [4]. Restricting the attention to the Masing models, the results given in Figure 2 allow for a synthetic description of 1dof systems. They show a softening behavior, with a resonance frequency which tends to that defined by the post elastic stiffness, and an unbounded response. The resonance peak is large and the response is very rich of superharmonics in a wide range of low frequencies. The asymptotic response is always periodic; for models with large hysteresis ($\alpha = 1$) only $1T$ periodic solutions occur (Figure 2a), while for small hysteresis ($\alpha < 1$) regions of coexisting solutions are found, $1T$ (Figure 2b) near the primary resonance and $1T$, $5T$, $7T$ in the subharmonic range.

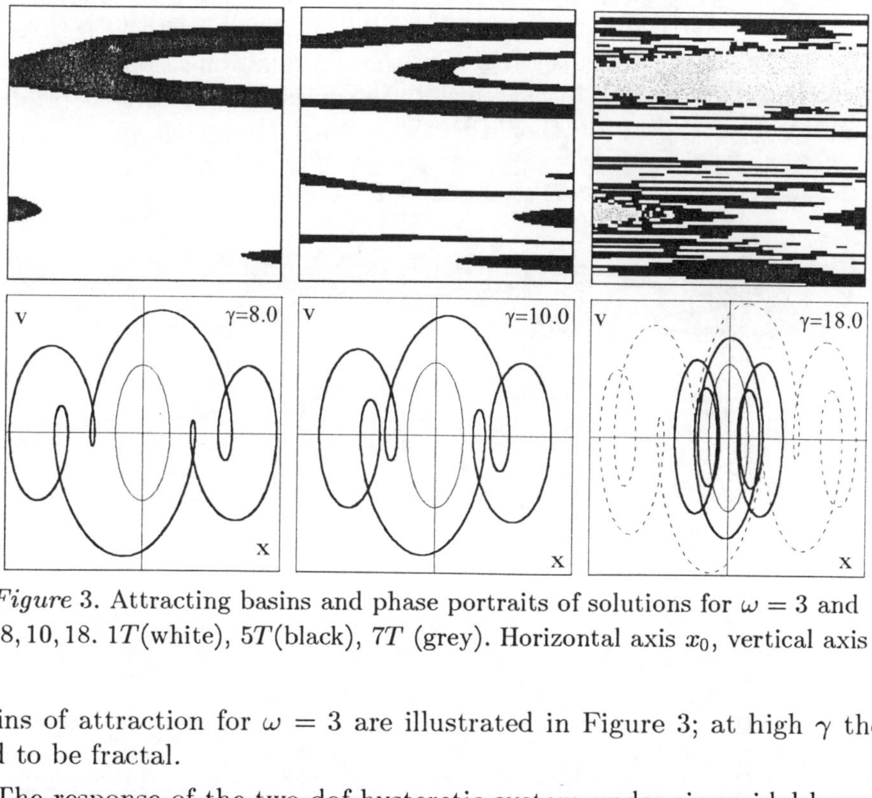

Figure 3. Attracting basins and phase portraits of solutions for $\omega = 3$ and $\gamma = 8, 10, 18$. $1T$(white), $5T$(black), $7T$ (grey). Horizontal axis x_0, vertical axis θ_0

Basins of attraction for $\omega = 3$ are illustrated in Figure 3; at high γ they tend to be fractal.

The response of the two dof hysteretic system under sinusoidal boundary acceleration is in general found to be periodic. Since the ratio of the small amplitude oscillation frequencies is three, there is an internal resonance conditions [6]. The frequency response curves (f.r.c.'s) of the relative velocities of the masses for model with complete hysteresis are single valued with a high number of superharmonics (Figure 4).

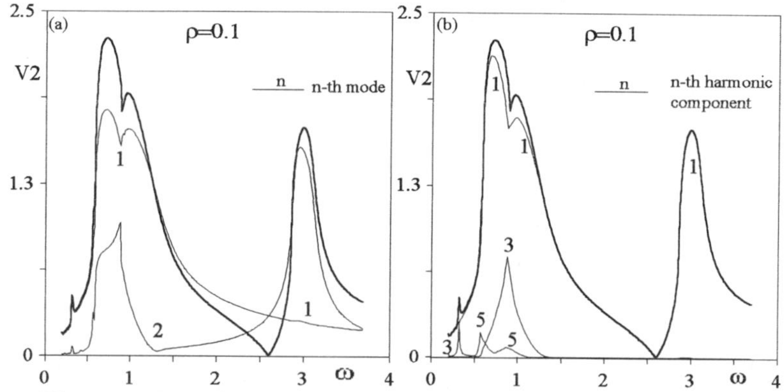

Figure 4. F.r.c.'s of velocity, modal and harmonic components.

The mode interaction is such that only around the first resonance both modes are involved, differently from nonlinear elastic systems[18]. A new resonance peak appears in the region of the combination resonance which is due to the second mode; by increasing the excitation intensity it produces a large resonance area together with the second mode peak [6].

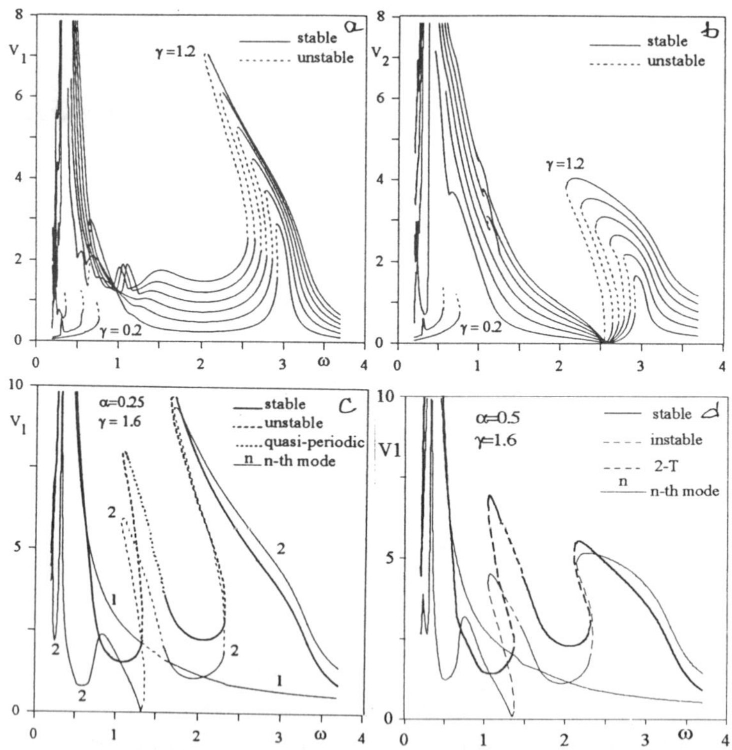

Figure 5. F.r.c's curves of velocity for low hysteresis. A general view

For models with reduced hysteresis (Figure 5), similar considerations of previous case hold, but the f.r.c.'s are no longer singlevalued and on the new peak in between the primary resonances nonstationary motions arise. This is the only example of nonperiodic motion, which was found for two different models (α= 0.25 and 0.50), in Figures 5c and 5d respectively.

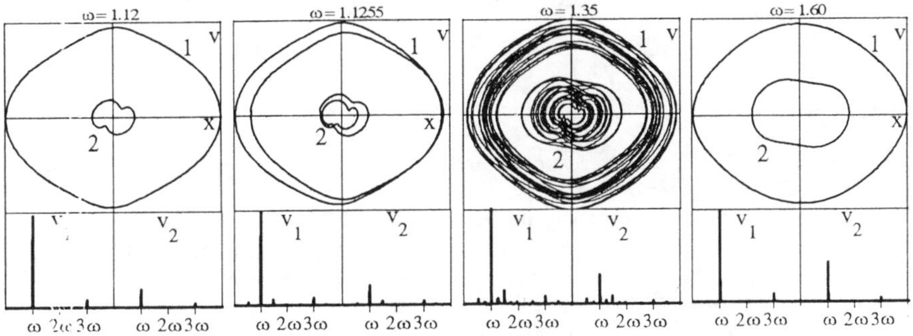

Figure 6. Phase portraits of quasi-periodic motions

By referring to the case of Figure 5c the trajectories on the phase plane are illustrated in Figure 6, passing from the periodic response to a nonstationary motion through a bifurcation and then regaining periodicity. For a slightly different control parameter value, in the quasi-periodic region a thin window of periodic oscillations occurs, as a consequence of a locking phenomenon which tunes the two new harmonics to be exactly a half and double the forcing frequency.

5. Conclusions

The description of the model as a dynamic system requires the introduction of internal variables that rises the dimension of the phase space. Thanks to its peculiarity the important class of Masing based models can be studied in a reduced dimension phase space.

The one dof oscillator is presented first, for a preliminary study of the basic behaviour. For full hysteresis only one $1T$ periodic response exists. For low hysteresis attracting basins on a two dimensional manifold displacement-phase, show that a rich variety of response is possible: resonant and nonresonant $1T$ periodic motions, $5T$ and $7T$ subharmonic responses have been found with large attracting basins.

For the two dof systems most aspects of the periodic response are studied looking at the frequency response curves. Multivaluedness appears, for the reduced hysteresis model both around the first and the second mode. It tends to disappear, for low frequencies, by increasing γ. The phenomena of nonlinear coupling strongly modifies the behaviour of the system. Depending on the yielding of the elements, new resonance peaks arise where quasi–periodic motion bifurcate with windows of $2T$ periodic motion, as a result of a frequency locking phenomenon.

Acknowledgements

This research has been partially supported by MURST 60 % 1995.

References

1. Capecchi D. (1991). Periodic response and stability of hysteretic oscillators. *Dynamics and Stability of Systems* **6**, 89-106.
2. Capecchi D. (1993). Asymptotic motions and stability of the elastoplastic oscillator studied via maps, *International Journal of Solids and Structures*, **30**, 3303-3314.
3. Capecchi D. and Masiani R. (1996), Stationary response of Masing hysteretic oscillators, *Chaos, Solitons & Fractals*, **7**, 1583-1600.
4. Capecchi D. and Vestroni F. (1990). Periodic response of hysteretic oscillators. *Int. J. Non-Linear Mechanics* **25**, 309-317.
5. Capecchi D. and Vestroni F. (1995). Asymptotic response of a two dof elastoplastic system under harmonic excitation. Internal resonance case. *Nonlinear Dynamics* **7**, 317-333.
6. Capecchi D., Masiani R. and F. Vestroni (1997). Periodic and non-periodic oscillations of a class of hysteretic two degree of freedom systems, Nonlinear Dynamics, **00**, 1-17.
7. Feng Z.C. and Li D.Z. (1997). Dynamics of a mechanical system with a shape memory alloy bar, *Journal of Intelligent Material Systems and Structures*, **7**, 399-410.
8. Fiedler L. and Nayfeh A. H. (1996), An experimental and theoretical investigation into the influence of hysteretic damping on the dynamic behaviour of a three–beam structure, *Sixth Conf. on Nonlinear Vibrations, Stability and Dynamics of Structures*, Blacksburg, Virginia.
9. Iwan W.D. (1965). The steady-state response of the double bilinear hysteretic model. *Journal of Applied Mechanics* **32**, 921-925.
10. Iwan W.D. (1965). The steady-state response of a two-degree-of-freedom bilinear hysteretic systems, *Journal of Applied Mechanics* **32**, 151-156.
11. Iwan, W.D. (1968). Response of multi-degree-of freedom yielding systems, *Journal of Engineering Mechanics Div. ASCE* **94** (EM2), 421.
12. Jennings P.C. (1963). Response of a general yielding structure, *J. Eng. Mechanics Division*, ASCE, **90**, 131-166. America **57**, 105-112.
13. Miller G.R. and Butler M.E. (1988). Periodic response of elastic-perfectly plastic SDOF oscillator *J. Eng. Mechanics*, ASCE **114**, 536-550.
14. Poddar B. Moon F.C. and Mukherjee S.(1988). Chaotic motion of an elastic plastic beam, *Journal of Applied Mechanics*, **55**, 185- 189.
15. Pratap P., Mukherjee S. and Moon F.C. (1994). Dynamic behaviour of a bilinear hysteretic elasto-plastic oscillator, part II: oscillations under periodic impulse forcing, *Journal of Sound and Vibration*, **172**(2), 339- 358.
16. Whiston G. S. (1987). Global dynamics of a vibroimpacting linear oscillator, *Journal of Sound and Vibration*, **118**(3), 395- 429.
17. Wong C. W., Ni Y.Q. and Lau S. L (1994). Steady-state oscillations of hysteretic differential model. Parts I and II, Journal of Engineering Mechanics, **120** , 2271-2325.
18. Bajkowski J. and Szemplinska- Stupnicka W. (1986). Internal resonance effects - simulation versus analytical method results, *Journal of Sound and Vibration* **104**(2), 259- 275.

MAPS, TRAPS, AND EQUILIBRIA FOR A FULLY DISSIPATIVE ELASTOPLASTIC OSCILLATOR

RUDRA PRATAP
India Institue of Science
Bangalore, India

AND

JOHN JUDGE
University of Michigan
Ann Arbor, MI, U.S.A.

Abstract. A simply supported elasto-plastic beam is subject to unforced damped vibration. It is modeled as a single degree of freedom oscillator with bilinear hysteresis and kinematic hardening. An iterative process is used to simulate the system's behavior and generate maps of maximum yield in either direction, for a variety of kinematic hardening coefficients and various degrees of damping. The maps are compared to show the effect of changing either of these parameters. Interesting regions of these maps and their meaning in reference to the model are discussed.

1. Introduction

Many materials exhibit elastoplastic behavior, that is, under small loads they are elastic, but subjected to loading beyond a certain point they respond plastically. When a beam of some elastoplastic material vibrates with enough energy to undergo plastic motion, there can sometimes be permanent plastic offset left after this energy has been dissipated. Pratap et al. [1] modeled the behavior of an elastoplastic beam under free vibration, and developed explicit maps for predicting the effects of the plastic cycles.

While the plastic motion itself dissipates energy, such a beam's vibration could also damped in a more classic sense. When a velocity-proportional damping term is added to the equations of motion, the solution becomes transcendental and the times at which the system switches from elastic

F. C. Moon (ed.),
IUTAM Symposium on New Applications of Nonlinear and Chaotic Dynamics in Mechanics, 213–220.

to plastic behavior and back cannot be solved for explicitly. (for a similar problem in piecewise elastic systems, see [2]). This simulation of the free responce of a damped elasto-plastic beam involves the use of an iterative process to find these times and thereby predict the behavior of the beam.

The model for the beam (Figure 1), consists of two massless, rigid links, of length l, with the mass of the beam, m, concentrated at the center [3, 4]. At this joint is an elastoplastic torsional spring to model the elastoplastic nature of the beam, and a rotational damper to model any sort of velocity proportional damping of the beam's vibration. It is assumed that, even under plastic deformation, the deflection y remains small enough that $\frac{1}{2}\alpha \approx y/l$.

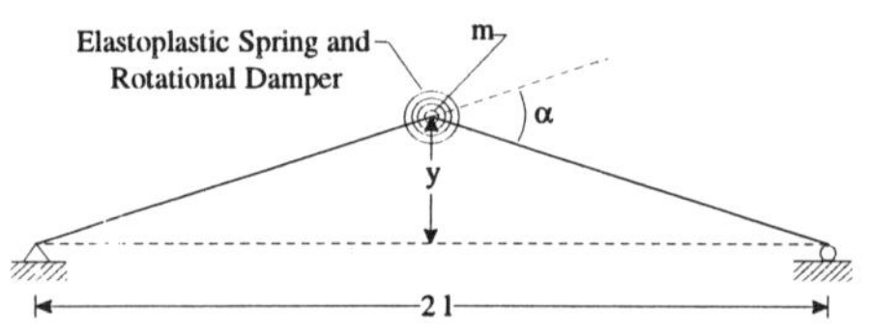

Figure 1. Model

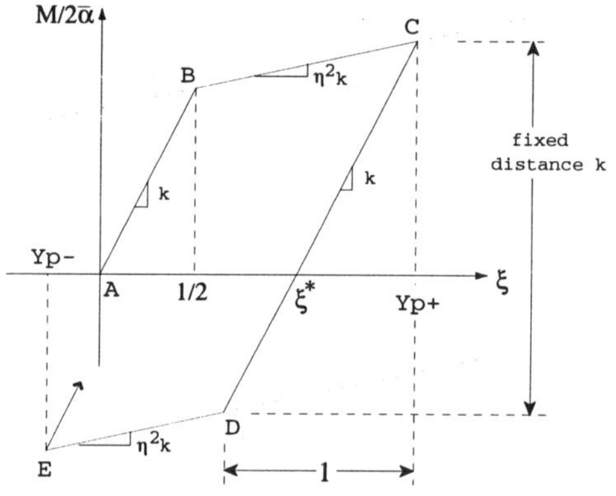

Figure 2. Spring Characteristic

Figure 2 shows the characterisitcs of the spring. Moment M is plotted against ξ, normalized displacement. The spring has spring constant k during elastic motion and $\eta^2 k$ during plastic motion. η^2 is the kinematic hardening coefficient, and is between 0 (perfect plasticity), and 1, for which there is no plastic region and the spring is perfectly elastic. The spring is modeled as having kinematic strain hardening, so that the difference in moment from one extreme of elastic motion to the other (C to D) is always the same [5]. The rotational damper provides for damping proportional to the rate of change of the angle α; the constant of this proportionality is b.

In any particular region of motion of the system (elastic or plastic motion of the spring) there is only one degree of freedom, and the equation of motion for the system is a single second order differential equation. This would indicate that there are two state variables, position and velocity. However, the constants in the equation of motion for each elastic region of motion depend on the state of the system at the end of the previous plastic region. Therefore, three variables are necessary to fully describe the state of the system: position, velocity, and a third variable to describe the amount of plastic offset in the system (and thus the equilibrium for elastic oscillations).

The non-dimensionalized equation of motion under free vibration is

$$\ddot{\xi} + 2\omega\zeta\dot{\xi} + \eta^2\omega^2\xi = c\omega^2 \tag{1}$$

where

$$\xi = y/l\bar{\alpha} \approx \tfrac{1}{2}\alpha/\bar{\alpha} \qquad \omega = \sqrt{4k/ml^2} \qquad \zeta = b/l\sqrt{km}$$

with η=1 and c=ξ^* in the elastic region,
and $c = -\frac{1}{2}(1-\eta^2)\text{sgn}(\dot{\xi})$ in the plastic region.

The three variables necessary to describe the state of the system are ξ, $\dot{\xi}$, and ξ^*, where $\xi^* = (\xi - \frac{1}{2}\text{sgn}(\dot{\xi}))(1-\eta^2)$ in the plastic region, and remains constant in the elastic region, where it is the equilibrium of the elastic oscillations.

2. Iterative Solution

Determining the behavior of the system during either type of motion is straightforward, but because there are both ξ and $\dot{\xi}$ terms in the equation of motion, time cannot be solved for explicitly. An iterative process based on Newton's method is used to determine the times at which the behavior changes from plastic to elastic or back. The system switches from plastic motion in one direction to elastic motion in the other whenever velocity, $\dot{\xi}$, reaches zero. It then switches from elastic motion to plastic motion in the same direction after ξ has changed by a unit amount (this is due to the kinematic strain hardening assumption and the fact the ξ is normalized so that the maximum difference between one extreme of elastic motion and the other is 1). Once the time of a behavior change is found, ξ and $\dot{\xi}$ at that time become initial conditions for the next motion, and, in the case where the system in beginning elastic motion, ξ^* at the time of the change becomes the equilibrium of the new elastic motion.

The permanent plastic offset is the ξ^* value when the system has finished all plastic motion. This occurs when ξ no longer changes by a full unit amount during an occurance of elastic behavior, and so no new plastic motion is begun. The various means by which ξ no longer changes a full unit amount are dicussed below in Section 3.

The maps shown in Figures 3 and 4 are the result of iterative simulations of the system's behavior. Y_{P+} and Y_{P-} are the extreme values of motion; the values of ξ at which $\dot{\xi}$ is zero and the behavior changes from plastic motion in the positive direction to elastic motion in the negative direction (Y_{P+}), or vice versa (Y_{P-}). The map allows prediction of Y_{P-} given Y_{P+}

then use of that value of Y_{p-} to predict the next Y_{p+}, and so on until the system no longer undergoes plastic deformation. It is then possible to find ξ^*, which, since any plastic motion is finished, is then the permanent plastic deformation. The size of the elastic region is always the same, so the value of ξ^* when the system switches from elastic to plastic behavior is either $Y_{p+} - \frac{1}{2}$ or $Y_{p-} + \frac{1}{2}$.

Each map has two curves: an upper curve which shows the next value of Y_{p-} given Y_{p+}, and a lower curve which shows the next value of Y_{p+} given Y_{p-}. The map is read by taking an initial known value of Y_{p+} or Y_{p-} and finding consecutive Y_p's by following a stairlike path between the two curves, upwards and to the left, until the line of slope 1 passing through (0.5, -0.5) is reached. Without damping ($\zeta = 0$), the two curves meet at (0.5, -0.5), so any intersection with this line will be at that point or outside the curves. When damping is introduced, the two curves move away from each other along this line, so that the stair-path could end either outside the curves or between them.

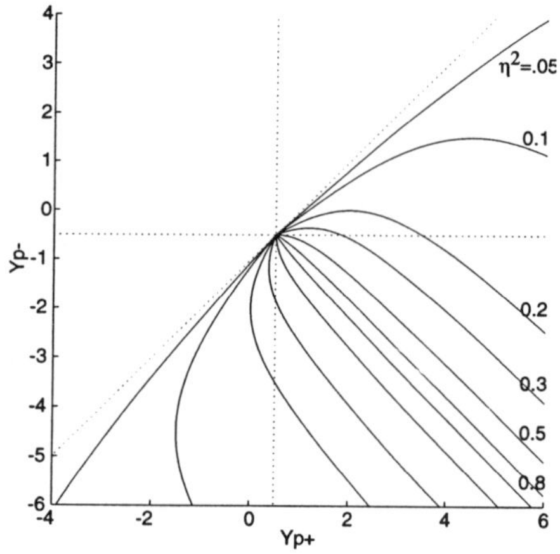

Figure 3. Map of Maximum Yield for $\zeta = 0$

3. Physical Interpretation

The line of slope 1 passing through (0.5, -0.5) can be called the *elastic oscillation line*. On this line, Y_{p+} and Y_{p-} are 1 unit apart, which is the amplitude of undamped elastic motion. Once the stair-path reaches this line, a beam can continue to undergo elastic osillations, but no further plastic motion will occur, since the amplitude of the elastic oscillations will be ≤ 1 (assuming $\zeta \geq 0$). Once its stair-path reaches this line, a damped beam will undergo decaying elastic oscillations about the ξ^* value determined by the most recent Y_p. Thus, the *elastic oscillation line* is the end of all stair paths, and marks the point at which no further plastic motion can

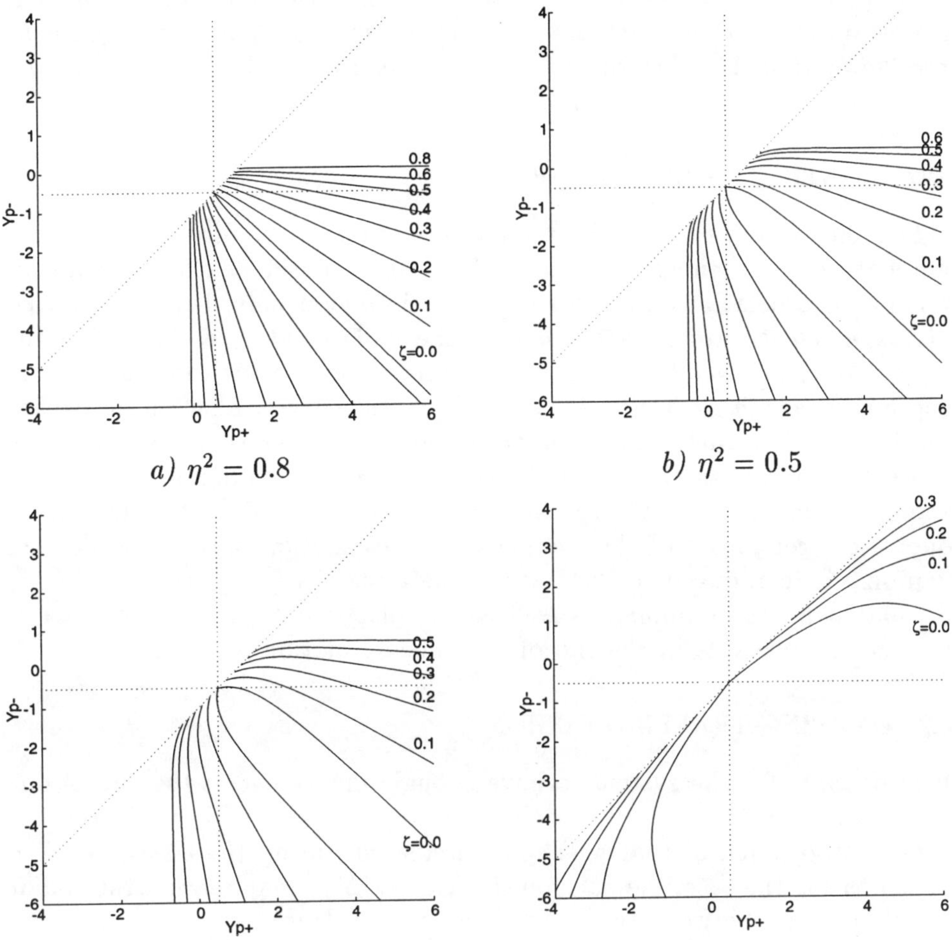

a) $\eta^2 = 0.8$ *b)* $\eta^2 = 0.5$

c) $\eta^2 = 0.4$ *d)* $\eta^2 = 0.1$

Figure 4. Maps of Maximum Yield

occur. The amount of permament plastic offset, ξ^*_{final} ($Y_{p+} - \frac{1}{2}$ or $Y_{p-} + \frac{1}{2}$), is zero at (0.5, -0.5) and increases as the distance from this point grows.

Figure 3 shows maps for various values of η^2, without any damping ($\zeta = 0$). As Pratap et al. [1] discussed, $\eta^2 = 0.5$ is a critical value. If $\eta^2 > 0.5$, all paths lead to (0.5,-0.5). No matter how much energy the vibration starts with, there is no net plastic deformation once the system reaches steady state (ξ^*_{final} is always zero). For $\eta^2 < 0.5$, the curves have regions to the left of Y_{p+}=0.5 and above Y_{p-}=-0.5. These areas, which result in permanent plastic offset ($\xi^*_{\text{final}} \neq 0$), are called *elastic trapping regions*, and are discussed in greater detail below.

Figure 4 shows maps for various amounts of damping, ζ, for four values of η^2 (0.1, 0.4, 0.5, and 0.8). As damping increases for constant η^2, the curves spread further apart and a gap opens between them along the *elastic oscillation line*. The portion of this line between the endpoints of the curves can be called a *damping trapping region*, also discussed below.

3.1. ELASTIC TRAPPING REGION

When the system reaches an extreme of motion in one direction and does not have enough energy left to reach the next onset of plastic behavior, remaining oscillations will be in the elastic region only. Pratap et al. [1] discussed *elastic trapping regions* for the undamped case. These are the regions of the curves on the maps in Figure 3 which allow the stair-path to end up on the *elastic oscillation line* outside the curves. That the stair-path reaches this line rather than the other curve indicates that the amount of energy remaining in the system at the most recent maximum yield is not enough to move the system elastically to the extent that it yields in the opposite direction, and the system continues motion only elastically. As damping is introduced to the system, these regions grow larger. Energy is dissipated by the damping as well as the plastic deformation, so there is less energy available at the end of each plastic motion.

3.2. DAMPING TRAPPING REGION

It is possible for the system to have enough energy left at the end of one plastic cycle to theoretically reach the next plastic cycle, but fail to do so because too much of that energy is dissipated during the elastic motion. When this is the case, the path on the Y_{p+} vs. Y_{p-} map reaches the *elastic oscillation line* between the ends of the curves. Once again, the vibration becomes limited to elastic oscillations, but in this case it is the damping alone which causes this rather than a combination of damping and plastic dissipation of energy.

3.3. CRITICAL DAMPING

Without any damping, both curves on any Y_{p+} vs. Y_{p-} map eventually reach a slope of -1 as $|Y_p|$ increases. As damping is increased, the curves not only spread apart, their slopes also diverge. The frequency of oscillation of the solution to the equation of motion in the plastic region is $\omega\sqrt{\eta^2 - \zeta^2}$. Thus, $\zeta = \eta$ is a critical value above which there are no oscillations. At that value, the curve for Y_{p-} as a function of Y_{p+} becomes horizontal for large values of $|Y_p|$, and the curve for Y_{p+} as a function of Y_{p-} becomes vertical. This is apparent when observing the curves for the largest values of ζ on Figure 4 (The phenomenon is not visible on Figure 4*d*, as it occurs at Y_p

values beyond the edge of the map). In the model, this means that the damping is just enough that the final maximum yield occurs when there is no energy stored in the spring, so that no subsequent oscillations of either the plastic or elastic variety occur.

3.4. INITIAL CONDITIONS

The Y_{p+} vs. Y_{p-} maps allow one to determine final position given the first maximum yield (first Y_{p+}). It is apparent from these maps that stronger initial conditions, which produce greater first maximum yields, will not necessarily result in more permanent plastic offset when the plastic oscillations are finished. The final plastic offset, ξ^*_{final}, is determined from either the last Y_{p+} or Y_{p-}, which will be within a certain range defined by the geometry of the map. Figure 5 shows the maximum amount of permanent plastic offset possible, given a certain combination of η^2 and ζ. While the exact amount of permament plastic offset is determined by the initial conditions, it cannot be outside a region determined only by the system parameters.

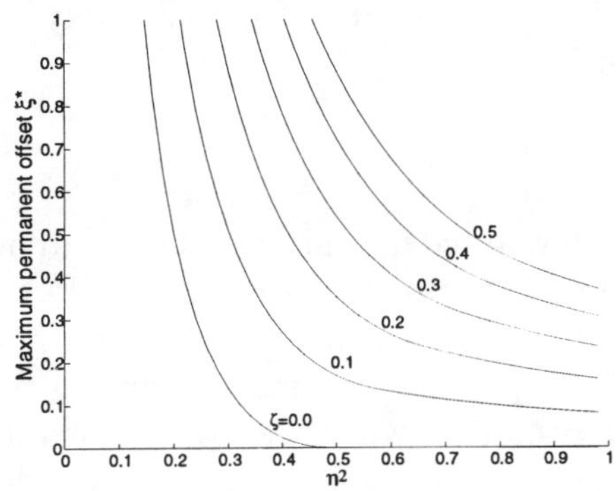

Figure 5. Maximum final offset (ξ^*) vs. η^2

Figure 6 shows, for two cases of η^2 (one less than 0.5 and one greater), which trapping region the motion ends up in, depending on ζ and intial speed $\dot{\xi}$. For simplicity's sake, zero initial displacement is assumed. The bottom region of both plots corresponds to initial conditions that do not impart enough energy to reach the first plastic yield.

For $\eta^2 < 0.5$, for which 0.4 is an example, undamped motion ends in the elastic trapping region, and as damping in introduced and increased, the damping trapping region appears and grows. The corresponding bands on Figure 6*a* move upwards to the right because as damping is increased, higher initial speeds are needed to achieve the same results. For $\eta^2 \geq 0.5$, all lightly damped motion ends up in the damping trapping region. Only when sufficient damping has been introduced do any initial conditions send the motion into an elastic trapping region.

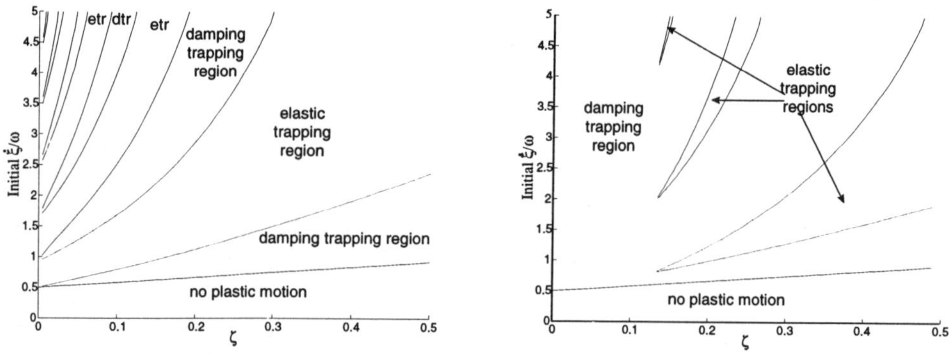

a) $\eta^2 = 0.4$ *b)* $\eta^2 = 0.6$

Figure 6. Mechanism responsible for final state given initial $\dot{\xi}$ and ζ

3.5. ACCURACY OF THE RESULTS

Since the results shown in the maps in Figures 3 and 4 are based upon the small angle approximation, they are only reasonably accurate for values of $|Y_p|$ less than $15°/2\bar{\alpha}$.

4. Conclusions

Maps such as those developed in this simulation allow one to take plasticity and damping into account when designing beams in structures and mechanisms. Beams can be analyzed to determine whether vibrations with enough energy to cause plastic deformation during oscillation will result in any permanent deformation when the oscillations have ceased, and the amount of any such permanent deformation can also be predicted.

Acknowledgements: The authors wish to thank Will Vodra whose term paper provided some of the derivations required in this work.

References

1. R. Pratap, S. Mukherjee, and F. C. Moon, "Limit cycles in an elastoplastic oscillator," *Physics Letters A*, Vol. 170, No. 5, pp 284-292, 1992.
2. S. W. Shaw and P. J. Holmes, "A periodically forced linear oscillator," *Journal of Sound and Vibration*, Vol. 90, No. 1, pp 129-155, 1983.
3. G. Ballio, "On the dynamic behavior of an elastoplastic oscillator (experimental investigation and theoretical developments)," *Meccanica*, Vol. 5(2), pp 85-97, 1970.
4. T. K. Caughey, "Sinusoidal excitation of a system with nilinear hysteresis," *ASME Journal of Applied Mechanics*, Vol. 27, pp 640-643, 1960.
5. A. Mendelson, *Plasticity*, The McMillan Company, NY, 1968.
6. R. Pratap, *Dynamics of a Bilinear Hysteretic Beam*, Ph.D. Thesis, Cornell University, Ithaca, NY, 1993.

IV. MACHINE SYSTEM DYNAMICS

The key lecture by Shaw et al. shows how concepts of nonlinear dynamics can be used in the design of vibration absorbers. The paper by Pfeiffer on chain dynamics shows how advanced simulation systems are able to capture complex multi-body dynamics in transmission systems.

Finally, Boedo's paper on hydro-dynamic rotor bearing dynamics provides some new evidence for chaos in rotor bearing systems.

All of these papers are close to industrial applications.

F. C. Moon (ed.),
IUTAM Symposium on New Applications of Nonlinear and Chaotic Dynamics in Mechanics, 221.

NONLINEAR DYNAMICS AND THE DESIGN OF TUNED PENDULUM VIBRATION ABSORBERS

S. W. SHAW, C.-P. CHAO AND C.-T. LEE
Department of Mechanical Engineering
Michigan State University
East Lansing, MI 48824, USA

Abstract

This paper presents a survey of recent results obtained from a series of analytical investigations into the nonlinear dynamic behavior of systems of centrifugal pendulum vibration absorbers. Issues related to the performance capabilities and limitations of existing and proposed designs are discussed. Specifically, we consider the performance acheived for various types of absorber paths, a subharmonic absorber, and the dynamic stability of systems composed of multiple absorber masses. A set of study summaries is offered for these topics, with details left to the references.

1. Introduction

There exist several methods for reducing troublesome torsional vibrations in rotating machinery, including the application of flywheels, tuned elastomeric dampers, and tuned pendulum absorbers. Flywheels are effective at all operating speeds, but add mass and inertia to the system, thereby reducing responsiveness. Elastomeric dampers are tuned to a particular resonance frequency of the system of interest and work quite well, but they are not effective over a range of operating speeds. Centrifugally driven pendulum vibration absorbers (CPVA's) offer vibration attenuation over a continuous range of rotation speeds. These devices have been in wide use since the 1940's and are commonly employed in helicopter rotors and light aircraft engines, and are finding more use in specialty applications, including the camshafts of certain heavy diesel engines and the crankshafts of some automotive racing engines (Albright, 1994).

A CPVA system is composed of a set of absorber masses, each of which is kinematically connected to a rotor in such a manner that its center of

F. C. Moon (ed.),
IUTAM Symposium on New Applications of Nonlinear and Chaotic Dynamics in Mechanics, 223–234.

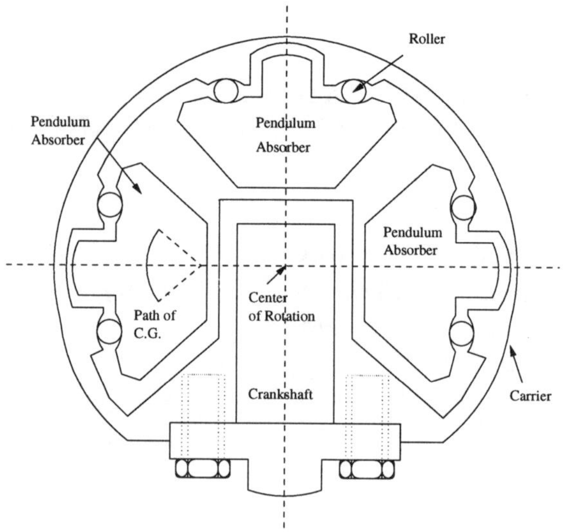

Figure 1. A schematic diagram of one possible arrangement of the rotor and absorber masses. This bifilar configuration supports the absorber masses by means of rollers that travel along machined surfaces cut into the absorber masses and the carrier. The carrier is fixed to the crankshaft at a location where a counterweight is typically found. The axis of rotation is perpendicular to the page.

mass is restricted to move along a prescribed path relative to the rotor. A configuration used in an experimental automotive application is schematically shown in Figure 1. The absorber masses use the centrifugal field, rather than a resilient element, in order to provide a restoring force, and this allows them to address vibrations at a given *order* of rotation, that is, a certain multiple of the rotation rate, rather than at a fixed frequency. The linear theory of CPVA's and some of their limitations have been known for several decades. The treatise by (Ker Wilson, 1968) describes the basic operation of CPVA's, presents several actual implementations, and offers some historical background.

This paper outlines results that have recently been obtained regarding the nonlinear dynamics of CPVA systems. The main objectives of this research have been the following:

- To point out limitations in absorber systems designed by conventional linear vibration analysis.
- To use these results to offer refined design guidelines that account for dynamic instabilities and nonlinear behavior.
- To generate new designs of absorber systems that exploit nonlinearities for improved system performance.

Here we present some highlights of the main results obtained and provide references where the reader can find detailed results. The paper is organized as follows. The remainder of this section covers a brief history of CPVA developments and an overview of why nonlinear effects are important for these systems; section 2 describes the dynamic model used in the analysis; section 3 deals with simple absorber systems consisting of one and two masses; section 4 briefly describes a special subharmonic absorber system; section 5 considers the dynamic stability and performance of general multi-mass absorber systems; and the paper closes with a description of some work in progress in section 6.

Since they are easily manufactured, virtually all existing CPVA systems use circular paths for the absorber masses, and the associated geometric parameters are selected so that the small amplitude dynamics are tuned as desired. However, such designs can fail at moderate amplitudes, due to the fact that the pendulum frequency changes as a function of its oscillation amplitude. This was known to Den Hartog (DenHartog, 1938), and Newland identified the possible catastrophic failure of circular-path CPVA's due to this mistuning, wherein a sudden jump in torsional vibration amplitude can occur as the torque level is increased (Newland, 1964). They both suggested a remedy for this problem: to build into the system a slight overtuning of the linearized pendulum frequency such that the absorber, while, perfectly tuned only at a single amplitude, will remain somewhat close to the desired tuning over a larger amplitude range. These tuning refinements allow devices to operate over a large amplitude range, but they ignore the fact that nonlinearities generate vibrations at higher harmonics which degrade performance and can excite structural resonances.

These simple nonlinear effects are crucial, and are among some of the several factors that make systems of tuned vibration absorbers fertile ground in which nonlinear phenomena can appear. While these effects can cause problems, they also offer the possibility that nonlinearities may play a useful role in generating new absorber configurations. These factors include the following:

- Absorbers are tuned to match the frequency of the applied fluctuating torque, and are therefore forced in a resonant manner.
- Absorbers maintain tuning at all rotation speeds and thus operate most effectively when they are lightly damped.
- Absorbers must be effective over a range of applied loads, and in conventional, linear-based designs this necessitates the use of a sufficient amount of absorber mass in order to maintain small amplitudes. Such added mass is highly undesirable.
- In many situations in which these absorbers are used, a system of several identical absorber masses is employed for dynamic balancing

and/or due to spatial restrictions around the rotor. These multi-mass systems inherently posses internal resonances, which raises questions about the stability of the desired synchronous motion of the absorbers.

- The nonlinear behavior of an absorber can be altered by varying the path followed by the absorber center of mass relative to the rotor. Thus, the possibility exists to design desirable nonlinear features into the system.

The final item on the above list has been the subject of investigation over the past twenty years. It is practically feasible since the commonly used bifilar suspensions of absorber masses make non-circular paths quite easy to realize in practice. In 1980, Madden proposed and implemented the use of *cycloidal* paths (Madden, 1980). These paths have favorable tuning behavior at large amplitudes — they stay quite close to perfectly tuned, with only a slight frequency increase as the amplitude is increased, thereby avoiding the catastrophic jump behavior. Madden presumably used the cycloidal path since it is known to be the solution of the tautochrone problem for a gravitational field, that is, it offers the path along which a point mass can slide without friction such that the resulting oscillations are independent of the amplitude of motion in a uniform force field. (Note that the solution of the tautochrone problem is identical to that of the more well-known brachristichrone problem (Denman, 1985)). Ir the mid-1980's Denman proposed that a certain epicycloid be used for the absorber path, as it is the exact tautochrone in the radially-dependent force field that is induced by constant-rate rotation. When tuned to frequency n, the motion of a tautochronic absorber provides a periodic torque which is capable of counteracting most of the order n harmonic of a disturbing torque, thereby reducing torsional oscillations at that order. This promising property of tautochronic paths launched analytical studies which explored the effectiveness of tautochronic absorbers (Denman, 1992; Cronin, 1992), and it also motivated an experimental study by Borowski and co-workers at the Ford Motor Company that considered absorbers riding on cycloidal and epicycloidal paths (Borowski *et al.*, 1991). In that investigation second-order cycloidal absorbers were experimentally tested and found to be very effective (92 %) in reducing the second-order torsional vibrations in an in-line, four-cylinder, four-stroke, 2.5L engine. However, fourth-order vibrations were significantly magnified, due to the harmonic generation mentioned above. The remedy offered for this problem was to use a combination of second and fourth order absorbers. A more recent development in terms of absorber paths is the subharmonic vibration absorber system proposed in (Shaw and Lee, 1995), in which a pair of identical absorbers tuned to one-half the order of the applied torque can totally eliminate a harmonic torque over a large amplitude range without generating higher-order harmonics. This promising system has yet to be experimentally tested.

2. Mathematical Model

The basis of the analytical work carried out by the authors to date is a nondimensionalized mathematical model for an idealized rotor/multi-absorber dynamic system. Details of its development can be found in (Shaw and Lee, 1995). It is based on the following assumptions and definitions:

- The rotor is rigid, rotates about a fixed axis O, has an orientation angle θ, and is subjected to an applied torque Γ.
- The N absorbers are taken to be point masses m_i that move along paths prescribed for the absorbers' centers of mass. For bifilar arrangements this is dynamically exact if one includes the moments of inertia of the absorbers about their respective centers of mass in the total moment of inertia of the rotor.
- The nondimensional arc-length position of the ith absorber is s_i and its path is specified by the function $x_i(s_i)$, which represents the square of the nondimentionalized distance from O to the location of m_i at s_i. An additional path function, defined for convenience, is $g_i(s_i) = \sqrt{x_i(s_i) - \frac{1}{4}\left(\frac{dx_i}{ds_i}(s_i)\right)^2}$.
- The disturbing torque is assumed to consist of a constant part (including load) and a harmonic fluctuating part of order n, as follows, $\Gamma = \Gamma_0 + \Gamma_\theta(\theta)$, where $\Gamma_\theta(\theta) = \Gamma_n \sin(n\theta)$. This represents n torque pulses per revolution; for example, $n = M/2$ for an M-cylinder, four-stroke engine.
- The damping between the ith absorber and the rotor is assumed to be an equivalent viscous damping with coefficient μ_{ai}. Resistance between the rotor and ground is also modeled as equivalent linear viscous damping with coefficient μ_0.
- The equations of motion are formulated with θ, rather than time, as the independent variable. This transforms the nonlinear term $\Gamma_\theta(\theta)$ into a periodic forcing term.
- The normalized rotor speed is given by y, where $y = 1$ is the desired steady-state.
- In terms of derivatives with respect to θ, denoted by a prime, $()'$, the rotor angular acceleration is given by yy'.
- The ratio of the moment of inertia of the ith absorber when positioned at its vertex to that of the rotor is given by b_i, a quantity that in practice is always much less than unity.

– Gravitational acceleration is extremely small when compared to rotational accelerations, and its effects are neglected.

The resulting equations of motion that describe the dynamics of the N absorbers and the rotor, respectively, are,

$$ys_i'' + [s_i' + g_i(s_i)]y' - \frac{1}{2}\frac{dx_i}{ds_i}(s_i)y = -\mu_{ai}s_i', \quad 1 \leq i \leq N, \quad (1a)$$

$$\sum_{i=1}^{N} b_i[\frac{dx_i}{ds_i}s_i'y^2 + x_i(s_i)yy' + g_i(s_i)s_i'yy' + g_i(s_i)s_i''y^2 + \frac{dg_i(s_i)}{ds_i}s_i'^2y^2]$$

$$+yy' = \sum_{i=1}^{N} b_i\mu_{ai}g_i(s_i)s_i'y - \mu_0 y + \Gamma_0 + \Gamma_n \sin(n\theta). \quad (1b)$$

A comment on the absorber paths is in order here, for which we consider a single absorber, $N = 1$, and drop the subscript i. The simplest way to show the nature of the absorber dynamics is to consider the case when the rotor speed is constant and there is no absorber damping. This is the case of a large flywheel attached to the system, so that the absorber moves under the action of the rotor, but the movement of the absorber does not affect the rotor. Specifically, take $y = 1$, $y' = 0$, and $\mu_a = 0$, in which case the equation for the absorber becomes a conservative oscillator in s, $s'' - \frac{1}{2}\frac{dx}{ds}(s) = 0$. The idea behind the tautochronic absorber is to select the path for which this equation becomes a linear harmonic oscillator, that is, by taking $x(s) = 1 - n^2s^2$. This path is a special epicycloid and *it renders the absorber motion purely harmonic with frequency n over the entire range of operation*, that is, until the absorber reaches the cusps on the corresponding epicycloidal path (where $g(s) = 0$). This design is optimal in eliminating the mistuning generated by large amplitudes. However, as seen from the equation of motion for the rotor, the harmonic motion of the absorber, $s(\theta) = A\sin(n\theta)$, induces a more complicated torque on the rotor containing many harmonics, even when $y = 1$ (more on this below). We now turn to some design issues related to simple CPVA systems.

3. Nonlinear Dynamics of Simple CPVA Systems

An ambitious analytical design process for selecting the paths for a CPVA system may be proposed as follows. Given an applied torque, determine an arrangement of absorbers that will result in minimizing torsional oscillations in the rotor. For example, one can use some measure of the angular acceleration of the rotor as an objective function in an optimization formulation, with the absorber masses and their paths in the role of design parameters. Posed in this manner the design of a general multi-absorber system is daunting, due to the sheer number of design variables and the complexity involved in finding periodic steady-state solutions of the underlying dynamic system. Furthermore, it will generally yield results of little practical importance, due to the fact that a given rotor system will be subjected to various combinations of harmonics. over different operating conditions. In practice, one designs absorber systems to address individual harmonics of the torque that are particularly bothersome. These harmonics are treated individually, and superposition is expected to provide the necessary additive quality to the overall design. Such a philosophy is adopted even for nonlinear designs, although the issue of nonlinear interactions looms large.

The first and simplest nonlinear design problem is that of attenuating vibrations due to harmonic order n using one absorber mass, or a set of masses constrained to move in a particular manner. Our approach to this problem has been to investigate the relationship between the path of the absorber, the torque it can generate, and the resulting angular acceleration of the rotor. This has produced some interesting and useful results. One approach of this type is rather brute-force — it attacks the problem by specifying particular forms for the absorber paths and testing their effectiveness by (approximately) determining the periodic steady-state response of the system over a range of paths and evaluating the results. For this approach one must have a systematic means of specifying the absorber path. There are many ways of doing this, and our choice has been guided by historical developments. A two-parameter family of paths has been systematically explored in which, roughly speaking, one parameter fixes the linear tuning and the other sets the sign and level of nonlinearity encountered along the path. Using a formulation suggested by Borowski and developed by Denman, these parameters can be independently varied, and this family of paths contains the most interesting and useful absorber paths, namely, circles, cycloids, and epicycloids (Denman, 1992). Perturbation techniques and simulations have been employed in systematic investigations of these paths, and the results show that the epicycloidal path is the most favorable for moderate torque ranges, and that the cycloidal and mistuned circular paths are also quite effective when torque amplitudes become large. In fact, by adding some mistuning to nearly any path, typically in the range of $5-10\%$, the absorber system is found to act in an essentially linear manner and still function as an absorber. Detailed results from these studies can be found in (Shaw *et al*, 1997).

Another approach along similar lines involved a classification of the (function) space of periodic torques that can be exactly addressed by a single absorber, or by a set of absorbers operating in some constrained manner, with *arbitrary* paths (Shaw and Lee, 1995). This investigation yielded a proof of the fact that a single absorber cannot exactly absorb a purely harmonic torque over a finite amplitude range, no matter how clever one is

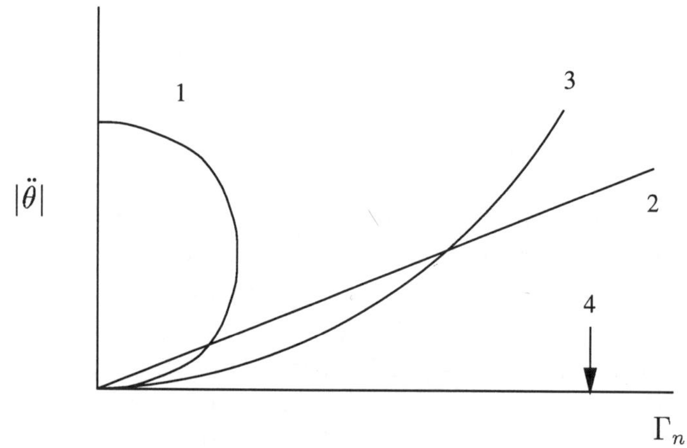

Figure 2. The qualitative features of the performance of various idealized absorber systems, depicted in terms of the amplitude of the rotor acceleration versus the amplitude of the fluctuating part of the applied torque. Curve 1: Perfectly tuned circular path; curve 2: slightly overtuned circular path; curve 3: perfectly tuned tautochronic epicycloidal path; curve 4: subharmonic absorber pair.

in designing the path, even in the ideal, undamped case. This is due to the kinematics of the system and the manner in which the absorber motion generates torque on the rotor. Furthermore, the undesirable torque harmonics generated at moderate and large amplitudes of motion for the tautochronic absorber system were quantified. The nature of these harmonics lead to the development of the subharmonic absorber system, as described below.

A schematic summary of the performance of these absorber systems is shown in Figure 2, which shows qualitative features for various idealized, undamped absorber systems. It shows the peak rotor acceleration versus the amplitude of the nth-order torque component. It is seen that the perfectly tuned circular path leads to disaster at a moderate torque level (curve 1), whereas mistuned circular paths behave essentially linearly over a large torque range (curve 2). The tautochronic epicycloid is seen to perform very well over a moderate torque range, but the amplification of higher harmonics limits its effectiveness at large torque amplitudes (curve 3). The subharmonic pair offers perfect performance, that is, zero angular acceleration, over this torque range (curve 4), although it requires more space to implement since the absorbers are tuned to order $n/2$ — a much shallower path than those tuned to order n.

An application study of these absorber systems was carried out wherein various configurations were tested via simulations on a detailed dynamic model of an internal combustion (IC) engine. The model considered was an in-line, four-stroke, four-cylinder engine that included realistic torques arising from the inertia of engine components, from gas pressure in the cylinder (based on experimental in-cylinder measurements), and from bearing resistance. These lead to a multi-harmonic applied torque that depends on $\dot{\theta}$ and $\ddot{\theta}$ in addition to θ, as well as on several inertial and geometrical parameters. It was found that systems of tautochronic absorbers tuned to various orders were the most robust in handling the variety of torques encountered under different operating conditions, although the subharmonic pair also worked very well (Lee and Shaw, 1995). A very intriguing observation made during the course of this study was that an identical pair of absorbers tuned to the dominant order of the applied torque and moving in exact unison could undergo a dynamic bifurcation at a moderate torque level, after which the absorbers' relative amplitudes and phases were shifted. It was observed that this instability actually *reduced* torsional vibration levels in the engine. This leads naturally to questions related to the dynamic stability and effectiveness of systems composed of multiple absorber masses, and this forms the subject of section 5.

4. The Subharmonic Absorber System

A tautochronic absorber tuned to order p, when moving in a harmonic manner on a rotor that is spinning at a constant speed, generates all odd order torque harmonics ($p, 3p, 5p, \ldots$), but only a single even order harmonic — the second ($2p$). Therefore, if an identical pair of absorbers move in an exactly out of phase manner, the net torque generated will be a pure harmonic of second order ($2p$). Thus, by tuning absorbers to one-half the order of the applied torque ($p = n/2$), a very attractive absorber system is realized in which no higher-order torque harmonics are generated. This system utilizes a global subharmonic response that is induced by Coriolis effects of the absorbers' motion, resulting in a system that works beautifully over a large torque range. This is an example where nonlinearity can improve system performance and where the proposed design cannot be imagined by simply extending ideas from linear dynamics. The general results of this research for undamped systems have been published in (Shaw and Lee, 1995), and a forthcoming paper offers the proof of dynamic stability and effectiveness of the subharmonic absorber pair when absorber damping is present (Lee *et al.*, 1997).

A more recent study has considered the robustness of the subharmonic absorber system to certain modeling uncertainties, particularly small variations in the absorber path. It was found that the absorber paths should be manufactured in a particular manner in order to achieve dynamic stability of the desired response and to avoid some potential subcritical bifurcations that lead to undesirable jumps in the response as torque levels vary. In particular, the absorber paths should be made as identical as possible, but the linearized frequency and the amplitude-dependent nonlinear frequency of the absorbers should both be very slightly overtuned (Chao and Shaw, 1997).

5. The Performance of Multi-Absorber Systems

When designing absorber systems for implementation, the total amount of absorber mass required is dictated by the torque levels encountered, but this mass must be distributed about and along the axis of rotation for purposes of balancing and to satisfy spatial constraints. It is implicitly assumed that these absorber masses will move in a unison manner during steady-state operation, thereby creating the dynamic effect of a single mass.

A study was undertaken of the dynamic stability of the unison response of a system of N tautochronic absorbers fitted to a rotor that is subjected to a harmonic torque. An investigation of the dynamic stability of this response turns out to be an example of a symmetric bifurcation problem. The symmetry of this system can be deduced directly from the absorbers' equations of motion for the case of N identical absorbers moving in a unison manner, whereupon the absorbers are completely interchangeable with one another. By allowing the N absorbers to move independently, it was found that the stability of the unison response is lost at a small torque level when the absorber damping is small (Chao *et al.*, 1997a). This result is of practical importance, since small damping is required for effective operation of CPVA's. It was also shown that the stable post-bifurcation response of the system has $N-1$ absorbers moving in unison, while one absorber moves at a different phase with larger amplitude. The character of this response is that torsional vibration levels are reduced when compared against the corresponding (unstable) unison motion, but the effective operating range is significantly reduced. This occurs since the one "out-of-step" absorber will reach the limits of its path at a lower torque level as compared with the unison response (Chao *et al.*, 1997b). This work has some interesting features, including the following: due to the tautochronic nature of the absorbers, results can be obtained for large amplitude absorber motions; the post-critical steady-state dynamics of the system lie on a stable invariant ellipse; the peak rotor acceleration is a constant on this ellipse; there are many possible steady-state solutions on the ellipse; but the only stable solution is the one mentioned above; and it leads to the smallest feasible torque range.

A similar study has also been carried out for multiple pairs of subharmonic absorbers, with encouraging conclusions. The desired solution for this case has the absorber masses moving in two synchronous groups of $N/2$ masses each, at the same amplitude with a phase difference of π between the groups. This response is found to be dynamically stable and can be made to be robust to variations in the absorber paths by proper selection of linear and nonlinear mistuning of the absorber path. This implies that systems composed of multiple subharmonic absorber pairs can be designed to perform as desired, that is, as predicted from the analysis of a single pair (Chao, 1997).

6. Work in Progress

These results have expanded our understanding of the operation of CPVA devices and have opened the door to potentially useful new CPVA configurations, but several interesting topics remain for study. The following analysis-based research topics are being pursued at this time: model extensions that include rotor flexibility and torque distribution along the shaft; model extensions that include the dynamic effects of the rollers which carry the absorber masses in typical bifilar arrangements and the stability of the unison response for systems of multiple, general-path absorbers, including the commonly-employed mistuned circular paths. In addition, the authors are aware of no studies of CPVA systems in which controlled systematic experiments are carried out over a range of operating conditions for different arrangements of absorbers. Such a program is currently being initiated by the lead author and Professor Alan Haddow of Michigan State University. A key feature of this investigation will be the evaluation of the subharmonic absorber system.

Acknowledgements

Portions of this work have been supported by the National Science Foundation. SWS is grateful to Victor Borowski of Ford Motor Company, Harry Denman of Wayne State University, and Don Cronin of the University of Missouri-Rolla for their continued interest in his work on CPVA's.

References

Albright, M., Crawford, T. and Speckhart, F., 1994 "Dynamic testing and evaluation of the torsional vibration absorber," *SAE Technical Paper Series 942519*, Vol. II, Engines and Drivetrains, P-288, pp. 185-192.

Ker Wilson, W., 1968, *Practical Solution of Torsional Vibration Problems*, Chapman and Hall Ltd, London, 3rd ed., Vol. IV, Ch. XXX.

Shaw, S. W. and Lee, C.-T., 1995, "On the Nonlinear Dynamics of Centrifugal Pendulum Vibration Absorbers," Ch. VII, pp. 247-309, *Smart Structures, Nonlinear Vibration, and Control*, Guran A. and Inman D. J., editors, Prentice Hall Inc., New Jersey.

Den Hartog, J. P., 1938, "Tuned Pendulums as Torsional Vibration Eliminators," *Stephen Timoshenko 60th Anniversary Volume*, The Macmillan Company, New York, pp. 17-26.

Newland, D. E., 1964, "Nonlinear Aspects of the Performance of Centrifugal Pendulum Vibration Absorbers," *ASME Journal of Engineering for Industry*, Vol. 86, pp. 257-263.

Madden, J. F., 1980, "Constant Frequency Bifilar Vibration Absorber," *United States Patent No. 4218187.*

Denman, H. H., 1985, "Remarks on brachistochrone-tautochrone problems," *American Journal of Physics*, Vol. 53, pp. 781-782.

Denman, H. H., 1992, "Tautochronic Bifilar Pendulum Torsion Absorbers for Reciprocating Engines," *Journal of Sound and Vibration*, Vol. 159, pp. 251-277.

Cronin, D. L., 1992, "Shake reduction in an automobile engine by means of crankshaft-mounted pendulums," *Mechanism and Machine Theory*, Vol. 27, pp. 517-533.

Borowski, V. J., Denman, H. H., Cronin, D. L., Shaw, S. W., Hanisko, J. P., Brooks, L. T., Milulec, D. A., Crum, W. B., and Anderson, M. P., 1991, "Reducing Vibration of Reciprocating Engines With Crankshaft Pendulum Vibration Absorbers," *SAE Technical Paper Series 911876.*

Lee, C.-T., Shaw, S. W. and Coppola, V. T., 1997, to appear, "A Subharmonic Vibration Absorber for Rotating Machinery," *ASME Journal of Vibration and Acoustics.*

Shaw, S. W., Garg, V. K. and Chao, C-P., 1997, "Attenuation of engine torsional vibrations using tuned pendulum absorbers," *SAE Noise and Vibration Conference and Exposition*, Vol. 2, pp. 713–722.

Lee, C.-T. and Shaw, S. W., 1995, "Torsional vibration reduction in internal combustion engines using centrifugal pendulums," *ASME Design Engineering Technical Conference*, DE-Vol. 84-1, Volume 3-Part A, pp. 487-492.

Lee, C.-T. and Shaw, S. W., 1997, "Nonlinear dynamic response of paired centrifugal pendulum vibration absorbers," *Journal of Sound and Vibration*, Vol. 203, No. 5, pp. 731-743.

Chao, C.-P. and Shaw S. W., 1997, to appear, "Effects of Imperfections and Mistunings on the Performance of the Subharmonic Vibration Absorber System," *Journal of Sound and Vibration.*

Chao, C.-P., Shaw, S. W. and Lee, C.-T., 1997a, "Stability of the Unison Response For a Rotating System with Multiple Centrifugal Pendulum Vibration Absorbers," *ASME Journal of Applied Mechanics*, Vol. 64, pp. 149-156.

Chao, C.-P., Lee, C.-T. and Shaw, S. W., 1997b, to appear, "Non-unison dynamics of multiple centrifugal pendulum vibration absorbers," *Journal of Sound and Vibration.*

Chao, C.-P., 1997, "The Performance of Multiple Pendulum Vibration Absorbers Applied to Rotating Systems." *Ph.D. Dissertation*, Department of Mechanical Engineering, Michigan State University.

DYNAMICAL BEHAVIOR OF CHAINS

FRIEDRICH PFEIFFER
Lehrstuhl B für Mechanik
TU-München, Boltzmannstr. 5
D-85748 Garching/Germany

Abstract: Chains are important machine components being applied in cars, in gears and in armoured vehicles. The dynamics of chains typically is characterized by free motion and by contact processes, which may include impacts and friction. Therefore modeling by multibody theory augmented by methods of contact mechanics represents an appropriate way to evaluate the dynamical chain behavior. Paper gives some examples.

1. Introduction

Typically chains consist of chain elements which might be connected by ideal joints, joints with backlash or by joints like journal bearing. Furthermore and considering Continuous Variable Transmission Systems (CVT) chains may be connected by rocker pins or by more complicated pin systems. Chains come into contact with guides and sprocket wheels, or they have contact with controllable sheaves. In the case of vehicles they have sprocket and ground contact. In all cases contacts start by nullifying relative distances between guides, sprockets, sheaves and some chain element, and contacts end by nullifying some force condition like normal forces or static friction forces. Contacts possess sliding friction or stick-slip transitions, there might be local separation and closure again. Altogether all types of contact mechanical phenomena may occur between chains and their controlling elements.

For a long time chains have been modeled as some kind of uniform strings (Fritzer, 1992; Yue, 1992), because the methods available could hardly deal with chains composed from single elements. Nevertheless polygonial effects have been considered already in a very early stage (Chen and Freudenstein; 1988, Mahaligham, 1958). The basic work of Binder (1956) describes the various possibilities of eigenfrequencies and parametric excitations. All these contributions did not regard contact processes between

F. C. Moon (ed.),
IUTAM Symposium on New Applications of Nonlinear and Chaotic Dynamics in Mechanics, 235–252.

chains and their controlling elements like sprockets, guides, sheaves. In an approximative way the paper (Naja and Marshek, 1983) analyzed the contact forces for a steady state behavior depending on the stress ratio between the tight and slack span. Nakanishi and Shabana (1994) developed very detailed chain models including contact forces for links, sprockets and idlers with special application to large-scale tracked vehicles.

At the author's institute a series of works have been performed dealing with contact problems in multibody systems (Fritzer, 1992; Glocker, 1995; Glocker and Pfeiffer, 1995; Pfeiffer, 1984; Pfeiffer and Glocker, 1996; Prestl, 1991; Seyfferth and Pfeiffer, 1992). These findings form a sound basis for analyzing chains.

2. Contact Dynamics

Multibody systems like chains with multiple unilateral contacts may be dynamically influenced by stick-slip phenomena and by impulsive processes. Such events are indicated by kinematical magnitudes like relative distances or relative velocities which become constraints when the contact event occurs. Examples are a closing contact or a transition from sliding to stiction. Constraints generate constraint forces which for an existing (active) contact situation like contact of two bodies or stiction serve as indicator for the duration of the event. For example, a contact separates when the normal constraint force changes sign, or stiction breaks down for static friction forces smaller than the tangential constraint forces. Designating with $\boldsymbol{g}_N \in \mathbb{R}^{n_N}$ the vector of relative distances in n_N contacts and with $\dot{\boldsymbol{g}}_H \in \mathbb{R}^{n_H}$ the vector of relative velocities in n_H contacts we may write the appropriate equations of motion in the form (Glocker and Pfeiffer, 1995)

$$\begin{gathered} \boldsymbol{M}\ddot{\boldsymbol{q}} - \boldsymbol{h} - \left(\boldsymbol{W}_N + \boldsymbol{W}_G \boldsymbol{\mu}_G \vdots \boldsymbol{W}_H\right) \cdot \begin{pmatrix} \boldsymbol{\lambda}_N \\ \boldsymbol{\lambda}_H \end{pmatrix} = \boldsymbol{0} \in \mathbb{R}^f, \\ \begin{pmatrix} \ddot{\boldsymbol{g}}_N \\ \ddot{\boldsymbol{g}}_H \end{pmatrix} = \begin{pmatrix} \boldsymbol{W}_N^T \\ \boldsymbol{W}_H^T \end{pmatrix} \ddot{\boldsymbol{q}} + \begin{pmatrix} \overline{\boldsymbol{w}}_N \\ \overline{\boldsymbol{w}}_H \end{pmatrix} \in \mathbb{R}^{n_N + n_H}, \end{gathered} \tag{1}$$

where the following abbreviations have been used: $\boldsymbol{M} \in \mathbb{R}^{f,f}$ mass matrix, $\boldsymbol{q} \in \mathbb{R}^f$ generalized coordinates, $\boldsymbol{h} \in \mathbb{R}^f$ all forces except constraint forces, $\boldsymbol{W}_N \in \mathbb{R}^{f,n_N}$, $\boldsymbol{W}_H \in \mathbb{R}^{f,n_H}$, $\boldsymbol{W}_G \in \mathbb{R}^{f,n_N-n_H}$ constraint matrices, $\overline{\boldsymbol{w}}_N \in \mathbb{R}^{n_N}$, $\overline{\boldsymbol{w}}_H \in \mathbb{R}^{n_H}$ constraint vectors mostly excitation terms, $\boldsymbol{\mu}_G = \text{diag}\{-\mu_i \text{sgn}\, \dot{g}_{Ti}\}$ sliding friction coefficient matrix, $\boldsymbol{\lambda}_N \in \mathbb{R}^{n_N}$, $\boldsymbol{\lambda}_H \in \mathbb{R}^{n_H}$ constraint forces (Lagrange vectors). The constraint equations for $\boldsymbol{g}_N$ and $\dot{\boldsymbol{g}}_H$ have been placed in second-order form. The different sets can conveniently be characterized in the following form:

$$\begin{aligned} I_G &= \{1,2,\cdots n_G\} \\ I_S &= \{i \in I_G \mid g_{Ni} = 0\} \text{ with } n_S \text{ elements} \\ I_N &= \{i \in I_S \mid \dot{g}_{Ni} = 0\} \text{ with } n_N \text{ elements} \\ I_H &= \{i \in I_H \mid \dot{g}_{Ti} = 0\} \text{ with } n_H \text{ elements} \end{aligned} \tag{2}$$

The multibody system under consideration possesses altogether n_G contacts. In n_S contacts the relative distance $g_{Ni} = 0$, but only in n_N contacts the relative velocity $\dot{g}_{Ni}$ is also zero. Only for such contacts stick-slip might happen, where stiction takes place in n_H contacts. Consequently, sliding friction developes only in $(n_N - n_H)$ contacts for $i \in I_N \backslash I_H$. Combining further

$$\ddot{\boldsymbol{g}} = \begin{pmatrix} \ddot{\boldsymbol{g}}_N \\ \ddot{\boldsymbol{g}}_H \end{pmatrix}, \ \boldsymbol{\lambda} = \begin{pmatrix} \boldsymbol{\lambda}_N \\ \boldsymbol{\lambda}_H \end{pmatrix}, \ \overline{\boldsymbol{w}} = \begin{pmatrix} \overline{\boldsymbol{w}}_N \\ \overline{\boldsymbol{w}}_H \end{pmatrix},$$

$$\boldsymbol{W} = \left(\boldsymbol{W}_N \vdots \boldsymbol{W}_H\right), \ \boldsymbol{N}_G = \left(\boldsymbol{W}_G \boldsymbol{\mu}_G \vdots \boldsymbol{0}\right), \tag{3}$$

we write eq. (1)

$$\begin{aligned} &\boldsymbol{M}\ddot{\boldsymbol{q}} - \boldsymbol{h} - (\boldsymbol{W} + \boldsymbol{N}_G)\,\boldsymbol{\lambda} = \boldsymbol{0} && \in \mathbb{R}^f \ , \\ &\ddot{\boldsymbol{g}} = \boldsymbol{W}^T \ddot{\boldsymbol{q}} + \overline{\boldsymbol{w}} && \in \mathbb{R}^{n_N + n_H} \ , \end{aligned} \tag{4}$$

which can be solved for $(\ddot{\boldsymbol{q}}, \boldsymbol{\lambda})$ for time instants where no transitions or separations occur. In this special case $\ddot{\boldsymbol{g}} = \boldsymbol{0}$, and we come out with

$$\left[\boldsymbol{W}^T \boldsymbol{M}^{-1} (\boldsymbol{W} + \boldsymbol{N}_G)\right] \boldsymbol{\lambda} = -\left(\boldsymbol{W}^T \boldsymbol{M}^{-1} \boldsymbol{h} + \overline{\boldsymbol{w}}\right) \tag{5}$$

which results in the unknown vector $\boldsymbol{\lambda}$. From the first equation (4) we then evaluate $\ddot{\boldsymbol{q}}$.

In cases of transitions or separations the sets (1) or (4) cannot be evaluated in a straight-forward way. In multiple contact situations every change of one contact only might influence all other contacts. It is not known beforehand what combination of constraints in all other contacts will follow such a change. We could start a combinatorial search for solving this problem, a more elegant way consists in a complementarity formulation (Lemke,

1970; Pfeiffer and Glocker, 1996) by applying a basic law of contact mechanics. It roughly says, that in mechanical contacts either magnitudes of relative kinematics are zero and the thus generated constraint forces are not zero, or vice versa. The product of these two groups of magnitudes is always zero.

To apply this for our chain problem we start with eq. (1), where the kinematical constraints are expressed on an acceleration level. The above law means for normal contacts

$$\ddot{\mathbf{g}}_N \geq \mathbf{0}, \quad \boldsymbol{\lambda}_N \geq \mathbf{0}, \quad \ddot{\boldsymbol{g}}_N^T \boldsymbol{\lambda}_N = 0 \ . \tag{6}$$

These are n_N inequalities for $(\ddot{\boldsymbol{g}}_N, \boldsymbol{\lambda}_N)$ and one complementarity condition. For stick-slip processes a complementarity form is much more complicated to establish, because we must consider very thouroughly the properties of the friction cone in each contact. For a contact i we have with $\ddot{\boldsymbol{g}}_H = \{\ddot{g}_{Ti}\}$ and $\boldsymbol{\lambda}_H = \{\lambda_{Ti}\}$:

$$\lambda_{Ni} \geq 0; \ |\lambda_{Ti}| \leq \mu_{0i}\lambda_{Ni} \ ;$$

$$\left.\begin{cases} |\lambda_{Ti}| < \mu_{0i}\lambda_{Ni} & \Rightarrow \ddot{g}_{Ti} = 0 \\ \lambda_{Ti} = +\mu_{0i}\lambda_{Ni} & \Rightarrow \ddot{g}_{Ti} \leq 0 \\ \lambda_{Ti} = -\mu_{0i}\lambda_{Ni} & \Rightarrow \ddot{g}_{Ti} \geq 0 \end{cases}\right\} ; \ i \in I_H \tag{7}$$

The inequalities (7) can be put in a standard complementarity form by a rather tedious decomposition procedure (Glocker and Pfeiffer, 1995; Pfeiffer and Glocker, 1996). Altogether we always can combine the equations (1), (6), (7) in a final set forming a Linear Complementarity Problem in its standard form (Lemke, 1970)

$$\begin{aligned} &\boldsymbol{y} = \boldsymbol{A}\boldsymbol{x} + \boldsymbol{b}, \ \boldsymbol{y} \geq \mathbf{0}, \ \boldsymbol{x} \geq \mathbf{0}, \ \boldsymbol{y}^T\boldsymbol{x} = 0 \, , \\ &\boldsymbol{y}, \ \boldsymbol{x} \in \mathbb{R}^{n_N + 4n_H} \ . \end{aligned} \tag{8}$$

It should be noted that for general spatial contact problems a formulation of a linear complementarity problem cannot be achieved. Spatial contact problems are connected with two tangential force directions in the plane of contact. As a consequence the direction of motion after a transition stick to slip, for example, depends on the geometric sum of two tangential forces thus generating a nonlinear complementarity form (Wösle and Pfeiffer, 1996).

For describing our chain problem we additionally need an impact analysis which will not be presented. All details about impacts with and without friction may be found in (Glocker, 1995; Pfeiffer and Glocker, 1996).

3. Chain and Chain Elements

3.1. GENERAL REMARKS

In modeling chains and chain elements we first must choose convenient coordinate frames. We start with an inertial coordinate base chosen in such a way that the geometry of the elements becomes as simple as possible, for example located in the centers of sprockets or of sheaves. We further fix at each element (chain element, sheave, sprocket, guides) a moving body-fixed coordinate system in a convenient point for instance in the mass-center. Sometimes it might be convenient to have a second body-related frame which is inertially fixed with one axis and rotates with a nominal speed, for example for shafts and wheels. Anyway, the system of coordinate frames must allow a complete and unambiguous geometrical description of all system elements.

Secondly, we must define a nominal motion, which is for chain dynamics nontrivial. We start with a prescribed (time-dependent or not) motion of the driving component, for instance one sprocket, or one pulley. But then we have to decide how a nominal motion of the chain system might look like. One could go back to the string-like model as developed earlier which is a reasonable basis. A more advanced nominal motion can be evaluated by taking into account the discrete chain structure and thus the polygonial effects but by not considering the detailed friction and impact phenomena. By experience of the last years this approach gives good results and, moreover, a sound basis for more refined investigations.

3.2. LINKS AND JOINTS

Typical links and joints for chains are very similar in different applications like roller chains in cars, rocker pin chains in CVT-gears or even chains for tracked vehicles. A typical chain element is shown in Fig. 1. Generally every element possesses six degrees of freedom though in most application it will be sufficient to consider only three degrees of freedom (two DOF translation, one DOF rotation).

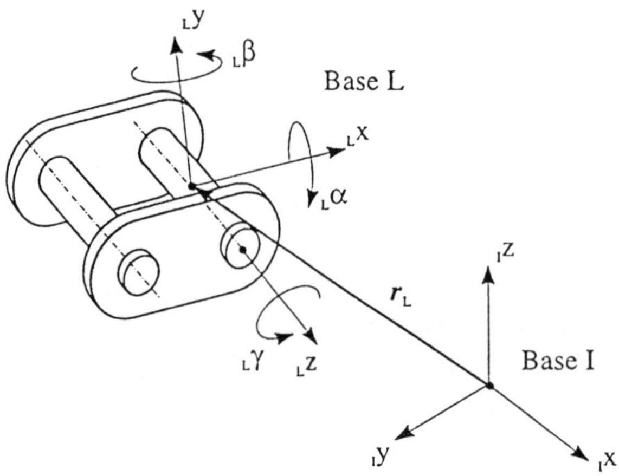

Figure 1. Typical chain link

From Figure 1 we get the coordinates of the link

$$_L\boldsymbol{r} = \begin{pmatrix} _Lx \\ _Ly \\ _Lz \end{pmatrix}, \quad _L\boldsymbol{\varphi} = \begin{pmatrix} _L\alpha \\ _L\beta \\ _L\gamma \end{pmatrix}, \quad _L\boldsymbol{q} = \begin{pmatrix} _L\boldsymbol{r} \\ _L\boldsymbol{\varphi} \end{pmatrix}, \tag{9}$$

which might be written down in any coordinate frame. For the roller chain- and CVT-models to follow we consider only the plane case ($_Lz = 0$, $_L\alpha = 0$, $_L\beta = 0$).

For the chain joints we may either use an ideal joint model reducing further the number of degrees of freedom to one namely $_L\boldsymbol{\varphi}$, and two additional geometrical constraints for $_L\boldsymbol{r}$, which then follows from a kinematical consideration. It is

$$_L\boldsymbol{\varphi} = \begin{pmatrix} 0 \\ 0 \\ _L\gamma \end{pmatrix}, \quad _L\boldsymbol{r} = {_L\boldsymbol{r}}({_L\boldsymbol{q}}) = \begin{pmatrix} _Lx \\ _Ly \\ 0 \end{pmatrix}. \tag{10}$$

Alternatively we may apply an approximation as a journal bearing with symmetric stiffness and damping matrices (Figure 2).

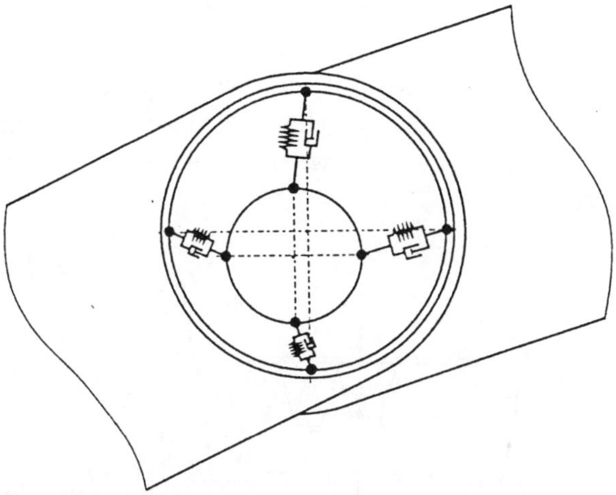

Figure 2. Chain connection as a journal bearing

3.3. CONTACT AND CONTROLING ELEMENTS

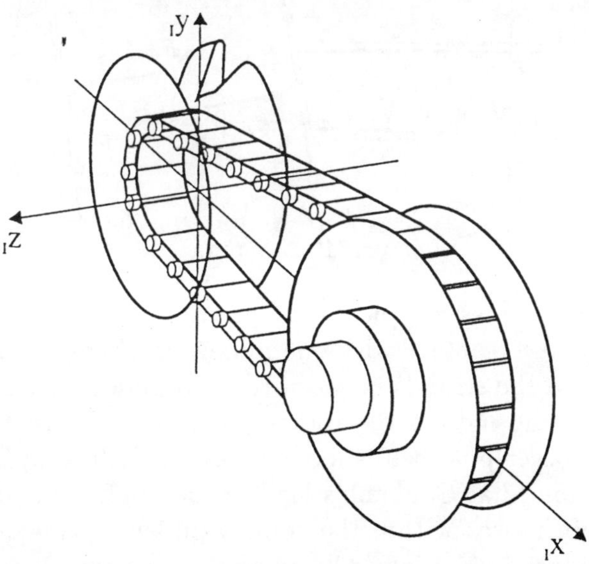

Figure 3. Typical CVT-configuration

CVT-chains are controlled by movable and non-movable sheaves, roller chains by sprockets and guides. The contact processes in the first case include impacts at the sheave entry and stick-slip phenomena during sheave contact. At the pulley exit we might have an impulsive effect when the chain rocker pin leave the pulley. In the second case of roller chains we do not observe stick-slip processes but more sliding friction effects. Instead we have more impulsive processes on the guides and when entering or leaving the sprockets. Separation effects with subsequent contacts might happen along the guides.

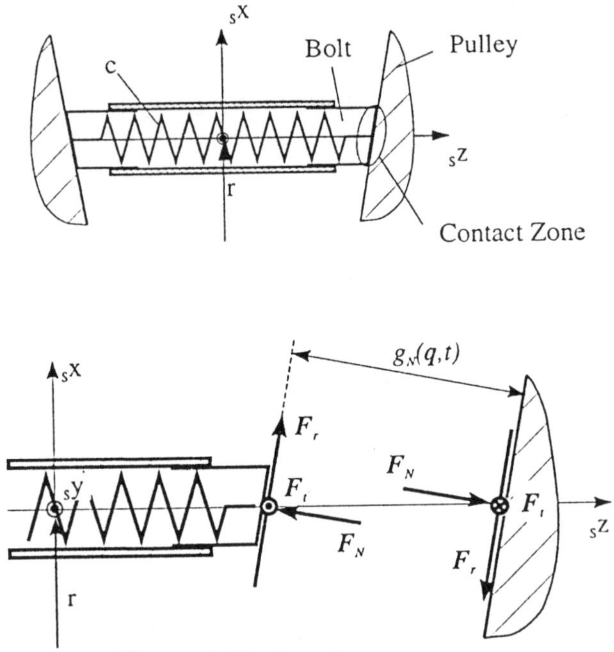

Figure 4. Rocker pin contacts

Figure 3 illustrates a typical configuration of a CVT-gear. The chain elements may have contact with one pulley, and they are moving without such contacts in the strands between the two pulleys. Pulley 1 is the driving wheel. During motion the rocker pin contacts with the pulley cones change. The rocker pin itself is modeled as one bolt with linear axial stiffness. We assume that Coulomb's law can be applied in the contact zone. The rocker pin moves within the conical pulley in two directions, radial and circumferential direction. This gives rise to a two-dimensional contact problem with nonlinear complementarity properties (Figure 4). With respect to Coulomb's law it turns out that computing time could in this case be reduced significantly by approximating the dry friction characteristic by a continuous function.

The relative normal distance $g_{Ni}(\boldsymbol{q}, t)$ and the relative tangential velocity $\dot{g}_{Ti}(\boldsymbol{q}, \dot{\boldsymbol{q}}, t)$ must be expressed in generalized coordinates and included in eqs. (1). The forces F_r, F_t in radial and circumferential directions depend on the constraint force vectors $(\boldsymbol{\lambda}_N, \boldsymbol{\lambda}_H)$. It should be noted that $g_{Ni}(\boldsymbol{q}, t)$ results from the approaching motion of the rocker pin as part of a strand element when approaching the pulley.

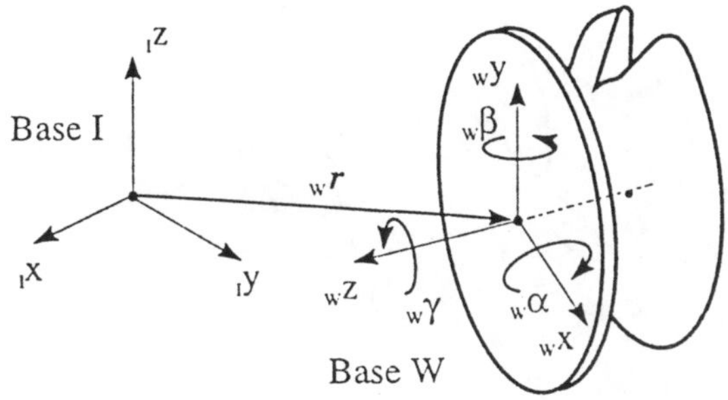

Figure 5. Wheel kinematics

The mechanical models for the pulleys are comparatively simple and very similar to those of sprockets. In both cases we have wheels with three degrees of freedom in a plane model and six degrees of freedom in a spatial model. Stiffness and damping of the wheel-bearings depend on their design, for instance roller or journal bearings. It is

$$
\begin{aligned}
{}_W\boldsymbol{r} &= \begin{pmatrix} {}_Wx \\ {}_Wy \\ {}_Wz \end{pmatrix}, \quad {}_W\boldsymbol{\varphi} = \begin{pmatrix} {}_W\alpha \\ {}_W\beta \\ {}_W\gamma \end{pmatrix}, \\
{}_W\boldsymbol{q} &= \begin{pmatrix} {}_W\boldsymbol{r} \\ {}_W\boldsymbol{\varphi} \end{pmatrix},
\end{aligned} \tag{11}
$$

where the index W stands for sprocket or sheave and the angles ${}_W\boldsymbol{\varphi}$ of deviation are small. In the plane case ${}_Wz = 0$, ${}_W\alpha = 0$, ${}_W\beta = 0$. It should be noted however that for more detailed investigations the elastic deformations of the pulley must be taken into account. This is a topic of ongoing research.

Coming back to the roller chains and considering the sprocket wheel we must take into account the external torques from the camshafts and the rotational vibrations of the crankshaft. In addition to the chain forces between the links contact forces act on the chain bolts and on the sprocket toothing (Figure 6). Particularly the impulsive forces due to the polygonial excitation, when a link enters the sprocket, influence the sprocket dynamics and the dynamics of the chain strands.

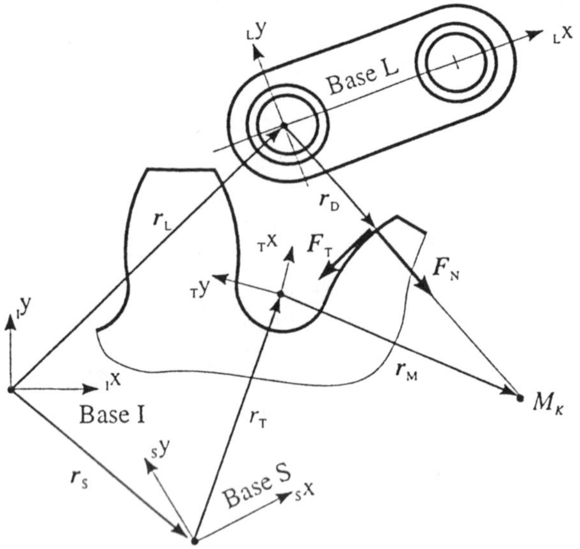

Figure 6. Contact forces between chain and sprocket

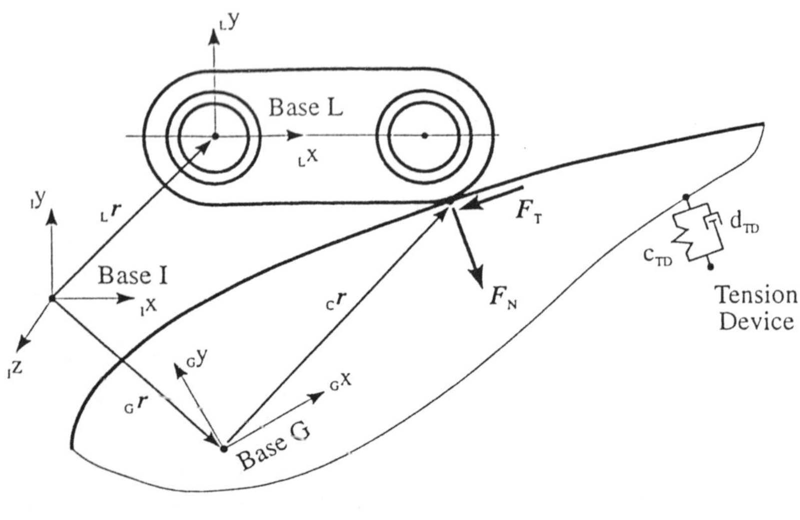

Figure 7. Guide-chain dynamics

Guides (Figure 7) are applied in chain drives to reduce the vibrations of the chain strands and to preload the chain with a definite stress, produced by some tension device. Therefore guides develop their own dynamics, so that the links might loose their contact to the guide. At each contact of the links normal and tangential forces act on the guide and the link plate. Again we have to calculate the impulsive forces when a contact between a link and a guide becomes active.

4. Examples

4.1. ROCKER PIN CHAINS IN CVT-GEARS

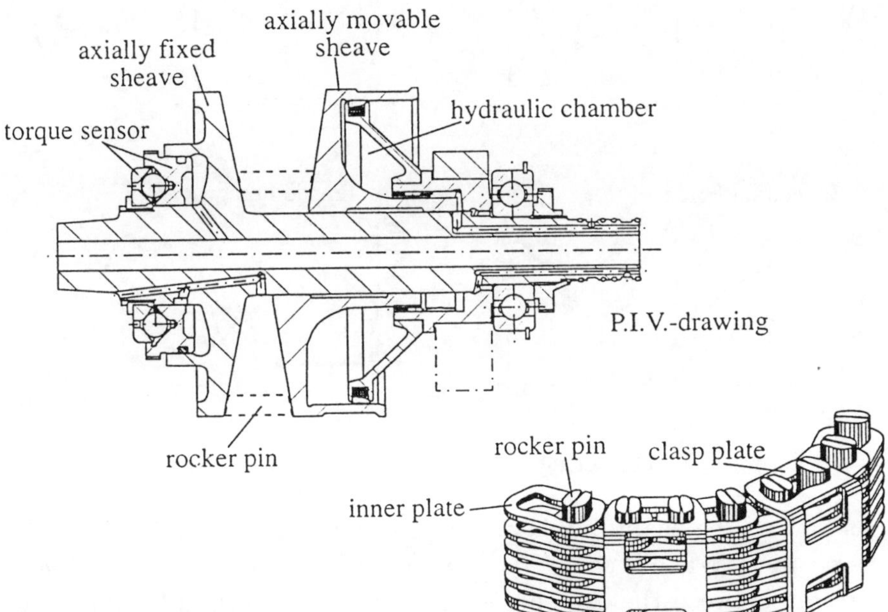

Figure 8. CVT chain drive components

Rocker Pin Chains represent one typical design in CVT-gears. Figure 8 shows the main elements, and their assemblage. The continuous variable properties are achieved by axially moving sheaves which are shifted hydraulically. This device applies also the necessary forces on the chain. The mechanical model takes into account the plane motion of the 63 links with 189 DOF, the driven pulley with one rotational DOF and a torsional excitation, the driving pulley with no DOF but a torsional excitation. The results are computed for a stationary operation with constant driving speed. Computing time for this case amounts about 8 hours on a SUN-workstation. Figure 9 shows the contact forces acting on a pair of rocker pins during one revolution as well as the tensile force in the corresponding chain link.

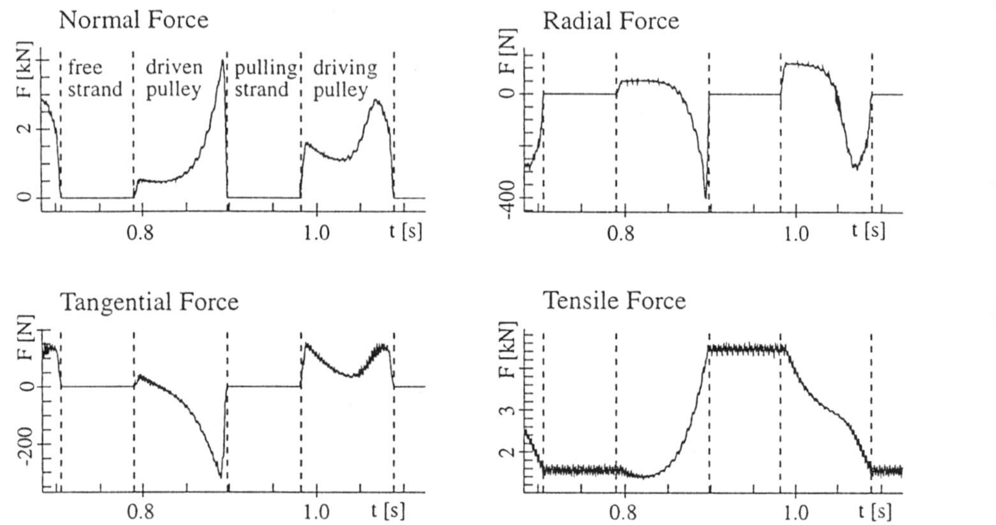

Figure 9. Forces acting on the rocker pins and in a chain link

As long as the chain link is part of a strand no contact forces work on its rocker pins. When it comes into contact with one of the pulleys the pins are pressed between the two sheaves and hence the normal force increases. Its amplitude depends on the geometry of the sheaves and the transmitted power. The frictional force is a function of this normal force and the relative velocity between the pulley and the pins. It is split into one radial and one circumferential component. The radial contact force coincides with a radial movement of the chain link generating power dissipation. In contrast to this the circumferential contact force causes the changes of the tensile force in the corresponding chain link, leading to different tensile force levels in the two strands which agree with the transmitted torque.

Due to the mechanical model the simulation provides an integrated value of the tensile force in the plates of a chain link, whereas measurement was performed for the tensile force in a clasp plate[1] (Figure 10 left).

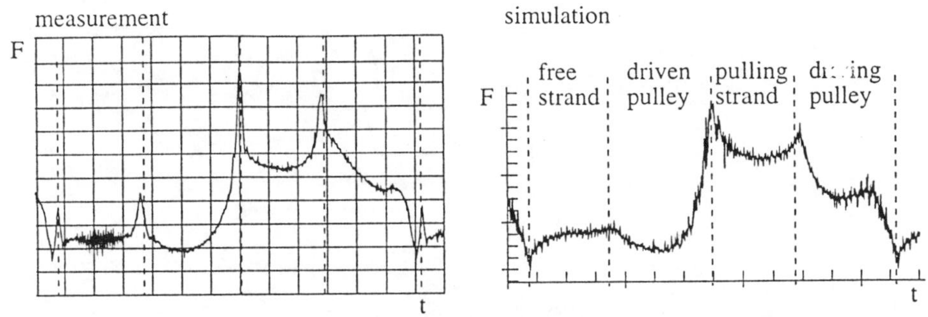

Figure 10. Tensile force in the clasp plate[1]

Therefore it is necessary to determine the distribution of the tensile force on the plates of the chain links. Using the results of the dynamic simulation shown in Figure 9 and modeling the pair of rocker pins as bending beams as well as the plates as linear springs we evaluated the graph of Figure 10 right for the clasp plate. The comparison of simulation and measurement confirms the mechanical model.

4.2. ROLLER CHAINS IN A CAR DRIVE SYSTEM

In combustion engines chain drives are applied as a driving unit for camshafts and injection pumps. The chain drive is composed of several sprockets (crankshaft, camshafts) and several guides. Figure 11 portrays a typical configuration, which is basis of the theoretical and experimental research.

The chain consists of 120 links with 360 DOF. The crankshaft (CrS) is excited by a rotational vibration. The angle dependent torques from the valve drive system as a result from measurement act on the camshafts (CS). On the right side of the chain drive the guide (G) is inertial fixed. On the other side the tension guide (TG) possesses one rotational DOF. The guide

[1] Measurements performed by G. Sauer and K.Th. Renius, Lehrstuhl für Landmaschinen, Technical University of Munich

is pressed against the chain by a hydraulic tension device (TD).

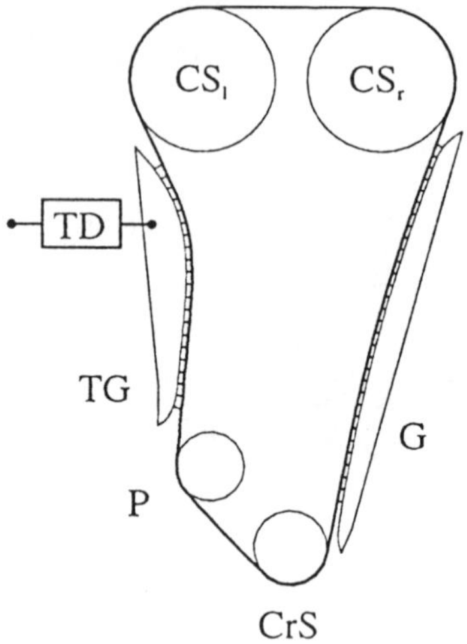

Figure 11. Configuration of a typical roller chain drive (BMW 318 is)

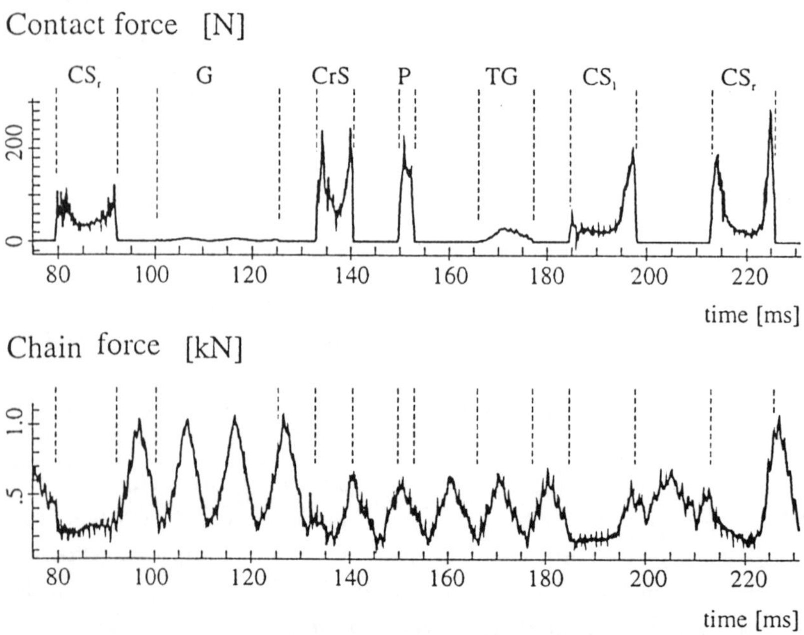

Figure 12. One cycle of a link

One cycle of a chain link is presented in Figure 12. The first graph shows the contact forces between the link and the sprockets or guides. Regarding the contact between a link and a sprocket (Naja and Marshek, 1983) the typical march of the contact force can be seen clearly. At the beginning and the end of the contact phase there is a peak in the contact force, due to the stress of the chain strands. These forces decrease rapidly when the link moves along the toothing of the sprocket. Because of the low curvature of the guide these contact forces are on a very low level.

The second graph presents the chain force of one link. The time history follows mainly the torque excitation of the valve drive unit. The effects of the polygonial excitation from the sprocket results in additional forces with small amplitudes. On the tight side of the chain drive the average forces are twice as large as the forces on the slack side.

Figure 13 portrays a comparison between measurements (left side) for the BMW 318 is - chain with corresponding simulations (right side). The speed of 4320 rpm lies around an internal resonance which is essentially generated by the dynamics of the hydraulic tension device in combination with other components. This resonance is well damped. The graphs at the top illustrate the amplitude response to a crankshaft-excitation for the tension guide motion on the left side (Figure 11).

The middle graphs depict the frequency response for the forces in the hydraulic tension device. In both cases the second engine order is dominant, whereas the polygonial frequency shows no significant effect. The bottom diagrams give a comparison of these tension device forces on a time scale. Measurements and simulations compare very well which has been also confirmed by many simulation and measurements for the whole speed range.

5. Summary

The dynamic of chains is investigated by applying multibody theory with unilateral contacts. The problem of multiple and non-decoupled contacts being typical for chains affords a special treatment of multibody systems including aspects of linear and nonlinear complementarity theories. The availability of such tools is an indispensable prerequisite for modeling complex systems like chains with their numerous impulsive and friction-induced contact processes. The related theory is presented and applied to two cases, the case of rocker pin chains in CVT-gears and the case of roller chains in car applications. The results and comparisons with measurements confirm the way of modeling such systems. Further research has to be invested in numerical and computer time problems.

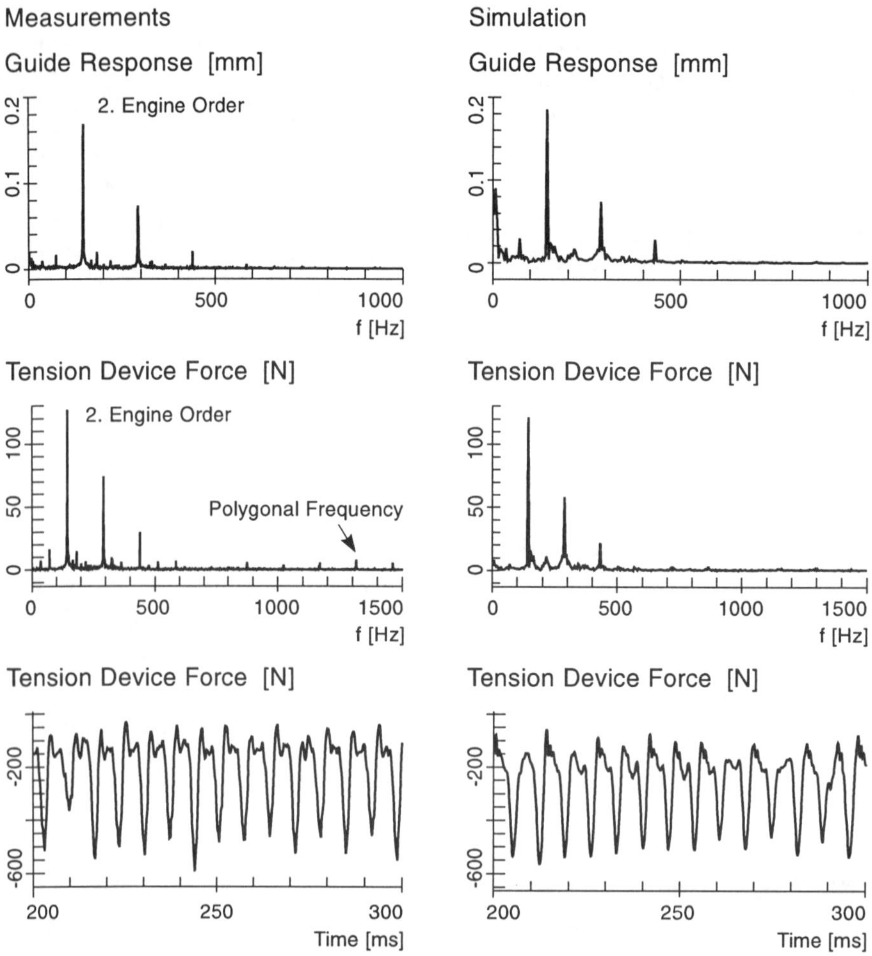

Figure 13. Theory and measurement for a 4-stroke-engine (BMW 318 is with 4320 rpm)

6. References

Binder, R.C. (1956) Mechanics of the Roller Chain Drive, Prentice Hall, Englewood Cliffs, New Jersey.

Chen, C.K. and Freudenstein, F. (1988) Towards a More Exact Kinematics of Roller Chain Drives, *Trans. ASME, Journal of Mechanisms, Transmissions and Automation in Design,* Vol.110, 169-275.

Fritz, P. and Pfeiffer, F. (1995) Dynamics of High Speed Roller Chain Drives, *ASME Conf. on Vibration and Noise,* Boston.

Fritzer, A. (1992) Nichtlineare Dynamik von Steuertrieben, VDI-Fortschrittsberichte, Reihe 11, Nr.176, VDI-Verlag, Düsseldorf.

Glocker, Ch. (1995) Dynamik von Starrkörpersystemen mit Reibung und Stößen, *VDI-Fortschrittsberichte*, Reihe 18, Nr. 182, VDI Verlag, Düsseldorf.

Glocker, Ch. and Pfeiffer, F. (1995) Multiple Impacts with Friction in Rigid Multibody Systems, *Nonlinear Dynamics,* Klumer Academic Publishers, Netherlands.

Lemke, C.E. (1970) Recent Results on Complementarity Problems, In: Rosen, J.B. et al. (eds.), *Nonlinear Programming.* Academic Press, New York, 349-384.

Mahaligham, S. (1958) Polygonial Action in Chain Drives, *J. Franklin Institute,* Vol.256, No.1, 23-28.

Naja, M.R. and Marshek, K.M. (1983) Analysis of the Sprocket Load Distribution, *Mechanism and Machine Theory,* Vol.18, No.5, 349-356.

Pfeiffer, F. (1984) Mechanische Systeme mit unstetigen Übergängen, *Ingenieur-Archiv,* Vol.54, 232-240.

Pfeiffer, F. and Glocker, Ch. (1996), Multibody Dynamics with unilateral Contacts, Nonlinear Science Series, John Wiley, New York.

Prestl, W. (1991) Zahnhämmern in Rädertrieben von Dieselmotoren, *VDI Fortschrittsberichte,* Reihe 11, Nr.145, VDI-Verlag, Düsseldorf.

Seyfferth, W. and Pfeiffer, F. (1992) Dynamics of Rigid and Flexible Part Mating with a Manipulator, *Proc. of IMACS Symp. MCTS,* 13-23.

Srnik, J. and Pfeiffer, F. (1994) Simulation of a CVT Chain Drive as a Multibody System with Variant Structure, *CISS-First Conference of International Simulation Societies,* Zürich.

Nakanishi, T. and Shabana, A.A. (1994) Contact Forces in the Nonlinear Dynamic Analysis of Tracked Vehicles, *International Journal for Numerical Methods in Engineering,* Vol.37, 1251-1275.

Wösle M. and Pfeiffer, F. (1996) Dynamics of Multibody Systems Containing Dependent Unilateral Constraints with Friction, *Journal of Vibration and Control,* 2:161-192, Sage Publications, Inc.

Yue, G. (1992) Belt Vibration Consideration Moving Contact and Parametric Excitation, *ASME - International Power Transmission and Gearing Conference,* DE-Vol.43-1, Vol.1, 311-318.

CHAOS IN OFFSET HYDRODYNAMIC ROTOR BEARINGS

S. BOEDO
Staff Engineer, Ithaca Technical Center
Borg-Warner Automotive, Ithaca, NY 14850 USA

Nonlinear diagnostic tools are applied to the dynamic analysis of the dual offset bearing over a representative set of design parameters employed in conventional rotordynamic applications. For particular values of gravity angle and dimensionless offset ratio, case studies reveal that improvements in bearing performance (as measured by cyclic-maximum segment eccentricity ratio) appear to be linked to observed period doubling and chaotic motion.

1. Introduction

Multi-lobe and tilting-pad bearing designs are often employed in rotordynamic machinery to alleviate half-speed whirl instability encountered with cylindrical journal bearings. An alternative and perhaps less costly design approach is to divide axially both journal and sleeve into two or more cylindrical segments with offset centerlines as shown in the exploded view of Figure 1, whereby each section retains circumferential symmetry. (Each cylindrical segment in Figure 1 is in fact complete.)

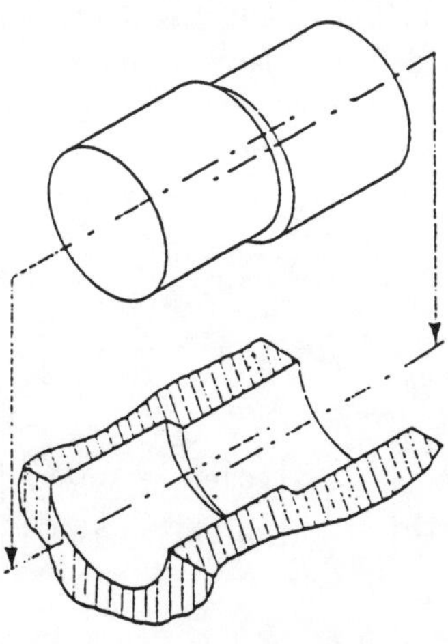

Figure 1. Dual offset bearing

F. C. Moon (ed.),
IUTAM Symposium on New Applications of Nonlinear and Chaotic Dynamics in Mechanics, 253–260.

Booker and Olikara (1984) found improved performance for offset bearings in which the load rotates at half shaft speed, the result of which prompted Booker and Govindachar (1984) to provide the first and apparently only study devoted to the stability of offset journal bearings for rotating machinery. However, their study was limited in the range of parameter space, and their conclusions related to the dynamic characteristics of the system remained essentially subjective.

This paper applies Poincaré maps, Lyapunov exponents, and frequency spectra to assess the dynamic behavior of offset bearings over a region of design parameter space known to produce undesirable period 2 limit cycles (half-speed whirl) in conventional bearings. As in Booker and Govindachar (1984), this study will be restricted to dual offset bearings shown in Figure 1 with a common offset applied to both journal and sleeve. In addition, the π-film (cavitating) short bearing lubrication model will be employed for each bearing segment. The use of the short bearing model allows for rapid qualitative assessment of bearing performance over a relatively large design space with minimal computational effort.

2. Problem Formulation

Much of what follows is based heavily on the analysis of Booker *et al.* (1982), Booker and Olikara (1984), and Booker and Govindachar (1984); a brief outline is presented here for completeness.

2.1 KINEMATIC RELATIONS

The geometry of the dual offset bearing is best understood by first considering separate descriptions of (disassembled) sleeve and journal as shown in Figures 2a and 2b, respectively. The sleeve is fixed in space and the journal rotates relative to the sleeve. In Figure 2a, the originally cylindrical sleeve is divided into two offset segments; each sleeve segment offset $\mathbf{a}_i$ is defined relative to fixed reference axes x,y. Similarly, in Figure 2b, the originally cylindrical journal is divided into two offset segments; each journal segment offset $\mathbf{b}_i$ is defined relative to reference axes x',y' which are fixed to and rotate with the journal at a constant angular velocity $\omega > 0$ about the sleeve z-axis.

Assembling journal and sleeve, Figure 2c shows that segment eccentricity $\mathbf{e}_i$ is related to journal reference position $\mathbf{e}_0$ by the kinematic loop

$$\mathbf{e}_i = \mathbf{e}_0 - \mathbf{a}_i + \mathbf{b}_i \tag{1a}$$

with sleeve-frame components

$$\begin{Bmatrix} e_i^x \\ e_i^y \end{Bmatrix} = \begin{Bmatrix} e_0^x \\ e_0^y \end{Bmatrix} - \begin{Bmatrix} a_i^x \\ a_i^y \end{Bmatrix} + \begin{bmatrix} \cos\omega t & -\sin\omega t \\ \sin\omega t & \cos\omega t \end{bmatrix} \begin{Bmatrix} b_i^{x'} \\ b_i^{y'} \end{Bmatrix} \tag{1b}$$

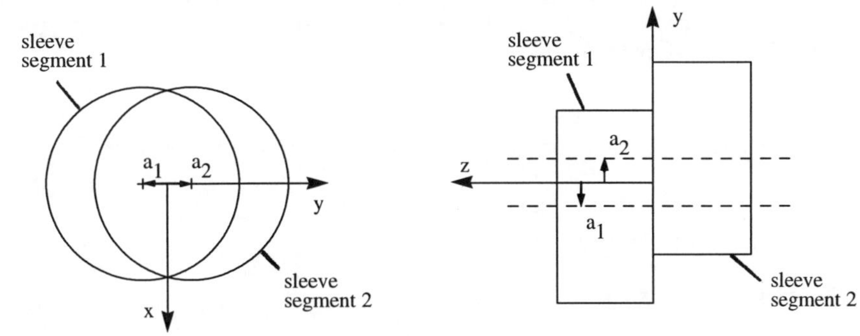

(a) sleeve geometry

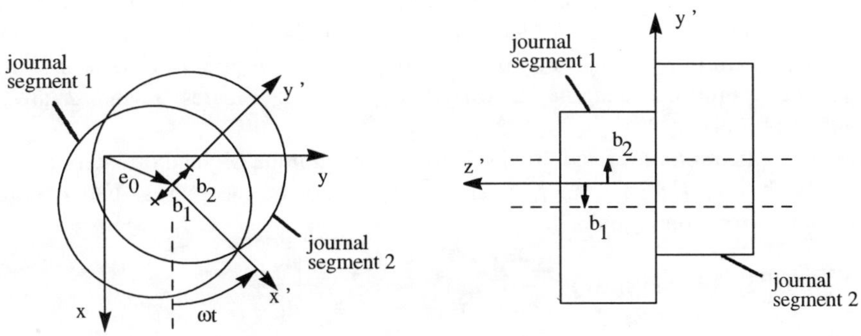

(b) journal geometry

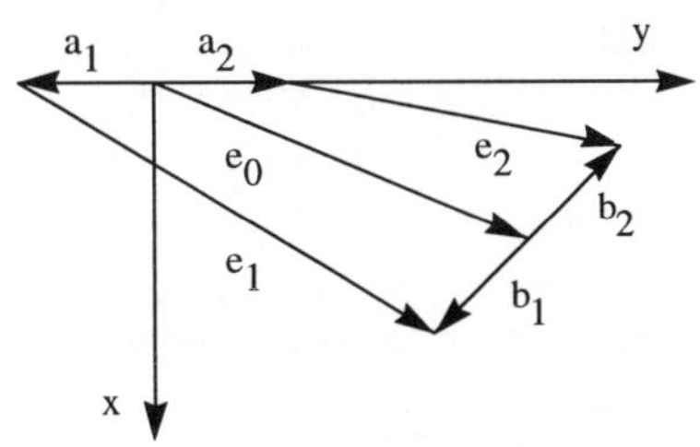

(c) kinematic relations

Figure 2. Geometry and kinematic relations

Thus, for non-zero journal offsets, the orientation of the journal relative to the sleeve is completely determined by specification of time t and *reference* position $\mathbf{e}_0$.

2.2 EQUATIONS OF MOTION

For the analysis which follows, each bearing segment has length L, diameter D, and radial clearance C, while segment offset components are given by

$$a_1^x = b_1^{x'} = a_2^x = b_2^{x'} = 0$$

$$a_1^y = b_1^{y'} = -\delta/2$$

$$a_2^y = b_2^{y'} = +\delta/2$$

where relative offset δ is defined over the range $0 \le \delta < C$. This application of identical segment offsets to both journal and sleeve in their respective reference frames results in the limiting condition $\delta < C$ required to assure full rotation without interference. (Journal and sleeve offsets depicted in Figures 1 and 2 are thus exaggerated for clarity.)

Upon application of a fixed gravity load with magnitude Mg and orientation angle β relative to the x,y sleeve frame, dynamic equilibrium of the journal under assumed steady journal rotation requires

$$Mg\cos\beta\,\mathbf{i} + Mg\sin\beta\,\mathbf{j} - \sum_{i=1}^{2}\mathbf{P}_i = M\frac{d^2\mathbf{e}_0}{dt^2} \tag{2}$$

in terms of segment film loads

$$\mathbf{P}_i = 2\mu L(R/C)^3 V_i \mathbf{W}_i$$

which, as described by Childs *et al.* (1977), depends upon segment squeeze velocity magnitude V_i and segment impedance $\mathbf{W}_i$. Segment impedance for short bearings is proportional to $(L/D)^2$ and is a function only of segment eccentricity ratio $\varepsilon_i \equiv \mathbf{e}_i/C$ and segment squeeze velocity vector orientation. The foregoing analysis also requires ambient pressure between the bearing segments which would be supplied in practice by a circumferential groove.

Introducing dimensionless time and dimensionless reference position

$$\tau = \omega t \qquad\qquad \varepsilon_0 = \mathbf{e}_0/C$$

dimensionless load, speed, and offset design parameters

$$F = \frac{Mg(C/R)^2 \ (D/L)^2}{\mu LD\omega} \qquad \sigma = \frac{\omega}{(g/C)^{1/2}} \qquad \bar{\delta} = \delta / C$$

and dimensionless state variable vector $\mathbf{z} \equiv [z_1, z_2, z_3, z_4]$ with components

$$z_1 \equiv \varepsilon_0^x \qquad z_2 \equiv \varepsilon_0^y \qquad z_3 \equiv \frac{d\varepsilon_0^x}{d\tau} \qquad z_4 \equiv \frac{d\varepsilon_0^y}{d\tau}$$

equation (2) can be rewritten in the functional form

$$\frac{d\mathbf{z}}{d\tau} = \mathbf{f}(\mathbf{z} \ ; \ F, \ \sigma) \qquad \bar{\delta} = 0 \tag{3a}$$

$$\frac{d\mathbf{z}}{d\tau} = \mathbf{f}(\mathbf{z} \ , \ \tau \ ; \ F, \sigma, \ \bar{\delta}, \ \beta) \qquad 0 < \bar{\delta} < 1 \tag{3b}$$

which includes conventional zero-offset cylindrical bearings. Specification of state initial conditions completes problem formulation.

3. Case Studies

Unlike conventional cylindrical rotors given in equation (3a), equation (3b) represents a non-autonomous system due to the incorporation of journal offset segments. Hence, a fixed equilibrium point does not exist for such offset bearings, and thus linearized small-motion stability analyses employed in representative studies by Holmes (1960) and Hollis and Taylor (1986) are not applicable here. Instead, offset bearing stability is investigated using nonlinear diagnostic tools based on the time history of state vector $\mathbf{z}$.

Fixing gravity orientation $\beta = 0$ and offset ratio $\bar{\delta} = 0.3$, we present some case studies which describe the behavior of offset bearings for load and speed parameters which lie near the well-known instability threshold locus of cylindrical bearings (Holmes, 1960). Each case is started with zero initial conditions and is run until the resulting attractor is established. Power spectral density per unit time is created from a fast-fourier transform (FFT) of dimensionless journal reference position $\varepsilon_0^{\hat{x}}(\tau)$ comprised of 2^{15} points at a sample rate corresponding to 0.25 degree of journal rotation. The largest Lyapunov exponent is estimated by repeated measurement of the divergence over time of randomly-selected trajectories of journal reference position which are initially close to the attractor (Moon, 1986).

Figure 3 shows that the offset bearing undergoes periodic motion synchronous with journal rotation for design parameters $F = 11$, $\sigma = 1$. (Circles mark position after each journal rotation.) This period 1 motion contrasts with the transient approach to a fixed equilibrium point found with the conventional cylindrical bearing. Yet, Table 1 shows

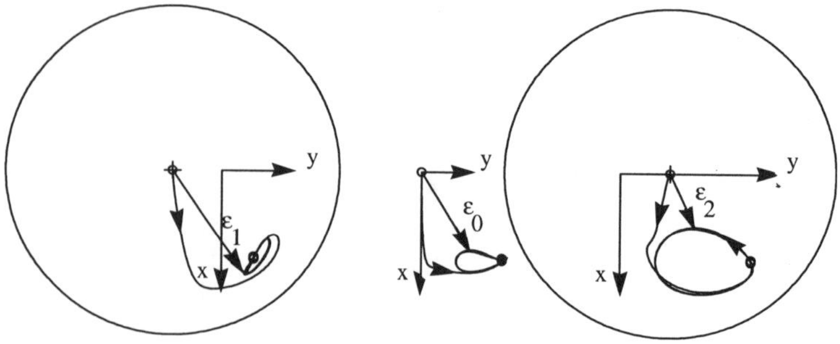

Figure 3. Period 1 motion: F = 11, $\sigma = 1$, $\beta = 0$, $\bar{\delta} = 0.3$

that the resulting cyclic-maximum segment eccentricity ratio for the offset bearing is only somewhat larger than the static eccentricity ratio for the cylindrical bearing.

Of particular interest is the behavior of offset bearings for load and speed parameters which can produce period 2 limit cycles in conventional bearings. Period 2 (half-speed) subharmonic journal motion is encountered in the offset bearing with load and speed parameters F = 11 and $\sigma = 4$. Adjusting the speed parameter to $\sigma = 5$, the resulting journal motion jumps to a period 4 subharmonic attractor. Further adjusting the speed parameter to $\sigma = 6$, Figures 4 and 5 show the appearance of chaotic journal motion as indicated by the fractal-like structure of the Poincaré projection coupled with a relatively broad frequency spectra and a converged positive Lyapunov exponent. The corresponding cylindrical bearing (with zero initial conditions) produces a large period 2 limit cycle for $\sigma = 4$ which continues to grow as the speed parameter is increased. Comparison of cyclic-maximum eccentricity ratio in Table 1 shows that the process of period doubling into chaos for offset bearings reduces eccentric journal motion and hence improves bearing performance over their cylindrical counterparts.

TABLE 1. Cyclic-maximum eccentricity ratio

F	σ	Cylindrical bearing, $\bar{\delta} = 0$		Offset bearing, $\beta = 0$, $\bar{\delta} = 0.3$ segment 1	segment 2	
11	1	0.718	fixed point	0.779	0.793	period 1
11	4	0.972	period 2	0.946	0.941	period 2
11	5	0.986	period 2	0.890	0.775	period 4
11	6	0.989	period 2	0.894	0.790	chaotic

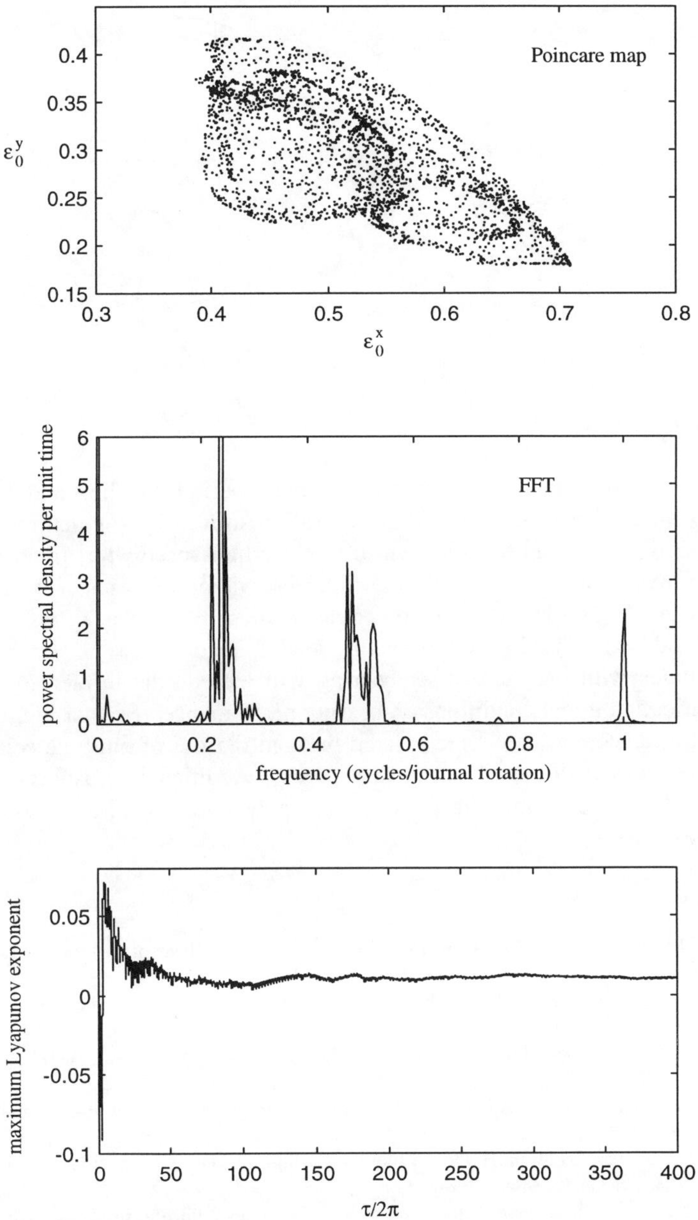

Figure 4. Chaotic motion: F = 11, $\sigma = 6$, $\beta = 0$, $\bar{\delta} = 0.3$

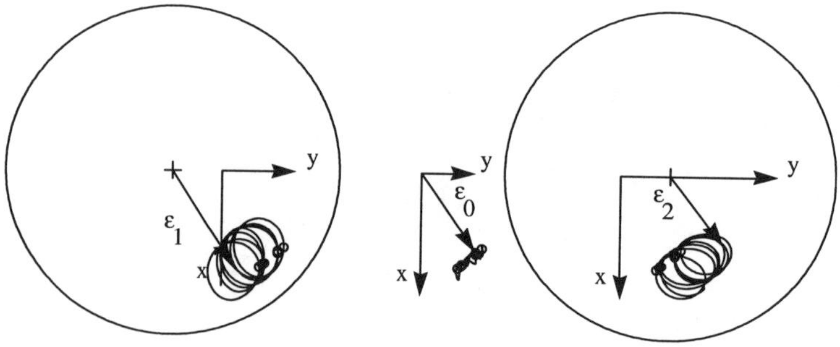

Figure 5. Chaotic motion: $F = 11$, $\sigma = 6$, $\beta = 0$, $\bar{\delta} = 0.3$

4. Conclusions

Case studies have shown that offset bearings, through period doubling and chaos, can provide improvements in bearing performance (as measured by segment eccentricity ratio) for load and speed conditions known to produce half-speed whirl in conventional cylindrical bearings. This is apparently the first observation of chaos with a balanced gravity load; chaos was observed in numerical studies by Brown *et al.* (1994) for cylindrical journal bearings subjected to rotating load imbalance.

Further studies with the dual offset bearing will include the influence of gravity angle, offset ratio, and initial conditions on bearing performance. Further improvements in bearing performance are also contemplated through the use of multiple segments or through variable journal/sleeve offset specifications. Additional measures of bearing performance such as maximum film pressure and power loss require inclusion in the design process.

5. References

Booker, J.F., Goenka, P.K., and van Leeuwen, H.J. (1982) Dynamic analysis of rocking journal bearings with multiple offset segments, *ASME J. Lub. Tech.* **104**, 478-490.

Booker, J.F. and Govindachar, S. (1984) Stability of offset journal bearing systems, in *Vibrations in Rotating Machinery*, Mechanical Engineering Publications Ltd., pp. 269-275.

Booker, J.F. and Olikara, P. (1984) Dynamics of offset bearings: parametric studies, *ASME J. Trib.* **106**, 352-359.

Brown, R.D., Addison, P., and Chan, A.H.C. (1994) Chaos in the unbalance response in journal bearings, *Nonlinear Dynamics* **5**, 421-432.

Childs, D., Moes, H., and van Leeuwen, H. (1977) Journal bearing impedance descriptions for rotordynamic applications, *ASME J. Lub. Tech.* **99**, 198-213.

Hollis, P. and Taylor, D.L. (1986) Hopf bifurcation to limit cycles in fluid film bearings, *ASME J. Trib.* **108**, 184-189.

Holmes, R. (1960) The vibration of a rigid shaft on short sleeve bearings, *J. Mech. Eng. Sci.* **2**, 337.

Moon, F.C. (1986) *Chaotic Vibrations*, John Wiley & Sons, New York.

V. IMPACT AND NONSMOOTH SYSTEMS

Dynamics problems with discontinuous force elements provide strong nonlinearities. The seven papers in this group explore a variety of problems. Theoretical methods are discussed by Balachandran et al., Hogan and Momer, Wiercigroch and Glocker. Numerical methods in impact problems are presently in the papers by Peterka and Van Campen et al. And Popp et al. present experimental results for a mechanical impact oscillator problem.

This area of research will likely remain a rich vein of problems in nonlinear dynamics, especially contact problems relating friction and impact between bodies.

F. C. Moon (ed.),
IUTAM Symposium on New Applications of Nonlinear and Chaotic Dynamics in Mechanics, 261.

DYNAMICS OF ELASTIC STRUCTURES SUBJECTED TO IMPACT EXCITATIONS

B. BALACHANDRAN
Assistant Professor
Department of Mechanical Engineering
University of Maryland
College Park, MD 20742, USA

M.-X. ZHAO
Graduate Research Assistant
Department of Mechanical Engineering
University of Maryland
College Park, MD 20742, USA

AND
Y.-Y. LI
Graduate Research Assistant
Department of Mechanical Engineering
University of Maryland
College Park, MD 20742, USA

Abstract.

A collection of results obtained through experimental and analytical investigations into the dynamics of thin-walled structures subjected to impact excitations is presented. A stainless steel cantilever beam with a tip mass is considered and this elastic structure is impacted close to the free end. The considered excitations include harmonic excitations and periodic excitations of the form $|Dcos(\Omega t)|$. The excitation amplitude and excitation frequency are used as control parameters. The experimental results are presented in the form of bifurcation diagrams and phase portraits. During harmonic impact excitations, period-doubled motions, incomplete period-doubling sequences, and aperiodic motions are observed. In addition to these responses, the responses observed during half-sine impact excitations

F. C. Moon (ed.),
IUTAM Symposium on New Applications of Nonlinear and Chaotic Dynamics in Mechanics, 263–272.

include period-three motions and modulated motions. A finite-dimensional model is developed in the analytical efforts through a Galerkin projection and numerical studies are conducted by using this model. The analytical results show many qualitative similarities with the experimental results. In the considered parameter ranges, nonlinear interactions are not observed. Some of the observed experimental responses appear to suggest the presence of grazing impacts. The implications of the current study for workpiece-tool interactions during milling of thin-walled structures are discussed.

1. Introduction

Dynamics of mechanical systems with impacts have quite been extensively studied in the literature previously [e.g., Wood and Byrne (1981,1982); Holmes (1982); Whiston (1987a,b); Shaw and Shaw (1989); Nordmark (1991); Moon (1992); Fang and Wickert (1994); Pfeiffer and Glocker (1996)]. Common assumptions made in some of these studies include the following: 1) the response amplitude is large compared to the amplitude of impactor motions, 2) the impacted oscillator's motion reverses after an impact, and 3) the impacted system is linear. To a certain extent, the validity of the first two assumptions has been addressed by Fang and Wickert (1994) in a setting where they considered the dynamics of an impacted elastic structure. Although the dynamics of systems subjected to harmonic impacts have been extensively studied, the same is not true for other types of periodic impacts.

In this work, the validity of all of the three assumptions stated above are examined in a setting where an elastic structure is considered. Furthermore, this work is motivated by a need to understand workpiece-tool interactions during milling of thin-walled structures. In recent work by Davies and Balachandran (1996), it was shown that when a low-immersion cut of a thin elastic structure is carried out by using a two-fluted cutter over a "small" span, one can effectively consider this problem to be a problem of an elastic structure subjected to a dynamic, half-sine transverse impact loading. Here, analytical and experimental investigations are conducted into the dynamics of an elastic structure subjected to harmonic as well as half-sine impact excitations. The rest of this article is organized as follows. In Sections 2 and 3, the experimental arrangement and analytical development are briefly discussed. Results are provided in the next section.

2. Experimental Arrangement

The experimental set up, which is depicted in Figure 1, is similar to that used by Fang and Wickert (1994). In this set up, a stainless steel can-

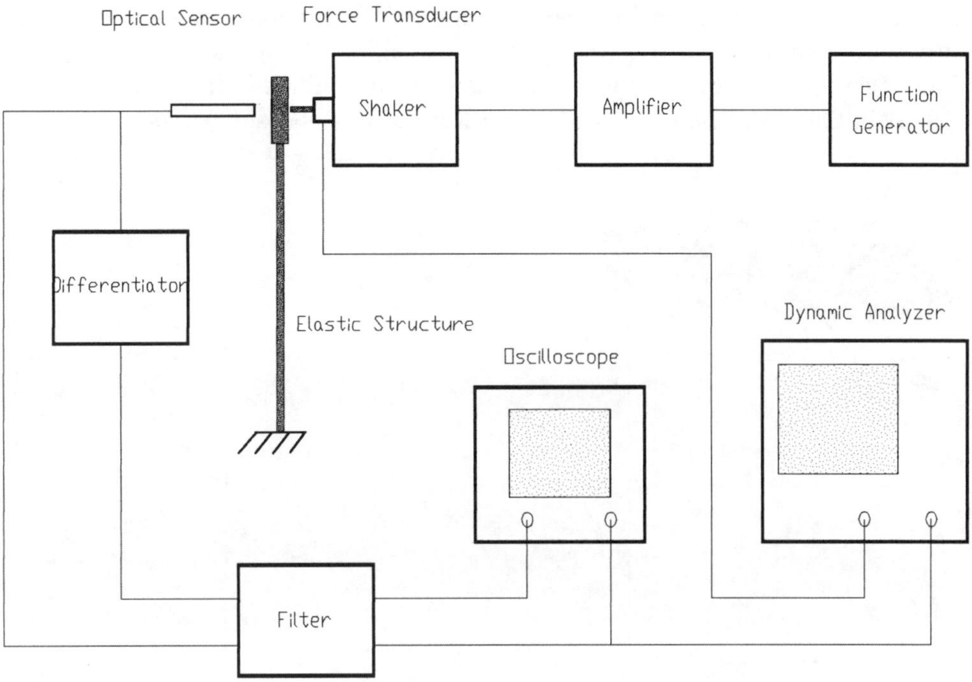

Figure 1. Experimental set up for impact studies

tilever structure with a tip mass is considered. From the cantilevered end, the beam has a length of about 184.2 mm. The beam width and beam thickness are 25.4 mm and 1.6 mm, respectively. On each side of the beam, at the top, a stainless steel block with the dimension 25.4 mm x 25.4 mm x 5.9 mm is located. The center of the tip mass is impacted by using a permanent magnet shaker, and the applied impact force is measured by using a piezoelectric force transducer. The beam is instrumented with strain gages at the base to get a measure of the motions of the free-end of the structure as well as to establish the initial contact position between the impactor and the structure. A noncontact fiber-optic sensor located as shown in Figure 1 is used to measure the structure's displacement at the free end. The first natural frequency of the structure is about 14.8 Hz, and the associated viscous damping factor is about 0.0028. The range of excitation frequencies considered in the experiments is restricted to be less than 150 Hz, which is below the second natural frequency of the structure.

3. Analytical Development

In Figure 2, the illustration used for modeling purposes is shown. Here, only planar motions of the structure are considered, and the steps used for deriving the equations follow along the lines of Balachandran and Nayfeh (1990). The Lagrangian L for the continuous system is constructed and it

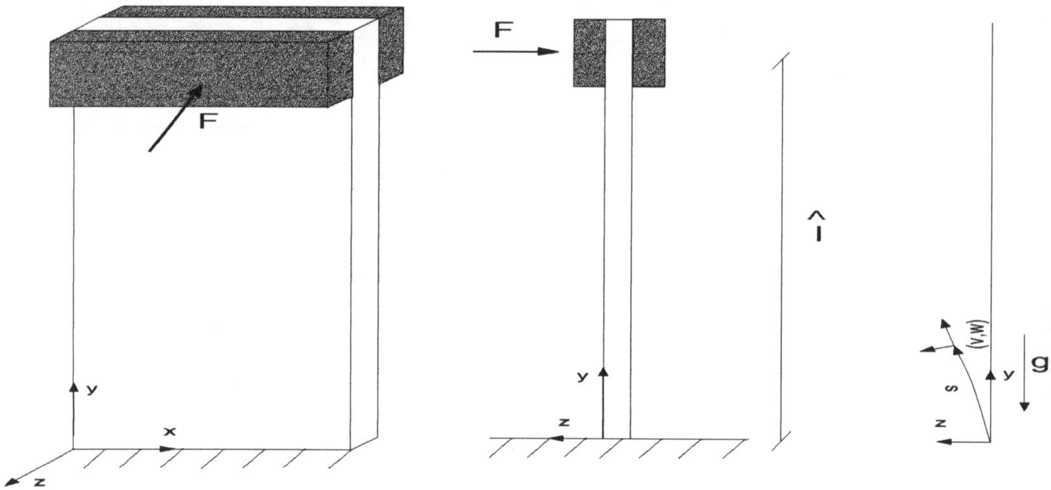

Figure 2. Modeling for impact studies

is augmented to account for the inextensionality constraint. The resulting equation is

$$L_{aug} = T - V - \frac{\lambda}{2}\int_0^{\hat{l}} [(1 + \frac{\partial v}{\partial s})^2 + (\frac{\partial w}{\partial s})^2 - 1] ds \qquad (1)$$

where T represents the kinetic energy of the continuous system, V represents the potential energy of the system, λ is a Lagrange multiplier, $v(s,t)$ is the axial displacement at spatial location s and time t, $w(s,t)$ is the transverse displacement, and $\hat{l}$ is the length from the catilevered end to the center of the tip mass.

After using extended Hamilton's principle, the governing equation obtained is of the form

$$\begin{aligned}\rho h\ddot{w} \quad &+EIw^{iv} - \rho gh[w' + (s-\hat{l})w''] + mgw'' \\ &+EI[w'(w'w'')']' - \tfrac{1}{2}mw'' \textstyle\int_0^{\hat{l}} \frac{\partial^2}{\partial t^2}(w'^2)ds \\ &+\tfrac{3}{2}mgw'^2w'' + \tfrac{1}{2}\rho hw' \frac{\partial}{\partial s}[\textstyle\int_{\hat{l}}^s \int_0^s \frac{\partial^2}{\partial t^2}(w'^2)d\hat{s}ds] \\ &+\tfrac{1}{2}\rho hw''[\textstyle\int_{\hat{l}}^s \int_0^s \frac{\partial^2}{\partial t^2}(w'^2)d\hat{s}ds] - \rho gh(\tfrac{1}{2}W'^3 + \tfrac{3}{2}(s-\hat{l})w'^2w'') = 0\end{aligned} \tag{2}$$

In equation (2), ρ is the mass density of the elastic structure, h is the thickness of the structure, EI is the flexural rigidity of the structure, and m is the tip mass. The impact force F, which is shown in Figure 2, is accounted for in the boundary conditions. These conditions without the nonlinear terms read as

$$\begin{aligned} w(s,t) &= 0 && at \quad s=0 \\ w'(s,t) &= 0 && at \quad s=0 \\ EIw''' + mgw' &= m\ddot{w} + F && at \quad s=\hat{l} \\ EIw'' + J_m\ddot{w}' &= 0 && at \quad s=\hat{l} \end{aligned} \tag{3}$$

A system similar to equation (2) was treated by Zavodney and Nayfeh (1989). Next, the transverse displacement $w(s,t)$ is written in the form $w(s,t) = q(t)\phi_1(s)$, where the spatial function $\phi_1(s)$ is chosen to satisfy the associated linear partial differential equation with constant coefficient terms and linear boundary conditions. Here, $\phi_1(s)$ is the first mode shape of the system. (The analytical predictions for the first and second natural frequencies are about 14.80 Hz and 167.6 Hz, respectively.) After carrying out a Galerkin projection and including dissipation, the resulting equation is

$$\begin{aligned} m_r\ddot{q_r} \quad +k_rq_r \quad &+c_r\dot{q_r} \\ +\alpha_1{q_r}^3 \quad &+\alpha_2[\dot{q_r}^2q_r + {q_r}^2\ddot{q_r}] = f_rF \end{aligned} \tag{4}$$

where $m_r = 2.2944$, $k_r = 1.9859 \times 10^4$, $c_r = 1.1625$, $\alpha_1 = -1.4289 \times 10^7$, $\alpha_2 = -242.1232$, and $f_r = -1.5405$. It is to be noted that the system has softening type of nonlinearities unlike the case in the study of Zavodney and Nayfeh (1989) and other related studies in the literature where beams with tip masses have been considered. Through numerical computations, it has been ascertained that if the tip mass is reduced to be about 10 grams or less, the resulting system has hardening type of nonlinearities. When the free-end displacement $w(s=\hat{l},t)$ is larger than or equal to the impactor displacement w_o, the forcing term is nonzero. Otherwise, it is zero. Other more complete conditions to define an impact are considered by Pfeiffer and Glocker (1996). Equation (3) is numerically integrated for different forcing parameter values to determine the numerical results.

4. Results and Discussion

In the experiments, the chosen Poincaré section is

$$\Sigma = \left\{(x, \dot{x}, t)\epsilon(R^1 \times R^1 \times R^1)|\dot{x} = 0, \ddot{x} < 0\right\} \tag{5}$$

where $x(t)$ is the displacement measured at the center of the tip mass by using the fiber-optic sensor. The bifurcation diagrams obtained during the forward and reverse sweeps of the excitation frequency, while holding the excitation amplitude F constant at 0.157 Newtons, are shown in Figures 3 and 4, respectively.

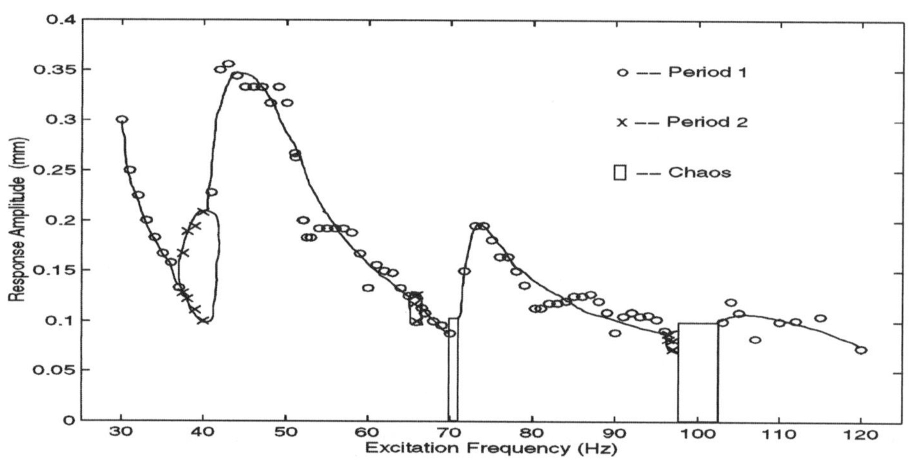

Figure 3. Bifurcation diagram on a Poincaré section: Forward sweep of excitation frequency during harmonic excitations

The features that can be discerned in Figure 3 include period-one motions, period-two motions, windows of aperiodic motions, and incomplete period-doubling bifurcation sequences. [Pointwise dimension calculations are indicative of fractal dimensions for the aperiodic motions. These motions are also characterized by a broadband spectral character. These features are suggestive of chaotic motions (Nayfeh and Balachandran, 1995)]. During the reverse sweep of the excitation frequency, similar features are observed. However, the character of the motion at an excitation frequency is not always the same during the forward and reverse sweeps. For instance,

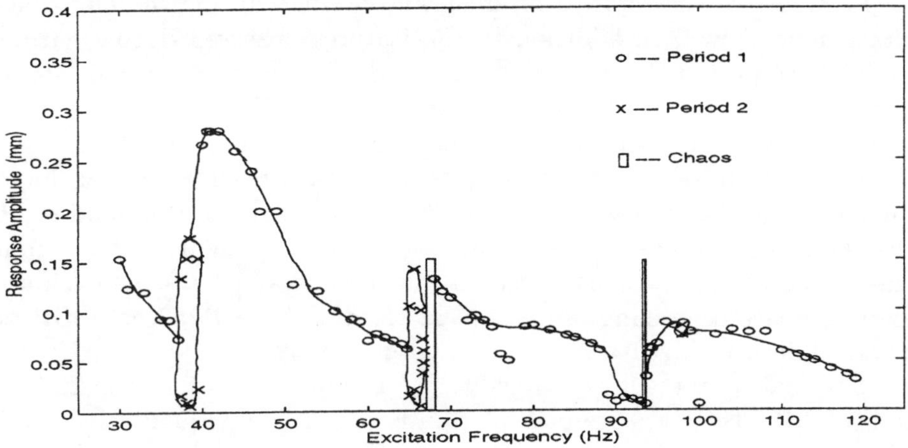

Figure 4. Bifurcation diagram on a Poincaré section: Reverse sweep of excitation frequency during harmonic excitations

at the excitation frequency of 70 Hz, aperiodic motions and period-one motions are observed during the forward and reverse sweeps, respectively.

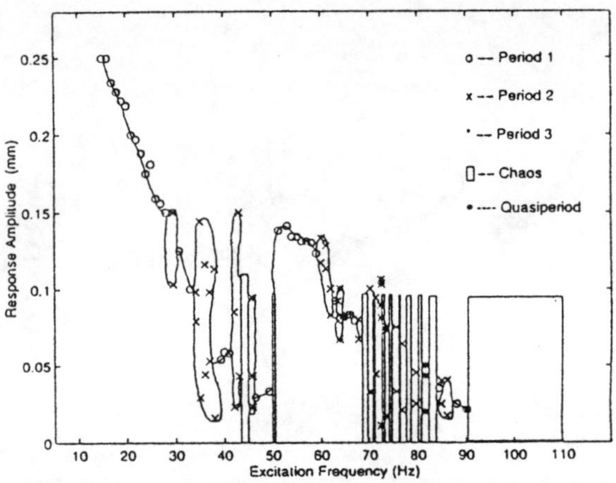

Figure 5. Bifurcation diagram on a Poincaré section: Forward sweep of excitation frequency during half-sine excitations

In Figure 5, the bifurcation diagram obtained during the half-sine excitations is shown. In addition to the features observed during harmonic excitations, period-three motions and modulated motions with dimensions of two are observed. Also, many more windows of aperiodic motions are observed. Also, in some of the experimental results obtained, it is noted that the motion of the structure does not always reverse after an impact. Furthermore, the displacement amplitude of the structure is found to be of the same order as the excitation amplitude; in particular, at low frequencies. Some of the motions observed during sine and half-sine excitations are suggestive of grazing impacts [e.g., Nordmark (1991); Nusse, Ott and, Yorke (1994); Chin, Ott, Nusse, and Grebogi (1995)].

The analytical results obtained for a harmonic excitation with an amplitude of 0.157 Newtons are presented in Figure 6. Many qualitative similarities between the bifurcation diagrams of Figures 3 and 6 can be discerned. The results obtained with the nonlinear terms in equation (3) are seen to agree better with the experimental results.

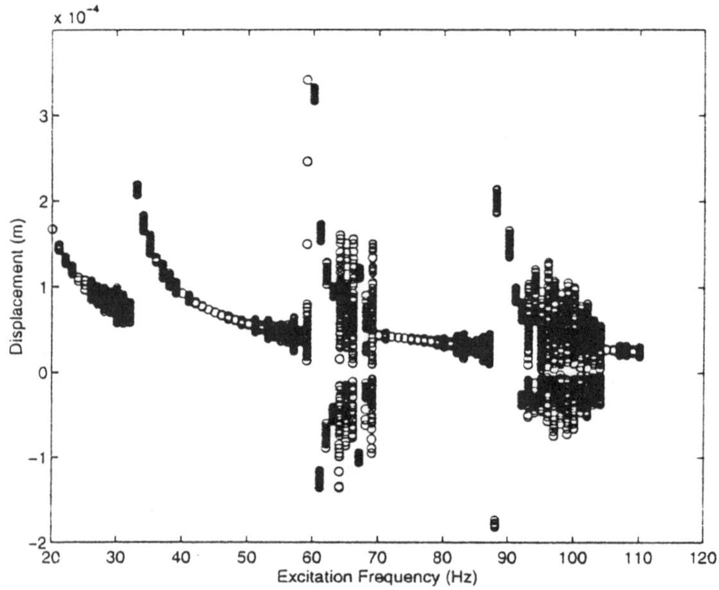

Figure 6. Bifurcation diagram on a Poincaré section: Forward sweep of excitation frequency during harmonic excitations

It is speculated that a good understanding of the dynamics of elastic structures subjected to impact excitations will be useful for manufacturing applications. Examples of two applications are milling of thin-walled structures (e.g., Davies and Balachandran, 1996) and micro-machining of alloy steels (e.g., Moriwaki, Shamoto, and Eguchi, 1996). In these applications, the cutting forces can take the form of impact forces. A good understanding of the dynamics of the impacted structure may help in controlling the workpiece and/or tool motions to achieve a good surface finish. The features of the responses observed during half-sine excitations may be important for cases where one carries out a low-immersion cut on a thin-walled structure by using a cutter with two identical flutes. Grazing impacts may also be useful in these cases. Furthermore, nonlinear interactions (e.g., Balachandran and Nayfeh, 1991) need to be considered when milling or cutting thin plate-like elastic structures.

5. Acknowledgements

Partial support received for this work from the U.S. Department of Commerce, National Institute for Standards and Technology through Grant No. 60NANB6D0052 is gratefully acknowledged. Dr. Alkan Donmez is the technical monitor for this contract. He is thanked for helpful suggestions.

6. References

Balachandran, B. and Nayfeh, A. H. (1990). Nonlinear Motions of Beam-Mass Structure, **Vol. 1, No. 1**, pp. 39–61.

Balachandran, B. and Nayfeh, A. H. (1991). Observations of Modal Interactions in Resonantly Forced Structures, *Nonlinear Dynamics*, **Vol. 2**,pp. 77–117.

Chin, W., Ott, E., Nusse, H. E., and Grebogi, C. (1995). Universal Behavior of Impact Oscillators Near Grazing Incidence, *Physics Letters A*, **Vol. 201**, pp. 197–204.

Davies, M. and Balachandran, B. (1996). "Nonlinear Oscillations in Milling of Thin-Walled Structures," *Journal of Sound and Vibration*, submitted for publication.

Fang, F. and Wickert, J. A. (1994). Response of a periodically driven impact oscillator, *Journal of Sound and Vibration*, **Vol. 170, No. 3**, pp. 397–409.

Pfeiffer, F. and Glocker, C.(1996). *Multibody Dynamics With Unilateral Contacts*, Wiley, New York.

Holmes, P. J. (1982). The Dynamics of Repeated Impacts with a Sinusoidally Vibrating Table, *Journal of Sound and Vibration*, **Vol. 84**, pp. 173–189.

Moon, F. C. (1992). *Chaotic and Fractal Dynamics: An Introduction for Applied Scientists and Engineers*, Wiley, New York.

Moriwaki, T., Shamoto, E., and Eguchi, D. (1996). Vibration Assisted Micro-Machining of Alloy Steel, in N. Narutaki, C. Dingchang, Y. Yamane, and A. Ochi (eds.), *Proceedings of the Third International Conference on Progress of Cutting and Grinding*, pp. 251–316.

Nayfeh, A. H., and Balachandran, B. (1995). *Applied Nonlinear Dynamics*, John Wiley & Sons, New York.

Nordmark, A. B. (1991). Non-Periodic Motion Caused by Grazing Incidence in an Impact Oscillator, *Journal of Sound and Vibration*, **Vol. 145, No. 2**, pp. 279–297.
Nusse, H. E. Ott, E., and Yorke, J. A. (1994). Border-Collision Bifurcations: An Explanation for Observed Bifurcation Phenomena, *Physical Review E*, **Vol. 49**, pp. 1073–1076.
Shaw, J. and Shaw, S. W. (1989). The Onset of Chaos in a Two-Degree-of-Freedom Impact Oscillator, *Journal of Applied Mechanics*, **Vol. 56**, pp. 168–174.
Whiston, G. S. (1987a). The Vibro-Impact Response of a Harmonically Excited and Preloaded One-Dimensional Linear Oscillator, *Journal of Sound and Vibration*, **Vol. 115, No. 2**, pp. 303–319.
Whiston, G. S. (1987b). Global Dynamics of a Vibro-Impacting Linear Oscillator, *Journal of Sound and Vibration*, **Vol. 118, No. 3**, pp. 395–429.
Wood, L. A. and Byrne, K. P. (1981). Analysis of a Random Repeated Impact Repeated Impact Process, *Journal of Sound and Vibration*, **Vol. 78, No. 3**, pp. 329–345.
Wood, L. A. and Byrne, K. P. (1982). Experimental Investigation of a Random Repeated Impact Process, *Journal of Sound and Vibration*, **Vol. 85, No. 1**, pp. 53–69.
Zavodney, L. D. and Nayfeh, A. H. (1989). The Non-Linear Response of a Slender Beam Carrying a Lumped Mass To a Principal Parametric Excitation: Theory and Experiment, *International Journal of Non-Linear Mechanics*, **Vol. 24, No. 2**, pp. 105–125.

A METHOD FOR FINDING ALL POSSIBLE PERIODIC ORBITS IN PIECEWISE CONTINUOUS MECHANICAL SYSTEMS OF ARBITRARY DIMENSION

S. J. HOGAN & M. E. HOMER
Department of Engineering Mathematics
Queen's Building
University Walk
Bristol BS8 1TR, United Kingdom

Abstract

Ideas from graph theory are used to systematically uncover all possible periodic orbits in piecewise continuous mechanical systems of arbitrary dimension. The method is illustrated by its application to one low dimensional system (a confined rocking block) and to one high dimensional system (a heat exchanger). In the latter case the number of possible periodic orbits is shown to be a Fibonacci sequence in the system dimension.

1. Introduction

The systems considered in this paper can be described by sets of ordinary differential equations of the form

$$\frac{dx}{dt} = f(x,t,p) \tag{1}$$

where x is a real vector, t is time, p is a real parameter vector. The real function f is piecewise continuous, that is, the phase space is divided into several adjacent regions, in each of which f has a different continuous form.

Systems such as these are studied in a variety of different areas of practical importance such as impact oscillators (Shaw & Holmes 1983), earthquake engineering (Housner 1963, Hogan 1989, 1992, 1994) and structural engineering (Doole & Hogan 1996). The theoretical approaches have been analytical (especially if f is linear in

F. C. Moon (ed.),
IUTAM Symposium on New Applications of Nonlinear and Chaotic Dynamics in Mechanics, 273–282.

each of its regions), numerical or more usually a combination of both. Chaotic solutions have been found for certain functions f over ranges of values of p. In addition, bifurcations between different solutions have been found. Not only are classical bifurcations possible but also a new type of bifurcation is seen when part of the system trajectory becomes tangent to a boundary that separates the different functional forms of f (e. g. Feigin 1970, Nordmark 1991). Path following computer codes can be used to follow these bifurcations (Doole & Hogan 1996).

In this paper we present a method for finding all possible periodic responses in piecewise continuous systems of arbitrary dimension which will always produce the full set of possible periodic solutions, even if this is extensive.

2. General Overview

A piecewise continuous system consists of several connected regions of phase space. During its evolution, a system trajectory may encounter a border between two regions, where it may alter one or more of its state vector values discontinuously. Subsequent evolution in the neighbouring region may occur with different dynamics until another border is reached.

Exactly how many different types of periodic solution are possible in a given piecewise continuous system and how can they be found? Any practical method would have to be both systematic and generalisable to more complicated problems of higher dimension. We now provide such a method, based on graph theory, which can fully answer both questions.

The main ideas of this paper are that:

1) the phase space borders of a piecewise continuous mechanical system can be interpreted as vertices of a directed graph of the system,

2) the phase space trajectory between two borders is then an edge of the graph between corresponding vertices,

3) the direction along an edge corresponds to increasing time,

4) a (proper) circuit of the graph (that is, starting at one point, visiting one or more vertices by traversing edges in the direction of increasing time, and then returning to the starting point) corresponds to a periodic orbit of the underlying piecewise continuous system.

In the next section, we give a few brief graph theory preliminar ies before illustrating the method on two different problems.

3. Graph theory preliminaries.

Following Biggs, Lloyd & Wilson (1976), a graph G = (V, E) is a collection of vertices V={A, B,....,D} and edges E={*a*, *b*,....,*h*}; see for example Figure 4. An edge *a* is *incident with* a vertex A if A is one of the ends of *a*. A *path* from (say) A to D is a sequence (say) A*a*B*b*C*d*D of alternating vertices and edges. If the initial and final vertices are the same, then the path is said to be a *circuit* of the graph. Circuits are crucial in this work as they are possible candidates for periodic orbits of the underlying piecewise continuous mechanical system, although we shall soon see that not all circuits can be periodic orbits. Our starting point is a well-known result from graph theory, that any circuit of a graph may be expressed as a combination of *fundamental circuits* of the graph (Biggs *et al.* 1976, Ch. 8). Circuits are independent if none of them can be expressed as the sum of any others. A maximal set of independent circuits is a set of fundamental circuits of the graph. Kirchhoff (1847) proved that any connected graph with N_v vertices and N_e edges has exactly $(N_e - N_v + 1)$ fundamental circuits. Veblen (1922) showed how to construct these circuits.

4. A low dimensional system.

In this section we consider the confined rocking block. This problem has been treated at length in Hogan (1994). The physical situation is illustrated in Figure 1. The block is mounted on a horizontal surface which is sufficiently rough to prevent sliding. When subject to a horizontal excitation, the block may rotate about O and O' and can impact the floor or either rigid wall. In suitably scaled co-ordinates the equations of motion can be written in the form

$$\frac{d^2x}{dt^2}+k\frac{dx}{dt}-x+1=F(t) \qquad (0<x<s) \qquad (2)$$

$$\frac{d^2x}{dt^2}+k\frac{dx}{dt}-x-1=F(t) \qquad (-s<x<0) \qquad (3)$$

Here $x(t)$ is the scaled angle of rotation of the block, s is the angular distance of the block from the wall, k is a damping coefficient and $F(t)$ is the forcing. Clearly this is in the form of equation (1) above and is hence a piecewise continuous system. The connected regions are the two sides of the phase plane; $(-s<x<0)$ and $(0<x<s)$. The borders are the boundary $x=0$ (the floor) and $x=\pm s$ (the walls), where energy is lost via impact rules, whose exact nature shall not concern us here.

As indicated earlier, our first step is to reduce the phase plane and its trajectories to a graph. We choose four vertices A, B, C and D as defined in Figure 2a and place them on our graph as shown in Figure 2b. [The origin x = y = 0 is highly singular: the only possible periodic orbit containing it is the trivial one.]

Our next step is to visit each vertex in the phase plane and examine where the system can evolve to from that vertex. This is illustrated in the phase plane in Figure 3. The full graph G_1 for the confined rocking block is shown in Figure 4. G_1 has exactly five fundamental circuits. We now use Veblen's (1922) method to find these circuits (see Hogan & Homer 1997 for full details). The results are shown in Figure 5. Any circuit in G_1 (and hence any periodic orbit of the underlying mechanical system) can be constructed from a suitable combination of F1-F5. However we can not interpret all the fundamental circuits themselves as periodic orbits. Instead we find a set of physically realisable circuits shown, as both graph and corresponding phase plane representations, in Figure 6. These physical circuits *do* correspond to possible periodic orbits of the mechanical system. P2 has already been observed in Hogan (1994). P1 & P5 are very similar to those analysed in Hogan (1992) and P4 was originally analysed in Hogan (1989). P3 and P6 have not been seen before. The deliberate lack of precision in defining vertices A - D can now be exploited to find other orbits in the system by simple composition of P1 - P6. For example a period-doubled P2 orbit is P2 + P2.

5. A high dimensional system.

In this section we consider a heat exchanger, illustrated in Figure 7. These systems typically consist of a large number (around 200) of thin pipes contained within an overall pressurised vessel. Hot fluid is pumped in one direction through the thin pipes, while cool fluid passes through the main vessel in the other direction. The fluid motion and local boiling cause the thin pipes to vibrate and impact both each other and the fixed spacers, leading to wear both along pipe lengths and around pipe circumferences. If allowed to proceed unchecked, pipework will fail and expensive plant shut-down will follow.

The vibrating pipes may be modelled most simply by a 'bouncing ball' system (similar to Holmes 1982), in which each mass oscillates parallel to the x - axis. Their equilibrium positions are spread along this axis. The masses can only impact nearest neighbours (see Figure 8). The displacement of the *j*th mass is $x_j(t)$.

The phase space diagram for particle *j* is shown in Figure 9. A' represents impact with pipe (*j*-1) and B' represents impact with pipe (*j*+1). No symmetry of A' and B' about the vertical axis is assumed. The graph G_2 for each pipe, as shown in Figure 10 for pipe *j*. G_2 has three fundamental circuits labelled *p*, *q*, *r* shown in Figure 11. All of them are physically realisable.

We now proceed to enumerate the simplest possible periodic orbits . Let $n_p(j)$ be the number of configurations in which pipe j has motion type p (similarly $n_q(j)$ and $n_r(j)$). Then:

$$n_p(j) = n_q(j\text{-}1) + n_r(j\text{-}1) \qquad (4)$$

$$n_q(j) = n_q(j\text{-}1) + n_r(j\text{-}1) \qquad (5)$$

$$n_r(j) = n_p(j\text{-}1) \qquad (6)$$

Now let the total number of configurations in a system with m pipes be N(m). Then

$$N(m) = n_p(m) + n_q(m) + n_r(m) \qquad (7)$$

Since $n_p(1) = n_q(1) = n_r(1) = 1$, we find from equations (5) - (8) that

$$N(1) = 3, \qquad N(2) = 5, \qquad N(j+1) = N(j) + N(j\text{-}1) \qquad (8)$$

Hence N(m) is a Fibonacci sequence i.e. N(3) = 8, N(4) = 13, N(5) = 21, N(6) =34 etc.

The case m = 3 is shown in Figure 12. With a few extra assumptions, it is possible to make progress on the problem of wear estimation in the heat exchanger. If each solution is equally likely to occur then, from Figure 12, we can see that the pipe in the middle sustains the most impacts (twelve in all), whereas its neighbours only endure eleven each and each wall is hit only five times. Hence the site of most likely wear is the central pipe and the time when failure is likely to occur can be estimated from the total number of impacts over a given time (assuming for example that it is known how many impacts cause failure). Even if, as is likely, each solution were not equally likely and impact wear were a function of impact velocity, then estimates can still be made with suitable weightings given to the solutions concerned.

6. Discussion

In this paper we have provided a method to find all possible periodic orbits in a piecewise continuous system. The strength of the method is that it can be used on any system of whatever size. It will always produce all the periodic orbits of a given topology and define the different types of mathematical problem which have to be solved. Even in a well studied problem with a low number of dimensions such as the confined rocking block, the method revealed solutions unknown to date. But it is in the area of high dimensional systems that we feel that method will prove indispensable. Not only can the method enumerate the orbits as we have already seen, but the different types of mathematical problem to be solved are now clear.

References

Biggs, N. L., Lloyd, E. K. & Wilson, R. J. 1976 *Graph theory 1736 - 1936.* Oxford: Clarendon Press.
Doole, S. H. & Hogan, S. J. 1996 A piecewise linear suspension bridge model: nonlinear dynamics and orbit continuation. *Dyn. & Stability of Systems* 11, 19-29
Feigin, M. I. 1970 Doubling of the oscillation period with C-bifurcations in piecewise continuous systems. *PMM J. Appl. Math. Mech.* 34, 861-869
Hogan, S. J. 1989 On the dynamics of rigid block motion under harmonic forcing. *Proc. Roy. Soc. Lond. A* 425, 441-476
Hogan, S. J. 1992. On the motion of a rigid block, tethered at one corner, under harmonic forcing. *Proc. Roy. Soc. Lond. A* 439, 35-45
Hogan, S. J. 1994 Rigid block dynamics confined between side-walls. *Phil. Trans. Roy. Soc. Lond. A* 347, 411-419
Hogan, S. J. & Homer 1997 A method for finding all possible periodic orbits in piecewise smooth dynamical systems of arbitrary dimension *Submitted to Proc. Roy. Soc. Lond. A.*
Holmes, P. J. 1982 The dynamics of repeated impacts with a sinusoidally vibrating table. *J. Sound Vib.* 84, 173-189
Housner, G. W. 1963 The behaviour of inverted pendulum structures during earthquakes. *Bull. Seism. Soc. Am.* 53, 403-417
Kirchhoff, G. R. 1847 Über die Ausflösung der Gleichungen, auf welche man bei der Untersuchung der linearen Vertheilung galvanischer Ströme geführt wird *Annalen der Physik und Chemie* 72, 497-508 (English part translation in Biggs *et al* (1976) pp 133-135)
Nordmark, A. B. 1991 Non-periodic motion caused by grazing incidence in an impact oscillator. *J. Sound Vib.* 145, 279-297
Shaw, S. W. & Holmes, P. J. 1983 A periodically forced piecewise linear oscillator *J. Sound Vib.* 90, 129-155
Veblen, O. 1922 Linear graphs *Analysis Situs*. (American Mathematical Society Colloquium Lectures 1916) (Part reproduction in Biggs *et al* (1976) pp 136-141)

Figure Legends

Figure 1. Sketch of confined rocking block

Figure 2. a) Definition of phase plane regions A, B, C and D. b) A, B, C and D as vertices of a graph.

Figure 3. Possible trajectories of confined rocking block.

Figure 4. Graph G_1 of confined rocking block.

Figure 5. Fundamental circuits F1 - F5.

Figure 6. Physically realisable fundamental circuits and their phase plane representations P1 - P6.

Figure 7. Sketch of heat exchanger.

Figure 8. Simple one space dimensional model of heat exchanger.

Figure 9. Possible trajectories in phase plane of pipe j in heat exchanger.

Figure 10. Graph G_2 of pipe j in heat exchanger.

Figure 11. Fundamental circuits p, q and r of G_2.

Figure 12. The eight simple periodic orbits in a three pipe heat exchanger.

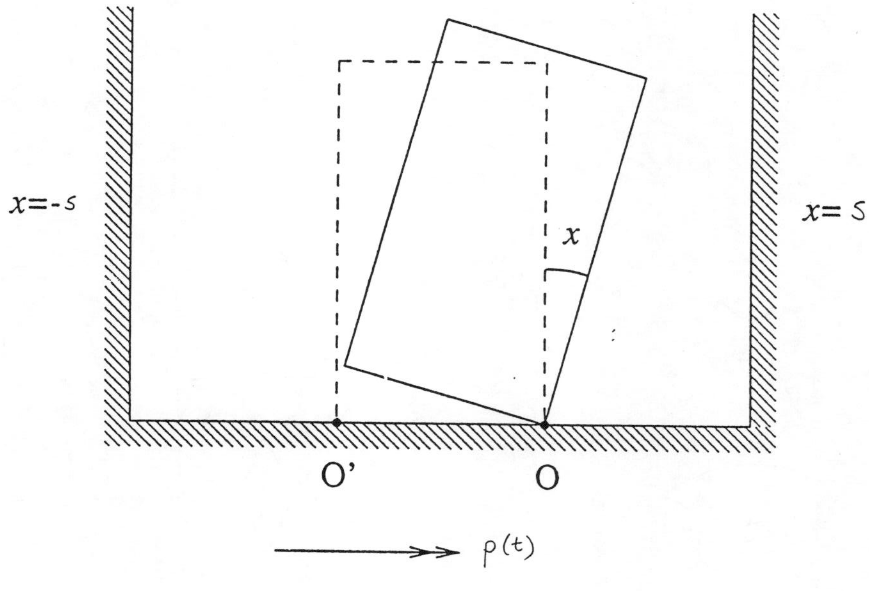

Figure 1

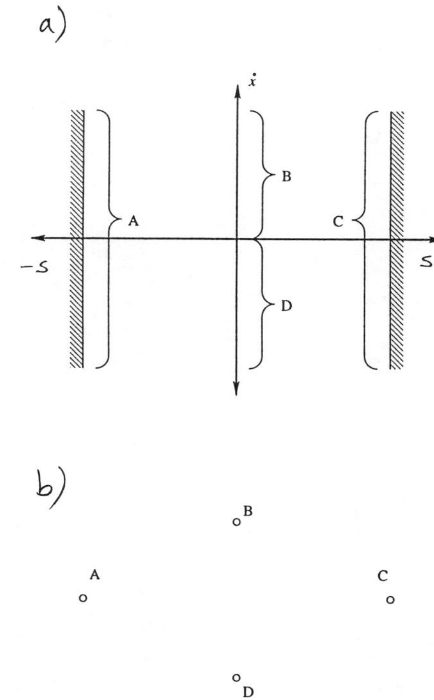

Figure 2

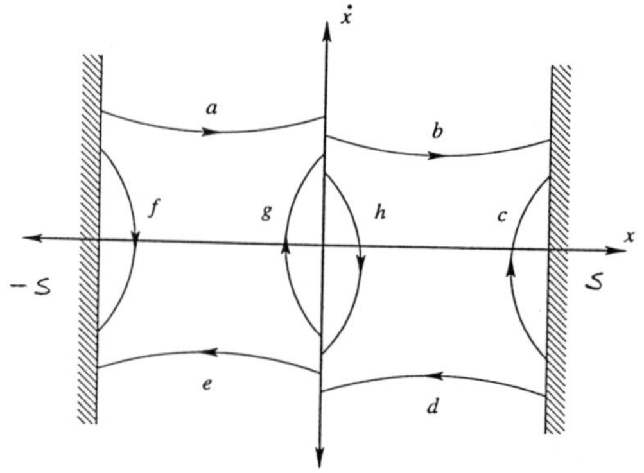

Figure 3

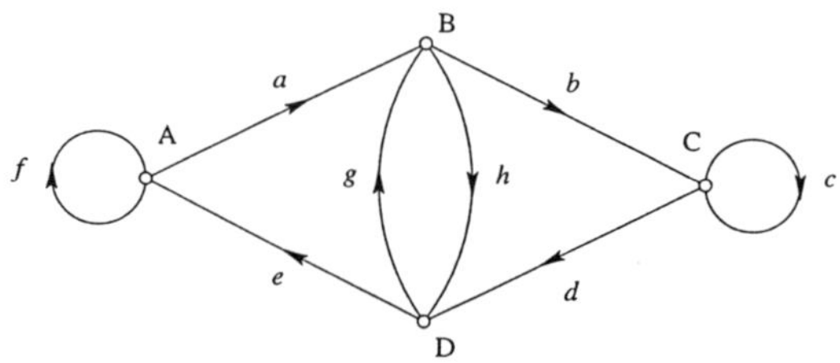

Figure 4

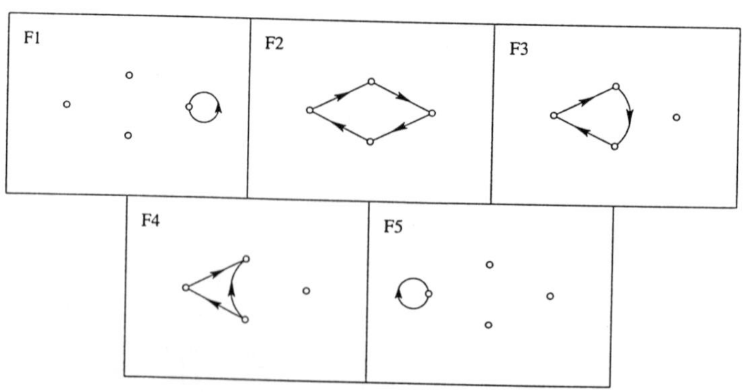

Figure 5

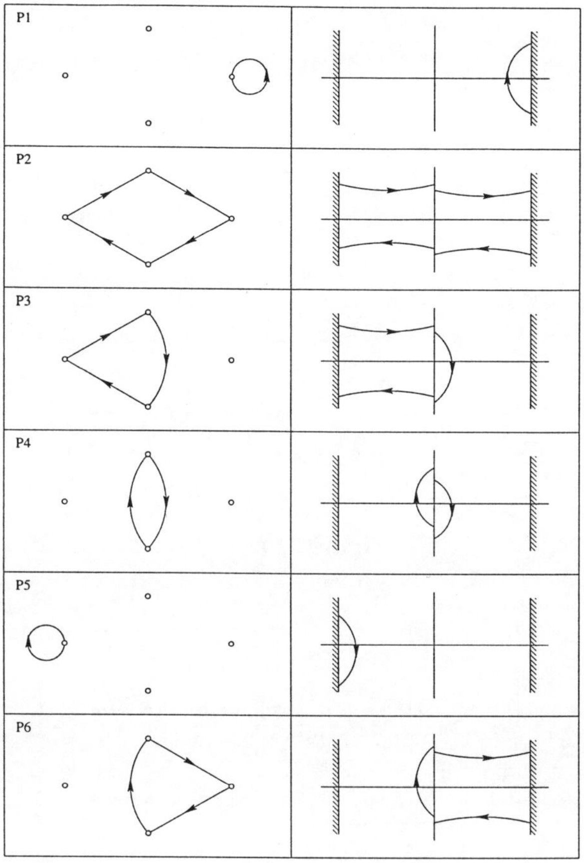

Figure 6

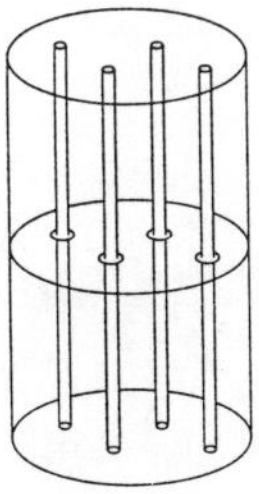

Figure 7

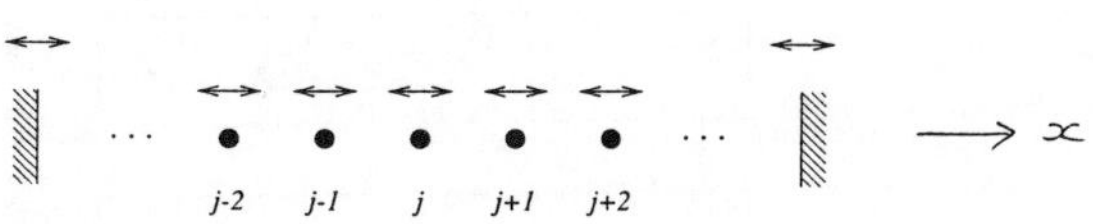

Figure 8

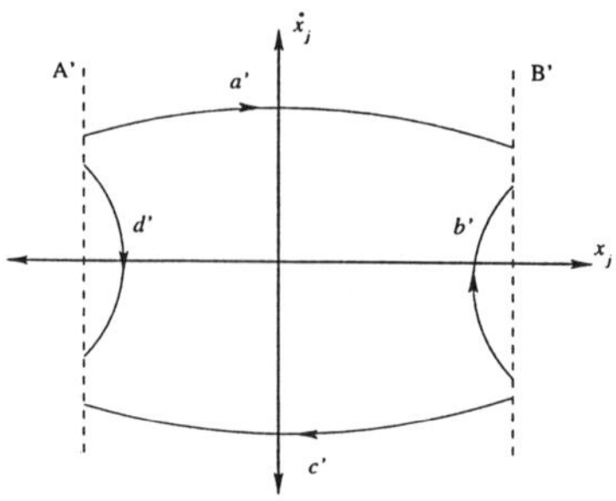

Figure 9

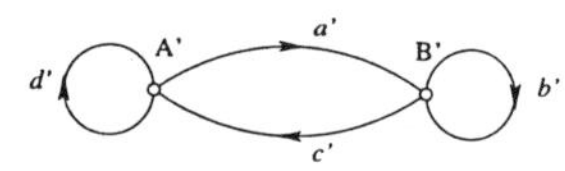

Figure 10

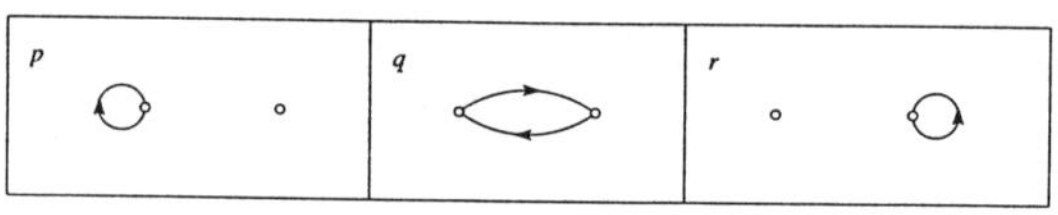

Figure 11

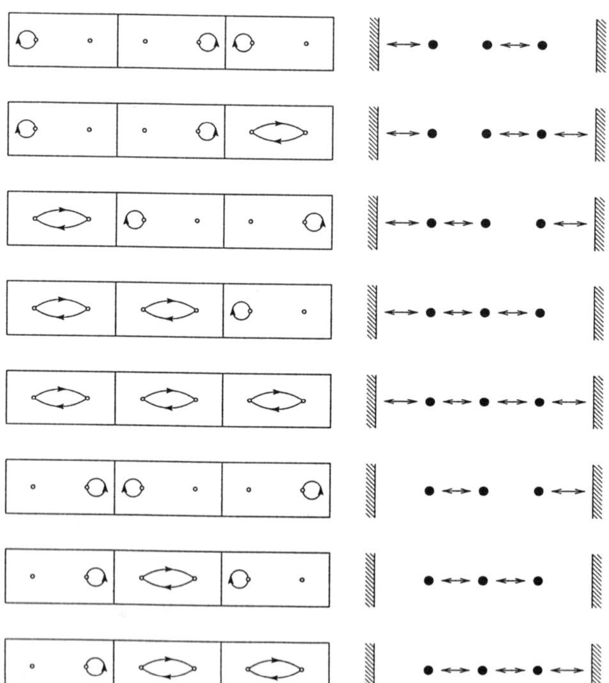

Figure 12

DYNAMICS OF THE IMPACT OSCILLATOR

F. PETERKA
Institute of Thermomechanics, Academy of Sciences of the Czech Republic
Dolejškova 5, 182 00 Prague 8, Czech Republic

1. Introduction

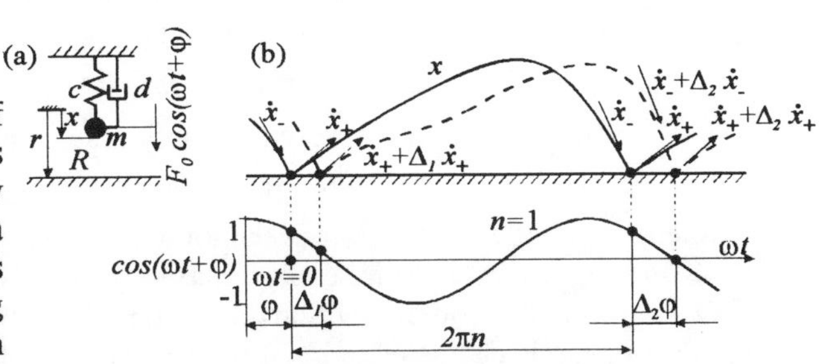

Figure 1. Scheme of the impact oscillator and its periodic motions $z = 1/n$

The impact oscillator is one of the simplest mechanical systems with the strong nonlinearity (Fig.1(a)), which is caused by a sudden change of stiffness during the contact of impacting bodies. When the time duration of the impact can be assumed negligibly short, then the Newton elementary theory of impact is used for the investigation of the system dynamics. This problem was investigated from many points of view. The paper introduces a short review of results and concentrates on the explaining the substance of four ways from periodic to chaotic impact motions, especially on the interrupted Feigenbaum cascade and the interrupted development of the saddle-node instability.

The history of the investigation of this system behaviour has been beginning since the middle of this century (e.g. Rusakov and Kharkhevich 1942) and it continued by theoretical analyses (e.g. Bespalova 1956), experiments (e.g. Sachse and Reum 1965) and analogue simulations (e.g. Irie *et al.*, 1974 or Peterka 1974b). The comprehensive review of the obtained results is published in the series of six papers *Laws of Impact Motion of Mechanical System with One Degree of Freedom* (Peterka 1974a,b), (Peterka and Vacík 1981), (Kotera and Peterka 1981), (Peterka and Kotera 1982) and (Kotera and Peterka 1984). Papers appeared in sixties and seventies are reviewed in (Peterka 1981).

The second wave of the interest in impact oscillators appeared from eighties in connection with the theory of bifurcations and with the application of numerical simulations (e.g. Shaw and Holmes 1983 or Nordmark 1991). New results obtained in the Institute of Thermomechanics were published in (Peterka and Čipera 1994), (Peterka 1996), (Peterka and Kotera 1996) and (Peterka and Szöllös 1997).

The response of this system on the periodic excitation is very heterogeneous. The simple periodic impact motions (Fig.1(b)) have been solved theoretically for the system without the viscous damping d (Fig.1(a)). The influence of the viscous and dry friction is investigated in (Peterka 1981).

F. C. Moon (ed.),
IUTAM Symposium on New Applications of Nonlinear and Chaotic Dynamics in Mechanics, 283–292.

The investigation of the periodic impact motion requires the stability and existence analysis. Regions of existence and stability of periodic impact motions in the space of system parameters result from such analysis. Regions are surrounded by the boundaries which correspond to the saddle-node (SN), period-doubling (PD) and grazing (G) bifurcations.

All these problems are explained in more detail in mentioned recent publications.

2. Short Outline of the Theoretical Analysis

The impact oscillator is composed from an elastically suspended mass m, which is excited by the harmonic force. The mass m impacts against the rigid stop with coefficient of restitution R. A clearance r is the distance of the rigid stop from the mass static position. It is assumed that the before-impact velocity $\dot{x}_-$ changes suddenly into the after-impact velocity $\dot{x}_+$ according to relation

$$\dot{x}_+ = -R\,\dot{x}_- . \tag{1}$$

The equation of the mass impactless motion has the form

$$m\ddot{x} + d\dot{x} + cx = F_0 \cos(\omega t + \varphi) , \tag{2}$$

where d - coefficient of viscous damping, c - spring stiffness, ω - excitation frequency, φ – phase of excitation force at the instant of impact.

One group of periodic motions with one impact, which repeats after n=1,2,... periods $T=2\pi/\omega$ is characterized by Fig.1(b). The stability of periodic impact motion is analyzed by the difference method. The periodic state of the system at the impact is perturbed by differences $\Delta_1\dot{x}_+$, $\Delta_1\varphi$. The motion stability is determined by the development of differences $\Delta_2\dot{x}_+$, $\Delta_2\varphi$ at the next impact. The relation between successive differences is assumed in the form

$$\Delta_2 = \beta\Delta_1 , \tag{3}$$

where β is the function of system parameters. The stability condition of the periodic impact motion is expressed by the relation $|\beta|<1$. Stability conditions and stability boundaries was derived for the assumed group of impact motions in the explicit form

$$s_{-1} < -\bar{r}\ \mathrm{sign}(K) < s_{+1} , \tag{4}$$

where $\bar{r} = rc/F_0$ - relative clearance, s_{+1} - SN stability boundary (β=1) and s_{-1} - PD stability boundary (β= -1).

Stability boundaries are following functions of the system parameters:

$$s_{+1} = g_0\sqrt{1+K^2}\ ;\ \ s_{-1} = g_0\frac{K^2+1-V}{\sqrt{(V-1)^2+K^2}} \tag{5}$$

where $g_0 = 1/|1-\eta^2|$; $\eta = \omega/\Omega$; $\Omega = \sqrt{c/m}$;

$$K = \frac{1+R}{1-R}\eta\,\mathrm{cotg}\frac{\pi n}{\eta}\ ;\ \ V = \frac{2(1+R^2)}{(1-R)^2\sin^2\pi n/\eta} .$$

A similar analysis was used for motions with two impacts per n excitation periods T (Kotera and Peterka 1981). The stability function (in the sense of mentioned function β) was named γ. Results of this important analysis, which was accomplished by Tadashi Kotera (Fukui University, Japan), will be used also in this paper.

3. Classification of Periodic Impact Motions

Impact motions can be classified by the quantity $z = p/n$, where p denotes the number of impacts in the motion period and n is the number of excitation period T in the motion period. The fundamental group is characterized by $n = 1$ and $p = 0, 1, 2, ...$ (including the impactless motion $p = 0$), i.e. $z = 0, 1, 2, 3, ...$.

The excitation frequency η and dimensionless clearance $\bar{r}$ substantially influence the type of the system motion, so the stability and existence regions of different impact motions are presented in the plane $(\eta, \bar{r})$. Summary pictures of such regions received by simulation methods are shown in Fig. 2 for $R = 0.6$, viscous damping $\tilde{d} = d/\sqrt{cm} = 0.1$ and the harmonic and rectangular excitation. Regions are defined by the stability boundaries s (PD and SN bifurcations) and by existence boundaries g (G bifurcation). Boundaries s, g mutually intersect and create two types of transtition regions: *hysteresis regions* (black regions in Fig. 2), where both neighbour z and z+1 motions can exist and they are stable and *beat motion regions* (white regions in Fig. 2), where neither of neighbour motions can exist and is stable and then subharmonic or chaotic impact motions appear (n>1). These motions also exhibit the hysteresis into the region of the fundamental group of impact motions (see dashed boundaries of z=1/2 motion regions in the frequency interval 2<η<3), especially for small viscous damping.

This law of impact motions has the general character. It was ascertained for different types of the periodic excitation and for different types of the motion damping (plastic and elastoplastic impacts or viscous and dry friction of the impactless motion).

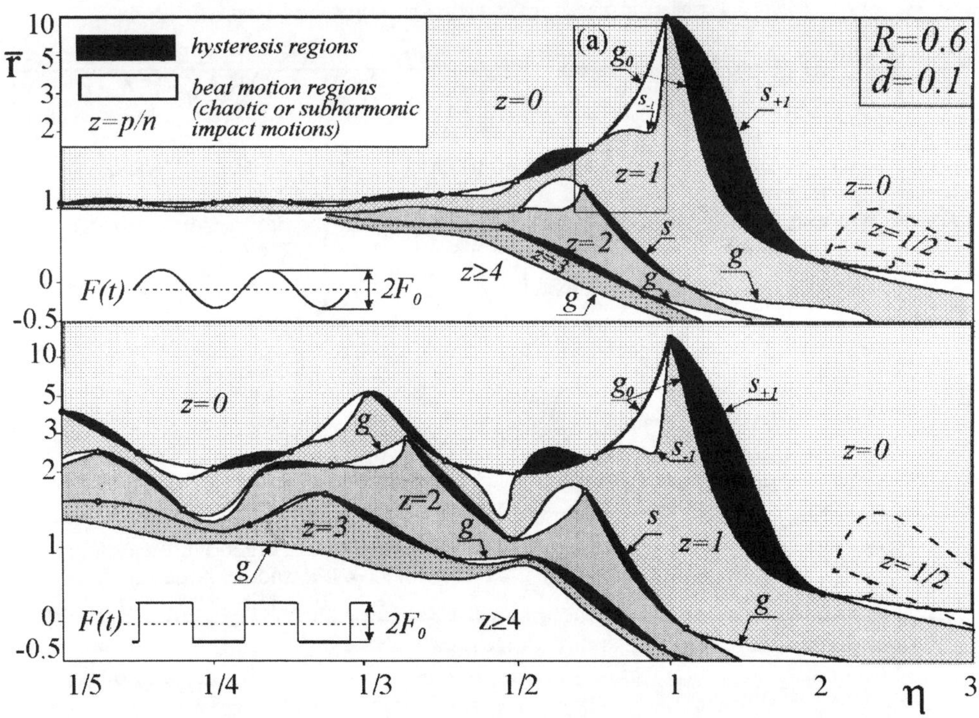

Figure 2. Regions of existence and stability of periodic motions for two types of the periodic excitation

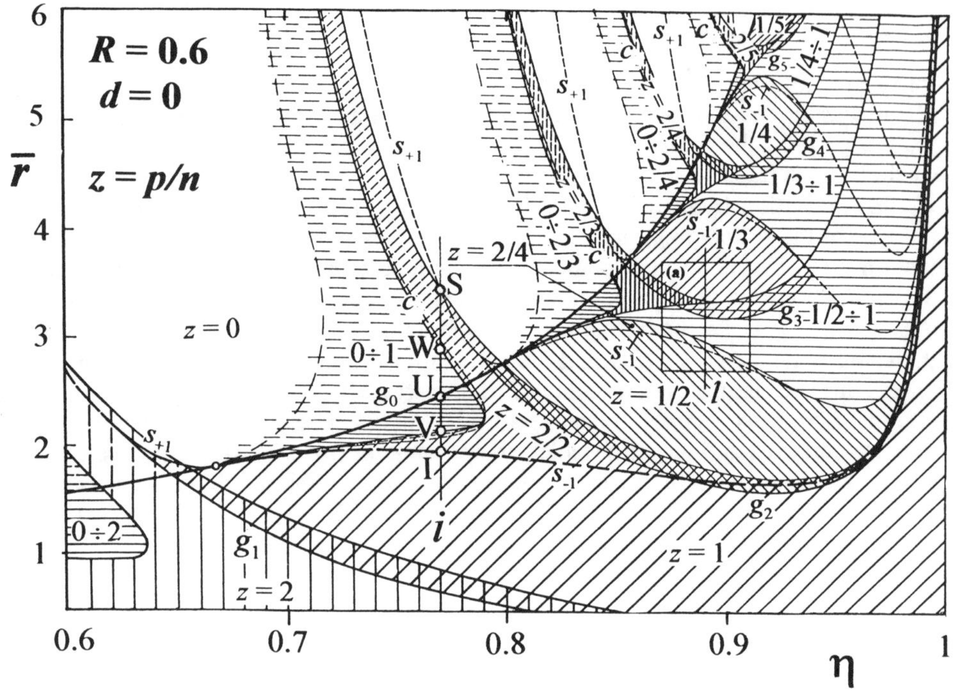

Figure 3. Periodic and chaotic impact motions in the beat motion region

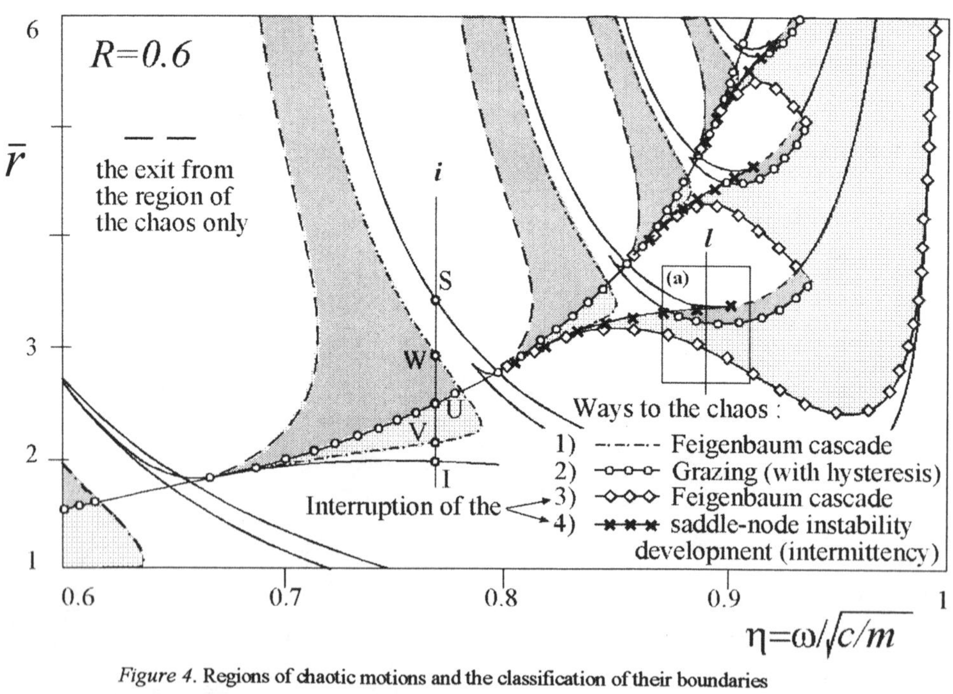

Figure 4. Regions of chaotic motions and the classification of their boundaries

4. Regions of Subharmonic and Chaotic Motions inside the Beat Motion Region

The structure of regions of periodic and chaotic impact motions in one beat motion region between impactless (z=0) and z=1 impact motion is shown in Fig.3, which is the detail (a) of Fig.2. Fig.3 corresponds to the system without the viscous damping (d= 0).

Regions of periodic and chaotic motions are oblique or vertically and horizontally hatched, respectively. Every region is characterized by the value of quantity z. The groups of theoretically solved motions z=1/n (Peterka 1974a,b) and z=2/n (Kotera and Peterka 1981) are indicated by series z=1, 1/2, 1/3, 1/4, 1/5 and z=2, 2/2, 2/3, 2/4, 2/5. Regions z=1/n are limited from above by stability boundaries s_{+1}, s_{-1}, and from below by existence boundaries g_1, g_2, g_3, .. . Existence boundaries g correspond to the grazing bifurcation, when the existence condition $x < r$ (Fig.1(b)) of the assumed periodic motion begins to be not met in the time interval nT between successive impacts. Boundaries g were ascertained numerically. Grazing bifurcation boundary g_0 (Fig.3) of the impactless motion (z=0) is expressed analytically (Eq. (5) for the harmonic excitation) and it is identical with the amplitude-frequency characteristics. The impactless motion exists over the boundaries g_0 ($\bar{r} > g_0$), but impact motions can exist also over the boundary g_0 in hysteresis regions hatched by dashed lines in Fig.3.

5. Ways from Periodic into Chaotic Impact Motions

Horizontally hatched regions of chaotic impact motions in Fig.3 are accented by dark and light shadows in Fig.4. Dark regions represent hysteresis regions in which periodic motions can also exist.

Five types of boundaries of chaotic impact motion regions are distinguished graphically (Fig.4). Four types of boundaries correspond to four different ways into the chaos from periodic motions and remaining type expresses only the exit from the chaos into the periodic motion. The origins of different ways into the chaos are derived from G, PD and SN bifurcations.

5.1 FEIGENBAUM'S PERIOD DOUBLING CASCADE

This way exists when the cascade of PD is not interrupted by the G bifurcation (additional impacts do not appear during the process of motion period doubling), i.e. the *value z remains unchanged* (Fig.5, z=1/1, 2/2, 4/4, 8/8). Such situation appears along boundaries c in Fig.3 or along dot-and-dashed boundaries in Fig.4 (see the boundary of z=2/2 region, passing through points V, W in Figs.3,4). The Feigenbaum cascade ends and

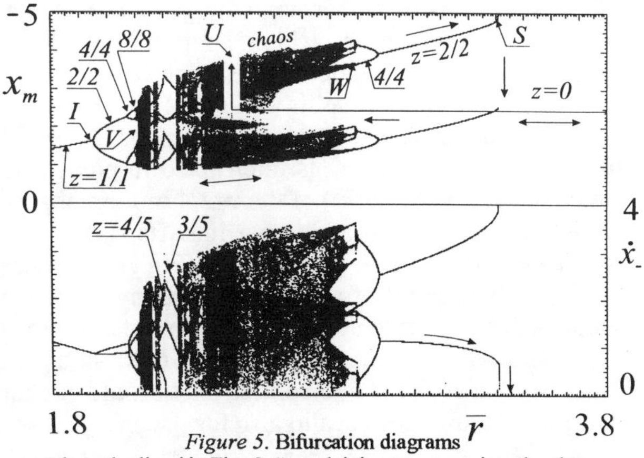

Figure 5. Bifurcation diagrams (along the line i in Figs.3,4) explaining two ways into the chaos: the Feigenbaum cascade and the grazing with hysteresis

the chaotic motion starts *without a hysteresis* on these boundaries. The behaviour of the system motion near these boundaries is shown on bifurcation diagrams in Fig.5, which correspond to the transit through the region z=0÷1 in Figs.3,4 along line i. Similar transitions into the chaos appear on boundaries c of z=2/3; 2/4; 2/5 impact motions. The chaotic motions z=0÷1, 0÷2/3, 0÷2/4, .. inside the horizontally hatched regions, left of boundaries c in Fig.3, are characterized by the appearance of additional impactless loops of motion in some periods T, so the quantity z of the chaotic motion decreases. There exist also windows of more complex periodic impact motions in these regions, e.g. motion z=4/5 or z=3/5 in Fig.5.

5.2 GRAZING BIFURCATION WITH HYSTERESIS

The grazing is characterized by the appearance of a new contact (the impact with zero before-impact velocity) in the motion period. Such motion is never stable, except the singular points (Peterka and Kotera 1996), and the motion of the system jumps either inwards a stable periodic impact motion region or inwards a region of chaotic motion (see segments of grazing bifurcation boundaries g_0, g_3, g_4, g_5 of motions z=0; 1/3; 1/4; 1/5 in Fig.3 or circled segments in Fig.4). These segments penetrate the horizontally hatched or shadowed regions of chaotic motions z = 0÷2; 0÷1; 0÷2/3; 0÷2/4; 0÷2/5; 1/2÷1; 1/3÷1; 1/4÷1 (see also points U on line i in Figs.3,4 and point L on line l in Figs.6,8). This transition *increases the value* $\boldsymbol{z}$ of the system motion and it exhibits the *significant hysteresis* (e.g. intervals U-W, U-V in Figs.3-5).

The characteristic feature of two remaining ways into the chaos is the existence of additional impacts, which appear during the development of the Feigenbaum cascade and during transition process after the saddle-node instability. Both ways will be explained using Fig.6 and bifuraction diagrams along line l, which crosses all types of bifurcation boundaries (grazing in points L, P, period-doubling in point N and saddle-node in points K, M in Fig.6).

5.3 THE INTERRUPTED FEIGENBAUM CASCADE

This way appears in subharmonic impact motions (e.g. z=1/n, z=2/n) and other types of periodic impact motion, which contain the impactless loop of the motion phase trajectory (e.g. z=1/2 motion; see point a on line l in Fig.6 and phase trajectory in Fig.7(a)). This motion splits (Fig.7(b)) on the PD bifurcation boundary (point N in Fig.6 on the dashed boundary s_{-1}; point N in bifurcation diagram Fig.8)). The increasing of the splitting is interrupted in point P (Fig.6), where the impactless loop with extreme $3b$ (Fig.7(b)) of the z=2/4 motion begins to touch the stop and the chaotic motion appears due to additional impacts (lower part of Fig.8). Its phase trajectories fill in the dark area in phase plane (Fig.7(c)). When additional impacts are avoided and impactless loops of the impact motion can penetrate the rigid stop (see trajectory z=2/4 with extreme $3b$ in Fig.7(c)), then the PD cascade can be accomplished.

It follows from the simulation (Fig.8) and theoretical (Fig.9(a)) analysis that velocities $\dot{x}_-$ of the stable splitted motion are higher than velocitites of the unstable nonsplitted motion. It is valid also for the energy losses

$$E_i = \tfrac{m}{2}(1 - R^2)\dot{x}_-^2 \tag{6}$$

caused by impacts in splitted and nonsplitted regime (bifurcation characteristics between points N, P in Figs.8,9). This is important phenomenon which explains the substance of this way into the chaos and it is documented by Fig.10. It shows the results of the evaluation of extremes x_e of the body motion and before-impact velocities $\dot{x}_-$ during 540 periods T. Extremes 1, 2, 3 correspond to the unstable nonsplitted motion (as in Fig.7(a)) and therefore its splitting grows up to the appearance of a new additional impact in the loop with extreme *3b*. The additional impacts (thick points in Fig.10) take away suddenly parts of kinetic energy (6), which causes the suppression of the splitting. The system is permanently in the transition regimes of the splitting increase and suppression, due to additional impacts. The instants and intensities of additional impacts are unpredictable and the response of this deterministic system is chaotic with the intermittency character. The presence of the *grazing bifurcation* in this way *increases the value* z, but the *chaotic motion does not exhibit the hysteresis.* This third way into the chaos appears along boundaries with squares in Fig.4.

5.4 INTERRUPTION OF THE SADDLE-NODE INSTABILITY DEVELOPMENT

Fourth way into the chaos was ascertained on the boundaries s_{+1} limitting the lower part of regions z=2/3, 2/4, 2/5, ... (vertically hatched regions in Fig.3; see also the boundary s_{+1} passing through points O,M,Q in Fig.6 and boundaries with crosses in Fig.4). Simulation and theoretical bifurcation diagrams along line l in Fig.6 are shown in Fig.8 and Fig.9(b), respectively.

The SN bifurcations appear on both boundaries s_{+1} (going through points M, K in Fig.6) of the z =2/3 motion, when the stability function γ_1 crosses the value +1 (Fig.9(b)). A jumps from unstable z=2/3 motion (Fig.11(a)) into z=1/3 motion (Fig.11(c)) should appear in points M, K (Figs.8, 9(b)), but only the jump in point K can be physically realized under assumed conditions of the impact motion.

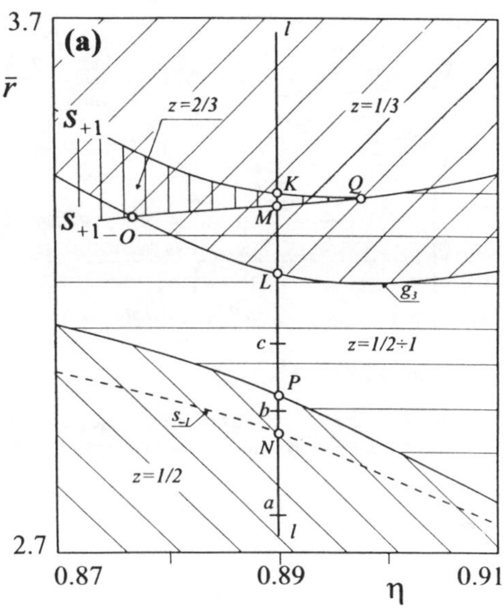

Figure 6. Enlarged detail (a) from Figs.3, 4

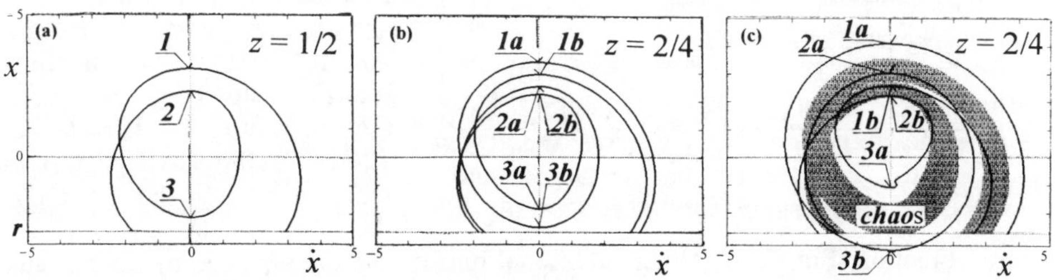

Figure 7. Explanation of the interrupted Feigenbaum cascade - third way into the chaos

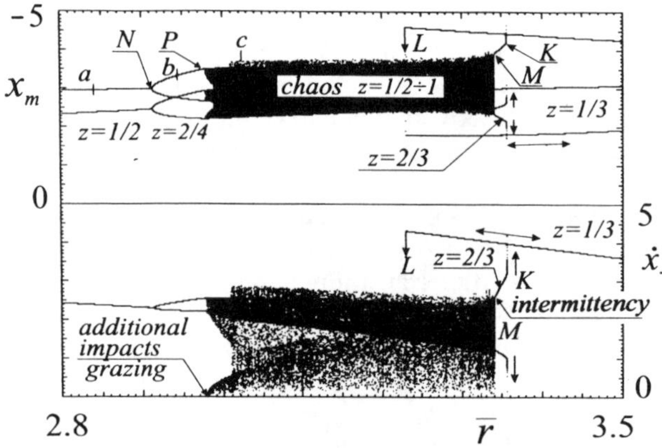

Figure 8. Bifurcation diagrams along line *l* in Fig.6

The z=2/3 impact motion near points M, K is described by Fig.9(b). Approaching to point K the before-impact velocity $(\dot{x}_-)_2$ of the weaker impact decreases up to the extinguishment and the velocity $(\dot{x}_-)_1$ of the stronger impact increases on the value of z=1/3 motion (Fig.8). On the contrary, approaching to point M, the weaker impact, indicated β in Fig.11(a), strengenths (point M_2 in Fig.9(b)) and the stronger

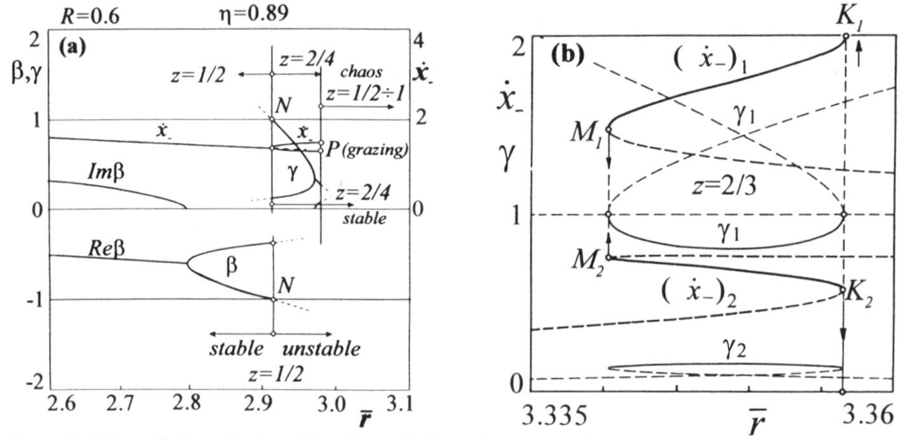

Figure 9. Theoretical results describing the periodic z=1/n (n=2) and z=2/n (n=3, 4) impact motions and their stability (β) and (γ) near interrupted PD cascade (a) and SN instability (b), along line *l* in Fig.6

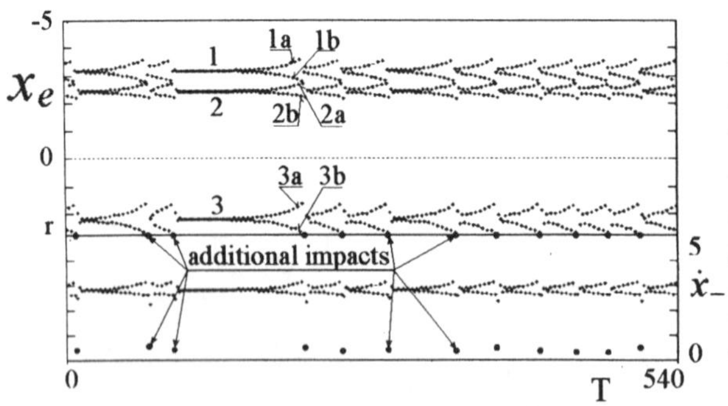

Figure 10. Explanation of interrupted PD cascade

impact indicated α weakens and vanishes. The sudden disappearance of the stronger impact causes the increase of the impactless loop extreme (point 4 in Fig.11(a)). An additional impact, *but not of the grazing character*, will appear and the system motion transits into the chaos (Figs. 8, 11(b)) of the intermittency character Fig. 12(a)). Additional impacts return more or or less the motion into the state of the unstable z=2/3 impact motion (Fig.12(a)). When additional impacts are omitted, e.g. by the repeatedly increasing clearance r about the difference Δr (Fig.13), then the jump from z=2/3 into

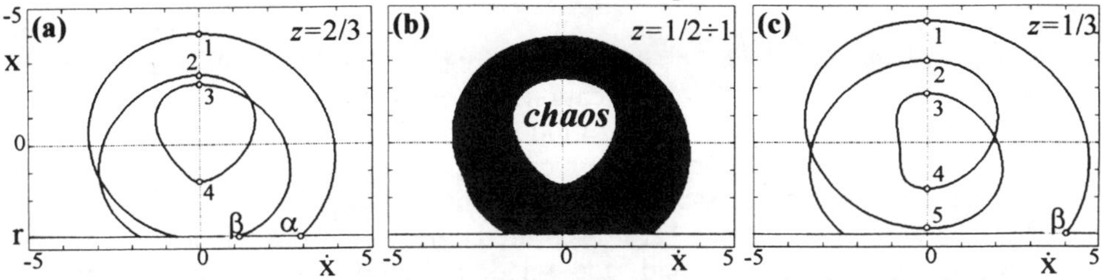

Figure 11. Explanation of the interrupted SN instability development - fourth way into the chaos

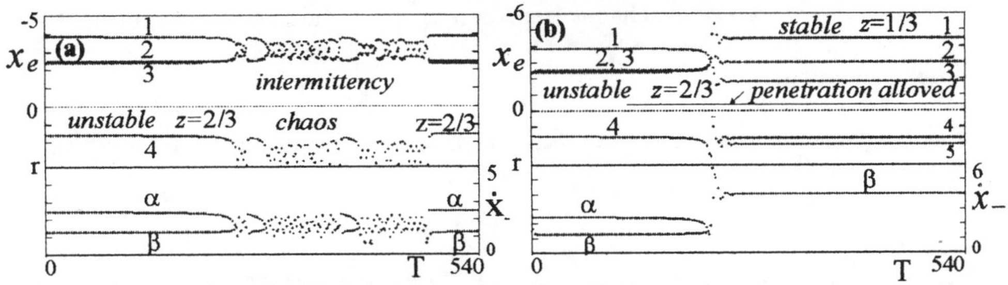

Figure 12. Development of the saddle-node instability up to the chaos - intermittency (a) and up to the periodic $z=1/3$ motion (b), when additional impacts are omitted

$z=1/3$ motion can be realized (Figs.11(c),12(b)) after the omission of only two addititonal impacts (Fig.13).

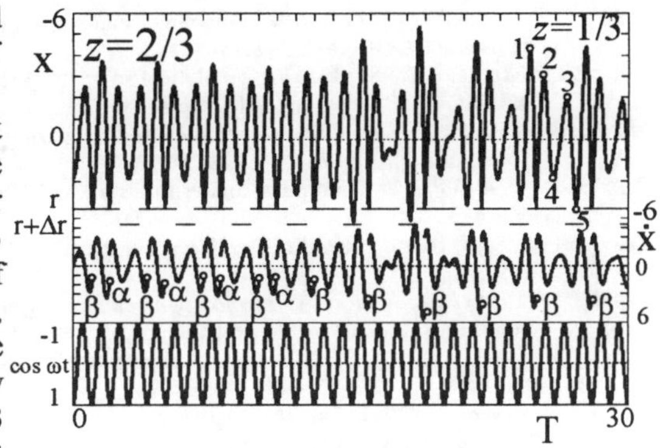

Figure 13. Transition from $z=2/3$ into $z=1/3$ motion, when additional impacts are omitted

Fig.13 shows the extinguishment of stonger impact α of the unstable $z=2/3$ motion. Formerly lower before-impact velocity β of $z=2/3$ motion stabilizes on the value β of $z=1/3$ motion (Figs.11(a),(c), 12(b), 13). It follows from Fig.12(a), that the energy level (6) of the intermittency chaos remains the same as for $z=2/3$ motion, but the energy level of $z=1/3$ motion (Fig.12(b)) is about 1.8 times higher than of unstable $z=2/3$ motion. Additional impacts play a similar role as in the interrupted Feigenbaum cascade. They take away suddenly parts of kinetic energy and return the system motion more or less into the state on the SN bifurcation boundary (s_{+1}). This state is unstable and the system is permanently in the development of SN instability and the recover into the state in the beginning of the SN instability. The instants and intensities of additional impacts are unpredictable and the response of this system is chaotic with the intermittency character.

Boundaries between periodic $z=2/n$ impact motions and the intermittencies are very sharp and *without the hysteresis*.

6. Conclusion

The laws of periodic and chaotic impact motions are briefly explained using the regions of existence and stability, bifurcation diagrams and motion trajectories. Four ways from periodic into chaotic impact motions are classified. The substance of two ways: a) Interruption of the Feigenbaum cascade and

b) Interruption of the saddle-node instability development,

caused by the additional impacts, are discovered. Additional impacts act, during the development of period-doubling and saddle-node instability, against the transition of the system impact motion on the periodic regime with higher energy level and they keep the system motion in the chaotic transition effects of the intermittency character.

Acknowledgement

The author gratefully acknowledge the financial supports of the Grant Agency of the Czech Republic (Grants No.101/94/0126 and No.101/97/0670).

References

Bespalova, L.V. (1956) On the theory of the vibro-impact system, Izv. AN SSSR, OTN, No..3,3-14 (in Russian).

Irie, T., Yamada, G., Matsuzaki, H. (1974) On the dynamic response of a vibro impact system to random force, *Bull. of the Faculty of Eng., Hokkaido University*, No.72, 13-24.

Kotera, T., Peterka, F. (1981) Part IV - Analytical solution of the (2/*n*)-impact motion and its stability, *Acta Technica CSAV*, No.6, 747-758.

Kotera, T., Peterka, F. (1984) Part VI - Analytical and analogue solution of the multi-impact motion and its stability, *Acta Technica CSAV*, No.3, 255-279.

Nordmark, A.B. (1991) Non-periodic motion caused by grazing incidence in an impact oscillator, *Journal of Sound and Vibration*, 145(2), 279-297.

Peterka, F. (1974a) Part I - Theoretical analysis of *n*-multiple (1/*n*)-impact motions, *Acta Technica CSAV*, No.4, 462-473.

Peterka, F. (1974b) Part II - Results of analogue computer modelling of the motion, *Acta Technica CSAV*, No.5, 569-580.

Peterka, F. (1981) *Introduction to vibration of mechanical systems with internal impacts*, ACADEMIA Publishing House, Prague, (in Czech).

Peterka, F., Vacík, J. (1981) Part III - Statistical characteristics of beat motions, *Acta Technica CSAV*, No.2, 161-184.

Peterka, F., Kotera, T. (1982) Part V - Regions of existence and stability and domains of attraction of different kinds of impact motion, *Acta Technica CSAV*, No.1, 92-117.

Peterka, F., Čipera, S. (1994) Chatic motions in mechanical systems with impacts, In *Proc. Chaos and nonlinear mechanics, World Scientific Series on Nonlinear Science*, Series B, Vol.4, 265-276.

Peterka, F., Kotera, T. (1996) Four ways from periodic to chaotic motion in the impact oscillator. *Machine Vibration*, 1996, 5, 71-81.

Peterka, F. (1996) Bifurcations and transition phenomena in an impact oscillator. *Chaos, Solitons & Fractals*, Vol.7, No. 10, 1635-1647.

Peterka, F., Szöllös, O. (1997) Dynamics of the opposed pile driver, In *Proc. IUTAM Symposium "Interaction between Dynamics and Control in Advanced Mechanical Systems"*, D.H. van Campen (ed.), Kluwer Academic Publishers, Solid Mechanics and its Applications, Vol.52, 271-278.

Rusakov, I.G., Kharkevich, A.A. (1942) Excited vibration of the system impacting against the stop, *Journal of Technical Physics* (ZhTF), Vol.12, No.11-12 (in Russian).

Sachse, R., Reum, G. (1965) Beiträge zur Theorie des Kurbel-Feder -Hammers (Teil I) - Der Kurbel-Feder Hammer als Ein-Masse System mit linearer Federkennlinie. *Wissenschaftliche Zeitschrift der Technischen Hochschule Karl-Marx-Stadt*, Jahrgang VII, Heft 2, 39-53.

Shaw, S.W., Holmes, P.J. (1983) A periodically forced piecewise linear oscillator. *J. of Sound and Vibration*, 90(1), 129-155.

NUMERICAL AND EXPERIMENTAL INVESTIGATION OF NONSMOOTH MECHANICAL SYSTEMS*

K. Popp, M. Oestreich, N. Hinrichs
Institute of Mechanics
University of Hannover
Appelstr. 11
30167 Hannover, Germany
e-mail: popp@ifm.uni-hannover.de

1. Introduction

During the past decade nonsmooth dynamical systems have become increasingly important in engineering and other applied sciences as nonsmooth effects are genuinely taken into account and are no longer smoothed out. Due to the lack of smoothness classical mathematical methods are applicable only to a limited amount and require extensions both for analytical and numerical methods. In the present paper the dynamics of nonsmooth oscillators with impact and friction are investigated numerically and by experiments.

Impact oscillators or related oscillators with piecewise linear restoring forces. They arise naturally in many applications such as impact print hammers, gear boxes and heat exchangers. In these examples the primary problems caused by the sucessive impacts are noise and wear.

Friction oscillators appear in everyday life as well as in engineering systems. Examples are bowed instruments, squealing wheels of tramways in narrow curves or grating brakes. In these examples self-sustained oscillations occur due to dry friction. In many engineering applications stick-slip vibrations are undesired and should be avoided since they debase precision of motion and safty of operation or create noise and wear.

Since impact and friction oscillators are nonlinear systems of sufficient complexity, also chaotic motions occur besides of regular behaviour. The bifurcational behaviour found numerically will be compared to experimental results in the present paper. In case of the friction oscillator, an extension of the model is necessary to fit the numerical simulations to experimental data.

2. Systems investigated

2.1. MODELS

The mechanical models used for the impact and friction oscillator are shown in Fig. 1. Both systems are described by mass m, damping b, spring stiffness c, dispalcement of

*Dedicated to Prof.Dr.Dr.h.c. Franz Ziegler on the occasion of his 60th birthday.

F. C. Moon (ed.),
IUTAM Symposium on New Applications of Nonlinear and Chaotic Dynamics in Mechanics, 293–302.

(a) 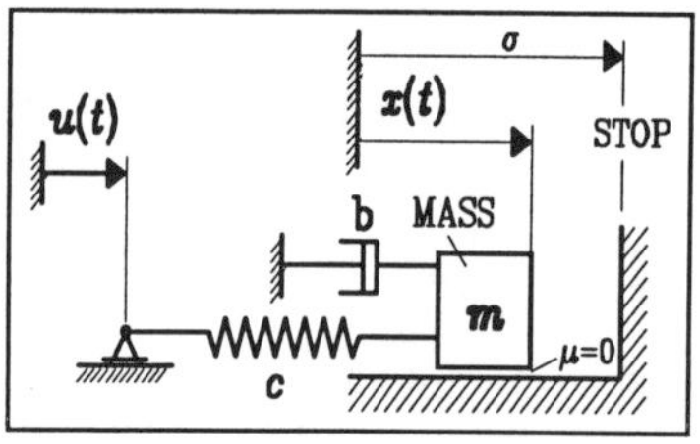

(b) 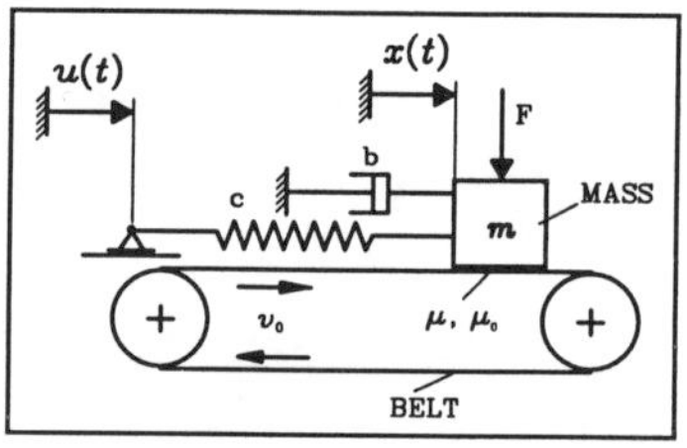

Figure 1. Investigated mechanical models: (a) impact oscillator, (b) friction oscillator.

mass x(t), external excitation $u(t)=u_0 \cos \Omega t$, ratio η of excitation frequency to eigenfrequency of the linear system and the normalized damping $D=b/(2\sqrt{cm})$. The system equation in normalized form is given by

$$x'' + 2Dx' + x = F_{NL} + u_0 \cos \Omega\tau, \tag{1}$$

For the impact oscillator during the free flight phase the nonlinear forces F_{NL} vanish. The impact with the obstacle is modeled by means of Newton's impact law, so infinite forces with zero duration lead to a sudden change of the velocity before (v_-) and after (v_+) the impact. Here $v_+= -e\, v_-$ holds, where e is the coefficient of restitution.
In case of the friction oscillator the linear oscillator is driven by a belt moving with constant velocity v_0. In the slip mode the nonlinear force F_{NL} depends on the friction coefficient μ and the normal force F_N, $F_{NL}= \mu(v_r \neq 0)\, F_N/c$. During the stick mode $F_{NL}=2Dx'+x-u_0 \cos \eta\tau < \mu(v_r=0)\, F_N/c$ holds. The dependence of the friction coefficient on the relative velocity $v_r=v_0-\omega_0 x'$ is modeled by means of three different friction characteristics

$$\text{I: } \mu_0=\mu(v_r\neq 0)=0.25, \qquad \text{II: } \mu_0=0.4,\ \mu(v_r\neq 0)=0.25,$$
$$\text{and III: } \mu_{III}(v_r)=\frac{0.3}{1+1.42|v_r|}+0.1+0.001\, v_r^2\ . \tag{2a}$$

For friction characteristic I, the friction coefficient does not change during stick and slip mode. Friction characteristic II is the well known friction characteristic of Coulomb-Amontouns and friction characteristic III is a spline approximation of a measured friction characteristic, cp. [4].

2.2. EXPERIMENTAL SET-UP

For the experimental investigation of the two systems a test stand has been built. Here, the linear oscillator is represented by a rotational pendulum supported by two springs (Fig. 2(a)). The rotational degree of freedom of the pendulum and its translational degree of freedom normal to the plane of rotation have been realized by means of air bearings providing small damping ($D<0.005$). The harmonic excitation is done by means of a magnetic excitation force F(t) (Fig. 2(b)) instead of the excitation u(t) for the

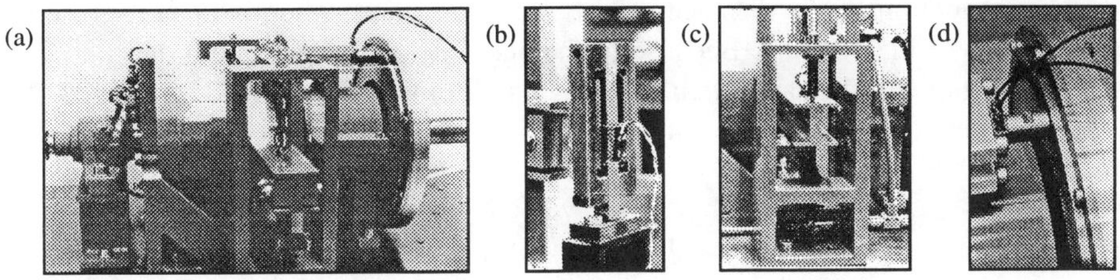

Figure 2. Experimental set-up: (a) linear oscillator, (b) magnetic excitation, (c) impact at stop, (d) friction contact.

mechanical model, so F(t)=cu(t) holds. For the realization of the impact oscillator the pendulum hits a sperical stop made of aluminum (Fig. 2(c)). In case of the friction oscillator the pendulum is pressed towards a disk by dead weights. The disk is driven with constant angular velocity. The nonlinear friction and normal forces are measured directly by a three component force transducer (Fig. 2(a), (d)). Displacement and velocity are sensed by a laser vibrometer. A more extended description of the test stand is given in Hinrichs et al. [5].

3. Numerical and experimental investigation

3.1. IMPACT OSCILLATOR

For the impact oscillator two different phases can be distinguished: free flight and impact. During free flight the motion is described by the superposition of the homogeneous and the particular solution of the linear equation of motion. During the time discrete events of the impacting bodies the velocity changes its sign and also the absolute value in case of e<1. Figure 3(a)-(c) show different types of experimentally measured phase trajectories, where the velocity of the pendulum, scaled in [mm/s], is

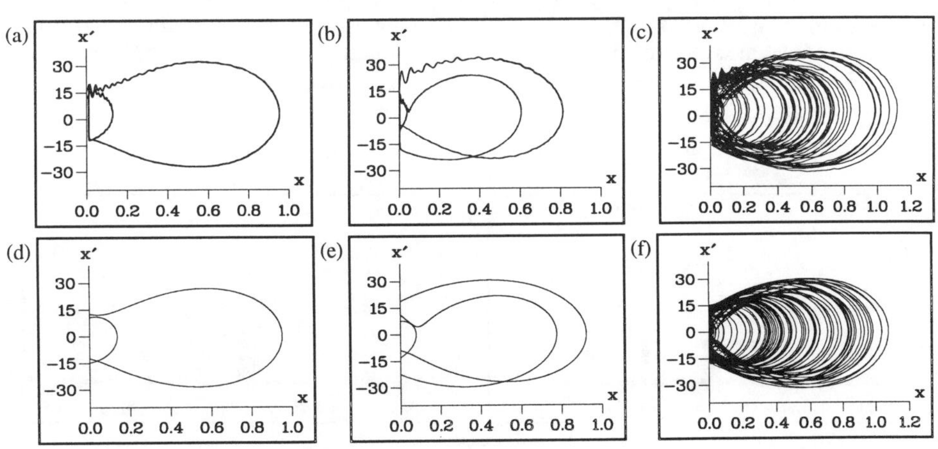

Figure 3. Different types of motions for σ=0.0, (a)-(c) Measuremnts, (d)-(f) Simulations with η=3.0 (a, d), η=2.76 (b, e), η=2.9 (c, f).

shown as a function of the displacement x, scaled in [mm]. The nonlinearity results in sudden jumps of the velocity. Because of the elasticity of the pendulum the time of the impacts does not vanish. Large impact forces also lead to higher mode oscillations of the pendulum immediately after the impact. The identified system parameters for the test stand are m=0.091 kg, b=0.017 kg/s, c=88.371 N/m and u_0=0.004 m. The coefficient of restitution e is measured by examining the velocity changes at impacts in the absence of forcing over a long time and results in e=0.86 for the experiments presented in this paper.
For η=3.0 the motion repeats periodically after two impacts and two forcing periods (Fig. 3(a)). For η=2.76 the motion repeats after three impacts and two forcing periods (Fig. 3(b)) and for η=2.9 the motion is apparantly chaotic (Fig. 3(c)).
The results of the simulations on the basis of the identified system parameters show a good agreement with the experimental ones, (Fig. 3(d)-(f)). Due to the simple model for the impact supposing zero impact time and rigid contact bodies, the trajectories differ during and immediatly after the impact but the periodicity of the solution and its amplitude can be predicted well by simulations.
A more global overview give bifurcation diagrams. The bifurcational behaviour depending on η can be obtained by slowly changing η in the experiment, allowing the system to reach an asymptotic state (typically after more than 120 seconds) which is a periodic or a chaotic attractor, and then recording the next impacts (up to 20 seconds). The initial state of the system was taken to be at equilibrium (x = v = 0). A similar diagram can also be obtained by direct numerical simulation. To record the data a Poincaré section is taken so that the values of x are recorded when v = 0. The results from measurements and simulations are given in Fig. 4(a), (b) for $2 < \eta < 4$. Several features of the measured and simulated bifurcation diagrams agree very well. In particular, large resonant peaks can be seen for the even values of η=2n, where the motion is periodic and repeats after one impact and n forcing periods. Around η=3 there is an interval of a period two motion which agrees with the analytic predictions made e.g. in [1]. For larger values of η there is a small chaotic window before a stable period one orbit appears which evolves toward the resonant peak at η=4. There are some differences between experiment and theory. In particular, the interval of the period two orbits centered at η=3 appears larger in the experiment than in the simulation. Furthermore, in the simulated bifurcation diagram there is a chaotic region at η=2.7, which cannot be found in the measurements. In this region the theory predicts coexisting

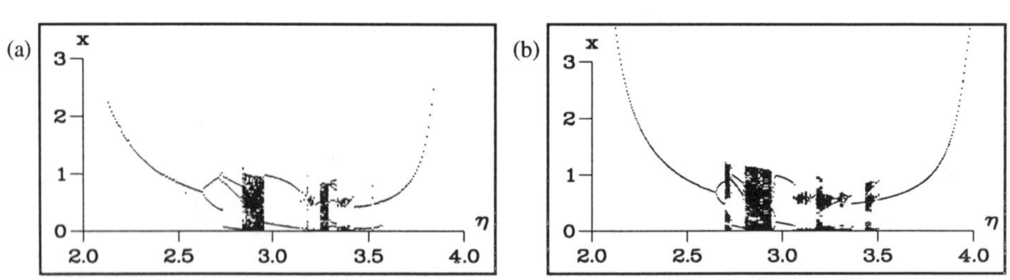

Figure 4. Bifurcation diagram showing (a) measurement: D=0.006, u_0=0.0043, e=0.86, σ=0 and (b) simulation.

solutions which depend upon the initial conditions, and these are not recorded in the measurements. Further bifurcation diagrams and poincaré-maps for simulations and measurements can be found in Hinrichs et al. [4]. Theoretical investigations of the impact oscillator can be found e.g. [7].

3.2. FRICTION OSCILLATOR

3.2.1. *Self excited vibrations*

In case of pure self excitation ($u_0 = 0$, $v_0 \neq 0$) one can observe stick-slip vibrations for the friction oscillator. Fig. 5(a) shows the phase portrait for friction characteristics III, where self excitation occurs due to the decreasing characteristic. As a typical feature of systems with friction the transients for all three characteristics decay very fast. As a result of the nonsmoothness, the integration is not invertible, so the system states on the stick line correspond to all sets of initial values.

The amount of energy transfered from the driven disk to the oscillator depends on the magnitude of the drop in friction between sticking and slipping. To get self excitation the energy transfered must be larger than the energy dissipated during one period of the oscillations. Measurements have shown, that the drop in friction for the materials steel-polyurethan is larger than that for the materials steel-aluminum, steel-brass and steel-bronze. So, this pair of materials leads to the largest limit cycle amplitude. Additionally, the contact area is treated with colophony. The measured phase curve for pure self excitation consists of a stick line representing constant velocity and the part of an ellipse during the slip mode. In contrast to the simulation (Fig. 5(a)), in the experiment the velocity during the stick mode oscillates slightly due to changes of the angular velocity of the driving shaft and small play in the gear box (Fig. 5(b)). For the transition from slip to stick there are small overshoots of the velocity signal, due to a stochastic component of the friction coefficient. So, in contrast to the simulations the system behaviour of the measurements is not periodic. In Fig. 5(c) the time dependent friction force is plotted. For the transition from stick to slip (t=0.9 [s]) the friction force jumps. The next jump occurs for the transition from slip to stick (t=1.0 [s]). This jump in the contact force results in decaying oscillations of the friction force. The period of these oscillations equals the period of the fluctuations in the velocity signal. Fig. 5(d) shows the measured friction characteristic that exhibits a drop for the transition from stick to slip.

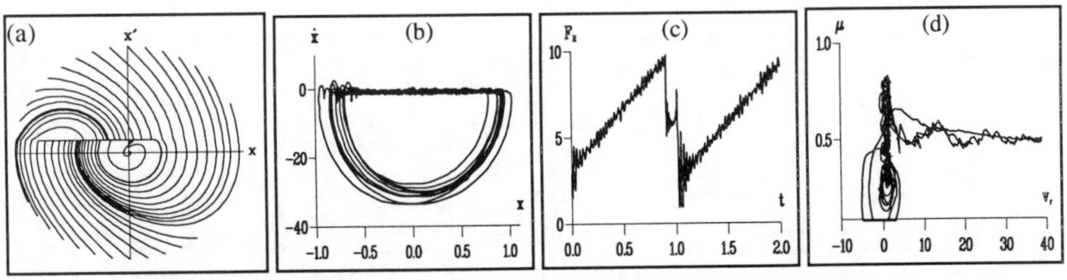

Figure 5. System behaviour for pure self excitation (materials: steel-polyurethan): (a) simulated phase plane plot (friction characteristic III), (b) measured phase plane plot, (c) measured friction force, (d) measured friction characteristic.

For pure self excitation there is no bifurcational behaviour. In order to investigate the bifurcational behaviour of the friction oscillator an external excitation is applied, which is done by means of a magnetic excitation.

3.2.2. *Forced vibrations*

The harmonically excited linear oscillator with a nonsmooth friction characteristic ($u_0 \neq 0$, $v_0 = 0$) exhibits qualitatively different types of motions. For small amplitudes u_0 of the excitation, $u_0 \leq \mu(v_r{=}0)\ F_N/c$, displacements in the range $-\mu(v_r{=}0)\ F_N/c + u_0 < x < \mu(v_r{=}0)\ F_N/c - u_0$ are stable equilibrium positions. For larger values of the excitation amplitude, depending on the bifurcation parameter η the friction oscillator shows motions without stop, motions with one stop, motions with two stops (Fig. 6(a)) and motions with four stops (Fig. 6(c)). This qualitative change of the system behaviour for small changes of η can be summarized by means of a bifurcation diagram, cp. Fig. 7(a), where the displacement x of the mass at the stops are shown. Obviously for $\eta > 0.55$ no regions of sticking exist. The number of stops increase with decreasing η. Further numerical investigations can be found in [3] and [13].

The results from the experiments shown in Fig. 6(b), (d) are qualitatively similar to the corresponding results gained from the simulations. For the sticking mass a high frequency oscillation can be observed in the velocity signal (Fig. 6(b), (d)).

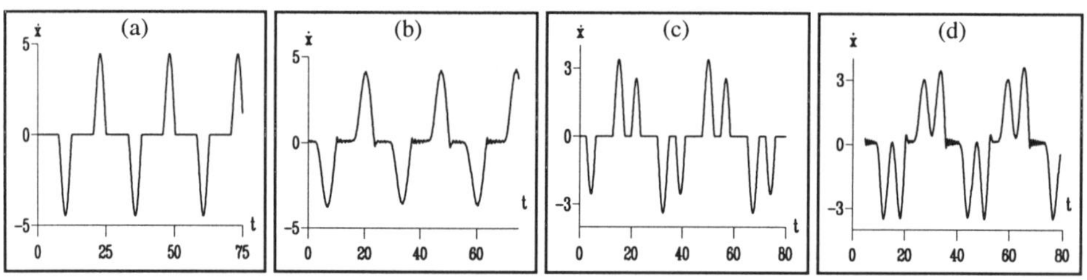

Figure 6. Friction oscillator with external excitation: (a) simulation, η=0.25 (friction characteristic II), (b) measurement, η=0.28, (c) simulation, η=0.20 (friction characteristic II), (d) measurement, η=0.18.

The corresponding measured bifurcation diagram under variation of the frequency ratio η is presented in Fig. 7(b). One can observe the bifurcation structure from solutions with one stop to solutions with multiple stops. A comparison with the bifurcation diagram

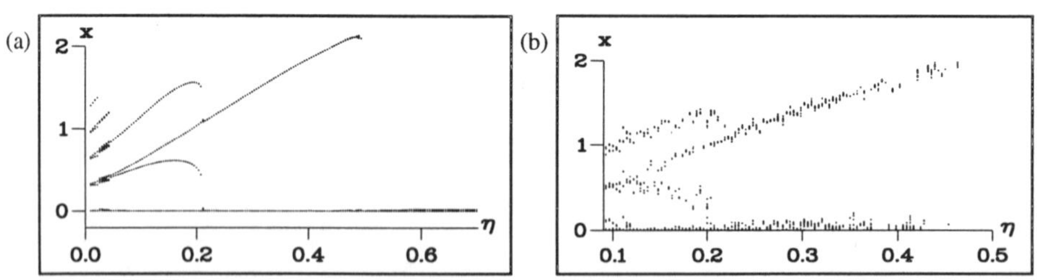

Figure 7. Bifurcation diagram showing the sticking points for the vibration with different frequency ratios η: (a) simulation, (b) measurement.

from simulations (Fig. 7(a)) shows again a good agreement. The only difference is that the stochasticity of the friction force results in a cloud of points in the experiment instead of a single point as in the simulations.

3.2.3. *Model extensions and results*

A comparison of the results from measurements and simulations shows that the global system behaviour of the friction oscillator, i.e. the amplitude and the period of the oscillations, can be predicted well by the simulations. However, the stochastic component of the friction coefficient has not been included in the simulations yet. Moreover, the fluctuations in the transition from slip to stick do not occur in the simulations. A more detailed insight in the dynamics in this regime is given in the following where also improved models are presented.

Stochastic friction model. The friction coefficient can be split up into a deterministic and a stochastic component:

$$\mu(v_r) = \mu_D(v_r) + \tilde{\mu}\,(v_r). \tag{2b}$$

With this improved model and a gaussian distribution of the stochastic component $\tilde{\mu}\,(v_r)$ (zero mean value and standard deviation σ=0.02) the system behaviour for pure self excitation comes close to the results from the measurement, as can be seen in Fig. 7(b), cp. also [5].

Model for the contact stiffness and contact damping. The model shown in Fig. 1(b) can be improved by adding a spring (stiffness c_R) and a damper (damping coefficient b_R), cp. Fig. 8 and [2]. The same model has been investigated in [15], however the contact stiffness and damping has been introduced for the regularisation of the nonsmooth problem.

If the absolute displacement of the single asperity in Fig. 8 is denoted by y, the system equation for the sliding block reads

$$m\ddot{x} + b\,\dot{x} + b_R\,(\dot{x} - \dot{y}\,) + c\,x + c_R\,(\,x - y) = cu\,, \tag{3}$$

and the equation of motion for the asperity is given by

$$b_R\,(\dot{y}\; - \dot{x}\,) + c_R\,(y - x) = F_R\,. \tag{4}$$

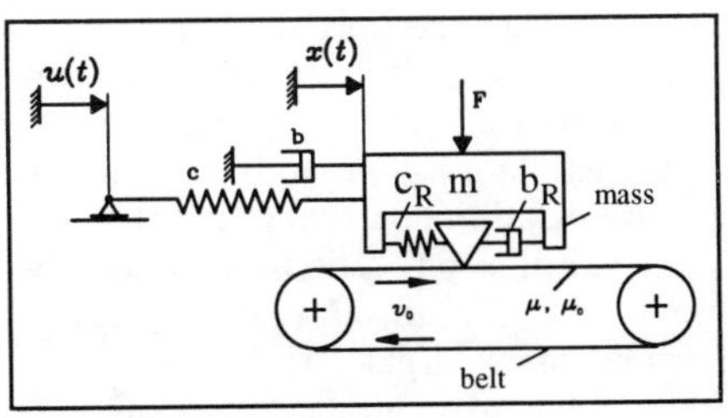

Figure 8. Improved model of the friction oscillator with contact damping b_R and contact stiffness c_R.

The transition point from stick to slip is reached for $F_R = \mu_0 F_N$, the slip mode ends for $\dot{y} = v_0$. The still unknown stiffness c_R and the damping b_R can be measured applying a small impuls to the mass, i.e. the pendulum in our experiments. (For the model Fig. 1(b) during the stick mode the displacement should equal to zero!) Fig. 9 shows the transient response of the systems for two different contact materials.

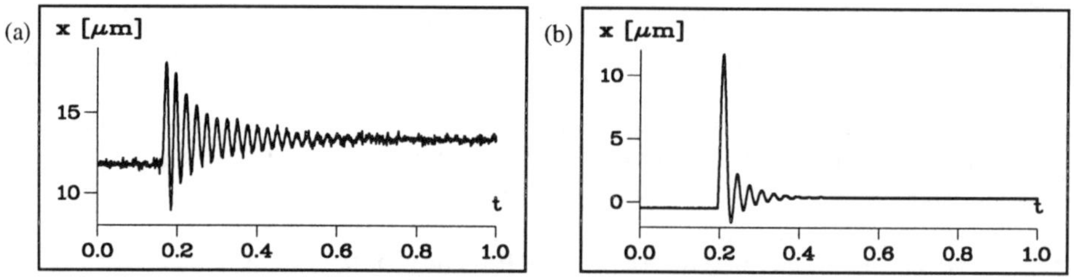

Figure 9. Free oscillations for impulse excitation of the sticking contact (materials: (a) steel-aluminum, (b) steel-polyurethan).

From the frequency of the resulting oscillations the contact stiffness can be evaluated, while the decay of the oscillations determines the damping. On the basis of the identified contact parameters simulations has been carried out for pure self excitation, cp. Fig. 10(c), (d). The results show the overshoots of the friction force and the velocity known from the measurements in Fig. 10(a),(b). The dynamics of the friction characteristic in the region of the transition from slip to stick can be predicted by means of this improved model.

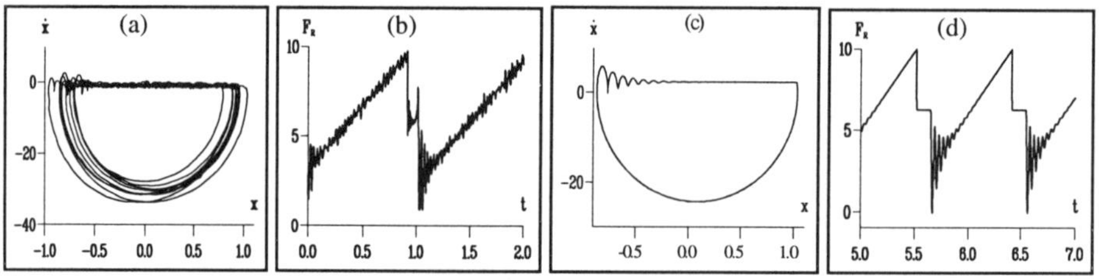

Figure 10. System response for pure self excitation (materials: steel-polyurethan): m=6.08 [kg], c=3956 [N/m], c_R=150000 [N/m], b_R=150 [Ns/m], μ_0=0.4, μ=0.25, F_N=25.0 [N], v0=0.0025 [m/s], (a), (b) measurements, (c), (d) simulations for the improved model with contact stiffness and damping.

In Fig. 11 (Fig. 12) the results from measurements and simulations for pure external excitation and the materials steel-polyurethan (steel-aluminum) are plotted . Fig. 11(a), (b) shows the experimental results for the system with pure external excitation and two stops during one period of the excitation. Due to the oscillations at the beginning of the stick phase, the transition of the friction coefficient takes place in the range of $-v_H < \dot{x} < v_H$ (here $v_H \approx 1.5$[mm/s]) instead of $\dot{x} = 0$. The transition shows a dynamic with limited stiffness and it can be shown that with an increase of the excitation frequency also v_H and the area of the hysteresis becomes larger, [5].

Due to the fact that the relative displacement and the relative velocity are sensed directly at the friction contact the dynamics of the pendulum cannot be responsible for the dynamics of the friction characteristic. A comparison of these results for the materials

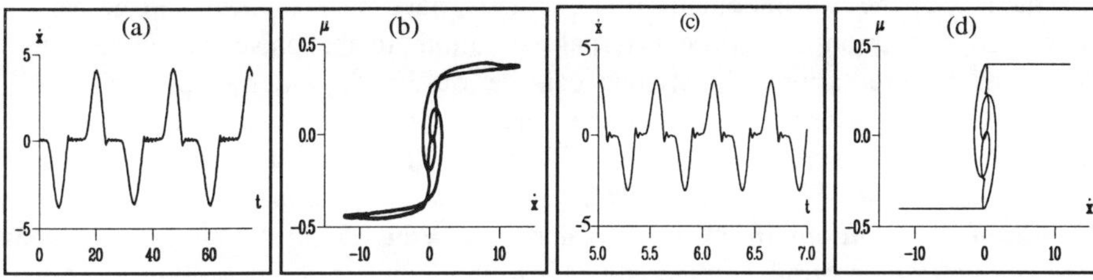

Figure 11. System response (velocity signal, friction characteristic) for pure external excitation (materials: steel-polyurethan): m=6.08 [kg], c=3956 [N/m], c_R=150000 [N/m], b_R=150 [Ns/m], μ_0=0.4, μ=0.25, F_N=25.0 [N], v0=0.0025 [m/s], Ω=11.5 [1/s] (a), (b) measurements, (c), (d) simulations for the improved model with contact stiffness and damping.

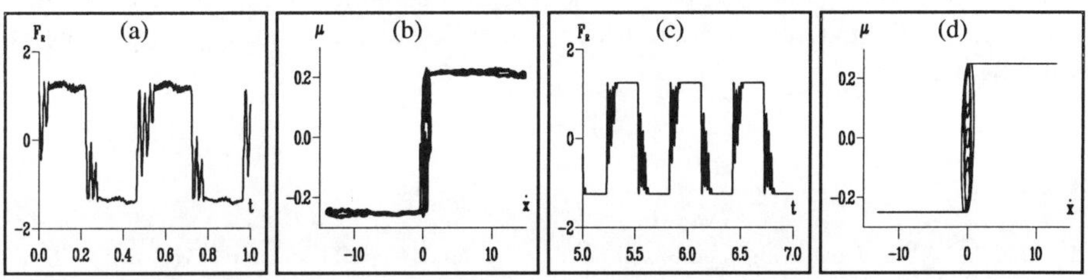

Figure 12. System response (velocity signal, friction characteristic) for pure external excitation (materials: steel-aluminum): m=6.08 [kg], c=3956 [N/m], c_R=150000 [N/m], b_R=150 [Ns/m], μ_0=0.4, μ=0.25, F_N=25.0 [N], v0=0.0025 [m/s], Ω=10.5 [1/s] (a), (b) measurements, (c), (d) simulations for the improved model with contact stiffness and damping.

steel-polyurethan with the experimental results for stiffer contact materials (steel-aluminum) shows that the frequency of the transients is larger in the latter case, see Fig. 12(b). As a consequence the hysteresis is much smaller for the stiffer material. Results similar to those in Fig. 12(b) can also be found in [2].
These phenomena cannot be predicted by the model Fig. 1(b), but by means of the introduced extensions of the model the observed phenomena can be shown.

4. Conclusions

The investigation of nonsmooth dynamical systems is presently a challange for the dynamisists and mathematicians, because many well known algorithms based on smoothness assumptions cannot be applied for systems with discontinuities. In the present paper the dynamical behaviour of two nonsmooth oscillators, an impact oscillator and a friction oscillator, has been investigated by experiments and compared to simulations.
In case of the impact oscillator, the application of Newton's impact law leads to a good agreement of measured and simulated phase curves and bifurcation diagrams. The higher mode oscillations and the deformation during the impacts do not affect the qualitative behaviour and the periodicity of the solution. However, an improved model requires the consideration of the elasticity of the impacting bodies.

The dynamics of the nonsmooth friction oscillator have been investigated numerically for pure self excitation and pure external excitation. In the presence of the external excitation rich bifurcational behaviour can be observed, which shows a sensitive dependence on the bifurcation parameters. The results for the two types of excitation have been verified experimentally. Also bifurcation diagrams have been measured for the friction oscillator with external excitation. For the prediction of the period and the amplitude of the solutions in the investigated parameter region the simple model of a friction oscillator is accurate enough and easy to implement for simulations. However, a more detailed look shows that stochastic features of the signal and oscillations at the nonsmooth transition points observed in the experiments cannot be predicted. Improved models comprise the required stochastic features and the dynamics in the transition regimes. The first one adds a stochastic component to the deterministic friction coefficient and the second one uses an additional tangential spring and damper in the contact area. A comparison of the improved models shows a good aggreement with the measurements.

5. References

1. Budd, C.J., Lee, G.: Double impact orbits of periodically forced impact oscillators, *Proc. R. Soc. Lond. A* **452** (1996), 2719-2750
2. Feeny, B. F., The effects of tangential contact stiffness on a harmonically forced friction oscillator, *Proc. ASME: IMECE: Elasto-impact and friction in dynamic systems* **90** (1996), 85-96
3. Den Hartog, J.P., Forced vibrations with Coulomb and viscous friction, *Trans. ASME*, **APM-53-9** (1931), 107-115
4. Hinrichs, N., Oestreich, M., and Popp, K.: Dynamics of oscillators with impact and friction, *Chaos, Solitons & Fractals*, **8**, 4 (1997), special issue: Guest editor T. Kapitaniak: Nonlinearities in applied engineering systems, 535-558
5. Hinrichs, N., Oestreich, M., and Popp, K., Friction induced vibrations: experiments, modelling and analysis, *Proc. of DETC*, Sacramento (1997), in press
6. Hinrichs, N., *Reibungsschwingungen mit Selbst- und Fremderregung: Experiment, Modellierung und Berechnung*, VDI-Verlag 11, 240, Düsseldorf: 1997
7. Oestreich, M., Hinrichs, N., Popp, K., and Budd, C.J., Analytical and experimental investigation of an impact oscillator, *Proc. of DETC*, Sacramento (1997), in press
8. Oestreich, M., Hinrichs, N., and Popp, K., Bifurcation and stability analysis for a non-smooth friction oscillator, *Archive of Applied Mechanics*, **66**, 5 (1996), 301-314
9. Oden, J.T., and Martins, J.A.C., Models and computational methods for dynamic friction phenomena, *Comp. Meth. in Appl. Mechs. and Engng.*, **52** (1985), 527-634
10. Persson, B.N.J., Theory of friction: The role of elasticity in boundary lubrication, *Phys. Rev. B*, **50**, 7 (1994), 4771-4786
11. Popp, K., Hinrichs, N., and Oestreich, M., Dynamical behaviour of a friction oscillator with simultaneous self and external excitation, Sãdhanã, **20**, Part. 2-4 (1995), 627-654
12. Popp, K., Hinrichs, N., and Oestreich, M., Analysis of a self excited friction oscillator with external excitation, In: Guran,A, Pfeiffer, F., and Popp, K. (Eds.): Dynamics with Friction: Modeling, Analysis and Experiment, World Scientific Publishing, Singapore, 1996, 1-35
13. Shaw, S.W., On the dynamic response of a system with dry friction,*J. Sound Vib.*, **108**, 2 (1986), 305-325.
14. Soom, A., and Kim, C.-H., The measurement of dynamic normal and frictional contact forces during sliding, *ASME:* 81-DET-40 (1981).
15. Vielsack, P.: Regularisierung des Haftzustandes bei Columbscher Reibung, *Z.angew. Math. Mech.*, **76**, 8 (1996), 439-446

STEADY-STATE BEHAVIOUR OF A SOLAR ARRAY SYSTEM WITH ELASTIC STOPS

D.H. VAN CAMPEN[1], R.H.B. FEY[2], F.P.H. VAN LIEMPT[3], A. DE KRAKER[1]
[1]*Eindhoven University of Technology, Department of Mechanical Engineering, PO Box 513, 5600 MB Eindhoven, The Netherlands*
[2]*TNO Building and Construction Research, Centre for Mechanical Engineering, PO Box 49, 2600 AA Delft, The Netherlands*
[3]*Fokker Space BV, PO Box 32070, 2303 DB Leiden, The Netherlands*

1. Introduction

In recent years a method was developed by the authors for efficient analysis of the long term behaviour of mechanical systems with local nonlinearities under periodic excitation. In this method the linear parts of the system are modelled using the finite element method. In order to keep the cpu-time for the nonlinear analyses acceptable, the number of degrees of freedom (dof) of the linear part of the system is reduced using a component mode synthesis (cms) technique. The cms technique used is based on free-interface eigenmodes and residual flexibility modes. Eigenmodes are kept up to a user-defined cut-off frequency. Subsequently, the reduced linear model is coupled to local nonlinearities, such as nonlinear springs and dampers, dry friction elements, backlash etc. The model obtained in this way is analysed using a nonlinear dynamics toolbox, which among others contains solvers for the calculation of periodic solutions and their stability and a path following method. The approach outlined above is described in Fey (1992) and Fey et al. (1996) and was integrated in the finite element package DIANA (1997). Until now, this approach was applied to rather academic, archetypal problems in order to verify its value. The approach turned out to be very successful: numerical results were compared with experimental results and a good correspondence was achieved (van de Vorst, 1996, van de Vorst et al., 1996a, van de Vorst et al., 1996b).

In this paper the dynamic behaviour of a real system is investigated: a solar array structure. These solar arrays, manufactured by Fokker Space BV, are connected to satellites. These arrays suffer from high vibrations during launch while mounted onto a satellite in folded position. The panel ends may strike each other if the excitation is too severe. This may cause damage to the structure. To prevent this, rubber snubbers are mounted at well chosen points of the structure. They act as elastic stops. The major advantage of this

F. C. Moon (ed.),
IUTAM Symposium on New Applications of Nonlinear and Chaotic Dynamics in Mechanics, 303–312.

solution is that no special adjustments are needed for the unfolding after launch, whereas the practical implementation is simple. The snubbers in the approximately linear folded solar array structure are local nonlinearities and will be modelled by means of one-sided springs, which are only active under pressure.
Fokker Space BV is interested in the steady-state behaviour of the folded system under periodic excitation with a constant acceleration amplitude of 2g perpendicular to the surface of the panels in the frequency range 20-100 Hz. DIANA (1997) was used to investigate the behaviour of the solar system. The main points of investigation are:

- Are snubbers needed to avoid contact between the panels?
- What is the influence of snubbers on the overall dynamic behaviour?
- Are there big differences between nonlinear dynamic analyses and linear dynamic analyses with snubbers modelled as linear springs?

2. System description and system model

Figure 1 gives a schematic view of the model of the folded system. Four primary z-levels are distinguished: level a, b, c and d at z=-33, 0, 48, and 82 mm respectively. The major parts of the model are named in Table 1 and are described below:

Part	Points	Part	Points
base	a1-a3, b1	edgemember 3	d2-d4
holddown 1	a2, c7, d7	joint yoke - panel 1	bc
holddown 2	a3, c7, c8	joint panel 1 - panel 2	cd
yoke	b1-b3	joint yoke - base	b1
panel 1	c1-c8	snubber 1	a1, b2 (ifcnod0, ifcnod1)
panel 2	d1-d8	snubber 2	c2, d2 (ifcnod2, ifcnod3)
edgemember 1	c1, c6, c5	snubber 3	c1, d1 (ifcnod4, ifcnod5)
edgemember 2	c2-c4	snubber 4	c6, d6 (ifcnod6, ifcnod7)

Table 1 Several parts of the structure.

- two solar panels. The panels have a thickness of 22.36 mm, whereas their mutual distance is 11.64 mm. The application of a lightweight sandwich structure for the solar panels leads to a very light structure in combination with high stiffness. With an estimate for the representative stiffness of such a composed material, a panel can be modelled with one layer of shell elements;
- two holddown and release systems. They clench the solar array to the body of the satellite, and try to maintain a constant distance between the panels. This is only achieved in the immediate vicinity of the holddowns, because the panels are flexible;
- one yoke. This arm forms the connection between the solar array and the operational satellite. The yoke is able to rotate the spread-out solar array to an optimal position with respect to the sun. During the considered launch this arm is folded along the panels;
- several joints, around which the panels and the yoke can rotate during unfolding (this process is no subject of investigation);

- eight snubbers, i.e. eight local nonlinearities, placed between the panels and between the satellite and the yoke;
- edgemembers distribute the stress smoothly at the point where a joint is connected to a panel. Moreover they add stiffness to the solar panels.

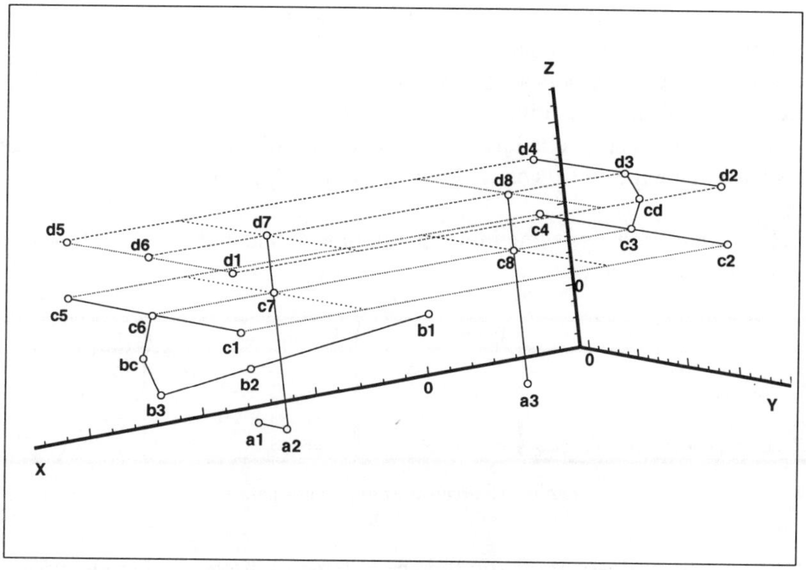

Figure 1 Schematic model of the solar array system.

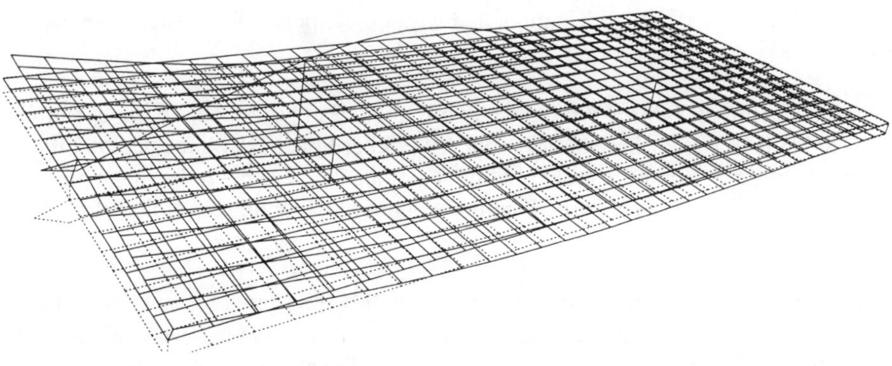

Figure 2 Lowest free-interface eigenmode (32.161 Hz) of the supported linear structure with snubbers with zero stiffness. Main motion: yoke swinging in x-z plane.

The following assumptions and idealisations are made:

- The system is excited by the spacecraft. This base excitation is in-phase at all base points (a1,a2,a3,b1; the corresponding nodes are tied) and perpendicular to the surface of the panels and has an acceleration amplitude of 2g.
- Because the system and the excitation are symmetric with respect to the plane y=0, only half the system is modelled.
- Snubbers and edgemembers are massless.
- There is no friction in the joints.
- Material damping is modelled by a modal damping of 3 percent for each linear mode and is based on experiments carried out by Fokker Space BV.

Table 2 shows constraints of some important points, to make clear which type of movement the model is restricted to.

	Points	Translations	Rotations
Symmetry edges	c4 c5, d4 d5	$u_y=0$	$\varphi_x=\varphi_z=0$
Part of base	a1-a3	$u_x=u_y=0$, $u_z=f(t)$	$\varphi_x=\varphi_y=\varphi_z=0$
Joint yoke-base	b1	$u_x=u_y=0$, $u_z=f(t)$	$\varphi_x=\varphi_z=0$
Connection holddowns-panel	c7, d7, c8, d8	$u_x=u_y=0$	

Table 2 Contraints at important points.

Four-node quadrilateral isoparametric flat shell elements are used to model the panels. Transverse shear stress is taken into account. The edgemembers are modelled with two-node 3D Thimoshenko beams. The yoke and the joints are also modelled with these beam elements. The total mass of the finite element model is 11.02 kg, which is half the mass of the real system because only one half is modelled. Each panel is modelled with 32x16=512 elements; the yoke is modelled with 20 elements. The structure model has 5894 dof.

The third column of Table 3 gives the eigenfrequencies of the model without snubbers supported at its base. Earlier studies, done by Fokker Space BV, indicated that the lowest eigenfrequency of the system without snubbers had to be in the range 30-40 Hz. Figure 2 displays the lowest corresponding eigenmode; the main motion is the vibration of the yoke in the plane y=0.

Later on results of nonlinear dynamic analyses will be presented for several snubber stiffnesses. Consider the ratio $\alpha = k_{snub}/k_j$, with k_{snub} the snubber stiffness, assuming that all snubbers are identical, and k_j the stiffness of the linear system at the position of ifcnod j; k_j follows from the reciprocal of the displacement, caused by a unit force on ifcnod j. This α is a measure for the degree of nonlinearity. Snubber stiffnesses of 10^4 N/m and 10^5 N/m will be considered. At snubber 3 (ifcnod 4 and 5) the highest ratio is reached, see Table 4 for the case $k_{snub}= 10^5$ N/m.

No.	Original model, base free	Original model, base supported	Reduced model, base supported (relative error %)
1	0.0003	32.161	32.161 (+0.000)
2	37.504	42.633	42.633 (+0.000)
3	42.734	75.552	75.552 (+0.000)
4	75.627	99.327	99.361 (+0.034)
5	103.43	123.63	123.64 (+0.008)
6	123.64	132.35	132.51 (+0.120)
7	132.68	135.39	156.80 (+15.81)
8	139.62	157.43	188.41 (+19.67)
9	157.56	164.47	194.27 (+18.11)
10	164.47	167.56	199.64 (+19.14)
11	167.59	186.22	224.09 (+20.33)
12	186.24	203.55	293.70 (+44.28)
13	204.01	207.54	447.44 (+115.5)
14	240.64	244.73	494.43 (+102.0)
15	244.79	245.25	

Table 3 Eigenfrequencies in Hz.

ifcnod j	k_j [N/m]	α[-]
1	4.5517E+04	2.197
2	3.8911E+04	2.570
3	3.8521E+04	2.596
4	3.4650E+04	2.886
5	3.2072E+04	3.118
6	8.7260E+04	1.146
7	7.0028E+04	1.428

Table 4 Values of α for $k_{snub}= 10^5$ N/m.

3. Reduction of the linear part of the finite element model

The number of dof of the linear model, i.e. the model without snubbers, will be reduced by application of a component mode synthesis method based on free-interface eigenmodes and residual flexibility modes (Fey 1992). The latter guarantee unaffected (quasi)-static load behaviour and are defined for dof which are loaded by (1) local nonlinearities, (2) prescribed motion or (3) external forces.
The model must be prepared to contain 4 snubbers, mounted between 8 nodes. These so-called interface nodes result in 8 residual flexibility modes since only the dof in z-direction are relevant. One of these dof, i.e. the z translation of the node at point a1 (see Table 1), will be subjected to a prescribed base excitation. The z-translations of the base nodes at the points a2, a3 and b1 are tied to this prescribed dof.
The number of dof of the reduced linear system now depends on the cut-off frequency f_c used for selecting the free-interface eigenmodes. As stated earlier, the frequency range of interest is 20-100 Hz. The second column of Table 3 shows the eigenfrequencies of the free-interface eigenmodes of the model without snubbers. Because the model is not supported at its base in this case the model has one rigid body mode: a translation in z-

direction. The choice of f_c is a compromise between long cpu times and accuracy of the solution in the coming nonlinear analyses. In general, f_c will be chosen higher than the highest excitation frequency to avoid as much as possible superharmonic resonances below f_c caused by artificial eigenfrequencies corresponding with the residual flexibility modes.

Selecting the cut-off frequency at 135 Hz, the linear model is reduced to 15 dof: 8 residual flexibility modes and 7 free-interface eigenmodes. As stated before, point a1 is subjected to a prescribed base motion in z-direction. So, the final reduced model has 14 dof. The eigenfrequencies of this model are presented in the last column of Table 3. If these eigenfrequencies are compared with those of the original linear model, we see a good correspondence up to the cut-off frequency.

4. Results of dynamic analyses

4.1 LINEAR DYNAMIC ANALYSES

Linear dynamic analyses have been carried out using the reduced model with 14 dof (f_c = 135 Hz). The four nonlinear elements have been replaced by linear two-sided springs. Frequency responses were determined for three cases: k_{snub}=0 (system without snubbers), k_{snub}=5.0 10^4 and k_{snub}=1.0 10^5 N/m, k_{snub} representing the linear stiffness of each spring. Figure 3 shows that for ifcnod1 with increasing snubber stiffnesses eigenfrequencies shift to higher values; displacement amplitudes decrease.

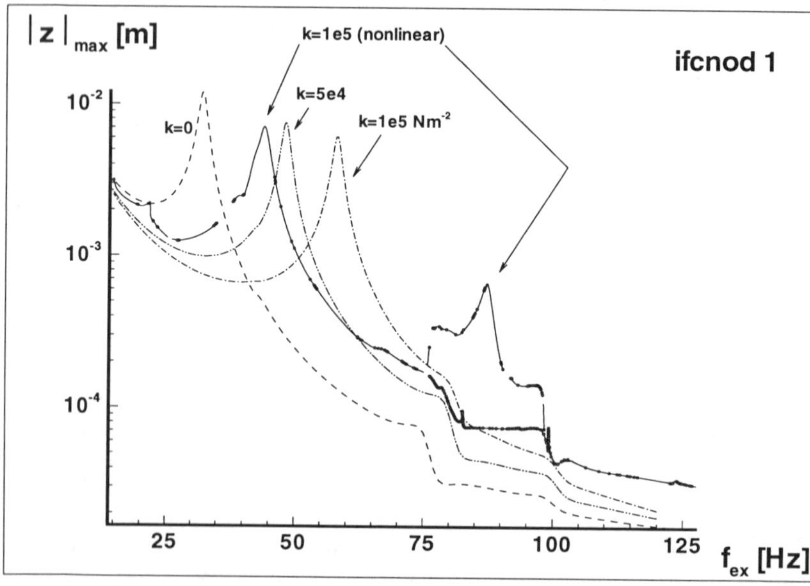

Figure 3 Three linear frequency responses and one nonlinear frequency response (strongly nonlinear case: k_{snub}= 10^5 N/m, small circles represent unstable periodic solutions).

4.2 NONLINEAR DYNAMIC ANALYSES

Again, the reduced 14 dof model will be used. Periodic solutions are calculated by solving a two-point boundary value problem using the finite difference method with an equidistant time grid. Path following is applied to follow branches of periodic solutions and Floquet theory is used to judge the local stability of the solutions and to detect bifurcation points.

4.2.1 Weakly nonlinear case: $k_{snub} = 10^4$ N/m

Results are shown in Figure 4: nonlinear dynamic analysis shows that all solutions are stable. The first harmonic resonance occurs at 34.2 Hz (32.2 Hz for the linear case without snubbers). The left inset of Figure 4 shows a 2nd superharmonic anti-resonance, originating from the 1st harmonic resonance peak. The right inset shows a superharmonic resonance caused by an artificial eigenfrequency (above f_c=135 Hz) originating from residual flexibility modes. Therefore, the occurence of this resonance has no physical relevancy. The response curves of ifcnod 2 and 3 are almost identical in the frequency range of interest: snubber 2 has little functionality. Near 34.2 Hz a small difference can be noticed between the response amplitudes.

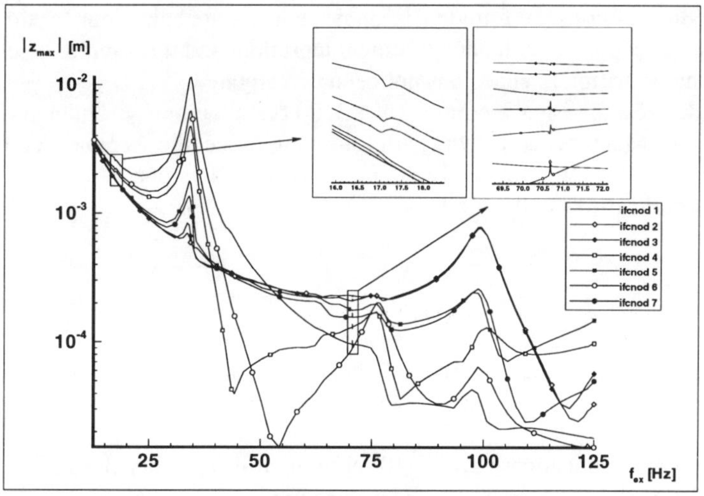

Figure 4 Frequency responses for the weakly nonlinear case k_{snub}=10^4 N/m.

4.2.2 Strongly nonlinear case: $k_{snub} = 10^5$ N/m

Let us consider the nonlinear frequency response of ifcnod 1 in Figure 3. A 2nd superharmonic resonance peak shows up at 21.9 Hz, caused by the first harmonic resonance

peak at 43.8 Hz. Furthermore, superharmonic resonances occur at 49.7, 62.0, 68.0, 82.4, 99.1 and 123 Hz, although not all are clearly visible.
The unstable region from 34.77 Hz to 34.83 Hz is initiated by a Neimark bifurcation. One may expect quasi-periodic behaviour. Numerical time integration at an excitation frequency of 34.8270 Hz (zero initial conditions) resulted in a Poincaré section with 17 points: apparently frequency locking on a 1/17 subharmonic occurs.
Within the range 74.94-98.26 Hz, a branch with stable and unstable 1/2 subharmonic solutions has been found, which bifurcates from the harmonic branch via flip bifurcations at the boundaries of this interval. The harmonic branch is unstable in this range. The ratio of the amplitudes on the 1/2 subharmonic and the harmonic branch reaches a maximum of 9.3 at 87.7 Hz.
Not all unstable regions on the 1/2 subharmonic branch have been investigated. But the unstable region from 95.51 Hz to 97.93 Hz is marked by two flip bifurcation points. In this interval 1/4 subharmonic solutions exist. Figure 5 zooms in on this branch for the displacement in z-direction of point c6, see Figure 1. The unstable region 96.808-97.018 Hz on the 1/4 subharmonic branch is marked by Neimark bifurcation points. At 96.808 Hz a Neimark bifurcation of a stable 1/4 subharmonic solution into a quasi-periodic torus is found. Figure 6 shows four closed curves in the Poincaré section. In terms of amplitude, the 1/4 subharmonic branch can hardly be distinguished from the 1/2 subharmonic solution branch.
At 99.136 Hz, Figure 5 displays a sharp peak. In the direct vicinity of this peak, no stable periodic solutions are found. The stable steady-state behaviour for this excitation frequency is investigated by means of time integration, and turns out to be chaotic, see Figure 7. The transition to chaos has not been investigated.
The frequency responses each contain over 3100 branch points, calculated in 46 path following runs. Many restarts of the path following procedure were needed because the solutions did not converge relatively often. This is probably due to the fact that there are four local nonlinearities in the model.

5. Conclusions

For the weakly nonlinear case no special phenomena are found: for each frequency within the frequency range of interest only one stable harmonic solution is found.

For the strongly nonlinear case $k_{snub} = 10^5$ N/m the following conclusions can be drawn:

- The analyses show very rich nonlinear dynamic behaviour. Neimark bifurcations of a stable 1/4 subharmonic solution into a quasi-periodic torus are found. Near 99.136 Hz the system reveals chaotic behaviour. In practice, especially the 1/2 subharmonic solution branch is important, because it spans such a wide frequency range (74.94-98.26 Hz). The difference in amplitude between this solution and the accompanying unstable harmonic solution reaches a maximum at 87.7 Hz, where the amplitude of the 1/2 subharmonic solution exceeds the amplitude of the corresponding harmonic solution by a factor 9.3.

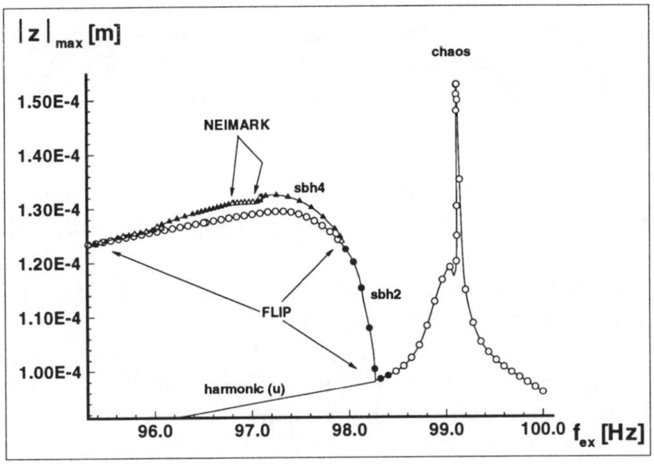

Figure 5 Detail of frequency response for point c6 (strongly nonlinear case $k_{snub}= 10^5$ N/m). Open symbols represent unstable periodic solutions.

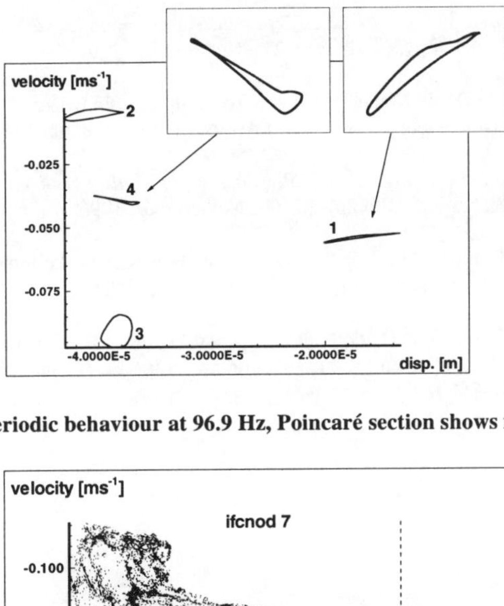

Figure 6 Quasi-periodic behaviour at 96.9 Hz, Poincaré section shows four closed curves.

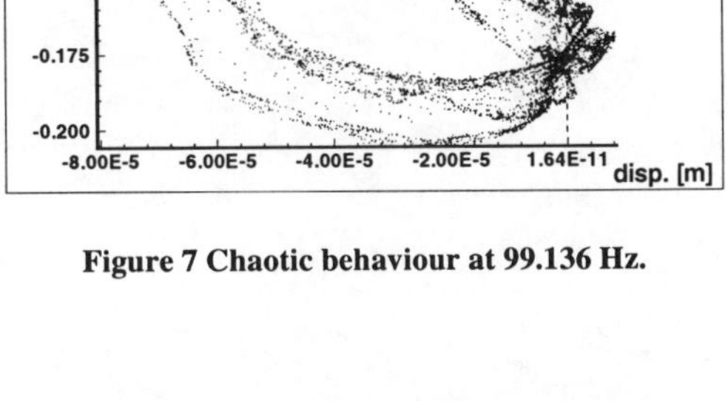

Figure 7 Chaotic behaviour at 99.136 Hz.

- A comparison of the results between nonlinear and linear dynamic analyses shows large differences in the frequency range 74.94-98.26 Hz, see Figure 3. Differences in amplitude of harmonic responses corresponding with three linear cases and the ½ subharmonic solutions calculated by means of nonlinear dynamic analysis can exceed a factor 10.
- Snubber 1 is most functional, because ifcnod 1 has the highest amplitude. Snubber 2 is least effective, and may be omitted because it hardly affects the calculated frequency responses.
- Using a prescribed base acceleration with a constant amplitude of 2g, the panels will not strike each other if no snubbers are applied.

DIANA (1997) appeared to be a very useful package for the analysis of steady-state dynamics of finite element models with local nonlinearities.

References

Diana, *User's Manual*, release 6.2. TNO Building and Construction Research, Delft, The Netherlands, 1997.

Fey, R.H.B., *Steady-state Behaviour of Reduced Dynamic Systems with Local Nonlinearities*, Ph.D. thesis, Eindhoven University of Technology, Eindhoven, The Netherlands, 1992.

Fey, R.H.B., van Campen, D.H., de Kraker , A., Long Term Structural Dynamics of Mechanical Systems With Local Nonlinearities, *Journal of Vibration and Acoustics*, Vol. 118, pp. 147-153, 1996.

van de Vorst, E.L.B., *Long Term Dynamics and Stabilization of Nonlinear Mechanical Systems*, Ph.D. thesis, Eindhoven University of Technology, Eindhoven, The Netherlands, 1996.

van de Vorst, E.L.B., van Campen, D.H., de Kraker , A., and Fey, R.H.B., Periodic Solutions of a Multi-dof Beam System with Impact, *Journal of Sound and Vibration*, Vol. 192, pp. 913-925, 1996a.

van de Vorst, E.L.B., van Campen, D.H., de Kraker , A., and Fey, R.H.B., Experimental Analysis of the Steady-State Behaviour of Beam Systems with Discontinuous Support, *Meccanica*, Kluwer Academic Publishers, Vol. 31, pp. 293-308, 1996b.

NONLINEAR DYNAMICS OF MECHANICAL SYSTEMS WITH DISCONTINUITIES

M. WIERCIGROCH
University of Aberdeen
Kings College, Department of Engineering
Aberdeen AB9 2UE, UK

Abstract - The paper outlines a general methodology for describing and solving dynamical systems with motion dependent discontinuities such as clearances, impacts, dry friction, etc. This approach assumes that the global solution is obtained by "gluing" together the local solutions at the connecting hypersufaces. An efficient numerical algorithm is presented and employed to analyse the dynamic interactions in the machine tool - cutting process system, which has five different discontinuities.

1. Introduction

Mathematical modelling of non-linear dynamical systems has become a powerful technique used by the engineering community to asses both qualitatively and quantitatively the dynamic systems responses. This has been utilised successfully in designing new or examining existing systems. There is an extensive library of mathematical tools, which covers almost completely linear dynamics, and an increasing number of different methods are available for non-linear dynamical systems. However, these tools are applicable mainly to continuous systems. As in many engineering applications we encounter either discontinuous or non-smoothed characteristics, there is a considerable need for a systematic methodology, which enables to solve such tasks. As well-known examples, one may point an oscillator with clearance analysed by Kreuzer *et al.* (1991), piecewise linear oscillator (Shaw and Holmes, 1983), jeffcott rotor with bearing clearances (Kim and Noah, 1990), systems with Coulomb friction (Popp and Stelter, 1990), and metal cutting processes (Grabec, 1988). All the above mentioned systems have been thoroughly investigated, however the analysis suffers from too specific approaches and difficulties in calculating some non-linear dynamics measures.

F. C. Moon (ed.),
IUTAM Symposium on New Applications of Nonlinear and Chaotic Dynamics in Mechanics, 313–322.

The discontinuous systems have been approximated either by oversimplified continuous models or mathematical types of discontinuities (McLeod, 1992), which hardly exist in engineering world. Therefore, the main goal of this paper is to show a new method of modelling systems with motion dependent discontinuities, which may supply a more accurate and adequate description for dynamics of such systems.

2. Modelling Of Discontinuous Systems

General methodology of describing and solving dynamic responses of a system with discontinuities presented here will be an extension of the work undertaken by Wiercigroch (1994a). This approach considers initially a dynamical system which is continuous in global hyperspace Ω, and can be described by the following first order differential equation

$$\frac{\partial \mathbf{x}}{\partial t} = \mathbf{f}(t; \mathbf{x}, \mathbf{p}), \tag{1}$$

where $\mathbf{x} = [x_1, x_2, ..., x_n]^T$ is the state space vector, $\mathbf{p} = [p_1, p_2, ..., p_m]^T$ is a vector of the system parameters, and $\mathbf{f}(\,) = [f_1, f_2, ..., f_n]^T$ is the vector function which is dependent upon the process being modelled. Then assume that the dynamical system (1) is continuous only in defined subspaces $\mathbf{X}_i$ of the global hyperspace Ω (which could be presented graphically as in Figure 1), therefore, the right hand side of Equation 1 may be written as follows

$$\forall_{t \in [t_1, t_2]} \ \forall_{x \in X_i} \ \mathbf{f}(t; \mathbf{x}, \mathbf{p}) = \mathbf{f}_i(t; \mathbf{x}, \mathbf{p}), \quad i \in N. \tag{2}$$

The global solution is obtained by "gluing" local solutions on the hypersurfaces $\Pi\mathbf{X}_{i,\,i+1}$ (where $i \in N$). As the dynamical system goes through these subspaces, a typical path of its suspected motion can be drawn. If any response of the investigated system is periodic, its trajectories repeat themselves in the state space, and its path forms a closed loop (see Figure 1). Otherwise an irregular motion (chaotic, unstable or quasiperiodic) would occur. When a hypersurface $\Pi\mathbf{X}_{i+1,\,i+2}$ is intersected by a trajectory emanating from the subspace X_{i+2} towards X_{i+1} for the k-th discontinuity crossing, the mapping $x^{(k)}_{(+)} \rightarrow x^{(k)}_{(-)}$ takes place. In order to solve the system (1) with piecewise continuous or discontinuous forcing functions (2) the precise value of the crossing time $t^{(k)}$ has to be evaluated since the response can be very sensitive to any inaccuracy of the computed solution on the above-mentioned hypersurface. As a consequence, a suitable switch function for the discontinuity location has then to be formulated, and a precise value determined when the discontinuity occurs. Such procedure has to be repeated for each time when a discontinuity is detected. This has been a base for a development of an effective numerical scheme providing a satisfactory solution for any dynamical system with motion dependent discontinuities. Some important aspects of this scheme will be discussed in a greater detail in the following section.

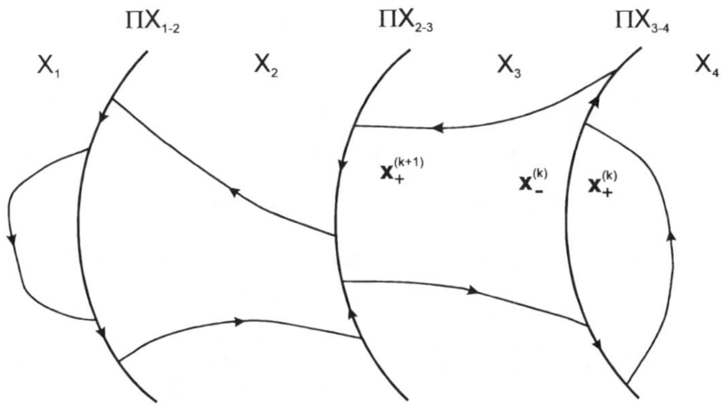

Figure 1. A conceptual model of a dynamical system with discontinuities

3. Numerical Algorithm

After transforming ODEs into algebraic form, in principle any numerical intergration solver can be used, but special attention must be paid to the error of solution. Either an exessively short step length or its variation has to be used in order to estimate a local truncation error to locate discontinuities sufficiently precisly. Also, as can be decipher from the proceed section, except for certain trivial or special cases, a closed form or approximate analytical solutions are not obtainable. Therefore, an approach has been made to devise a numerical algorithm which can provide solutions in a general case. It follows the argument of mathematical modelling explained in the previous section. In practical terms, the solution for any nonlinear system with motion dependent discontinuities is obtained using the classical fourth-order Runge-Kutta method with two extra procedures, namely

(i) the discontinuity detection and

(ii) the calculation of the precise time value when the discontinuity occurs.

The detection procedure calculates values of an appropriate switch function (see Equation 3) each time step to evaluate whether its value at the beginning and end of each time being considered straddle zero or indicate the presence of a local extremum. After the discontinuity is detected, the system is govern by a different set of equations and mapping (4) takes place.

$$\Phi_k(t; \mathbf{x}) = 0, \tag{3}$$

$$\aleph : \mathbf{f}_{i,-}(t;\mathbf{x},\mathbf{p}) \rightarrow \mathbf{f}_{i+1,+}(t;\mathbf{x},\mathbf{p}), \qquad \mathbf{x}_{-}^{(k)} \rightarrow \mathbf{x}_{+}^{(k)} \tag{4}$$

A flow diagram incorporating the discontinuity detection procedure into a numerical integration scheme is shown in Figure 2, where decission is taken by evaluation of the product of two values of the switch function calculated for the subsequent time steps.

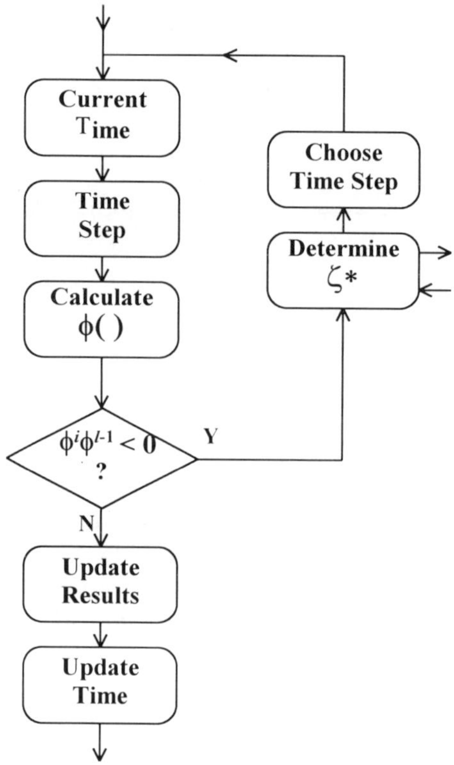

Figure 2. The first step in the proposed algorithm of numerical integration of ODEs containing motions dependent discontinuities.

Once the discontinuity is detected, an interpolation scheme is used to locate a precise value of the time when the discontinuity occurs. Figure 3 shows a flow chart which was designed to be most effective for handling any displacement (e.g. clearances) and velocity (e.g. dry friction) discontinuities. The major shortcoming of this algorithm is that criteria for choosing an appropriate procedure to be used (e.g. bisection or inverse interpolation), cannot be generated automatically.

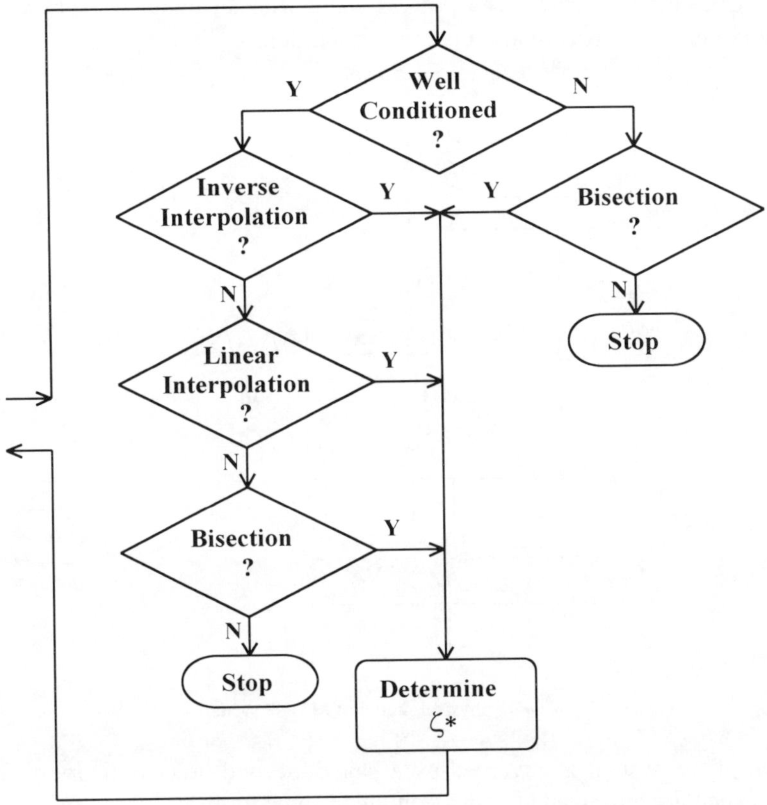

Figure 3. The second step in the proposed algorithm of numerical integration of ODEs containing motions dependent discontinuities.

4. An Example

As an example to test the introduced methodology, the dynamic interactions in the machine tool - cutting process systems (MT-CP) are investigated. The systems has five different discontinuities (Wiercigroch, 1997) of both the displacement and velocity types, and it generates rich dynamic responses. A short description of MT-CP physical model for the orthogonal metal cutting and its dynamic behaviour is presented below.

The machine tool performing the orthogonal cutting is assumed to be a planar oscillator with linear springs and dashpots, as depicted in Figure 4. This model has been successfully used for intermittent cutting initially by Grabec (1988), and recently refined by Wiercigroch (1997). The process starts with an initial velocity v_0 and an initial depth of cut h_0 which are also the initial conditions.

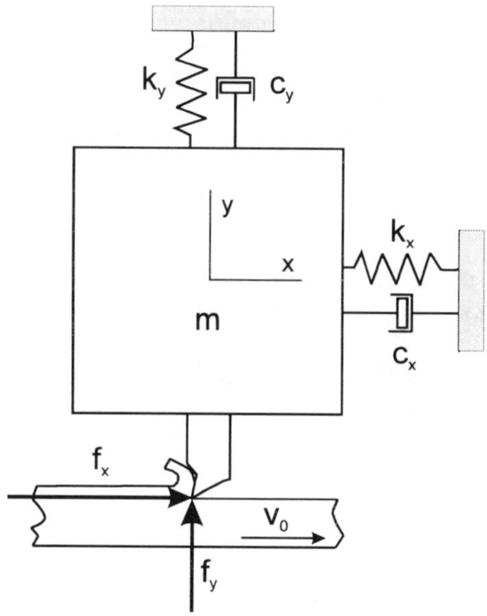

Figure 4. A physical model of MT-CP system

The motion of the system is governed by a pair of second order ordinary differential equations, which are presented here in a nondimensional form

$$x'' + 2\xi_x x' + x = f_x(x, y, x', y') \tag{5}$$

$$y'' + 2\xi_y \sqrt{\alpha}\, y' + \alpha\, y = f_y(x, y, x', y') \tag{6}$$

The viscous damping in the system is represented by ξ_x and ξ_y, and a stiffness ratio between x and y direction denoted by α. The cutting forces $f_x()$ and $f_y()$ are dependent upon the stochastic properties of the workpiece

$$f_x(x, y, x', y') = f_0 h H(v_r)\, (a_1(v_r - 1)^2 + 1)\; H(h) \tag{7}$$

$$f_y(x, y, x', y') = \chi\, (v_r, v_f, h)\; f_x(x, y, x', y') \tag{8}$$

where

$$
\begin{aligned}
\chi(\ ) = \chi_0(a_2(v_f-1)^2+1)(a_3(h-1)^2+1)H(f_x)\,\mathrm{sgn}(v_f) \\
v_r = v_0 - x',\ v_f = v_0 - py',\ h = h_0 - y,\ p = p_0(a_4(v_r-1)^2+1)
\end{aligned}
\tag{9}
$$

a_1 to a_4 are the cutting process constants, p is the shear deformation coefficient, $H(\ )$ represents Heaviside function, and $\chi(\ )$ is a variable friction coefficient.

Although one cannot deny usefulness and elegance of approximate solutions, in this particular case, a more extensive analysis cannot be accomplished without numerical simulation. Therefore, Eqs. (5) and (6) are transformed to the system of four first order differential equations, which can be written as follows

$$
\begin{aligned}
&x_1' = x_2, \\
&x_2' = -x_1 - 2\xi_x x_2 + f_{x2}(x_2, x_3, x_4), \\
&x_3' = x_4, \\
&x_4' = -\alpha\, x_3 - 2\xi_y \sqrt{\alpha}\, x_4 + f_{x4}(x_2, x_3, x_4),
\end{aligned}
\tag{10}
$$

where

$$
\begin{aligned}
&f_{x2}(x_2, x_3, x_4) = f_x(y, x', y'), \\
&f_{x4}(x_2, x_3, x_4) = f_y(y, x', y').
\end{aligned}
\qquad ....(11)
$$

The expressions for the cutting forces (7) and (8) have five different discontinuities, which can be labelled to one of two groups: either to the continuos discontinuity (unsmooth function), *DiscC* or discontinuous discontinuity, *DiscD*. This classification was used to further develop the explained above integration scheme (Wiercigroch, 1997). The discontinuity *DiscC* is a product of linear and Heaviside function, where *DiscD* is a straightforward *sign* function.

The MT-CP system has been thoroughly numerically investigated by Wiercigroch (1994b), and this paper gives only a small sample of the most interesting results. Since the adopted model of the cutting process allows the tool to leave the material being cut, the intermittent cutting process may occur. This is confirmed by the intermittency of the cutting force components (Figures 5a and b), which coincides with an erratic profile of the surface profile depicted in Figure 5c.

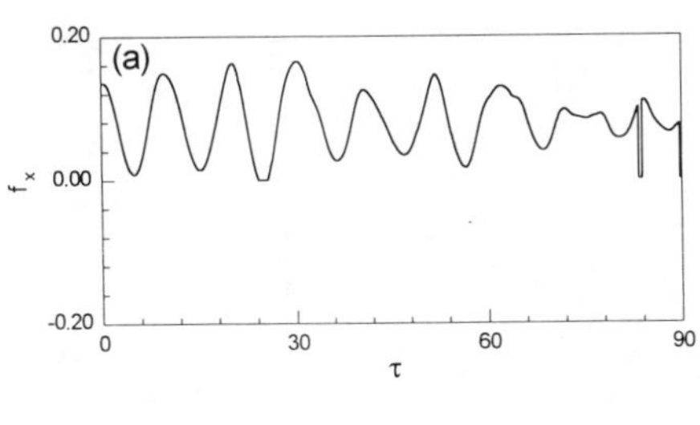

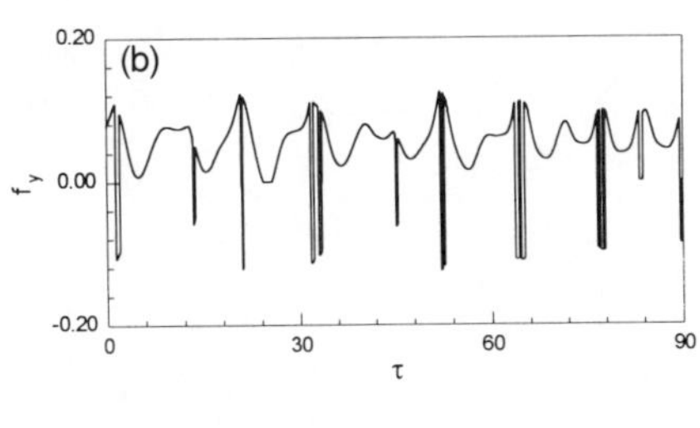

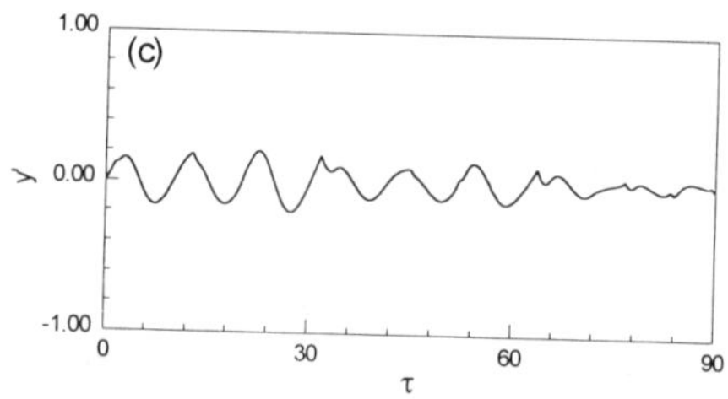

Figure 5. Time histories of (a) horizontal cutting force, (b) vertical cutting force and (c) surface profile

The investigated system is described by a twelve-parameter vector $\mathbf{p} = [\alpha, \xi_x, \xi_y, \mu_0, c_1, c_0, c_3, c_4, q_0, h_0, R_0, v_0]^T$. The results presented here were obtained by fixing the values of the following parameters, i.e., $\mu_0 = 0.1$, $c_1 = 0.3$, $c_2 = 0.7$, $c_3 = 1.5$, $c_4 = 1.2$, $R_0 = 2.2$ and $v_0 = 0.5$. The bifurcation diagrams constructed for $\alpha = 4$ show an another example of an unusual system's behaviour, that is *unidirectional bifurcation*, where the considered system bifurcates in the x direction and is completely stable in the y, where f_0 is between 0.24 and 0.54 despite the fact the equations of motions are coupled. This might be explained as well as a *shift* of the critical point for the x and y directions, i.e., the values for the x and y are equal 2.04 and 2.38 respectively. For the x direction the system starts with two bifurcation *period of doubling* type, and then vibrates chaotically. For the y direction, the system after crossing the critical value oscillates with period 4, and then becomes chaotic.

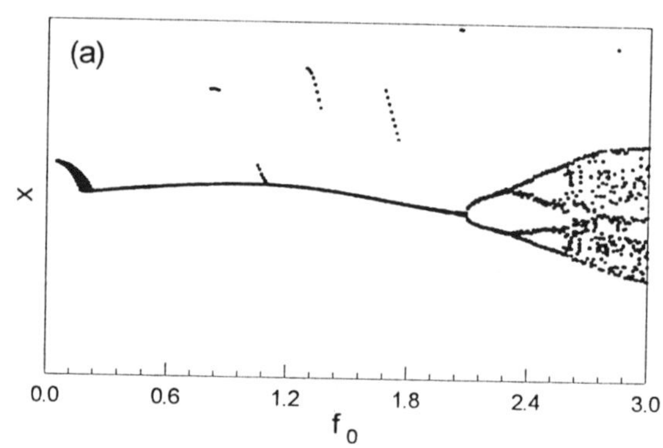

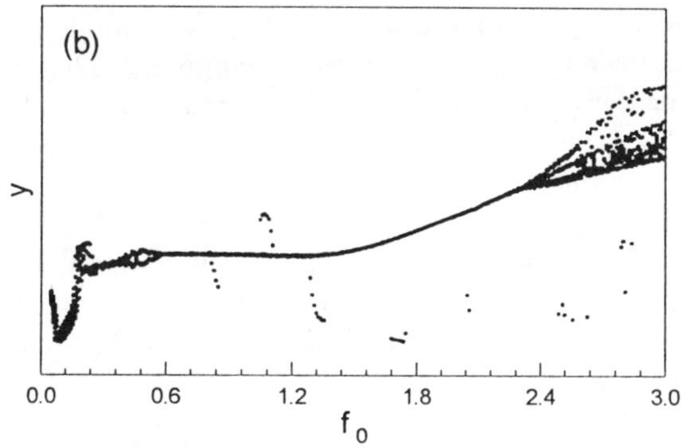

Figure 6. Bifurcation diagrams (a) $x = f(f_0)$ and (b) $y = g(f_0)$

5. Concluding Remarks

The paper presents general methodology for describing and solving dynamical systems with motion dependent discontinuities such as clearances, impacts, dry friction, etc. This approach is based on the assumption that global solution is obtained by "gluing" together the local solutions at the connecting hypersufaces, which can be 'a priori' mathematically described.

General numerical algorithm implementing the introduced mathematical modelling of systems with discontinuities was proposed. Essentially, the discontinuity problems are handled in two stages: detection of the discontinuity and then a precise calculation of the time when it occurs. The algorithm was tested on a simple model of the machine tool - cutting process system, which has five different discontinuities.

The elastic, dissipative and inertial properties of the machine tool structure, tool and the workpiece are represented by a planar oscillator, which is excited by the cutting forces. It is assumed that the relationship between the cutting forces and the chip geometry, namely the cutting process characteristics is captured by the intermittent orthogonal cutting process. To make the process more realistic a dry friction force acting on the cutting edge was added. The undertaken analysis demonstrates a complex dynamic behaviour, which is manifested by an existence of periodic, quasi-periodic, subharmonic and chaotic motion (see Wiercigroch, 1997). It was found out that some of the bifurcation diagrams cannot be classified into standard routes to chaos, however, crisis type transition to chaos is dominating. Two new phenomena were detected, which are unique for the non-linear dynamical systems. The first one is called unidirectional bifurcation, i.e., the system is stable in the one direction and unstable in the another. The deaths and births of periodic solutions is the second one, where the system oscillates between the stable and asymptotically stable states as the damping ratio is increasing. In general, it may be stated that dynamic responses of the system can be controlled most effectively by a mean of the cutting force modulus. Most importantly, contrary to classical machine tool theory it was been observed that for certain values of the parameter vector the amplitude of vibration can decrease with an increase of the cutting force, which can be used as a design information to improve the productivity and accuracy of the machine tools.

Acknowledgment. The support provided to the author by the British-Dutch Joint Research Programme under the grant JRP388 is deeply appreciated. The hospitality of the Mechanical Engineering Department at the Eindhoven University of Technology is also gratefully acknowledged.

6. References

Grabec, I. (1988) Chaotic dynamics in the cutting process, *Int. J. Mach. Tools Manufact.* **28**(1), 19-32.

Kim, Y.S. and Noah, S.T. (1990) Bifurcation analysis for a modified jeffcott rotor with bearing clearances, *Nonlinear Dynamics* **1**, 221-241.

Kreuzer, E., Kleczka, M. and Schaub, S. (1991) Chaotic dynamics of a single oscillator: A pictorial introduction, *Chaos Solitons and Fractals* **1**, 439-456.

McLeod, B.J. (1992) Smoothing of Stokes discontinuities, *Proc. R. Soc. Lond. A* **437**, 343-354.

Popp, K. and Stelter, P. (1990) Stick-slip vibrations and chaos. *Proc. R. Soc. Lond. A* **332**, 1477-1480.

Shaw, S.W. and Holmes, P.J. (1983) A periodically forced linear piecewise oscillator, *Sound and Vibration* **90**, 129-155.

Wiercigroch, M. (1994a) A note on switch function for stick-slip phenomenon, *J. Sound and Vibration* **175**(5), 700-704.

Wiercigroch, M. (1994b) *Dynamics of Discrete Mechanical Systems with Discontinuities*, Scientific Press of Silesian Technical University, Gliwice.

Wiercigroch, M. (1996) On modelling discontinuities in dynamic systems, *Machine Vibration* **5**, 112-119.

Wiercigroch, M. (1997) Chaotic vibration of a simple model of the machine tool-cutting process system, *Trans. ASME, JVA* **119** (in press).

DISPLACEMENT POTENTIALS IN NON-SMOOTH DYNAMICS

CH. GLOCKER
Lehrstuhl B für Mechanik
Technische Universität München
D-85747 Garching
Germany

Abstract. The paper treats the evaluation of the accelerations in rigid multibody systems which are subjected to displacement dependent set-valued force interactions. The interaction laws are represented by non-smooth displacement potentials and derived through generalized differentiation. The resulting multifunctions contain the cases of smooth force characteristics, bilateral constraints, as well as combinations of them like unilateral constraints or prestressed springs with play. Impacts are excluded. A generalization of the classical principles of d'Alembert, Jourdain, and Gauss in terms of hemi-variational inequalities is given. A strictly convex minimization problem depending on the unknown accelerations is stated, known in classical mechanics as the Principle of Least Constraints. The theory is applied to unilaterally constrained systems.

1. Introduction

In classical mechanics the concept of virtual work is used in order to classify displacement-dependent forces. Forces are called "active" or "applied forces" if they produce virtual work, otherwise they are called "passive" or "constraint forces". Both types are treated completely different with respect to formulation and evaluation: Applied forces are continuous functions of the displacements and can therefore be directly evaluated when the displacements are known. Passive forces originate from bilateral constraints and can not be represented by such functions. Usually certain displacements are prescribed, but the forces are arbitrary. The evaluation of the passive forces is done by either inverting the corresponding force characteristics

F. C. Moon (ed.),
IUTAM Symposium on New Applications of Nonlinear and Chaotic Dynamics in Mechanics, 323–330.

which yield additional equations depending on the displacements known as constraints, or by choosing appropriate reduced sets of new coordinates, called "minimal coordinates", such that the constraints are automatically satisfied and thus eliminated from the equations of motion. This approach makes sense as long as the above mentioned classification applies to all forces within a multibody system. However, there are forces of intermediate type acting in some regions as applied forces, in other regions as constraints. Usually one becomes aware of them when dealing with dry friction, unilateral constraints, or impact effects. This type of forces leads to hemivariational inequalities which are variational expressions for differential inclusions obtained by generalized differentiation of certain (nonsmooth) superpotentials. In classical mechanics only small attention is paid to this case. Even in modern literature only a few works have appeared concerning inequality constraints and nondifferentiable energy functions. We refer especially to the papers [1], [4], [6] and to the few references given there. The aim of this paper is to introduce general interaction laws including both, active type forces and constraint type forces, and to give a new interpretation of d'Alembert's Principle of Virtual Work which leads to a strictly convex optimization problem in terms of the unknown accelerations of the system. As an example the theory will be applied to systems with arbitrary unilateral constraints.

2. The Principles of d'Alembert, Jourdain, and Gauss

In this section we define the Principles of d'Alembert, Jourdain, and Gauss for holonomic scleronomic multibody systems, and we show that the Principle of Gauss is equivalent to a strictly convex optimization problem which corresponds to the classical Principle of Least Constraints.

We consider a dynamical system consisting of a certain number of rigid bodies. The generalized displacements of the system may be described by a coordinate vector $q \in \Re^f$ which depends on time t. The velocities are assumed to be functions of bounded variations [4], [5], i.e. their left and right limits $\dot{q}^-$ and $\dot{q}^+$ exist at every point t, the set of points at which $\dot{q}$ is discontinuous is at most countable, the displacements are therefore absolutely continuous, and the differential measure of $\dot{q}$ consists of three parts: The Lebesgue measurable part which is absolutely continuous, the atomic part which is concentrated on the countable set of discontinuities of $\dot{q}$, and a part with support on sets with Hausdorff dimensions between 0 and 1. Neglecting the latter and splitting off the atomic part one may state the equilibrium of the linear and angular momenta and forces which yields

$$M(q)\,\ddot{q}^+ - h(q, \dot{q}^+) = f^+ \tag{1}$$

with a symmetric positive definite mass matrix M and a vector h consisting of the gyroscopical accelerations. In order to obtain a complete description of the dynamics of the system we must introduce some force laws which connect the forces f^+ and the displacements q. We assume these interaction laws to be derived from a *displacement potential* $V(q)$ through generalized differentiation [6], and we will equip it with the following properties:

Let $V : \Re^f \to \Re \cup \{+\infty\}$ be a (non-smooth, non-convex) lower semicontinuous (l.s.c.) function with effective domain C such that its epigraph epi V consists of regular points only, i.e. the tangent cone $T_{\text{epi}\,V}$ and the contingent cone $K_{\text{epi}\,V}$ to the set epi V coincide for every point $(q, V(q)) \in \text{epi}\,V$. With that assumption we have excluded the so-called "reintrant corners" [3], the physical meaning of which being not yet well understood in multibody dynamics. The potentials $V(q)$ considered here thus "share the properties of nonsmooth convex and smooth nonconvex functions" rather than being general which, however, seems to be sufficient at the present time with respect to application problems. We will now assume force laws of the form

$$-f^+ \in \bar{\partial} V(q) \tag{2}$$

where $\bar{\partial} V(q)$ denotes the generalized gradient of V at q. The generalized gradient $\bar{\partial} V(q)$ is a convex set [8]. Roughly speaking, it consists of the convex hull of the gradients $\lim_{i\to\infty} \nabla V(q_i)$ of all sequences $\{q_i\}$ approaching q for $i \to \infty$. If we are interested in the accelerations from the past ($\ddot{q}^-$) we have to consider, instead of (1), the equation $M(q)\,\ddot{q}^- - h(q, \dot{q}^-) = f^-$ together with the force law $-f^- \in \bar{\partial} V(q)$. Note that both, $-f^-$ and $-f^+$ belong to the same set $\bar{\partial} V(q)$ but might be different. They coincide, for example, if $\bar{\partial} V(q)$ consists of one element only, i.e. for example in the smooth case where we have $-f^+ = -f^- = \nabla V(q)$.

By using the force law (2) we are able to take into account "smooth" interaction forces in the classical sense, for example springs with nonlinear characteristics, as well as unilateral and additional (this also means all) smooth bilateral constraints, and certain set-valued interactions like prestressed springs with plays. Every kind of viscous and dry friction is excluded as well as nonholonomic constraints. They demand a representation via velocity potentials which are not considered here. Combining (1) and (2) we get the differential inclusions of the system [1]

$$-(M\ddot{q}^+ - h) \in \bar{\partial} V(q). \tag{3}$$

It is known from non-smooth analysis that inclusion (2) can also be expressed by means of a hemivariational inequality, i.e.

$$-f^+ \in \bar{\partial} V(q) \;\Leftrightarrow\; V^\uparrow(q, q^\star - q) \geq \langle -f^+, q^\star - q \rangle, \quad q \in C,\ \forall q^\star, \tag{4}$$

where $V^{\uparrow}(q, q^{\star} - q)$ denotes the generalized directional derivative of V at q in the direction $(q^{\star} - q)$, and C is the set of the admissible values of q, see [8]. Together with (1) we obtain the expression

$$-\langle M\ddot{q}^{+} - h, q^{\star} - q\rangle \leq V^{\uparrow}(q, q^{\star} - q), \quad q \in C, \ \forall q^{\star} \tag{5}$$

which holds for every fixed time t and which we will call the *Principle of d'Alembert* [1]. Note that the variations $(q^{\star} - q)$ are not restricted to any subset of $\Re^{f}$, thus values $V^{\uparrow}(q, q^{\star} - q) = +\infty$ which occur when q is at the boundary of C are allowed. The physical dimension of eq. (5) is that of "work" [Nm]. Indeed, we will see in Section 3 when dealing with unilateral constraints that eq. (5) leads to Fourier's classical principle that the virtual work produced by constraint forces is always greater than or equal to zero.

Although providing a complete description of the dynamics of the system, equations (3) and (5) are too general for further investigations. Particularly, there is no direct way to obtain the accelerations $\ddot{q}^{+}$ if the displacements q and velocities $\dot{q}^{+}$ are given. However, one can show that d'Alembert's principle (5) is equivalent to a variational expression in terms of velocities which yields an expression with the physical dimension of a power [Nm/s]. This approach is also based on potential functions. We define the function

$$\Phi : p \rightarrow V^{\uparrow}(q, p) \tag{6}$$

and call it the *velocity potential* of the multibody system. Note that the velocity potential Φ is nothing more than the generalized directional derivative of V at a point q considered as a function of the direction p. The velocity potential is a convex, l.s.c., positively homogeneous, and subadditive function, and obviously dom $\Phi = T_C(q)$ for every fixed q, where $T_C(q)$ denotes the tangent cone to C at q. We define the *Principle of Jourdain* for every fixed time t to be

$$-\langle M\ddot{q}^{+} - h, \dot{q}^{\diamond} - \dot{q}^{+}\rangle \leq \Phi^{\uparrow}(\dot{q}^{+}, \dot{q}^{\diamond} - \dot{q}^{+}), \quad q \in C, \ \dot{q}^{+} \in T_C(q), \ \forall \dot{q}^{\diamond}. \tag{7}$$

The magnitude $\dot{q}^{\diamond}$ denotes just any arbitrary velocity in order to express that the inequality in eq. (7) has to be valid for every direction $(\dot{q}^{\diamond} - \dot{q}^{+})$. Like in classical mechanics $\dot{q}^{\diamond}$ has to be understood as the (right) time derivative of any arbitrary trajectory starting at time t in point q, i.e. there are sequences $\{q^{\diamond}(t_n)\} \subset \{q^{\star}(t_n)\}$ approaching q asymptotic to the half-line emanating from q in the direction $\dot{q}^{\diamond}$ when $t_n \rightarrow t$ for $n \rightarrow \infty$, [1], [9]. Analogously to (3) a differential inclusion being equivalent to Jourdains Principle (7) can be stated

$$-(M\ddot{q}^{+} - h) \in \partial\Phi(\dot{q}^{+}), \tag{8}$$

where due to the convexity of Φ one may use the subdifferential from Convex Analysis ∂ instead of the generalized gradient $\bar{\partial}$, [9].

In order to define a variational principle on acceleration level we proceed in the same manner as above. We introduce the *acceleration potential* to be the generalized directional derivative of the velocity potential as a function of its direction

$$\Psi : p \rightarrow \Phi^{\uparrow}(\dot{q}^{+}, p) \tag{9}$$

and notice that $\Psi(p)$ is also convex, l.s.c., positively homogeneous, and subadditive, and its effective domain is given by dom $\Psi = T_{T_C(q)}(\dot{q}^{+})$ for every fixed q, $\dot{q}^{+}$. We define the *Principle of Gauss* for every fixed time t to be

$$\begin{aligned} -\langle M\ddot{q}^{+} - h, \ddot{q}^{\times} - \ddot{q}^{+}\rangle \leq \Psi^{\uparrow}(\ddot{q}^{+}, \ddot{q}^{\times} - \ddot{q}^{+}), \\ q \in C,\ \dot{q}^{+} \in T_C(q),\ \ddot{q}^{+} \in T_{T_C(q)}(\dot{q}^{+}),\ \forall \ddot{q}^{\times} \end{aligned} \tag{10}$$

which has the physical dimension of a power per time [Nm/s^2], and where the term $\ddot{q}^{\times}$ has to be understood in the same sense as $\dot{q}^{\diamond}$ above. As in (8) we obtain the differential inclusion

$$-(M\ddot{q}^{+} - h) \in \partial\Psi(\ddot{q}^{+}) \tag{11}$$

which is an equivalent representation of (10).

Equation (11) constitutes the necessary and sufficient optimality conditions of a strictly convex optimization problem [9] which is obvious by rewriting it in the form

$$0 \in M\ddot{q}^{+} - h + \partial\Psi(\ddot{q}^{+}). \tag{12}$$

The accelerations $\ddot{q}^{+}$ are hence the optimal solutions of the program

$$\ddot{q}^{+} = \arg\min \{f(\ddot{q}^{+})\}; \quad f(\ddot{q}^{+}) = \tfrac{1}{2}\langle M\ddot{q}^{+}, \ddot{q}^{+}\rangle - \langle h, \ddot{q}^{+}\rangle + \Psi(\ddot{q}^{+}) \tag{13}$$

which is called in classical mechanics the *Principle of Least Constraints*. The cost function f is strictly convex because M is a symmetric and positive definite matrix and the acceleration potential Ψ is convex. Moreover, since f is strictly convex, the optimal solutions $\ddot{q}^{+}$ are unique.

In [1] it has been proved that the principles of d'Alembert, Jourdain, and Gauss are equivalent. One important property of such systems may be expressed via the inequalities

$$-\langle M\ddot{q}^{+} - h, \ddot{\eta}^{\times}\rangle \leq \Psi^{\uparrow}(\ddot{q}^{+}, \ddot{\eta}^{\times}) \leq \Phi^{\uparrow}(\dot{q}^{+}, \ddot{\eta}^{\times}) \leq V^{\uparrow}(q, \ddot{\eta}^{\times}), \quad \forall \ddot{\eta}^{\times}, \tag{14}$$

or, by rewriting (14) in terms of the corresponding differential inclusions (3), (8), (11), via

$$-(M\ddot{q}^{+} - h) \in \partial\Psi(\ddot{q}^{+}) \subseteq \partial\Phi(\dot{q}^{+}) \subseteq \bar{\partial}V(q). \tag{15}$$

The advantages of introducing acceleration potentials and solving Gauss' principle instead of d'Alembert's principle are multiple: First of all we were able to proceed from a non-convex displacement potential to a convex and hence much better behaved acceleration potential which is positively homogeneous in addition. Furthermore the number of inequalities describing the sets (15) becomes smaller, because $\partial\Psi(\ddot{q}^+)$ is contained in $\bar{\partial}V(q)$ (15). Sometimes it may happen that $\partial\Psi(\ddot{q}^+)$ consists of one element only whereas $\bar{\partial}V(q)$ is really "big." For systems allowing the formulation of the generalized gradients via linear complementarity conditions the dimension of the matrix of the corresponding Linear Complementarity Problem is directly related to the dimensions of the sets in (15). Finally, by the Principle of Gauss we achieved a representation of the dynamics equations which enables a direct access to the right accelerations of the system. Moreover, these accelerations are uniquely determined by the corresponding strictly convex optimization problem (13).

3. Unilateral Constraints

It is well known by now that unilateral constraints may be taken into account by indicator functions. We consider a multibody system, the displacements of which being subjected to geometrical restrictions $q \in \operatorname{dom} V = C$. The indicator function I_C of a set C is defined by

$$I_C(q) = \begin{cases} 0 & \text{if } q \in C \\ +\infty & \text{if } q \notin C \end{cases} \tag{16}$$

and has some important connections to the normal and the tangential cone of C. In [8] it is shown that

$$\bar{\partial} I_C(q) = N_C(q) \quad \text{and} \quad I_C^{\uparrow}(q,p) = I_{T_C(q)}(p), \tag{17}$$

where $N_C(q)$ is the normal cone to C at q defined by $N_C(q) = \{x \mid \langle x, y\rangle \leq 0,\ \forall y \in T_C(q)\}$. We assume that only forces resulting from unilateral constraints act on the multibody system. In this case the displacement potential $V(q)$ is

$$V(q) = I_C(q). \tag{18}$$

In order to derive the velocity potential (6) and the acceleration potential (9) we just have to apply equation (17). We obtain

$$\Phi(p) = I_C^{\uparrow}(q,p) = I_{T_C(q)}(p), \quad \Psi(p) = I_{T_C(q)}^{\uparrow}(\dot{q}^+,p) = I_{T_{T_C(q)}(\dot{q}^+)}(p). \tag{19}$$

With the help of (17) we get immediately from (18) and (19) the three sets $\bar{\partial}V(q)$, $\partial\Phi(\dot{q}^+)$, and $\partial\Psi(\ddot{q}^+)$. They are

$$\bar{\partial}V(q) = N_C(q), \quad \partial\Phi(\dot{q}^+) = N_{T_C(q)}(\dot{q}^+), \quad \partial\Psi(\ddot{q}^+) = N_{T_{T_C(q)}(\dot{q}^+)}(\ddot{q}^+), \tag{20}$$

and the differential inclusion (15) becomes

$$-(M\ddot{q}^+ - h) \in N_{T_{T_C(q)}(\dot{q}^+)}(\ddot{q}^+) \subseteq N_{T_C(q)}(\dot{q}^+) \subseteq N_C(q). \tag{21}$$

At that point, note the polarity between the normal and the tangential cone. Due to (21) we obviously have

$$T_C(q) \subseteq T_{T_C(q)}(\dot{q}^+) \subseteq T_{T_{T_C(q)}(\dot{q}^+)}(\ddot{q}^+). \tag{22}$$

The tangential cones increase from the displacement level to the acceleration level in the same manner as the corresponding normal cones decrease. One may say that the accelerations are in some sense less constrained than the displacements, a fact completely different from bilaterally constrained motion. Using the potential functions in (18), (19) the three principles (5), (7), and (10) become

$$\begin{aligned}
&-\langle M\ddot{q}^+ - h, q^\star - q\rangle \le I_C^\uparrow(q, q^\star - q), \quad \forall q^\star, q \in C \\
&-\langle M\ddot{q}^+ - h, \dot{q}^\diamond - \dot{q}^+\rangle \le I_{T_C(q)}^\uparrow(\dot{q}^+, \dot{q}^\diamond - \dot{q}^+), \quad \forall \dot{q}^\diamond, q \in C, \dot{q}^+ \in T_C(q) \\
&-\langle M\ddot{q}^+ - h, \ddot{q}^\times - \ddot{q}^+\rangle \le I_{T_{T_C(q)}(\dot{q}^+)}^\uparrow(\ddot{q}^+, \ddot{q}^\times - \ddot{q}^+), \\
&\qquad\qquad \forall \ddot{q}^\times, q \in C, \dot{q}^+ \in T_C(q), \ddot{q}^+ \in T_{T_C(q)}(\dot{q}^+).
\end{aligned} \tag{23}$$

We apply again equation (17) to the right-hand sides of the inequalities in (23) and choose the variations of the displacements, velocities, and accelerations such that they belong to the sets specified by the resulting indicator functions. The values of the indicators are then equal to zero and (23) becomes

$$\begin{aligned}
&-\langle M\ddot{q}^+ - h, q^\star - q\rangle \le 0, \quad \forall (q^\star - q) \in T_C(q), q \in C \\
&-\langle M\ddot{q}^+ - h, \dot{q}^\diamond - \dot{q}^+\rangle \le 0, \quad \forall (\dot{q}^\diamond - \dot{q}^+) \in T_{T_C(q)}(\dot{q}^+), q \in C, \dot{q}^+ \in T_C(q) \\
&-\langle M\ddot{q}^+ - h, \ddot{q}^\times - \ddot{q}^+\rangle \le 0, \quad \forall (\ddot{q}^\times - \ddot{q}^+) \in T_{T_{T_C(q)}(\dot{q}^+)}(\ddot{q}^+), \\
&\qquad\qquad q \in C, \dot{q}^+ \in T_C(q), \ddot{q}^+ \in T_{T_C(q)}(\dot{q}^+).
\end{aligned} \tag{24}$$

Setting $M\ddot{q}^+ - h = f^+$ (cf. eq. (1)) we see that all scalar virtual expressions such as virtual work or virtual power produced by forces from unilateral constraints are always greater than or equal to zero. The Principle of d'Alembert for unilateral constraints was stated in 1821 by Fourier and can also be found in [2], [3]. Moreau's work is completely based on Jourdain's principle in the second line of (24) but contains additionally impacts and Coulomb friction [4], [5]. The Principle of Gauss as stated in the third equation of (24) can be found in [2] and is often heuristically used in multibody dynamics, cf. for example [7]. Taking the acceleration potential (19) we may state the Principle of Least Constraints (13),

$$\ddot{q}^+ = \arg\min \{f(\ddot{\bar{q}}^+)\}; \quad f(\ddot{q}^+) = \tfrac{1}{2}\langle M\ddot{q}^+, \ddot{q}^+\rangle - \langle h, \ddot{q}^+\rangle + I_{T_{T_C(q)}(\dot{q}^+)}(\ddot{q}^+) \tag{25}$$

which is a strictly convex program with inequality constraints, i.e. find $\ddot{q}^+$ such that $\frac{1}{2}\langle M\ddot{q}^+, \ddot{q}^+\rangle - \langle h, \ddot{q}^+\rangle$ becomes minimal under the restriction that $\ddot{q}^+$ belongs to the closed convex cone $T_{T_{C(q)}}(\dot{q}^+)$.

4. Conclusion

The topic of the presented paper may be described as follows: When considering the velocities of a multibody system as functions of bounded variation the accelerations of the system exist almost everywhere. In this case there must be a way to determine their values directly from the equations of motion. For that purpose we considered the equations of motion of a system under the influence of forces which were derived by generalized differentiation from a displacement potential. The corresponding differential inclusion was reformulated in terms of a hemivariational inequality; this is the Principle of d'Alembert. Following classical mechanics we also introduced a variational principle on the velocity and on the acceleration level, i.e. the Principles of Jourdain and Gauss, respectively. We also showed that unilateral constraints were covered by the chosen approach. The presented theory can be extended to non-integrable potential functions on velocity level in order to take into account additional forces resulting from viscous damping, dry friction, and nonholonomic constraints. At a first view these forces seem to be of completely different nature, only sharing the property of non-integrability of some related functions. It is a remarkable fact that all of them can be treated by the unified concept of using velocity potentials.

References

[1] Glocker, Ch.: The Principles of d'Alembert, Jourdain, and Gauss in Nonsmooth Dynamics. Part I: Scleronomic Multibody Systems, *ZAMM*, to appear.

[2] Glocker, Ch.: *Dynamik von Starrkörpersystemen mit Reibung und Stössen,* VDI-Fortschrittb. Mechanik/Bruchmechanik, Reihe 18, Nr. 182, VDI-Verlag, Düsseldorf, 1995.

[3] May, H., and Panagiotopoulos, P.D.: F.H. Clarke's Generalized Gradient and Fourier's Principle, *ZAMM* **65 (2)** (1985), 125–126.

[4] Moreau, J.J.: Unilateral Contact and Dry Friction in Finite Freedom Dynamics, *Non-Smooth Mechanics and Applications,* CISM Courses and Lectures, Vol. 302, Springer Verlag, Wien, 1988.

[5] Moreau, J.J.: Bounded Variation in Time, *Topics in Nonsmooth Mechanics,* Birkhäuser Verlag, Basel, 1988.

[6] Panagiotopoulos, P.D.: Ungleichungsprobleme und Differentialinklusionen in der Analytischen Mechanik, *Annual Public. School of Technology,* Aristotle University Thessaloniki, Greece, Vol. Θ' (1982), 100 – 139.

[7] Pfeiffer, F., and Glocker, Ch.: *Multibody Dynamics with Unilateral Contacts,* John Wiley & Sons, New York, 1996.

[8] Rockafellar, R.T.: Generalized Directional Derivatives and Subgradients of Non-convex Functions, *Can. J. Math.* **XXXII** (1980), 257–280.

[9] Rockafellar, R.T.: *Convex Analysis,* Princeton University Press, Princeton, New Jersey, 1972.

VI. CIRCUITS, CONTROL, CARDIAC MODELLING

The control of chaos using feedback in a double pendulum (Schiehlen) provided a very interesting demonstration at the symposium and interested readers can learn how to build such a demo from this paper. The experimental paper by Virgin showed new experimental techniques to measure transient nonlinear behavior.

Matsumoto and Nishi give a nice discussion of chaos control and synchronization in an electric circuit. The paper by Englebrecht et al. explores dynamic cardiac modelling.

F. C. Moon (ed.),
IUTAM Symposium on New Applications of Nonlinear and Chaotic Dynamics in Mechanics, 331.

DRIVEN NONLINEAR OSCILLATORS FOR MODELING CARDIAC PHENOMENA

J. ENGELBRECHT[1,2], R.v. HERTZEN[3], O. KONGAS[1]
[1]*Institute of Cybernetics, Akadeemia 21, EE0026 Tallinn, Estonia*
[2]*Estonian Academy of Sciences, Kohtu 6, EE0001 Tallinn, Estonia*
[3]*Helsinki University of Technology, SF02150 Espoo, Finland*

Abstract

A model of cardiac transmembrane action potential excitation and propagation is presented. The model consists of strongly asymmetric van der Pol oscillators and novel quiescent nerve pulse equations modeling cardiac pacemakers and conducting tissues, respectively. A detailed analysis of bifurcation structures of driven nerve pulse equations is given. The rich variety of calculated arrhythmias and conduction blocks agrees well with measured behavior of animal cardiac Purkinje fibers.

1. Introduction

Driven oscillators form the backbone of nonlinear dynamics. Physical phenomena modeled by such oscillators may be different but the general theory and methods are applicable for a wide range of problems that is of a special importance in interdisciplinary areas.

Many biological rhythmic processes can be modeled by nonlinear differential equations exhibiting limit cycle behavior, among them are also the cardiac rhythms. The cardiac rhythms are governed by self-oscillating pacemakers and quiescent, excitable conducting system (for detailed description, see Glass et al 1991). The sino-atrial (SA) node, being the primary pacemaker, and the atrioventricular (AV) node, a secondary pacemaker, can be modeled as a pair of coupled relaxation oscillators (van der Pol and van der Mark 1928, Katholi et al 1977, West et al 1985). Under normal conditions, the intrinsically faster SA node appears to entrain the slower secondary pacemaker resulting in a one to one phase-locking (Goldberger et al 1985). The AV node, on the other hand, acts as a drive for the conducting system - His bundle and Purkinje fibers - considered as non-pacemaking excitable media. Purkinje fibers conduct the electrical activation to the ventricular subendocardium that leads to a contraction of the atrial muscle.

This complicated physiological system, based on cell energetics, ion currents, electrical activation, contraction, mechanical stresses, etc., has been studied for a long time (Glass et al 1991) and mathematical modeling of these phenomena is nowadays closely related to the methods of nonlinear dynamics (Jalife 1990). In this work, based

F. C. Moon (ed.),
IUTAM Symposium on New Applications of Nonlinear and Chaotic Dynamics in Mechanics, 333–342.

on our earlier results (Engelbrecht 1981, 1991, von Hertzen and Kongas 1996), the pacemakers of the heart in an entrained mode are modeled by strongly asymmetric van der Pol oscillators and the cardiac conducting tissues (His-Purkinje network) are modeled by the nerve pulse equation (NPE) derived by Engelbrecht (1981). The novelty is in the analysis of the NPE that despite of its simplicity exhibits rich dynamics similar to the results of experiments (Chialvo et al 1990).

The paper is organize as follows. We first consider the structured model of heart and the model for nerve pulse propagation (the NPE). The main emphasis is put on the analysis of driven NPE's where the bifurcation structures and bifurcation diagrams of periodically driven NPE's are presented. The discussion shows the rich variety of calculated arrhythmias and conduction blocks agreeing with experiments.

2. Basic models

2.1. STRUCTURAL ELEMENTS

As described above, the myocardium is preceded by four main elements of heart related to the electrical pulse excitation and conduction - two oscillating nodes and the conduction system carrying and distributing the action potential. Following the structured model elements (Fig. 1), we propose a mathematical model which consists of coupled equations corresponding to each of these elements (Hertzen and Kongas 1996). For the nodes we use strongly asymmetric van der Pol-type oscillators and for the His-Purkinje network - Liénard-type NPE's (Engelbrecht 1991).

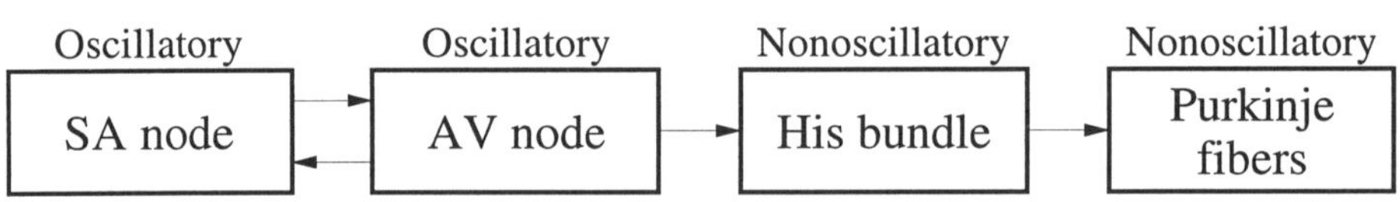

Figure 1. Structured model of heart

2.2. BASIC EQUATIONS

Here we put emphasis on the dynamics between atrium and ventricle. For this reason we model the SA and AV nodes in a 1:1 entrained mode which acts as a drive for the His-Purkinje network. The general asymmetric van der Pol equation is used for the action potential u

$$\ddot{u} + q_u \left(u - u_1\right)\left(u - u_2\right)\dot{u} + k_u u = 0 \tag{1}$$

with standard notations and $u_1 u_2 < 0$ in order to guarantee the existence of the limit cycle. The parameters q_u, u_1, u_2 and k_u are chosen to get a "unit solution" with the unit amplitude and the unit angular frequency (Hertzen and Kongas 1996). It should be noted, however, that the average value of this solution is zero as immediately follows from the properties of the van der Pol equation. According to Liebert (1991), the average value of the real AV action potential, measured from the corresponding resting state, is, however, evidently different from zero. This means that the drive $G(\tau)$ should have a constant bias C

$$G(\tau) = I\ u(\omega\tau) + C, \tag{2}$$

where I is the amplitude.

The NPE for modeling His-Purkinje network is derived from the hyperbolic telegraph equations (Lieberstein, 1967) in their full form in order to avoid a sensitivity of the stationary profile (Engelbrecht 1991). For the membrane current, a FitzHugh-Nagumo type cubic expression is used with one recovery (internal) variable. The evolution equation (Engelbrecht 1981, 1991) governing the transmembrane action potential v is then

$$\frac{\partial^2 v}{\partial\xi\partial X} + f(v)\frac{\partial v}{\partial\xi} + av = 0, \quad f(v) = a_0 + a_1 v + a_2 v^2, \tag{3}$$

where $\xi = c_0 t - X$, a_0, a_1, a_2, and a are constants. Here c_0 is the propagation velocity without ion currents. Equation (3) describes the space-time dynamics of a pulse and the corresponding stationary profile is governed by

$$\ddot{v} + f(v)\dot{v} + (a / \theta)v = 0\,, \tag{4}$$

where dot denotes the differentiation with respect to $\tau = X + \theta\xi$, θ = const. It is shown (Engelbrecht 1991) that eq. (4) exhibits a threshold, a possible amplification of the initial excitation and a formation of a stationary profile with a characteristic refraction part. This is the sought NPE and later we prefer writing eq. (4) in the following form

$$\ddot{v} + q_v(v - v_1)(v - v_2)\dot{v} - v = 0 \tag{5}$$

with q_v, v_1, v_2 constants. What makes eq. (5) different from eq. (1) is that $v_1 v_2 > 0$ which is the basic condition for the NPE becoming a quiescent excitable nerve pulse equation. Note also that the harmonically driven NPE is equivalent to the Bonhoffer-van der Pol equation except for a missing Duffing-type cubic term. A single stimulus acting on the NPE in its resting state will be either attenuated or amplified resulting in a sub-threshold (small stimulus) or supra-threshold (larger stimulus) response. The minimum threshold impulse (velocity kick) needed to create a supra-threshold response is analyzed by Hertzen and Kongas (1996) who have shown that the NPE under consideration behaves as a supernormal neuron.

3. Numerical simulation

3.1. DRIVING BY A TRAIN OF DIRAC DELTA SPIKES

The NPE driven periodically by a train of Dirac delta spikes is of a special practical interest since many experimental transmembrane action potential measurements are performed using cell simulation by short current pulses (Chialvo and Jalife 1987, Chialvo et al 1990). The emphasis is on a detailed study of the bifurcation diagrams of the response and on the bifurcation map in the drive frequency-amplitude plane. The model NPE under consideration is

$$\ddot{v} + q_v (v - v_1)(v - v_2)\dot{v} + v = I \sum_{n=0}^{\infty} \delta(\tau - nT), \tag{6}$$

where I accounts for the strength and $T = 2\pi/\omega$ for the basic period of the stimulus, and $\delta(\cdot)$ is the Dirac delta function.

The overall bifurcation structure of eq. (6) displaying the bifurcation lines of period one solution in the (I, ω)-control plane is shown in Fig. 2.

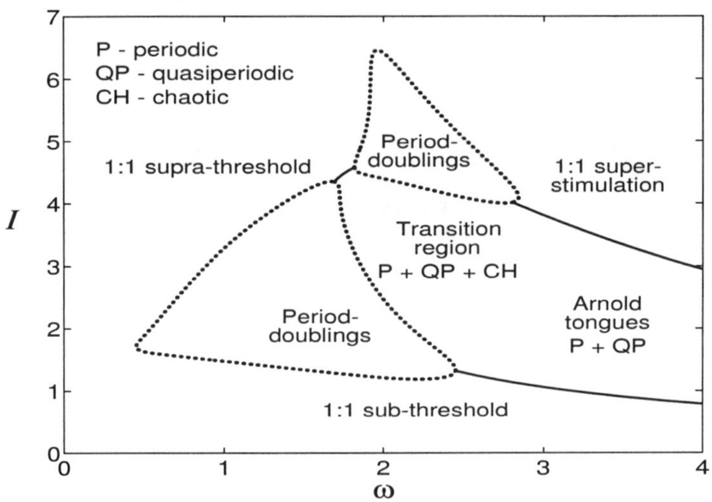

Figure 2. The skeleton of the bifurcation structure of eq. 6.
Dotted lines - period doublings, solid lines - Neimark-Sacker bifurcations.

A detailed study of bifurcations up to period six solutions revealed the classic Farey organization and Arnold tongue structure for phase locking zones originating from the Neimark-Sacker lines within the quasiperiodic region, period doubling cascades and chaos originating from the two "butterfly wings", and a complicated transition region between these main domains. A more detailed bifurcation structure within the zone $i \leq \omega \leq 3$ is shown in Fig. 3.

In Fig. 3, Neimark-Sacker bifurcations are labeled by NS and saddle-node and period doubling bifurcations by $sn(m,n)$ and $pd(m,n)$, respectively, where m and n denote torsion and period numbers (Mettin et al 1993). The distinction between period doubling and saddle-node bifurcations is easily made on the basis of the general winding number (GWN). Hertzen and Kongas (1996) have analyzed this case in detail. Here we stress some special features only needed for the overall picture. The $sn(m,n)$ lines may appear very close to $pd(m,n)$ lines (Fig. 3) and may be connected to the NS lines. From the viewpoint of arrhythmias, the $pd(1,2)$ lines which intervene the NS line, arise due to the 2:1 conduction block, which is primarily caused by supernormality of the NPE under considerations. The cuts of bifurcation planes for fixed I's show clearly the difference between small drive amplitudes, amplitudes around the firing threshold I_{th} and large amplitudes. For small drive amplitudes, for increasing ω the stable period one attractor experiences a NS bifurcation leading to the quasiperiodic behavior. For a stimulus, slightly less than the threshold, a period doubling cascade and chaotic region precede the quasiperiodic regime and replace the NS bifurcation. For large amplitudes, the bifurcation diagram displays a sequence of alternating quasiperiodic regimes and phase-locked states.

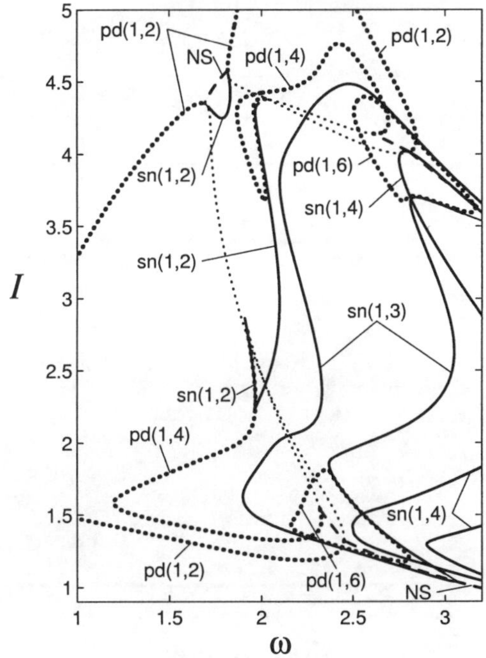

Figure 3. Detailed bifurcation structure around the transition region

3.2. ROBUSTNESS OF THE BIFURCATION STRUCTURE

The tissue of the cardiac conducting system in the intact heart and in preparations for electrophysiological experiments meets very different driving conditions. This suggests the need for the analysis of the influence of drive type. Here three different driving functions are used: the periodic train of Dirac delta spikes (delta drive), sinus drive and unit drive (see Sec. 2.2). The 2D diagrams showing the response periodicities corresponding to these driving conditions are shown inf Figs. 3 and 4. For sinus and unit drives, the driving amplitudes and biases are simultaneously adjusted to make both mean values and amplitudes of first harmonics equal to the corresponding values of delta drive. This gives the r.h.s.

$$G(\tau) = \frac{I_k\omega}{\pi H_k} u_k(\omega\tau) + \frac{I_k\omega}{2\pi} \tag{7}$$

for sinus (subscript $k = s$) and unit (subscript $k = u$) drives, respectively. The values $H_s = 1$ and $H_u = 0.499$ are the corresponding amplitudes of the first harmonics.

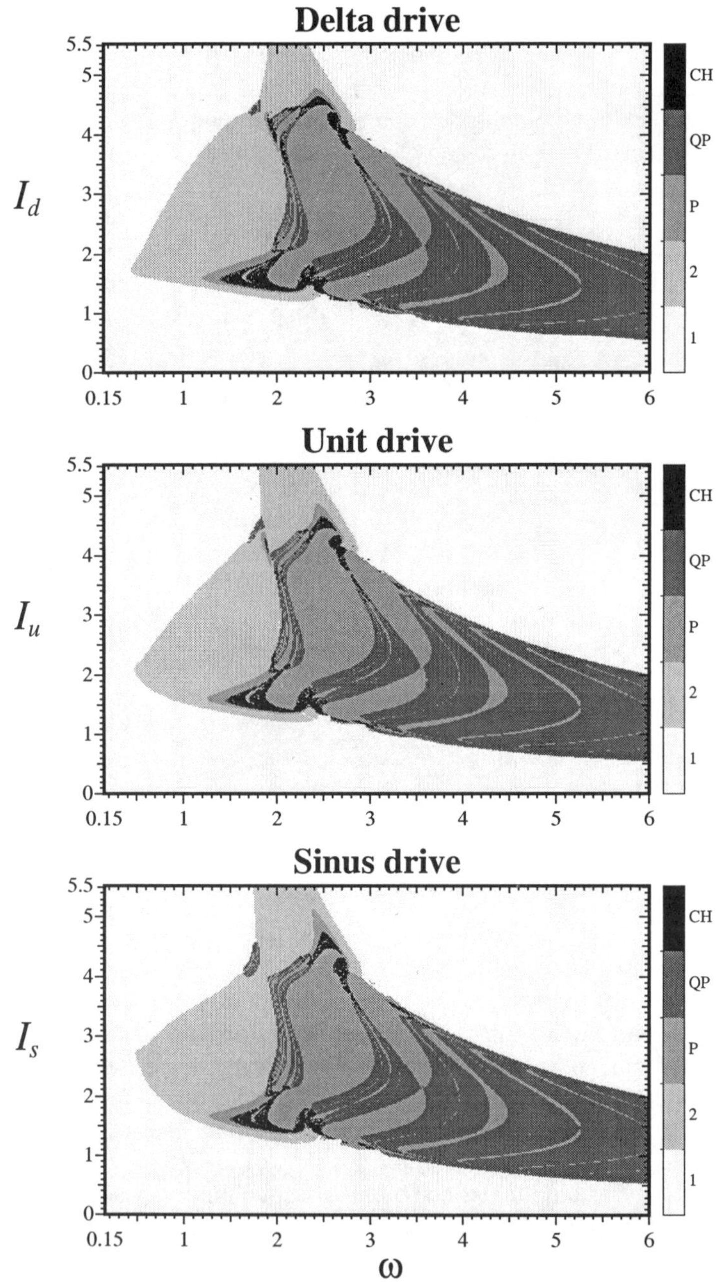

Figure 4. Bifurcation diagrams for various drive types.
Gray tone labels indicate the type of solution: 1 - period one, 2 - period two, P - periodicities higher than two, QP - quasiperiodicity, CH - chaos; subscripts *d, u, s* denote the intensities of the drives.

Figure 4 shows the amazing robustness of the bifurcation structure of the NPE with respect to different drive types. Minor differences occur for low frequency domain involving slight distortions of periodic regions. This indicates that the characteristic features of the drive are well contained in bias and first harmonic. The NPE has its intrinsic self-oscillations frequency and for increasing ω, obviously, the influence of higher harmonics decays quickly, i.e., they become "well out of resonance". This suggests that almost any kind of periodic drive can be used without affecting the bifurcation structure.

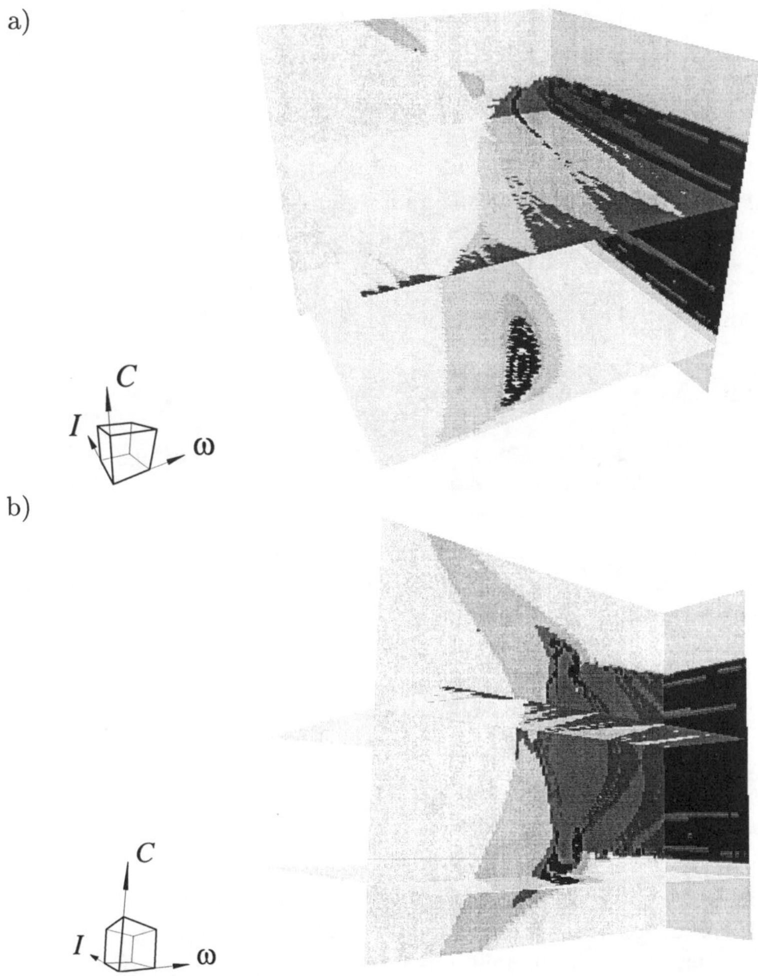

Figure 5. Two views at the 3D bifurcation diagram of the NPE. The gray tones correspond approximately to the tones in Fig. 4. Box dimensions: I = 0.1-15.0, ω = 0.15-4.0, C = 0.0-2.4; a) planes I = 15.0, ω = 4.0, C = 0.4, C = 1.5, and b) planes I = 4.0, ω = 4.0, C = 0.4, C = 1.3.

3.3. DRIVING BY A BIASED UNIT GENERATOR

In this section we present the 3D bifurcation diagram of the NPE with respect to driving frequency ω, driving amplitude I and bias C. The unit drive is used here, i.e. the r.h.s. of the NPE is taken from eq. (2). This gives us a direct possibility to study the asymmetry of the resulting diagram with respect to bias C.

It can be shown that if the driving function $u_k(\cdot)$ possesses the inversion symmetry property - $u_k(\tau + T/2) = u_k(\tau)$, where T is the period of $u_k(\tau)$, then the 3D bifurcation diagram is symmetric.

The results are shown in Fig. 5. Analytically it can be shown that the Arnold tubes (3D Arnold tongues) start at the plane $I = 0$, widening with increasing I. For $I = 0$, the NS bifurcation (actually a Hopf bifurcation for this particular case) occurs at planes $C = u_1 = 0.5$ and $C = u_2 = 1.9$. The complicated bifurcation structure with Arnold tubes remains between these two planes for all values of I and ω. Merely the period doubling 'wings' extend beyond these planes (cf. Fig. 2).

Careful exploration of the 3D bifurcation structure reveals weak asymmetry indicating that the corresponding structure for sinus drive would be very similar. The bifurcation structure shown in Figs. 3, 4 can easily be recognized.

4. Discussion

4.1. COMPARISON WITH EXPERIMENTS

Low dimensional chaotic behavior in sheep and dog cardiac Purkinje tissues has been demonstrated by Chialvo et al (1990), where period doubling bifurcations of the transmembrane action potential amplitude were shown to precede the irregular action potential behavior. For a 1.5 × threshold drive amplitude and increasing drive frequency, the period doubling bifurcations associated with changes in the stimulus:response locking were manifested in the sequence 1:1 → 2:2, 2:1 → 4:2, 3:1 → 6:2 and 4:1 → 8:2 until irregular activity occurred at very brief cycle lengths. Careful exploration also revealed irregular dynamics between the 4:2 and 3:1 states. For a 2.6 × threshold drive amplitude, the corresponding sequence was 1:1 → 2:2 → 4:4 → irregular dynamics. The consistency with our results (Fig. 3) is good.

4.2. COEXISTING ATTRACTORS

The well-known coexisting functioning modes of the heart are the reentry tachycardia and the normal mode. The switching between these modes occurs usually due to ectopic beats. One of the important tasks of artificial pacemakers is to drive the fibrillating heart back to the normal mode of functioning.

Analysis of the NPE shows that the heart is a very stable system for normal rates and for higher rates the complicated rhythms can appear (expected result). In addition, the detailed analysis shows the presence of coexisting attractors (CAs). Extensive number of papers deals with analysis and modeling of Wenckebach-type brady-arrhythmias arising within or below the AV node. Most of them tend to ascribe the alternating Wenckebach periodicities to the multilevel block that is also clinically reported (Castellanos et al. 1993). The change in Wenckeback periodicities (e.g. from 2:1 to 3:1 entrainment) is usually achieved with change in driving frequency. However, our studies suggest that several stable Wenckebach periodicities could be produced by a single level of a block within or below the AV node without changing the driving frequency. The possibility for CAs is not earlier reported in case of bradycardias.

To demonstrate the presence of CAs, we paced the NPE with delta drive using the driving parameters $I = 1.41$ and $\omega = 2.14$. We exerted an irregular beat (IB), an extra pulse exactly in the middle of two contiguous regular beats. After some evolution of the system we exerted another IB in the same phase as the first one. Figure 6 shows the result of this simulation. Before the first IB was exerted, the system exhibited 3:1 entrainment. The first IB kicked it to the chaotic mode and the second IB brought it back to 3:1 mode. The point (ω, I) used in this experiment is located both inside the Arnold tongue three and within the fully developed period doubling cascade of the lower 'wing' (see Fig. 3). As the subharmonic cascade within the tongue three intersects transversely the cascade within the lower 'wing', there exists a wide spectrum of different CAs. In clinical situations, this type of an arrhythmic phenomenon could be observed in patients with atrial fibrillation together with fast AV conduction, irregular ectopy and a single level block within His bundle. However, in view of the significant fluctuations existing in parameters describing the human heartbeat, unambiguous identification of chaos in clinical recordings with frequent ectopy will be difficult (Courtemanche et al. 1989).

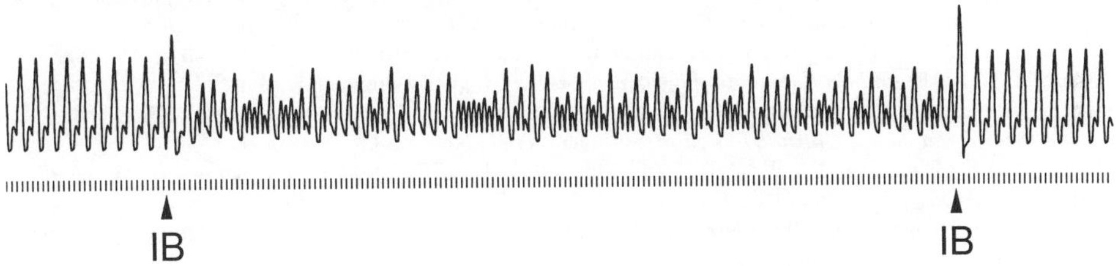

Figure 6. Coexisting attractors possessed by delta-driven NPE.
Irregular beats (IB) kick the system from one stable attractor to another.

The idea that human disease may sometimes be associated with bifurcations in the dynamics of living organisms was elaborated by Mackey and Glass (1977) with the notion of 'dynamical disease' to denote abnormal dynamics in physiological systems associated with changes in sysem parameters. Our results suggest that sudden change in cardiac functioning indicates the possibility for CAs rather than the change in 'dynamical disease'. Moreover, some 'dynamical diseases' may have more than one representative form in ECG recordings.

4.3. EXTENSIONS

We have demonstrated in section 3.2 that the characteristic features of the drive for the NPE are well contained in bias and first harmonic, whereas the higher harmonics have minor influence to the bifurcation structure. This suggests that from both the analytical and computational viewpoints the optimal drive for the bifurcation analysis would be a biased sine. By a linear change of variables, the sinusoidally driven NPE can be transformed into the van der Pol form

$$\ddot{x}+\varepsilon\left(x^{2}-1\right)\dot{x}+x=I\sin\omega t+C \tag{8}$$

(Hertzen and Kongas 1996). This indicates the need for 4D bifurcation analysis of the van der Pol equation in the space of (ε, I, ω, C). The encouraging point should be that then all the major characteristics of the bifurcation structure are involved.

Acknowledgements. This study is supported by the grant from the Estonian Science Foundation. J.E. would like to thank the LOC of Chaos 97 for financial support to attend the symposium

References

Castellanos, A., Interian, A., Jr., Cox, M.M. and Myerburg, R.J. (1993) Alternating Wenckebach periods and allied arrhythmias. *Pace* **16**, 2285-2300.

Chialvo, D.R. and Jalife, J. (1987) Non-linear dynamics of cardiac excitation and impulse propagation, *Nature* **330**, 749-752.

Chialvo, D.R., Gilmour, R.F. Jr and Jalife J. (1990) Low dimensional chaos in cardiac tissue. *Nature* **343**, 653-657.

Courtemanche, M., Glass, L., Rosengarten, M.D. and Goldberger, A.L. (1989) Beyond pure parasystole: promises and problems in modeling complex arrhythmias. *Am. J. Physiol.* **257(2)**, H693-H706.

Engelbrecht, J. (1981) On theory of pulse transmission in a nerve fibre, *Proc. R. Soc. Lond.* **A375**, 195-209.

Engelbrecht, J. (1991) *An Introduction to Asymmetric Solitary Waves*, Longman, Harlow.

Glass, L.; Hunter, P.J. and McCulloch, A. (Eds.) (1991) *Theory of Heart: Biomechanics, Biophysics, and Nonlinear Dynamics of Cardiac Function*, Springer-Verlag, New York.

Goldberger, A.L., Bhargava, V., West, B.J. and Mandell, A.J. (1985) Nonlinear dynamics of the heartbeat, *Physica* **17D** , 207-214.

Hertzen, R.v. and Kongas, O. (1996) Nonlinear dynamics of cardiac action potential oscillators, in L. Pust, F. Peterka (eds.), *Proc. Euromech 2nd European Nonlinear Oscillation Conf.*, Acad. Sci., Prague, vol 1, pp. 23-32.

Jalife, J. (ed.) (1990) *Mathematical Approaches to Cardiac Arrhytmias*, Annals, New York Acad. Sci., 591.

Katholi, C.R., Urthaler, F., Macy, J., Jr. and James, T.N. (1977) A mathematical model of automaticity in the sinus node and AV junction based on weakly coupled relaxation oscillators, *Comp. Biomed. Res.* **10**, 529-543.

Lieberstein, H.M. (1967) On the Hodgkin-Huxley partial differential equation, *Math. Biosci.* **1**, 45-69.

Liebert, W. (1991) *Chaos und Herzdynamik.* Verlag Harri Deutsch, Frankfurt am Main.

Mockey, M.C. and Glass, L. (1977) Oscillation and chaos in physiological control system. *Science* **197**, 287-289.

Mettin, R., Parlitz, U. and Lauterborn, W. (1993) Bifurcation structure of the driven van der Pol oscillator, *Int. J. Bifurcation and Chaos* **3**, 1529-1555.

Van der Pol, B. and van der Mark, J. (1928) The heartbeat considered as a relaxation oscillation and an electrical model of the heart, *Phil. Mag. Suppl.* **6**, 763-775.

West, B.J., Goldberger, A.L., Rovner, G. and Bhargava, V. (1985) Nonlinear dynamics of the heartbeat, *Physica* **17D**, 198-206.

CHAOS, SYNCHRONIZATION AND BIFURCATIONS IN A DRIVEN R-L-DIODE CIRCUIT

T. MATSUMOTO AND M. NISHI
Department of Electrical, Electronics and Computer Engineering, Waseda University, Shinjuku, Tokyo, 169 JAPAN

1. Introduction

Consider the R-L-Diode circuit given in Fig. 1 driven by the sinusoidal voltage source $E\sin(2\pi ft)$. The dynamics is described by

$$\begin{aligned} \frac{dq}{dt} &= i - g(f(q)) \\ L\frac{di}{dt} &= -Ri - f(q) + E_b + E\sin(2\pi ft) \end{aligned} \tag{1}$$

where q is the charge stored in the (parasitic) capacitor of the diode, i is the current through the circuit, R and L are the series resistor and the inductor, and E_b is a DC bias. Function $g(\cdot)$ is the well-known exponential characteristic of a diode, whereas $f(\cdot)$ represents the nonlinear characteristic of the capacitive part of the diode. Analytical expression is rather complicated [7]. Figure 2 shows a typical graph of $f(\cdot)$.

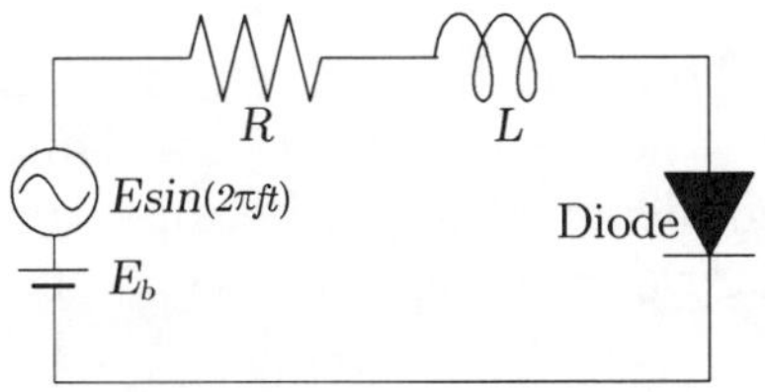

Figure 1. R-L-Diode circuit driven by a sinusoidal voltage source.

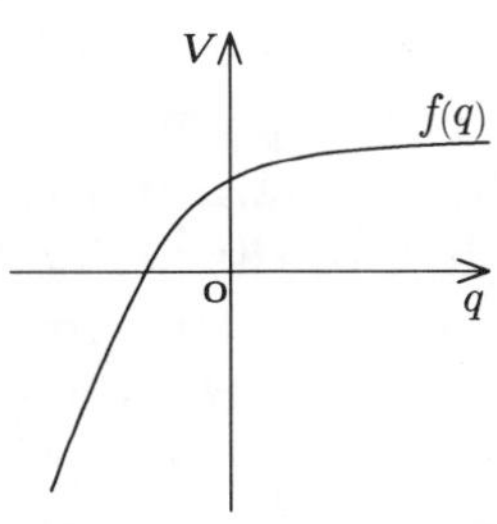

Figure 2. q-v characteristic.

F. C. Moon (ed.),
IUTAM Symposium on New Applications of Nonlinear and Chaotic Dynamics in Mechanics, 343–352.

In a series of papers, [1]–[10], we have been studying various aspects of this circuit; bifurcations, chaotic attractor formation, synchronization and masking among others. Due to its simplicity and naturalness, this circuit has been extensively studied by engineers as well as physicists [11]–[19]. It is not surprising to see that other systems including gear meshing [17] and compliant off shore structure [18] are described by equations similar to (1), because there is no artificially synthesized element in this circuit.

2. Chaos

One of the first questions is: What is responsible for the chaotic behavior when the circuit is so simple and natural? The answer is the nonlinearity of the parasitic capacitor associated with the diode [1]. The next natural question is: How are the chaotic attractors formed? The answer is that a significant difference in the slopes of $f(\cdot)$ in the regions $q \geq 0$ and $q < 0$ gives rise to significant difference in the vector fields in each of the two regions. This yields "folding mechanism" which in turn is responsible for the chaotic attractor formation [2]. Figure 3 shows a schematic picture of how a topological horseshoe can be formed in this system where simplifications are made; (i) Smooth nonlinearity $f(\cdot)$ is replaced by a piecewise linear one;

$$f(q) = \begin{cases} \frac{1}{C_d}q + E_0, & q \geq 0 \\ \frac{1}{C_j}q + E_0, & q < 0 \end{cases}, \quad C_j \ll C_d \tag{2}$$

(ii) nonlinearity $g(\cdot)$ is ignored, and (iii) $E \sin \omega t$ is replaced by $E sgn(\sin \omega t)$. These simplifications imply that the solution can be represented as compositions of four linear flows $\phi_1^t, \cdots, \phi_4^t$ on $\mathbb{R}^2$ depending on $q \geq 0$ or $q < 0$, and $sgn(\sin \omega t) > 0$ or $sgn(\sin \omega t) > 0$. In Fig. 3, the upper plane corresponds to $sgn(\sin \omega t) > 0$, where curve Σ represents the solution passing through the origin. Other curves with arrow show the flows. Observe the significant speed difference for $q \geq 0$ and $q < 0$ which is attributable to $C_j \ll C_d$. Pick an initial rectangle which has a non empty intersection with Σ. Due to the significant speed difference in the i - direction, the initial rectangle is bent and stretched. This takes place during the half period where $sgn(\sin \omega t) > 0$. This is followed by the deformation of the bent object shown in the lower plane where $sgn(\sin \omega t) < 0$. During this second half of the period, nothing significant happens because the object in question does not cross the boundary $q = 0$ and because the flow ϕ_3^t is approximately a rotation, thereby gives rise to a topological horseshoe.

This is only a schematic picture with rather drastic simplifications. A recent result [10] demonstrates that one can show the formation of a topological horseshoe with verified error bounds where more accurate model is used. Observe that existence of a topological horseshoe implies that an

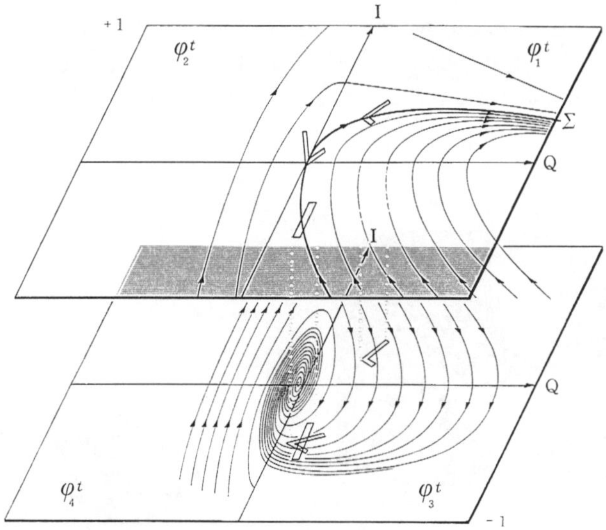

Figure 3. topological horseshoe.

extremely complicated behavior is embedded in the dynamics because a topological horseshoe map acting on its invariant set is semi-conjugate to (σ, S_2), the shift map σ on two symbols S_2 so that
(i) There are countably many periodic points with different periods in S_2;
(ii) The set of all periodic points is dense in S_2;
(iii) σ is topologically transitive. Namely, there exists a point $c \in S_2$ whose orbit is dense in S_2;
(iv) σ is expansive, that is, there exists a positive number ϵ such that any two points which stay within the distance ϵ under all forward and backward iterates of σ must be identical;
(v) S_2 is homeomorphic to a Cantor set.

3. Bifurcations

Figure 4(a) shows a typical global bifurcation diagram. The horizontal axis is the amplitude E of the voltage source, and the vertical axis is the current of the circuit sampled at a particular phase of the voltage source. The lower part of Fig. 4(a) explains corresponding bifurcation mechanisms. A question naturally arises: Why does the period of the large periodic window increases exactly by one? This is answered in [2].

At the lower E and the lower f values, qualitatively different bifurcations are observed. Namely, instead of the increase of the period by one, a repeated appearance of period one attractors are observed sandwiched between chaotic bands (Fig. 5). It was found that the "multi folding" mechanism [6] is responsible for this phenomenon. The multi folding is in turn due to the fact that with lower E and f values, the trajectory spends a much longer time in a region (of the state space) where the vector field is fast so that the trajectory rotates more frequently than with larger E and f values.

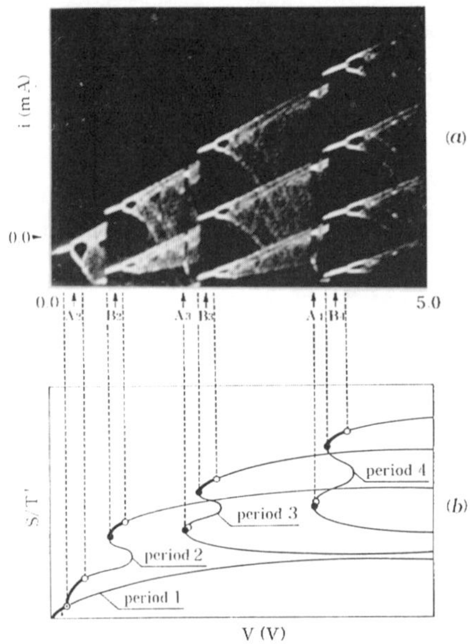

Figure 4. A global one parameter bifurcation diagram.(from [7])

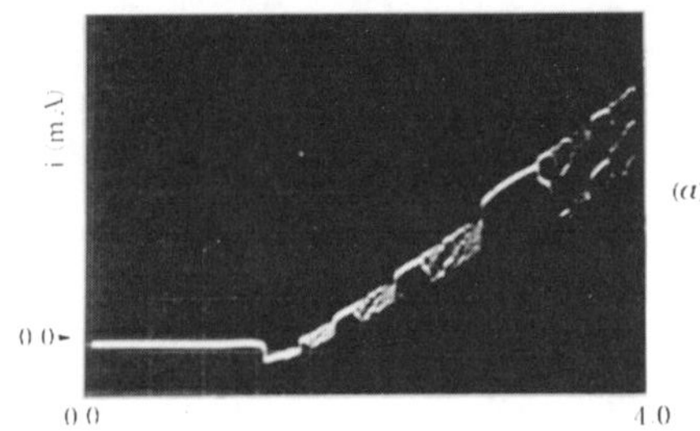

Figure 5. Another one parameter bifurcation diagram with a different setting.(from [7])

Based on these results, global bifurcation structure has been revealed [7] (Fig. 6). This is drawn in $\mathbb{R}^3$ where one coordinate is f, the driving frequency, another is E, the driving voltage amplitude, and $(S/T)'$, which represents state variable information. An important aspect of this global bifurcation diagram is the fact that periodic orbits have *sheet structure.*

4. Synchronization / Masking

The chaotic synchronization of Pecora and Carroll [19] has several interesting features including potential applications. The scheme consists of two basic ideas. One is an appropriate decomposition of a nonlinear dynamical system into subsystems. The other is the stability concept of the subsystems as generalized to chaotic systems. If each subsystem has negative conditional Lyapunov exponents, then a chaotic synchronization is possible. Figure 7 shows a synchronization scheme which is described by

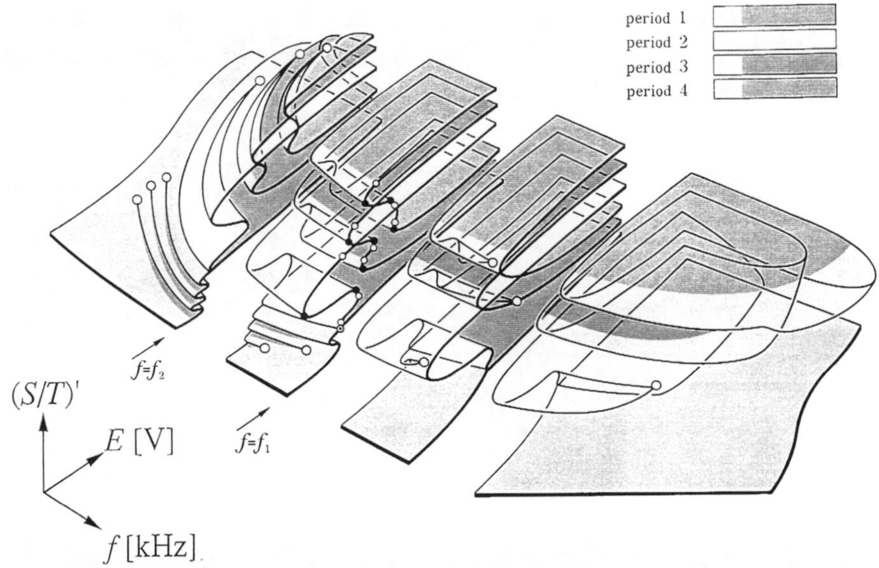

Figure 6. Sheet structure in global bifurcation diagram (from [7]).

$$\frac{dq}{dt} = i - g(f(q)) \tag{M1}$$

$$L\frac{di}{dt} = -Ri - f(q) + e(t) \tag{M2}$$

$$\frac{dq_s}{dt} = i - g(f(q_s)) \tag{S1}$$

$$L\frac{di_s}{dt} = -Ri_s - f(q_s) + e(t). \tag{S2}$$

Observe that the first slave system (S1) is driven by i, the current of Master system (M1), (M2). The following gives a theoretical justification for synchronization.

Proposition Assume

A1. $f(\cdot)$ and $g(\cdot)$ are uniformly increasing, i.e., there are positive numbers α_1, α_2, β_1, β_2 with

$$0 < \alpha_1 \leq \frac{f(q) - f(q')}{q - q'} \leq \alpha_2, \quad q \neq q'$$

$$0 < \beta_1 \leq \frac{g(v) - g(v')}{v - v'} \leq \beta_2. \quad v \neq v'$$

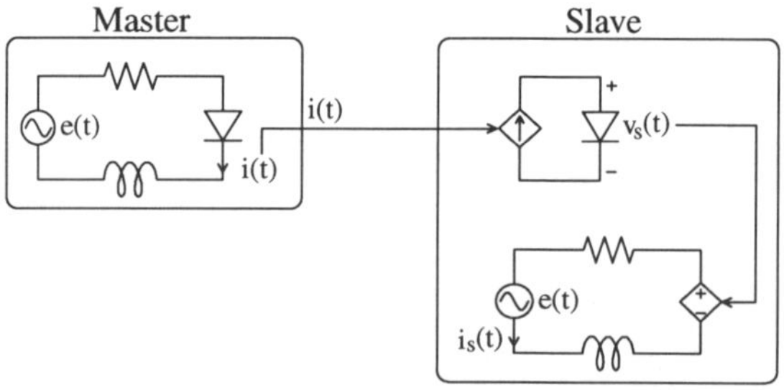

Figure 7. Block diagram of Master-Slave System.

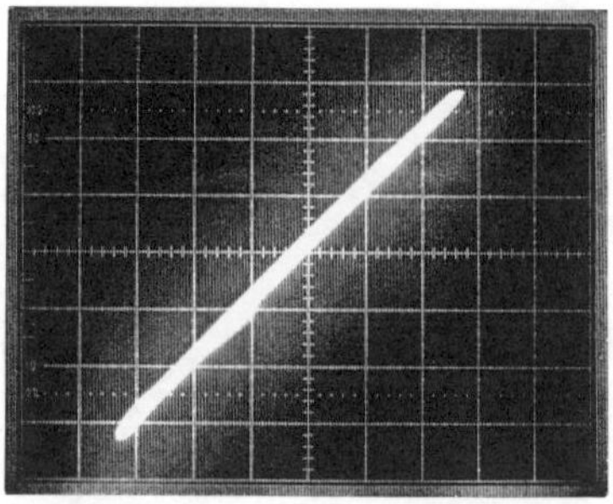

Figure 8. i_{Master} vs. i_{Slave}

A2. $e(t)$ is continuous and $|e(t)| \leq B$.

Then given arbitrary $(q(0), i(0))$ and $(q_s(0), i_s(0))$,

$$\lim_{t \to \infty} ||(q(t), i(t)) - (q_s(t), i_s(t))|| = 0.$$

A proof is given in the Appendix. No conditional Lyapunov exponents are used. It should be noted that by definition, conditional Lyapunov exponents are difficult to establish because they demand statistical treatment.

This circuit appears to be one of the very few real physical systems where the chaotic synchronization is confirmed both experimentally and theoretically.

Figure 9 shows that if one replaces the receiver diode by a different type from the transmitter diode, then synchronization is not possible.

Chaotic masking is also confirmed experimentally where an information barring signal, e.g., music is masked by a chaotic waveform from the circuit and transmitted to a receiver and appropriate circuits "decode" the original signal. Figure 10 shows the masking scheme, where $m(t)$ represents information signal where $max|m(t)|$ is approximately $1/50$ of $max|i(t)|$, the amplitude of the circuit current.

Figure 10 shows the masking scheme whereas Fig. 11 gives experimental data. Figure 11 (a) and (b) compare $i(t)$ with $j(t) := j(t) + m(t)$ where m(t) is a music signal. No difference is discernible. In fact the sounds of $i(t)$ and j(t) exhibit typical chaotic sounds. Figure 12 (a) and (b) show the original music signal $m(t)$ and the decoded signal $\hat{m}(t)$. Although there are some distortions, original music signal is essentially recovered.

It should be noted, however, that no theoretical justifications are achieved for chaotic masking in this circuit as well as in other systems.

References

1. T. Matsumoto, L. O. Chua, and S. Tanaka, Simplest chaotic nonautonomous circuit, Phys. Rev. A, vol.30, no.2, pp.1155–1157, August 1984,

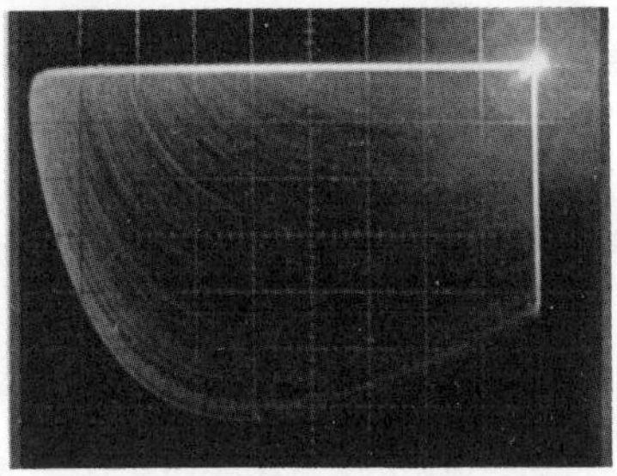

Figure 9. different type diode.

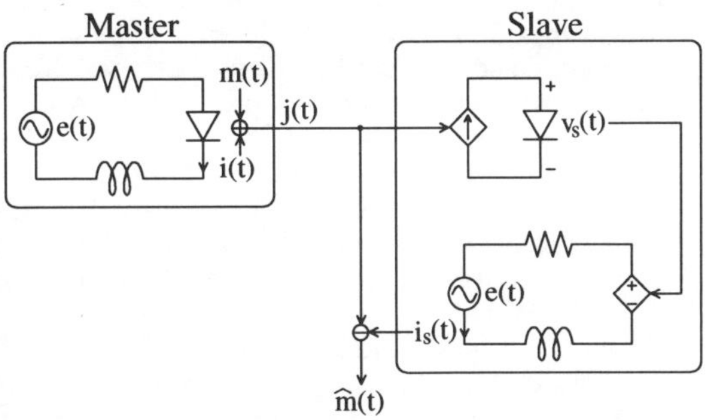

Figure 10. Block diagram of Chaotic Masking.

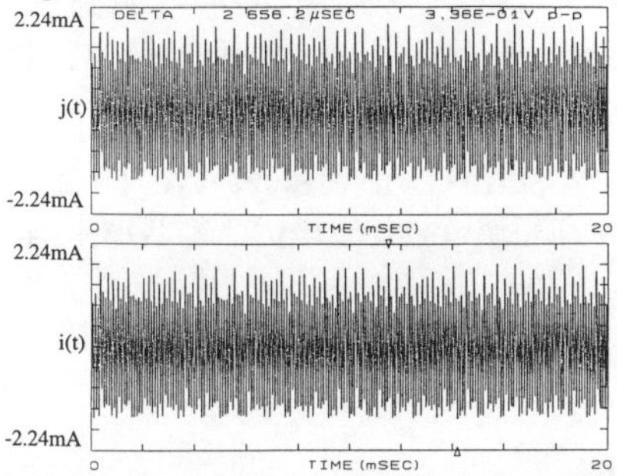

Figure 11. $i(t)$ and $j(t)$.

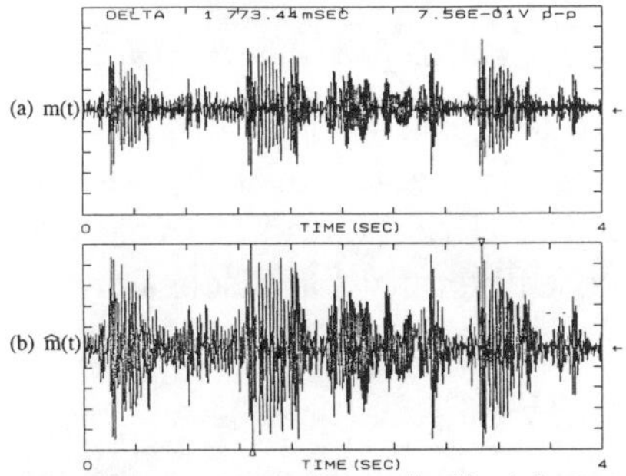

Figure 12. Chaotic masking (from [10]). (a) Original music signal $m(t)$.(b) "Decoded" signal $\hat{m}(t)$.

2. S. Tanaka, T. Matsumoto, and L. O. Chua, Bifurcation scenario in a driven R-L-Diode circuit, Physica D, vol.28, no.3, pp.317–344, October 1987.
3. S. Tanaka, T. Matsumoto, J. Noguchi, and L. O. Chua, Multi-Folding: Alternative appearance of period-one attractors and chaotic attractors in a driven R-L-Diode circuit, Phys. Lett. 157A, no.1, pp.37–43, July 1991.
4. S. Tanaka, J. Noguchi, S. Higuchi, and T. Matsumoto, A repeated appearance of period-1 attractor in a driven R-L-Diode circuit: Experimental and theoretical bifurcation analysis, IEICE Trans., vol.E-74, no.6, pp.1406–1413, June 1991.
5. S. Higuchi, S. Tanaka, M. Komuro and T. Matsumoto, in *Advanced Series in Dynamical Systems*, edited by H. Kawakami (World Scientific, Singapore, 1991), Vol. 10, pp. 119–138.
6. T. Matsumoto, M. Komuro, H. Kokubu, and R. Tokunaga, *Bifurcations: Sights, Sounds and Mathematics* (Springer-Verlag, Tokyo, 1993).
7. S. Tanaka, S. Higuchi, and T. Matsumoto, Sheet Structure in Global Bifurcations of a Driven R-L-Diode Circuit, Phys. Rev. E, vol.54, no.6, pp.6014–6028, August 1984,
8. Y. Kurimoto and T. Matsumoto, Chaotic Synchronization of R-L-Diode Circuit, IEICE National Convention, 44 (1994).
9. T. Matsumoto and M. Nishi, The Pecora-Carroll Synchronization in a Driven R-L-Diode Circuit, Proc. IEEJ Workshop on Chaos, Dec (1995).
10. N. Takeuchi and T. Matsumoto, Topological Horseshoe in a Driven R-L-Diode Circuit, Proc. IEEJ Workshop on Chaos, to appear.
11. P. S. Linsay, Period doubling and chaotic behavior in a driven anharmonic oscillator,

Phys. Rev. Lett. vol.47, no.19, pp.1349–1352, November 1981.
12. J. Testa, J. Perez, and C. Jeffries, Evidence for universal chaotic behavior of a driven nonlinear oscillator, Phys. Rev. Lett, vol.48, no.11, pp.714–717, March 1982.
13. S. D. Bronson, D. Dewey, and P. S. Linsay, Self-replicating attractor of a driven semiconductor oscillator, Phys. Rev. A, vol.28, no.2, pp.1201–1203, August 1983.
14. T. Klinker, W. M. Ilse, and W. Lauterborn, Period doubling and chaotic behavior in a driven Toda oscillator, Phys. Lett. 101A, no.8, pp.371–375, April 1984.
15. J. Mevissen, R. Seal, and L. Waters, Chaos: Period-2 windows and splitting, Phys. Rev. A, vol.32, no.5, pp.2990–2992, November1985.
16. J. H. Baxter, M. F. Bocko, and D. H. Douglass, Behavior of a nonlinear resonator driven at subharmonic frequencies, Phys. Rev. A **41**, 619 (1990).
17. M. Kuroda, T. Hikawa, and M. Nakai, Trans. JSME **61**, 815 (1995).
18. J. M. T. Thompson, Complex dynamics of compliant off-shore structure, Proc. R. Soc. London, A, vol.387, no.1793, pp.407–427, June 1983.
19. L. M. Pecora and T. L. Carroll, Driving systems with chaotic signals Phys. Rev. A, vol.44, no.4, pp.2374–2383, August 1991.
20. J. Szarski, *Differential Inequalities*, Polish Scientific Publisher, Warszawa, (1967).

Acknowledgement: This work is supported in part by the Japanese Ministry of Education and the Tokutei-Kadai of Waseda University. We thank S. Tanaka of Hitachi, and H. Kokubu of Kyoto University, and M. Komuro of Teikyo University for discussions.

Appendix

Step 1.

It follows from *Step 2* and *Step 3* below that q and q_s are uniformly bounded so that

$$V_1(q, q_s) := \frac{1}{2}(q - q_s)^2 \tag{A1}$$

is well defined. It follows from A.1 that $q \neq q_s$ implies $f(q) \neq f(q_s)$ and

$$\begin{aligned} \frac{dV_1(q, q_s)}{dt} &= (q - q_s)(-g(f(q)) + i + g(f(q_s)) - i) \\ &= -(q - q_s)(g(f(q)) - g(f(q_s))) \\ &= -(q - q_s)^2 \frac{g(f(q)) - g(f(q_s))}{f(q) - f(q_s)} \frac{f(q) - f(q_s)}{q - q_s} \\ &\leq -\beta_1 \alpha_1 (q - q_s)^2 = -2\beta_1 \alpha_1 V_1(q, q_s). \end{aligned} \tag{A 2}$$

An elementary result of differential inequality [20] implies

$$V_1(t) \leq e^{-2\alpha_1 \beta_1 t} V_1(0) \tag{A3}$$

and hence

$$\lim_{t \to \infty} |q(t) - q_s(t)| = 0. \tag{A4}$$

It follows from *Step 2* and *Step 3* below that

$$V_2 := i - i_s \tag{A5}$$

is also well defined. Note that

$$\begin{aligned} \frac{dV_2}{dt} &= \frac{1}{L}(-Ri - f(q) + e(t) + Ri_s + f(q_s) - e(t)) \\ &= -\frac{R}{L}V_2 - \frac{1}{L}(f(q) - f(q_s)). \end{aligned}$$

Using A.1, one obtains

$$-\alpha_2|q-q_s| \leq -(f(q)-f(q_s)) \leq \alpha_2|q-q_s|$$

so that

$$-\frac{R}{L}V_2 - \frac{\alpha_2}{L}|q-q_s| \quad \leq \quad \frac{dV_2}{dt} \leq -\frac{R}{L}V_2 + \frac{\alpha_2}{L}|q-q_s|. \tag{A 6}$$

Combining the right hand side of (A 6) with (A 1) and (A 3), one obtains

$$\frac{dV_2}{dt} \leq -\frac{R}{L}V_2 + \frac{\alpha_2}{L}\sqrt{2e^{-2\alpha_1\beta_1 t}V_1(0)}.$$

It follows from this differential inequality that

$$\begin{aligned} V_2(t) &\leq e^{-\frac{R}{L}t}V_2(0) + \int_0^t \frac{\alpha_2}{L}\sqrt{2V_1(0)}e^{-\frac{R}{L}(t-\tau)}e^{-\alpha_1\beta_1\tau}d\tau \\ &= e^{-\frac{R}{L}t}V_2(0) + \frac{\alpha_2}{L}\sqrt{2V_1(0)}e^{-\frac{R}{L}t}\int_0^t e^{(\frac{R}{L}-\alpha_1\beta_1)\tau}d\tau \\ &= e^{-\frac{R}{L}t}V_2(0) + \frac{\frac{\alpha_2}{L_1}\sqrt{2V_1(0)}}{\frac{R}{L}-\alpha_1\beta_1}(e^{\alpha_1\beta_1 t} - e^{-\frac{R}{L}t}) \end{aligned}$$

provided that $R/L \neq \alpha_1\beta_1$. If $R/L = \alpha_1\beta_1$,

$$\begin{aligned} V_2(t) &\leq e^{-\frac{R}{L}t}V_2(0) + \frac{\alpha_2}{L}\sqrt{2V_1(0)}e^{-\frac{R}{L}t}\int_0^t d\tau \\ &= e^{-\frac{R}{L}t}V_2(0) + \frac{\alpha_2}{L}\sqrt{2V_1(0)}te^{-\frac{R}{L}t}. \end{aligned}$$

A similar argument applied to the left hand inequality of (A 6) yields

$$V_2(t) \geq e^{-\frac{R}{L}t}V_2(0) - \frac{\frac{\alpha_2}{L}\sqrt{2V_1(0)}}{\frac{R}{L}-\alpha_1\beta_1}(e^{-\alpha_1\beta_1 t} - e^{-\frac{R}{L}t}), \qquad \frac{R}{L} \neq \alpha_1\beta_1$$

$$V_2(t) \geq e^{-\frac{R}{L}t}V_2(0) - \frac{\alpha_2}{L}\sqrt{2V_1(0)}te^{-\frac{R}{L}t}, \qquad \frac{R}{L} = \alpha_1\beta_1.$$

These inequalities imply

$$\lim_{t\to\infty}|i(t) - i_s(t)| = 0.$$

Step 2. (uniform boundedness of (q, i))

Assumption A1 implies that the vector field (M1), (M2) is Lipschitz so that there is a unique local solution for an arbitrary initial condition. In order to demonstrate the uniform boundedness of (q, i), consider

$$E(q,i) := \int_{q(0)}^q f(u)du + \frac{1}{2}Li^2 \tag{A 7}$$

which is the total energy stored in the capacitor and the inductor. It follows from A.1 that

$$\lim_{||(q,i)||\to\infty} E(q,i) = \infty. \tag{A 8}$$

Assumption A.1 also implies that

$$A := \{(q,i)|f(q)g(f(q)) \leq \frac{B^2}{4R}, |i| \leq \frac{B}{R}\} \tag{A 9}$$

is bounded or empty so that it suffices to show that

$$\frac{dE(q,i)}{dt} < 0, \quad (q,i) \in \mathbf{R}^2 - A. \tag{A 10}$$

Note that

$$\begin{aligned} \frac{dE(q,i)}{dt} &= f(q)\frac{dq}{dt} + Li\frac{di}{dt} \\ &= f(q)(-g(f(q)) + i) - Ri^2 - if(q) + ie(t) \\ &= -(f(q)g(f(q)) + Ri^2) + ie(t). \end{aligned} \tag{A 11}$$

The first term represents the power dissipated by the resistive part of the circuit while the second term represents the power supplied by the voltage source. Since

$$\min_{|i|}(R|i|^2 - B|i|) = -\frac{B^2}{4R}.$$

one sees that for an arbitrary i,

$$\frac{dE(q,i)}{dt} \leq -f(q)g(f(q)) + \frac{B^2}{4R} \tag{A 12}$$

Therefore, for q satisfying

$$f(q)g(f(q)) > \frac{B^2}{4R}$$

the right hand side of (A 12) is negative, and hence (A 10) follows.

Step 3. (uniform boundedness of (q_s, i_s))

Given a uniformly bounded i of (M1), (M2), one needs to show that (q_s, i_s) is also uniformly bounded. Let $E_s(q_s, i_s)$ be the energy defined in a manner similar to (A 7). Then

$$\begin{aligned} \frac{dE_s(q_s,i_s)}{dt} &= f(q_s)(-g(f(q_s)) + i(t)) - Ri_s^2 - i_s f(q_s) + i_s e(t) \\ &= -f(q_s)(g(f(q_s)) + i_s - i(t)) - Ri_s^2 + i_s e(t) \\ &:= -P(q_s, i_s). \end{aligned} \tag{A 13}$$

Due to the fact that (S1) is driven by i, the term $i_s - i$ in (A 13) does not cancel as it does in (A 11). Note, however, that

$$P(q_s, i_s) \leq f(q_s)g(f(q_s)) + i_s f(q_s) + I|f(q_s)| + Ri_s^2 + B|i_s|$$

which follows from A.1, where I is a bound on $|i(t)|$. This implies that

$$\lim_{||q_s, i_s|| \to \infty} P(q_s, i_s) = \infty$$

and hence

$$A_s := \{(q_s, i_s)|P(q_s, i_s) \leq 0\}$$

is bounded or empty. Therefore

$$\frac{dE_s(q_s,i_s)}{dt} < 0, \qquad (q_s, i_s) \in \mathbf{R}^2 - A_s$$

which yields

$$\lim_{||(q_s,i_s) \to \infty||} E_s(q_s, i_s) = \infty.$$

TRANSIENT GLOBAL BEHAVIOR IN NONLINEAR EXPERIMENTAL OSCILLATORS

L.N. VIRGIN, M.D. TODD, C.J. BEGLEY, S.T. TRICKEY AND E.H. DOWELL
School of Engineering
Duke University, Durham, NC 27708-0300

Abstract

This paper is concerned with the determination of basin boundaries in experimental nonlinear oscillators. Global transient behavior has received considerable attention in numerical studies but is relatively unexplored from an experimental perspective, despite the fact that a global view of transient behavior provides a much more complete description of the dynamics of a system than a traditional concentration on steady-state behavior. Three different physical systems are studied: two mechanical and one electrical.

1. Introduction

Perhaps the most dominant characteristic of dissipative dynamical systems is the contraction of the vector field onto attracting sets, i.e., initiated transients asymptote to certain special solutions as time tends to infinity. A typical forced nonlinear oscillator may possess co-existing attractors thus providing a dependence on initial conditions, which in some cases may be extremely sensitive [9]. Relatively little attention has been paid to the determination of basins of attraction in experimental mechanics, mainly due to the difficulty of accurately prescribing initial conditions [3].

In this paper three archetypal experimental systems are investigated with a view to examining their basins of attraction. The determination of the basins of attraction and global transient behavior is approached in a similar way for each of these systems. The systems studied are: a double-well potential gravity-driven Duffing system [6, 16]; a piecewise linear impacting system including dry friction [1, 2]; and an electric circuit modeling a driven Duffing system [17].

Each of the above systems can be modeled as a single degree of freedom oscillator with a periodic input. As such the periodic attractors in the

F. C. Moon (ed.),
IUTAM Symposium on New Applications of Nonlinear and Chaotic Dynamics in Mechanics, 353–362.

resulting three-dimensional (cylindrical) phase space can be assessed by their penetration of a surface of section defined by a prescribed forcing phase. In this way regions of initial position and initial velocity describe domains in the plane which map to the various fixed points, and how long they take to get there.

2. Stochastic Interrogation

A systematic method for investigating the basins of attraction in an experimental context has recently been developed [3]. In this approach an initial condition map is built up by imparting bursts of stochasticity to perturb a steady-state. The subsequent decay of these generated transients is then monitored and the whole process repeated in an automated fashion.

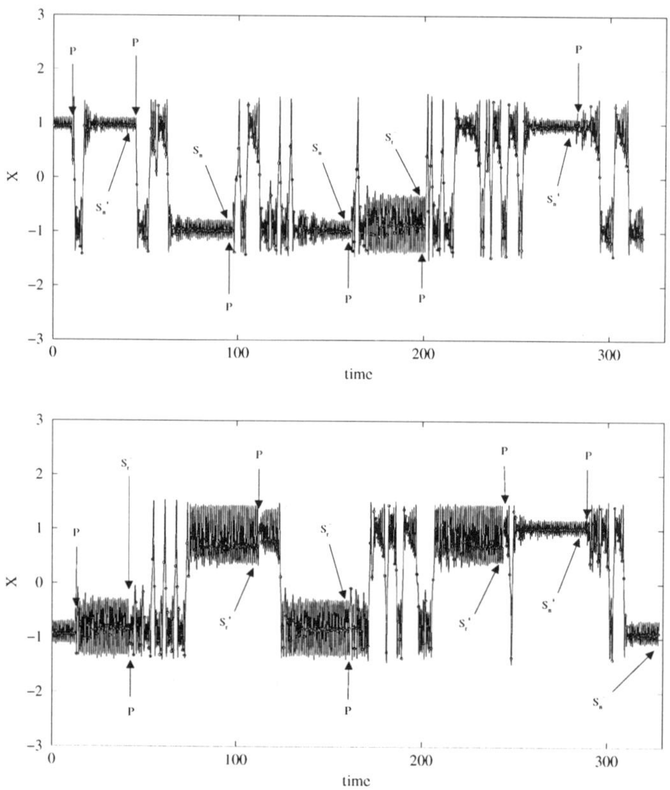

Figure 1. Typical stochastic interrogation time series showing perturbations of the attractors.

As an illustration of this procedure consider the typical occasionally perturbed time series of Fig. 1. Here, there are four co-existing period-one attractors. The system is initially exhibiting steady-state motion when a perturbation (P) is applied, causing a relatively large transient which eventually settles back down on to the original attractor. The process is repeated and this time the transient leads to an alternative small amplitude motion. A collage of initial conditions can thus be built up effectively from a single time series. There are a number of other practical considerations when using this approach including the optimal size of perturbations and their stochastic distribution.

3. A Twin-Well Mechanical Oscillator

The double-well oscillator is one of the archetypal systems in nonlinear dynamics. Its experimental realization has been achieved using a buckled beam [7], and more recently used as the basis for a roller-coaster type mechanical system [6]. The potential energy function has two minima separated by a hilltop and thus provides the cencept of a particle moving in a gravitational field. Details of the experimental realization of this arrangement can be found in [6], including a derivation of the actual equation of motion:

$$\begin{aligned}(1+\alpha^2X^2(X^2-1)^2)(X''+2\zeta X')+\frac{\mu}{2\alpha}sgn(X')\\ +\alpha^2X'^2X(X^2-1)(3X^2-1)\\ +\frac{1}{2}X(X^2-1)=F\Omega^2 sin(\Omega\tau+\phi),\end{aligned} \tag{1}$$

where X is a nondimensional horizontal projection. Provided α is relatively small [13], this equation is a reasonable apporoximation to Duffing's equation with the addition of a Coulomb damping term and transmissible excitation. In this paper the following parameters are fixed at their experimentally measured values: $\alpha^2 = 0.653$, $\zeta = 0.002$, $\mu = 0.02$, $F = 0.03$, $\Omega = 0.89$ and $\phi = 0$.

In the presence of a small horizontal harmonic base excitation with the parameter values indicated the asymptotic behavior is dominated by periodic attractors. When the system is forced close to its natural frequency (with a sufficiently large magnitude) the classic hysteresis is seen together with the accompanying resonant amplitude jumps. Since this phenomenon occurs within each of the two wells it is possible to locate a robust regime of parameter space where there are four coexisting attractors. Unstable fixed points and manifolds play a central role in determining long-term behavior and a variety of subtle (local and global) bifurcational behavior has been detailed in careful numerical simulations [8]. In this section of the paper a typical initial condition study is used as a basis of comparison between numerical and experimental data.

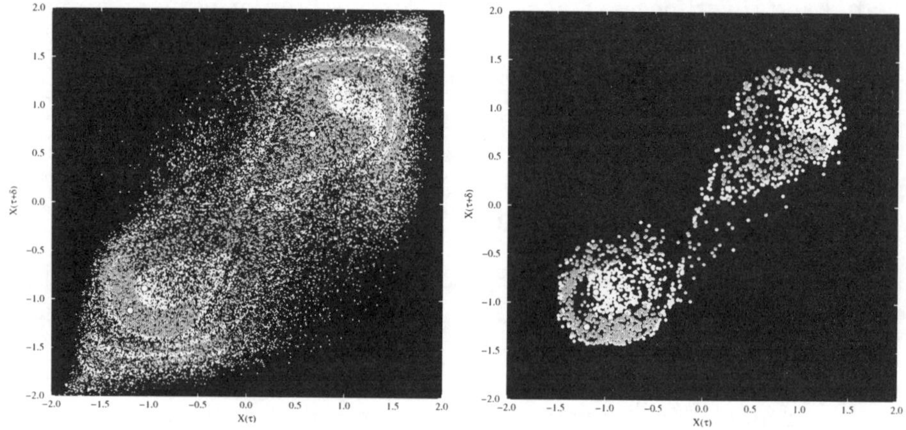

Figure 2. Basins of attraction for the Duffing system with F = 0.03, $\Omega = 0.89$ and $\phi = 0$. Four attractors compete and the initial state is based on time-lag embedding (1/4 cycle delay). (a) Numerical integration using a set of randomly generated starts, (b) experimental results using stochastic interrogation.

The stochastic interrogation technique described in the previous section is used to generate a random set of initial conditions for a given set of parameter values (see above). Figure 2 shows the basin boundaries within the resonant hysteresis. In Fig. 2(a) a randomly-generated set of 50,000 initial conditions based on simulation is shown in terms of time-delay coordinates. The equivalent experimental results are shown in Fig. 2(b) for 4,000 perturbations. The white data points are the set of initial conditions that lead to the non-resonant periodic solution, and the gray data points lead to the resonant oscillations (in either well). The fixed points are indicated by the white circles. These regions can be further divided according to which well the solution appears in, and it can then be observed that these basins are smooth, in agreement with the Melnikov criterion.

It is worth mentioned that in the numerical generation of Fig. 2(a) a shooting procedure was used to match the initial conditions in terms of a quarter-cycle position delay and the equivalent conventional initial conditions used in numerical simulation. Other studies of this system in relation to indeterminate bifurcations have shown the presence of fractal basin boundaries with the possibility of chaotic transients, although noise plays an increasing role in the experimental data acquisition [14, 16]. Determination of the onset of fractal basin boundaries is one of the ingredients of indeterminate post saddle-node jumps in the response.

4. A Piecewise Linear System

Impact oscillators have been the focus of many studies [12]. This is also the case with dry friction systems [4]. Here, these two piecewise linear characteristics are considered together in the context of a single-degree-of-freedom mechanical oscillator consisting of a rotational inertia, harmonically forced through a base spring, and subject to impact at critical angles, as shown in Figure 3.

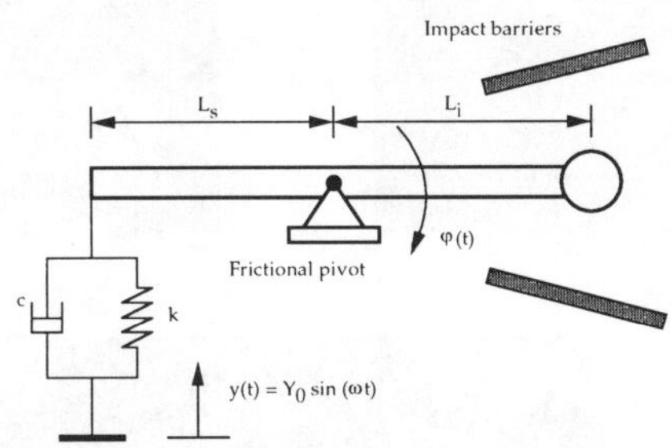

Figure 3. A schematic of the frictional-impacting system. The forcing is transmitted through a spring-damper. Dry friction occurs at the pivot and motion is constrained by the impact barriers.

Friction is assumed to be present at the pivot. For small frictional force, a summation of moments about the pivot and appropriate nondimensionalization provides the equation of motion:

$$\ddot{\theta} + 2h\dot{\theta} + \theta + f(\dot{\theta}) + g(\theta) = 2h\gamma\cos(\gamma\tau) + \sin(\gamma\tau), \qquad (2)$$

where $f(\dot{\theta})$ and $g(\theta)$ represent nonlinear forces due to friction and impact respectively. They are both piecewise linear and are discussed in more detail in [2], together with experimental details. It is worth mentioning that rather than using the time-lag embedding coordinates as in the previous section, here the position and velocity were measured independently.

The ratio of the driving frequency to natural frequency is held fixed at $\gamma = \omega/\omega_n$ =1.4; the normalized friction force, f = 0.0505 (based on a simple Coulomb law); the viscous damping ratio, h = 0.02; and the (asymmetric) left and right impact locations, σ_L = -1.574 and σ_R = 1.36 (nondimensionalized with respect to the distance from the pivot and forcing magnitude). For this set of parameter values there are three co-existing periodic attractors and, where appropriate, their mirror images. The three periodic attractors coexisting at this specific set of parameters values are: (a) a small amplitude non-impacting (linear) solution with a fixed point at (-0.202, -1.285); (b) a period-1 double impacting solution with a fixed point at (-0.602, 1.377); and (c) a period-2 triple impacting solution with a fixed points at (-1.33, 0.0.077) and (0.108, 2.066), where the fixed point locations are based on the results of the experiment.

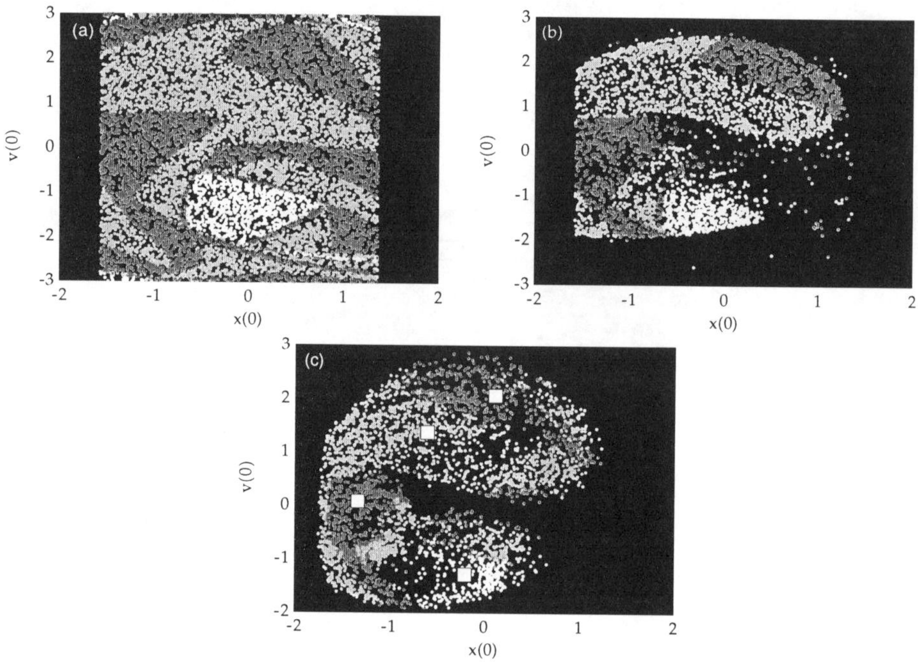

Figure 4. Basins of attraction for the impact system. (a) Numerical integration from a randomly generated set of starts, (b) adjusted numerical integration with a 1.5 cycle delay before initial conditions are acquired, (c) experimental results based on stochastic interrogation.

The initial condition map at this set of parameter values (Fig. 4(a)) is determined numerically by generating 10,000 random initial conditions and labeling them according to their subsequent long term behavior as (a) white; (b) light gray; (c) dark gray, for the attractors (a), (b) and (c) in the previous apargraph. Here, the axis labels $x(0) \equiv \theta(0)$ and $v(0) \equiv \dot{\theta}(0)$. The corresponding experimental result using stochastic interrogation to generate 4,000 initial conditions is shown in Fig. 4(c) using the same gray shading and with the fixed points labeled as white squares. Apart from the unvisited region the correlation is quite good. Since there is a finite time delay in switching off the stochastic forcing mix it turns out that there tends to be typically a one and a half cycle lag before the initial condition can be acquired. This suggests that it is not possible to generate initial conditions in areas where transients exit very quickly. To confirm this scenario a 1.5

cycle time-lag was built in to the numerical algorithm and the results are shown in Fig. 4(b). Clearly initial conditions move somewhat rapidly away from a central region as well as from regions of high initial velocity. The uneven nature of phase space contraction is not very surprising given the severe nonlinearity provided by the impact condition.

5. A Nonlinear Electric Circuit

Ordinary differential equations are often represented in a block diagram form such as in Fig. 5, which is a schematic representation of:

$$\ddot{x} + 2\zeta\omega_n\dot{x} - (\omega_n^2/2)(x - x^3) = A\cos(\omega t). \tag{3}$$

The circuit used in this study was designed by literally translating the block diagram into basic circuit components consisting of resistors, capacitors, operational amplifiers and multiplier chips [10].

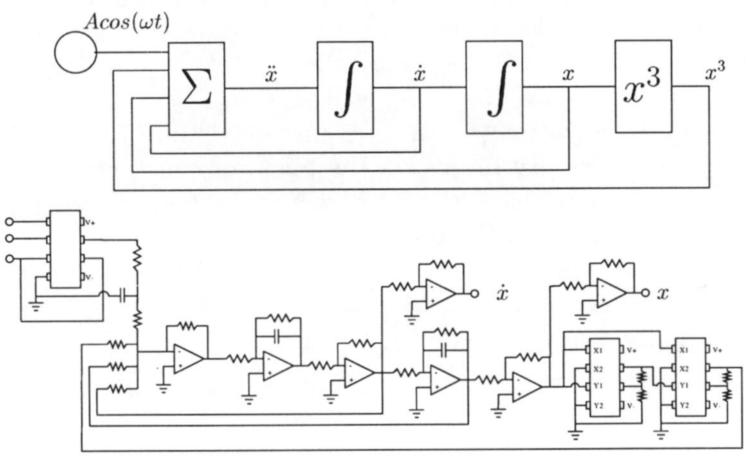

Figure 5. The electric circuit shown as a (a) block diagram, (b) circuit diagram.

In the circuit, input from a function generator is modulated using a voltage controlled amplifier constructed from a multiplier chip as shown at the far left of the circuit in Fig. 5(b). The forcing input is then added to the position, velocity, and the cubic term through the use of a summing amplifier. The sum, representative of the acceleration, is then integrated using a low pass filter. Rescaling the output signal, now representative of the velocity, using an inverting amplifier serves two purposes. First, the signal is returned to a nominal magnitude within the operational range of the circuit components as well as data acquisition system and second, the signal is given the correct sign sense for the acceleration summation. A similar process follows when integrating the velocity to obtain the position. The most important section of the circuit is composed of the two multiplier chips at the far right of Fig. 5(b). Together they constitute the cubic term in Duffing's Eq. (3). The circuit was tuned to have a linear natural frequency (f_n) in the vicinity of 10 Hz (due to the double-well nature of the potential this value represents an average of the two natural frequencies) and stable equilibrium solutions at ± 1 Volt. A standard circuit analysis book, such as [5], provides a good review of integration, summing and gain amplification circuits.

The system parameters were chosen in the region where four stable readily distinguishable solutions coexist: a low and high amplitude oscillation in both potential wells of the system, analogous to the mechanical oscillator examined in the first part of this paper. Again stochastic interrogation was used to perturb the system about the baseline parameter values. The forcing amplitude and frequency were assigned random values and held constant over a forcing cycle with new values assigned at each interrogation forcing cycle. An additional randomizing element is added by varying the number of interrogation forcing cycles. The data collection algorithm simply involves turning on the pseudo-random forcing described above then resetting to the nominal system parameters and acquiring the initial condition when the phase space trajectory intersects the plane of constant forcing phase. The initial condition is assigned a basin when the trajectory converges to within a tolerance of one of the predetermined steady state solutions. This choice of interrogation input is validated by the results seen in Fig. 6 where the initial conditions are well distributed in phase space.

For the frequency value in Fig. 6(a) there are just two coexisting attractors with the dark gray shading representing the small amplitude solution in the left well, and the light gray shading corresponding to the motion in the other well. Not surprizingly, just because an initial condition is generated in one well it does not necessarily lead to the attractor in that well, indeed a trajectory may traverse across the hilltop a number of times before the transient has decayed. Fig 6(b) corresponds to a frequency within the main resonant hysteresis. The shading *now* corresponds to light gray for both small amplitude solutions and dark gray corresponds to both large amplitude solutions. Again this data can be broken down in to the four basins, the boundaries of which are smooth. The white symbols correspond to the location of the fixed points. Furthermore the duration of the transients is found to broadly correspond to the distance from the initial conditions, although the long duration transients give some indication of the location of unstable fixed points and their manifolds. This may have some use in the control of chaos [11] and identifying bottlenecking [15].

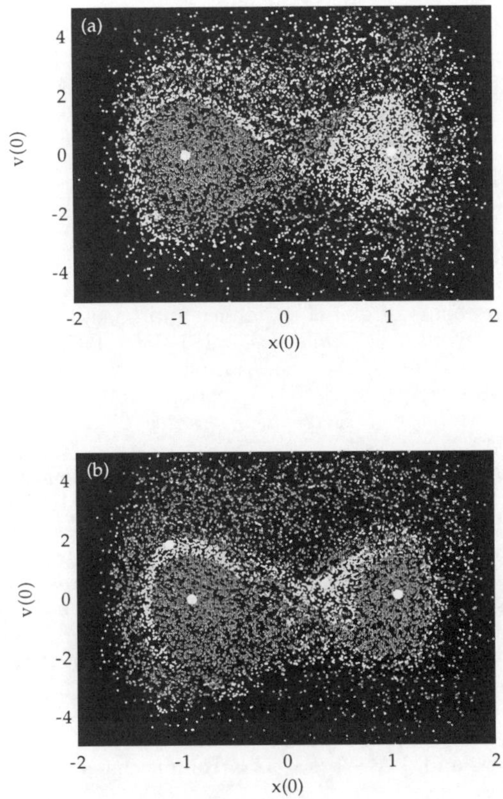

Figure 6. Initial condition maps obtained from the circuit: (a) two attractors at $\omega = 6.5$ Hz, (b) four attractors at $\omega = 8.3$ Hz.

6. Conclusions

This paper has shown the determination of basins of attraction for three distinct experimental systems using the method of stochastic interrogation. Initial condition maps are generated which illustate the relative dominance of co-existing periodic attractors for typical parameter ranges. In the examples shown, parameter values have been intentionally chosen where smooth basin boundaries exist, mainly for validation purposes. Although not included in the current paper, the data also contains information relating to how long the initial conditions take to reach a periodic attractor (to a given tolerance). This is an especially interesting question when fractal basin boundaries are encountered since chaotic transients can have arbitrary durations. Clearly this general approach can play a key role in identifying global bifurcations in experiments. Although global issues have received quite intensive study from a numerical standpoint, very few experimental studies have been conducted. Consideration of global behavior adds robustness to the understanding of the general properties of a dynamic system.

References

1. P. V. Bayly and L. N. Virgin. An experimental study of an impacting pendulum. *Journal of Sound and Vibration*, 164:364–374, 1994.
2. C. J. Begley and L .N. Virgin. A geometric interpretation of grazing bifurcations in an impact oscillator (in preparation).
3. J. P. Cusumano and B. W. Kimble. A stochastic interrogation method for experimental measurements of global dynamics and basin evolution: application to a two-well oscillator. *Nonlinear Dynamics*, 8:213–235, 1995.
4. B.F. Feeny and F.C. Moon. Autocorrelation on symbol dynamics for a chaotic dry-friction oscillator. *Physics Letters A*, 141:397–400, 1989.
5. L. R. Fortney. *Principles of Electronics: Analog & Digital.* Harcourt Brace Jovanovich, Inc., 1987.
6. J. A. Gottwald, L. N. Virgin, and E. H. Dowell. Experimental mimicry of Duffing's equation. *Journal of Sound and Vibration*, 158:447–467, 1992.
7. P.J. Holmes and F.C. Moon. Strange attractors and chaos in nonlinear mechanics. *Journal of Applied Mechanics*, 50:1021–1032, 1983.
8. A.L. Katz and E.H. Dowell. From single well chaos to cross well chaos: A detailed explanation in terms of manifold intersections. *International Journal of Bifurcation and Chaos*, 4:933–941, 1994.
9. G. X. Li and F. C. Moon. Fractal basin boundaries in a two-degree-of-freedom nonlinear oscillator. *Nonlinear Dynamics*, 1:209–219, 1990.
10. T. Matsumoto, L.O. Chua, and S. Tanaka. Simplest chaotic nonautonomous circuit. *Physical Review A*, 30:1155–1157, 1984.
11. E. Ott, C. Grebogi, and J.A. Yorke. Controlling chaos. *Physical Review Letters*, 64:1196–1199, 1990.
12. S. W. Shaw. The dynamics of a harmonically excited system having rigid amplitude constraints, parts 1 and 2. *Journal of Applied Mechanics*, 52:453–464, 1985.
13. S. W. Shaw and A. G. Haddow. On 'roller-coaster' experiments for nonlinear oscillators. *Nonlinear Dynamics*, 3:375–384, 1992.
14. M.S. Soliman and J.M.T. Thompson. Basin organization prior to a tangled saddle-node bifurcation. *International Journal of Bifurcation and Chaos*, 1:107–118, 1991.
15. S.H. Strogatz. *Nonlinear Dynamics and Chaos.* Addison-Wesley, 1994.
16. M.D. Todd and L.N. Virgin. An experimental verification of basin metamorphoses in a nonlinear mechanical system. *International Journal of Bifurcation and Chaos*, 7, 1997.
17. S.T. Trickey and L.N. Virgin. Experimental studies of a nonlinear electric circuit (in preparation).

CONTROL OF CHAOS FOR PENDULUM SYSTEMS

W. O. SCHIEHLEN
Institute B of Mechanics, University of Stuttgart,
D-70550 Stuttgart, Germany,
email: wos@mechb.uni-stuttgart.de

1. Introduction

A mathematical pendulum as well as a one-body mechanical pendulum is restricted to periodic motions. A two-body pendulum shows during the transition from oscillations in the gravity field with low energy to oscillations in the centrifugal field with high energy chaotic behaviour. A three-body pendulum with suitable inertia properties may perform already in the gravity field chaotic motions what is generally true for pendulum systems, see e.g. Rott [3]. In theory, pendula represent conservative mechanical systems. Therefore, in simulation conservative pendula show stationary, irregular motion but they do not feature strange attractors. In experiments, due to damping and friction pendula are in nonstationary transition from some initial state to equilibrium and strange attractors are missing, too. By adding a control device mechanical pendula may be converted to mechatronic pendula representing a very rich dynamic behaviour including chaos and strange attractors.

An experimental pendulum apparatus has been described by Rott and Smith [4]. The fundamental phenomen with mechatronic pendula were presented by Schiehlen and Meinke [6].

2. Modelling of a Three-Body Pendulum

A three-body pendulum with an inner parametric resonance consists of three rigid bodies, Fig. 1. For modelling the mulitbody system approach is used and the motion of the mechanical device has to be controlled to overcome the energy dissipation by damping and friction.

2.1. MULTIBODY SYSTEM DYNAMICS

The three-body pendulum is composed of the main pendulum K1 and two equal auxiliary pendula K2 and K3.

F. C. Moon (ed.),
IUTAM Symposium on New Applications of Nonlinear and Chaotic Dynamics in Mechanics, 363–370.

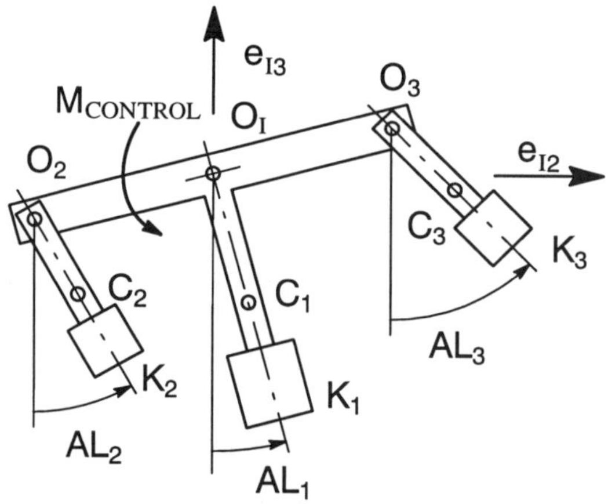

Figure 1. Three-body pendulum

The masses are M_i, the moments of inertia I_i, the damping coefficients of the bearings D_i, $i = 1(1)3$, GE is the gravitational acceleration and $M_{CONTROL}$ means the control torque. The pendulum is symmetric with respect to the geometrical parameters: distance O_IC_1 is A, distances O_2C_2, O_3C_3 are B and distances O_IO_2, O_IO_3 are C. The pendulum has $f = 3$ degrees of freedom, the generalized coordinates are introduced as AL_i, $i = 1(1)3$.

The equations of motion are symbolically computed by the program NEWEUL, see Leister [2] and read as

$$\boldsymbol{M}(\boldsymbol{y},t)\,\ddot{\boldsymbol{y}} + \boldsymbol{k}(\boldsymbol{y},\dot{\boldsymbol{y}},t) = \boldsymbol{q}(\boldsymbol{y},\dot{\boldsymbol{y}},t)\ , \tag{1}$$

The elements of the inertia matrix $\boldsymbol{M}$ are denoted as $RM(l,m)$, the generalized force vectors $\boldsymbol{k}$ and $\boldsymbol{q}$ are written as $K(l)$, $Q(l)$, l, $m = 1(1)3$. The generalized applied force vector $\boldsymbol{q}$ includes gravity, damping and control. The vector of the generalized coordinates is specified as

$$\boldsymbol{y}(t) = [AL_1\ AL_2\ AL_3]^T \tag{2}$$

The matrices and vectors of the equations of motion are presented in Fig. 2.

2.2. CONTROL OF MOTION

Due to the inner resonance a properly designed inherent conservative three-body pendulum shows chaotic behaviour. The only task of the control device is to compensate the energy dissipation. For this purpose controlled torque thrusts may be used.

```
Inertia matrix
  RM(1,1)=A**2*M1+C**2*M2+M3*C**2+I1.
  RM(1,2)=B*C*M2*SIN(AL1-AL2).
  RM(1,3)=-M3*B*C*SIN(AL1-AL3).
  RM(2,2)=B**2*M2+I2.
  RM(2,3)=0.
  RM(3,3)=M3*B**2+I3.
Vector of generalized gyroscopic forces
  K(1)=-B*C*M2*AL2P**2*COS(AL1-AL2)
      +M3*B*C*AL3P**2*COS(AL1-AL3).
  K(2)=B*C*M2*AL1P**2*COS(AL1-AL2).
  K(3)=-M3*B*C*AL1P**2*COS(AL1-AL3).
Vector of generalized applied forces
  Q(1)=-GE*A*M1*SIN(AL1)+C*M2*GE
      *COS(AL1)-M3*C*GE*COS(AL1)
      -D1*AL1P+D2*AL2P-D2*AL1P
      +D3*AL3P-D3*AL1P+L1+MCONTROL.
  Q(2)=-B*M2*GE*SIN(AL2)-D2*AL2P
      +D2*AL1P.
  Q(3)=-M3*B*GE*SIN(AL3)-D3*AL3P
      +D3*AL1P.
```

Figure 2. Equations of motion

The control law has to provide a mechanism for self-excitation which can be achieved by additional logical curcuits as shown in Fig. 3, where the 6×1–state vector

$$\boldsymbol{x}(t) = [\boldsymbol{y}^T \dot{\boldsymbol{y}}^T]^T = [\alpha_1 \alpha_2 \alpha_3 \dot{\alpha_1} \dot{\alpha_2} \dot{\alpha_3}]^T \tag{3}$$

is introduced in modified notation. The control torque, here denoted as M, is most of the time zero, only for short time intervals a constant torque $M_{CONTROL}$ is applied. In summary, the control law reads as

$$M = M(\alpha_1) \tag{4}$$

i.e. only the position α_1 of the main pendulum has to be measured.

3. Nonlinear Chaotic Dynmaics

The chaotic motion of the three-body pendulum follows from six strongly coupled, highly nonlinear state equations. Therefore, the mathematical analysis is restricted to a mapping, some Fourier analyses and a computation of Ljapunov exponents. Further, an experimental verification will be discussed.

3.1. MATHEMATICAL ANALYSIS

The inner resonance condition of the three-body pendulum is easily found by linearization of the conservative pendulum. Using the parameters $M_2 = M_3$,

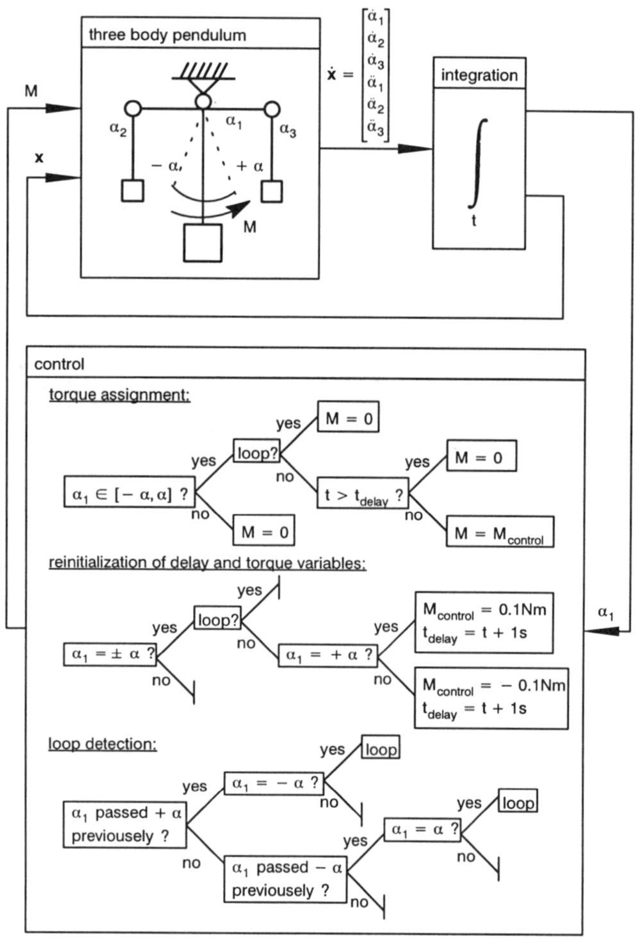

Figure 3. Logical curcuits for the control law

$I_2 = I_3$, $D_i = 0$, $M = 0$, $i = 1(1)3$, the initial conditions $\dot{\alpha}_i = 0$, $i = 1(1)3$ and $\alpha_{10} << 1$, $\alpha_{20} < 1$, $\alpha_{30} < 1$ it follows

$$\ddot{\alpha_1} + [(\alpha_{20} - \alpha_{30}) \cos \omega t + \Omega^2]\alpha_1 = 0 \ , \tag{5}$$

$$\ddot{\alpha_2} + \omega^2 \alpha_2 = 0 \ , \tag{6}$$

$$\ddot{\alpha_3} + \omega^2 \alpha_3 = 0 \ . \tag{7}$$

Equation (5), well-known as Matthieu differential equation, features parameter resonance for $\alpha_{20} \neq \alpha_{30}$ and the frequency condition

$$\omega = 2\Omega \ . \tag{8}$$

For a conservative pendulum, $D_i < 0$, trajectories and FFT plots are shown in Ref. [7]. However, these results can not be experimentally verified due to damping and friction.

The controlled mechatronic three-body pendulum has been analysed using the following parameters:

$$\begin{aligned}
&M_1 = 0,613kg \\
&M_2 = M_3 = 0,209kg \\
&I_1 = 0,0223kgm^2 \\
&I_2 = I_3 = 0,00324kgm^2 \\
&A = 0,107m \\
&B = 0,114m \\
&C = 0,180m \\
&D_1 = 0,0015Nms \\
&D_2 = 0,000055Nms \\
&D_3 = 0,00011Nms \\
&M_{control} = 0,1Nm \\
&\alpha = 21grad \\
&t_{delay} = 1s
\end{aligned}$$

The mapping of the motion of the main pendulum is shown in Fig. 4 subject to the condition of the $\alpha_2 = 0$, $\dot{\alpha}_2 = 10$, $\alpha_3 = 0$, $\dot{\alpha}_3 = 10$. The mapping of a higher dimensional state-space to a twodimensional plane is also discussed by Schaub [5]. It turns out that there exists an attractor in the neighborhood of the separatrix of a mathematical pendulum.

A stationary steady state motion of the main pendulum is shown in Fig. 5, the trajectory is not periodic at all. This follows from the FFT analysis of the trajectories, too, Fig. 6.

There is one periodic component for each pendulum, for the main pendulum $f_1 \approx 0,5Hz$ and for both auxiliary penudula $f_{2,3} \approx 0,8Hz$.

These frequencies are related to the frequencies of small osciallations which are found to be $f_1 \approx 0,62Hz$ and $f_{2,3} \approx 1Hz$. As known from the theory of nonlinear oscillations, the frequencies are decreasing for larger amplitudes of pendula.

The computation of the Ljapunov exponents resulted in the following steady-state values

$$\sigma_1 = 1.215 \qquad \sigma_2 = 0.588 \qquad \sigma_3 = 0.043$$
$$\sigma_4 = -0.074 \qquad \sigma_5 = -0.606 \qquad \sigma_6 = -1.232$$

Details on the numerical method are available from Kreuzer and Lagemann [1]. It clearly turns out, due to two positive Ljapunov exponents, that the mechatronic pendulum system shows irregular, chaotic behaviour.

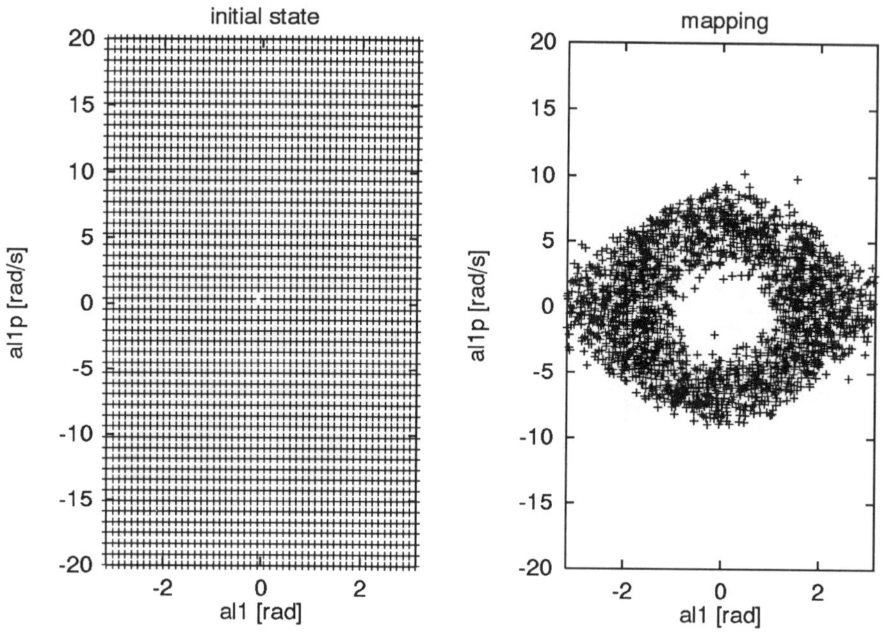

Figure 4. Mapping of the main pendulum motion

3.2. EXPERIMENTAL VERIFICATION

A mechatronic three-body pendulum has been designed and set up at the Insitute B of Mechanics of the University of Stuttgart with the kind support of the Institute of Mechanics, University of Hannover. The control device consists of an inductive sensor, a digital controller and an electric drive. Further, an initialisation of the motion from the equilibrium condition was implemented. The artistical design of the three-body pendulum is shown in Fig. 7. By observation it is obvious, that the three-body pendulum features a stationary chaotic motion which is completely unpredictable. It turns out that the mechatronic pendulum is a robust system, it is not required to adjust the parameters of the system accurately due to the selfexcitation implemented in the control law.

4. Conclusions

Steady-state chaotic motions of pendulum systems can be achieved by digital control. Then, the energy dissipation by damping and friction is compensated by electric energy. The resulting differential equations of motion of mechatronic pendula are strongly nonlinear. However, the fundamental frequencies are related to the linearized equations valid for small oscillations only. From

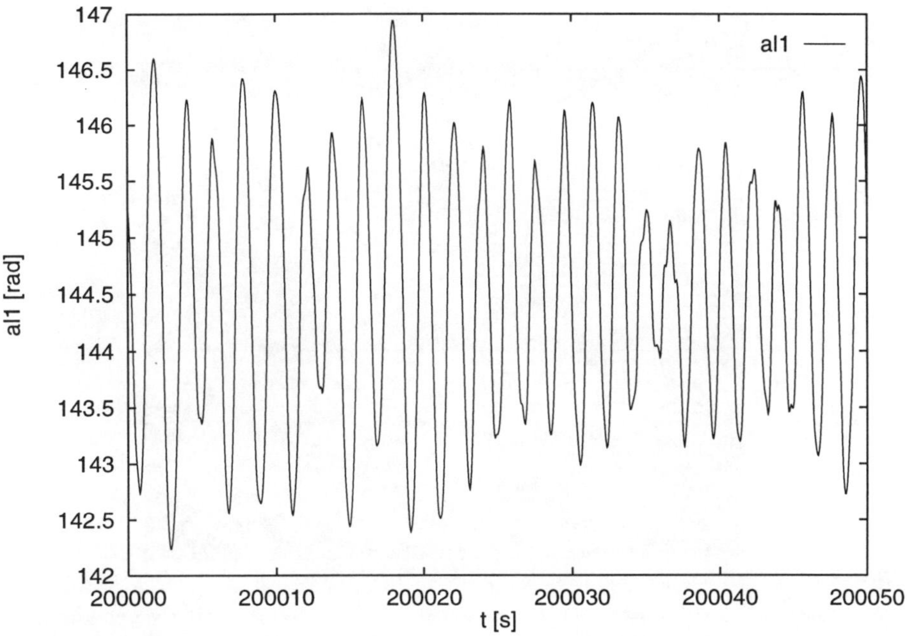

Figure 5. Stationary steady state motion of the main pendulum

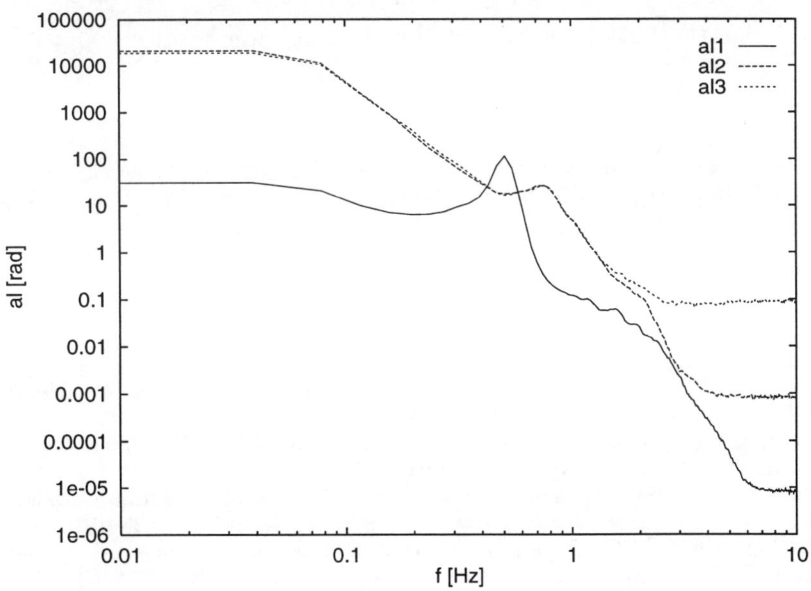

Figure 6. FFT analysis of the trajectories

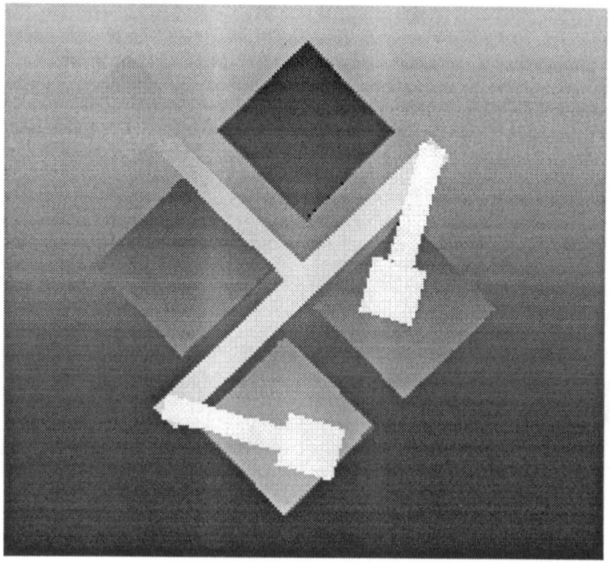

Figure 7. Experiments with a artistical design of the three-body pendulum

the observation of the experiments and the mathematical signal analysis of the simulations it can be concluded that even for very large time intervals irregular chaotic motions must not turn into perodic motions.

5. Acknowledgement

The author thanks Dr. Andreas Rükgauer, Rainer Dürr, M.Sc. and Bodo Lagemann, M.Sc. for the support with the mathematical analysis of the pendulum system investigated.

References

1. Kreuzer, E., Lagemann, B.: Computing Lyapunov Exponents for Retarded Dynamical Systems. In: *Proc. of the 3rd Polish-German Workshop "Dynamical Problems in Mechanical Systems"*, pp. 269-280, 1993.
2. Leister, G.: *Programmsystem NEWEUL'92.* Anleitung AN-32. University of Stuttgart, Institute B of Mechanics, Stuttgart, 1993.
3. Rott, N.: A Multiple Pendulum for the Demonstration of Nonlinear Coupling. *Zeitschrift für Angewandte Mathematik in der Physik* **21**, pp. 570-582, 1970.
4. Rott, D.P., Smith, J.C.: *Pendulum Principle Demonstration Apparatus.* United States Patent No. 5, 145,378, 1992.
5. Schaub, S.: *Erweiterte Zellabbildung zur globalen Untersuchung nichtlinearer dynamischer Systeme.* Ph.D.-Thesis, University of Stuttgart, Stuttgart, 1997.
6. Schiehlen, W., Meinke, P.: Chaospendel. In: *Proc. Nichtlneare Dynamik.* Akademischer Verlag, Munich, to appear.
7. Schiehlen, W.: Nonlinear Phenoma in Multibody Systems. *Delft Progress Report* **10**, No. 2, pp. 105-120, 1985.

VII. VEHICLE AND SHIP DYNAMICS

Ships, walking, railways and shimmying wheels are all covered in these five papers. Coleman et al. presents a model and experimental evidence for stable passive walking machines. Stépán presents a key lecture on theoretical models for shimmy dynamics in wheels. A similar theoretical problem is found in the paper by O'Reilly and Varadi who discuss pulling a luggage cart on wheels.

Jensen et al. provide a theoretical view of nonlinear rail models and Spyrou and Bishop extend earlier work on ship dynamics to multiple degree of freedom motions.

Despite the ubiquitous nature of vehicles, ship and animal locomotion, the nonlinear dynamics problems in these systems are not completely understood and should provide the new researcher with a rich vein of research to mine.

F. C. Moon (ed.),
IUTAM Symposium on New Applications of Nonlinear and Chaotic Dynamics in Mechanics, 371.

DELAY, NONLINEAR OSCILLATIONS AND SHIMMYING WHEELS

G. STÉPÁN
Department of Applied Mechanics
Technical University of Budapest, H-1521 Budapest, Hungary

1. Introduction

The lateral vibration of towed wheels – the so-called shimmy – is a well-known classical example of self-excited nonlinear oscillations. The developments of more and more sophisticated wheel suspension systems and tires in vehicles, as well as the appearance of semi-trailers, caravans, high–speed motorcycles and other large towed vehicle systems require a continuous analysis of this phenomenon. The paper calls the attention for the complexity of the modeling issues and for that of the nonlinear dynamics of the phenomenon strictly from the view-point of the wheel only. This means that the complex dynamics of the vehicle itself is simplified and only those key elements of the mechanical models are considered and analyzed which are related to the wheel–ground connection. The resulting finite and infinite dimensional dynamical systems exhibit a great variety of nonlinear oscillations such as stable and unstable periodic and quasi-periodic oscillations, chaotic and transient chaotic behavior. The infinite dimensional phase space is due to the presence of a special kind of memory, or time delay in the system which is related to the deformed tire contact surface keeping a record of the past motion of the wheel.

The common meaning of the word 'shimmy' is a dance back to the early thirties of this century. This shows that the phenomenon named by the same word, shimmy, has been studied for several decades already (the earliest scientific study the author could find is that of Schlippe *et al.* (1941). However, the phenomenon has not been fully explored. This is due to two important reasons. One problem is caused by the fact that the vehicle itself is a complex dynamical system having important nonlinearities and many degrees of freedom (DOF) serving several low-frequency linear vibration modes which may all be important components of the several types of dynamical behavior at different running speeds and conditions. This part of the problem is somewhat resolved by the appearance of multi-body dynamics and good commercial computer codes. The other part of the problem is related to the wheel-ground contact. One of the most challenging tasks of continuum mechanics is the contact problems which need long computation time even in stationary cases and when the most advanced finite element programs are used

F. C. Moon (ed.),
IUTAM Symposium on New Applications of Nonlinear and Chaotic Dynamics in Mechanics, 373–386.

(see Kalker (1990) for railway wheels and Böhm *et al.* (1989) for tires). The dynamic contact problems need special codes, tremendous computational effort and still, there are no analytical examples and results to check the calculations which are available for a couple of parameter settings only.

In this study, the simplest possible mechanical structures are considered, with the lowest number of mechanical DOF which still exhibit shimmy motion. On the other hand, several wheel models are considered, and these models are analyzed from nonlinear vibration view-point.

The models have 4 important elements: the vehicle, the king pin, the caster, and the wheel. In accordance with the above goals, the vehicle is always modeled as a rigid body running straight with a constant speed v. Clearly, the towed wheel may have a straight stationary running if all the mechanical parts are rigid. Shimmy may occur if at least one of the king pin, the caster, and the wheel is elastic. Two groups of the basic models can be classified this way.

If the wheel is rigid then the king pin (and/or the caster) is considered to be elastic. These non-holonomic models may describe realistic systems at low speed, like the trolleys at supermarkets or airports (see Plaut, 1996). Also, when long semi-trailers are modeled, an elastic king-pin may have an important effect on the dynamics. The often chaotic and/or transient chaotic 'dance' of these wheels can be experienced on these simple trolleys where the king pin is loose (see Stépán, 1991). The control of this motion has been studied recently via feedback linearization by Goodwine *et al.* (1997).

If the wheel is elastic then shimmy may occur even with rigid caster and king pin. These models are realistic for pneumatic tires in the case of short caster length. However, the description of the contact of the elastic wheel and rigid ground is much more complicated than the case of the elastic king pin. For engineering applications in the middle range of the towing speed, the quasi-stationary idea of the lateral creep force and moment helps to simplify this problem and to achieve reasonable agreement with experiments (see Pacejka, 1988). These models use a holonomic mechanical model. As successful studies, the paper of Scheidl *et al.* (1985) can be mentioned here for tractor-semi-trailer systems, the analysis of [Sharp] for motorcycles, and that of Fratile *et al.* (1995) for caravans. Although, the creep force idea can strongly be criticized from nonlinear dynamics point of view, the conclusions of these studies are really useful and important: they give advice how to pack in a caravan not to increase its moment of inertia about its vertical axis, etc. However, the most complex phenomena which can be studied by these models are the different structures of stable and unstable limit cycles only.

In the case of the elastic wheel models of airplane front landing gears, Schlippe (1941) realized in the late thirties already the presence of a past effect in the system. The wheel contact region behaves like a memory in the system, it records the past position and orientation of the shimmying wheel. First, this memory effect was modeled by using the assumption that the contact line is straight. This resulted a delay differential equation with discrete delays. Since the theory of delay differential equations was missing from the mathematical literature at that time, the mathematical analysis suffered of several simplifications.

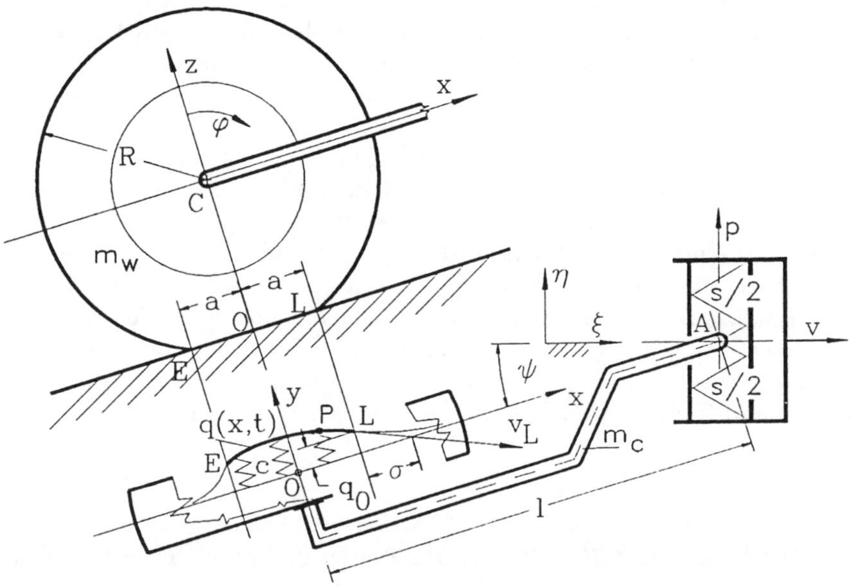

Figure 1: Model of a towed wheel.

In the final section of this paper, those models are established and analyzed which describe this memory effect more precisely, using the contact line deformation as a distributed parameter in the system. The resulting coupled partial and integral differential equation is transformed into a general functional differential equation which describes the memory effect. Stability analysis and nonlinear oscillations are presented based on the existing mathematical theory of these infinite dimensional systems.

2. Mechanical model

The mechanical model shown in Figure 1 has 4 parts. The rigid vehicle runs with a constant velocity **v** in the ξ direction providing a non-stationary geometric constraint for the structure. Note that the system is not conservative due to this constraint. The king pin at the point A has its axis in the vertical direction z, and it is attached to the vehicle via a spring of overall stiffness s. The qualitative effect of the viscous damping of this support will also be mentioned, but it is neglected in the analytical calculations. The rigid caster of mass m_c has a length l. This length and the towing speed v will serve as the two bifurcation parameters. The homogeneous wheel of mass m_w has a radius R. Its central part is assumed to be rigid. The tire is elastic, modeled as a thin stretched string (see Pacejka, 1988). This string is characterized by the relaxation length σ which depends on the tension in the string and the specific lateral stiffness c of the tire. The circumferential elasticity of the tire is neglected. The half-contact-length is denoted by a, and the wheel center has a constant distance $b = \sqrt{R^2 - a^2}$ from the ground.

If the above geometrical constraints are considered only, the following general coordinates are chosen: p is the lateral displacement of the king pin, ψ is the angle of rotation of the caster, φ is that of the wheel, and $q(x)$ is the lateral displacement of the tire contact points. The scalar coordinates p, ψ, φ as well as the function $q(x)$ depend on the time t, of course. The lateral shape q of the deformed tire is described in the moving coordinate system (x, y, z) attached to the caster.

The wheel can roll, slide, or creep. It rolls if all the wheel-ground contact points P have zero velocity, it slides if all of them has non-zero velocity. We speak about creep if some parts of the contact line q stick to the ground, some others slip. The static and dynamic coefficients μ_s, μ_d of friction are used to describe constraints and friction forces between the wheel and the ground.

3. Rigid tire

If the tire has no lateral elasticity, that is $c \to \infty$, the contact region is reduced to a single contact point $P \equiv O$, $q(0,t) \equiv 0$, $a = 0$, $\sigma = 0$. Then the system is non-holonomic with the kinematical constraint $|\mathbf{v}_P| = 0$. This results a system of two first order scalar ordinary differential equations for the general coordinates p, ψ, φ in accordance with the formula

$$\mathbf{v}_P = \begin{pmatrix} v_{Px} \\ v_{Py} \end{pmatrix} = \begin{pmatrix} v\cos\psi + \dot{p}\sin\psi - R\dot{\varphi} \\ -v\sin\psi + \dot{q}\cos\psi - l\dot{\psi} \end{pmatrix} . \tag{1}$$

Thus, the system can be described by a single quasi-velocity which can be the angular velocity $\Omega = \dot{\psi}$ of the caster. The Appell-Gibbs equation coupled to the above definition of the quasi-velocity and the kinematical constraints (1) provides the four dimensional system of first order scalar equations

$$\begin{aligned}
\dot{\psi} &= \Omega\,, \\
\dot{\Omega} &= -\frac{\frac{v}{l}\left(\frac{1}{\cos^2\psi} - \frac{1}{2} + \frac{3m_w}{2m_c}\tan^2\psi\right)\Omega + \frac{s}{lm_c}p + \left(1 + \frac{3m_w}{2m_c}\right)\frac{\tan\psi}{\cos\psi}\Omega^2}{\left(\frac{1}{3} + \tan^2\psi\right)\cos\psi + \frac{m_w}{4m_c}\left(\frac{R^2}{l^2}\cos\psi + 6\tan^2\psi\cos\psi\right)}\,, \\
\dot{p} &= v\tan\psi + \frac{\Omega l}{\cos\psi}\,, \\
\dot{\varphi} &= \frac{v + \Omega l\sin\psi}{R\cos\psi}\,.
\end{aligned} \tag{2}$$

Since φ is a cyclic coordinate, it does not appear in the first 3 equations, and the trajectories of the rolling system can uniquely be represented in the three dimensional subspace $\psi, \dot{\psi}, p$ of the caster angle and angular velocity and the king pin lateral displacement.

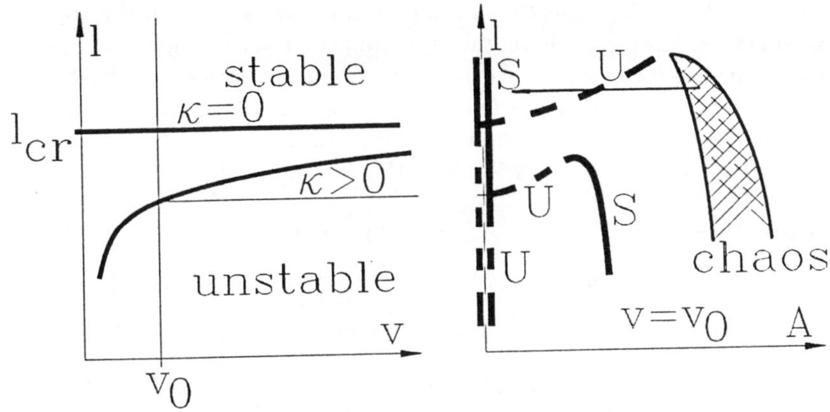

Figure 2: Stability charts and bifurcation diagrams.

The linear stability analysis of the straight rolling (the trivial solution of (2)) gives a critical value of the caster length:

$$l > l_{cr} = \sqrt{\frac{3m_w}{2m_c}} R \,. \tag{3}$$

Figure 2 shows the corresponding stability chart. The stability limit at zero viscous damping is represented by the line at $\kappa = 0$. In this case, the speed of towing does not appear in the stability condition. However, it does appear when $\kappa > 0$. The corresponding stability limit is the result of a lengthy calculation when the viscous damping is included in the mathematical model. Thus, short caster length and increasing speed destabilizes the towed wheel.
At the critical value of the caster length, the Hopf bifurcation calculation (see details in Stépán, 1991) based on the third degree truncation of (2) presents a subcritical case, that is there exists an unstable limit cycle around the asymptotically stable zero solution for some values of $l - l_{cr} > 0$. This is also presented by the bifurcation diagram of Figure 2, where A is the amplitude of the oscillation.

The numerical simulation of (2) does not show any attractor outside this limit cycle. In fact, the trajectories run to some singular surfaces at $\psi = \pm\pi/2$. Clearly, the real structure will not roll anymore when the vibrations are so violent. The wheel is going to slide when the static friction is not able to provide the necessary constraining force for rolling. Consequently, the above equations (2) of rolling describe the system only for that region of the phase space where the friction force F at the contact point P satisfies the condition

$$F \leq \mu_s N = \mu_s (m_c/2 + m_w) g \,. \tag{4}$$

This condition can be represented in the phase space as codimension 1 surfaces.

Within these surfaces the trajectories are described by (2). When the trajectories hit these surfaces, they switch to the dynamics of the sliding wheel. The corresponding equations of motion are the Lagrangian equations of the holonomic system where the kinematical constraint $|\mathbf{v}_P| = 0$ is released. The 3 dimensional system of the second order equations in the 6 dimensional phase space of the general coordinates and velocities assumes the form:

$$\begin{pmatrix} \left(\frac{1}{3}m_c + \left(1+\frac{R^2}{4l^2}\right)m_w\right)l^2 & -\left(\frac{1}{2}m_c + m_w\right)l\cos\psi & 0 \\ -\left(\frac{1}{2}m_c + m_w\right)l\cos\psi & m_c + m_w & 0 \\ 0 & 0 & \frac{1}{2}m_w R^2 \end{pmatrix} \begin{pmatrix} \ddot{\psi} \\ \ddot{p} \\ \ddot{\varphi} \end{pmatrix} =$$

$$\begin{pmatrix} \frac{v_{Px}}{\sqrt{v_{Px}^2+v_{Py}^2}}l \\ -\left(\frac{1}{2}m_c + m_w\right)l\dot{\psi}^2\sin\psi - sp - \frac{v_{Px}\sin\psi + v_{Py}\cos\psi}{\sqrt{v_{Px}^2+v_{Py}^2}} \\ \frac{v_{Py}}{\sqrt{v_{Px}^2+v_{Py}^2}}R \end{pmatrix} \mu_d(m_c/2 + m_w)g \qquad (5)$$

where the coordinates of the contact point velocity $v_{Px,Py}$ has to be substituted from (1).

The numerical simulation of the equations (2) and (5) and the precise check of the changes between the two dynamics are difficult. The general methods of Glocker *et al.* (1995) are needed if the system is more complicated and consists of several bodies which may all slide or stick. The above case is somewhat simpler, although orbitally stable periodic motion as well as chaotic or transient chaotic motion can also be experienced outside the unstable limit cycle for different values of the parameters (see the bifurcation diagrams in Figure 2).

Figure 3 explains a case when transient chaos occurs (for simulation results and parameter values see Stépán, 1991). The plane $\psi, \dot{\psi}$ represents the 4 dimensional phase space of rolling. The thick trajectories start running outwards around the unstable limit cycle presented by a thin ellipse in the figure. When the trajectory hits the condition of rolling calculated from (4) and approximated by two planes here at $\pm\dot{\psi}_{sl}$, the wheel starts sliding. The speed $|\mathbf{v}_P|$ of the contact point represents the next 2 dimension in the phase space of Figure 3. When the thick trajectory changes to a thin one and leaves the space of rolling, it follows the dynamics of (5) in the 6 dimensional phase space of sliding. This dynamics is strongly dissipative due to the sliding friction.

The trajectories enter the space of rolling several times. When they enter it within the conditions of rolling, and 'outside' the unstable limit cycle, they start spiraling outwards again, and the wheel shows a chaotic dance. This may disappear after a stochastically varying time of transient period when the trajectory arrives back to the space of rolling within the direct domain of attraction of the zero solution, and the straight running of the wheel is approached asymptotically. The above structure of the trajectories can also be followed by an approximate 1 dimensional Poincare section of the trajectories considered along the ψ axis. This

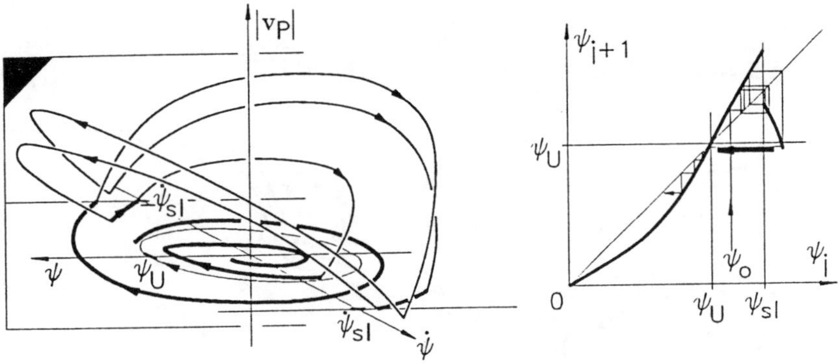

Figure 3: Phase space structure and discrete map.

cartoon of the transient chaotic wheel motion is presented on the discrete map of Figure 3.

Note, that the above described transient chaotic motion occurs for certain parameter values only, when the static friction is about 2 times greater than the dynamic one. In other cases chaotic motion or just simpler periodic motions may coexist with the asymptotically stable straight running and they are 'separated' by the unstable periodic motion.

4. Elastic tire and creep force

If the tire is elastic, there is a contact region of length $2a$ between the wheel and the ground (see Figure 4). In the meantime, the king pin is assumed to be rigid, that is $s \to \infty$. It can be shown that there is always a region at the rear end E of the contact line where the contact points slide, although the contact points at the other part of the wheel may still stick to the ground. This phenomenon is called creep.

In case of a constant (!) drift angle, the calculation of the resultant force and couple of the distributed lateral (partly dynamic and partly static) friction forces is presented by Pacejka (1988). Since the drift angle ψ is fixed in this calculation, $\mathbf{v}_L = \mathbf{v}$, the shape of the contact line can analytically be determined, and it depends on the lateral deformation $q_0 = q(a,.)$ of the leading edge L only. Thus, the lateral creep force F_y and moment M_{Oz} can be presented in the form

$$F_y(q_0) = \begin{cases} 22.2\frac{ca^2}{\sigma}q_0\,, & |q_0| < q_{0sl} = 0.045\frac{\sigma}{ca^2}\mu_d N \\ \mu_d N \,\mathrm{sign}\, q_0\,, & |q_0| \geq q_{0sl} \end{cases}, \tag{6}$$

$$M_{Oz}(q_0) = \begin{cases} 0.16\mu_d N a \sin\left(69.8\frac{ca^2}{\sigma}\frac{1}{\mu_d N}q_0\right), & |q_0| < q_{0sl} \\ 0\,, & |q_0| \geq q_{0sl} \end{cases}, \tag{7}$$

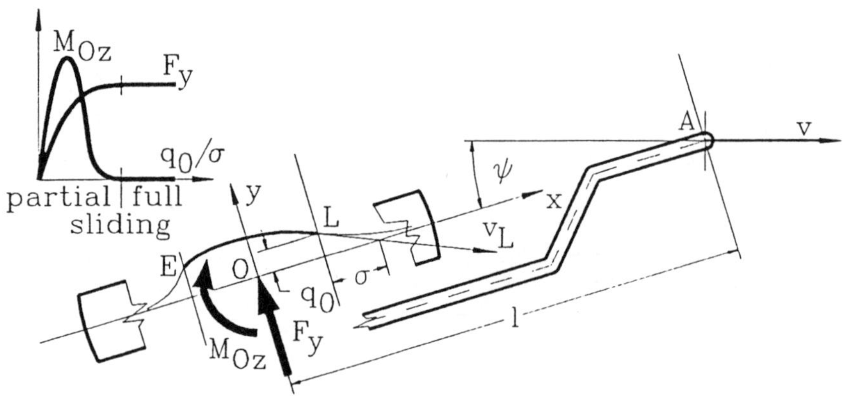

Figure 4: Creep force and moment.

where $\mu_d N = \mu_d(m_c/2 + m_w)g$ is the overall friction force when the wheel fully slides, that is when $|q_O| > q_{Osl}$.

Now, this stationary creep force and moment is used instead of the real kinematic constraint along the contact line to describe the dynamics of the towed wheel. In the three dimensional phase space $\psi, \dot{\psi}, q_O$ of the caster angle, angular velocity and the lateral displacement of the leading edge L, the equations of motion assume the simple form

$$I_A \ddot{\psi} = -l F_y(q_O) - M_{Oz}(q_O)\,, \tag{8}$$

$$\dot{q}_O = v \sin\psi + (l - a)\dot{\psi} - \frac{v}{\sigma} q_O \cos\psi\,, \tag{9}$$

where I_A is the mass moment of inertia of the wheel and the caster with respect to the z axis at the king pin A. The first equation (8) describes a simple holonomic system, while the second equation (9) still preserves a kinematical constraint with respect to the leading edge L and also a 'no kink at L' condition introduced by Pacejka (1988).

Although the above equations are the results of a complicated simplification of the wheel-ground contact, the linear stability analysis of the zero solution of (8,9) with (6,7) gives a similar condition to that of the case of the rigid tire in (3):

$$l > l_{cr} = a + \sigma\,. \tag{10}$$

Moreover, the Hopf bifurcation analysis of the 3rd degree truncation of (8,9) with (6,7) presents a subcritical case again (for details see Stèpàn, 1992), thus both the stability chart and the bifurcation diagram of Figure 2 are valid for this model, too. However, there is no chaotic attractor or transient chaos outside the unstable limit cycle. An orbitally asymptotically stable limit cycle exists, however, when the viscous damping at the tire-ground contact is also included in the model (see the $\kappa > 0$ case in the bifurcation diagram of Figure 2). In this case, the speed v of towing shows up in the linear stability condition, just as in the case of the rigid tire.

As mentioned in the introduction, the inclusion of a stationary creep force in the dynamic equations simplifies the mathematical model tremendously (compare equations (2,5) to (8,9)), the nonlinearities are partly put into the nonlinear creep force and moment characteristics, but in the meantime, most parts of the complex dynamics are lost. Even some linear stability effects are missing in the above models. For example, experiments show that the stability of the straight rolling is regained sometimes with increasing speed, then lost again, etc. These issues are discussed in the subsequent section.

5. The memory effect of the elastic tire

In this section, the condition of rolling is enforced for the whole contact line between the wheel and the ground. Although it is true for large values of static coefficient μ_s of friction only, and never true in the vicinity of the rear edge E, this model still serves important and new nonlinear oscillation phenomena. The main advantage of this model is, that it takes into account the actual and dynamically varying shape $q(x,t)$ of the contact region.

To simplify the calculations, we restrict ourselves to the case of rigid carcass, and the elasticity of the tread elements are modeled only. This means that the relaxation length is negligible, that is $\sigma = 0$, and the lateral displacement $q(a,t)$ of the leading edge L is also zero as follows from (9).

The equations of motion with respect to the scalar ψ and the function $q(x,.)$ assume the form

$$I_A\ddot{\psi}(t) = -c\int_{-a}^{a}(l-x)q(x,t)dx \tag{11}$$

$$\dot{q}(x,t) = v\sin\psi(t) + (l-x)\dot{\psi}(t) + q'(x,t)\big(v\cos\psi(t) - q(x,t)\dot{\psi}(t)\big)\,, \tag{12}$$

$$x \in [-a,a]\,,\ t \in [t_0,\infty)\,, \quad \text{and} \quad q(a,t) = 0\,. \tag{13}$$

where prime denotes the derivative with respect to the moving coordinate x attached to the caster. In these equations, the partial differential equation (PDE) (12) with the boundary condition (13) describes the kinematical constraint $|\mathbf{v}_P| = 0$ for any point P along the contact line in accordance with Figure 1. This system of integral and partial differential equations describes the dynamics in an infinite dimensional phase space.

As mentioned in the Introduction, the contact line keeps a history of the wheel past positions and orientations while the contact points stick to the ground, and these past values determine the actual lateral force system which appear in the right hand side of the integral-differential equation (11). This can be interpreted mathematically in the following way.

The PDE part (12) of the equations has a traveling wave-like solution. This can be constructed by giving the absolute coordinates (ξ,η) of the contact point

P as

$$\xi(x,t) = vt - (l-x)\cos\psi(t) - q(x,t)\sin\psi(t)\,,$$
$$\eta(x,t) = -(l-x)\sin\psi(t) + q(x,t)\cos\psi(t)\,.$$

Actually, the time derivatives of these equations also lead to the above PDE (12). This time, however, we are going to change the space variable x to the time delay τ which is the time needed for the leading edge L at $x = a$ to travel backwards (relative to the caster) to the actual point P. This means that

$$\xi(x,t) = \xi(a,t-\tau)\,,$$
$$\eta(x,t) = \eta(a,t-\tau)\,.$$

From the above expressions, simple algebra gives

$$l - x = v\tau\cos\psi(t) + (l-a)\cos(\psi(t) - \psi(t-\tau))\,, \tag{14}$$
$$q(x,t) = v\tau\sin\psi(t) + (l-a)\sin(\psi(t) - \psi(t-\tau))\,, \tag{15}$$
$$\frac{\mathrm{d}x}{\mathrm{d}\tau} = -v\cos\psi(t) + (l-a)\dot{\psi}(t-\tau)\sin(\psi(t) - \psi(t-\tau))\,. \tag{16}$$

Note that a linear approximation provides explicit expression for the connection of the delay and space coordinate in the physically obvious form

$$\tau \approx \frac{a-x}{v}\,.$$

Introduce the dimensionless bifurcation parameters L and V for the caster length and the speed of towing by:

$$L = \frac{l}{a}\,, \quad T = \frac{2a}{v}\,, \quad V = \frac{1}{\alpha T}\,,$$

where $\alpha^2 = 2ac(l^2 + a^2/3)/I_A$ is the square of the natural angular frequency of the system when its speed is zero.

With the help of formula (15), $q(x,t)$ can be eliminated from (11), and the variable x of integration is changed to τ via formulae (14,16). The use of the dimensionless time $t := t/T$ and the new dimensionless variable $\vartheta = -\tau/T$ of integration results in the 3rd degree truncated form of a nonlinear delay-differential equation (DDE)

$$V^2\ddot{\psi}(t) + \psi(t) - \frac{7}{6}\psi^3(t) =$$

$$\frac{L-1}{L^2+1/3}\left(\int_{-1}^{0}(L-1-2\vartheta)\psi(t+\vartheta)\mathrm{d}\vartheta + f(\psi_t) + g(\psi_t,\dot{\psi}_t)\right)\,. \tag{17}$$

where the odd fúnctions f and g are calculated as functions of ψ_t defined by

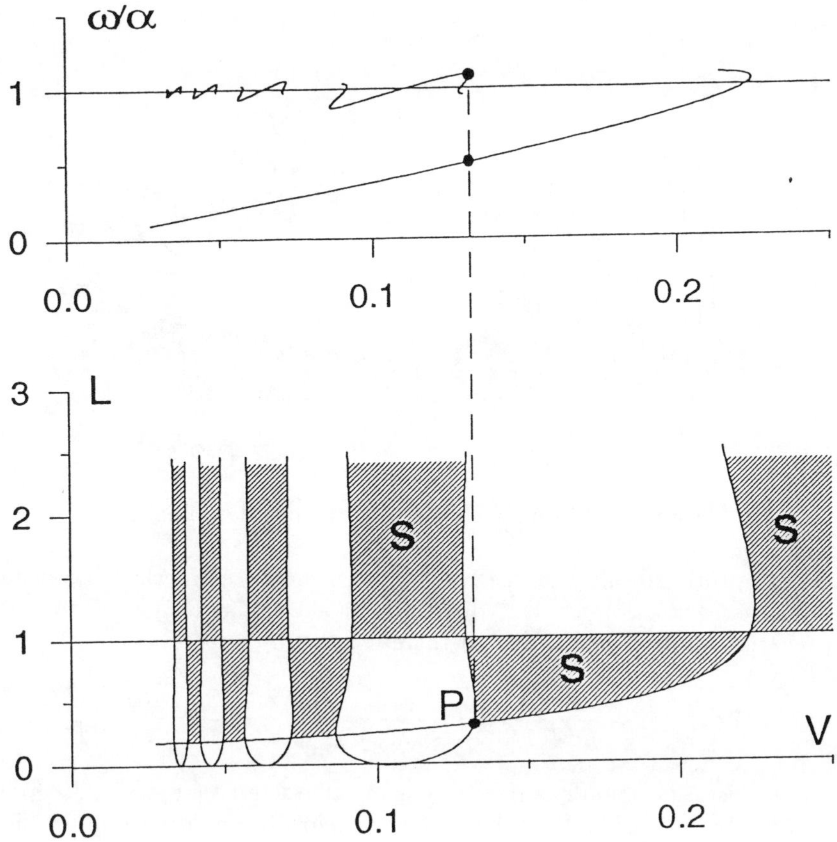

Figure 5: Stability chart and vibration frequencies.

$\psi_t(\vartheta) = \psi(t+\vartheta)\,,\ \vartheta \in [-1,0]$:

$$\begin{aligned}
f(\psi_t) &= -\tfrac{5}{2}\psi^2(t)\int_{-1}^{0}(L-1-2\vartheta)\psi(t+\vartheta)\mathrm{d}\vartheta \\
&\quad +2\psi(t)\int_{-1}^{0}(L-1-\vartheta)\psi^2(t+\vartheta)\mathrm{d}\vartheta \\
&\quad -\tfrac{2}{3}\int_{-1}^{0}(L-1-\tfrac{1}{2}\vartheta)\psi^3(t+\vartheta)\mathrm{d}\vartheta\,, \\
g(\psi_t,\dot\psi_t) &= \tfrac{1}{2}\psi^2(t)\int_{-1}^{0}(L-1-2\vartheta)^2\dot\psi(t+\vartheta)\mathrm{d}\vartheta \\
&\quad -\psi(t)\int_{-1}^{0}(L-1-\vartheta)(L-1-2\vartheta)\psi(t+\vartheta)\dot\psi(t+\vartheta)\mathrm{d}\vartheta \\
&\quad +\tfrac{1}{2}(L-1)\int_{-1}^{0}(L-1-2\vartheta)\psi^2(t+\vartheta)\dot\psi(t+\vartheta)\mathrm{d}\vartheta\,.
\end{aligned}$$

Its only unknown is the scalar ψ, but due to the presence of the integration of its past values on $[t-1,t]$, the phase space is still infinite dimensional: it is the space of the continuous functions ψ_t on $[-1,0]$. For more details of the theory see e.g. Kuang (1993).

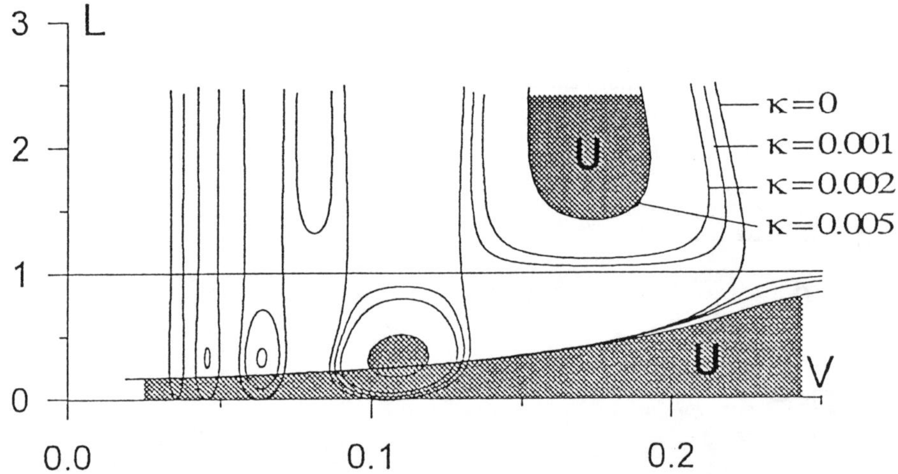

Figure 6: Stability charts with viscous damping.

The asymptotic stability of the zero solution of (17) is checked by means of the characteristic function associated with the linear part of (17) when the $\psi(t) = K\exp(\lambda t)$ trial solution is substituted there:

$$D(\lambda) = V^2\lambda^2 + 1 - \frac{L-1}{L^2+1/3}\left(L\frac{1-e^{-\lambda}}{\lambda} - \frac{1+e^{-\lambda}}{\lambda} + 2\frac{1-e^{-\lambda}}{\lambda^2}\right) \qquad (18)$$

This characteristic function has infinitely many characteristic roots in accordance with the infinite dimensional nature of the problem. The necessary and sufficient condition of asymptotic stability in the linear systems is the same as in the case of ODEs, that is $\mathrm{Re}\lambda_j < 0,\ j = 1, 2, \ldots$.

The method presented in Theorem 2.19 in Stépán (1989) gives the stability chart of Figure 5 where a complex structure of stability domains appear. The stability limit at $L = 1 \Leftrightarrow l = a$ also appears just as in the case of the creep force model in (10) with the actual assumption $\sigma = 0$, but there are also many alternating borders of stability with Hopf bifurcations along them. The corresponding vibration frequencies are also presented above the stability chart.

This model presents several (infinitely many) bifurcation points as the running speed v is decreased. However, if viscous damping is added to the model at the tire, the number of these bifurcation points become finite, at great values of the damping only a couple of them survives. This is shown in Figure 6 where the $\kappa > 0$ stability limits are the result of a long calculation not discussed here.

There is also another important nonlinear oscillation phenomenon which appears in this delay model of the tire. At the parameter values V=0.1319, L=0.3049 of the point P in the stability chart of Figure 5, quasiperiodic oscillations appear with 2 vibration frequencies: one is somewhat more than the natural frequency α of the structure at $v = 0$, the other one is at about the half of it.

The codimension 1 and 2 Hopf bifurcation analysis requires a long and tedious algebraic calculation using the nonlinear functions f, g in (17), and it serves the structure of stable and unstable limit cycles and tori embedded in a 2 or 4 dimensional center manifold of the infinite dimensional phase space. Actually, the calculations are very similar to those of Campbell (1995) and Stépán (1995) for models in population dynamics and robotics.

At this point we refer only to a simple and useful application of Theorem 3.24 in Stépán (1989) or in Stépán (1997). The zero solution of the delay-differential equation

$$\ddot{x}(t) + c_0 x(t) = c_1 \int_{-1}^{0} x(t+\theta)d\theta \tag{19}$$

is asymptotically stable if and only if

$$0 < c_1 < c_0 \quad \text{and} \quad c_0 \neq 4j^2\pi^2 \;\; j = 1, 2, \ldots . \tag{20}$$

The linear part of the delay model (17) is the same as (19) when the caster length $L \to \infty$. In this case $c_0 = c_1 = 1/V^2$. Clearly, the first stability condition of (20) is just violated: there exist a $\lambda = 0$ characteristic root in this case. The meaning of this root is obvious from the physical model, since the position of the wheel in η direction is neutral in the case of infinitely long caster. The second conditions of (20) give those critical speeds where self-excited vibrations are likely to occur at finite caster length which are still greater than a:

$$V = \frac{1}{2\pi}(\approx 0.159), \frac{1}{4\pi}, \frac{1}{6\pi}, \ldots .$$

This means that the stability limits of Figure 5 tend to these values as $L \to \infty$, and the corresponding unstable regions shrink and then disappear, since the shimmying wheel is always stable for very long caster apart of some critical speeds. At these critical speeds one condition of the Hopf Bifurcation Theorem is not satisfied, namely the critical characteristic roots do not cross the imaginary axis with nonzero derivative with respect to the bifurcation parameter.

5. Conclusions

Three basic models of the shimmying wheels have been considered and analyzed here from nonlinear oscillations point of view. The model of rigid tires may exhibit chaotic and transient chaotic motion. The creep force models describe the partially sliding and rolling elastic tire with the help of a stationary tire model fitted in the dynamic model. This model gives some simple and useful results for engineering design, but does not present most of the complex behavior of the shimmying wheels. The more precise modeling of the dynamic contact problem of the tire and the ground may include the memory effect of the tire. This effect appeared in the early models of the shimmying wheel (see Schlippe, 1941). These models must be reconsidered again, and the recent developments of mathematics both in bifurcation theory and in functional analysis of infinite dimensional systems provide much better understanding of the complex motion of the elastic wheels of motorcycles, cars, trucks, airplanes.

Acknowledgments

This research was partially supported by the Danish Council for Scientific and Industrial Research Grant No. 16-4786.M, the Hungarian Scientific Research Foundation OTKA Grant No. 11524, the US-Hungarian Science and Technology Program Grant No. 336, and the Greek-Hungarian Technology Program. The author thanks Professor Hans Pacejka and Professor Hans True for the useful discussions.

References

Böhm, F., Kollatz, M., (1989) *Some Theoretical Models for Computation of Tire Nonuniformities*, Fortschrittberichte VDI, **12.124**, VDI Verlag, Berlin.

Campbell, S. A., Bélair, J., Ohira, T., and Milton, J. (1995) Complex dynamics and multistability in a damped oscillator with delayed negative feedback, *Journal of Dynamics and Differential Equations* **7**, 213-236.

Fratila, D., Darling, J. (1995) Improvements to car/caravan handling and high speed stability through computer simulation, in *Proceedings of the ASME International Mechanical Engineering Congress* (San Francisco, 1995) **95-WA/MET-9** pp. 1-15.

Glocker, Ch., Pfeiffer, F. (1995) Multi impacts with friction in rigid multibody systems, Nonlinear Dynamics 7 (1995) 471-497.

Goodwine, B., Stèpàn, G. (1997) Controlling unstable rolling phenomena, *Journal of Vibration and Control*, accepted for publication.

Guckenheimer, J. and Holmes, P. (1986) *Nonlinear Oscillations, Dynamical Systems, and Bifurcations of Vector Fields*, Springer Verlag, New York.

Kuang, Y. (1993) *Delay Differential Equations*, Academic Press, Boston.

Kalker, J. J. (1990) Wheel-rail rolling contact, *Wear* **144**, 243-261.

Pacejka, H. (1988) *Modelling of the pneumatic tyre and its impact on vehicle dynamic behaviour*, Research report **i72B**, TU Delft.

Plaut, R. H. (1996) Rocking instability of a pulled suitcase with two wheels, *Acta Mechanica* **117**, 165-179.

Scheidl, R., Stribersky, A., Troger, H., Zeman, K. (1985) Nonlinear stability behaviour of a tracktor-semitrailer in downhill motion, in *Proceedings of the 9th IAVSD Symposium* (Linköping, 1985) pp. 509-522.

Schlippe, B. v., Dietrich, R. (1941) Shimmying of a pneumatic wheel, *Lilienthal-Gesellschaft für Luftfahrtforschung*, Bericht **140**, translated for the AAF in 1947 by Meyer & Company, pp. 125-160.

Sharp, R. S., Jones, C. J. (1980) A comparison of tyre represetations in a simple wheel shimmy problem, *Vehicle System Dynamics* **9**, 45-57.

Stépán, G. (1989) *Retarded Dynamical Systems*, Longman, London.

Stépán, G. (1991) Chaotic motion of wheels, *Vehicle System Dynamics* **20**, 341-351.

Stépán, G. (1992) Nonlinear modelling of shimmying wheels, in P. Christiansen (ed.), *Future Directions of Nonlinear Dynamics in Physical and Biological Systems*, NATO ASI series, Plenum, New York, pp. 169-172.

Stépán, G., Haller, G. (1995) Quasiperiodic oscillations in robot dynamics, *Nonlinear Dynamics* **8**, 513-528.

Stépán, G. (1997) Delay-differential equation models for machine tool chatter, F. C. Moon (ed.), *Nonlinear Dynamics of Material Precessing and Manufacturing*, Wiley, N.Y.

SYMMETRY, GENERIC BIFURCATIONS, AND MODE INTERACTION IN NONLINEAR RAILWAY DYNAMICS

CARSTEN NORDSTRØM JENSEN
ph.d. student
ES-Consult Ltd, Staktoften 20, DK-2950 Vedbæk, Denmark, e-mail: escons@inet.uni-c.dk

MARTIN GOLUBITSKY
professor
University of Houston, Department of Mathematics, TX 77204-3476, USA, e-mail: mg@uh.edu

AND

HANS TRUE
professor
The Technical University of Denmark, Department of Mathematical Modelling, Bldg 321, DK-2800 Lyngby, Denmark, e-mail: ht@imm.dtu.dk

1. Introduction

In this article we investigate some dynamical features of the Cooperrider bogie. A modern railway passenger car has a car body supported at each end by a carriage or *bogie* with a relatively short wheel base. The suspension systems are placed in the bogies and between the bogies and the car body. Since the guiding forces from the rails act on the wheelsets in the bogies, the bogies play an important role in the dynamics of the vehicle motion. The mathematical model is presented at the end of this section.

In Section 2 we outline the mathematical theory related to our investigations of this model and we use the symmetry of the model to deduce the form that generic bifurcations from symmetric periodic solutions will have. In contrast to systems without symmetry, period-doubling bifurcations are not generic, but symmetry breaking pitchfork bifurcations are. This remark was noted previously by Swift and Wiesenfeld and by Fiedler. If two critical eigenvalues are close in the control parameter space, different types of secondary bifurcations may occur. This type of degeneracy is called *mode interaction*. In our model equations we find a mode interaction between a saddle-node and a pitchfork bifurcation. This mode

F. C. Moon (ed.),
IUTAM Symposium on New Applications of Nonlinear and Chaotic Dynamics in Mechanics, 387–396.

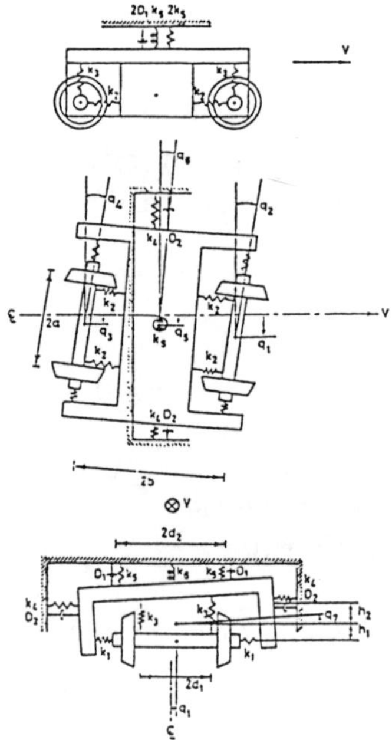

Figure 1. Conventional Bogie Model.

interaction was studied by Dangelmayr and Armbruster and explains a number of features in the bifurcation diagram of the Cooperrider model.

The results of our numerical investigations are presented in Section 3. The mathematical theory verifies the numerical results. The stringent mathematical theory suggested a generic interpretation of the numerical results and led to the discovery of the correct splitting and the correct sequence of the bifurcations of the periodic attractor. It also helped the authors to look for and find a solution that had not been found before due to its instability and/or small basin of attraction. Section 4 contains the conclusions.

1.1. DYNAMICAL MODEL

The Cooperrider model was developed as a model for a conventional passenger car bogie with two axles. We assume that all parts except the suspension elements are rigid and that the suspension elements all have linear characteristics. Furthermore, we assume that the vertical displacements are so small that the equations for the vertical and horizontal motions are uncoupled and consider only the lateral motion.

The model of the conventional bogie is a multibody system. The bogie frame can rotate without friction in a bearing in the floor of the car body. It is supported on two wheelsets, through springs and dampers as shown in Figure 1. The bogie model has seven degrees of freedom: lateral and yaw motion for each wheelset and the bogie frame, and roll motion of the bogie frame. In a coordinate system moving with constant speed v along the track center line, the variables are denoted

$q_1, \ldots, q_7$, see Figure 1. The speed v is chosen as the control parameter, and all other parameter values are kept constant in these investigations.

The bogie runs on a straight, horizontal, perfect track. The profile of the rail surface is an arc of a circle, and the wheels have a conical profile, with inner flange. The nonlinearities in the system stem from the creep-creep force relation in the ideal contact point between each wheel and the rail and from the flange force.

The Vermeulen-Johnson creep force law relates the resulting creep force $F_R(\xi_R)$ to the resulting creep $\xi_R = \sqrt{\left(\frac{\xi_x}{\Psi_1}\right)^2 + \left(\frac{\xi_y}{\Phi}\right)^2}$, where $\xi_{xf} = \frac{\dot{q}_1}{v} - q_2$ and $\xi_{xr} = \frac{\dot{q}_3}{v} - q_4$ are the front and rear lateral creepages, and $\xi_{yf} = \frac{a\dot{q}_2}{v} + \frac{\delta q_1}{r_0}$ and $\xi_{yr} = \frac{a\dot{q}_4}{v} + \frac{\delta q_3}{r_0}$ are the front and rear longitudinal creepages. In this model $a = 0.716m$ is half the track gauge, $\delta = 0.05$ is the contact angle, and $r_0 = 0.4572m$ is the centered rolling radius of the wheel. Hertz theory is used to calculate the contact area between a wheel and the rail with the coefficient of adhesion $\mu = 0.15$ and the normal force N given by $\mu N = 10kN$. The constant G is the shear modulus and with a_e and b_e as the semiaxes of the contact ellipse, $G\pi a_e b_e = 6.563MN$. Setting $u = \frac{G\pi a_e b_e}{\mu N}\xi_R$ it follows that

$$\frac{F_R}{\mu N} = \begin{cases} u - \frac{1}{3}u|u| + \frac{1}{27}u^3 & u < 3 \\ 1 & u \geq 3 \end{cases}$$

which defines F_R. Vermeulen and Johnson show that the lateral and longitudinal creep forces are $F_x = \frac{\xi_x}{\Psi_1}\frac{F_R}{\xi_R}$ and $F_y = \frac{\xi_y}{\Phi}\frac{F_R}{\xi_R}$, where $\Phi = 0.60252$ and $\Psi_1 = 0.54219$ are weight factors. The flange force F_T is modelled as a stiff nonlinear spring

$$F_T(u) = \begin{cases} k_0(u - \eta) & \eta < u \\ 0 & -\eta \leq u \leq \eta \\ k_0(u + \eta) & -\eta > u \end{cases}$$

where we have used $k_0 = 14.60MN$, and $\eta = 0.0091m$.

The equations of motion for the system give seven coupled nonlinear second-order differential equations:

$$\begin{array}{lllllllllll}
m_w\ddot{q}_1 & +A_1 & & & & & & & +2F_{xf} & +F_T(q_1) & = 0 \\
I_{wy}\ddot{q}_2 & & & +A_3 & & & & & +2aF_{yf} & + & = 0 \\
m_w\ddot{q}_3 & & +A_2 & & & & & & +2F_{xr} & +F_T(q_3) & = 0 \\
I_{wy}\ddot{q}_4 & & & & +A_4 & & & & +2aF_{yr} & + & = 0 \\
m_f\ddot{q}_5 & -A_1 & -A_2 & & & +A_5 & & & & & = 0 \\
I_{fy}\ddot{q}_6 & -bA_1 & +A_2 & -A_3 & -A_4 & & +A_6 & & & & = 0 \\
I_{fr}\ddot{q}_7 & -h_1A_1 & -h_1A_2 & & & +h_2A_5 & & +A_7 & & & = 0
\end{array}$$

where $A_1 = 2k_1(q_1 - q_5 - bq_6 - h_1q_7)$, $A_2 = 2k_1(q_3 - q_5 - bq_6 - h_1q_7)$, $A_3 = 2k_2d_1^2(q_2 - q_6)$, $A_4 = 2k_2d_1^2(q_4 - q_6)$, $A_5 = 2D_2(\dot{q}_5 - h_2\dot{q}_7) + 2k_4(q_5 - h_2q_7)$, $A_6 = k_6q_6$ and $A_7 = 2D_1d_2^2\dot{q}_7 + 2k_5d_2^2q_7 + 4k_3d_1^2q_7$. The front and rear, lateral and longitudinal creep forces resulting from the creepage between rails and wheels are F_{xf},F_{yf} and F_{xr},F_{yr}. The mass and moment of inertia of the axles are $m_w = 1022kg$ and $I_{wy} = 678kgm^{-2}$. The mass and moment of inertia (yaw direction) of the bogie frame are $m_f = 2918kg$ and $I_{fy} = 6780kgm^{-2}$ while the moment of inertia in the roll direction is $I_{fr} = 6780kgm^{-2}$. Other spring and damper parameters are

$k_1 = k_4 = 1.823MN/m$, $k_2 = k_3 = 3.646MN/m$, $k_5 = 0.3333MN/m$, $k_6 = 2.710MN \cdot m$, $D_1 = 20.0\ kNs/m$ and $D_2 = 29.2kNs/m$ (see Figure 1). The remaining constants $b = 1.074m$, $h_1 = 0.0762m$, $h_2 = 0.6584m$, $d_1 = 0.620m$ and $d_2 = 0.680m$ are geometrical quantities (see Figure 1).

For $n = 1, \ldots, 7$ we define $x_{2n-1} = q_n$ and $x_{2n} = \dot{q}_n$ and obtain an autonomous system of 14 coupled first-order differential equations with the speed v as control parameter. Abstractly, this system is:

$$\dot{x} = F(x, v), \tag{1}$$

where $x \in \mathbb{R}^{14}$, $v \in \mathbb{R}^+$ and $F : \mathbb{R}^{14} \times \mathbb{R}^+ \to \mathbb{R}^{14}$.

A 14×14 matrix γ is a symmetry of this system of differential equations if $F(\gamma x, v) = \gamma F(x, v)$ for all $x \in \mathbb{R}^{14}$ and $v \in \mathbb{R}^+$. The set of all symmetries of F is a group, which we denote by Γ.

Let I_{14} be the 14×14 identity matrix. It is easy to verify that $-I_{14}$ is a symmetry of F, since all of the terms in F are odd in x. It follows that $\Gamma = \{I_{14}, -I_{14}\} \cong \mathbb{Z}_2$ is a symmetry group for F. It is also true that Γ is the only nontrivial symmetry group of F, and F is said to be $\mathbb{Z}_2$-symmetric .

2. Theory

In this section we present the mathematical background for the numerical results discussed in the next section. In Subsection 2.1 we deduce the generic bifurcations of a system that is slightly more general than (1), and in Subsection 2.2 we discuss mode interactions in our system.

2.1. GENERIC BIFURCATIONS OF SYMMETRIC PERIODIC SOLUTIONS

We consider a k-parameter family of n-dimensional systems of first order differential equations

$$\dot{x} = F(x, \mu) \tag{2}$$

where $x \in \mathbb{R}^n$, $\mu \in \mathbb{R}^k$ and $F : \mathbb{R}^n \times \mathbb{R}^k \to \mathbb{R}^n$ has symmetry group $\Gamma = \mathbb{Z}_2 = \{\pm I_n\}$. For convenience we set $\gamma = -I_n$.

We study the generic bifurcations of symmetric periodic solutions of (2). Let $c(t)$ be a periodic solution of F with period T, $c(t+T) = c(t)$. The periodic solution c is *symmetric* with respect to γ when

$$c(t + \frac{T}{2}) = \gamma c(t). \tag{3}$$

Let Σ_1 be a transverse section to the periodic orbit c, and let $P : \Sigma_1 \to \Sigma_1$ be the Poincaré map. See Figure 2. We now discuss the restrictions placed on P by the symmetry γ.

Let Σ_2 be the section given by $\Sigma_2 = \gamma\Sigma_1$, and define the mappings given by the flow: $\Theta : \Sigma_1 \to \Sigma_2$ and $\Psi : \Sigma_2 \to \Sigma_1$. Let z be a point in the section Σ_2. From the symmetry we have

$$\Psi(z) = \gamma\Theta(\gamma z), \tag{4}$$

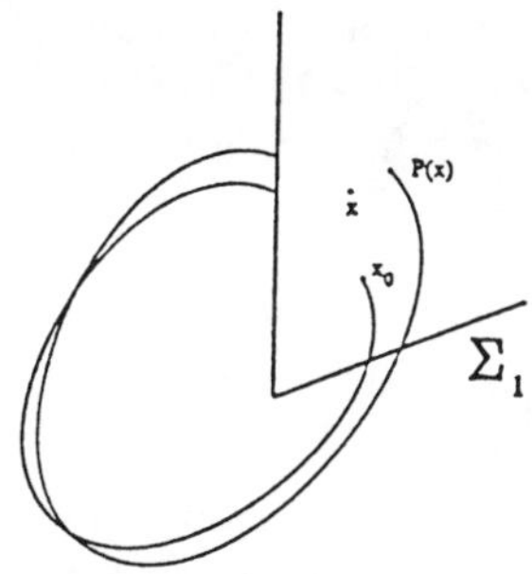

Figure 2. Poincaré Section of the Flow.

which for a given point x in the section Σ_1 yields the relation (5):

$$P(x) = \Psi \circ \Theta(x) = \Psi(\Theta(x)) = \gamma\Theta(\gamma\Theta(x)) = (\gamma\Theta)^2(x). \tag{5}$$

We define the map $Q : \Sigma_1 \to \Sigma_1$ by $Q = \gamma\Theta$ and obtain

$$P = (\gamma\Theta)^2 = Q^2. \tag{6}$$

Let x_0 be the intersection of the symmetric periodic solution c with the section Σ_1 and let V be an $(n-1)$-dimensional subspace of $\mathbb{R}^n$ that is transverse to $\dot{c}(t_0)$ where $c(t_0) = x_0$. Write $Q : (x_0 + V) \times \mathbb{R}^k \to (x_0 + V)$, and change coordinates to obtain the map $Q : V \times \mathbb{R}^k \to V$. By considering generic bifurcations of the map Q we deduce the generic bifurcations of the Poincaré map P and thereby the generic bifurcations from the symmetric periodic solution c.

Note that symmetric periodic solutions of F correspond to fixed points of Q. For suppose that z is the intersection of a symmetric periodic solution of F and the section Σ_1. Symmetry implies that

$$Q(z) = \gamma\Theta(z) = z. \tag{7}$$

Thus, a symmetric periodic solution of F corresponds to a fixed point of Q. Next let y be the intersection of a asymmetric periodic solution of F and the section Σ_1. Since the solution is asymmetric and periodic we have

$$Q(y) = \gamma\Theta(y) \neq y \quad \text{and} \quad Q^2(y) = P(y) = y.$$

Thus an asymmetric periodic solution of F is a period two point of the map Q.

The generic bifurcations of Q and their effect on P are:

- A saddle-node bifurcation of the map Q leads to a saddle-node for the Poincaré map $P = Q^2$ and, therefore, also a saddle-node bifurcation of the periodic solution of (2).
- A Hopf bifurcation of the map Q corresponds (generically) to a Hopf bifurcation of the periodic solution of F leading to either stable or unstable quasiperiodic motion in the vicinity.

- A period-doubling bifurcation of the map Q leads to a pitchfork bifurcation of the Poincaré map $P = Q^2$ and, thereby, a pitchfork bifurcation of the flow or a symmetry breaking bifurcation.

Note that period-doubling bifurcations of symmetric periodic solutions are *not* generic.

2.2. MODE INTERACTION

The eigenspaces associated with the Jacobian matrix are often called *modes*. When varying one control parameter we expect to have only one critical mode at a time. By varying another parameter, multiple critical modes are possible. Near parameter values at which there are multiple critical modes different types of secondary bifurcations may occur. These secondary solutions are created by nonlinear interactions of the two modes and are called *mode interactions*.

Suppose that a saddle-node point and a period-doubling point for the map Q occur simultaneously at the origin (by varying two parameters at a time). To study this mode interaction and its unfolding, we use a method for analyzing period doubled states that was originally developed by Vanderbauwhede . Define the mapping $R : V \times V \times \mathbb{R}^k \to V \times V$ given by

$$R(\sigma, \rho, \mu) = (Q(\sigma, \mu) - \rho, Q(\rho, \mu) - \sigma) \tag{8}$$

for all $\sigma, \rho \in V$ and $\mu \in \mathbb{R}^k$.

Note that zeroes of R correspond to either fixed points or period two points of Q. Thus, it is possible to study the bifurcations of fixed points and period two points of Q by studying the bifurcation of zeroes of R, and the bifurcation of zeroes has been well studied.

To look for bifurcation of zeroes of R, we need to find zero eigenvalues of the linearization L of the map R evaluated at the origin. Assume that v is an eigenvector of the linearization of Q corresponding to the critical eigenvalue $+1$ (the saddle-node) and that w is an eigenvector of the linearization of Q corresponding to the critical eigenvalue -1 (the period-doubling). We find $\ker L = \operatorname{span}\left(\begin{pmatrix} v \\ v \end{pmatrix}, \begin{pmatrix} w \\ -w \end{pmatrix}\right)$ and a Lyapunov-Schmidt reduction leads to a dynamical system in a two dimensional space that is tangential to $\ker L$ at the origin. Now we determine the characteristics of such a system. Let κ act on V^2 by

$$\kappa(\sigma, \rho) = (\rho, \sigma)$$

for all $\sigma, \rho \in V$. Note that κ is a symmetry of R since

$$\begin{aligned} \kappa R(\sigma, \rho, \mu) &= \kappa(Q(\sigma, \mu) - \rho, Q(\rho, \mu) - \sigma) \\ &= (Q(\rho, \mu) - \sigma, Q(\sigma, \mu) - \rho) \\ &= R(\rho, \sigma, \mu) \qquad\qquad = R\kappa(\sigma, \rho, \mu). \end{aligned}$$

Observe that κ acts on (v, v) and $(w, -w)$ by

$$\begin{aligned} &\kappa(v, v) = (v, v) \\ &\kappa(w, -w) = (-w, w) = -(w, -w). \end{aligned}$$

Thus we can determine the action of κ on $\ker L$. We write the vectors in $\ker L$ as

$$x(v,v) + y(w,-w);$$

we can then identify $\ker L \cong \mathbb{R}^2$ using the coordinates $(x,y) \in \mathbb{R}^2$. Since

$$\kappa(x(v,v) + y(w,-w)) = x(v,v) - y(w,-w),$$

the action of κ on $\mathbb{R}^2 \cong \ker L$ is given by

$$\kappa(x,y) = (x,-y).$$

Since the Lyapunov-Schmidt reduction respects symmetry, we find that the zeroes of R are parametrized by the zeroes of a map $T : \mathbb{R}^2 \times \mathbb{R}^k \to \mathbb{R}^2$ that satisfies $T(\kappa(x,y),\mu) = \kappa T(x,y,\mu)$. Thus we have a system in two state variables with $\mathbb{Z}_2$-symmetry, $(x,y) \to (x,-y)$. The bifurcations in this case have been studied by Dangelmayr and Armbruster and they are described in Golubitsky et al., Chapter XIX, §§2-3. In the next section, we present an example of such a mode interaction that is also found in our bogie model.

3. Results

In this section we discuss those results from our numerical investigations of the mathematical model of the railway bogie presented in Section 1 that can be analysed by the mathematical tools presented in Section 2. Other interesting features of the mathematical model have been presented elsewhere by Jensen and True. The main numerical tool for the analysis of this model is the continuation routine PATH developed by Kaas-Petersen.

It is easily seen that the fixed point $x_1 = \ldots = x_{14} = 0$ is an equilibrium solution for all v. For low speeds the solution is asymptotically stable, but at $v = 65.2m/s$ the solution loses stability in a Hopf bifurcation. The bifurcating periodic solution is symmetric with respect to the symmetry group presented in Section 1. We have managed to follow the symmetric periodic solution as it undergoes five saddle-node bifurcations, three pitchfork bifurcations and one Hopf bifurcation. This is illustrated in Figures 3 and 4. (Stable periodic solutions are depicted by solid lines and unstable periodic solutions are depicted by dotted lines).

The final bifurcation diagram is shown on Figure 5. It is noticeable that in the region $115m/s < v < 147m/s$ the only stable periodic solutions are the asymmetric ones bifurcating from the symmetric periodic solution at P_1. The asymmetric periodic solutions bifurcate subcritically, turn around in a saddle-node, gain stability in a Hopf bifurcation at $v = 112.59\ m/s$ and remain stable up to $v = 203.33\ m/s$ where they lose stability in another Hopf bifurcation. The symmetric periodic solution is stable in the region $147.59m/s < v < 181.73m/s$ but, as found by Galvanetto et al. , the symmetric periodic solution in this region has much smaller basin of attraction than the asymmetric periodic solutions. Thus, in the region $115m/s < v < 200\ m/s$, the asymmetric periodic solutions are *dominant*.

In Figure 6 we see the details of the final bifurcation diagram in the speed interval $109.5m/s < v < 110m/s$ showing an example of mode interaction. The

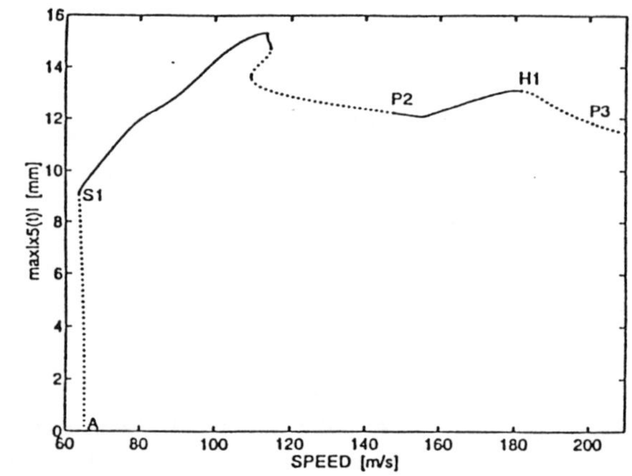

Figure 3. The symmetric periodic solutions with its bifurcations.

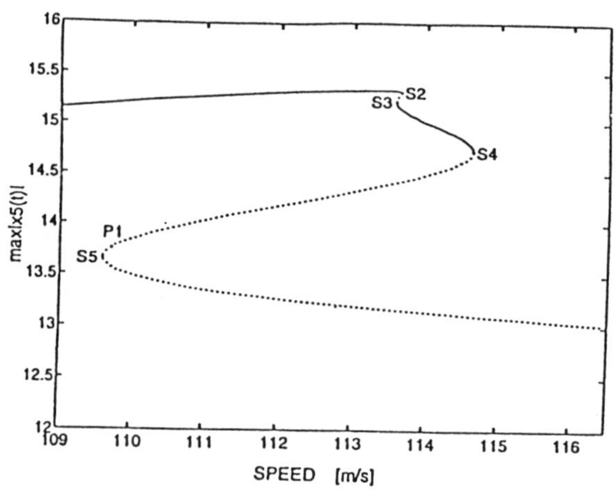

Figure 4. Blow up of Figure 3.

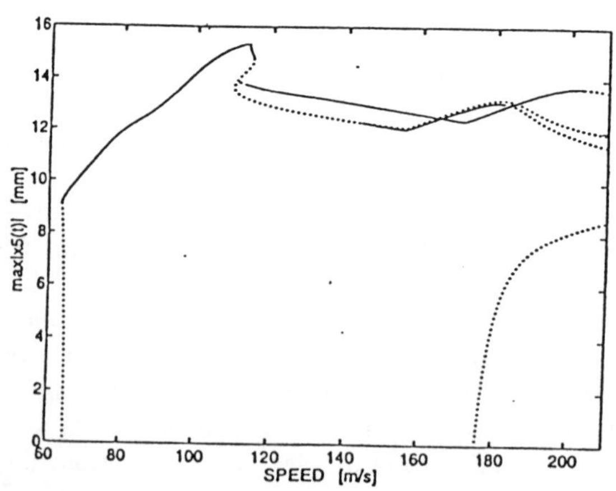

Figure 5. Final bifurcation diagram.

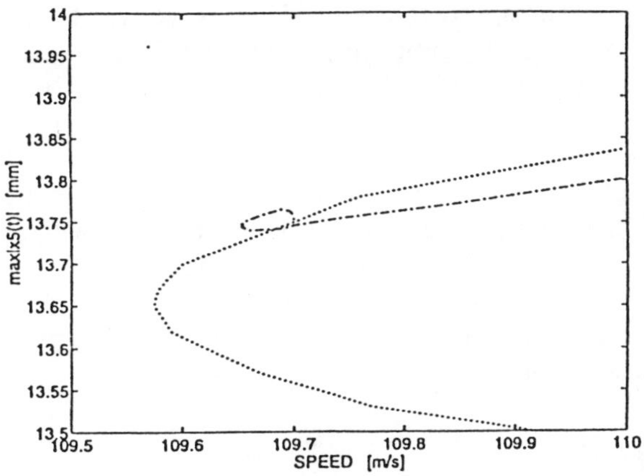

Figure 6. Blow up of Figure 5.

figure shows the region near the pitchfork bifurcation P_1 and the saddle-node bifurcation S_5. The dotted curve in figure 6 is the symmetric periodic solution and the dash-dotted curve are the bifurcating asymmetric periodic solutions. The saddle-node point and the pitchfork point are very close in parameter space. This is exactly the situation we considered in Section 2.2. We find that our Figure 6 is identical to Golubitsky et al, figure XIX.3.5 and we conclude that in the vicinity of the mode interaction, the dynamics of the our system can locally be described by the two state variable system with the normal form $(x^2 + y^4 - \beta y^2 - \lambda, -(x - \alpha)y)$.

4. Conclusion

We have considered the Cooperrider model of a railway bogie and we have analysed the bifurcations using symmetry groups. We find that the model is $\mathbb{Z}_2$-symmetric.

We then develop the generic bifurcations of symmetric periodic solutions of a system on that form, considering the Poincaré map P and a map Q with the property $P = Q^2$. The numerical bifurcation analysis of our system has revealed many bifurcations, all of them generic in the sense described in Section 2.

Subsection 2.2 deals with the theory for the dynamics of the system near the parameter values where bifurcation points of low codimension coincide. We consider the form of the mode interaction in such a situation. The numerical bifurcation analysis of our system reveals a very complicated region where a mode interaction as described in Section 2.2 takes place. This mode interaction yields the information necessary to complete the bifurcation diagram of our bogie model. The splitting and the correct sequence of the bifurcations in the complicated region was found after studies of the stringent mathematical theory for symmetry and bifurcations. The theory verifies the numerical results and helped the authors to refine the numerical investigations until only generic bifurcations appeared in the bogie model. Furthermore, these refined investigations revealed the continuation of the symmetric periodic solution. This solution is unstable for a large speed range but gains stability in the speed range 147.59 $m/s < v <$ 181.73 m/s. In this speed range, however, the asymmetric periodic solutions are dominant, and the symmetric periodic solution had not been found in earlier investigations.

From the mode interaction in the complicated region, the theory in Golubitsky et al, Chapter XIX, §§2-3 also predicts the Hopf bifurcation in the bogie system and the presence of a quasiperiodic solution bifurcating from the asymmetric periodic solution. This quasiperiodic attractor, its development and its symmetry characteristics have been studied in Jensen and True.

Asymmetric wear of railway wheelsets do occur in real life. We suggest that this lopsided wear may be related to a symmetry breaking pitchfork bifurcation as described in this paper.

5. Acknowledgements

This work was supported by Otto Mønsted's Foundation Grant 93-33-1233 and The Danish Council for Scientific and Industrial Research Grant n. 16-4786.M. The research of MG was supported in part by NSF Grant DMS-9403624. The Dept. of Mathematics of the University of Houston is kindly acknowledged for hosting CNJ during a part of this work and for the use of its computing facilities.

References

N.K. Cooperrider. The hunting behaviour of conventional railway trucks. *J. Engineering in Industry*, 94:752–762, 1972.

G. Dangelmayr and D. Armbruster. Classification of $\mathbb{Z}_2$-equivariant imperfect bifurcations of corank 2. *Proc. London Math. Soc.*, 46:517–546, 1983.

G. Dangelmayr and D. Armbruster. Classification of $\mathbb{Z}_2$-equivariant imperfect bifurcations of corank 2. *Proc. London Math. Soc.*, 46:517–546, 1983.

Ugo Galvanetto et al. Optimal axle distance of a railway bogie. *Intern. J. Bifur. & Chaos*, To Appear.

M. Golubitsky, I. Stewart, and D.G. Schaeffer. *Singularities and Groups in Bifurcation Theory*, volume 2. Springer Verlag, 1988.

C.N. Jensen and H. True. On a new route to chaos in railway dynamics. *Nonlin. Dynam.*, To Appear.

C. Kaas-Petersen. *PATH — User's guide.* Dept Appl Math Studies & Centre Nonlin Stud, Univ Leeds, 1989.

J.W. Swift and K. Wiesenfeld. Suppression of period doubling in symmetric systems. *Phys. Rev. Lett.*, 52:9:705–708, 1984.

H. True. Some recent developments in nonlinear railway vehicle dynamics. In E. Kreuzer and G. Schmidt, editors, *1st European Nonlinear Oscillations Conference, Proceedings of the International Conference in Hamburg*, pages 129–148, Berlin, August 1993. Akademie Verlag.

P.J. Vermeulen and K.L. Johnson. Contact of nonspherical elastic bodies transmitting tangential forces. *J. Appl. Mech.*, 31:338–340, 1964.

A TRAVELER'S WOES: SOME PERSPECTIVES FROM DYNAMICAL SYSTEMS

OLIVER M. O'REILLY AND PETER C. VARADI

Department of Mechanical Engineering,
University of California at Berkeley,
Berkeley, CA 94720-1740, USA

Abstract. Some features of vehicular dynamics are explored in the context of a simple model for a two-wheeled suitcase. Despite the simplicity of the model, we find some peculiar phenomena. These include an uncountable infinity of heteroclinic orbits, bifurcations in the presence of reversible symmetries and complex dynamics.

1. Introduction

Two-wheeled suitcases have become increasingly common in recent years. For many travelers, these suitcases pitch, yaw, rock and fall in the most undesirable situations. In a recent paper, Plaut [11] examined some of the features of the rocking instabilities. Motivated by his work, we analyze a simpler model of a suitcase in order to explain its yaw and pitch dynamics. Despite the simplicity of the model, we find a system with a rich dynamical behavior. Our analysis is a precursor to a more detailed examination of rocking. It is also a starting point for examining maneuvering instabilities in trollies and articulated vehicles.

Here, we examine the dynamics of the suitcase from the perspective of non-holonomically constrained, reversible mechanical systems. There is a vast Russian literature on this subject. The interested reader is referred to Karapetyan and Rumyantsev [7], Neimark and Fufaev [4], and Tkhai [13]. Some examples of work in this area pertaining to vehicle dynamics include O'Reilly and Tongue [10], Troger and Steindl [14], and Verbitskii and Lobas [15, 16].

F. C. Moon (ed.),
IUTAM Symposium on New Applications of Nonlinear and Chaotic Dynamics in Mechanics, 397–406.

2. A Model for the Dynamics of a Suitcase

We begin by developing a simple model for the dynamics of a suitcase. Here, the suitcase is modeled as a rigid body which is being pulled. It has two wheels mounted at its rear which roll without slipping on a horizontal surface (cf. Figure 1). The final results of this section include a set of four first-order ordinary differential equations governing its motion. In our derivation we will use the treatment of rigid body dynamics proposed by Casey [3] which uses a convenient tensor notation.

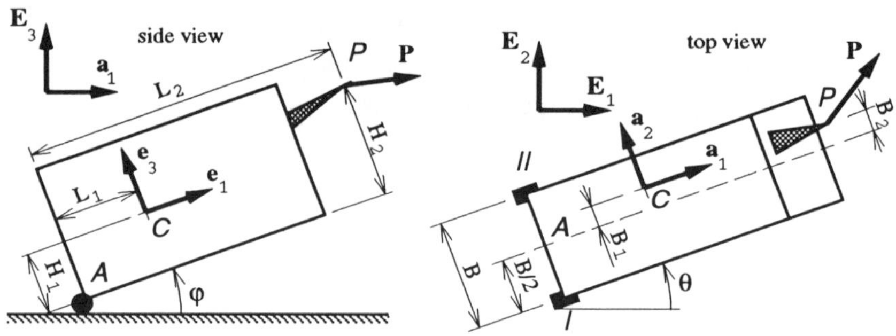

Figure 1. Definitions of the suitcase's parameters.

The vectors $\mathbf{E}_i$ $(i = 1,2,3)$ denote a fixed orthonormal basis for Euclidean three-space. Further, the vector $\mathbf{E}_3$ is assumed to be normal to the horizontal ground plane on which the suitcase moves. Another orthonormal basis $\{\mathbf{e}_i\}$ is also used in the sequel. This basis co-rotates with the suitcase such that each $\mathbf{e}_i$ is aligned with one of the principal directions of the inertia tensor $\mathbf{J} = \sum_{i=1}^{3} \lambda_i \mathbf{e}_i \otimes \mathbf{e}_i$ of the suitcase. Here, $\otimes$ denotes the usual tensor product and λ_i are the principal moments of inertia. We recall that these two bases are related by the rotation tensor $\mathbf{Q}$ of the suitcase: $\mathbf{e}_i = \mathbf{Q}\mathbf{E}_i$. To parametrize $\mathbf{Q}$, we use three Euler angles: the yawing angle $\theta \in [0, 2\pi)$, the pitching angle $\varphi \in [0, 2\pi)$ and the rocking angle $\alpha \in (-\frac{\pi}{2}, \frac{\pi}{2})$. Using these angles, the angular velocity vector $\boldsymbol{\omega}$ of the suitcase has the representation

$$\boldsymbol{\omega} = -\frac{1}{2}\,\varepsilon\left[\dot{\mathbf{Q}}\mathbf{Q}^T\right] = \dot{\theta}\,\mathbf{E}_3 + \dot{\alpha}\,\mathbf{a}_1 + (-\dot{\varphi})\,\mathbf{e}_2 \quad . \tag{1}$$

Here, ε is the third-order permutation tensor and the superposed dot denotes the time derivative. For later convenience, we have defined an orthonormal basis $\{\mathbf{a}_i\}$:

$$\mathbf{a}_1 = \cos\theta\,\mathbf{E}_1 + \sin\theta\,\mathbf{E}_2 \ , \ \mathbf{a}_2 = -\sin\theta\,\mathbf{E}_1 + \cos\theta\,\mathbf{E}_2 \ , \ \mathbf{a}_3 = \mathbf{E}_3 \ . \tag{2}$$

In the sequel, it is assumed that the principal axis $\mathbf{e}_2$ is parallel to the axis connecting the centers of the wheels.

The vectors $\mathbf{r}_C$, $\mathbf{r}_A$ and $\mathbf{r}_P$ denote the position of the center of mass of the suitcase, the midpoint of the rear wheel axis and the point of application of the pulling force $\mathbf{P}$, respectively. We define the velocities $\mathbf{v}_C = \dot{\mathbf{r}}_C$ and $\mathbf{v}_A = \dot{\mathbf{r}}_A$.

We now consider the three constraints on the motion of the suitcase. First, both wheels are assumed to remain in contact with the ground:

$$\mathbf{v}_A \cdot \mathbf{E}_3 = 0 \quad , \quad \alpha = 0 \quad . \tag{3}$$

The latter constraint implies that we are not considering the rocking instabilities discussed by Plaut [11][1]. Second, it is assumed that the (massless) wheels roll without slipping:

$$\mathbf{v}_A \cdot \mathbf{a}_2 = 0 \quad . \tag{4}$$

It is important to note that we will assume that no part of the suitcase other than the wheels comes in contact with the ground. Hence, we have the unilateral constraints $\varphi \in (\varphi_1, \varphi_2)$, where φ_1 and φ_2 are functions of the suitcase's geometry. If these are violated then our model is inapplicable as it does not include the additional reaction forces exerted by the ground plane.

Imposing the constraints (3) and using equations (1) and (2), the relative position vectors $\mathbf{r}_C - \mathbf{r}_A$ and $\mathbf{r}_P - \mathbf{r}_A$ are

$$\begin{aligned} \mathbf{r}_C - \mathbf{r}_A &= X_1\,\mathbf{a}_1 + B_1\,\mathbf{a}_2 + Z_1\,\mathbf{a}_3 \quad , \\ \mathbf{r}_P - \mathbf{r}_A &= X_2\,\mathbf{a}_1 + B_2\,\mathbf{a}_2 + Z_2\,\mathbf{a}_3 \quad . \end{aligned} \tag{5}$$

Here, we have defined the relative coordinates X_α and Z_α ($\alpha = 1, 2$):

$$\begin{aligned} X_\alpha &= L_\alpha \cos\varphi - H_\alpha \sin\varphi \quad , \quad \dot{X}_\alpha = -\dot{\varphi} Z_\alpha \quad , \\ Z_\alpha &= L_\alpha \sin\varphi + H_\alpha \cos\varphi \quad , \quad \dot{Z}_\alpha = \dot{\varphi} X_\alpha \quad . \end{aligned} \tag{6}$$

The geometrical parameters L_α, B_α and H_α are shown in Figure 1. We also define the speeds $u_\mathcal{I} = \mathbf{v}_\mathcal{I} \cdot \mathbf{a}_1$, $\mathcal{I} = A, C$. It follows from the no-slip condition (4) that

$$\mathbf{v}_A = u_A\,\mathbf{a}_1 \quad , \quad u_C = u_A - B_1\dot{\theta} - Z_1\,\dot{\varphi} \quad . \tag{7}$$

The suitcase, which has mass m, is subject to a gravitational body force $-mg\,\mathbf{E}_3$ and a pulling force $\mathbf{P} = \sum_{i=1}^{3} P_i\,\mathbf{a}_i$ which act at C and P, respectively. The lateral reaction force that enforces the constraint (4) is

[1]Specifically, Plaut considers rocking instabilities when φ and θ are constrained to be constant, but α and u_C are varied. He also includes the inertia and spin angles of the wheels which we have neglected.

$R\mathbf{a}_2$ and it acts along the wheel axes. The normal reaction forces which enforce the constraints (3) are $N_\alpha \mathbf{E}_3$, and they act on the wheels I and II, respectively.

The reduced equations of motion are obtained by eliminating the constraint forces and constraints from the balances of linear and angular momenta:

$$\frac{d\hat{u}_C}{d\tau} - \hat{X}_1\omega^2 = \hat{P}_1 \quad ,$$

$$\hat{\lambda}_2\frac{d^2\varphi}{d\tau^2} + \hat{X}_1\frac{d}{d\tau}\left[\hat{X}_1\frac{d\varphi}{d\tau}\right] + (\hat{\lambda}_3 - \hat{\lambda}_1)\omega^2 \sin\varphi\cos\varphi = \hat{P}_1(\hat{Z}_1 - \hat{Z}_2) + \hat{P}_3\hat{X}_2 - \hat{X}_1 \quad ,$$

$$\frac{d}{d\tau}\left[(\hat{\lambda}_1 \sin^2\varphi + \hat{\lambda}_3\cos^2\varphi + \hat{X}_1^2)\omega\right] + \hat{X}_1\omega\left(\hat{u}_C + \hat{Z}_1\frac{d\varphi}{d\tau}\right) =$$
$$\hat{P}_2\hat{X}_2 - \hat{P}_1(\hat{B}_2 - \hat{B}_1) \quad . \quad (8)$$

In writing these equations, we have used the dimensionless quantities

$$\tau = \sqrt{\frac{g}{L_1}}\,t \quad , \quad \hat{\mathbf{P}} = \frac{\mathbf{P}}{mg} \quad , \quad \hat{u}_C = \frac{u_C}{\sqrt{gL_1}} \quad , \quad \omega = \frac{d\theta}{d\tau} \quad ,$$

$$\hat{X}_\alpha = \frac{X_\alpha}{L_1} \,,\; \hat{Z}_\alpha = \frac{Z_\alpha}{L_1} \,,\; \hat{B}_\alpha = \frac{B_\alpha}{L_1} \,,\; \hat{\lambda}_i = \frac{\lambda_i}{mL_1^2} \quad . \quad (9)$$

For brevity's sake, we refrain from writing the explicit expressions for the constraint forces which may be calculated as functions of the motion. These expressions are needed to conclude if the wheels have lifted off the ground.

It is of interest to use the work-energy theorem to ascertain if an energy is conserved. A standard calculation shows that

$$\frac{dE}{dt} = -\dot{\mathbf{P}}\cdot\mathbf{r}_P \quad , \quad E = T + mg\,\mathbf{E}_3\cdot\mathbf{r}_C - \mathbf{P}\cdot\mathbf{r}_P \quad . \quad (10)$$

Here, the kinetic and total energies of the suitcase are denoted by T and E, respectively.

The reduced equations of motion (8) show no explicit dependence on the orientation θ (which is to be expected on the grounds of symmetry), nor do they depend explicitly on $\mathbf{r}_C$. Also, as expected, (8) reduce to the equations of motion of a three-wheeled cart discussed by O'Reilly and Tongue [10] when $\mathbf{P} = \mathbf{0}$ and the additional constraint $\varphi = \varphi_1$ is imposed. This additional constraint entails the introduction of a constraint moment in $(8)_2$. The resulting equations are identical to those governing the sled which were discussed by Carathéodory [2] and Fufaev [4].

We will henceforth use only dimensionless quantities and omit the superscripted hat.

3. Reversibility and Steady Motions

We recall that a dynamical system $\dot{\mathbf{x}} = \mathbf{f}(\mathbf{x})$ is reversible if there exists a linear transformation $\mathbf{R} = \mathbf{R}^{-1}$ such that $\mathbf{Rf}(\mathbf{Rx}) = -\mathbf{f}(\mathbf{x})$. The equations of motion (8) constitute a reversible dynamical system. In particular, they are invariant under the following transformation[2]:

$$(u_C, \varphi, \frac{d\varphi}{d\tau}, \omega, \tau) \longmapsto (-u_C, \varphi, -\frac{d\varphi}{d\tau}, -\omega, -\tau) \quad . \tag{11}$$

In this section, we will examine the steady motions and their stability in light of this reversibility.

A steady motion of the suitcase corresponds to an equilibrium of the reduced equations of motion (8):

$$u_C = u_{C0} \quad , \quad \varphi = \varphi_0 \quad , \quad \omega = \omega_0 \quad , \tag{12}$$

where u_{C0}, φ_0 and ω_0 are constants. From (8), the following three conditions must be satisfied:

$$-X_{10}\,\omega_0^2 = P_{10} \quad , \quad X_{10}\,\omega_0\,u_{C0} = P_{20}\,X_{20} - P_{10}(B_2 - B_1) \quad ,$$

$$\omega_0^2\,((\lambda_3 - \lambda_1)\sin\varphi_0\cos\varphi_0 + X_{10}\,(Z_{10} - Z_{20})) + X_{10} = P_{30}\,X_{20} \quad . \tag{13}$$

The added subscript 0 here indicates that the functions are evaluated at the equilibrium. We observe that if $X_{20} = 0$, then (13) are independent of P_{20} and P_{30}.

If one linearizes the reduced equations (8) about an equilibrium, one finds that the trace I_1 of the Jacobian is

$$I_1 = \frac{-X_{10}u_{C0}}{\lambda_1\sin^2\varphi_0 + \lambda_3\cos^2\varphi_0 + X_{10}^2} \quad . \tag{14}$$

It follows that the sum of the eigenvalues is not necessarily zero. If $X_{10}u_{C0} < 0$, then a steady motion is necessarily unstable. On the other hand, it is possible to have an asymptotically stable steady motion even though there is no dissipation in our model[3].

[2]When $\mathbf{P} = P_{30}\mathbf{E}_3$ is constant, then this agrees with the results of Tkhai [13], who showed that a Chaplygin system is generically reversible under an involution of the form (11). For further details on Chaplygin systems, we refer the reader to Neimark and Fufaev [8].

[3]Although there are several similarities between reversible and Hamiltonian dynamical systems (cf. O'Reilly, Malhotra and Namachchivaya [9] and Sevyruk [12]), there also exist several distinctions. For example, Hermans [6] noted that another reversible dynamical system, the rattleback (or wobblestone), has asymptotically stable steady motions.

From the point of view of a traveler, the most interesting steady motions arise when $\omega_0 = 0$. In this case, (13) reduce to

$$P_{10} = 0 \quad , \quad P_{20}\, X_{20} = 0 \quad , \quad X_{10} = P_{30}\, X_{20} \quad . \tag{15}$$

Assuming that $X_{20} \neq 0$, P_{20} must also vanish. It follows that these motions can be sustained by $\mathbf{P} = P_{30}\, \mathbf{E}_3$. Further, if one has a steady motion where $u_C = u_{C0}, \varphi = \varphi_0$ and $\omega_0 = 0$, then, from (11), its counterpart $u_C = -u_{C0}, \varphi = \varphi_0$ and $\omega_0 = 0$, also exists. The linearized equations in the neighborhood of the former equilibrium are

$$\frac{d}{d\tau}\begin{bmatrix} \varphi - \varphi_0 \\ \frac{d\varphi}{d\tau} \\ u_C - u_{C0} \\ \omega \end{bmatrix} = \begin{bmatrix} 0 & 1 & 0 & 0 \\ J_1 & 0 & 0 & 0 \\ 0 & 0 & 0 & 0 \\ 0 & 0 & 0 & I_1 \end{bmatrix}\begin{bmatrix} \varphi - \varphi_0 \\ \frac{d\varphi}{d\tau} \\ u_C - u_{C0} \\ \omega \end{bmatrix} \quad , \tag{16}$$

where $J_1 = \frac{-P_{30}\, Z_{20} + Z_{10}}{\lambda_2 + X_{10}^2}$. The eigenvalues are $\{\pm\sqrt{J_1}, 0, I_1\}$. It should be noted how the last eigenvalue depends on the sign of u_{C0}; this is a consequence of (11) and has obvious implications for linear stability results.

Due to the non-hyperbolic nature of these equilibria, we proceeded to calculate the center manifolds for the equilibria. There are two distinct cases to consider. In the first case, J_1 is negative or zero, and the center manifold is $\omega = h(\varphi - \varphi_0, \frac{d\varphi}{d\tau}, u_C - u_{C0})$. Examining the partial differential equation for h, we found that $h = 0$ was an exact solution. In the second case, J_1 is positive, and the center manifold is one-dimensional. Here, we found that the equations for the center manifold were $\omega = 0$, $\varphi - \varphi_0 = 0$ and $\frac{d\varphi}{d\tau} = 0$. It follows that $u_C = u_{C0}$ on the center manifold of an equilibrium. Clearly we are dealing with a degenerate case here[4].

4. Dynamics on the Invariant Plane

We henceforth restrict our attention to the case $\mathbf{P} = P_{30}\, \mathbf{E}_3$. It follows from (8) that, if $\omega = 0$, then the φ-$\frac{d\varphi}{d\tau}$ plane is invariant. From our previous discussion on center manifolds, it suffices to examine the dynamics on this plane to determine non-linear stability. Of further interest is the case where $\lambda_1 = \lambda_3 = \lambda$, and the φ-$\frac{d\varphi}{d\tau}$ plane is independent of ω and u_C.

For all of the aforementioned cases, the equations (8) restricted to the invariant plane are

$$\lambda_2 \frac{d^2\varphi}{d\tau^2} + X_1 \frac{d}{d\tau}\left[X_1 \frac{d\varphi}{d\tau}\right] = P_{30}\, X_2 - X_1 \quad . \tag{17}$$

[4]The uniqueness of these results was confirmed to fourth order using the approximate solution method discussed by Guckenheimer and Holmes [5].

This dynamical system has a first integral (cf. (10)):

$$E(\varphi, \frac{d\varphi}{d\tau}) = \frac{1}{2}\left(\lambda_2 + X_1^2\right)\left(\frac{d\varphi}{d\tau}\right)^2 - P_{30}Z_2 + Z_1 \quad . \tag{18}$$

Clearly, (17) is, in principle, integrable by quadrature.

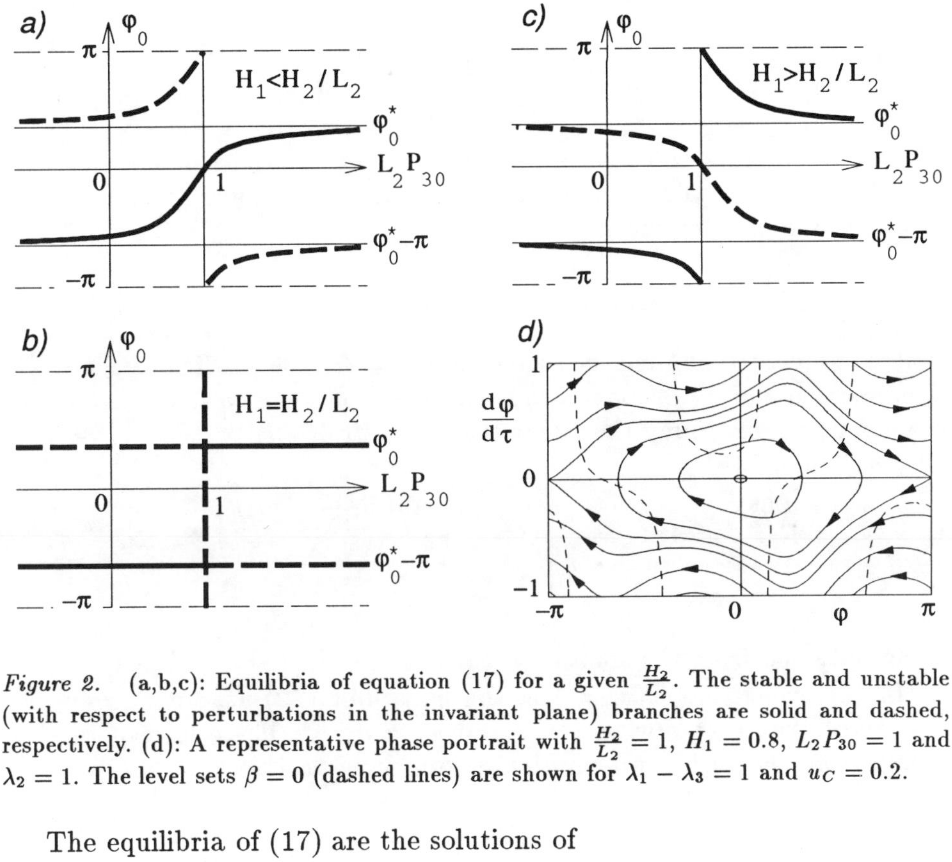

Figure 2. (a,b,c): Equilibria of equation (17) for a given $\frac{H_2}{L_2}$. The stable and unstable (with respect to perturbations in the invariant plane) branches are solid and dashed, respectively. (d): A representative phase portrait with $\frac{H_2}{L_2} = 1$, $H_1 = 0.8$, $L_2P_{30} = 1$ and $\lambda_2 = 1$. The level sets $\beta = 0$ (dashed lines) are shown for $\lambda_1 - \lambda_3 = 1$ and $u_C = 0.2$.

The equilibria of (17) are the solutions of

$$(P_{30}H_2 - H_1)\sin\varphi_0 = (P_{30}L_2 - 1)\cos\varphi_0 \quad . \tag{19}$$

These solutions can be conveniently discussed by considering the three cases which arise when H_1 is smaller, equal or greater than $\frac{H_2}{L_2}$ (cf. Figure 2(a-c)). If one imagines a line connecting A to P in the side view of Figure 1, then these three cases correspond to the center of mass C lying below, on, or above this line, respectively. The branches of equilibria asymptote to either $\varphi_0 = \arctan\left(\frac{L_2}{H_2}\right) = \varphi_0^* < \frac{\pi}{2}$ or $\varphi_0 = \varphi_0^* - \pi$ as $|P_{30}| \longrightarrow \infty$. The stability of an equilibrium is inferred from the following equation:

$$-(P_{30}L_2 - 1)\sin\varphi_0 < (P_{30}H_2 - H_1)\cos\varphi_0 \quad . \tag{20}$$

Clearly, (local) bifurcations only occur when $H_1 = \frac{H_2}{L_2}$. For completeness, a typical phase portrait is displayed in Figure 2(d).

It is of interest to examine the behavior of orbits in the neighborhood of the invariant plane. To this end, we consider the boundary layer where $|\omega| \ll 1$. From (8), it follows that $\varphi(\tau)$ is governed by (17), that u_C is constant, and

$$\frac{d\omega}{d\tau} = -\beta\,\omega \quad , \quad \beta = \frac{(\lambda_1 - \lambda_3)\frac{d\varphi}{d\tau}\sin 2\varphi + X_1(u_C - Z_1\frac{d\varphi}{d\tau})}{\lambda_1 \sin^2\varphi + \lambda_3\cos^2\varphi + X_1^2} \quad . \tag{21}$$

The level sets of $\beta = 0$ separate regions on the invariant plane where the orbits are ejected and injected into the plane (cf. Figure 2(d)). We shall return to this matter shortly.

5. A Multitude of Heteroclinic Orbits in a Suitcase

It is of interest to examine the case where $\lambda_1 = \lambda_3 = \lambda$ and $\mathbf{P} = P_{30}\,\mathbf{E}_3$. The dynamics on the φ-$\frac{d\varphi}{d\tau}$ plane are then independent of ω and u_C. Further, the equations governing the latter states are, from (8),

$$\frac{du_C}{d\tau} = X_1\omega^2 \quad , \quad \frac{d\omega}{d\tau} = -\left(\frac{X_1(Z_1\frac{d\varphi}{d\tau} + u_C)}{\lambda + X_1^2}\right)\omega \quad . \tag{22}$$

where $X_1 = X_1(\varphi(\tau))$, $Z_1 = Z_1(\varphi(\tau))$ and $\frac{d\varphi}{d\tau} = \frac{d\varphi}{d\tau}(\tau)$, these functions being obtained from the solutions of (17).

In the event that $\varphi(\tau) = \varphi_0$, the equations (22) have a uncountable infinity of fixed points with $\omega = 0$ and $u_C = \pm u_{C0}$. These fixed points are pairwise connected by two elliptical heteroclinic orbits:

$$u_C^2 + \omega^2(\lambda_2 + X_{10}^2) = u_{C0}^2 \quad . \tag{23}$$

Indeed, the entire phase space of (22) is foliated by these orbits, and the phase portrait is qualitatively identical to that shown in Figure 2 of O'Reilly and Tongue [10]. Paralleling the work of Carathéodory [2], one can calculate the change in θ as one of these orbits is traversed:

$$\Delta\theta = \pm\left(\sqrt{\frac{\lambda_2 + X_{10}^2}{X_{10}^2}}\right)\pi \quad . \tag{24}$$

In addition, the motion of the suitcase on the horizontal plane, can be easily inferred from figures in Bahar [1] and Carathéodory [2].

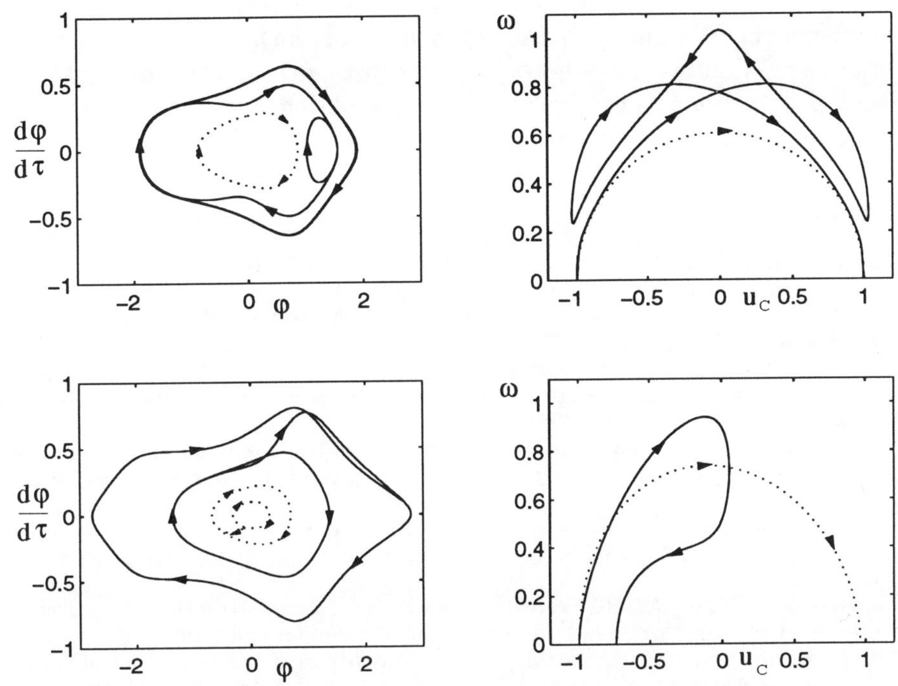

Figure 3. **Orbits connecting periodic orbits of the invariant plane with the same period (top) and different periods (bottom). The parameters are $H_1 = 0.8$, $H_2 = L_2 = 2$, $P_{30} = 0.5$, $\lambda_1 = 1.5$, $\lambda_2 = 1$, $\lambda_3 = 0.5$. The initial conditions are $\varphi(0) = 0$, $u(0) = -1$, $\omega(0) = 10^{-3}$. In addition, for the top figures, $\frac{d\varphi}{d\tau}(0) = 0.51069$ (solid), $\frac{d\varphi}{d\tau}(0) = 0.2641$ (dotted), and for the bottom figures, $\frac{d\varphi}{d\tau}(0) = 0.4$ (solid), $\frac{d\varphi}{d\tau}(0) = 0.1$ (dotted).**

6. Concluding Remarks

When $\lambda_1 \neq \lambda_3$ and $\mathbf{P} = P_{30}\,\mathbf{E}_3$, then the heteroclinic connections mentioned earlier will be altered. Our numerical simulations indicate that almost every orbit asymptotes to $\omega = 0$ as $|\tau| \to \infty$ (cf. Figure 3). That is to say, these orbits connect periodic orbits in invariant planes with different values of u_C. The injection and ejection dynamics appear to be governed by the function β of equation (21). The projections of the orbits in the ω–u_C plane can be extremely tangled. It is also possible to start close to a periodic orbit where $\frac{d\varphi}{d\tau}$ is positive and end up close to an orbit where $\frac{d\varphi}{d\tau}$ is negative. This is reminiscent of the behavior observed in a wobblestone. On the other hand, by starting off the invariant plane, but close to the homoclinic orbit in the invariant plane, one observes what appears to be non-integrable behavior. The mechanisms for these complex dynamics seems to be the breaking of the heteroclinic and homoclinic orbits. However, we have been unable to explain these dynamics using Shilnikov–type mechanisms (cf. Wiggins [17] and references therein).

One issue that we have ignored is the presence of the unilateral constraints $\varphi \in (\varphi_1, \varphi_2)$. These can be accommodated by introducing an impact model and truncating the phase space.

References

1. L. Y. Bahar, On a non–holonomic problem proposed by Greenwood, *International Journal of Non-Linear Mechanics* **28** 1993, 169-186.
2. C. Carathéodory, Der Schlitten, *Zeitschrift für angewandte Mathematik und Mechanik (ZAMM)* **13** 1933, 71-76.
3. J. Casey, On the advantages of a geometrical viewpoint in the derivation of Lagrange's equations for a rigid continuum, *Journal of Applied Mathematics and Physics (ZAMP)* **46** 1995, S805-S847.
4. N. A. Fufaev, On the possibility of realizing a non-holonomic constraint by means of viscous friction, *Journal of Applied Mathematics and Mechanics (PMM)* **28** 1964, 630-632.
5. J. Guckenheimer and P. Holmes, *Nonlinear Oscillations, Dynamical Systems, and Bifurcations of Vector Fields*, 4th corrected printing, Springer-Verlag, New York, 1993.
6. J. Hermans, *Rolling Rigid Bodies With and Without Symmetries*, Thesis, Universitet Utrecht, Faculteit Wiskinde en Informatica, The Netherlands, 1996.
7. A. K. Karapetyan and V. V. Rumyantsev, Stability of conservative and dissipative systems. In *Applied Mechanics: Soviet Reviews* **1** 1990, 3-144, G. Mikhailov and V. Z. Parton, eds, Hemisphere, New York.
8. Ju. I. Neimark and N. A. Fufaev, Dynamics of nonholonomic systems, translated from Russian by J. R. Barbour, *Translations of Mathematical Monographs* **33** 1972, American Mathematical Society, Providence, Rhode Island.
9. O. M. O'Reilly, N. K. Malhotra and N. S. Namachchivaya, Some aspects of destabilization in reversible dynamical systems with application to follower forces, *Nonlinear Dynamics* **10** 1996, 63-87.
10. O. M. O'Reilly and B. H. Tongue, Some comments on vehicle instability due to brake lockup, *Journal of Sound and Vibration* **194** 1996, 760-764.
11. R. H. Plaut, Rocking instability of a pulled suitcase with two wheels, *Acta Mechanica* **117** 1996, 165-179.
12. M. Sevryuk, Lower dimensional tori in reversible systems, *Chaos* **1** 1991, 160-167.
13. V. N. Tkhai, The reversibility of mechanical systems, *Journal of Applied Mathematics and Mechanics (PMM)* **55** 1991, 461-468.
14. H. Troger and A. Steindl, *Nonlinear Stability and Bifurcation Theory: An Introduction for Engineers and Applied Scientists*, Springer-Verlag, Vienna, 1991.
15. V. G. Verbitskii and L. G. Lobas, Bifurcations of steady states in systems with rolling under constant force perturbations, *Journal of Applied Mathematics and Mechanics (PMM)* **58** 1994, 933-939.
16. V. G. Verbitskii and L. G. Lobas, Real bifurcations of two-unit systems with rolling, *Journal of Applied Mathematics and Mechanics (PMM)* **60** 1996, 413-419.
17. S. Wiggins, *Introduction to Applied Nonlinear Dynamical Systems and Chaos*, 3rd corrected printing, Springer-Verlag, New York, 1996.

STABILITY AND CHAOS IN PASSIVE-DYNAMIC LOCOMOTION

M.J. COLEMAN, M. GARCIA, A. L. RUINA AND J. S. CAMP
Department of Theoretical and Applied Mechanics
Cornell University, Ithaca, NY 14853-7501

AND

A. CHATTERJEE
Engineering Science and Mechanics
Penn State University, University Park, PA 16802

1. Introduction

Human locomotion is a complicated process, controlled and actuated by the neuro-muscular system. Tad McGeer [8], however, studied bipedal walking and completely neglected the neuro-muscular system in his models. Some of his machines, powered only by gravity, can walk stably and somewhat anthropomorphically down shallow slopes. We have continued study of the dynamics of McGeer-like physical and mathematical biped models that have little or no actuation or control. This paper summarizes some of our results.

These *passive-dynamic* walking mechanisms are built of hinged bodies that make collisional and rolling contact with the ground at the foot. In models with knees, the lower leg, or *shank*, is prevented from hyperextending (swinging too far ahead of the upper leg, or *thigh*) by means of angular stops at the knees. Thus, the kneed walkers have an internal rotational collision. In our modeling, we assume that all collisions are instantaneous and non-bouncing (plastic).

Following McGeer, our analysis is built around simulation of a single walking step. One step, or cycle of motion, starts at an arbitrary point, say just after a foot collision. A cycle then includes the motion between foot collisions, as well as the discontinuities at the next foot collision. The cycle of motion is represented mathematically by a return map, termed the 'stride-function' by McGeer, that maps the state of the system from just after one heel-strike to just after the next. Fixed points of the map correspond to period-one motion cycles, or period-one 'gaits' of the model.

F. C. Moon (ed.),
IUTAM Symposium on New Applications of Nonlinear and Chaotic Dynamics in Mechanics, 407–416.

Gait stability can be determined by calculating (most often numerically) the eigenvalues of the linearization of the map at the fixed points (see [3] for a detailed description of the modeling and analysis procedures).

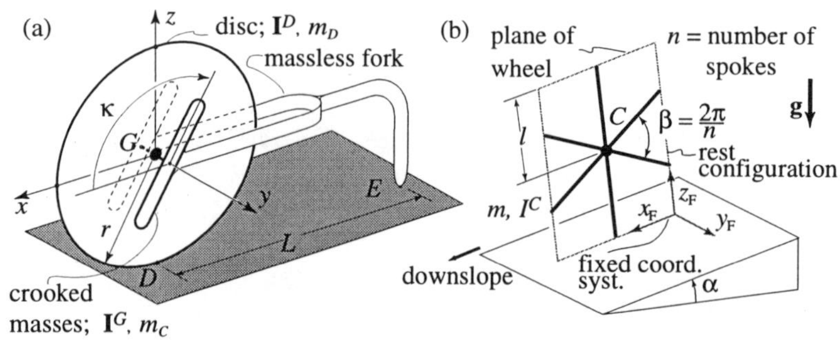

Figure 1. The parameters and orientation variables for (a) a uniform rolling disk with oblique masses added and (b) a rimless spoked wheel.

Using this scheme, we have studied several passive-dynamic models of increasing complexity, progressing from rolling wheels to 2D straight-legged and kneed models to 3D straight-legged models, each of which is described below. We also describe a simple barely-controlled powering scheme for a 2D straight-leg walker, which produces stable gait on level ground.

2. Rolling Wheels in 2D and 3D

Perhaps the simplest passive-dynamic system to study, that has some features in common with walking, is the rimless spoked wheel, or rolling polygon, confined to 2D [8]. The 2D rimless wheel has a stable limit cycle motion whose eigenvalues and associated global basins of attraction we have completely determined analytically [3]. The primary lesson of the rimless wheel in 2D is that speed regulation comes from a balance of collisional dissipation, which is proportional to speed squared, and gravitational work, which is proportional to speed.

Next, we studied a 3D rolling disk with oblique masses added [3] (see Figure 1). The masses can bank and steer with the disk but cannot roll (or pitch) with it. The purpose of this investigation was to study the effects of mass distribution on stability. The oblique masses, if adjusted properly, change the stability of the uniform rolling disk, a conservative nonholonomic system, from *neutrally stable* to *asymptotically stable* [10]. This result suggests that mass distribution may affect side-to-side balance in more complicated walking models.

Finally, we studied the 3D rimless wheel (see Figure 1) and found analytically, for many spokes and small slope angles, the stability eigenvalues for steady 'rolling' motions [1]. The 3D rimless wheel is a piecewise-conservative-holonomic (but globally non-conservative and nonholonomic) system with intermittent, dissipative impacts – features that are shared by some more realistic human walking models. The rolling rimless wheel can also exhibit asymptotic stability when perturbed from a downhill limit cycle, even when its mass distribution corresponds to that of a disk that does *not* have asymptotic stability. Thus, the intermittent collisions can play a role in side-to-side balance.

Despite the useful lessons from rolling models, they are not anthropomorphic walkers. They cannot fall down forwards or backwards, and they lack swinging legs.

3. Straight-Legged 2D Walkers

The next-simplest class of models live in 2D and consist of two linked swinging legs and point-feet [8, 6, 3, 5]. Asymptotically-stable walking motions of these models exist for a variety of parameters. The simplest such 'point-foot' straight-legged model has a huge hip mass and tiny masses (relatively infinitesimal) at its point-feet (see Figure 2).

A typical plot of the stance-leg and swing-leg angles is shown over one stable step in Figure 2. This model exhibits two steady walking motions, or period-one gaits, all the way to $\gamma \to 0^+$ as shown in Figure 3. At these gaits, the stance angle (and step length) are proportional to $\gamma^{1/3}$. Figure 4 shows how stable limping (period-two) and apparently-chaotic 'staggering' gaits appear as the slope angle is increased.

That this machine can walk on arbitrarily small slopes means that, by some reasonable measures, it is capable of near-perfectly-efficient gait (zero+ energy cost per unit distance of transport). At small slopes, the gravitational power used by this model in downhill walking is proportional to the fourth power of the walking speed. This result gives insight into achieving similar efficiency in more complicated models (e.g. models with knees, circular feet, and/or more general mass distribution). The power scaling depends, in part, on the infinitesimally small feet. With finite-mass feet, there are two modes of energy loss at heelstrike: one due to deflection of the hip mass, and one due to dissipation of the foot's kinetic energy in a plastic collision with the ground. Preliminary studies of point-foot models with finite foot mass show that the long-period gaits retain the same scaling laws at small slopes, while the short-period gaits do not.

4. More General 2D Walkers

We have reproduced and extended McGeer's results for more general 2D walkers with knees[8]: Figure 5 shows the state variables and parameters for

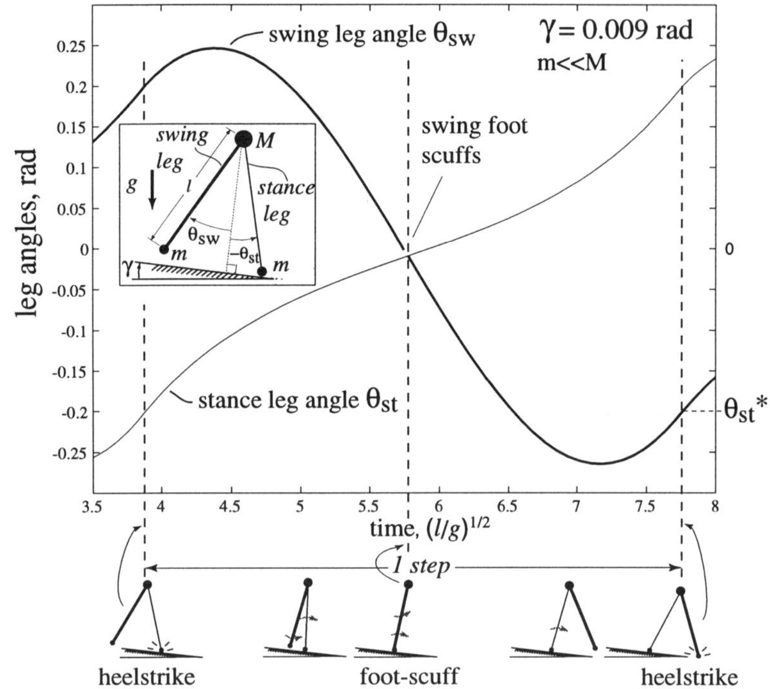

Figure 2. The simplest walking model and one of its typical passive walking steps. The inset schematic describes the variables and parameters that we use. In the cartoon below the graph, the new stance leg (lighter line) has just made contact with the ramp (left-most picture). The swing leg (heavier line) swings until the next heel-strike (right-most picture). We ignore foot-scuffing of the swing leg, allowing its foot to pass through the floor. Leg angles versus time are shown over one step at a gait cycle. Leg lines are drawn with different weights to correspond to heavy-line leg of the cartoon below the graph. Heelstrike returns the system to its initial conditions. A perturbation analysis predicts $\theta_{st}{}^* \approx C_1\gamma^{1/3} + C_2\gamma$, where $\theta_{st}{}^*$ is the stance angle at a fixed point (see Garcia,*et al.* [5]).

the kneed model and Figure 6 shows a typical gait cycle and its qualitative comparison to experimental data.

Certain conditions on the mass distribution are necessary for general 2D kneed and straight-legged walkers to achieve walking at arbitrarily small slope angles. These 'balanced' models follow similar scaling laws as their simpler point-foot, straight-legged cousins. We have also found period doubling and chaos for these kneed walkers.

5. 3D Walking

Finally, we move our point-foot model into 3D [3]. McGeer [9] and Fowble and Kuo [4] began studies of three-dimensional passive walking mechanisms, finding only unstable periodic motions. We have built a simple two-

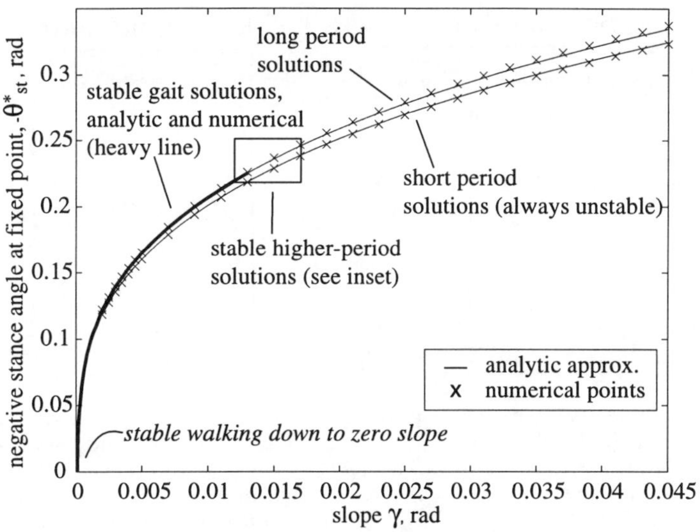

Figure 3. Comparison of numerical and analytic predictions for point-foot stance angle at fixed point as a function of slope. The box is shown expanded in Figure 4.

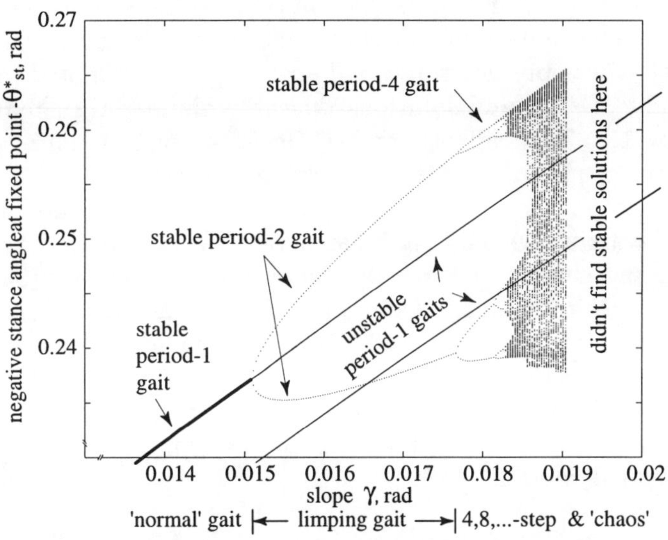

Figure 4. Period doubling of stable walking motions, inset from Figure 3. Period doubling occurs when one of the map eigenvalues for a period-n walking cycle passes through -1. Unstable period-one cycles are shown for reference. Dotted lines represent stable cycles while solid lines represent unstable ones. No persistent walking was found at slopes much steeper than 0.019 radians.

legged Tinkertoy® model that walks passively, apparently stably, down gentle slopes [2] (see Figure 7).

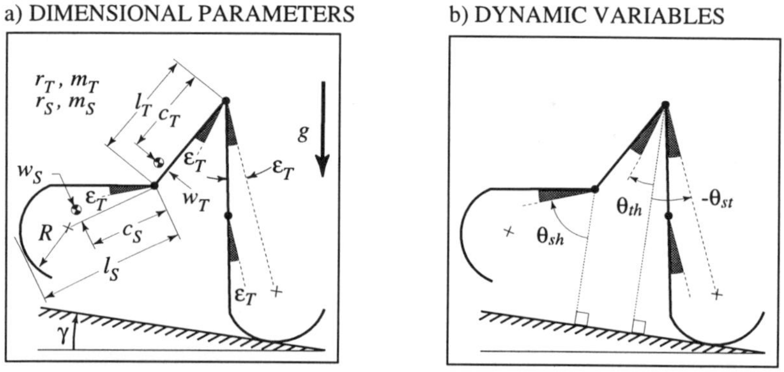

Figure 5. McGeer's kneed walking model. Shown above are (a) model parameters, and (b) dynamic variables. Radii of gyration and masses of thigh and shank are denoted by r_T, m_T, r_S, and m_S, respectively. The foot is a circular arc centered at the "+". ϵ_T is defined to be the angle between the stance thigh and the line connecting the hip to the foot center. Dynamic variable values θ_{st}, θ_{th}, and θ_{sh} are measured from ground-normal to lines offset by ϵ_T from their respective segments. A stop (not shown) at each knee prevents hyperextension of either knee.

The configuration and mass distribution of the legs of the Tinkertoy® model were suggested by numerical simulations of a simpler 3D model (see Figure 8) that was predicted to be almost-stable. The model predicts near-stable steady 3D walking solutions (the maximum return map eigenvalue is $|\sigma|_{max} \approx 1.15$ with all others $\sigma \leq 1$) for very low center-of-mass and lateral center-of-mass location comparable to the leg length. As the lateral center-of-mass position get very large, the model predicts something like 'tight-rope' walking with a long balance bar: the step period and length get very small, and the maximum map eigenvalue modulus approaches 1 (neutral stability) asymptotically from above (see Figure 10). More detailed 3D modeling is currently in progress. This walking mechanism joins a small list of passive mechanical devices free to move in three dimensions but without fast spinning parts, that are statically unstable, yet can be dynamically, asymptotically stable. Figure 9 shows typical 3D periodic behavior predicted by the model.

6. Powered 'Passive' Walking

Once power is added to our passive devices, they are, of course, no longer uncontrolled in the pure sense. Nevertheless, as shown by McGeer [7], a stable passive-dynamic model is a good basis for simple 'open-loop' powering schemes. Figure 11 shows the configuration and simulated walking cycle for a powered 2D point-foot-like model. The torque is provided by a constant-voltage DC motor at the stance-ankle. The ankle is locked during the passive mode.

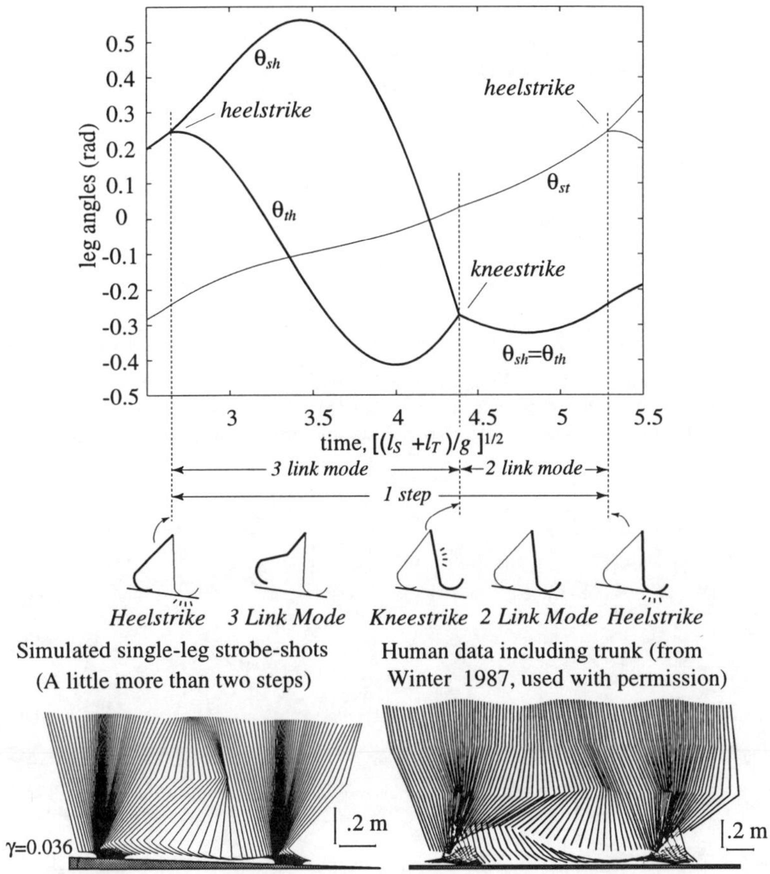

Figure 6. Simulated gait cycle (ours, similar to McGeer's). Angles of leg segments are shown from just before one heelstrike to just after the next heelstrike in a stable gait of the walker in Figure 5. The heavy line on the graph corresponds to the motion of the heavy-line leg on the small cartoon under the graph. At the start of the step, this is the stance leg, but it becomes the swing leg just after the first heelstrike. The strobe-like picture of the walker on the bottom left, created from the simulated gait cycle in the graph, shows the anthropomorphic nature of the gait. The stroboscopic picture on the lower right was generated from experimental data from [11]. The parameters values used, from a working physical model in our lab, are: l_t = 0.35m, w_t = 0m, m_t = 2.345kg, r_t = 0.099m, c_t = 0.091m, l_s = 0.46m, w_s = 0.025m, m_s = 1.013kg, r_s = 0.197m, c_s = 0.17m, R = 0.2m, γ = 0.036rad, g = 9.81 m/s^2, ϵ_T = 0.097rad.

7. Conclusions

The human-like and complicated motions of McGeer-like passive dynamic devices studied by ourselves and others imply that coordination in locomotion may be largely governed by pure mechanics. It has yet to be determined whether or not these models have medically-useful lessons to teach us, and whether or not they are a good spring-board for biomechanical or robotic

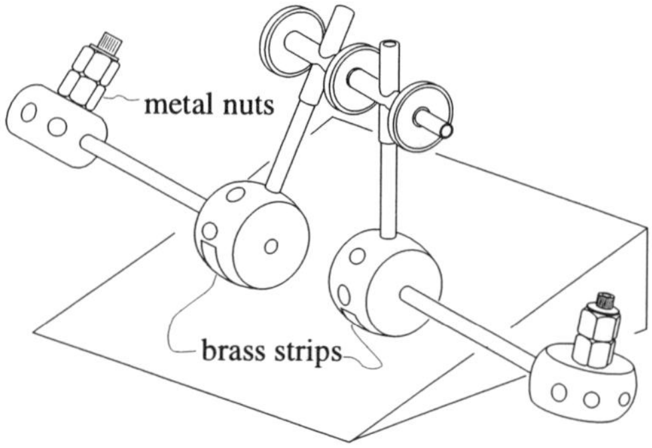

Figure 7. A drawing of our 3D Tinkertoy® walking model. The center-of-mass of the device is above the centers of the wheel-like feet and behind the leg axes. The metal nuts for weight and the brass strips to round the foot bottoms are fastened with black electrical tape.

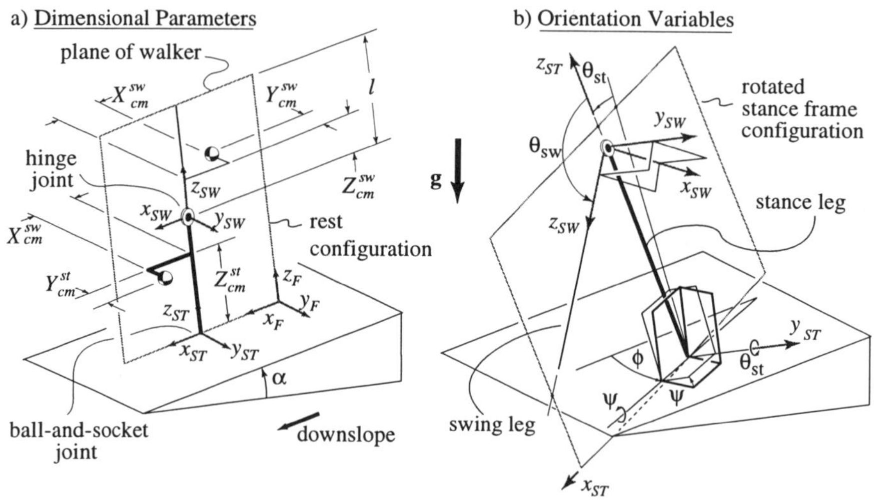

Figure 8. The orientation variables and parameters for the 3D straight-legged point-foot walking model. Each leg has mass M, moment of inertia matrix $\mathbf{I}^{cm}$, and length l.

models that incorporate neuro-muscular elements or mechanical actuators.

References

1. M. Coleman, A. Chatterjee, and A. Ruina. Motions of a rimless spoked wheel: A simple 3D system with impacts. *Dynamics and Stability of Systems*, 1997. In press.
2. M. Coleman and A. Ruina. A Tinkertoy® model that walks. *Physical Review Letters*, 1997. accepted for publication.
3. M. J. Coleman. *A Stability Study of a Three-dimensional Passive-dynamic Model of Human Gait.* PhD thesis, Cornell University, Ithaca, NY, 1997. In preparation.
4. J. V. Fowble and A. D. Kuo. Stability and control of passive locomotion in 3-D.

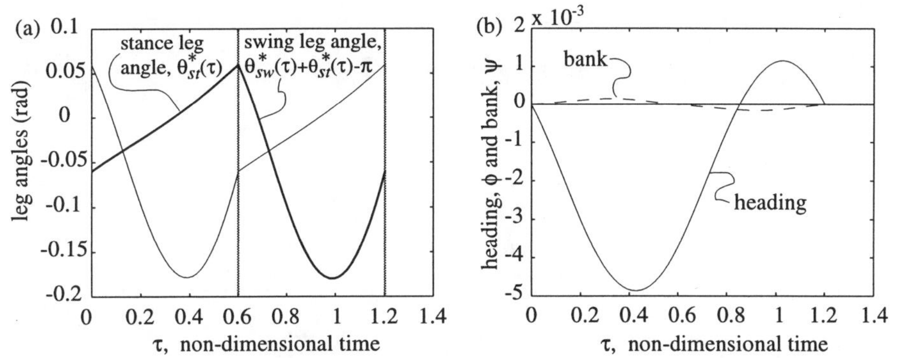

Figure 9. Simple 3-D simulations. Typical periodic gait cycle behavior over two steps for $I_{xx} = 0.5577$, $I_{yy} = 0.00021$, $I_{zz} = 0.5579$, $I_{xy} = 0.0000$, $I_{xz} = 0.0$, $I_{yz} = 0.0$, $\alpha = 0.0037$,$x = 0.0$, $y = 0.2706$, and $z = 0.9270$. The fixed point for this case is $\mathbf{q}^* = \{0.0000\ 0.000008\ -0.0597\ 3.2610\ -0.0132\ 0.00051\ 0.1866\ -0.8523\}^T$; the maximum eigenvalue is $|\sigma_{max}| = 2.58$ and the non-dimensional step period is $\tau^* = 1.2031$. (a) The periodic gait cycle leg angles are very similar to those for 2D walking. The heavy line on the graph corresponds to the motion of the heavy-lined leg in Figure 8. At start-of-step, this is the stance leg, but at a foot-collision, it becomes the swing leg. The instants of foot-strike are denoted by the light gray lines. (b) The plots show the relationship of the heading and the bank angle of the walker over two steps.

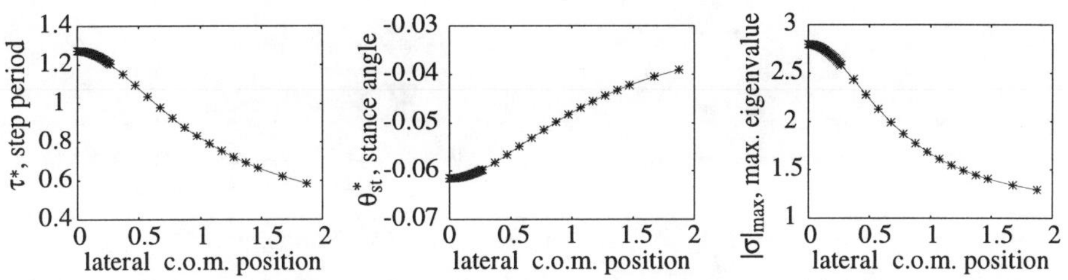

Figure 10. Point-foot 3D simulation. The effect of lateral c.o.m. mass position on the limit cycle period, stance angle, and maximum eigenvalue modulus. Note that the simple simulation does *not* predict stability whereas the more complex physical model *is* stable.

In *Biomechanics and Neural Control of Movement*, pages 28–29, Mount Sterling, Ohio, 1996. Engineering Foundation Conferences.

5. M. Garcia, A. Chatterjee, A. Ruina, and M. J. Coleman. The simplest walking model: Stability, complexity, and scaling. *ASME Journal of Biomechanical Engineering*, 1997. In press.
6. A. Goswami, B. Thuilot, and B. Espiau. Compass-like biped robot, part I: Stability and bifurcation of passive gaits. Rapport de recherche 2996, Unité de recherche INRIA Rhône-Alpes, St. Martin, France, October 1996.
7. T. McGeer. Dynamics and control of bipedal locomotion. *Progress in Robotics and Intelligent Systems*, 1990.
8. T. McGeer. Passive dynamic walking. *The International Journal of Robotics Research*, 9(2):62–82, April 1990.
9. T. McGeer. Passive dynamic catalogue. Technical report, Aurora Flight Sciences Corporation, 1991.
10. J. Papadopolous. personal communication, 1996.

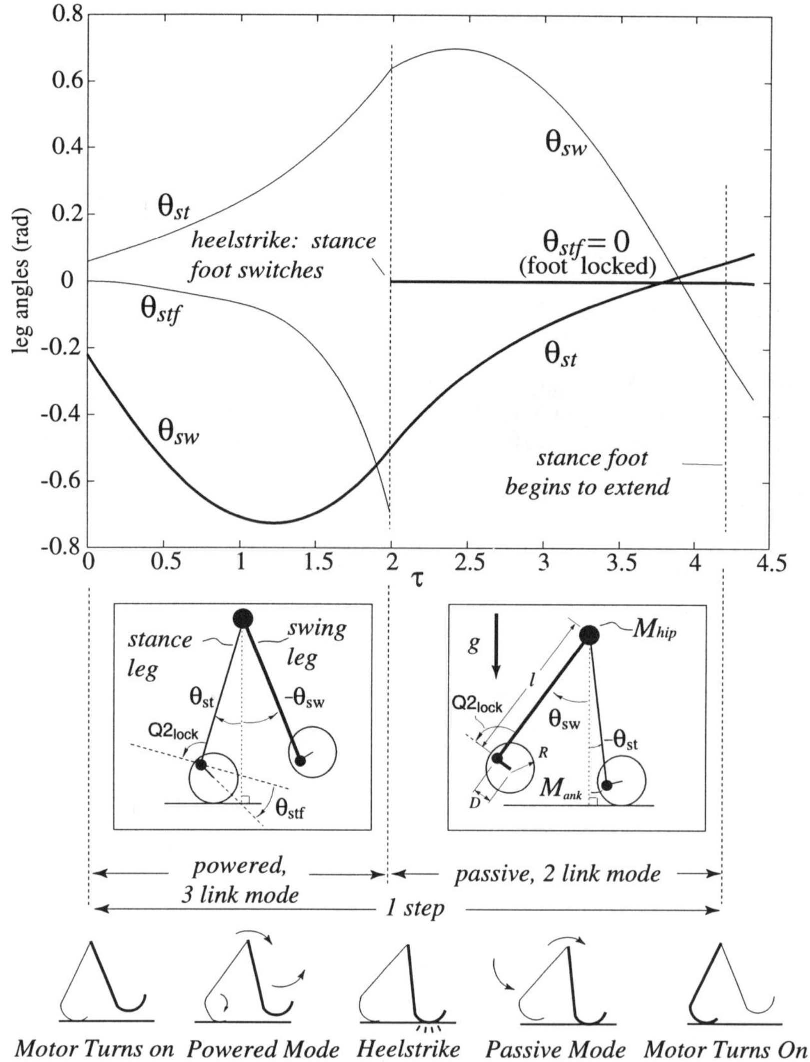

Figure 11. The state of the powered 2D walker versus time over one stable gait cycle. The two inset boxes show the parameters and orientation variables of the 2D powered walker gait cycle. Parameters are as follows: $M_{ank} = 1, M_{hip} = 1000, l = 1, R = 0, D = 0.05, Q2_{lock} = 3\pi/4$, and $Q3_{switch} = 2.8625$. DC motor characteristics are $\omega_{no-load} = 100$ and $T_{stall} = 725$. The actuation begins when $Q3 = Q3_{switch}$, where $Q3 = \pi + \theta_{sw} - \theta_{st}$ (the swing leg angle measured from the projection of the stance leg).

11. D. A. Winter. *The Biomechanics and Motor Control of Human Gait.* University of Waterloo Press, Waterloo, Ontario, 1987.

ON THE MULTI-DEGREE OF FREEDOM, NONLINEAR DYNAMICS OF SHIP MOTIONS WITH APPLICATION TO THE BROACHING PROBLEM

K.J. Spyrou and S.R. Bishop
Centre for Nonlinear Dynamics and its Applications
University College London
Gower Street, London WC1E 6BT, UK

1. Introduction

Over the last few decades there has been substantial progress in the theoretical prediction of ship motions and there is now a solid body of knowledge which, under the banner of seakeeping theory, can be utilized in ship design [1-3]. However, due to the intractable nature of the hydrodynamic problem posed by the motion of a body at the water-air interface, much of this knowledge is based on linearized dynamics. Seakeeping theory is useful for assessing the average performance degradation in rough seas and for deducing certain operability limitations on the basis of the anticipated magnitude of wave-induced motions or accelerations, at various locations along the ship hull. It can even predict the occurrence of certain large amplitude phenomena such as bottom slamming, deck wetness and propeller emergence. These go some way towards addressing the important problem of ship survivability in extreme sea-states. However, when the response contains sufficient nonlinearity this approach cannot offer satisfactory insight into the dynamics.

The roll restoring function is perhaps the best known source of nonlinearity in ship motions and indeed it is the one that generates the possibility of capsize. The capsize problem has been studied to considerable depth recently [4-7]. Instability phenomena can also arise however in the other degrees of freedom although this is less frequently acknowledged in the literature. Typical examples are, directional instability which is a feature of the sway and yaw motions, surf-riding that involves surge and porpoising for high-speed planing craft in pitch. An additional matter of concern is the fact that coupling mechanisms can give rise to safety-critical internal transfers of energy engaging directions of motion that appear at first sight to be unrelated. Internal resonances have been predicted for the pitch or heave coupled with roll in longitudinal seas [8] and in heave coupled with roll in beam seas [9]. In general, however, little has been done so far in exploring mechanisms that involve more than one degrees of freedom.

2. Broaching: Description of the problem

An effort aimed at unveiling and classifying the instability mechanisms that can arise when steep waves approach a ship from various directions is currently underway. Presently, attention is focused on a phenomenon that has puzzled naval architects for several years, namely an instability on the horizontal plane known as broaching. It is potentially encountered when steep waves approach a ship from behind. Broaching is the sudden loss of heading by an actively steered ship, that gives rise to quick build-up of significant deviation from the desired course. In the "post-critical" stage the interest is largely on

F. C. Moon (ed.),
IUTAM Symposium on New Applications of Nonlinear and Chaotic Dynamics in Mechanics, 417–424.

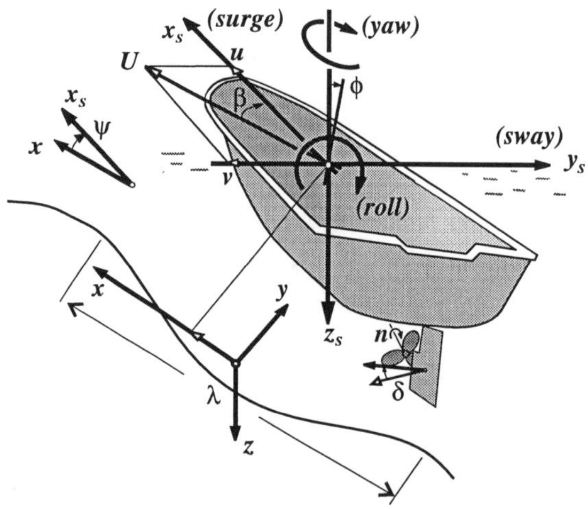

Fig. 1 : Basic variable and parameter definitions

roll motion because broaching sometimes results in capsize. In Fig.1 we introduce the terminology that will be used in the present article.

Broaching is often associated with ship speeds near to the wave celerity. Accordingly we first consider the surge response of a ship under a gradual increase of speed in a following sea. Fig. 2 shows schematically the main changes in ship response. The parameter varied is the rate of rotation of the propeller but it is generally preferable to present the information on the basis of another parameter that has a more universal meaning, the nominal Froude number *Fn*. This is basically a nondimensional version of the corresponding steady speed in still-water. For low *Fn* there is only the ordinary periodic response [plane (a) of Fig. 2] at the frequency of encounter between the ship and the wave (the frequency of encounter is defined as $\omega_e = k[c - U\cos(\psi - \beta)]$ where k, c are respectively wave number and celerity and U is the speed of the ship). However, as the wave celerity is approached (which means that ω_e tends to 0) the response becomes asymmetric [plane (c)]. At the same time an alternative, stationary type of behaviour can emerge [to the right of plane (b)], where the ship is carried forward by a single wave! This condition is called surf-riding [10 -11] and is characterized by two speed thresholds: The first is where stationary behaviour becomes possible due to static balance between the longitudinal wave force and the hydrodynamic resistance of the ship [plane (b)]. The second represents the disappearance of the periodic motion that creates global attraction towards surf-riding [see planes (c) and (d)]. This happens due to a homoclinic saddle connection as the periodic orbit collides with the unstable fixed point [11] that is physically located in the vicinity of the wave crest.

The underlying dynamics of surge resemble those of a pendulum with constant torque. The periodic motion of the ship as is overtaken by the waves corresponds to full rotations whereas surf-riding represents the equilibrium state. Nonlinearity exists primarily due to sinusoidal type restoring in terms of the relative position of the ship on the wave. Nonlinearity may also influence the dynamics through damping due to the nonlinear nature of the resistance curve of a ship.

It has been suspected for long time that surf-riding may play the role of precursor of broaching especially when the heading of the ship does not coincide with the direction of propagation of the incident wave (an angle between the 10 and 30 deg is thought to be the most dangerous) [12]. Consideration of nonzero heading angle introduces however degrees of freedom in addition to surge, thus leading to a multi-degree system. New nonlinear effects come into play, through position-dependent wave loads in sway and yaw. Nonlinearities with an origin in the hydrodynamic reaction due to the motion of the ship in these directions can also arise if the ship's yaw rate or drift become significant. These are

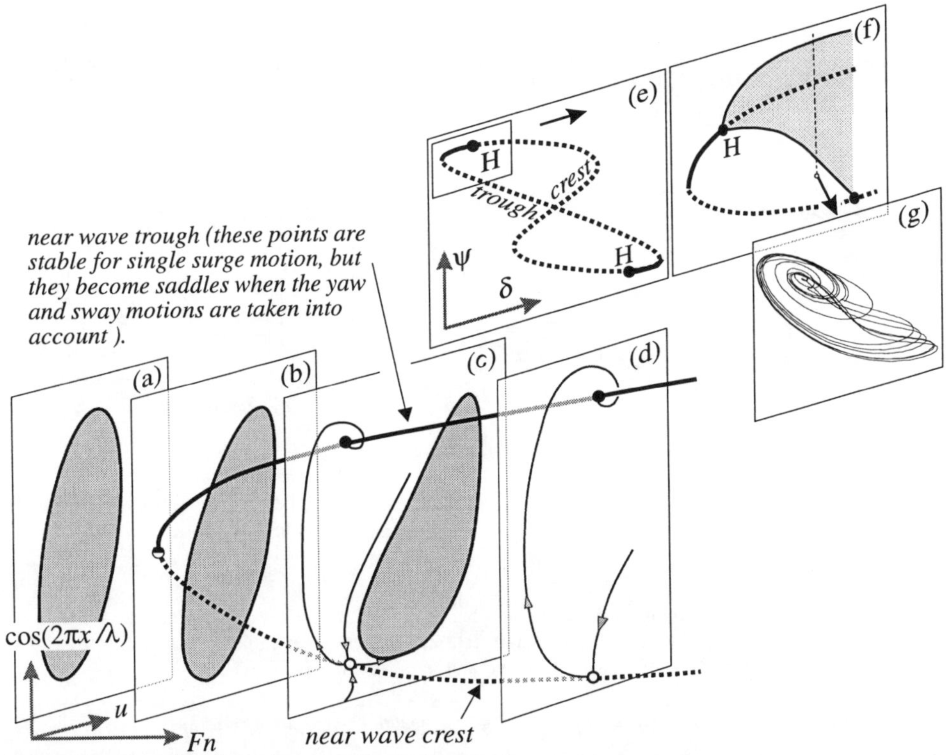

Fig. 2 : The main stages of surf-riding, from its onset to its final disappearance.

damping-like effects (the so called added masses/moments of inertia have been shown to be primarily linear). Sway and yaw motions can further induce roll since the lateral resistance acts generally off the roll centre of rotation. Roll couples back to the sway/yaw pair because the lateral resistance of the inclined hull is generally different from that of the upright. These new effects introduce additional possibilities for capsize.

Finally, it is pointed out that the behaviour of the moving ship is governed by the hydrodynamics not only of the hull, but rather of the hull-propeller-rudder combination. It is essential, for example, to take into account that the rudder acts as a lifting surface placed in the slipstream of the propeller as well as in the wake of the drifting and/or turning hull.

3. Mathematical model

In order to assemble all the above considerations into a mathematical model we assume the ship to be a rigid body and we apply Newton's second law for a non-inertial system that moves with the ship. The equations of motion in surge, sway, yaw and roll (assuming the existence of static equilibrium in heave and pitch) are as follows, Fig 1 :

surge : $$m\,(\dot{u} - r\,v - x_s^{(G)}\,r^2 + z_s^{(G)}\,p\,r) = X_{(H)} + X_{(W)} + X_{(P)} + X_{(R)}$$

sway : $$m\,(\dot{v} + r\,u + x_s^{(G)}\,\dot{r} - z_s^{(G)}\,\dot{p}) = Y_{(H)} + Y_{(W)} + Y_{(P)} + Y_{(R)}$$

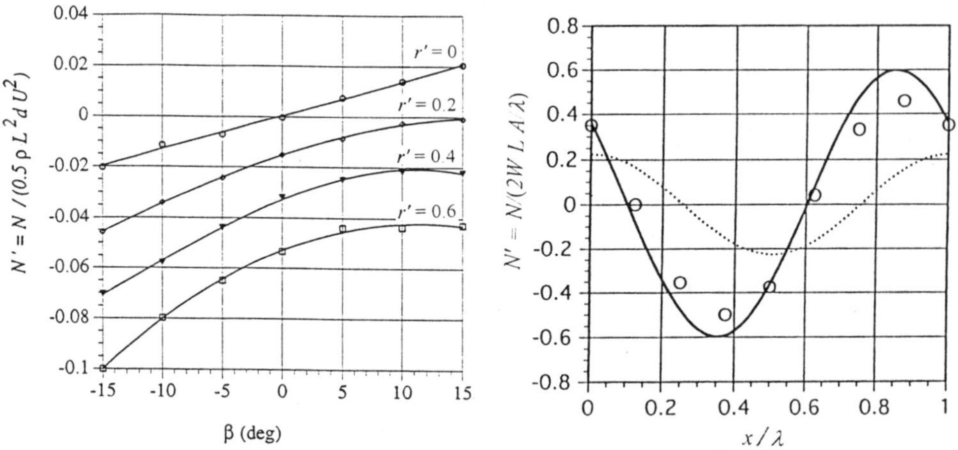

N' = -0.0739947 r' - 0.0819427 v' - 0.0802578 $r'|r'|\,v'$ -
-0.487212 $r'\,v'|v'|$ - 0.0219821 $r'|r'|$

(a) (b)

Fig 3 : Experimentally measured moments in yaw based on a 2.3m model of a 34.5m purse-seiner.
(a) Hydrodynamic reaction (experimental data denoted by symbols) and polynomial fit.
(b) Wave moment: Comparison of theory and experiment [14].

$$\text{yaw:}\quad I_z \dot{r} + m\, x_s^{(G)} (\dot{v} + r\,u) = N^{(H)} + N^{(W)} + N^{(P)} + N^{(R)}$$

$$\text{roll:}\quad I_x \dot{p} - m\, z_s^{(G)} (\dot{v} + r\,u) = K^{(H)} + K^{(W)} + K^{(P)} + K^{(R)}$$

The letters X, Y, N and K denote the external loads on the ship, respectively in surge, sway, yaw and roll. The superscripts $^{(H)}$, $^{(W)}$, $^{(P)}$ and $^{(R)}$ represent fluid reaction, the wave, the propeller, and the rudder. Also, u, v are surge and sway velocities and r, p are yaw and roll angular velocities; m is ship mass, I_x, I_z are roll and yaw mass moments of inertia and $x_s^{(G)}$, $z_s^{(G)}$ denote the longitudinal and vertical position of the ship's centre of gravity.

The hydrodynamic reaction on the hull normally includes contributions from ideal potential flow, hull lifting, cross-flow and separation effects. Neglecting the frequency dependence (related with wave-making of the hull and viscous effects) that, in the time domain, would lead to a convolution-type representation, it is quite common to express the excitations in polynomial form, retaining linear accleration terms and up to second or third-order velocity terms [11]. The coefficients of the acceleration terms can be theoretically predicted while for the velocity terms it is sometimes possible to use existing regression type formulae. Alternatively, the forces can be measured experimentally through 'captive' physical model tests and then the coefficients can be derived through a 2-variable curve fitting method. In Fig. 3a is shown such an example for the damping moment in yaw.

The wave excitation should be decomposed into contributions based on the "Froude-Krylov" assumption (where the pressure field around the ship is only due to the characteristics of the incident wave) and diffraction (the modification of the pressure field due to the presence of the hull). In yaw for example, the Froude -Krylov excitation, is:

$$N^{(W)} = \rho\, g\, A\, k \sin\psi \int_L a(\psi, x_S)\, e^{-k\, d(x_s)}\, x_S\, C(x_s) \sin k(x + x_S \cos\psi)\, dx_S$$

where ρ, g are water density and acceleration of gravity; A is wave amplitude; L is ship length; $C(x_S)$ is the area of a ship cross section lying at x_S and with draught $d(x_s)$. Also, $a(\psi, x_s)$ is a parameter assumed here equal to 1.0. In Fig. 3b we give a comparison between theory and experiment for the wave yaw moment. A discussion about the contribution of diffraction, also taken into account in our calculations, is given in [14].

As is usual, the propeller force is expressed in terms of the so-called thrust coefficient which, in turn, is written as a second-order polynomial of the so-called speed of advance. As for the rudder forces/moments, these are expressed in terms of the so called rudder normal force F_N and the rudder angle δ, as for example in yaw:

$$N^{(R)} = -[1+a_H\,(x_s^{(H)}/x_s^{(R)})]\,x_s^{(R)}\,F_N\cos\delta$$

It should be noted that the rudder normal force depends on the rudder inflow velocity vector. Also, a_H is the so-called rudder-to-hull interaction coefficient which accounts basically for an additional force acting on the hull, at a distance $x_s^{(H)}$ from the middle of the ship. Finally, $x_s^{(R)}$ indicates the location of the rudder.

The rudder is controlled with a linear law that is based on heading angle and rate:

$$\dot{\delta} = t_\delta[-\delta - a_\psi(\psi-\psi_r) - a_r r]$$

where t_δ is the inverse time constant of the steering engine; a_ψ , a_r are respectively the so-called proportional and differential gain; and ψ_r is the desired heading of the ship.

4. Key nonlinear dynamical features of surf-riding at nonzero relative heading angle

As is perhaps obvious, due to the multi-dimensional nature of our system we can visualize behaviour only through 2-d (or possibly 3-d) projections. A number of new features emerge in the dynamics of surf-riding if the waves are allowed to meet the ship also obliquely from behind ("quartering sea"). Some of these are shown on the planes (e), (f) and (g) of Fig. 2. Firstly, by applying fixed-point continuation it has been found that the states of surf-riding lie, in general, on a closed curve in state space [11]. This means that the pair of surf-riding points that are shown to exist to the right of plane (b) in Fig. 2, represent in fact the intersection of this curve with the ψ=0 plane. The range of headings close to the exactly following sea for which surf-riding can arise is however relatively narrow. Numerical studies were carried out both for a rudder-fixed and for an automatically controlled ship, based on models of a 7.14m-fishing vessel and a 34.5m-purse seiner. When the rudder is fixed at preselected angles, surf-riding is in principle unstable, with the exception of two very short and symmetrically located regions, at relatively large heading, which begin with a saddle node and end with a Hopf bifurcation. This is shown in Fig.2 [plane (e)] through a projection in terms of the parameter δ and the state variable ψ.

The oscillations that emanate from the Hopf bifurcations can lead sometimes to chaotic behaviour [plane (g)]. The chaotic domain is very short in terms of the control parameter and it arises through the period-doubling route. Details may be found in [15]. Unfortunately, the experimental verification of such behaviour through free-running, radio-controlled model tests somehow surpasses the current level of technology in this field. However it is worth mentioning that oscillatory surf-riding has been observed by Kan during model experiments [10] (at that time no theoretical explanation was supplied). The oscillations that are created at the Hopf bifurcation end abruptly with a new homoclinic connection. How this happens is shown on plane (f).

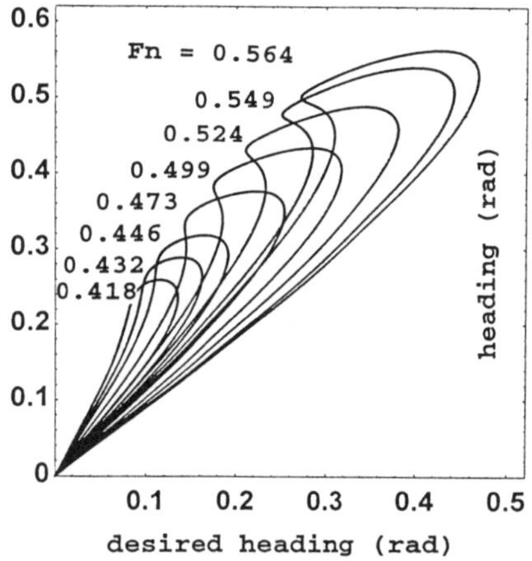

Fig. 4 : Evolution of surf-riding with automatic control for increasing Fn : a_ψ=3, a_r=3, t_δ=1, H/λ =1/20, λ/L=2.0 (H=2A).

If the ship is actively controlled, the desired heading ψ_r should replace the rudder angle δ as the varied parameter, while δ should be incorporated into the state variables. Then, depending on the autopilot gains, stable surf-riding is possible in the vicinity of the wave trough, even at $\psi_r = 0$, Fig. 4. The proportional gain a_ψ controls the location of the saddle node that determines the region of stability. More information about the role of the gains for surf-riding is given in [11].

5. Global behaviour

Broaching can be regarded as a dynamic change of state instigated by the variation of a certain control parameter. Therefore, transient dynamics rather than steady-states dictate the fortune of the ship. The control vector of our system includes a number of ship-based parameters, such as the propeller rate and the desired heading which can be varied at will, as well as exogenous controls that represent the sea state, such as the wave height and length. To facilitate our discussion we shall classify ship behaviour in quartering seas into four categories, as following: (1) surf-riding, (2) Oscillatory-type motion with a clearly defined heading and the waves running before or ahead of the ship (non-zero encounter frequency condition), (3) broaching and (4) capsize. Types (3) and (4) are non-ordinary responses and, obviously, as motion patterns of the ship that could play the role of initial conditions it makes sense to consider only types (1) or (2). However, the parameter change will cause a transition towards any of the four presented types of behaviour, thus creating a rather unusual "multiple-effect" scenario [13]. An additional problem to be dealt with is the multidimensional character of our system. For this we have assumed that the ship, just before the change of parameter setting, was in steady-state.

In Fig. 5 is presented an application of this approach, focusing on the problem of escape from surf-riding. The ship is assumed initially to be lying at a fixed point with its nominal Froude number 'tuned' on the wave celerity (Fn=0.56 if λ/L=2.0) and with some desired heading angle ψ_r in the range where stable surf-riding can exist. The locus of the possible initial conditions is traced in advance using a path-following technique. Then the nominal Froude number is suddenly reduced to $(Fn)_{low}$. The pair $[\psi_r, (Fn)_{low}]$ will be thus our plane of control parameters. It should be pointed out however that, as the initial conditions lie on fixed points, the state and control variables are directly connected. For example from the autopilot equation of Section 3 is easily derived that, for $d\delta/dt = r = 0$, it will be $\psi - \psi_r = -\delta/a_\psi$. This means that we can use ψ [or indeed any of the other constituents of the state vector ($u, v, r, p, \phi, \psi, x, \delta$)] instead of ψ_r, and so we can study the organization of the domains of the four types of behaviour, on the plane $[\psi, (Fn)_{low}]$ which seems to be more relevant.

In Fig. 5 the domain of surf-riding can be seen occupying the upper part of the graph while below it is located the domain of periodic motion. This appears to be however in

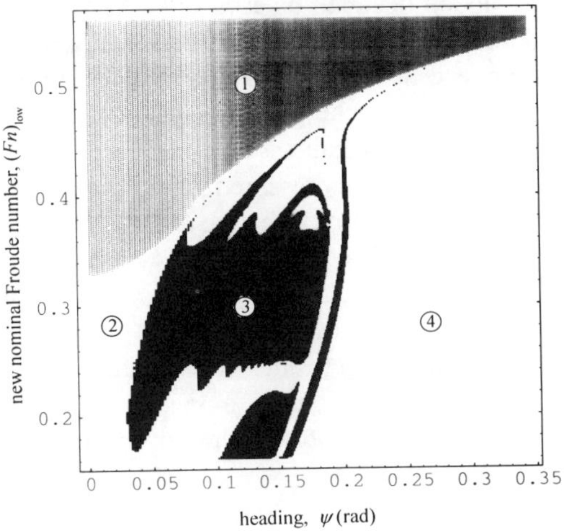

Fig. 5 : The organization corresponding to the escape from surf-riding problem.

competition with the broaching domain. With increasing heading the broaching domain comes very near to developing a direct interface with surf-riding. A boundary of a qualitatively different nature arises due to an intrusion of capsize into the broaching domain. At larger heading capsize dominates. Further studies are currently underway about the character of the boundaries. Quantification of the tendency of a ship for broaching is possible, by measuring relative areas and their change under the effect of selected parameters, in a similar fashion to the "integrity curves" concept of Thompson [4].

6. Broaching with low speed

There have been several reports based on observations (see for example [16]), that broaching can also occur for speeds that are considerably lower than the wave celerity and that, in fact, there is no need for a ship to go through surf-riding. In this case broaching should arise directly from the periodic motion.

Our numerical study of steady periodic responses for the purse-seiner model, considering a range of speeds below the first surf-riding threshold, has shown that in extreme waves, with gradual increase of the desired heading angle a supercritical flip is encountered. Further increase of the control parameter paves the way for a sudden jump to resonance from a fold [17]. The resonant motion is an oscillation that often tends to extend beyond the [-35, 35] range of feasible rudder angles. The wave steepness required to cause this mechanism is however higher than the one through surf-riding. It is likely that with proper parameter variation we can go through the codimension-2 coalescence of the supercritical flip and fold that will give rise to a subcritical flip. There, the change of state will be even more sudden and unexpected. This matter is currently under investigation. It is interesting that in the type of broaching that is discussed here, the divergence from the initial state is of oscillatory type and this agrees very well with observations about a "cumulative-type" character of broaching at lower speeds.

Studies have been carried out also concerning the effect of the autopilot gains for this particular type of broaching [17]. Through proper gain values' selection, especially of the differential one which is the most influential, it is possible to increase the range of headings of safe ship operation.

7. Concluding remarks

The study of broaching from a nonlinear dynamics perspective has helped to understand what causes this phenomenon. However further studies will be needed in a number of areas. Firstly, pitch and heave must be taken into account in a more rational way, particularly when motions away from the zero frequency of encounter condition are studied. Also the effect of frequency dependence of fluid reaction needs to be assessed. At the same time, simplified models that can capture the dominant features of response must be derived in order to improve our understanding of the phenomena at a fundamental level. In this direction we have been successful in showing that the broaching mechanism at low speed is basically a case of parametric instability, [17]. Thirdly, there is a compelling need to carry out extensive free running model tests in order to verify the presented theory. Lastly, the new information must be brought into a form that would allow effective use of it by naval architects during the design stage of a ship.

8. Acknowledgements

The experimental results of Fig. 3 and some of the numerical results were obtained during the first author's stay at the National Research Institute of Fisheries Engineering of Japan. Also, the first author acknowledges his current financial support by the European Union with a Human Capital and Mobility Fellowship.

9. References

1. Price W.G.. (ed.) : *The Dynamics of Ships*, The Royal Society, London (1991).
2. Lloyd, J.R.M.: *Seakeeping: Ship Behaviour in Rough Weather*, Ellis Horwood Series in Marine Technology, Chichester (1989).
3. Faltinsen, O.: *Sea Loads on Ships and Offshore Structures*, Ocean Technology Series, Cambridge University Press, Cambridge (1993).
4. Thompson, J.M.T.: Designing against capsize in beam seas : Recent advances and new insights, *Applied Mechanics Reviews*, **50**, 5 (1997), 307-325.
5. Nayfeh, A.H. and Sanchez, N.E.: Stability and complicated rolling responses of ships in regular beam seas, *International Shipbuilding Progress*, **37** (1990), 331-352.
6. Kan, M.: Chaotic capsizing, in *Proceedings*, The 20th ITTC Seakeeping Committee, Osaka, (1992), 155-180.
7. Falzarano, J.M., Shaw, S.W., and Troesch, A.: Applications of global methods for analyzing dynamical systems to ship rolling motion and capsizing, *International Journal of Bifurcation and Chaos*, **2**, 1 (1992), 101-115.
8. Nayfeh, A.H.: Undesirable roll characteristics of ships in regular seas, *Journal of Ship Research*, **32**, 2 (1988), 92-100.
9. Thompson, J.M.T. and De Souza J.R.: Suppression of escape by resonant modal interactions: shell vibration and heave-roll capsize, *Proc. R. Soc. Lond. A*, **452** (1996), 2527-2550
10. Kan, M.: Surging of large amplitude and surf-riding of ships in following seas, *Selected Papers in Naval Architecture and Ocean Engineering*, The Society of Naval Architects of Japan, Tokyo, **28**, (1990).
11. Spyrou, K.J.: Dynamic instability in quartering seas: The behaviour of a ship during broaching, *Journal of Ship Research*, **40**, 1 (1996), 46-59.
12. Fuwa, T., Sugai, K., Yoshino, T. and Yamamoto, T.: An experimental study of broaching of a small high speed boat, *Papers of Ship Research Institute*, Tokyo, **66** (1982), 1-43.
13. Spyrou, K.J.: Dynamic instability in quartering seas - Part II: analysis of ship roll and capsize for broaching, *Journal of Ship Research*, **40**, 4 (1996), 326-336.
14. Spyrou, K.J. and Umeda, N.: From surf-riding to loss of control and capsize: A model of dynamic behaviour of ships in following/quartering seas, in *Proceedings*, 6th International Conference of Practical Design of Ships and Mobile Units (PRADS), (1995), Seoul, 494-505.
15. Spyrou, K.J.: Homoclinic connections and period doublings of a ship advancing in quartering waves. *CHAOS*, **6**, 2 (1996), 209-218.
16. Conolly, J.E.: Stability and control in waves: A survey of the problem, *Journal of Mechanical Engineering Science*, Supplementary issue, **14**, 7 (1972), 186-193.
17. Spyrou, K.J.: Dynamic instability in quartering seas-Part III: Nonlinear effects on periodic motions, *Journal of Ship Research*, **41**, 3 (1997) (in press).

VIII. FLUID-ELASTIC PROBLEMS

This group of nine papers was the largest group covering topics such as aeroelasticity, rotorcraft, aircraft stability and hydrodynamic systems. The key lecture by Dowell et al. summarized the nonlinear dynamics of areoelastic systems and provides many references. Macmillan and Thompson presented a view of how nonlinear dynamics can be employed in the aircraft industry.

Heat exchanger tube dynamics are examined in the papers of Païdoussis et al. and Thothadri and Moon.

Readers can learn how to build a hydrodynamic, chaotic water wheel demonstration in the paper by Kriebel et al. These chaotic demos are now in many museums and educational institutions.

F. C. Moon (ed.),
IUTAM Symposium on New Applications of Nonlinear and Chaotic Dynamics in Mechanics, 425.

NONLINEAR DYNAMICS OF AEROELASTIC SYSTEMS[1]

E. H. Dowell, J. A. Jones Professor and Dean of the School of Engineering
L. N. Virgin, Associate Professor
D.M. Tang, Research Associate
M.D. Conner, Research Assistant
Duke University, Durham, North Carolina 27708-0300

Abstract

This paper is an overview of a research program for nonlinear aeroelasticity of aircraft and rotorcraft in forward flight conducted at Duke University and supported by the Air Force Office of Scientific Research and the Army Research Office. Theoretical and experimental investigations of aircraft and rotor aeroelastic stability and nonlinear response are described. Flexible airfoil and rotor blade models with geometrical and freeplay structural nonlinearities and the ONERA stall aerodynamic model are considered. Analytical and numerical solutions of nonlinear mathematical models and experimental methods for flutter and forced response in a wind tunnel environment are discussed. Correlations between theoretical and experimental results are presented and some significant conclusions are drawn. It is hoped that this paper will help provide improved fundamental understanding of the nonlinear aeroelastic behavior of aircraft and rotorcraft.

This work was supported by the Air Force Office of Scientific Research under Grant F49620-97-1-0063 and the Army Research Office under Grant DAAL03- 87-K-0023; Major Brian Sanders and Dr. Gary Anderson are the technical monitors.

1. Introduction

Aeroelasticity in aircraft and rotorcraft is concerned with the dynamic interaction between the air stream and the elasticity of the aircraft or rotorcraft structure. The air stream cannot only provide sufficient energy through the aircraft or rotor system to support flight in air, but it is also a potential source of aeroelastic instability such as flutter as well as forced vibration. An acceptable structural design should avoid or ameliorate such dynamic response. However, it is not easy to reach this goal. There are many linear and nonlinear stability and forced response phenomena that need to be studied. During the last thirty years numerous papers

[1] An earlier version of this paper was presented as a plenary lecture at the International Forum on Aeroelasticity and Structural Dynamics,June 17-20, 1997, Rome, Italy

F. C. Moon (ed.),
IUTAM Symposium on New Applications of Nonlinear and Chaotic Dynamics in Mechanics, 427–449.

and reports on linear aeroelasticity in aircraft and rotorcraft have been published. Here the focus in on nonlinear effects.

Some readers will recall a seminal article entitled, "The Engineer Grapples with Nonlinear Problems," written by Theodore Von Karman [1] many decades ago. Today, engineers continue to grapple with nonlinear aeroelasticity in aircraft and rotorcraft, but on the whole more successfully due to work of numerous individuals. This paper is an overview of a research program for nonlinear aeroelasticity of aircraft and rotorcraft in forward flight conducted by the authors and their colleagues. The emphasis is on physical understanding and describing the nonlinear aeroelastic phenomena theoretically and experimentally. The goal is not to model any particular practical aircraft or rotorcraft, but rather to develop accurate models of the several possible phenomena that may be verified by experiments. Various aspects of aircraft and rotor aeroelastic stability and nonlinear response are included in this program and described in this paper including the following topics.

1. The study of the subject will be motivated by considering the most significant physical nonlinearities that occur in aeroelasticity of aircraft and rotorcraft.
2. The consequences of nonlinearity for constructing mathematical models and analytical solution procedures will be discussed.
3. A comparison and discussion of experimental and theoretical results will be presented.

For a discussion of nonlinear aeroelasticity in a more general context see [2,3]. Two of the authors have reviewed the work on rotorcraft previously [4] and hence here the emphasis will be primarily on more recent results for an aircraft model. However we will first briefly summarize the key results from our rotorcraft studies in Part I of this paper before turning to the aircraft model in Part II.

Part I: ROTORCRAFT

I-1. Physical Nonlinearities in Aeroelasticity of Rotorcraft

The aeroelastic behavior of a rotor system is most complex. Structural nonlinearity and stall aerodynamics are dominant and the deformations are typically of the order of the blade chord or span. The possible combinations of structural motions (bending in the lift and drag directions and twisting) and flow field conditions (hover and high speed forward flight) are several. Here a few basic nonlinearities and combinations that we have studied in our work will be addressed.

I-1.1. STRUCTURAL NONLINEARITIES:

I-1.1.1. *Geometrical Structural Nonlinearity*

Hodges and Dowell [5] derived a set of nonlinear equations of motion for the elastic bending and torsion of twisted nonuniform hingeless rotor blades. A moderate deflection theory was developed and the geometrically nonlinear terms up to the second order were taken into account in terms of an ordering scheme. Subsequently, the theoretical computations discussed by DaSilva and Hodges [6,7] indicated that when geometrically nonlinear terms up to third order are retained, the stability characteristics of the hingeless blade are only slightly affected (12% or less) as compared to the results when terms up to only second order retained. Friedmann [8] has emphasized that from the view-point of cost and accuracy, improvements in the aerodynamic model are more effective than the retention of third and higher order geometrically nonlinear structural terms. The stability and structural response are more sensitive to the improved aerodynamic model than to the geometrically nonlinear structural terms. In our work, only the second order geometrically nonlinear terms, $\phi w''''$, $\phi v''''$ and $w''v''$, are included in equations and the cross-section warping terms and other higher order terms are neglected, see Figure.1. v, w and ϕ are the chordwise, flapping bending deflections and twist about deformed elastic axis, respectively. It is noted that the higher order interactive terms included in the effective angle of attack, α, due to the structural motion deformation, ϕ, v, w, and the aerodynamic coefficients are retained.

I-1.1.2. *Freeplay and Parabolic Pitch Stiffness Nonlinearites*

A freeplay pitch stiffness nonlinearity of the rotor system may exist in a helicopter control system due to a loose hinge, or linkage, or possible joint slippage in the system. Because there is an initial pitch angle of the blade, in general, the blade is subjected to a preload moment; hence a static deviation angle , ϕ_p, exists and the freeplay stiffness curve is not symmetric. Also a parabolic pitch stiffness nonlinearity may be found in light helicopters with a direct rod link control system without a power actuator. Typical restoring moment versus rotation characteristics for our experimental model are shown in Figure.2. Curve (a) is for the nominal linear pitch spring stiffness; curve (b) for the parabolic or cubic stiffness and curve (c) for the freeplay stiffness with a preload.

I-1.2. STALL AERODYNAMIC NONLINEARITY:

A high angle of attack is usually required in normal operation for an advanced helicopter with high load and high forward speed. If the angle of attack of the blade sections becomes large enough, dynamic stall of the fluid flow may occur. Typically, this occurs on the retreating blade. When the blade sections undergo repeated excursions into stall, a large hysteresis in the fluid force and moment behavior can arise and lead to a negative pitch damping. Under appropriate conditions, it can excite the blade torsion mode at its natural frequency and lead to stall flutter. Dynamic stall and stall flutter are quite complex as is seen from wind tunnel and flight tests.

There are several aerodynamic models that can be used to predict the dynamic stall behavior of rotor airfoils. The ONERA dynamic stall model developed by Tran and Petot [9] seems to be a particularly useful model for dynamic aeroelastic analysis. This model describes dynamic stall in terms of differential equations that describe the lift, drag and moment coefficients of a blade section element. In our work we have chosen the ONERA dynamic stall model for a systematic investigation of helicopter blade stall flutter and forced response. The effects of large angle of attack and a reversed flow region will be taken into account. For more details, see references [10-15].

For linear aeroelastic behavior in rotorcraft , the two- dimensional incompressible Theodorsen aerodynamic model [16], or Roger's transfer function [17] (in fact, this is an alternative formulation of the Theodorsen model) is frequently used. The unsteady aerodynamic coefficients are expressed only at discrete values of reduced frequency, or the coefficients of transfer function are calculated by a best fit of the Theodorsen function. In fact, there is a connection between the Theodorsen and ONERA aerodynamic models for a given airfoil. The Theodorsen aerodynamic model is a special case of the ONERA model when the effective angle of attack is less than the stall angle of attack. When transforming the linear part of the ONERA model for small angles of attack into the Laplace domain, we obtain an unsteady aerodynamic transfer function with a simple pole similar to Roger's formulation.

I-2. Mathematical Consequences of Nonlinearity

Formulation of the complete aeroelastic equations of motion requires a combination of structural, aerodynamic and inertia terms. The structural and aerodynamic model have been described in the previous section. The inertia loads are obtained in a direct manner by using D'Alembert's principle or by using variational methods. The complexity of a mathematical model may be determined by whether the model is linear or nonlinear and the number of degrees of freedom (DOF) and the number of parameters. Obviously, the fewer the number of needed DOF and the number of parameters, the less complex the model. It is also clear that linear models are simpler than nonlinear ones. However it may be worthwhile to note a fundamental blurring of the distinction between parameters and degrees of freedom that occurs when the model is nonlinear. For a linear system the long time, steady state solution is independent of the initial conditions, though it does depend on model parameters. However for nonlinear midels the long term, steady state solution may depend upon initial conditions. Thus in a nonlinear model, initial conditions become possible parameters.

After formulation of the equations of motion, a general solution approach is to carry out a structural spatial discretization. For a flexible rotor blade with bending-torsional coupling, the spatial discretization often consists of applying the well known Galerkin method based upon the free vibration modes of the rotating blade. If N modes are used for each of the flapwise, chordwise and twist directions, there are $3N$ structural DOF and $6N$ DOF in the state space.

The ONERA dynamic stall model is based on a blade airfoil element. It is applied to an isolated blade motion in hover and forward flight which satisfies the force and momnet equilibrium trim requirement. A simple assumption is that the blade is divided into several spanwise aerodynamic sections, say NN, and the ONERA model is applied to each section both for the rigid or flexible blade . Generally, there are three aerodynamic degrees of freedom, $C_{za}, C_{z\gamma}$ and C_{zb} in the state space for each aerodynamic section, and for each aerodynamic coefficient, lift coefficient (C_l), drag coefficient (C_d), and pitch moment coefficient (C_m). The total DOF of the aerodynamic model is thus $3 \times 3 \times NN$.

When either structural nonlinearity or aerodynamic nonlinearity is included in the mathematical model, it may be difficult to obtain explicit equations of motion. The blade motions are functions of the aerodynamic coefficients and, in turn, the aerodynamic coefficients depend upon the blade motion. Therefore, an implicit solution approach is frequently applied to the equations of nonlinear aeroelasticity which mandates iterative solutions or requires considerable additional algebraic effort. For the pitch stiffness nonlinearities, the additional complexity of the algebraic procedure is significant, because the pitch stiffness characteristics change with the blade motion. Therefore, the critical flutter airspeed and response is a function of the amplitude of blade oscillation. A linearization technique, e.g. the harmonic balance method, is often used to develop an equivalent linear pitch stiffness which is represented as an algebraic equation and then added into the original nonlinear ordinary differential equations to provide a more readily soluble mathematical model. For details, see reference [12].

It is important to note that for a rotor in forward flight, the nonlinear equations have periodic parameter excitation terms which appear in explicit form in the damping and stiffness terms of the equations. When we solve these nonlinear equations, the first step is to use a perturbation procedure to obtain a set of linear equations for small motion about an equilibrium position. The solutions of the (nonlinear) equilibrium position are contained in a set of nonlinear algebraic equations which are solved by an iteration procedure such as the Newton- Raphson method. The second step is to solve the linear, perturbation equations with the periodic coefficients using Floquet theory [19].

Of course, the full nonlinear equations can also be solved by the nonlinear periodic shooting method [18],[19] or by direct numerical time marching integration.

By contrast for an airfoil plus control surface model, there are only three structural degrees of freedom (or six states) plus two additional states from the simplified but accurate aerodynamic model. See the discussion in Part II.

I-3. Experimental Methods for Flutter and Forced Response

All flutter and forced response tests were performed in the Duke University low speed wind tunnel. The wind tunnel is a closed circuit tunnel with a test section of 2.3x1.75 ft^2 and a length of 5 ft. The maximum air speed attainable is 293 ft/sec. An overall view of the wind tunnel test model, measurement and excitation systems is represented in this section.

I-3.1. EXPERIMENTAL MODELS

The rotor blade experimental model includes two parts: a helicopter blade section and a root support mechanism.

There are two blade model configurations used in our experiment. One is a rigid blade model which is rectangular and untwisted. The blade is constructed from a stiff aluminum beam with mass per unit length uniformly distributed along the blade span and a wood styrofoam fairing covering the entire chord and span. The other model is a flexible blade which is rectangular, untwisted and flexible in the flap, lag and torsional directions. The blade is constructed from a flexible aluminum beam with eight light wood styrofoam fairing elements covering the entire chord and span which provide the aerodynamic contour of the blade.

The root mechanism is mounted to a very heavy support frame which is attached to the ground. A root support mechanism allows the blade to have rigid body flap and pitch motion degrees of freedom. The two degrees of freedom are independent of each other. The pitch axis of the blade is supported on upper and lower bearings in a blade root socket and is free to move in the pitch direction. The pitch spring uses a steel wire spring material. The steel wire is inserted tightly into a slot of the blade root socket. According to the different constraint condition at the support of the pitch spring in the experimental model, three different pitch stiffness characteristics were obtained. When the support at the end of pitch spring is allowed to have rotation and slip motions without transverse motion, the linear pitch stiffness is obtained. When the support point has a transverse gap, the freeplay structural nonlinearity is obtained, and the magnitude of the deadspace can be adjusted. When the support point is tightly clamped (both transverse motion and rotation at this support point are equal to zero), the pitch stiffness has a hard nonlinear spring characteristic with a parabolic or cubic moment-rotation relationship. The flap axis of the blade is supported on two precision ball bearings and is free to move in the flap direction. The natural frequencies of flap and pitch motion can be adjusted by moving the support position of the flap springs and pitch spring (wire spring), respectively. The effective mean pitch angle can be adjusted by rotating the root support mechanism as shown in Figure.3.

I-3.2. MEASUREMENT AND EXCITATION METHODS

Strain gages were used for measurement of flap and elastic torsional deflections. A micro-accelerometer was installed at the tip of the blade for measuring the chordwise response. The pitch and flap angular displacements at the blade root was measured by the rotational velocity/displacement transducers, RVDT. The digitized response data could be graphically displayed either on-line or off-line as a time history, phase plane plot, FFT, PSD, or Poincare map.

Two excitation methods are used. One is a mechanical base pitch excitation where the driving frequency and amplitude can be adjusted. In this system, only a pure single harmonic excitation in the pitch direction is provided. The other excitation is an aerodynamic gust which is described in [4] and [15].

I-4. Theoretical and Experimental Results

I-4.1 PARABOLIC PITCH STIFFNESS NONLINEARITY

Figure 4 shows the limit cycle flutter amplitude and oscillation frequency vs airspeed for the parabolic torsional stiffness nonlinearity. Figure 4a gives the flutter amplitude, and 4b the flutter frequency. In the figure, the theoretical curve is indicated by the solid line, and the experimental results are indicated by the symbol ∘. The agreement is generally good. Both theory and experiment show that the limit cycle flutter amplitude includes many higher harmonic components in addition to the dominant flutter oscillation frequency, and the magnitude of those higher frequency components increases as airspeed U increases. The dominant oscillation frequency (flutter frequency) is 14.5 Hz, but higher order frequency components of two and three times the dominant frequency also appear in this limit cycle response due to the pitch structural nonlinearity.

For forced response behavior, the motions are periodic. Figure 5 shows the frequency response curve of the pitch motion for $U = 15$ m/s and $\theta_s = 1^0$. The theoretical curve is constructed from the harmonic balance, Newton-Raphson algorithm. The linear frequency response curve is also drawn in the same figure. For comparison the time integration, numerical solutions are plotted in this figure as well. Two stable solutions are found for a range of base excitation frequency and different initial conditions. For example, when $\omega = 25Hz$, the amplitude is equal to 0.0373 for $\phi_0(0) < 0.24$, and 0.154 for $\phi_0(0) \geq 0.24$. The limit cycle amplitude and equilibrium position may vary with the initial conditions. It is noted that the pitch motion is dominant as a consequence of the pitch only excitation and a substantial flap aerodynamic damping. Experimental data are also plotted in Figure 5 for confirmation of the theoretical prediction. There is a weak jump phenomenon from $\omega = 23Hz$ to $24Hz$. Because of safety limitations of the model configuration, we could not obtain the larger pitch responses experimentally that were predicted theoretically. Also the higher frequencies were not accessible to our experimental model, i.e. $\omega \geq 28Hz$.

I-4.2. FREEPLAY PITCH STIFFNESS NONLINEARITY

Figure 6 shows air stream or flow speed vs limit cycle flutter amplitude response normalized by ϕ_a, μ_r, for a magnitude of deadspace in pitch of $\phi_a = 1.5^0$, and an initial pitch angle, $\theta_0 = 0^0$. In the figure, the theoretical curve is indicated by the solid line and dotted line. The solid line denotes a stable limit cycle vs airspeed, while the dotted line denotes an unstable limit cycle. These two states vary with the airspeed and, upon their coalescence, bifurcation occurs as confirmed by the theory of nonlinear oscillations. The experimental results are indicated by the symbols $\bullet$, $\odot$ and $\triangle$. The symbol $\bullet$ denotes the motion is chaotic, $\odot$ denotes the motion is convergent to a static equilibrium position, and $\triangle$ denotes the motion tends to a limit cycle oscillation. Also, the results obtained from the time integration are plotted in these figures. The symbol $\otimes$ denotes the motion is chaotic and the symbol $\square$ denotes the motion tends to a limit cycle oscillation or tends to be convergent to a static equilibrium position. The agreement of theory with experiment is good.

A chaotic oscillation is observed in a narrow airspeed range and for certain initial conditions. It occurs in an amplitude-sensitive region. Figure 7 shows the chaotic oscillation time history of the pitch motion for $\theta_0 = 0^0$ and $U = 12.42m/s$. In the figure, (a) is from the theory, and (b) is from the experiment. An evident behavior is the global large oscillation around all three rest positions, $\phi = 0, \phi_{p+,-} = \pm 1.25^0$, with a small fluctuation around the two static equilibrium positions, $\phi_{p+,-}$. The pitch motion is more dominant as compared with the flap motion. The motion is similar to that of a buckled beam under large harmonic excitation [3].

The limit cycle and chaotic motions may be quite sensitive to the initial condition and airspeed. As an experimental example of the system sensitivity to the initial condition, in Figure 6 there are two values of μ_r for U=11.5 m/s. When the initial pitch angle, $\phi_0(0) = 0.5^0$, the motion converges to a static equilibrium position ($\phi = 1.15^0$). When the $\phi_0(0) > 1.5^0$, the motion tends to a limit cycle oscillation. Corresponding to the above example, the theoretical prediction also exhibits this feature As an another example, when a small airspeed increment or a large external disturbance is added to a chaotic motion, the motion will tend to a limit cycle oscillation. When the airspeed increases from 12.42 to 12.8 m/s, the motion becomes a limit cycle oscillation, evolving from a steady chaotic motion. Similar results have been obtained for flexible blades and gust excitation; these effects are discussed in [4].

I-5. Conclusions

A variety of structural and aerodynamic nonlinearities occur in rotorcraft which may lead to complex dynamic responses including chaos. Mathematical models have been developed to describe these responses and their accuracy confirmed by experiment.

Part II: AIRCRAFT

II-1. Physical Nonlinearities in Aircraft

Nonlinear stall aerodynamics may also be an issue for aircraft as can nonlinear transonic flow effects at high speeds. These topics are still under active study and much work remains to be done. Here we concentrate on the effects of freeplay in an airfoil plus control surface model (aileron) with plunge, pitch and flap degrees of freedom.

The potentially serious effect of a freeplay nonlinearity on the response of airfoils is well established [20-22], especially within the context of wear and maintenance where a degradation of moving joints is inevitable. Given the relatively complex nature of the fluid-structure interaction for an airfoil containing a control surface freeplay nonlinearity it is inevitable that any theoretical study must rely heavily on numerical methods. Historically a limited number of approximate analytical techniques have been developed to incorporate piece-wise linearity into the analysis primarily based on harmonic balance [21], but it will be seen that the assumption of periodic response is often inappropriate.

The aeroelastic section under study is shown in Figure 8(a). The degrees of freedom are: plunge, h; pitch, α and flap, β. The freeplay is present at the axis of rotation of the aileron or control surface at the point of attachment to the airfoil. The system is modeled as a number of connected linear sub-systems which switch abruptly due to a discrete change in stiffness. This is shown schematically in Figure 8(b) where the flap experiences no restoring force between certain contact angles. However, once the critical angle δ is exceeded then the system experiences a linear restoring force since spring contact is made. Further details of the mathematical modeling can be found in [23-27], consequently only an overview is given here.

Previous experimental research in this area includes the classical work of Theodorsen and Garrick [28, 29] as well as the more recent work of Tang and Dowell, e.g. [30]. The Tang/Dowell studies had highlighted the possibility of finite-amplitude limit cycle oscillations prior to the onset of exponentially growing flutter. The present study confirms the presence of quite complex nonlinear dynamics, for a relatively large range of flow velocity in both the numerical and experimental studies.

This type of investigation is aided considerably by the recent advances provided by dynamical systems theory [31, 32]. Such an approach has been used in related studies [33-35] in aeroelasticity. An innovative aspect of this research concerns quasi-periodic characterization and the *prediction* of changing limit cycle behavior, which, although previously applied to other simpler mechanical systems [36], is considered here for the first time for an aeroelastic system. For further background material the reader is referred to [23, 37].

II-2. Theoretical Model

The following nondimensional equations of motion describe the mathematical model:

$$\ddot{\alpha} r_\alpha^2 + \ddot{\beta}(r_\alpha^2 + (c-a)x_\beta) + \ddot{h} x_\alpha + \alpha r_\alpha^2 \omega_\alpha^2 = M_\alpha \tag{1}$$

$$\ddot{\alpha}(r_\beta^2 + (c-a)x_\beta) + \ddot{\beta} r_\beta^2 + \ddot{h} x_\beta + \beta r_\beta^2 \omega_\beta^2 = M_\beta \tag{2}$$

$$\ddot{\alpha} x_\alpha + \ddot{\beta} x_\beta + \ddot{h} + h\omega_h^2 = L. \tag{3}$$

The aerodynamic modeling is contained in the aerodynamic moments and lift, M_α, M_β and L, and is based on two-dimensional, incompressible, potential flow theory. The x, r, and ω parameters are the reduced distance, radii of gyration and natural frequencies in the various degrees of freedom: they are held fixed and appropriately nondimensionalized. More details of this modeling can be found in [27]. The value of (c - a) represents the distance between the flap axis of rotation and the overall center of rotation of the airfoil as shown in Figure 8(a). In the results section uppercase letters are used for normalized quantities, where the positions are nondimensionalized with respect to the freeplay range [27].

This three-degree-of-freedom mechanical or structural model can be cast in state-space form [24] to give a six dimensional phase space. Two augmented states are added because of the Jones' approximation to the Wagnor indicial function [27] for the aerodynamic modeling. Due to the sudden change in stiffness resulting from the freeplay condition, the state-space model switches between the relevant linear regions [27] as shown in Figure 8(b). A small amount of structural damping was included in the modeling based on a linear viscous assumption and an in-vacuo modal response as determined from experiment.

The central issue is then the following. Suppose all the parameters of the model are held fixed and the flow rate, U, is slowly increased from zero. In the absence of any kind of freeplay nonlinearity the linear response is very well characterized and the stability of equilibrium is lost at the critical flutter speed leading to unbounded oscillations. When the freeplay is present a variety of pre-flutter oscillatory behavior is observed which may be quite remarkably complex. It is the onset of limit cycle behavior and the transition between one type of motion and another that is considered in this paper.

II-3. Numerical Simulation

A 4th-order Runge-Kutta algorithm is used to time march the solution incorporating an accurate location of the switching points when contact is made. The importance of accurately locating this stiffness change has been noted by a number of authors, and in this study use is made of the approach suggested by Hénon [38] whereby a simple change of independent variable can be used to locate the discrete jump in phase space. This is useful because it reduces computation time, improves accuracy, and in a nonlinear system with complex, possibly chaotic, behavior any errors in inaccurately locating the sudden change in stiffness may have a cumulative effect on the long term response.

A number of other numerical techniques are used to characterize the periodicity of limit cycle behavior. This is primarily achieved by constructing a Poincaré section of the trajectory and then investigating stability properties via the behavior of small perturbations. The decay or otherwise of induced transients is used as a stability measure through the calculation of characteristic eigenvalues. It will be seen that certain trends can be detected before instability occurs. Fast Fourier transforms are also obtained and some interesting quasi-periodic behavior is observed. Although much of this behavior is highly complex, it is a interesting to observe certain generic features often encountered in simpler nonlinear dynamical systems.

II-4. Experimental Overview

Some experiments were conducted in the low speed wind tunnel at Duke University. Only a brief description is given here: further details of the experiments can be found in [27] and [37]. The three-degree-of-freedom airfoil was constructed such that varying degrees of freeplay could be incorporated into the model. The NACA 0012 experimental model section is shown in Figure 8. The main wing has a

19 cm chord and 52 cm span and is attached (using two pairs of micro-bearings with pins) to a flap with a 6.35 cm chord and a 52 cm span. The wing is made of an aluminum alloy circular spar beam with a diameter of 2.54 cm and a wall thickness of 0.32 cm, and acts as the pitch axis running through 14 pieces of airfoil plate. A 0.254 mm thick aluminum sheet provides the aerodynamic contour of the wing. Additionally an aluminum tube is incorporated to enable adjustment of the center of gravity. The flap control surface is constructed in a similar manner. Freeplay is incorporated into the model by means of leaf springs freely moving between rigid contact blocks. Upon encountering the blocks, the leaf springs bend to provide the nominal elastic stiffness. The model is supported outside of the wind tunnel at the top and bottom. The main wing supports consist of bi-cantilever beams and allowance is made for plunge motion which is effectively independent from the pitch degree-of-freedom. The characteristic Reynolds number was on the order of 0.5×10^6. The measurement signals in the three degrees of freedom were acquired using RVDT's and the LabVIEW data acquisition and analysis software incorporating a NB-MIO-16 board. The structural inertia, stiffness, damping and frequency data were then measured according to the procedure described in [27]. Initial testing was conducted on the linear system, i.e., with the freeplay clamped down. Results from this study can be found in [37] and showed excellent agreement between theory and experiment.

II-5. Response

Given the freeplay configuration it is relatively easy to conduct preliminary linear testing by reducing the freeplay range to zero. Consider the results shown in Figure 9. Here, the flow rate is increased from zero until the onset of linear flutter (at 23.9 m/s) and the numerical root locus shows the expected movement of the eigenvalues in 9(a). An alternative plot is shown in 9(b) where the imaginary parts of the eigenvalues are evolved as a function of flow rate U. The table in Figure 9(c) shows a good correlation between numerical and experimental characteristics for the linear system.

For the fully nonlinear system three specific freeplay configurations were chosen corresponding to nominal angular gaps or freeplay ranges of: 2.3, 3.66, and 4.24 degrees. Theory suggests the response normalized by the angular freeplay range will be universal as shown in Figure 10. Hence attention is focussed on typical results using gap 3. Starting from the rest state the airflow velocity is gradually increased and after allowing for the decay of transients the displacement in each of the degrees-of-freedom was measured. A similar procedure was followed in the numerical integration. Numerical responses were obtained for all flow rates: numerical data points correspond to periodic behavior and the lines indicate an average amplitude based on RMS. A summary of results is given in Figures 10(a) to 10(d). Four distinct regions of pre-flutter behavior can be observed in this figure. First, up to about 20% of the flutter velocity there is a static response. Here any initial disturbance dies out. For a flow rate slightly greater than 20% there is a discrete jump to a low frequency limit cycle.

The flow velocity for the onset of this limit cycle may be calculated by considering the linear system. For zero flap stiffness, $\omega_\beta \equiv 0$, linear theory predicts a flutter instability for U/U_f greater than 0.36 for any set of initial conditions. However it is seen that a limit cycle occurs for U/U_f as low as 0.19. This limit cycle behavior has a simple and interesting physical explanation as follows, as has been confirmed by nonlinear numerical time simulations and also harmonic balance or equivalent linearization calculations.

With freeplay, the flap frequency is really a function of the amplitude of response, varying between zero when the motion is small with in the freeplay range and increasing to the nominal flap frequency when the motion is large extending well beyond the freeplay range. Now <u>linear</u> theory predicts the flutter velocity has a minimum for some value of flap frequency. If that flap frequency is between zero and the nominal value, then flutter may occur for velocities above the minimum flutter velocity and, indeed, the minimum flutter velocity for the present example is 0.19. From <u>nonlinear</u> theory, one may determine the magnitude of the initial disturbance required to initiate such a flutter limit cycle. It turns out that the magnitude of the initial disturbance required to initiate the limit cycle is smallest in the plunge degree of freedom. In fact, the value is quite small in physical terms, $h/2\delta b > 0.12$. For typical δ, the amplitude of the plunge initial displacement required to initiate the limit cycle is no more than 1% of the airfoil half chord, b. Thus in our experiments, we able to apply a small disturbance in the plunge degree of freedom to readily excite the limit cycle.

If the nominal flap frequency is decreased, theory suggests a larger initial disturbance is required to excite the limit cycle below the flutter velocity predicted by linear theory for zero flap frequency. This observation explains the difference in results between those reported for our experimental model and those described by Lacabanne and Humbert in [40] for an experimental model with a flap frequency well below the pitch frequency.

Since the freeplay is in the flap motion, attention is now focused on the response behavior in this degree of freedom. This is shown as a (dimensional) time series in Figure 11(a), where the numerical and experimental time series are slightly phase shifted for clarity. One second of data is shown in each plot. This type of behavior then gives way to some more complicated periodic and non-periodic behavior at about one third of the linear flutter speed. This *transition* region is characterized by oscillatory behavior that, although having roughly the same energy (in terms of RMS) as the periodic response, has a much greater variety of frequency content and appears significantly more noisy. The region between the vertical dashed lines in Figure 10 contains this interesting transitional behavior and will be discussed at greater length later in the paper.

For flow rates just below 50% of the linear flutter speed a subharmonic response occurs. This is shown as a time series in Figure 11(b) and was found to be robust in the sense that this response was the only outcome for a variety of initial conditions (studied numerically and experimentally) and arbitrary disturbances.

At about 55% of the linear flutter speed there is another abrupt change in the system response. The low frequency limit cycle becomes unstable (rather suddenly) and a high frequency response occurs. This is also characterized by a sudden drop in the plunge amplitude. The flap amplitude remains relatively constant. The flap time series for this type of response is shown in Figure 11(c) at 73% of the linear flutter velocity. Tests were not conducted for flow rates greater than 93% of the linear flutter velocity in order to prevent the possibility of damage to the experimental model. It is seen that the numerical and experimental results show good correlation.

In order to take a closer look at these various responses a more detailed numerical study was conducted. Consider the response at 21% of the linear flutter speed. Figure 12 shows the (dimensional) results of a numerical simulation with all three degrees of freedom shown. In Figure 12(a,c,e) a phase projection (i.e., position versus velocity) illustrates the nature of this response. This is also confirmed by the power spectrum in Figure 12(f): it is by no means simple harmonic. Figure 13 shows a similar set of (nondimensional) numerical results but now at a nondimensional flow rate of 49% of critical. The motion is now aperiodic and the broadband nature of the power spectrum (Figure 12(f)) suggests a chaotic response. At a comparable flow rate the experimental model exhibited what appeared to be an increased amount of noise [27].

II-6. Stability of the Limit Cycles

The stability of these limit cycles can be assessed in the following manner. A system oscillating in a steady-state by definition passes through a point in the phase space once every period of the motion. If this stable cycle is perturbed then the induced transient will return to the limit cycle. An unstable cycle will be characterized by the movement of transients away from the fixed point. If the size of the perturbations is small then the local transient dynamics can be described within a linear framework (i.e. a finite difference approximation to the Jacobian of the dynamical system), from which characteristic multipliers can be extracted. In this way the stability of a limit cycle can be assessed in an analogous manner to the stability of an equilibrium point, e.g., the logarithmic decrement has often been used in linear flutter estimates. Since the Poincaré sampling is a discrete mapping from one period of motion to the next the condition for stability is that the eigenvalues must be less than one in magnitude. Further details of this approach and the typical instability mechanisms can be found in [27, 36].

Using this approach consider the evolution of stability properties over a range of flow rates. Figure 14(a) shows a root locus of the characteristic multipliers over the flow range from 19% to 35% of critical. Alternative views are provided in Figures 14(b) and 14(c). Each state variable has its own eigenvalue and the approach to a magnitude of unity can be clearly seen for a complex conjugate pair. This corresponds to the transition from periodic to quasi-periodic behavior, which can be seen at about 33% of the critical flow velocity in Figure 10. This is typically the result of at least two incommensurate frequencies influencing the motion. Hence the behavior is not periodic and may look quite random as a time series. However the Poincaré section tends to exhibit a closed curve as the orbit gradually passes along a (toroidal) surface in the phase space which is effectively cut by the Poincaré section [31, 32]. This effect is not dissimilar to the familiar beating effect in some linear systems with adjacent natural frequencies. Other stability transitions can be followed in this way.

Returning to the range of flow rates referred to as the transition region it was found that the oscillations in this regime typically did not correspond to strictly periodic behavior. The results reported in Figure 10 were based on a RMS measure of the amplitudes and hence tend to disguise this fact. However within the transition region (for both the experimental and numerical results) marked by the vertical dashed lines a variety of subharmonic, quasi-periodic and even chaotic behavior was observed [31, 32]. This is primarily based on the numerical results although experimentally it was clear that this same type of qualitative behavior occurred, but due to noise and the sheer complexity of the response it could not be observed as clearly as in the numerical simulations. A typical set of results is shown in Figure 8 as pitch Poincaré projections. 'A' is the nondimensional α normalized by the freeplay gap, 2 δ. A harmonic response would have produced just a single point in this projection. When quasi-periodic behavior appears in a nonlinear dynamical system it will often undergo a series of generic bifurcations as a system parameter is varied. Here, the flow rate is changed (very slightly) and a splitting of the quasi-periodicity into two distinct branches is seen. On further (extremely slight) changes in the flow rate the Poincaré projection has broken up into a classic chaotic geometry. This has some interesting features in its power spectrum [39].

II-7. Conclusions

In the presence of a region of freeplay in the flap motion of a three-degree-of-freedom typical airfoil section, a variety of pre-flutter oscillatory (limit cycle) behavior is observed. Experimental results show good agreement with numerical simulations despite the relatively simple modeling approach used. The character and stability of the dynamic responses prior to the complete loss of stability at the flutter boundary are assessed using techniques from dynamical systems theory. It is interesting to note that this characteristic behavior can be expected to be present whenever significant nonlinearities are present in an aeroelastic system.

EPILOGUE

The advances in computational and experimental laboratory capability combined with new understanding of nonlinear dynamical systems has led to more complex and more accurate mathematical models of aircraft and rotorcraft. These models have been verified by experiment and now offer a new and better approach to analysis and design.

References

[1] Von Karman, Theodore (1940), *The Engineer Grapples with Nonlinear Problems,* Bulletin of the American Mathematical Society, Vol. 46.

[2] Dowell, E.H., et al(1995), *A Modern Course in Aeroelasticity,* Kluwer.

[3] Dowell, E.H., and Ilgamov, M.(1988), " Studies in Nonlinear Aeroelasticity ," Springer-Verlag.

[4] Tang, D.M., and Dowell, E.H.(1993), Nonlinear Aeroelasticity in Rotorcraft, *Mathematical and Computer Modeling* , **8** (3/4), pp 157-184.

[5] Hodges, D.H., and Dowell, E.H.(1974), Nonlinear Equations of Motion for the Elastic Bending and Torsion of Twisted Nonuniform Rotor Blades, NASA TN D-7818.

[6] Crespo DaSilva, M.R.M., and Hodges, D.H.(1986), Nonlinear Flexure and Torsion of Rotating Beams, with Application to Helicopter Rotor Blades, I. Formulation, *Vertica,* **10** (2), pp.151-169.

[7] Crespo DaSilva, M.R.M., and Hodges, D.H.(1986), Nonlinear Flexure and Torsion of Rotating Beams, with Application to Helicopter Rotor Blades, II. Results for Hover, *Vertica,* **10** (2) pp.171-186.

[8] Friedmann Peretz P.(1990), Rotary-Wing Aeroelasticity with Application to VTOL Vehicles, AIAA-90-1115-CP, pp.1624-1670.

[9] Tran, C.T., and Petot, D.(1981) , Semi-Empirical Model for the Dynamic Stall of Airfoils in View to the Application to the Calculation of Responses of a Helicopter Blade in Forward Flight, *Vertica* **5**, pp. 35-53.

[10] Tang, D.M., and Dowell, E.H.(1992), Chaotic Stall Response of a Helicopter in Forward Flight , *Journal of Fluids and Structures,* **6** (3), pp. 311-335.

[11] Tang, D.M., and Dowell, E.H.(1992), Flutter and Stall Response of Helicopter Blade with Structural Nonlinearity, *Journal of Aircraft* ,**29** (5), pp. 953-960.

[12] Tang, D.M., and Dowell, E.H.(1993), Comparison of Theory and Experiment for Nonlinear Flutter and Stall Response of a Helicopter Blade, *Journal of Sound and Vibration* , **162** (1), pp.1-26.

[13] Tang, D.M., and Dowell, E.H.(1993), Experimental and Theoretical Study for Nonlinear Aeroelastic Behavior of a Flexible Rotor Blade, *AIAA J.*, **31** (6), pp. 1133-1142.

[14] Tang, D.M. and Dowell, E.H.(1996), " Nonlinear Response of a Non-rotating Rotor Blade to a Periodic Gust, *Journal of Fluids and Structures*, **10** (3), pp. 721-742.

[15] Tang, D.M., Paul G.A. Cizmas and Dowell, E.H.(1996), Experiments and Analysis for A Gust Generator in A Wind Tunnel, *Journal of Aircraft* ,**33** (1), pp.139-148.

[16] Fung Y. C.(1955), *An Introduction to The Theory of Aeroelasticity,* John Wiley and Sons, Inc.

[17] Rogers, J. P.(1984), Applications of an Analytic Stall Model to Time-History and Eigenvalue Analysis of Rotor Blades , *Journal of the American Helicopter Society* **29**, pp.25-33.

[18] Friedmann Peretz P.(1986), Numerical Method for Determining the Stability and Response of Periodic Systems with Applications to Helicopter Rotor Dynamics and Aeroelasticity, *Computers and Mathematics with Applications,* **12A** (1), pp. 131-148.

[19] Peters, D.A., and Izadpahan, A.P.(1981), Helicopter Trim by Periodic Shooting with Newton-Raphson Iteration, Proceedings 37th Annual Forum of the American Helicopter Society, pp.217-226.

[20] Woolston, D.S., Runyan, H.L. and Andrews, R.E.(1957), An investigation of effects of certain types of structural nonlinearities on wing and control surface flutter, *Journal of Aeronautical Sciences*, **24**, pp.57.

[21] Shen, S.F.(1959), An approximate analysis of nonlinear flutter problems, *Journal of Aeronautical Sciences*, **26**, pp.25.

[22] Breitbach, E.(1977), Effects of structural nonlinearities on aircraft vibration and flutter. *AGARD Tech. Rep. 665.*

[23] Lin, W.B. and Cheng, W.H.(1993), Nonlinear flutter of loaded lifting surfaces (I)&(II). *Journal of the Chinese Society of Mechanical Engineers*, **14**, pp.446.

[24] Conner, M.D., Virgin, L.N. and Dowell, E.H.(1996), A note on accurate numerical integration of state-space models for aeroelastic systems with freeplay. *AIAA Journal*, **34**, pp.2202.

[25] Edwards, J.W., Ashley, H. and Breakwell, J.V.(1979), Unsteady aerodynamic modeling for arbitrary motions, *AIAA Journal*, **17**, pp.365.

[26] Bisplinghoff, R.L. and Ashley, H.(1962), *Principles of Aeroelasticity.* Dover.

[27] Conner, M.D.(1996), Nonlinear aeroelasticity of an airfoil section with control surface freeplay. PhD Dissertation, Duke University.

[28] Theodorsen, T. and Garrick, I.E.(1940), Mechanism of flutter, a theoretical and experimental investigation of the flutter problem. *NACA Report No. 685.*

[29] Theodorsen, T. and Garrick, I.E.(1940), Flutter, calculations in three degrees of freedom. *NACA Report No. 741.*

[30] Tang, D.M. and Dowell, E.H.(1995), Damping Prediction for Stalled Rotor Flap-Lag Stability with Experimental Correlation. *Journal of the American Helicopter Society*, **40**, pp.79.

[31] Moon, F.C.(1992), *Chaotic and Fractal Dynamics*, Wiley: New York.

[32] Thompson, J.M.T. and Stewart, H.B.(1986), *Nonlinear Dynamics and Chaos*, Wiley : London.

[33] Hauenstein, A.J., Zara, J.A., Eversman, W. and Qumei, I.(1992), Chaotic and nonlinear dynamic response of aerosurfaces with structural nonlinearities, *AIAA-92-2547-CP.*

[34] Price, S.J., Alighanbari, H. and Lee, B.H.K.(1994), Post-instability behavior of a two-dimensional airfoil with a structural nonlinearity, *Journal of Aircraft*, **31**, pp.1395.

[35] Virgin, L.N. and Dowell, E.H.(1992), Nonlinear aeroelasticity and chaos, in *Computational Nonlinear Mechanics in Aerospace Engineering*, AIAA.

[36] Murphy, K.D., Bayly, P.V., Virgin, L.N. and Gottwald, J.A.(1994), Measuring the stability of periodic attractors using perturbation-induced transients: Applications to two nonlinear oscillators. *Journal of Sound and Vibration*, **172**, pp.85.

[37] Conner, M.C., Tang, D.M., Dowell, E.H. and Virgin, L.N.(1997), Nonlinear behavior of a typical airfoil section with control surface freeplay: A numerical and experimental study. *Journal of Fluids and Structures*, **11**, pp.89.

[38] Hénon, M.(1982), On the numerical computation of Poincaré maps, *Physica D*, **5**, pp.412.

[39] Grebogi, C., Ott, E., Pelikan, S. and Yorke, J.A.(1984), Strange attractors that are not chaotic. *Physica D*, **13**, pp.261.

[40] Lacabanne, M. and Humbert, M. (1997), An Experimental Analysis of the Aeroelastic Behavior with a Freeplay in a Control Surface. *Proceedings of the International Forum on Aeroelasticity and Structural Dynamics,* June 17-20, Rome, Italy, pp. 239.

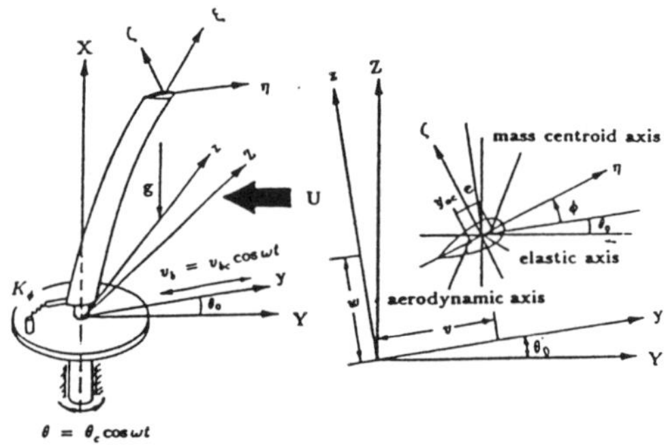

Figure 1: Physical representation of a flexible rotor blade

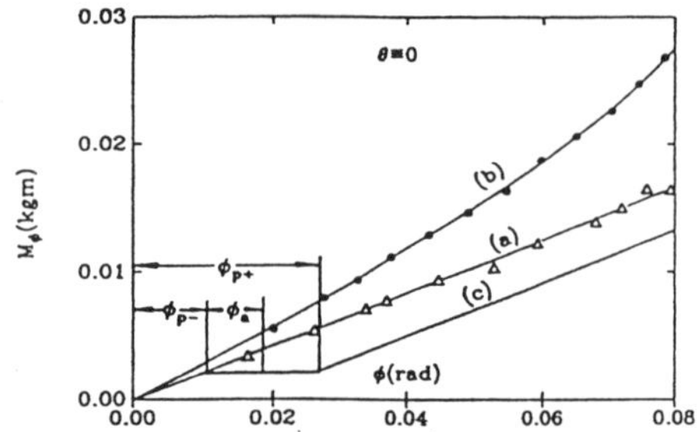

Figure 2: Static torque vs twist angle for experimental model.

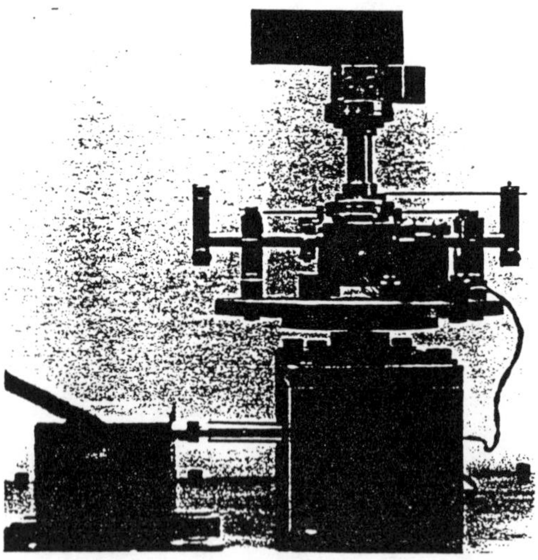

Figure 3: Photograph of root support mechanism of experimental model.

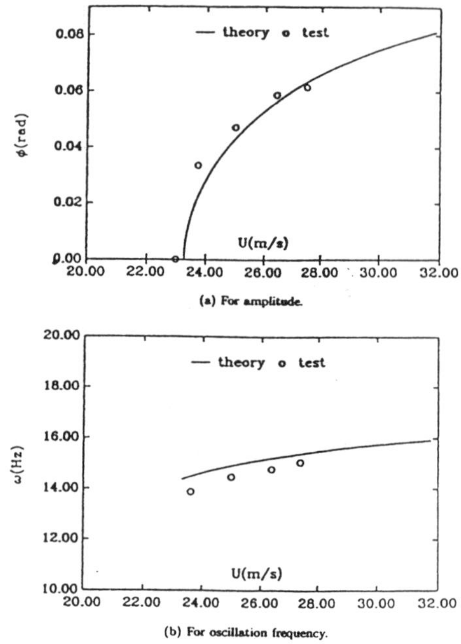

Figure 4(a)(b): Limit cycle flutter amplitude vs airspeed with parabolic pitch stiffness nonlinearity, (a) for amplitude, (b) for oscillation frequency.

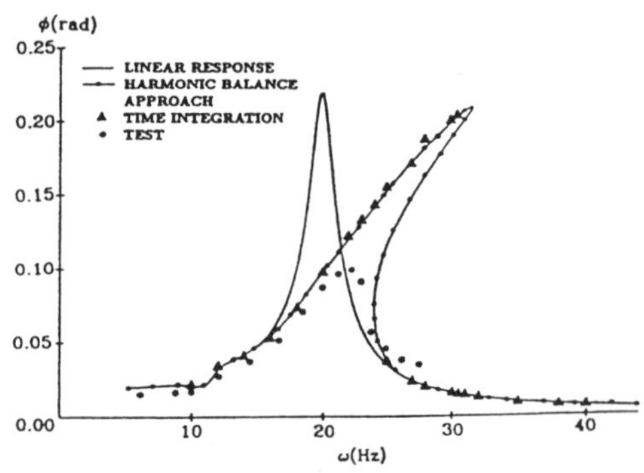

Figure 5: Frequency response behavior with parabolic pitch stiffness nonlinearity.

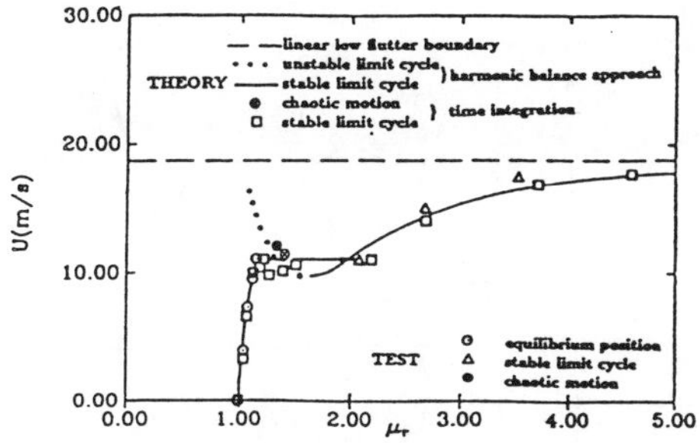

Figure 6: Flow speed vs limit cycle flutter amplitude ratio, μ_r, with freeplay nonlinearity for $\theta_0 = 0^0$.

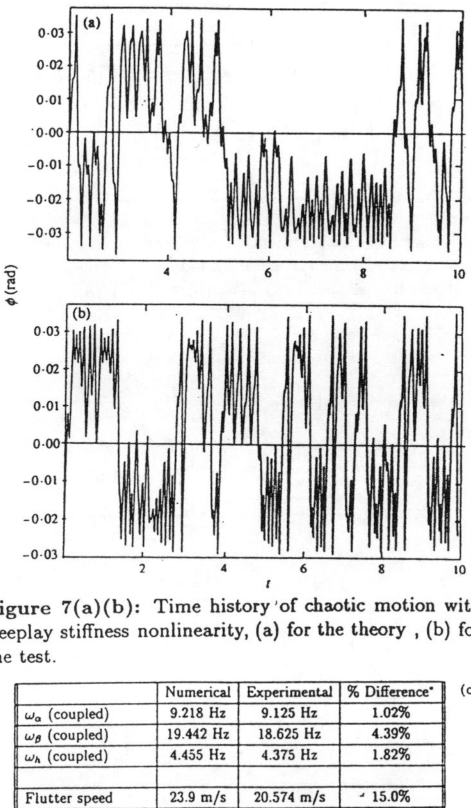

Figure 7(a)(b): Time history of chaotic motion with freeplay stiffness nonlinearity, (a) for the theory, (b) for the test.

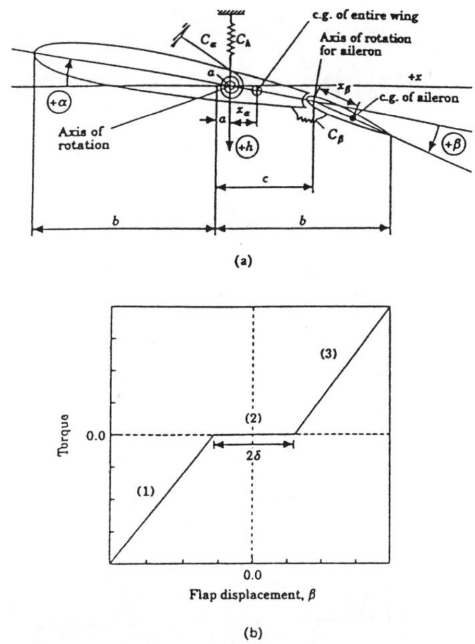

Figure 8: (a) Schematic of the aeroelastic typical section with control surface, (b) structural restoring moment for the control surface (flap).

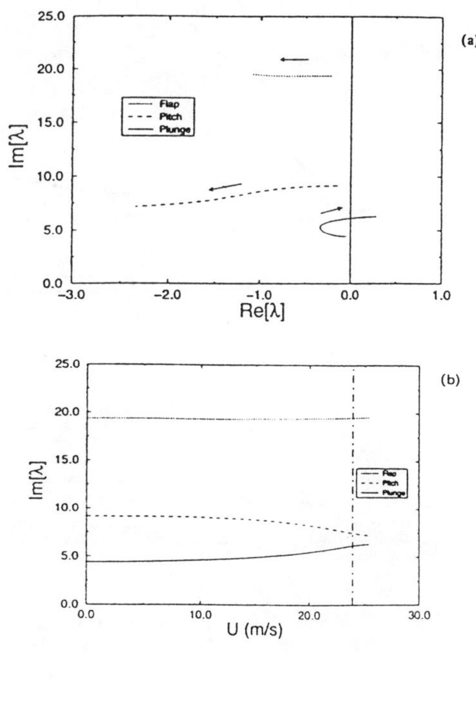

(c)

	Numerical	Experimental	% Difference*
ω_α (coupled)	9.218 Hz	9.125 Hz	1.02%
ω_β (coupled)	19.442 Hz	18.625 Hz	4.39%
ω_h (coupled)	4.455 Hz	4.375 Hz	1.82%
Flutter speed	23.9 m/s	20.574 m/s	- 15.0%
(Reduced velocity)	(3.249)	(2.825)	(15.0%)
Flutter frequency	6.112 Hz	5.47 Hz	11.74%

Figure 9: (a) Root-locus plot for the nominal linear system (numerical) as the freestream increases from 0 to 25 m/s, (b) Variation of the frequencies with freestream velocity. The vertical dashed line represents the linear flutter boundary, (c) A numerical-experimental comparison of dynamic characteristics.

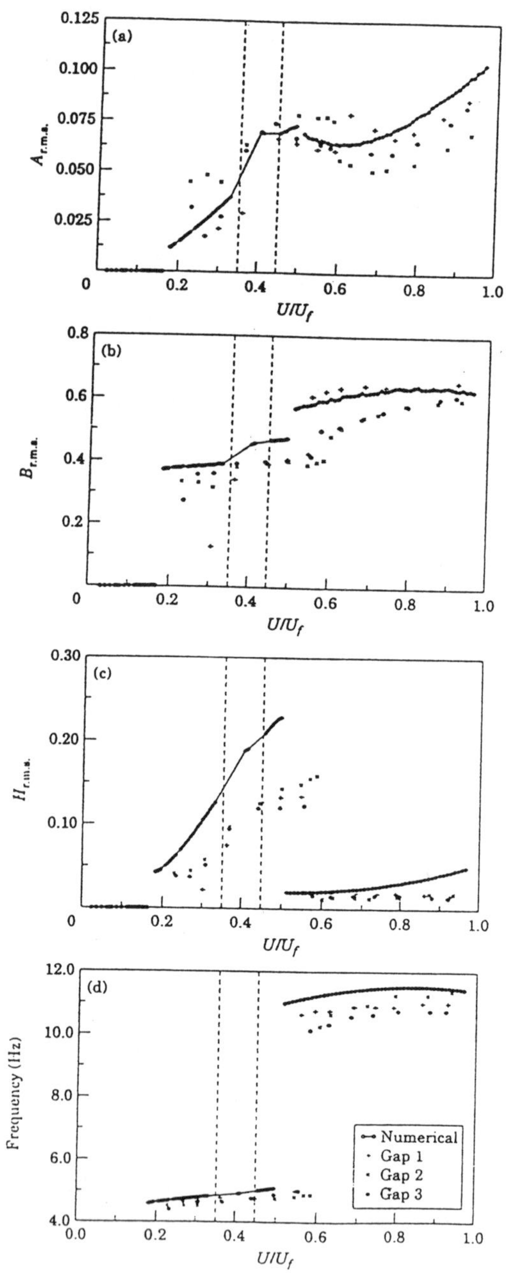

Figure 10: Numerical and experimental normalized steady-state RMS amplitude for (a) pitch, (b) flap, (c) plunge, and (d) frequency.

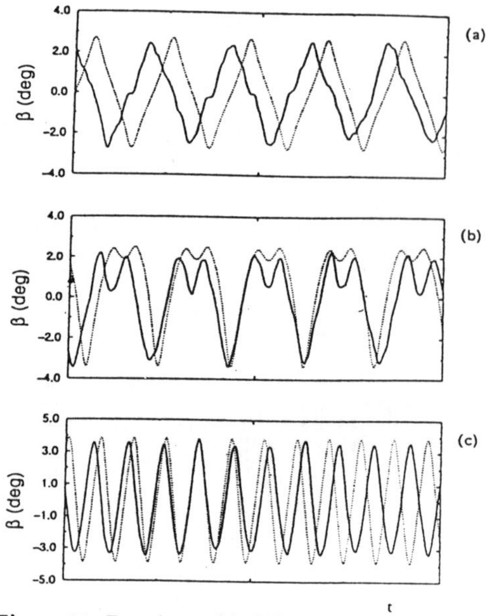

Figure 11: Experimental (solid) and numerical (dashed) one-second time series for flap at various fractions of the linear flutter speed (a) 27%, (b) 49%, (c) 73%.

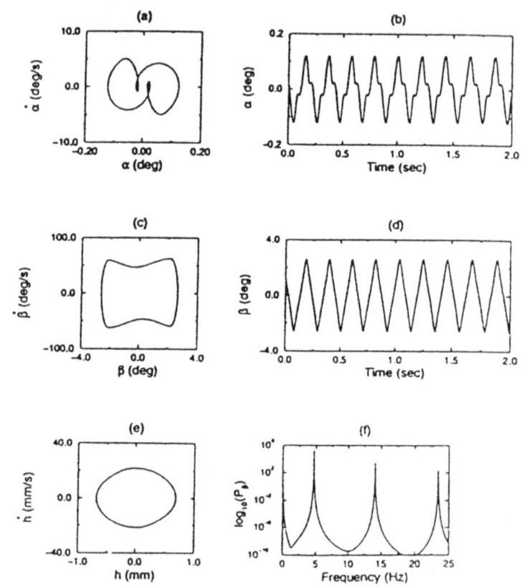

Figure 12: Phase projections (a,c and e), dimensional time series (b and d) and power spectrum (f), based on numerical simulation at 21%.

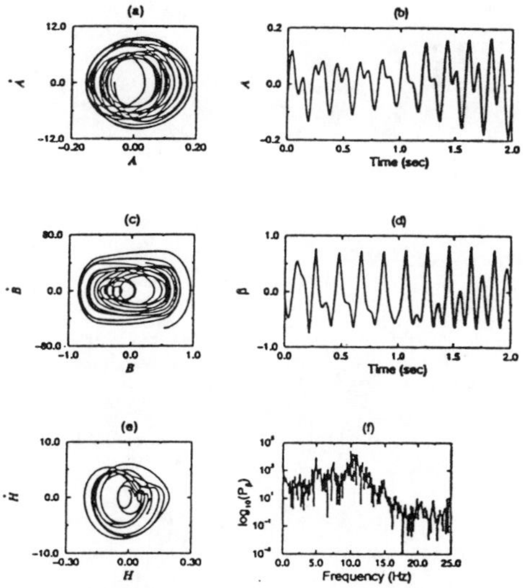

Figure 13: Phase projections (a,c and e), dimensional time series (b and d), and power spectrum (f), based on numerical simulation at 49%.

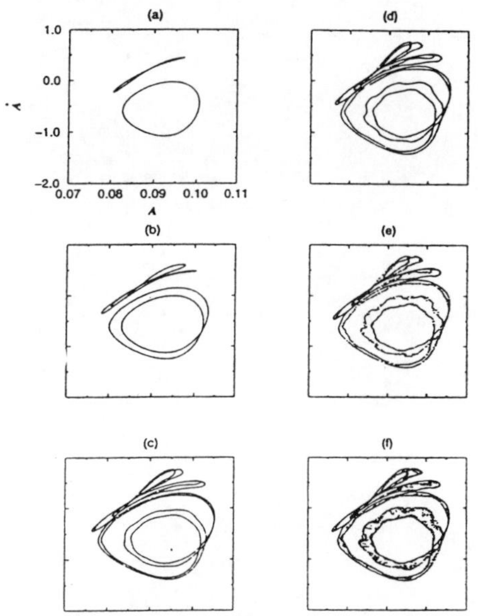

Figure 15: Poincaré sections in pitch for various freestream velocities: (a) 38.36%, (b) 38.42%, (c) 38.46%, (d) 38.47%, (e) 38.473%, (f) 38.48%.

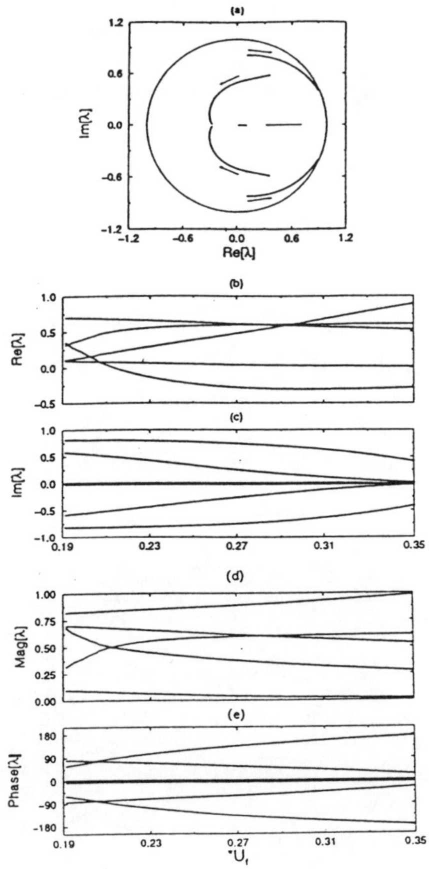

Figure 14: Progression of the Jacobiam matrix eigenvalues as U increases from 19% to 35%, (a) complex plane, (b) real and imaginary components, (c) amplitude and phase.

STABILIZATION OF NONLINEAR HYDROELASTIC STRUCTURES VIA RANDOM PARAMETRIC EXCITATION

R. A. IBRAHIM AND M. HIJAWI
Wayne State University
Department of Mechanical Engineering
Detroit, MI 48202, USA

1. Introduction

The problem of stabilizing dynamical nonlinear systems by introducing random fluctuations to the control parameter has been of great interest to physicists, engineers and mathematicians (Ibrahim, 1997). For nonlinear one dimensional systems with widely separated time scales, Graham and Schenzle (1982) showed the possibility of shifting the bifurcation point by a broad-band multiplicative noise. The classical problem of stabilizing the inverted position of a simple pendulum during vertical support oscillation was considered by Stoker (1950), Bogdanoff and Citron (1965), Valeev (1971), and many others. The unstable equilibrium position can be stabilized if one imposes a sinusoidal support motion with appropriate amplitude and frequency. Bogdanoff and Citron (1965) observed experimentally that it was not possible to stabilize the inverted pendulum with Gaussian random noise, regardless of the excitation spectral shape. However, it was possible to stabilize the pendulum for brief intervals with Gaussian random noise clipped at one sigma interval.

Sethna and Orey (1980) examined the possibility of stabilizing the pendulum by allowing the support motion to be sample functions of a stochastic process with continuous spectrum. By using the averaging method they showed that the pendulum can indeed be stable in the inverted position when the support motion is stationary in the wide sense and has a continuous spectrum. This result was confirmed by Prussing (1981), who proved that the inverted pendulum can be stabilized in terms of the first and second moments if the support motion is a physical white noise. However, if the excitation is a mathematical white noise, the pendulum cannot be stabilized, as proved by Nevel'son and Khasminskii (1966) and by Nakamizo and Sawaragi (1972). Khasminskii (1980) constructed a two-dimensional unstable system that can be stabilized by introducing two independent white noise sources. Arnold (1979) presented another class of unstable deterministic system which can be stabilized by applying a single nonwhite random noise.

The present work deals with the stabilization and response transition of nonlinear hydroelastic structures subjected to random hydrodynamic excitations. A unified

F. C. Moon (ed.),
IUTAM Symposium on New Applications of Nonlinear and Chaotic Dynamics in Mechanics, 451–460.

second-order stochastic averaging approach based on the Stratonovich-Khasminskii limit theorem is developed to predict the response pdf, stochastic stability, phase transition (known as noise-induced transition), and stabilization by multiplicative noise in one treatment. It is known that first-order stochastic averaging fails to account for inertia and stiffness non-linearities because their effect is lost during the averaging procedure. The effect of such nonlinearities can only be determined by performing a second-order averaging. Second-order averaging has not been widely used among dynamicists apparently because it requires tedious mathematical manipulations. The second-order averaging results are found in good agreement with those estimated by Monte Carlo simulation. Under pure parametric random excitation, the stochastic averaging and Monte Carlo simulation predict the on-off intermittency phenomenon near bifurcation point, in addition to stochastic bifurcation.

2. Statement of the Problem

The dynamic analysis of ocean structures involves the interaction of inertia forces, elastic forces, and hydrodynamic forces. Basically, the hydrodynamic forces induced on structural elements are nonlinear and a combination of inertia and drag contributions. The hydrodynamic forces or the time variation of the system parameters are random in nature. Figure 1 shows an elastic beam of length L and carries a mass M at the top. The beam is subjected to two types of excitations. The first force is due to hydrodynamic loading while the second is parametric due to two vertical wave loads which can take place as a result of the heaving motion of the top mass as it responds to the seaway. The equation of motion of the first bending mode takes the form (Hijawi, 1996)

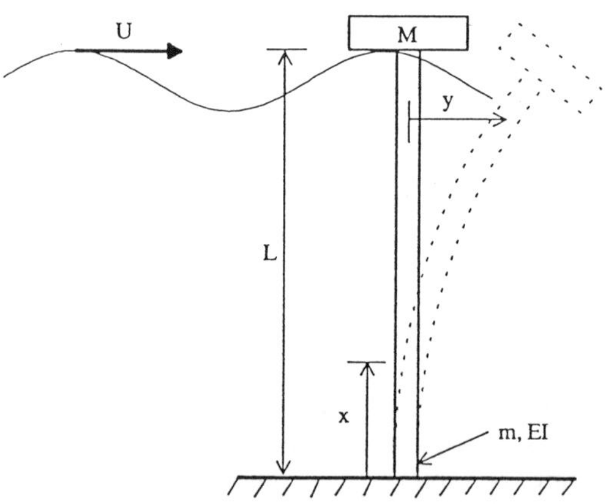

Figure 1. Schematic diagram of an elastic structure subjected to ocean waves.

$$Y'' + Y + \varepsilon(\zeta Y' + c_1 Y'^2 Y + c_2 Y^3 - \lambda_1 q_0|q_0| - \lambda_2 \mathrm{sgn}(q_0) Y'^2 + \lambda_3 |q_0| Y') + \sqrt{\varepsilon} Y \tilde{W}(\tau) = \sqrt{\varepsilon}\lambda_4 q_0' \quad (1)$$

where Y is a dimensionless deflection of the beam at its free end, a prime denotes differentiation with respect to the non-dimensional time $\tau = \omega_n t$, ω_n is the first mode natural frequency, $\varepsilon = D/L$, D is a the characteristic width of the structure, $\varepsilon\zeta$ is the linear damping factor, c_1 and c_2 are the coefficients of nonlinear inertia and curvature, respectively, λ_i, i=1,2,3, are the coefficients of nonlinear hydrodynamic drag, λ_4 is the coefficient of inertia hydrodynamic force and $\tilde{W}(\tau)$ is a dimensionless white random process representing the parametric excitation and $q_0=u/\omega_n D$ is a dimensionless fluid wave velocity. The velocity q_0 is represented by the output of the second-order shaping filter

$$q_0'' + 2\omega_f \zeta_f q_0' + \omega_f^2 q_0 = \xi(\tau) \tag{2}$$

where $\omega_f=\Omega_f/\omega_n$ is a dimensionless filter frequency defined as the ratio of the filter frequency to the system natural frequency and ζ_f is the filter damping ratio and $\xi(\tau)$ is a zero mean white noise process of intensity ν. The power spectral density and mean square of the process $q_0(\tau)$ are, respectively

$$S_{q_0}(\omega) = \frac{\nu}{2\pi} \frac{1}{\left(\omega_f^2 - \omega^2\right)^2 + \left(2\omega_f \omega \zeta_f\right)^2}, \quad S_{q_0'}(\omega) = \omega^2 S_{q_0}(\omega), \quad E\left[q_0^2\right] = \frac{\nu}{4\,\omega_f^3\,\zeta_f} \tag{3}$$

The average value $\langle |q_0| \rangle$ can be calculated by using the pdf of the linear filter (2)

$$\langle |q_0| \rangle = \frac{2}{\sqrt{2\pi}\,\sigma} \int_0^\infty |q_0| \exp\left(\frac{-q_0^2}{2\sigma^2}\right) dq_0 = \sqrt{\frac{\nu}{2\pi \zeta_f \omega_f^3}} \tag{4}$$

where σ is the variance given by $\sigma = E[q_0^2]$. More realistic shaping filters can be used to model the wave spectrum. However, the second-order filter (2) is adopted in the present analysis for the sake of simplification.

3. Second-Order Stochastic Averaging

The solution of equation (1) has a similar form to the linear solution (ε=0), but both amplitude $A(\tau)$ and phase $\varphi(\tau)=\tau+\theta(\tau)$ are slowly varying with time, i.e.,

$$Y = A\cos\varphi, \qquad Y' = -A\sin\varphi \tag{5}$$

introducing (5) into equation (1) and following the procedure of stochastic averaging method the following standard amplitude and phase equations are obtained

$$\begin{Bmatrix} A' \\ \theta' \end{Bmatrix} = \varepsilon \begin{Bmatrix} f_1(A,\theta,\tau) \\ f_2(A,\theta,\tau) \end{Bmatrix} + \sqrt{\varepsilon} \begin{bmatrix} g_{11}(A,\theta,\tau) & g_{12}(A,\theta,\tau) \\ g_{21}(A,\theta,\tau) & g_{22}(A,\theta,\tau) \end{bmatrix} \begin{Bmatrix} q_0'(\tau) \\ \tilde{W}(\tau) \end{Bmatrix} \tag{6}$$

where f_1 and f_2 stand for drift terms, while the functions g_{ij} are associated with diffusion terms. The nonlinear terms in f_1 and f_2 will contain numerous products of sine and cosine functions with phase angle $\varphi(\tau)$. These terms may be expanded into series of sine and cosine functions at the multiple phase angle $n\varphi$, n=0,2, 4,... . We can therefore eliminate the oscillatory effects and simplify the equations of motion by introducing the near-identity transformation

$$A(\tau) = \bar{A}(\tau) + \varepsilon u(\bar{A},\bar{\theta},\tau), \qquad \theta(\tau) = \bar{\theta}(\tau) + \varepsilon v(\bar{A},\bar{\theta},\tau) \tag{7}$$

where $\bar{A}(\tau), \bar{\varphi}(\tau) = \bar{\theta}(\tau) + \tau$, and $\bar{\theta}(\tau)$ stand for non-oscillatory amplitude, phase angle, and phase shift, respectively. Differentiating relations (7), equating each result with the corresponding drift functions in (6) adding the contribution of the diffusion terms as outlined in Hijawi, et al., (1997) yields the average equation for the amplitude $\bar{A}'$

$$\bar{A}' = C_0 \frac{1}{\bar{A}} + C_1 \bar{A} + C_3 \bar{A}^3 + \sqrt{2\,C_0 + D_2 \bar{A}^2}\; W_1(\tau) \tag{8}$$

where C_i and D_2 are coefficients which depend on the system parameters and excitation statistics.

Introducing the Hamiltonian $\bar{H} = \bar{A}^2$, and differentiating $\bar{H}$ according to the Ito formula gives the averaged equation

$$\bar{H}' = 4\,C_0 + \left(2\,C_1 + D_2\right)\bar{H} + 2\,C_3\,\bar{H}^2 + \sqrt{8\,C_0\,\bar{H} + 4\,D_2\,\bar{H}^2}\; W_1(\tau) \tag{9}$$

The stationary response pdf of (9) is

$$p(\bar{H}) = K_N \left(2\,C_0 + D_2\,\bar{H}\right)^{\kappa} \exp\left(\frac{C_3}{D_2}\,\bar{H}\right) \tag{10}$$

where $\kappa = -\left(\frac{3}{2} + 2\frac{C_3\,C_0}{D_2^2} - \frac{C_1}{D_2}\right)$ and the normalization constant K_N is

$$K_N = \frac{(-C_3)^{\kappa+1}\exp(2C_3C_0/D_2^2)}{D_2^{2\kappa+1}\Gamma(\kappa+1, -2\frac{C_3C_0}{D_2^2})}, \qquad \Gamma(a,x) = \int_x^{\infty} e^{-t}\, t^{a-1}\, dt \tag{11}$$

$\Gamma(a,x)$ is the incomplete gamma function which is monotonic and (for $a>1$) rises from "near-zero" to "near-unity" in a range of x centered on about a-1, and of width about $\sqrt{a}$. Note that the response pdf (10) includes the effects of inertia and curvature nonlinearities of the structure and the effects of two components of the hydrodynamic forces $\lambda_3|q_0|Y'$ and $\sqrt{\varepsilon}\lambda_4 q_0'$. However, it does not capture the effect of the two hydrodynamic drag components $-\lambda_1 q_0|q_0|$ and $-\lambda_2 \mathrm{sgn}(q_0)Y'^2$. These two terms only have a localized effect (in time) and do not contribute significantly to the average behavior of the system over a long period of time. The normalization is valid provided $C_3 < 0$ or

$$\left(c_1 - 5\,c_2\right) < 0 \tag{12}$$

In the absence of hydrodynamic forces the averaged equation is

$$\bar{H}' = \left(2\,C_1^* + D_2^*\right)\bar{H} + 2\,C_3^*\,\bar{H}^2 + \sqrt{4\,D_2^*\,\bar{H}^2}\;W_1(\tau) \tag{13}$$

where $C_1^* = \varepsilon\,\{-\frac{1}{2}\,\zeta + \frac{3}{8}\,\pi\,S_{\tilde{W}}(2)\}$, $C_3^* = \varepsilon^2\,\{\,\frac{1}{32}\left(c_1 - 5\,c_2\right)\left[\zeta + \pi\,S_{\tilde{W}}(2)\right]\}$, $D_2^* = \varepsilon\frac{1}{4}\pi S_{\tilde{W}}(2)$.

The corresponding stationary pdf is

$$p(\bar{H}) = K_N^*\,\bar{H}^{\left(\frac{C_1^*}{D_2^*} - \frac{3}{2}\right)}\exp\!\left(\frac{C_3^*}{D_2^*}\bar{H}\right), \quad \text{where } K_N^* = \left(-\,\frac{C_3^*}{D_2^*}\right)^{\frac{C_1^*}{D_2^*} - \frac{1}{2}} \Big/ \Gamma\!\left(\frac{C_1^*}{D_2^*} - \frac{1}{2}\right) \tag{14}$$

The normalization constant K_N^* is valid only under the two conditions

$$\text{(i)}\;\; \frac{C_1^*}{D_2^*} - \frac{1}{2} > 0, \qquad\qquad \text{(ii)}\;\; C_3^* < 0 \tag{15}$$

In terms of the system parameters the first condition gives

$$\frac{\pi S_{\tilde{W}}(2)}{\zeta} > 2 \tag{16a}$$

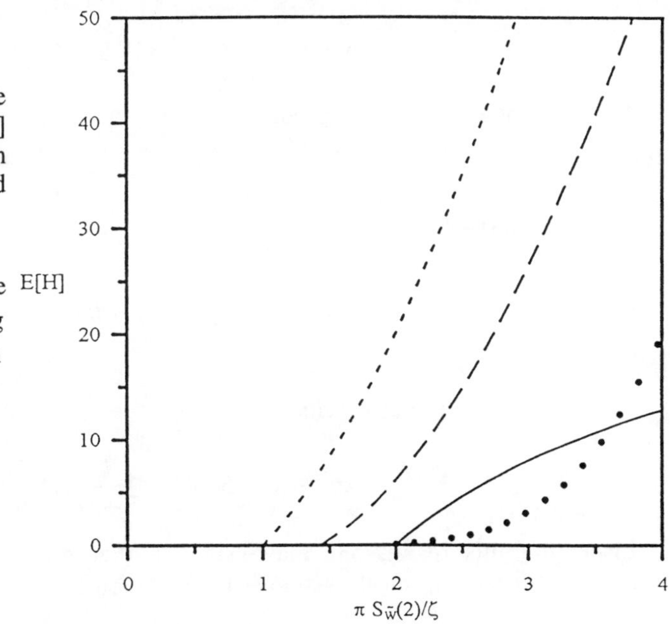

Figure 2 Dependence of the response mean energy E[H] on the parametric excitation level for $\varepsilon = 0.005$, $\zeta = 4$, and $c_2 = 9$.

- - - Gaussian Closure

— —— non-Gaussian closure

—— second-order averaging

• • • Monte Carlo simulation

Similarly, the second condition gives

$$\left(c_1 - 5\,c_2\right) < 0 \tag{16b}$$

Finally, the expression of the mean value of the response energy is

$$E\left[\bar{H}\right] = -\frac{D_2^*}{C_3^*}\Gamma\left(\frac{C_1^*}{D_2^*} + \frac{1}{2}\right)/\Gamma\left(\frac{C_1^*}{D_2^*} - \frac{1}{2}\right) \tag{17}$$

The dependence of the response mean square E[H] on the parametric excitation level is shown in Figure 2 by the solid curve. Figure 2 also shows the mean energy level as predicted by Gaussian and non-Gaussian closures and Monte Carlo simulation. It is seen that there is a critical excitation level, given by relation (16a), above which the response mean square branches to a non-zero value. This level corresponds to the well-known ensemble stochastic stability condition of dynamic systems parametrically excited by a white noise. The bifurcation point predicted by second-order averaging is in good agreement with the one estimated by the Monte Carlo simulation (shown by solid small circles).

4. Noise-Induced Transition And Stabilization Problem

Noise-induced transition implies a change in the number of extrema in the response pdf. This transition will be examined by considering the averaged equation of the response amplitude by setting the external excitation to zero in equation (8)

$$\bar{A}' = C_1^*\,\bar{A} + C_3^*\,\bar{A}^3 + \sqrt{D_2^*\,\bar{A}^2}\;W_1(\tau) \tag{18}$$

The corresponding stationary pdf is

$$p(\bar{A}) = K_N^*\,\bar{A}^{\,2(\frac{C_1^*}{D_2^*}-1)}\exp\left(\frac{C_3^*}{D_2^*}\bar{A}^2\right)\,, \qquad K_N^* = 2\left(-\frac{C_3^*}{D_2^*}\right)^{\frac{C_1^*}{D_2^*}-\frac{1}{2}}/\Gamma\left(\frac{C_1^*}{D_2^*}-\frac{1}{2}\right) \tag{19}$$

In terms of system parameters the response pdf takes the form

$$p(\bar{A}) = 2\left(-\frac{1}{8}\varepsilon(c_1 - 5c_2)\left(1 + \frac{\zeta}{\pi\,S_{\tilde{W}}(2)}\right)\right)^{(1-\frac{2\zeta}{\pi\,S_{\tilde{W}}(2)})}\bar{A}^{(1-\frac{4\zeta}{\pi\,S_{\tilde{W}}(2)})}$$

$$\times\exp\left(\frac{1}{8}\varepsilon(c_1 - 5c_2)\left(1 + \frac{\zeta}{\pi\,S_{\tilde{W}}(2)}\right)\bar{A}^2\right)/\Gamma\left(1 - \frac{2\zeta}{\pi S_{\tilde{W}}(2)}\right) \tag{20}$$

Inspecting this result, one may establish the following three different response regimes, depending on the parametric excitation level $\pi S_{\tilde{W}}(2)/\zeta$:

1. Zero motion given for all excitation levels defined by the range $0<\pi S_{\tilde{W}}(2)/\zeta<2$.

2. *Partially developed random motion (or on-off intermittency)* takes place for all excitation levels within the range $2<\pi S_{\tilde{W}}(2)/\zeta<4$. Within this range the peak of the response pdf occurs at zero response amplitude, as detected by the exponent of $\bar{A}$. The time evolution of the response will experience a sequence of zero and non-zero response periods (Ibrahim, 1997).
3. *Fully developed random motion* occurs for all excitation levels exceeding 4, i.e., when $\pi S_{\tilde{W}}(2)/\zeta>4$. This regime is characterized by continuous random motion and the peak of the response pdf will be shifted from zero amplitude.

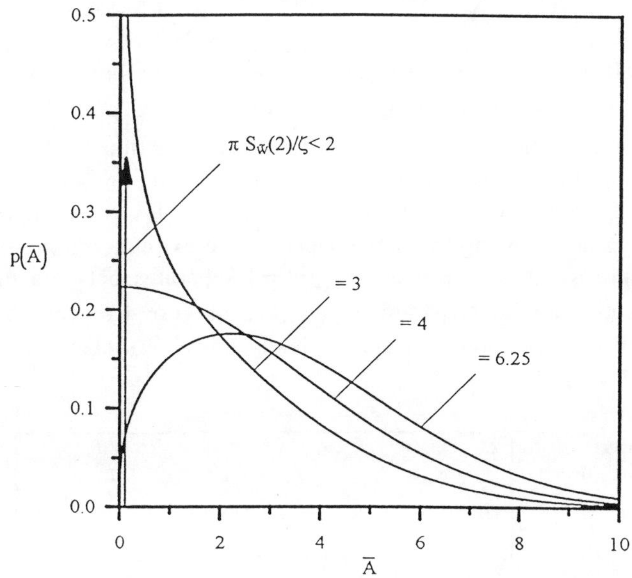

Figure 3 Response amplitude pdf for different values of excitation level

The response pdf $p(\bar{A})$ is shown in Figure 3 for different parametric excitation levels $\pi S_{\tilde{W}}(2)/\zeta$. Note that both on-off intermittency and noise-induced transition cannot be predicted by estimating the response moments because the moments eliminate a lot of information about the state of the system. The appropriate indicator of a transition is the extrema of the pdf which is obtained from the condition $dp(\bar{A})/d\bar{A}=0$. The on-off intermittency can only be uncovered by using Monte Carlo simulation or experimental tests (Ibrahim, 1991). The extrema of the response stationary pdf, are determined from the condition

$$a_1(\bar{A})-\frac{1}{2}\frac{d\left(b_{11}(\bar{A})\right)}{d\bar{A}}=0 \tag{21}$$

Substituting for $a_1(\bar{A})=C_1^*\,\bar{A}+C_3^*\,\bar{A}^3$ and $b_{11}(\bar{A})=D_2^*\,\bar{A}^2$, (21) becomes

$$C_3^* \bar{A}^3 - \bar{A}(D_2^* - C_1^*) = 0 \tag{22}$$

This equation has three solutions given by the roots

$$\bar{A}_1 = 0, \; \bar{A}_{2,3} = \pm\sqrt{(D_2^* - C_1^*)/C_3^*} \tag{23a}$$

In terms of system and excitation parameters these roots are

$$\bar{A}_1 = 0, \quad \bar{A}_{2,3} = \pm\sqrt{4(4 - \frac{\pi S_{\tilde{W}}(2)}{\zeta})/\{\varepsilon(c_1 - 5c_2)[1 + \frac{\pi S_{\tilde{W}}(2)}{\zeta}]\}} \tag{23b}$$

For $(c_1 - 5c_2) > 0$, there is only one peak of the response pdf at $\bar{A}_1 = 0$ as long as $\pi S_{\tilde{W}}(2)/\zeta < 4$. Above this level a transition of the response pdf takes place and the peak is determined by the second root of (23b). A transition to the peak $\bar{A}_2$ takes place due to the multiplicative noise and its value depends on whether the net value of the nonlinear inertia and stiffness expression $(c_1 - 5c_2)$ is greater or less than zero. A bifurcation diagram showing the dependence of the extrema on the excitation level is shown in Figure 4. It is seen that $\pi S_{\tilde{W}}(2)/\zeta = 4$ separates between the extrema of positive and negative nonlinear parameter $(c_1 - 5c_2)$. Figure 4 reveals the stabilization effect of the multiplicative noise on the originally unstable system when $(c_1 - 5c_2)$ only assumes positive values.

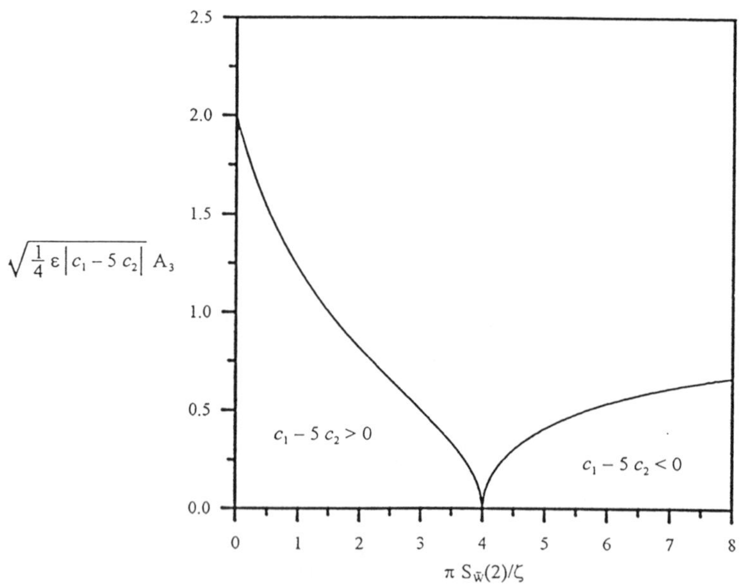

Figure 4. Bifurcation diagram showing stabilization for soft nonlinear system

The Monte Carlo simulation time history records are processed to estimate the pdf and power spectra. Figure 5 shows typical pdf plots for excitation levels

$\pi S_{\tilde{W}}(2)/\zeta$ =1.875, 3.0, 4.0, and 6.25, respectively. The response pdf for the zero response motion is a delta Dirac function, while the one corresponding to intermittency displays a peak at zero amplitude where the sharp delta peak is broadened. For a fully developed motion the response pdf is non-Gaussian whose peak at zero amplitude is reduced and is broadened over large amplitudes at the tails. For excitation level 4 there is a transition in the response pdf characterized by a zero slope at A=0. This feature was also predicted analytically. As the excitation level increases the response will have a peak at A=6.0 which is in agreement with second order averaging as given by relation (20). The estimated response power spectra corresponding to different values of excitation levels have peaks which increase as the excitation level increases and are shifted to the right.

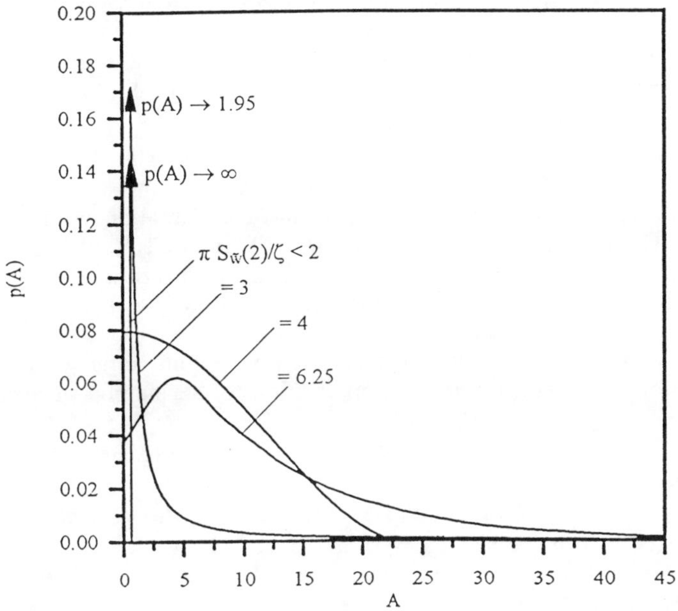

Figure 5. Response amplitude pdf as estimated by Monte Carlo simulation

5. Conclusions

Unstable dynamical systems dominated by nonlinear inertia can be stabilized by imposing parametric random excitation. The excitation acts as a source of positive stiffness. If the system is dominated by nonlinear stiffness, the parametric excitation will destabilize the system. This has been demonstrated using the second-order stochastic averaging method. The method is capable in predicting stochastic phenomena associated with noise-induced transition such as stochastic stability, on-off intermittency, and transition of the peak of response pdf.

Acknowledgment: This research is supported by a grant from ONR under research grant no. N000149310936

References

Arnold, L., 1979, "A New Example of an Unstable System Being Stabilized by Random Parameter Noise," *Match* **7**, 133-140.

Baxter, G.K., 1971, ,*The Non-linear Response of Mechanical Systems to Parametric Random Excitation*, Ph.D. Thesis, University of Syracuse.

Bogdanoff, J. L. and Citron, S., J., 1965, "Experiments with an Inverted Pendulum Subjected to Random Parametric Excitation," *J. Acoust. Soc. Amer.* **38**, 447-452.

Graham, R. and Schenzle, A., 1982, "Stabilization by Multiplicative Noise," *Phys. Rev. A* **26**(3), 1676-1685.

Hijawi, M., Moshchuk, N. and Ibrahim, R. A., 1997, "Unified Second-Order Stochastic Averaging Approach," *ASME J. Appl. Mech.* **64**, 281-291.

Horsthemke, W. and Lefever, R., 1989, "Noise-Induced Transition," in *Noise in Nonlinear Dynamical Systems, Vol. 2: Theory of Noise Induced Processes in Special Applications,* Chapter 8, F. Moss and P. V. E. McClintock, editors, Cambridge University Press, Cambridge.

Ibrahim, R. A., 1991, "Nonlinear Random Vibration: Experimental Results," *ASME Applied Mechanics Reviews* **44**(10), 423-446.

Ibrahim, R. A., 1997, "Stabilization and Stochastic Bifurcation with Application to Nonlinear Ocean Structures," in Stochastically Excited Nonlinear Systems, M. Shlesinger and T. Swean (eds.), World Scientific, (in Press).

Khasminskii, R. Z., 1980, Stochastic Stability of Differential Equations, Sijthoff & Noordhoof, The Netherlands.

Mitchell, R. R., (1972), "Stability of the inverted pendulum subjected to almost periodic and stochastic base motion: an application to the method of averaging," *Int. J. Nonlinear Mechanics* **7**, 101-123.

Nakamizo, T. and Sawaragi, Y., 1972, "Analytical Study on n-th Order Linear System with Stochastic Coefficients," Proceedings of the IUTAM Symposium on *Stability of Stochastic Dynamical Systems, Lecture Notes in Mathematics* **294**, R. F. Curtain (ed.) Springer-Verlag, Berlin, 173-185.

Nevel'son, M. B. and Khasminskii, R. Z., 1966, Stability of a Linear System with Random Perturbations of Its Parameters," *Applied Mathematics and Mechanics (PMM)* **30**, 487-493.

Prussing, J. E., (1981), "Stabilization of an unstable linear system by parametric white noise," *ASME Journal of Applied Mechanics* **48**, 198-199.

Schmidt, G., 1981, "Vibrations caused by simultaneous random forced and parametric excitations," *Zeitschrift fur Angewandte Mathematik und Mechanik (ZAMM)* **60**, 409-419.

Sethna, P. R. and Orey, S., (1980), "Some asymptotic results for a class of stochastic systems with parametric excitations," *International Journal of Nonlinear Mech*anics **15**, 431-441.

Stoker, J. J., 1950, *Nonlinear Vibrations in Mechanical and Electrical Systems*, Wiley & Sons, Mew York.

Valeev, K. G., 1971, "Dynamic Stabilization of Unstable Systems," *Izv. An SSSR. Mekhanika Tverdogo Tela* **16**(4), 9-17.

AIRCRAFT STABILITY AND CONTROL:
Bifurcation Analysis In The Design Process?

F.B.J. MACMILLEN
Bombardier Inc., Canadair
Montreal, Canada

J.M.T. THOMPSON
University College London
London, UK

1 Abstract

Since 1977 [27] various studies have been conducted to analyse the dynamics of aircraft using continuation schemes and bifurcation analysis, [1-5,7-29]. The technique has not, however, been adopted as an engineering tool in the industry. Having performed a review of previous work the authors have applied bifurcation analysis to an industrial scale aircraft model [26]. The technique is shown to locate and identify key changes in the dynamic behaviour of the aircraft as its flight controls are varied. The issues facing the aircraft industry in implementing this technique are raised and a set of software improvements suggested. With these improvements the technique could be made into an engineering tool with important benefits to industry.

2 Background to Aircraft Dynamics and Modelling

An aircraft has 6 degrees-of-freedom, giving a 12 state dynamic problem, 4 of these states (the spatial position of the aircraft and its heading angle) do not effect the dynamic behaviour of interest. The remaining 8 phase space variables are (Figure 1):

3 rotational rates:

p roll rate (about the longitudinal axis)

q pitch rate (about the transverse or lateral axis)

r yaw rate (about the directional or vertical axis)

3 states relating the aircraft to the velocity vector:

V total aircraft velocity

α angle-of-attack: angle between the symmetric (longitudinal) component of velocity and the aircraft's longitudinal axis.

β angle-of-sideslip: angle between the plane of symmetry and the velocity vector.

2 Euler angles describing the aircraft orientation relative to the gravity vector:

θ pitch attitude

ϕ roll attitude (bank angle)

F. C. Moon (ed.),
IUTAM Symposium on New Applications of Nonlinear and Chaotic Dynamics in Mechanics, 461–470.

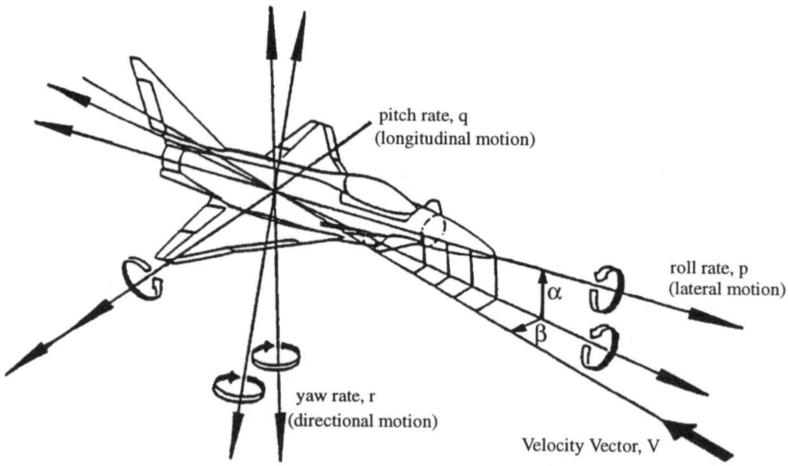

Figure 1. Aircraft axes and notation

The rigid-body equations of motion are nonlinear due to inertial effects. In addition the equations include the total aerodynamic forces and moments acting on the aircraft. The aerodynamic forces and moments are typically generated from a nonlinear model in the form of multiple, multi-dimensional interpolation tables. These tables are usually constructed from empirical data. Adding a control system adds to the number of states of the problem. In addition a control system may result in degenerate solutions (an infinite number of solutions for the same control parameter setting), [1,5,15,17,28].

Usually an aircraft will fly in a stable equilibrium state corresponding to steady level flight, or a steady banked turn. In typical steady turns, values of α will be moderate (0 to 20 deg), values of β will be small, and p, q, and r will be steady. Other equilibrium states usually exist: a 'stall' in which the airflow over the wing is detached and turbulent is an equilibrium state characterised by large α (>20 deg) and low speed. A 'spin' may also be an equilibrium state. A 'spin' is a multi-axis rotation of an aircraft with a stalled wing. Spins may exhibit various characteristics but typically are characterised by excessive yaw rates and moderate to very high angles-of-attack (30- 90 deg). Bifurcation analysis has often been used to analyse spins, [2-4,7,13,16,17,21,29].

In combat aircraft the most common periodic (limit cycle) behaviour is 'wing rock'. This consists of a lateral oscillation with a period of about 2-3 seconds. The oscillation may develop at moderate angles-of-attack (before stall), and is an undesirable characteristic. Bifurcation analysis has also been used to analyse wing rock, [2,5,8,15,16,20,28,29]. Spins may also be oscillatory in nature.

3 The 'Harrier' Model

The British Aerospace/McDonnell Douglas Harrier II has been chosen for this study. It is known to exhibit a wide range of dynamic behaviour and a complete nonlinear simulation model is available. In the current study the stability augmentation system has not been modelled, but the auto-flap system has been included. The aerodynamic model for the Harrier consists of over half a million data points contained in 56 to 130 tables (depending on configuration) of 2, 3 and 4 dimensions. This is representative of the complexity of aircraft models used in the aircraft industry. The AUTO94 code used for continuation is unable to accommodate linear interpolation of these tables, hence a tensioned spline algorithm has been written to interpolate 1 and 2 dimensional tables. The model therefore has been simplified by limiting tables to 2 independent variables: neglecting second order effects. All analysis has been performed at a single Centre of Gravity (CG) and inertial condition. In practise CG, mass or inertia may be varied or be used as a continuation parameter.

4 AUTO94

The AUTO94 continuation and bifurcation analysis software [6] has been chosen for its availability, zero cost, compatibility with existing hardware and software, and the comprehensiveness of its algorithms. Certain software changes had to be made to the AUTO94 code in order to run with models containing more than 5 dimensions. A standard Newton scheme has been used to obtain starting solutions. AUTO94 is a generic program and the user must set 36 parameters to define the problem and the operation of AUTO94. The following parameters are of particular importance:

DSMAX: the maximum allowable pseudo-arclength step size requires setting to ensure convergence with acceptable speed. Values ranged from 0.5 to 0.01. An adaptive mesh and step size was used for all runs.

ITMX: this sets the number of Newton iterations for accurate location of bifurcations, it was set high at 50.

NTST: for periodic orbits NTST sets the number of mesh points for the collocation scheme used. NTST must be as small as possible for speed, but if set too low resulted in the continuation scheme reversing direction at bifurcation points. A value of 50 has usually been used, a value of 20 usually being too low.

5 Bifurcation Analysis of the Harrier Model

All results presented are with the horizontal stabilitor (μ), the longitudinal (pitch) control, as the continuation (control) parameter. All other controls have been held constant (ailerons, rudder, thrust, etc.). On the bifurcation diagrams solid lines denote stable equilibria, dashed lines unstable equilibria. Solid circles denote stable periodic solutions, open circles unstable periodic solutions. Supercritical Hopf bifurcations are identified by solid squares.

5.1 MAIN EQUILIBRIUM BRANCH

The first bifurcation diagram, Figure 2, is from the most complete model analysed. It shows the main equilibrium branch which corresponds to the normal flight regime of the aircraft. Here 2 of the 8 state variables are shown. As the pilot moves the stabilitor (μ) to negative deflections the angle-of-attack (α) increases. At $\alpha = 8$ - 10 deg is an unstable region bounded by fold bifurcations. This corresponds to a small aerodynamic pitch instability in the basic aircraft model.

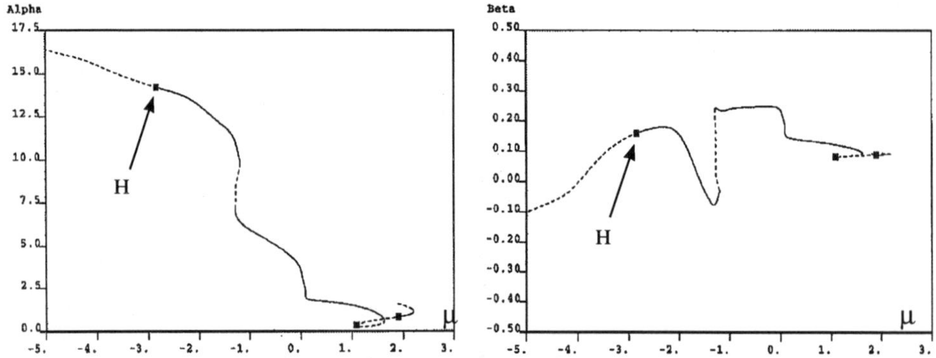

Figure 2. The angle-of-attack (α), & angle-of-sideslip (β) vs. stabilitor: main equilibrium branch

Angle-of-sideslip (β) is small, as expected in normal flight. It is non-zero however, due to small aerodynamic asymmetries included in the model.

The Harrier aircraft has a spiral mode which is stable in the dynamical system sense, (but unstable in the terminology of flight dynamicists): ie. left to its own devices the aircraft will enter a steady downward spiral. Hence the three rotational rates p, q and r are all non-zero but under 10 deg/s in the stable region of Figure 2.

The main equilibrium branch becomes unstable at low α by means of a fold, and at moderate α by a supercritical Hopf bifurcation (H). Further analysis concentrates on the dynamics at the higher angles-of-attack: more commonly reached during aggressive manoeuvering.

5.2 PERIODIC SOLUTIONS

The following of the periodic orbit arising from the Hopf bifurcation at moderate α has initially been performed on a simplified model. This simplified model gives the same qualitative dynamics as the more complex model in the vicinity of the Hopf bifurcation. The supercritical Hopf (H) yields a stable limit cycle which almost immediately undergoes a couple of cyclic folds to give a stable periodic solution (X). These cyclic folds result in an effective jump to an oscillatory state as the pilot increases angle-of-attack by decreasing μ. The maximum values of two state variables in this stable periodic solution are shown below (Figure 3):

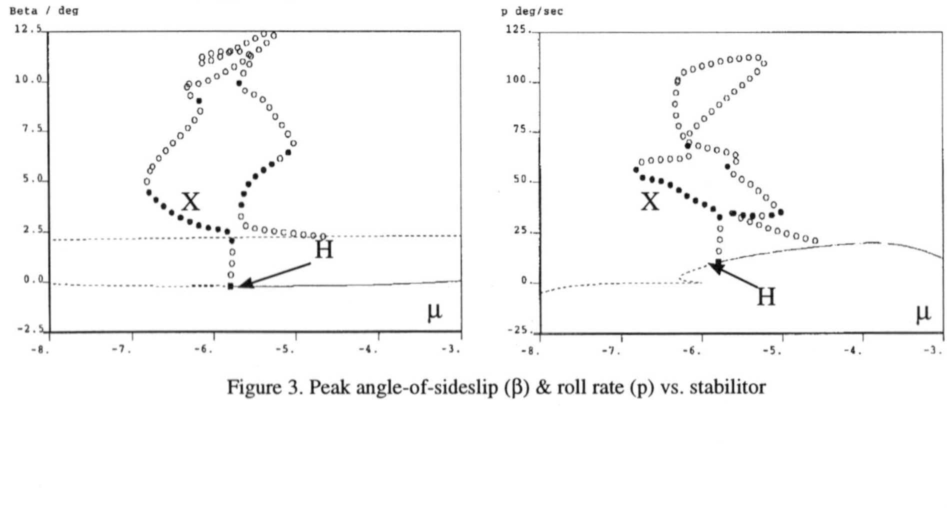

Figure 3. Peak angle-of-sideslip (β) & roll rate (p) vs. stabilitor

In the simplified model the jump occurs at about α=17 deg. The resulting oscillation is a lateral-directional mode characterised by high roll rates of 30 to 50 deg/s. The period is 2.7 seconds. Such a motion is known as 'wing rock'. A time history of the wing rock development for the same two state variables is shown in Figure 4. Harriers can exhibit such 'wing-rock' behaviour in wind-up turns in which the pilot progressively increases angle-of-attack. Whilst the α from flight tests at which onset of wing-rock occurs, agrees well with the above analysis such wing-rock is not usually sustained in flight and comparison of steady-state characteristics is more difficult. However, wing-rock develops when the complex eigenvalues corresponding to a 'dutch roll' mode become unstable. The frequency of dutch roll on Harriers is typically 2-3 seconds: this agrees well with the time period of the wing rock. Indeed it is through bifurcation analysis that the link between dutch roll and wing rock was conclusively proven.[18]

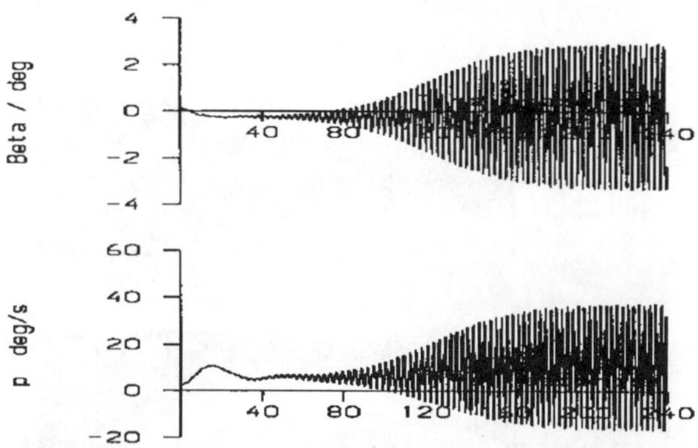

Figure 4. Angle-of-sideslip (β) & roll rate (p) versus time in seconds, following the Hopf

The above example demonstrates how continuation and bifurcation analysis may be used to find important changes in an aircraft's dynamic behaviour, and focus simulation effort. In return the simulation demonstrates how slowly (or otherwise) the dynamic behaviour changes during the transient motion.

5.3 TOROIDAL SOLUTIONS

The above results are from a simplified aerodynamic model. The final model analysed contained all major aerodynamic terms and the auto-flap control system which is usually engaged in flight. Following the resulting periodic orbit with this more comprehensive model reveals more complex behaviour.

Again a pair of cyclic folds yields a jump to stable wing-rock (X), the magnitude of oscillation increasing as μ is decreased. At μ = -3.7 deg (α = 15.5 deg) a cyclic fold (F) gives an unstable branch which eventually doubles back via another cyclic fold to give a new stable periodic branch (Y) but with much increased rotational rates, sideslip and velocity (Figure 5). With the loss of a locally stable solution at μ = -3.7 deg the aircraft will jump to the new periodic solution (Y) as the control is moved past -3.7 deg. This jump is shown in a simulation time history (Figure 6), in which the control μ has been varied slowly.

In the AUTO software it is not currently possible to record the minimum value of each state in a periodic orbit easily. However, for asymmetric motions, as is the case here, the minimum state value may have greater magnitude and importance than the maximum value.

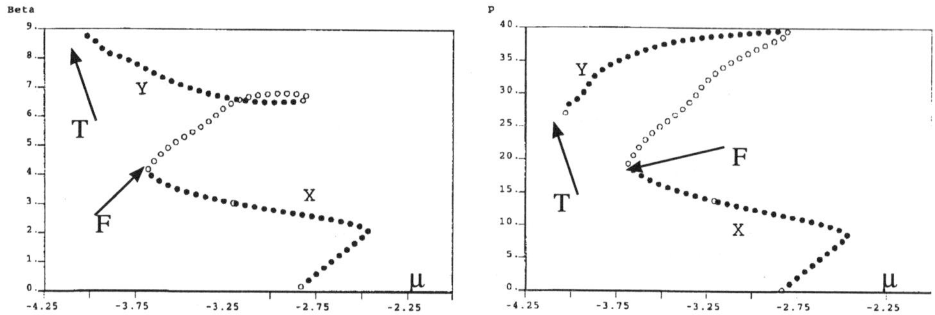

Figure 5. Peak angle-of-sideslip (β) & roll rate (p) for multiple periodic solutions

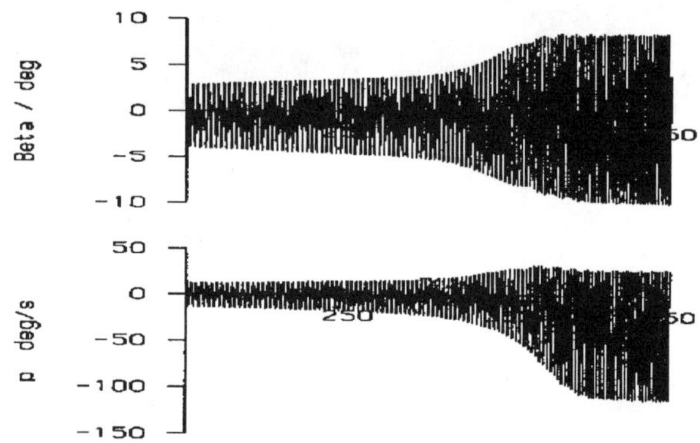

Figure 6. Angle-of-sideslip (β) & roll rate (p) versus time in seconds: jump between periodic solutions

If μ (and α) is increased further the periodic solution becomes unstable - via a secondary Hopf bifurcation (T) to a torus. This behaviour contains two periods, one very close to the wing-rock, 2.7 seconds, and one considerably slower, 19.2 seconds, (Figure 7). Such toroidal behaviour has been demonstrated before in aircraft models [9]. Plots of phase space projections illustrate the toroidal nature of the motion (Figure 8).

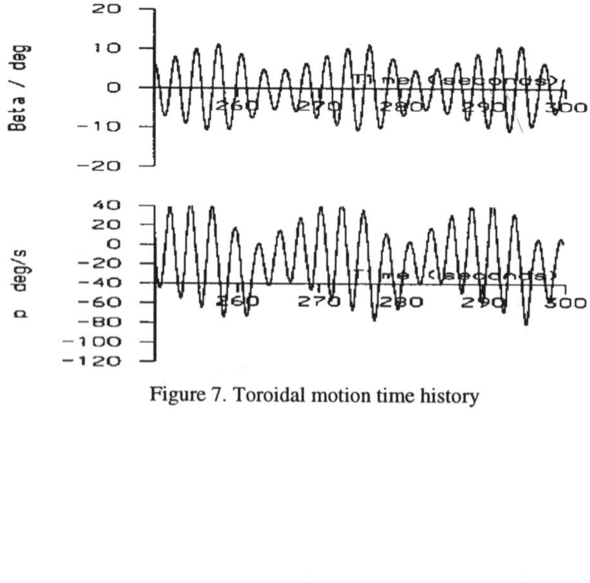

Figure 7. Toroidal motion time history

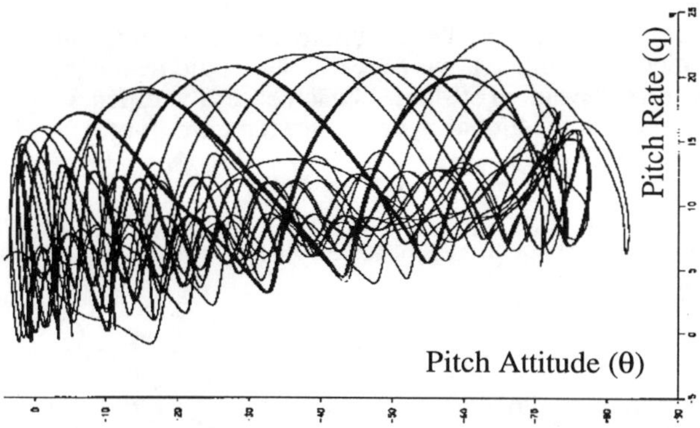

Figure 8. Toroidal motion phase projection

6 Benefits

In all studies of this application, continuation and bifurcation analysis has been shown to highlight areas of interest for simulation analysis. It provides a more global picture than simulation: one which sets the context in which changes in dynamic behaviour may be explained and understood. Other studies have begun to demonstrate the technique's potential for control law design. To what extent therefore would the technique benefit industry?

To clear an aircraft for flight test requires obtaining as close to a global and complete picture of the aircraft's dynamics as possible: and ensuring that this behaviour is acceptable. Combat aircraft fly in a large phase space. Flight condition is dependent upon altitude, Mach number, CG, inertia, and external stores configuration. There are multiple control surfaces which may be moved in almost any combination. Flight clearance is currently performed through repeated nonlinear simulation. It may typically represent about 60% of the total aerodynamic cost of a project. Over 40 man-years may be spent on flight clearance of a combat aircraft. Hence a reduction of say 30% in this process time would save in excess of 10 man-years per project. Could bifurcation analysis offer such a saving?

The software may be developed so as to use the same aircraft model as the simulation tool: thereby eliminating additional setting-up costs. One may therefore consider the ability of the technique to reduce the amount of simulation, and analysis, required. Continuation of equilibrium branches provides insight as to where bifurcations occur and the qualitative dynamics to be expected. Such a knowledge could potentially save of the order of 30% of simulation time. However, such a saving is dependent upon certain improvements to the software used.

Application may also be found in flight control system design. Control law design is almost exclusively performed using linear design methods: whether classical or modern control techniques. Problems often arise when control laws are implemented within a nonlinear model. Nonlinear simulation is the principal tool for analysing nonlinear control systems. Bifurcation analysis may be applied to both preliminary control law layout and to detailed/corrective design. In the first of these applications studies have focused on using bifurcation analysis in comparing the use of control effectors or feed-back schemes when multiple options exist [23,24,29]. The aim is to compare the effect of different schemes on the stability of the aircraft: ie. its bifurcational behaviour.

In the second potential application the technique may be used to aid design of a control law to address a specific problem area. An example would be a wing-rock suppression system [29], or a spin-prevention system. In such cases the efficacy of the potential scheme, and its impact on the global dynamics of the aircraft may be checked.

7 Developments

The following amounts to a specification for an 'ideal' software tool which would be suited to use in industry.

7.1 SOFTWARE CONTROL

If the tool is to be used for flight clearance of aircraft it must be both validated and configuration controlled.

7.2 CONTINUATION SCHEME

The pseudo-arclength continuation method used in AUTO94 is unsuitable. A robust continuation method is required which converges for non-smooth systems: that is for models with linearly interpolated aerodynamics and discontinuities such as rate and amplitude limits. Such schemes are being incorporated into other software packages (KRIT), and have been used successfully on aircraft models, [1,10]. Ultimately the software should be compatible with the same aircraft model which is used for nonlinear simulation.

7.3 SPEED

For this study AUTO94 was run on an SGi Iris. This gave acceptable performance for equilibria paths but was too slow for periodic solutions. A typical periodic branch took 2-4 hours to compute. Considering that a complete aircraft model may be 2 to 3 times as large, run times are 'over-night'. By far the greatest computational time (67%) was spent in interpolation of aerodynamic tables. A further 8% was spent on pivoting and Gaussian elimination. Speed may be increased by being able to use linear interpolation and optimising the search/interpolation routines. The interpolation task is also highly suited to parallel processing, and this could achieve 2 orders of magnitude improvement in speed. Such a speed increase is needed as the software is best used interactively.

7.4 COMPREHENSIVENESS

AUTO94 is good in terms of its ability to identify all bifurcations of interest; its restart, branch switching and bifurcation following capabilities are all required. Some improvements would include: recording maxima and minima of periodic orbits in all state variables in a single continuation sweep; the ability to increase the dimensions of the model must be straight forward; the ability to control the output of information is important; and the versatility of the graphical output needs improving.

7.5 EASE OF USE

The graphical user interface (GUI) of AUTO94 is of benefit. However the program, by reason of its generality, is difficult to set up and use on a new application, without prior experience. Software developed for industry would require parameter settings/guidelines specific to the application.

7.6 BASINS OF ATTRACTION

Knowing basin boundaries would complete the picture given by bifurcation analysis. There are two issues however:

1. How easily could knowledge of basin boundaries be used to aid simulation/flight clearance, or control law design?

2. How are the boundaries to be calculated?

The first issue is the dominant: for an un-augmented aircraft it would be possible to look at projections in all 8 state variables and rapidly determine in which states the boundaries are closest to the equilibrium. This would give a 'manoeuvre' boundary within which the aircraft would not jump to a different steady state solution. For an aircraft with control augmentation the number of states increases rapidly and the picture may become too complex.

Basins of attraction may be determined via 'carpet bombing' techniques [10] but this is computationaly inefficient. Further research is required to look at ways in which boundaries may be rapidly estimated to sufficient accuracy.

7.7 EXPERIENCE

Two decades of experience in this application have been gained, but this experience needs to develop in two ways:

1) More experience is required on the application to large aircraft models with complex control systems within an industrial environment. A control system increases dimensions and complexity and gives rise to additional problems: multiple solutions for the same control parameters; speed of computation; finding all solution branches.

2) Engineers in industry often face a terminology gap in the application of modern non-linear dynamics theory applied to aircraft dynamics. The technique has to be both explained and demonstrated to be beneficial, for it to be adopted by industry.

8 Conclusions

The authors have found that bifurcation analysis provides an insight into an aircraft's dynamic characteristics which is both different from, but complementary to, nonlinear simulation. The technique has clear abilities to trace out all the trim states of an aircraft, and locate and identify qualitative changes in its dynamics. These abilities may be used to reduce the amount of simulation time required to clear an aircraft for flight. They may also be useful aids to control law design. Due to the large cost of both these aspects of an aircraft project there is an incentive to develop the techniques and software in order to provide a beneficial tool for industry. After two decades, it is time for the industry to have an input into future development of the technique and its software, and ultimately to assess the technique, and hopefully benefit from it.

9 References

1) Avanzini, G., Matteis, Guido de.; Bifurcation Analysis of a Highly Augmented Aircraft Model, *AIAA-96-3367-CP*; August 1996

2) Barth, T.J., Planeaux, P., High Angle-of-Attack Dynamic Behaviour of a Model of High Performance Fighter Aircraft, *AIAA-88-4368*, 1988

3) Baumann, D.D., F-15B High Angle-of-Attack Phenomena and Spin Prediction Using Bifurcation Analysis, M.Sc. thesis, AFIT/GAE/ENY/89D-01 891200
4) Carroll, J.V., Mehra, R.D., Bifurcation Analysis of Nonlinear Aircraft Dynamics, *J. Guidance*, Vol 5, 1982; pp529-536
5) Davison, M.T., An Examination of Wing Rock for the F-15, M.Sc. Thesis, AFIT/GAE/ENY/92m-01 9202000
6) Doedel, E.J., Kernevez, J.P., AUTO: Software for Continuation and Bifurcation Problems in Ordinary Differential Equations, Applied Mathematics Report, CALTECH, September 1994
7) Gao H., He, X.D., Zhou Z.Q., The Study of Global Stability and Sensitivity Analysis of High Performance Aircraft at High Angles-of-Attack, *ICAS/AIAA*, 1988
8) Gao H., Wang, Z., Zhang, S.G., A Study of Wing Rock, *ICAS, 1990, Proceedings*. Vol 2.
9) Goman, M.G., Zagaynov, G.I., Bifurcation analysis of Critical Aircraft Regimes, *ICAS-80-4.2.1*, 1984
10) Goman, M.G., Khramtsovsky, A.V., A Review of Bifurcation Theory Methods in Nonlinear Flight Dynamics Problems. Application to HIRM high angle-of-attack dynamics, Contract report to Flight Systems Dept. Defense Research Agency, Bedford UK, ASF/2017/E
11) Guicheteau, Ph., Bifurcation Theory in Flight Dynamics: An Application to Real Combat Aircraft, *ICAS-90-5.10.4*, 1990
12) Guicheteau, Ph., Stability Analysis through Bifurcation Theory (1&2), 2 papers in *Nonlinear Dynamics and Chaos*, AGARD-LS-191, June 1993
13) Guicheteau, Ph., Nonlinear Flight Dynamics, in *Nonlinear Dynamics and Chaos,* AGARD-LS-191, June 1993
14) Hawkins, C.A., Application of Bifurcation and Catastrophe Theories to Near Stall Flight Mechanics, M.Sc. Thesis, AFIT/CR/NR 86-54T, 1985
15) Jahnke, C.C., Culick, F.E.C., Application of Bifurcation Theory to Nonlinear Aircraft Dynamics, *AIAA 88-4372*, Aug 1988
16) Jahnke, C.C., Application of Dynamics Systems Theory to Nonlinear Aircraft Dynamics, Ph.D., CALTECH,CA, 1990
17) Jahnke, C.C., Culick, F.E.C., Application of Bifurcation Theory to the High Angle-of-Attack Dynamics of the F-14, *J. of Aircraft*, Vol 31, No 1, Jan-Feb 1994
18) Liebst, B.S., Nolan R.C., A simplified wing rock prediction method, *AIAA CP 93-3662*, Aug 9-11, 1993
19) Littleboy, D.M., Smith. P.R., Closed Loop Analysis of a Simple Aircraft Model Using Bifurcation Methods, *AIAA-96-3366-CP*, 1996
20) Lowenberg, M.H., Application of the Bifurcation Analysis Technique to Non-linear Aircraft Dynamics, M.Sc., University of Witwatersrand, Johannesburg, 1991
21) Lowenberg, M.H., Non-linear oscillations and chaotic Behaviour in Aircraft Dynamics, Third SA Aeronautical Engineering Conference, Pretoria, August 1991
22) Lowenberg, M.H., Optimising the Use of Multiple Control Effectors using Bifurcation Analysis, *AIAA-94-3456-CP*, 1994
23) Lowenberg, M.H., Bifurcation Analysis for Aircraft Dynamics, Dept. of Aerospace Engineering Report No 495, University of Bristol, UK, June 1994
24) Lowenberg, M.H., Stability and Controllability of Sustained Flight Manoeuvres, *AIAA-96-3422-CP*, 1996
25) Macmillen, F.B.J., Application of Bifurcation Analysis to Flight Dynamics, Centre for Nonlinear Dynamics Report, University College London, April 1995
26) Macmillen, F.B.J., Continuation and Bifurcation Analysis of Flight Dynamics for Combat Aircraft: an assessment for the aircraft industry, M.Sc., Centre for Nonlinear Dynamics, University College London, UK, Sept. 1996
27) Mehra, R.D., Kessel, W.C., Carroll, J.V., Global Stability and Control Analysis of Aircraft at High Angles-of-Attack, Annual Technical Reports 1-3, ONR-CR215-248-1-3, Scientific System. USA, 1977, 1978, 1979
28) Nolan, R.C., Wing Rock Prediction Method for a High Performance Fighter Aircraft, AFIT/GAE/ENY/92J-02; 920500
29) Planeuax, J.B., Beck, J.A., Baumann, D.D., Bifurcation Analysis of a Modern Fighter Aircraft with Control Augmentation, *AIAA-90-2836*, August 1990

10 Acknowledgments

This work was supported by, and performed in conjunction with, British Aerospace (Defence), Military Aircraft Division, UK. Writing and presentation of this work has been supported by Bombardier Inc. Canadair.

GLOBAL DYNAMICS IN GYROSCOPIC AND AEROELASTIC SYSTEMS

N. SRI NAMACHCHIVAYA AND R. MCDONALD
Nonlinear Systems Group
Department of Aeronautical and Astronautical Engineering
University of Illinois at Urbana-Champaign
104 S. Wright Street, Urbana, IL 61801

1. Introduction

In this paper, we study certain gyroscopic and aeroelastic systems near resonances, subject to internal damping, symmetry-breaking and time periodic perturbations. The interplay between these three effects results in a variety of local and global bifurcations. This paper deals with obtaining new bifurcation results for fairly general nonlinear gyroscopic and aeroelastic systems using the method of normal forms, global bifurcation techniques, and various other dynamical systems tools. While deriving these equations of motion, many in the past have neglected some nonlinear terms as insignificant which, in fact, may completely change the bifurcation behavior. In both gyroscopic and aeroelastic systems it is the nonlinear dissipative terms that can change the behavior, and it is essential to carefully model these terms in the physical problems. For general nonlinear gyroscopic and aeroelastic systems, we present these effects as various bifurcation results due to symmetry-breaking imperfections. In the case of the gyroscopic systems the symmetry is broken by the addition of imperfections while for general aeroelastic systems, the reversible symmetry is broken by the addition of unsteady aerodynamic terms. We also determine parameter regions where homoclinic foci exist, which may give rise to Silnikov phenomena, depending on the linearized eigenvalues. In a mechanical system, such phenomena may lead to large amplitude oscillations leading to catastrophic failure or persistent small amplitude motions leading to structural fatigue. In this paper we shall present two important classes of problems for which some global methods can be applied to detect such behavior.

F. C. Moon (ed.),
IUTAM Symposium on New Applications of Nonlinear and Chaotic Dynamics in Mechanics, 471–482.

2. Gyroscopic System

A gyroscopic system, in the absence of damping, is Hamiltonian, and the spectrum of its linearization at an equilibrium is symmetric with respect to both real and imaginary axes. The necessary condition for stability is that the spectrum lies entirely on the imaginary axis. In generic situations, simple eigenvalues remain on the imaginary axis under parameter perturbations. However, multiple (coincident) eigenvalues split and leave the imaginary axis. Generic eigenvalue movements are modified by the presence of symmetries, such as spatial reflections or rotations. Instabilities in families of symmetric Hamiltonian systems are affected by their symmetry type.

Almost all mechanical and structural systems have inherent imperfections which break the symmetry and can give rise to both local and global changes in the dynamics of the system. Such symmetry-breaking causes a pair of eigenvalues to split at the origin, causing instability. We study the combined effects of symmetry-breaking imperfections, damping, and forcing, all of which are weak, on the nonlinear dynamics of a gyroscopic system. Thus, we treat the problem as a nearly Hamiltonian system, making use of the many results available for Hamiltonian systems.

2.1. EQUATIONS OF MOTION AND SYMMETRY

For most of the gyroscopic problems encountered in engineering, such as rotating shafts, supported pipes conveying fluid, etc., the discretized equations of motion in the neighborhood of an equilibrium configuration can be written in the form

$$\ddot{q} + 2G\dot{q} + Kq + \nabla U(q) + \zeta[\Lambda\dot{q} + F(q,\dot{q})] + \mathcal{F}(q,\dot{q},t) = 0, \tag{1}$$

where $q \in \mathbf{R}^d$ $(d \geq 2)$ is a generalized coordinate, and the $d \times d$ matrices G, K, and Λ represent gyroscopic, linear stiffness, and linear damping terms respectively. The function $U(q)$ is a general polynomial which contains higher-order nonlinearities in the potential energy, with coefficients that can also depend on loading parameters. The vector $\mathcal{F}$ is periodic in t and represents parametric perturbations. $F_i(q,\dot{q})$, $(i = 1,2)$ is a general polynomial linear in $\dot{q}$, representing nonlinear dissipation. In the absence of dissipation $(\zeta = 0)$, equation (1) represents a conservative gyroscopic system, while $\zeta > 0$ corresponds to the gyroscopic system with damping.

Introducing the generalized momentum in the usual way, the Hamiltonian equation for (1) can be written in terms of $x = (q,p) \in \mathbf{R}^{2d}$ as

$$\dot{x} = Ax + JDH_1(x) + JDH_t(x) - \zeta[Bx + f(x)], \tag{2}$$

where

$$A = \begin{bmatrix} -G & I \\ -C & -G \end{bmatrix}, \quad B = \begin{bmatrix} O & O \\ -\Lambda G & \Lambda \end{bmatrix}, \quad J = \begin{bmatrix} O & I \\ -I & O \end{bmatrix}.$$

The $d \times d$ matrices C, O and I are, respectively, $C = K - G^2$, the null matrix, and the identity matrix. The nonlinear conservative terms are given by $JDH_1(x)$ while the linear parametric terms are represented by $JDH_t(x)$, where $H_1(q,p) = U(q)$ and $H_t(q,p) = \mathcal{U}(q,t)$. In the types of applications that we are concerned with in this paper, the nonlinearities stem from the potential energy, such as strain energy due to stretching and bending of elastic elements and periodic perturbations stem from variation of axial loads and rotaion speeds. However, there are many examples where $H_1(q,p)$, $H_t(x)$ are also functions of the momentum vector $p \in \mathbf{R}^d$. To this end, we assume a general form of H_1 and H_t throughout the analysis. The nonlinear dissipative terms are contained in $f(x)$. When $\zeta = 0$, the Hamiltonian formulation of the nonlinear gyroscopic system, taking advantage of canonical transformations, reduces a significant amount of algebra and provides a better insight into the mathematical structure of the problem.

2.2. BEHAVIOR OF EIGENVALUES

First, we consider the linear, autonomous Hamiltonian system, studied by Nagata and Sri Namachchivaya [6]. In the absence of damping, the system is Hamiltonian and the governing linearized equations reduce to

$$\dot{x} = JSx = Ax, \tag{3}$$

and we note that A depends on system parameters. Since we are interested in the stability of the trivial solution, the four eigenvalues of A can be written as

$$\pm\, i\omega_i = \pm i \sqrt{\left[\Omega^2 + \frac{1}{2}(\overline{\omega}_1^2 + \overline{\omega}_2^2)\right] \mp \sqrt{2\Omega^2\left(\overline{\omega}_1^2 + \overline{\omega}_2^2\right) + \frac{1}{4}\left(\overline{\omega}_1^2 - \overline{\omega}_2^2\right)^2}}. \tag{4}$$

For the linear system with S^1 symmetry, i.e. $\overline{\omega} = \overline{\omega}_1 = \overline{\omega}_2$, the eigenvalues reduce to

$$\pm\, i(\overline{\omega} - \Omega)\,, \quad \pm i(\overline{\omega} + \Omega) \tag{5}$$

and depending on the system parameters, the symmetric problem can have one of three critical cases: (1) a double zero eigenvalue at $\overline{\omega} = \Omega$, (2) a pair of purely imaginary eigenvalues of multiplicity two (i.e., 1:1 resonance) $\pm i\Omega$, $\pm i\Omega$ at $\overline{\omega} = 0$, or (3) four zero eigenvalues at $\overline{\omega} = \Omega = 0$. However, in this paper we only address the double zero eigenvalue case. To simplify the analysis and capture the essential dynamics near the double zero eigenvalue, we introduce a small bifurcation parameter δ by $\Omega = \Omega_c + \delta = \omega + \delta$ where $\Omega_c = \omega$ is the critical value of Ω for the symmetric problem. In addition, we introduce the symmetry-breaking parameter β (chosen to simplify calculation of the linear normal form) to measure the deviation of the system from the symmetric case (see Nagata and Sri Namchchivaya [6]). With these parameters, the linear matrix $\tilde{A}$ has eigenvalues

$$\pm\sqrt{\beta^2 - \delta^2}, \qquad \pm i\sqrt{(2\Omega - \delta)^2 + \beta^2}$$

Note that the system is stable for $\delta < -\beta$, unstable for $-\beta < \delta < \beta$, and restabilized for $\delta > \beta$. Including the nonlinear terms, dissipative terms, and the periodic terms and applying various nonlinear transformations as in Nagata and Sri Namachchivaya [6] and McDonald and Sri Namachchivaya [4] the normal form for a damped, forced gyroscopic system for the subharmonic resonance case is as follows:

$$\dot{z}_1 = \frac{1}{4\Omega}\left\{\left(\frac{\delta\zeta\lambda_0}{2\Omega} - i\delta\right) z_1 + \left(-\frac{\beta\zeta\lambda_0}{2\Omega} + i\beta\right)\overline{z}_1 \right. \tag{6}$$
$$\left. + \left[i\hat{A}_0 + \zeta\left(\Re(A) - \frac{\lambda_0\hat{E}_0}{4\Omega}\right)\right] z_1^2\overline{z}_1 + \left[i\hat{B}_0 + \zeta\left(\Re(B_1) - \frac{\lambda_0\hat{F}_0}{2\Omega}\right)\right] z_1 z_2\overline{z}_2\right\}$$

$$\dot{z}_2 = \frac{1}{4\Omega}\left\{\left[-\zeta\lambda_0 + i\left(2\Omega\lambda - \delta\right)\right] z_2 + \left[\beta h\left(iq_i - q_r\right)\right]\overline{z}_2 \right. \tag{7}$$
$$\left. + \left[i\hat{B}_0 + \zeta\left(\Re(B_2) + \frac{\lambda_0\hat{E}_0}{2\Omega}\right)\right] z_1 z_2\overline{z}_1 + \left[i\hat{C}_0 + \zeta\left(\Re(C) + \frac{\lambda_0\hat{F}_0}{2\Omega}\right)\right] z_2^2\overline{z}_2\right\}$$

where λ is a detuning of the forcing frequency from a critical frequency ω_0.

2.3. DYNAMICS OF THE DISSIPATIVE SYSTEM

In this section, we study the dynamics of (6-7) in the absence of forcing. The z_1-plane $\{z_2 = 0\}$ is invariant and exponentially attracting in a neighborhood of the origin, so we need only study the dynamics restricted to the z_1-plane, which are described by

$$\dot{z}_1 = \left(\frac{\delta\zeta\lambda_0}{2\Omega} - i\delta\right) z_1 + \left(-\frac{\beta\zeta\lambda_0}{2\Omega} + i\beta\right)\overline{z}_1 + \left[i\hat{A} + \zeta\left(\operatorname{Re} A - \frac{\lambda_0\hat{E}}{4\Omega}\right)\right] z_1^2\overline{z}_1.$$

We assume $\hat{A} \neq 0$. If $\beta > 0$, then by rescaling $z_1 \to \left(\frac{\beta}{|\hat{A}|}\right)^{1/2} z_1$, $\zeta \to \frac{2\Omega}{\lambda_0}\zeta$, $\delta \to \beta\delta$, and $t \to \frac{t}{\beta}$, and introducing action angle coordinates $I \geq 0$ and $\theta \in S^1$ by $z_1 = \sqrt{I}e^{i\theta}$ we get

$$\dot{I} = (\delta\zeta - \zeta\cos 2\theta + \sin 2\theta)I + 2\zeta\alpha I^2, \quad \dot{\theta} = -\delta + \zeta\sin 2\theta + \cos 2\theta + \gamma I. \tag{8}$$

where

$$\gamma = \operatorname{sgn}(\hat{A}) = \pm 1, \quad \alpha = \frac{1}{|\hat{A}|}\left(\frac{2\Omega \operatorname{Re} A}{\lambda_0} - \frac{\hat{E}}{2}\right).$$

The lowest order damping terms make the I-plane exponentially attracting. However, the effect of damping on the system restricted to the I-plane is weak, and higher order damping terms are required to obtain generic and structurally stable dynamics.

Treating δ as a bifurcation parameter and linearizing about the fixed point at the origin, we find that the origin is a sink for $\delta < -1$, a saddle for $-1 < \delta < 1$, and a source for $\delta > 1$. Thus we expect bifurcations to occur at $\delta = -1$ and $\delta = 1$. To check this, we look for nontrivial fixed points. Solving for nontrivial fixed points of (8), and eliminating θ, we get

$$\delta^2 - 2(1+\zeta^2)^{-1}(\gamma - \zeta^2\alpha)\delta I + (1+\zeta^2)^{-1}(1+\zeta^2\alpha^2)I^2 = \beta^2, \tag{9}$$

which, for fixed positive ζ, is the equation of an ellipse in the δI-plane. We are concerned with the portion satisfying $I \geq 0$. The ellipse of nontrivial fixed points intersects the axis $I = 0$ at $\delta = -1$ and $\delta = 1$, giving pitchfork bifurcation points. If $\gamma = +1$, then the solutions bifurcate supercritically from $I = 0$ at both $\delta = -1$ and $\delta = 1$, while the bifurcations are both subcritical for $\gamma = -1$. For δ near -1, there is a pair of bifurcating solutions with I near 0 and θ near $\pi/2$ and $3\pi/2$, while for δ near 1, there is another pair of bifurcating solutions with I near 0 and θ near 0 and π. Solving (9) for I, we obtain

$$I = \frac{\delta(\gamma - \zeta^2\alpha) \pm \sqrt{\Delta}}{1+\zeta^2\alpha^2}, \quad \Delta = (1+\zeta^2\alpha^2)(1+\zeta^2) - \delta^2\zeta^2(\gamma+\alpha)^2$$

and Δ is positive for all sufficiently small ζ.

Linearizing (8) about the nontrivial fixed points, we find that the determinant and trace of the linearization are given by

$$Det = \pm I\sqrt{\Delta}, \quad Tr = 2\zeta(\delta + 2\alpha I).$$

Thus, if $\gamma = +1$, the pair of solutions bifurcating from $I = 0$ at $\delta = -1$ are sinks, at least near the bifurcation point, and the pair of solutions bifurcating at $\delta = 1$ are saddles, and conversely if $\gamma = -1$.q Secondary Hopf bifurcations ($Tr = 0$, $Det > 0$) can occur from the top branch of the ellipse (9) if it intersects the line $I = -\delta/2\alpha$ for some $I > 0$. For example, if $\gamma = +1$ then secondary Hopf bifurcations occur when $2\alpha > -(1+\zeta^2\alpha)/(1-\zeta^2\alpha^2) = -1 + \mathcal{O}(|\zeta|^2)$. We discuss the results in terms of the branching of solutions along one-parameter paths, as δ is increased while $\alpha < 0$ is fixed. For small values of $|\alpha|$, we consider the one-parameter

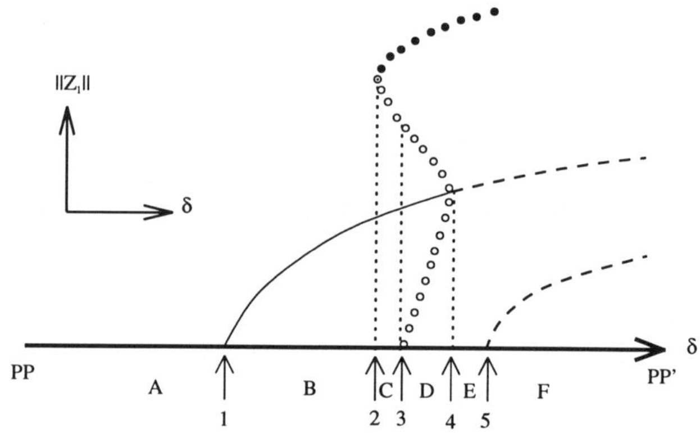

Figure 1. One-parameter branching diagram for $\gamma = 1$ and $\alpha_c = 0.33 < \alpha < 0$.

bifurcation diagram. For $\delta < -1$, the origin is asymptotically stable, i.e., a sink. At $\delta = -1$, indicated by the point 1 on the δ-axis in Figure 1, the origin undergoes a pitchfork bifurcation. At point 2 on the δ-axis, a semi-stable large periodic orbit exists. As δ is increased through the value corresponding to the point 2, there is a saddle-node bifurcation of periodic orbits. For δ just beyond the point 2, there is a pair of large periodic orbits with unstable inner orbit and a stable outer orbit. While the outer stable orbit remains qualitatively the same, the inner unstable orbit shrinks to form a homoclinic orbit at the point 3 on the δ axis. Further increase of δ produces a pair of unstable small periodic orbits around the nontrivial fixed points while the large stable periodic orbit remains qualitatively the same. The small pair of unstable periodic orbits ceases to exist at the point 4 on the δ-axis,and this represents the secondary Hopf bifurcation point obtained in the local analysis. Similarly, we can find a local bifurcation diagram for larger values of $|\alpha|$, as shown in Nagata and Sri Namachchivaya [6].

2.4. DYNAMICS OF THE FORCED SYSTEM: LOCAL ANALYSIS

The normal form equations for a damped, forced gyroscopic system for the subharmonic resonance case are given in (6-7): Since the damping in the second mode is much greater than that in the first mode, we can show that the z_1-plane ($z_2 = 0$) is invariant, and exponentially attracting. Thus, since there is no forcing in the first mode, the local analysis is the same as for the autonomous case, studied in section 2.3.

2.5. DYNAMICS OF THE FORCED SYSTEM: GLOBAL ANALYSIS

When the damping and forcing are both scaled to order ϵ, an integrable structure is obtained, and a global analysis may be done [8]. In this case, the equations of motion, in action-angle coordinates, can be written as:

$$\dot{I}_1 = \frac{\beta}{2\Omega} I_1 \cos 2\phi_1, \quad \dot{\phi}_1 = -\frac{\delta}{4\Omega} + \frac{\beta}{4\Omega}\cos 2\phi_1 + \frac{\hat{A}_0}{4\Omega} I_1 + \frac{\hat{B}_0}{4\Omega} I_2 \quad (10)$$

$$\dot{I}_2 = \epsilon \left\{ \frac{\beta h}{2\Omega} I_2 \left(-q_r \cos 2\phi_2 + q_i \sin 2\phi_2\right) - \frac{\zeta \lambda_0}{2\Omega} \right\} \quad (11)$$

$$\dot{\phi}_2 = \frac{\lambda}{2} - \frac{\delta}{4\Omega} + \frac{\hat{B}_0}{4\Omega} I_1 + \frac{\hat{C}_0}{4\Omega} I_2 + \epsilon \left\{ \frac{\beta h}{4\Omega} \left(q_i \cos 2\phi_2 + q_r \sin 2\phi_2\right) \right\} \quad (12)$$

Equations (10) may also be written in terms of rectangular coordinates (x_1, x_2) to avoid singularities in some calculations.

The unperturbed system possesses, for $\max(0, \frac{\delta-\beta}{\hat{B}_0}) < I_2 < \frac{\delta+\beta}{\hat{B}_0}$, a 2-dimensional normally hyperbolic invariant manifold $\mathcal{M}_0$, given by $\mathcal{M}_0 = \{(x, I_2, \phi_2) | x = \tilde{x}_0(I_2), I_2 \in \mathcal{I}, 0 \leq \phi_2 < 2\pi\}$ where $\tilde{x}_0(I_2)$ is a saddle point in the x-plane, and a 3-dimensional homoclinic manifold Γ, given by $\Gamma = \left\{x^h(t, I_2), I_2, \phi_2(t, I_2, (\phi_2)_0)\right\}$, where $x^h(t, I_2)$ is a homoclinic orbit in the x-plane. This homoclinic orbit is given by

$$\phi_1(t) = -\tan^{-1}\left[a \tanh(bt)\right] \qquad I_1(t) = \gamma_1 - \frac{\beta}{\hat{A}_0} \cos\left[2\tan^{-1}\left(a \tanh(bt)\right)\right]$$

$$\phi_2(t) = (\phi_2)_0 + \omega_A t + \nu_A \tan^{-1}\left[\frac{\coth(bt)}{a}\right]$$

where the constants are given by:

$$a = \sqrt{-\frac{\delta - \beta - \hat{B}_0 I_2}{\delta + \beta - \hat{B}_0 I_2}} \qquad b = \frac{\sqrt{-(\delta - \beta - \hat{B}_0 I_2)(\delta + \beta - \hat{B}_0 I_2)}}{4\Omega}$$

$$\gamma_1 = \frac{1}{\hat{A}_0}\left(\delta - \hat{B}_0 I_2\right) \qquad \omega_A = \frac{2\Omega\lambda - \delta + \hat{C}_0 I_2}{4\Omega} \qquad \nu_A = \frac{2\hat{B}_0}{A_0}$$

The unperturbed dynamics on the manifold $\mathcal{M}_0$ are given by the $\dot{I}_2$ and $\dot{\phi}_2$ equations, evaluated at the saddle $I_1 = 0$, and $\epsilon = 0$, i.e.

$$\dot{I}_2 = 0 \qquad \dot{\phi}_2 = \frac{\lambda}{2} - \frac{\delta}{4\Omega} + \frac{\hat{C}_0}{4\Omega} I_2 \quad (13)$$

A resonance is said to occur in the dynamics on $\mathcal{M}_0$ for a value of I_2 such that $\dot{\phi}_2 = 0$. Thus, a resonance occurs when $I_2 = I_2^r = \frac{\delta - 2\Omega\lambda}{\hat{C}_0}$

For the non-resonant case, the fixed point of the averaged system on $\mathcal{M}_0$ has a fixed point only at $I_2 = 0$, which is on the boundary of the manifold $\mathcal{M}_0$. This violates an assumption of the persistence theorem, since points on the boundary of $\mathcal{M}_0$ may flow out of the perturbed manifold $\mathcal{M}_\epsilon$.

However, we can obtain results when $I_2 = I_2^r$ as in Kovačič and Wiggins [2]. In this case, the manifold $\mathcal{M}_0$ persists, and we can examine the dynamics on the perturbed manifold $\mathcal{M}_\epsilon$ by rescaling to slow time near the resonance, i.e. $I_2 = I_2^r + \sqrt{\epsilon}\eta$, $\tau = \sqrt{\epsilon}t$. In these coordinates, the unperturbed system on $\mathcal{M}_0$ has center fixed points at $q_0 = (0, (\phi_2)_c)$ and saddle fixed points at $p_0 = (0, (\phi_2)_s)$, where

$$(\phi_2)_c = \frac{1}{2}\left(\phi_q + n\pi + \cos^{-1}\delta_1\right) \qquad (\phi_2)_s = \frac{1}{2}\left(\phi_q + n\pi - \cos^{-1}\delta_1\right),$$

$\delta_1 = \frac{\zeta\lambda_0}{\beta h q}$ and $\cos\phi_q = \frac{q_r}{q}$. Using Bendixson's criteria on the perturbed system, we can determine that no closed orbits exist in the perturbed system, and that the center fixed point perturbs to a hyperbolic sink q_ϵ in the $\eta - \phi_2$ plane, with the homoclinic orbit of the unperturbed system as an approximation to the domain of attraction of the sink.

We now find conditions under which the hyperbolic sink is connected to itself by a homoclinic orbit. This homoclinic orbit exists if the unstable manifold of q_ϵ (transverse to the $\eta - \phi_2$ plane) transversely intersects the stable manifold of the annulus $\mathcal{A}_\epsilon$ centered at I_2^r, and if this orbit returns to the domain of attraction of the sink q_ϵ. Whether or not the orbit returns to this domain is determined by the phase shift, $\Delta\phi_2$, calculated as

$$\Delta\phi_2 = \phi_2(\infty, I_2^r, (\phi_2)_0) - \phi_2(-\infty, I_2^r, (\phi_2)_0) = \frac{4B_0}{A_0}\tan^{-1}\frac{1}{a} \tag{14}$$

The intersection of the unstable manifold of q_ϵ with the stable manifold of the annular resonance region is determined by the Melnikov function. Simple zeros of the Melnikov integral define parameter values at which these two manifolds intersect transversely. The Melnikov integral is then

$$M(I_2^r, (\phi_2)_c) = -\frac{\beta h q}{4\Omega} I_2^r \left[\delta_c(\cos\Delta\phi_2 - 1) - \sqrt{1-\delta_c^2}\sin\Delta\phi_2 - 2\delta_1\Delta\phi_2\right)$$

where

$$\delta_c = -\frac{2\delta_1 q_r q_i}{q^2} + \sqrt{1-\delta_1^2}\left(\frac{q_i^2 - q_r^2}{q^2}\right)$$

Figure 2 shows a plot of the zeros of the Melnikov function as a function of $0 < \delta_1 < 1$, $-1 < \delta_c < 1$, and $\Delta\phi_2$

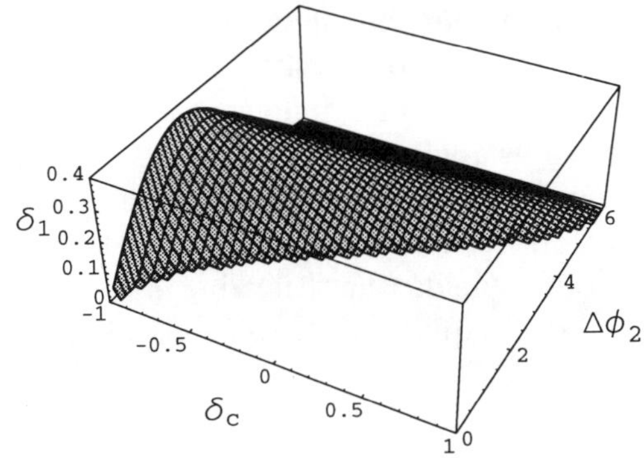

Figure 2. Zeros of Melnikov function as function of δ_1, δ_c, and $\Delta\phi_2$

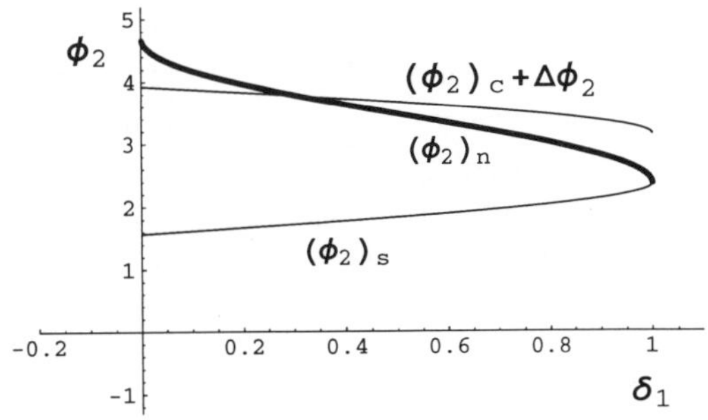

Figure 3. Phase condition for $\Delta\phi = 5.0$, $\phi_q = \frac{\pi}{2}$ is satisfied for $\delta_1 < .304$

Finally, to ensure that the orbit returns to the domain of attraction of q_ϵ, a phase condition must be satisfied. Assuming $\phi_2(-\infty) = (\phi_2)_c$, and $(\phi_2)_n$ is the "nose" of the homoclinic orbit, the phase condition is given by

$$(\phi_2)_s < (\phi_2)_c + \Delta\phi_2 < (\phi_2)_n$$

For $\Delta\phi = 5$, a typical value for the phase shift, and $\phi_q = \frac{\pi}{2}$ ($q_r = 0$) we have plotted this phase condition as a function of δ_1 in Figure 3. In this case, the phase condition is only satisfied for $\delta_1 < .304$

We can interpret these results for the case of a rotating shaft. The manifold $\mathcal{M}_0$ represents for each value of I_2, a periodic orbit, i.e. a pure whirling motion of the shaft, with some whirling frequency which is, in general, unequal to the rotation frequency of the shaft. The homoclinic focus thus consists of two distinct motions: (1) a quick excursion away from this steady whirling motion, and (2) an asymptotic approach back to the whirling motion. The existence of this homoclinic focus, as well as conditions on the eigenvalues of the linearization of the perturbed vector field about the sink q_ϵ, are sufficient to show the existence of chaotic dynamics in this problem, according to the Silnikov theorem [8].

3. Aeroelastic Problem

It is our intention to study a fairly general model for the panel flutter problem. We take the general partial differential equations of motion in terms of the displacements u, v and w as in Dowell [1], assuming a two-dimensional, isotropic flat plate with nonlinear coupling between in-plane stretching and out-of-plane bending. The supersonic flow over the panel generates the pressure difference Δp with respect to the cavity below (which we assume is deep enough not to have a significant influence on the panel response). For inviscid high supersonic flow regimes, a third-order, quasi-steady piston theory approximation can be derived as an estimate of Δp. Following Lee and Sri Namachchivaya[3], applying Galerkin's method to the resulting partial differential equations yields the following system of nonlinear second order ordinary differential equations (*two mode approximation*):

$$\ddot{z} + Kz + F(z) + [G^{\epsilon}\dot{z} + F^{\epsilon}(z)] = 0 \tag{15}$$

where

$$K = \begin{bmatrix} \omega_1^2 & -\gamma\lambda \\ \gamma\lambda & \omega_2^2 \end{bmatrix}, \qquad G = \begin{bmatrix} \left(\frac{\mu\lambda}{M}\right)^{\frac{1}{2}} & 0 \\ 0 & \left(\frac{\mu\lambda}{M}\right)^{\frac{1}{2}} \end{bmatrix}$$

$$\begin{aligned} \gamma &= \frac{2m_1 m_2(1-(-1)^{m_1+m_2})}{m_2^2 - m_1^2}, \qquad \Gamma = -R_x \\ \omega_1^2 &= \pi^4(m_1^2 + n^2r^2)^2 - m_1^2\pi^2\Gamma, \quad \omega_2^2 = \pi^4(m_2^2 + n^2r^2)^2 - m_2^2\pi^2\Gamma \end{aligned}$$

In several respects, Eq.(15) contains terms that have not been considered in previous works. Since the term $\sqrt{\mu\lambda/M}$ is small (i.e. $O(\epsilon)$) compared to the other coefficients, Eq.(15) is separated into two parts. The linear and nonlinear terms $Kz + F(z)$ contain some aerodynamic terms which are of $O(1)$ and all of the terms that stem from the nonlinear panel. The $O(\epsilon)$ aerodynamic terms are given by $G^{\epsilon}\dot{z} + F^{\epsilon}(z)$. This separation of order is essential to the subsequent analysis. We can rewrite the above second order equations as four first order equations in the state variables $x_1 = z_1$, $x_2 = \dot{z}_1$, $x_3 = z_2$, $x_4 = \dot{z}_2$ as

$$\dot{x}(t) = Ax + F(x) + F^{\epsilon}(x) \tag{16}$$

where

$$A = \begin{bmatrix} 0 & 1 & 0 & 0 \\ -\omega_1^2 & 0 & \gamma\lambda & 0 \\ 0 & 0 & 0 & 1 \\ -\gamma\lambda & 0 & -\omega_2^2 & 0 \end{bmatrix}$$

where the usual normal form notation is used as in Sri Namachchivaya *et al.* [7] and the nonlinear coefficients $f_{i,jklm}$, $f^{\epsilon}_{i,jklm}$ are given in [3] in terms of various system parameters. Global analysis may be possible in the presence of certain symmetries, which reduce the dimensionality of the system and thus simplify the analysis. Reversible dynamical systems are invariant under a certain spatial transformation accompanied by a reversal of time. More precisely, consider an n-dimensional autonomous dynamical system, $\dot{x} = f(x)$, where $x \in \Re^n$ and f is C^∞. This dynamical system is defined to be reversible if there exists a transformation of the state variables, R, such that $R = R^{-1}$ and the equations are invariant under $x \to Rx$ and $t \to -t$. Reversibility of a dynamical system is entirely dependent on the existence of a transformation R which satisfies the above mentioned conditions. For the panel flutter problem, the first order system (given by Eq.(16)) with $\epsilon = 0$, i.e., the corresponding nondissipative system is reversible under the transformation

$$R = \begin{bmatrix} 1 & 0 & 0 & 0 \\ 0 & -1 & 0 & 0 \\ 0 & 0 & 1 & 0 \\ 0 & 0 & 0 & -1 \end{bmatrix}$$

Here we have $AR + RA = 0$, and $RF(x) + F(Rx) = 0$ These properties enable us to reduce the original system considerably thus making the analytical analysis possible. The primary task is to determine the various instabilities that take place as the parameters are varied. The stability of the trivial solution can be completely specified by examining the eigenvalues of the linear operator A. The equilibrium has a pair of purely imaginary eigenvalues of multiplicity two (i.e., 1:1 resonance)

$$\rho_{1,2} = \rho_{3,4} = \pm i\sqrt{\frac{1}{2}\left(\omega_1^2 + \omega_2^2\right)} \tag{17}$$

when $\lambda_c = \frac{\omega_1^2 - \omega_2^2}{2\gamma}$ and we only present the results pertaining to this case. The local bifurcation results can be summarized in terms of two nonlinear terms, π_3 and d_3 representing the nonlinear reversible and dissipative unsteady aerodynamic terms respectively, and the three unfolding parameters μ_1, μ_2, and α. In the analysis the effects of the reversible (μ_1) and non-reversible unfolding parameters (α, μ_2) which arise from linear dissipative effects have been incorporated

Theorem 1 Non-degenerate Case *Assuming $\pi_3 d_3 \alpha \neq 0$, there are zero, one or two periodic solutions:*

$$\rho_c = \frac{-\left(\mu_2 d_3 + 2\alpha^2\pi_3\right)}{d_3^2} \pm \frac{2\alpha\sqrt{D}}{d_3^2}, \quad u_c = -\alpha, \quad v_c = \frac{\pi_3\alpha \mp \sqrt{D}}{d_3} \tag{18}$$

where

$$D = d_3\left(\pi_3\mu_2 - d_3\mu_1\right) + \alpha^2\left(\pi_3^2 + d_3^2\right)$$

Along the parabola

$$\mathcal{B}: \quad \mu_2^2 + 4\alpha^2\mu_1 - 4\alpha^4 = 0 \tag{19}$$

the trivial solution loses stability through a Hopf bifurcation giving rise to one or two periodic solutions. The nontrivial periodic solution undergoes a saddle node bifurcation along the line

$$\mathcal{L}: \quad d_3\left(\pi_3\mu_2 - d_3\mu_1\right) + \alpha^2\left(\pi_3^2 + d_3^2\right) = 0 \tag{20}$$

This line divides the parameter space into regions of zero periodic solutions and two periodic solutions. Finally the periodic solutions undergo secondary Hopf bifurcation along

$$\mathcal{H}: \quad 2\alpha\left(3D + \pi_3 d_3\mu_2\right) \mp d_3\mu_2\sqrt{D} \mp 14\alpha^2\pi_3\sqrt{D} + 4\alpha^3\left(d_3^2 + 2\pi_3^2\right) = 0 \tag{21}$$

The stability of the bifurcating 2-torus solutions are given by the standard Hopf formula.

Acknowledgments

The authors acknowledge the support of the Air Force Office of Scientific Research through grant 96-1-0265 monitored by Capt. Brian Sanders.

References

1. Dowell, E. H., and Ilgamov, M., (1988) *Studies in Nonlinear Aeroelasticity*, Springer-Verlag, New York,
2. Kovačič, G. and Wiggins, S. (1992) Orbits Homoclinic to Resonances, with an Application to Chaos in a Model of the Forced and Damped Sine-Gordon Equation *Physica D*, **57**, pp. 185-225.
3. Lee, A., and Sri Namachchivaya, N., (1997) Nonlinear Analysis of Aeroelastic Systems *International Forum on Aeroelasticity and Structural Dynamics , Rome, Italy*, **III**, pp. 371-378.
4. McDonald, R. and Sri Namachchivaya, N. (1997) Global Bifurcations in Periodically Perturbed Gyroscopic Systems with Application to Rotating Shafts *Journal of Chaos, Solitons and Fractals*, **8**(4), pp. 613-636.
5. McDonald, R., Murdock, J. and Namachchivaya, N. (Submitted for publication) Normal Forms for Nonlinear Hamiltonian Systems with Weak Periodic Perturbations.
6. Nagata, W. and Sri Namachchivaya, N. (1997) Bifurcations in Gyroscopic Systems with an Application to Rotating Shafts *Proceedings of the Royal Society: Series A* (to appear)
7. Sri Namachchivaya, N., Doyle, M. M., Langford, W. F., and Evans, N. W., (1994) Normal Form for Generalized Hopf Bifurcation with Non-semisimple 1:1 Resonance *Journal of Applied Mathematics and Physics (ZAMP)*, **45**, pp. 312-335.
8. Wiggins, S. (1988) *Global Bifurcations and Chaos: Analytical Methods*. Springer-Verlag, New York.

CHAOTIC OSCILLATIONS OF A LOOSELY SUPPORTED TUBE IN A HEAT-EXCHANGER ARRAY IN CROSS-FLOW

M.P. PAÏDOUSSIS, S.J. PRICE AND N.W. MUREITHI †
Department of Mechanical Engineering, McGill University
Montreal, Québec, H3A 2K6, Canada

1. Introduction

At sufficiently high flows, tubes in tube-in-shell heat exchangers (incl. steam generators and condensers) are subject to 'fluidelastic instability'. This self-excited flutter phenomenon has caused damages in the $\$10^8$ range. Typically, the tubes are inverted-U shaped and are positively supported only at the ends of the U, and generally loosely supported elsewhere by baffle plates (TSPs) or flat 'antivibration bars' (AVBs) — the looseness being for ease of assembly, to allow thermal expansion, or because of imperfections. Thus, theoretically, a tube may be supported at A, B, C,..., K, and hence, be designed against fluidelastic instability for spans AB, BC, etc., while in practice, if support B is loose, span AC may easily become fluidelastically unstable; the oscillation then grows till impact with the loose support occurs.

It has been shown experimentally for so-called 'rotated triangular' tube arrays, one of the typical tube layout geometries, that the onset of fluidelastic instability is sensibly the same whether (i) all tubes are flexible or flexibly mounted, or (ii) only one is, while the rest are rigid and immobile. Hence, a set of experimental-theoretical studies have been undertaken of the dynamics of a flexibly mounted, loosely supported tube in an otherwise rigid array in cross-flow.

Four studies have been conducted at McGill: (a) for planar motion of the tube, to model impacting with AVBs, revealing a period-doubling route to chaos [1]; (b) a two-degree-of-freedom (2-d.o.f.) model of the motion in the TSP plane, in the confines of a circular baffle hole [2]; (c) a higher-order model of a flexible tube interacting with the baffle [3,4]. Studies (a) and (b) are theoretical, while (c) is combined experimental-theoretical. The presentation here stresses (b) and (c). The problem has also been studied by others; e.g., Axisa *et al.* [5] and Cai & Chen [6].

2. A simple 2-d.o.f. model

In the simplest possible model, based on the observation that the fluidelastic vibration is in a single flexural model of the tube, a modal approach is taken in which the motion

†Presently with the University of Kobe, Kobe, Japan.

F. C. Moon (ed.),
IUTAM Symposium on New Applications of Nonlinear and Chaotic Dynamics in Mechanics, 483–492.

is modelled as if it were 2-D; i.e., as if the tube were flexibly supported and vibrating as a rigid-rod, normal to the tube hole (Figure 1). The equations of motion then are

$$m\ell\ddot{x} + c\dot{x} + kx = F_{xf} + F_{xi}\,, \qquad m\ell\ddot{y} + c\dot{y} + ky = F_{yf} + F_{yi}\,, \tag{1}$$

where F_{xf} and F_{yf} are the fluid-dynamic forces, and F_{xi} and F_{yi} are the impact forces with the TSP.

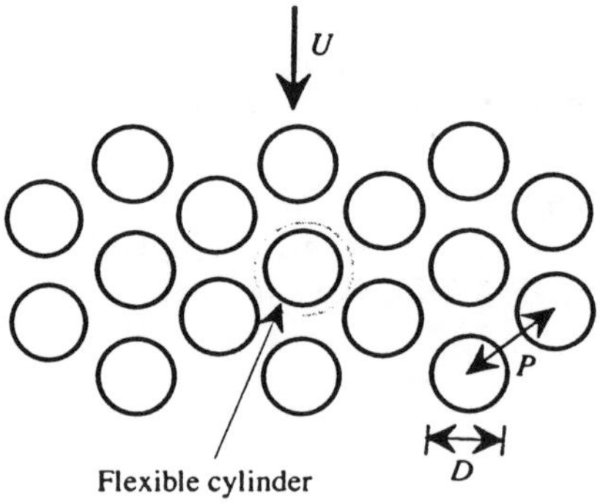

Figure 1: Cross-sectional view of the tube array.

According to the Price & Païdoussis model [7], $F_{xf} = \frac{1}{2}\rho(U_r/a)^2 D\ell\,[C_L \sin\alpha + C_D \times \cos\alpha]$ and $F_{yf} = \frac{1}{2}\rho(U_r/a)^2 D\ell\,[C_L\cos\alpha - C_D\sin\alpha]$, where $U_r = [(U-\dot{x})^2 + \dot{y}^2]^{1/2}$ is the approach flow velocity, corrected for cylinder motion, U being the velocity far upstream; U_r/a is the approach velocity in the array, with $a = T/(T - \frac{1}{2}D)$, $T = \frac{1}{2}\sqrt{3}\,P$; $\alpha = \sin^{-1}(\dot{y}/U_r)$ is the motion-induced angle of attack; ℓ is the tube length and D the diameter; C_L and C_D are empirically determined lift and drag coefficients, available in maps [2]. Defining $\tilde{m} = m/\rho D^2$, $V = U/\omega_0 D$, $\tilde{x} = x/D$, $\tilde{y} = y/D$ and $(\ )' = \mathrm{d}(\ )/\mathrm{d}\tau$, $\tau = \omega_0 t$, we have the dimensionless equations

$$\tilde{x}'' + \tilde{c}\tilde{x}' + \tilde{x} = \tilde{F}_{xf} + \tilde{F}_{xi}, \quad \tilde{y}'' + \tilde{c}\tilde{y}' + \tilde{y} = \tilde{F}_{yf} + \tilde{F}_{yi}\,; \tag{2}$$

$$\tilde{F}_{xf} = (2\tilde{m}a^2)^{-1/2}\left\{(V-\tilde{x}')^2 + \tilde{y}'^2\right\}^{1/2}\,[\tilde{y}'\,C_L + (V-\tilde{x}')C_D]\,, \tag{3}$$

and similarly for $\tilde{F}_{yf}$. In this quasi-steady analysis, it is supposed that there is a time delay $\Delta\tau$ between motions of the tube and the full development of forces thereon, so that

$$C_L(\tau) = C_L\left(\tilde{x}(\tau-\Delta\tau), \tilde{y}(\tau-\Delta\tau)\right), \; C_D(\tau) = C_D\left(\tilde{x}(\tau-\Delta\tau), \tilde{y}(\tau-\Delta\tau)\right)\,, \tag{4}$$

where $\Delta\tau = \tilde{\mu}\omega D/U$, $\tilde{\mu} = \mathcal{O}(1)$; in this study, $\tilde{\mu} = 1$ always.

The nonlinear forces $\tilde{F}_{xi}$ and $\tilde{F}_{yi}$ in equation (2) are not explicitly calculated. Instead, they are accounted for via (i) an impact model for the radial motion and (ii) a Coulomb friction model for the transverse motion of the tube relative to the

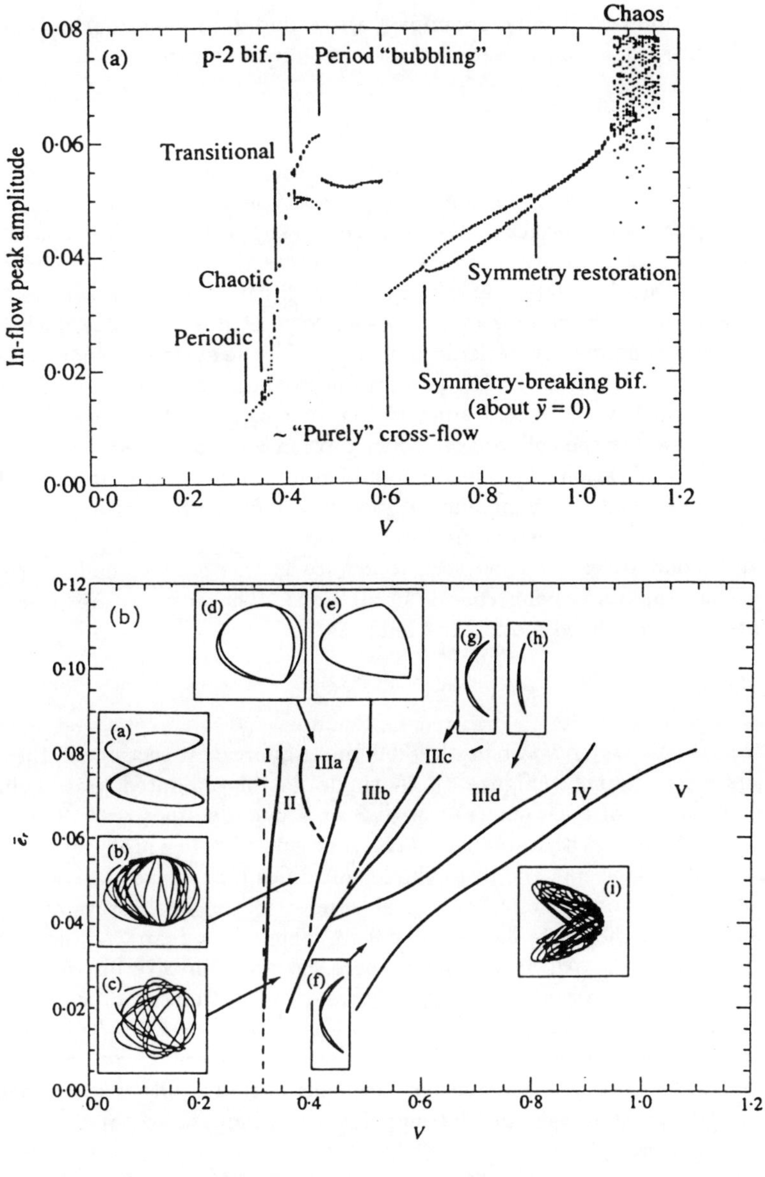

Figure 2: (a) Bifurcation diagram for in-flow ($\tilde{x}$) peak amplitude for $\tilde{m} = 10$, $\delta = 0.05$, $\tilde{e}_r = 0.08$; (b) bifurcation map in terms of V and $\tilde{e}_r$, showing the different motions in the ($\tilde{x}$, $\tilde{y}$) plane — according to the 2-d.o.f. model.

support. If the tube velocities before impact are u_r and u_t in the radial and tangential (to the hole) directions, and v_r and v_t those after impact, then

$$\tfrac{1}{2}m\,v_r^2 = e^2\left(\tfrac{1}{2}m\,u_r^2\right)\;;\quad v_t = u_t - \mu_{fr}(1+e)u_r\,,\qquad v_t > 0\,, \tag{5}$$

where e is the coefficient of restitution and μ_{fr} is the Coulomb friction coefficient.

The equations are integrated numerically, and typical results are summarized in Figure 2, where $\tilde{e}_r$ = (clearance gap)/D; these are supported by time traces, PSDs, orbital diagrams and Poincaré maps. As V is increased, there is a narrow region of weakly chaotic vibrations [at $V \simeq 0.35$ in Figure 2(a)]; they are associated with a chaotic component in the in-flow motion — while cross-flow motion, which is 10-25 times larger, is predominantly periodic. For higher V, the system becomes effectively periodic, undergoing many bifurcations before becoming subject to generalized chaos at $V \simeq 1$ — but at lower V if $\tilde{e}_r$ is smaller [Figure 2(b)]. The bifurcation at $V \simeq 0.6$ in Figure 2(a) involves the collapse of in-flow motion and the onset of figure-of-eight mainly cross-flow vibration. This is accompanied by a jump in the in-flow vibration frequency — practically *very* significant, because a higher impacting frequency is directly related to higher tube-material wear rates.

Two questions arise: (i) how close to reality is this simple model? (ii) what are the mechanisms underlying chaotic oscillation? The answers are provided via experiments, a more refined model, and analysis.

3. Experiments

The test apparatus is mounted in the 260 mm square test-section of the Kempf & Remmers water tunnel. (Figure 3). A single flexibly-mounted brass cylinder is located in the midst of a 7-row array, with 8 or 9 cylinders per row. The cylinder diameter $D = 12.7$ mm, with a smaller-diameter extension below providing the flexible mounting. At its upper end, the test cylinder protudes into a box-like compartment; the 'impact piece' can impact on a cylindrical 'support piece', in air or water. Different gaps were tested, in the range $0.07 < \tilde{e}_r < 0.23$, where $\tilde{e}_r$ = (gap)/D; also, different support-piece materials. All tests reported here, however, involve brass-on-brass in-air impacting. The in-air natural frequency and log decrement of vibration were $\omega = 38.3$ rad/s and $\delta = 0.01$; the mass ratio was $m/\rho D^2 = 1.87$. Tube motion in two orthogonal directions was sensed by an Optron non-contacting optical motion follower, providing both velocity and displacement outputs. Other instrumentation included a digital oscilloscope, an FFT signal analyzer, an analog tape recorder, and an on-line computer.

Typical results are shown in Figure 4 for $V = 1.17\,V_c$, $1.47\,V_c$ and $1.91\,V_c$. In agreement with the simple model, cross-flow motion is almost periodic and is not presented. In Figure 4(a), for V a little higher than the fluidelastic instability threshold (V_c), the motion involves single-sided impacting and is quasiperiodic,

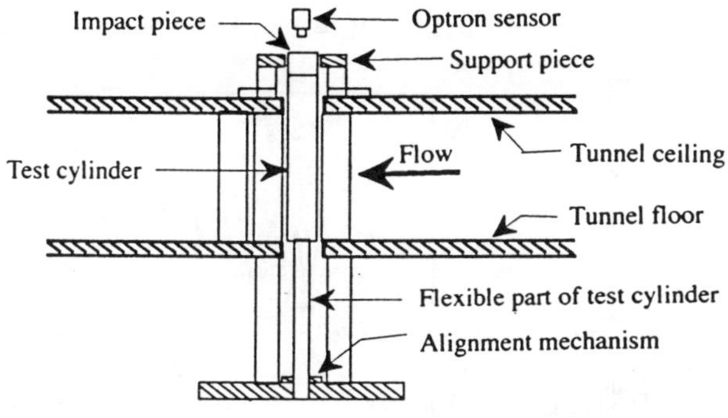

Figure 3: Schematic of the model in the water tunnel.

the second frequency representing the time interval between impacts; however, this time interval appears to be random, introducing a chaotic component in the response.

For $V = 1.24\,V_c$, the response involves double-sided impacting and is essentially periodic. This breaks down at $V = 1.43\,V_c$: the impacting response intermittently loses stability, breaking down into complex orbiting motion [see Figure 4(b)]. The in-flow motion exhibits bursts of amplitude growth during the orbiting phase, which are apparently random, again introducing a chaotic component in the vibration. The time interval between bursts becomes progressively shorter and, for $V = 1.69\,V_c$, periodicity has vanished completely [see Figure 4(c)].

The experimental bifurcation diagram (Figure 5) is in qualitative agreement with the 2-d.o.f. model. In particular, we have low-V weak chaos and high-V generalized chaos, both with aspects of intermittency. This intermittency has also been reported in field observations: "quiet phases", interspersed with "violent vibration".

4. A higher-dimensional model

The strategy here was to (i) introduce enough refinement in the model to represent the physics of the experimental system, and thus be able (ii) to compare quantitatively with the experiment and with the 2-d.o.f. model. This was done by (a) effecting a 5-mode Galerkin approximation of the cantilevered beam system of Figure 3 in each direction of motion, and introducing (b) a trilinear model for impact, with Coulomb friction for sliding, (c) nonlinear impact damping [8,9] and (d) squeeze-film damping in the tube-hole annulus [10]. The $\tilde{x}$-equation in this case is

$$\gamma\,\tilde{p}_i'' + 2\tilde{\zeta}_i\,\tilde{p}_i' + (\lambda_i/\lambda_1)^4\,\tilde{p}_i = \frac{1}{2\tilde{m}a^2}\int_o^1 \left[(V-\tilde{x}')^2 + (\tilde{y}')^2\right]^{1/2} \left(C_L\,\tilde{y}' + (V-\tilde{x}')\,C_D\right)\phi_i\,\mathrm{d}\tilde{s}$$
$$+\phi_i(\tilde{s}_p)\left\{\tilde{K}_s(\tilde{r}-\tilde{e}_r)^{\xi} + \tilde{r}'\left[\tilde{c}_s\,\tilde{r}^{\xi} + \tilde{\beta}\left(\frac{1}{1-\tilde{r}/\tilde{e}_r}\right)\right]\right\}(\mu_{fr}\,\sin\theta - \cos\theta), \qquad (6)$$

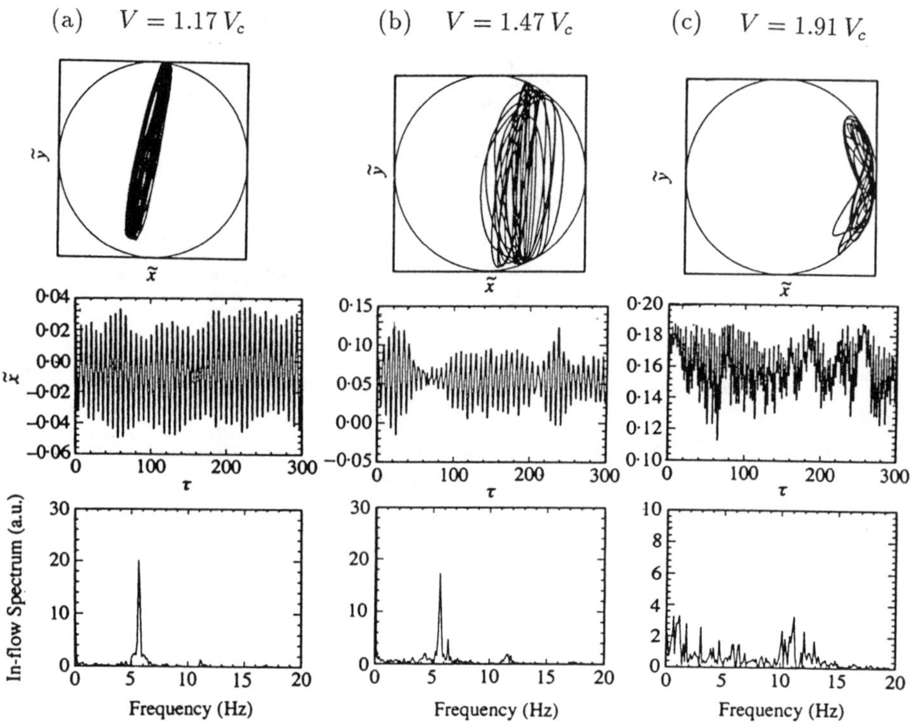

Figure 4: Typical experimental results of orbital motion, in-flow time trace and PSD.

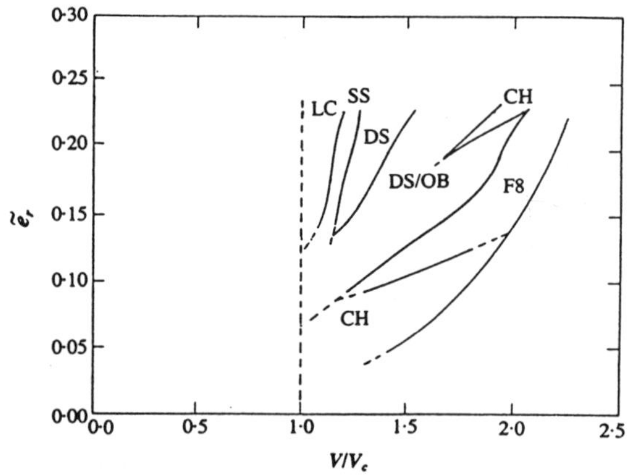

Figure 5: Experimental bifurcation diagram. LC: limit cycle; SS and DS single- and double-sided impacting; F8: figure-of-eight (weakly chaotic); DS/OB: double-sided/orbiting (intermittently chaotic); CH: generalized chaotic.

$i = 1, 2, \ldots, N$; just to show the structure, where some of the expressions (e.g., for $\tilde{K}_s$ and $\tilde{c}_s$) are themselves complex — see ref.[3].

Typical agreement with with experiment is demonstrated in Figure 6 and Table 1. It is seen that quantitative agreement of some of the main bifurcations is reasonable. Furthermore, the key qualitative aspects of the experimental response are captured by the model (not detailed here).

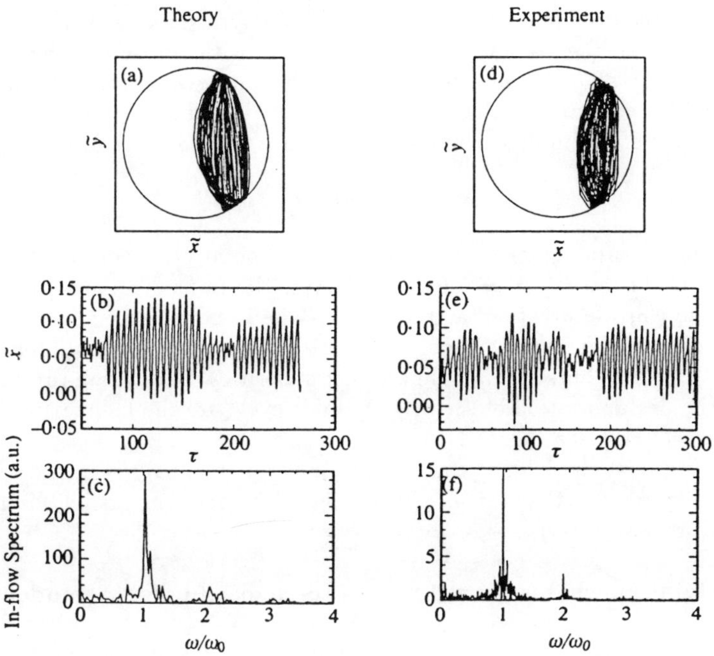

Figure 6: Typical comparison of theory (higher-N model) at $V = 1.76\,V_c$ with experiment at $V = 1.62\,V_c$

Table 1: Comparison of theoretically predicted bifurcation velocities (higher-N model) with experimental measurements; $m/\rho D^2 = 1.87$, $\delta = 0.01$

Transition	V_{th}/V_c	V_{exp}/V_c	Error (%)
Hopf bifurcation	0.86	1.00	-14
Double-sided impacting	1.24	1.20	3
Intermittency	1.52	1.41	7.8
Chaos	1.86	1.70	9.4
P-2 motion	2.10	1.95	7.2

5. Underlying mechanisms and routes to chaos: 2-d.o.f. model

The higher-dimensional model has been shown to be fairly reliable in predicting the experimentally observed dynamics. The remaining question is: what is the predicted route to chaos?

Chaotic motion near the onset of impacting results from destabilization of the limit cycle after the Hopf bifurcation. For this non-analytical model, a natural starting point is to examine the stability of the limit cycle with impacting, by examining the Poincaré map. Thus, the intersection of the phase-space trajectories with a hypersurface $\sum$ is considered, where $\{\sum | \dot{x} = 0,\ y > 0\}$, and a map relating successive extremum values of the in-flow displacement, X, are viewed, in the form

$$X_{n+1} = G(X_n;\ V)\,. \qquad (7)$$

Figure 7(a) shows this map for $V = 0.96\,V_c$; the fixed point has an eigenvalue $|\lambda_p| < 1$, and hence it is stable. For $V = 1.01\,V_c$, however, the in-flow response has suffered a subharmonic (flip) bifurcation, and $\lambda_p = -1$. The in-flow response is no longer at double the cross-flow frequency, though this 2:1 relationship appears intermittently, just before 'turbulent' bursts. The subharmonic subcritical bifurcation (unstable limit cycle) and the destabilization of the original limit cycle are the hallmarks of *type III intermittency* [11, 12]; see Figure 7(b), where $|\lambda_p| > 1$.

Considering next the high-V chaos, and pursuing the analysis in the same way, it is seen in Figure 7(c), that this is a *type I intermittency* [12], arising via a saddle-node bifurcation ($\lambda_r = +1$).

6. Underlying mechanisms and routes to chaos: higher-N model

In this case, low-V chaos is found to be associated with the *switching mechanism* [13]. Without going into details, two kinds of behaviour have been discovered to exist, again displayed as 1-D maps [Figure 7(d,e)]. No stable fixed point exists, but there is a trapping region — the union of the two types of map — within which the trajectories 'live'.

The transition to high-V chaos occurs, as for the 2-d.o.f. model, by *type I intermittency*, but a little differently. Here, a subharmonic bifurcation in the in-flow motion precedes intermittency, which then occurs via destabilization of this period-2 motion [Figure 7(f)].

7. Conclusion

Chaotic regions in the vibration of heat exchanger tubes have been determined, and they have been found to be mainly associated with intermittency and/or switching — both near the onset of impacting and in a higher flow range. Some quantitative comparisons of the underlying chaotic attractors via fractal-dimension and saddle-

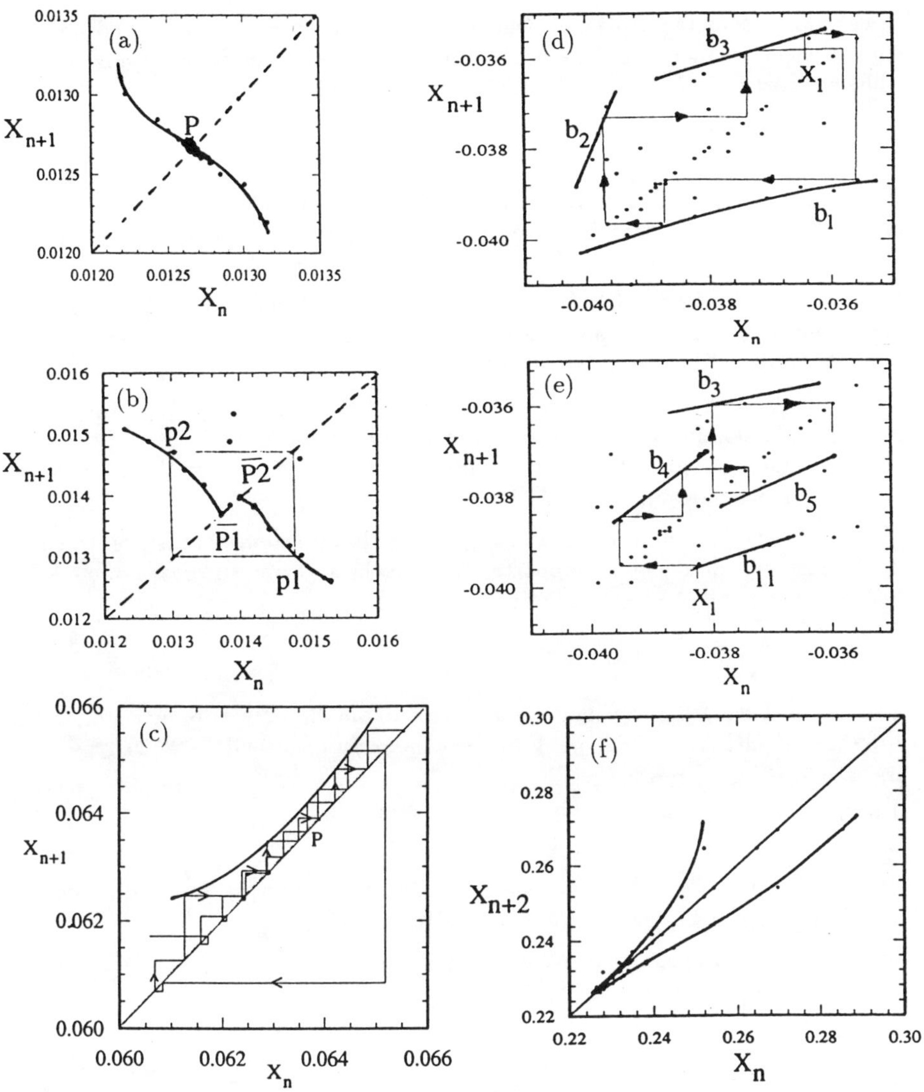

Figure 7: The *2-d.o.f. model*: (a) marginally stable system at $V = 0.96\,V_c$, and (b) for $V = 1.01\,V_c$ (type III intermittency); (c) type I intermittency at $V = 3.14\,V_c$. The *higher-N model*: (d,e) the two maps involved in the switching mechanism at $V = 1.01\,V_c$; (f) type I intermittency from period-2 motion at $V = 2\,V_c$.

Throughout this paper, V_c is the *experimental* critical value. Thus, $V = 0.96V_c$ corresponds to $1.09V_{c,th}$ where $V_{c,th}$ is the *theoretical* critical value; and so on.

orbit calculations show good agreement with observations — not presented here for brevity. Hence, the model may be used with confidence for engineering purposes.

The support by NSERC and FCAR is gratefully acknowledged.

References

1. Païdoussis, M.P. & Li, G.X.: Cross-flow-induced chaotic vibrations of heat exchanger tubes impacting on loose supports, *J. Sound and Vibration* **152** (1992), 305-326.
2. Païdoussis, M.P., Price, S.J. & Mureithi, W.N.: Nonlinear and chaotic dynamics of a two-degree-of-freedom analytical model for a rotated triangular array in cross-flow, *J. Fluids and Structures* **7** (1993), 497-520.
3. Mureithi, N.W., Price, S.J. & Païdoussis, M.P.: The post-Hopf-bifurcation response of a loosely supported cylinder in an array subjected to cross-flow. Part I: experimental results. Part II: theoretical model and comparison with experiments, *J. Fluids and Structures* **8** (1994), 833-852 and 853-876.
4. Mureithi, N.W., Païdoussis, M.P. & Price, S.J.: Intermittency transition to chaos in the response of a loosely supported cylinder in an array in cross-flow, *Chaos, Solitons & Fractals* **5** (1995), 847-867.
5. Axisa, F., Antunes, J. & Villard, B.: Overview of numerical methods for predicting flow-induced vibration, *ASME J. Pressure Vessel Technology* **110** (1988), 6-14.
6. Cai, Y. & Chen, S.S.: Chaotic vibrations of nonlinearly supported tubes in crossflow, *ASME J. of Pressure Vessel Technology* **115** (1993), 128-134.
7. Price, S.J. & Païdoussis, M.P.: A single-flexible-cylinder analysis of the fluidelastic instability of an array of flexible cylinders in cross-flow, *ASME J. of Fluids Engineering* **108** (1986), 193-199.
8. Hunt, K. & Crossley, F.: Coefficient of restitution interpreted as damping in vibroimpacting, *J. Applied Mechanics* **42** (1975), 440-445.
9. Goldsmith, W.: *Impact: The Theory and Physical Behaviour of Colliding Solids*, Edward Arnold, London.
10. Kim, B.S., Pettigrew, M.J. & Tromp, J.H.: Vibration damping of heat exchanger tubes in liquids: effects of support parameters, *J. Fluids and Structures* **2** (1988), 593-614.
11. Pomeau, Y. & Manneville, P.: Intermittent transition to turbulence in dissipative dynamical systems, *Communications in Mathematical Physics* **74** (1980), 189-197.
12. Bergé, P., Pomeau, Y. & Vidal, C.: *Order within Chaos*, John Wiley, New York.
13. Pikovsky, A.S. & Rabinovitch, M.I.: Stochastic oscillations in physical systems, *Physica D* **2** (1981), 8-24.

STABILIZATION OF PERIODIC FLAP-LAG DYNAMICS IN ROTORCRAFT

J.M. SCHMITT
Princeton University
Princeton, NJ

AND

P.V. BAYLY AND D.A. PETERS
Washington University
St. Louis, MO

Abstract. An approach developed to stabilize periodic orbits in chaotic attractors is generalized and applied to non-chaotic flap-lag instability in helicopter rotor blades. Periodic flapping and lead-lag oscillations occur in rotor blades during forward flight; the governing equations are nonlinear with periodic coefficients. Oscillations become unstable as the advance ratio of the helicopter increases. Stabilization may be achieved by control of the mean pitch angle of the blade once per period according to a discrete control law. The control law is applied to the Poincaré map which governs samples of the system obtained once per period. The controller stabilizes but does not attempt to change underlying periodic orbits. This approach is particularly well-suited to systems with periodic coefficients (such as rotorcraft) since the discrete version of the system is time-invariant.

1. Introduction

Rotorcraft dynamics are dominated by periodic orbits of varying stability. For example, the flap-lag dynamics of helicopter rotor blades can become unstable as forward flight speed is increased at a given rotor speed [1]. The equations governing flap-lag dynamics in forward flight are nonlinear and contain strongly time-varying coefficients [1].

Presently helicopter designs incorporate extra components such as mechanical lag dampers to improve stability. "Smart" blade materials, as well

F. C. Moon (ed.),
IUTAM Symposium on New Applications of Nonlinear and Chaotic Dynamics in Mechanics, 493–502.

as active servoflaps, are currently being considered for use in rotorcraft [2]. Control of rotor blade instability is complicated by three factors, however: (1) the steady state is intrinsically periodic; (2) coefficients are also periodic; and (3) behavior is nonlinear. These complications are addressed by the use of the Poincaré map which describes the behavior of samples of the system taken once per blade-passing period. The behavior of a periodic-coefficient, continuous system near a periodic orbit is thus reduced to the dynamics of an equivalent discrete, time-invariant system near a fixed point.

The use of a linearized Poincaré map to stabilize periodic orbits was first proposed by Ott, Grebogi, and Yorke [3] (OGY) for the purpose of controlling chaos. Since the original algorithm was described it has been modified by several investigators [4, 5, 6]. and validated in experiments [6, 7, 8]. However, for a number of possible reasons [9] benefits of these methods have not been demonstrated in practical mechanical or aerospace systems.

In this paper, stabilization of periodic orbits is investigated in an aeroelastic system that does not exhibit chaos: a helicopter rotor blade in forward flight. We extend the basic procedure presented by Ott, Grebogi, and Yorke [3] in the following ways: (1) The control law is optimized, taking into account the control effort; (2) Unstable orbits possessing two complex unstable eigenvalues are controlled; (3) Control is implemented without knowledge of fixed point location; (4) Constraints on the performance in steady state ("trim" constraints) are satisfied; (5) Control strategies to cope with (i) lack of knowledge of state variables and (ii) delay in implementation of control are used.

2. Rotor blade flap-lag dynamics

The rotor blade model was developed by Peters [1]. In this model, a thin rigid blade rotates at a specified speed about a fixed hub. The blade is connected to the hub by root hinges with linear torsional springs. The governing equations of the flap-lag system are derived in a rotating coordinate system (Figure 1a). The coordinate system is defined with the Z axis oriented along the rotor shaft. The $X - Y$ axes are rotated about the Z axis with an angular displacement $\psi = \Omega t$.

The position of a single blade with no hinge offsets is uniquely determined by three angles: the flap angle (β), the lag angle (ζ), and the pitch angle (θ). Dimensionless airloads on the blade in the flap and lag directions, F_β and F_ζ, are determined through quasi-steady airfoil theory (see the Appendix). The resulting nondimensional nonlinear equations for cou-

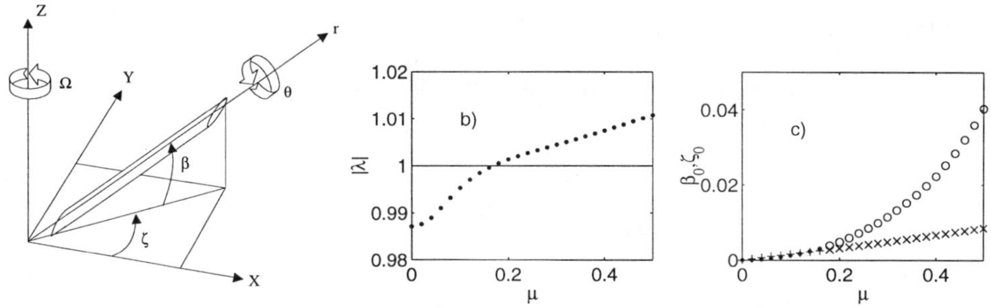

Figure 1. a) The rotor blade coordinate system and generalized coordinates β (flap angle) and ζ (lag angle). The pitch angle of the blade is θ. The $X-Y$ axes rotate about the Z-axis with angular velocity Ω. b) The magnitude of the largest eigenvalue of the Floquet transition matrix plotted vs advance ratio μ. c) Amplitudes of flap ($\cdot$ stable, o unstable) and lag (+ stable, $\times$ unstable) displacement as a function of μ.

pled flap-lag motion may be written as [1]

$$\ddot{\beta} + \sin\beta\cos\beta(1+\dot{\zeta})^2 + \omega_\beta^2\beta = \int_0^1 F_\beta r\, dr \tag{1}$$

$$\cos^2\beta\ddot{\zeta} - 2\sin\beta\cos\beta(1+\dot{\zeta})\dot{\beta} + \omega_\zeta^2\zeta = \cos\beta\int_0^1 F_\zeta r\, dr. \tag{2}$$

Dots $(\dot{\ })$ indicate differentiation with respect to $\psi = \Omega t$. The parameters ω_β and ω_ζ are the dimensionless natural frequencies in the flap and lag directions of the non-rotating system [1]. The pitch angle is prescribed to be $\theta = \theta_0 + \theta_s\sin\psi + \theta_c\cos\psi$, where θ_0, θ_s, and θ_c are input parameters. The pitch angle and the advance ratio μ (the ratio of forward flight speed to rotor tip speed) affect motion through the aerodynamic forces on the blade, which also depend on β and ζ (see the Appendix). All other parameters are fixed at values given in the Appendix.

Trimmed solutions of these equations are found from a Newton-Raphson technique applied to Runge-Kutta simulations [10]. Initial conditions and values for θ_0, θ_s, and θ_c are found that (1) provide periodic motion, (2) suppress the first harmonic component of β, and (3) achieve a specified coefficient of thrust; this combination is known as moment trim [1].

As the advance ratio μ is increased, stability of the trimmed solution is lost. The magnitude of the largest eigenvalue of the Floquet matrix exceeds unity for $\mu > 0.17$ (Figure 1b). The amplitude of oscillations also increases (Figure 1c), but remains reasonably small even at large values of μ. If these unstable orbits could be stabilized, the range of useful performance of the system could be significantly increased.

3. Discrete control methods

The general approach used here is motivated by the OGY approach to stabilizing unstable periodic orbits in chaotic nonlinear systems [3, 6]. Several concepts have been modified or generalized to apply control to the flap-lag system. The control technique is applicable to continuous dynamical systems of the form $\dot{\mathbf{x}} = \mathbf{f}(\mathbf{x}, t; \eta)$. In the flap-lag system the state-vector $\mathbf{x}$ is the vector $[\beta, \dot{\beta}, \zeta, \dot{\zeta}]^T$; η is a control parameter, which is taken to be the mean blade pitch angle θ_0. A discrete Poincaré map is obtained that describes the dynamics of the system sampled once every period, T: $\mathbf{x}_{n+1} = \mathbf{g}(\mathbf{x}_n; \eta)$. A periodic orbit in the continuous system ($\mathbf{x}(t+T) = \mathbf{x}(t)$) becomes a fixed point of the associated Poincaré map ($\mathbf{x}_{n+1} = \mathbf{x}_n$). Also, periodic coefficients in the original system are eliminated.

3.1. BACKGROUND: OGY CONTROL

The original OGY approach [3] follows the following argument. Near a fixed point, the Poincaré map may be linearized and represented by a matrix equation:

$$\xi_{n+1} = A\,\xi_n + \mathbf{h}\,\delta\eta, \tag{3}$$

where ξ_n is the location in state space relative to the fixed point: $\xi_n = \mathbf{x}_n - \mathbf{x}_f$. The characteristic matrix A in this approximation is the Floquet transition matrix. The vector $\mathbf{h}$ represents the effect of a small change in the parameter η on the location of ξ_{n+1}.

Suppose (as in reference [3]) that the matrix A has one unstable eigenvalue λ_u. Let ξ_n^u be the component of ξ_n in the direction of the unstable contravariant eigenvector: $\xi_n^u = \mathbf{f}_u^T \xi_n$. Then, at each iteration the control parameter is perturbed in proportion to the unstable component: $\delta\eta_n = \alpha \mathbf{f}_u^T \xi_n$. Stabilization is achieved by directing ξ_{n+1} onto the stable manifold, which occurs if $\alpha = -\lambda_{\dot{u}}/(\mathbf{h}^T\,\mathbf{f}_u)$.

3.2. GENERALIZED DISCRETE CONTROL METHODS

Several shortcomings exist in the basic OGY procedure, if we wish to apply it to a general non-chaotic system. First, the algorithm requires knowledge of the location of the fixed point. Second, performance criteria other than stability must often be achieved. Third, instability due to pairs of complex conjugate eigenvalues penetrating the unit circle is not considered. Fourth, the cost of the prescribed control effort may be excessive. Finally, changes in the control parameter may not be instantaneous, and must often be based on incomplete or imprecise knowledge of the system.

3.2.1. *Control of the difference map*

To overcome lack of knowledge of the fixed point, we use an idea presented in reference [6] and implement control based on the difference between consecutive iterates of the Poincaré map. Subtraction from Equation 3 the equation for the previous iteration of the map, it can be seen that

$$\mathbf{d}_{n+1} = A\,\mathbf{d}_n + \mathbf{h}\,e_n, \tag{4}$$

where $\mathbf{d}_n = \mathbf{x}_n - \mathbf{x}_{n-1}$ and $e_n = \delta\eta_n - \delta\eta_{n-1}$. The fixed point is no longer explicitly included in the linearized map. The stability of the map in Equation 4 is still governed by the eigenvalues of the Floquet matrix A.

3.2.2. *Pole placement and optimal control*

By the application of state feedback, the difference in the control parameter from cycle to cycle (e_n) can be made a function of the difference between successive iterates of the Poincaré map. Specifically, a linear control law can be expressed as $e_n = \mathbf{k}^T\mathbf{d}_n$, where $\mathbf{k}$ is a gain vector. The control problem is then reduced to finding the gain vector $\mathbf{k}$ so the eigenvalues of $(A - \mathbf{h}\mathbf{k}^T)$ are inside the unit circle.

The choice of gain proposed by Ott, Grebogi, and Yorke [3] is only one of an infinity of possible values. It is the gain which leads to the *most stable* controlled system. However, if control effort is not free, it is probably not the best choice of gain. By introducing a cost function $J = \sum_{n=1}^{\infty} \mathbf{d}_n^T Q \mathbf{d}_n + Re_n^2$, one can find a set of gains which minimizes a logical combination of transient dynamics and control effort [11]. In this paper $Q = nI$ and $R = 1$ are chosen for several values of n. The gain vector $\mathbf{k}$ which minimizes J is found from the matrix Riccati equation [11].

3.2.3. *Trim*

In rotorcraft applications certain performance or trim criteria (other than stability) must be met. For example, a rotorcraft should maintain a specified thrust. The trim problem is posed as follows: select values for K parameters which will drive an N-dimensional system to a steady state satisfying K constraints. The chosen performance measure (average thrust, for example) depends on the state variables of the system and on the control parameters. Close to the trimmed condition we can write the linear approximation

$$\mathbf{p}_{n+1} = J_{px}\,\mathbf{x}_n + J_{p\eta}\,\eta_n + \mathbf{b} \tag{5}$$

where the vector $\mathbf{p}_n$ contains all the trim errors for the nth period. Subtracting the equation governing the previous iteration to eliminate the unknown constant vector $\mathbf{b}$, and combining with Equation 4, we obtain

$$\begin{bmatrix} \mathbf{d}_{n+1} \\ \mathbf{p}_{n+1} \end{bmatrix} = \begin{bmatrix} A & 0 \\ J_{px} & I \end{bmatrix} \begin{bmatrix} \mathbf{d}_n \\ \mathbf{p}_n \end{bmatrix} + \begin{bmatrix} \mathbf{h} \\ J_{p\eta} \end{bmatrix} \mathbf{e}_n. \tag{6}$$

where $\mathbf{d}_n = \mathbf{x}_n - \mathbf{x}_{n-1}$ and $\mathbf{e}_n = \eta_n - \eta_{n-1}$. The system is now represented compactly in the same form as Equation 4.

3.3. PRACTICAL CONCERNS AND NECESSARY ENHANCEMENTS

Limited measurements and state estimation In many situations, not all state variables are accessible for measurement. A discrete observer is used to control the system when only the lead-lag state variables are measured. The output (measured) variables are expressed as $\mathbf{y}_n = C\mathbf{d}_n$. A dynamic estimate of the entire state can be formed from the input and output of the true system [11]:

$$\hat{\mathbf{d}}_{n+1} = A_c\hat{\mathbf{d}}_n + \mathbf{L}y_n + \mathbf{h}_c e_n, \tag{7}$$

where $A_c = A - \mathbf{L}CA$ and $\mathbf{h}_c = \mathbf{h} - \mathbf{L}C\mathbf{h}$. If $\mathbf{L}$ is chosen so that the eigenvalues of A_c are all small magnitude, then the estimate will converge rapidly to the true state.

Actuator dynamics Physical actuators do not change state instantly. To investigate the effects of a finite actuation delay, we model the actuator as a first-order system with a specified time constant: $\dot{\theta}_0 = (\tilde{\theta}_0 - \theta_0)/\tau_\theta$. The commanded pitch position, $\tilde{\theta}_0$ could change instantly, however. The order of the system is increased by one state variable, and the commanded pitch angle becomes the control input.

Model and parameter identification No mathematical model acn exactly describe a physical system. Accordingly, we attempt to stabilize the flap-lag system (a) with gains computed for erroneous parameters, and (b) with control laws developed from models fitted to simulated noisy data, without use of the equations of motion. All methods are tested in the presence of white noise measurement errors and random disturbances of the simulated system.

4. Results

4.1. CONTROL OF THE IDEAL SYSTEM INCLUDING TRIM

To stabilize unstable periodic orbits, the eigenvalues of the Floquet matrix must be moved within the unit circle. The eigenvalues of the Floquet matrix A, for the unstable orbit at $\mu = 0.4$, are slightly outside of the unit circle, as illustrated in Figure 2. An optimal gain vector $\mathbf{k}$ is calculated to minimize the cost function J defined above. The eigenvalues of the Floquet matrix of the stabilized system, $(A - \mathbf{h}\mathbf{k}^T)$ are moved significantly inside the unit circle by this algorithm, as shown in Figure 2.

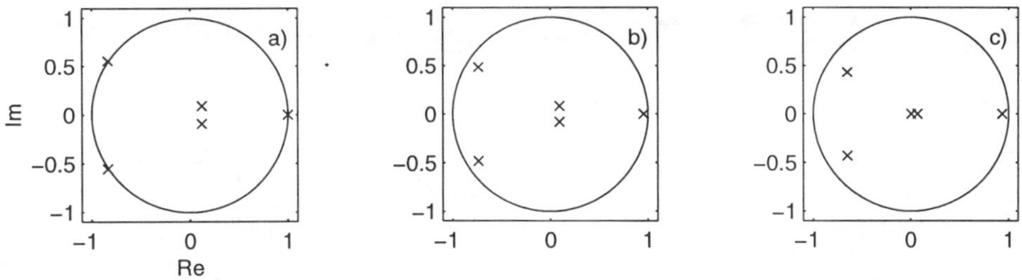

Figure 2. Eigenvalues of the Floquet matrix at an advance ratio of $\mu = 0.4$ for (a) the uncontrolled system, (b) the optimally controlled system with weighting matrices $Q = I$ and $R = 1$, and (c) the optimally controlled system with $Q = 100I$ and $R = 1$.

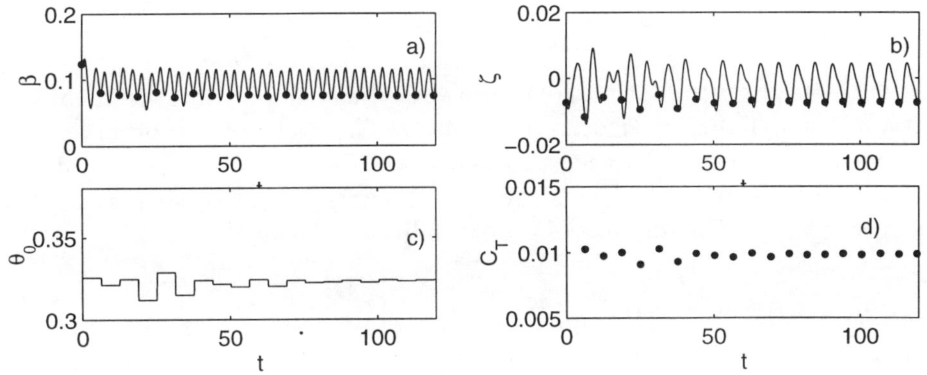

Figure 3. Response of the controlled flap-lag system to an initial perturbation (0.05 radians in β) from the periodic orbit. Time series of the flap (β) and lag (ζ) displacements, the control parameter (mean pitch angle, θ_0), and the coefficient of thrust, C_T are shown.

The ability to control unstable orbits in the flap-lag system is demonstrated in simulations of the full nonlinear system (Figure 3). Simulations are started near an unstable periodic orbit and control is then applied. The oscillations gradually approach the stabilized periodic orbit. The required deviation in mean pitch angle has a maximum value near 0.01 and decays with time.

4.2. CONTROL OF IMPERFECT SYSTEMS

Noise and measurement error In the next results, random disturbances of amplitude 5% of the RMS amplitude of each state variable are added to the corresponding variable every period. The results of a controlled simulation are illustrated in Figure 4; control is turned off after 20 periods to show

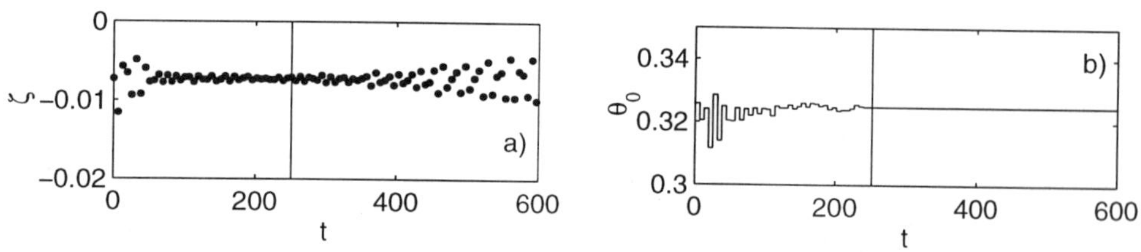

Figure 4. Response of the controlled flap-lag system to an initial perturbation from the periodic orbit and additive random noise (5% of RMS amplitude of each state variable). Discrete points from the lag (ζ) response, and the mean pitch angle (θ_0) are shown. Control is turned off after 40 periods (to the right of the vertical line).

the underlying instability. While the deviation in the pitch angle decreases significantly after the initial perturbation, the control effort will fail to decay to zero.

Observer-based control Control is implemented with only knowledge of the flap angle and flapping velocity (β and $\dot{\beta}$). The poles of the observer are chosen to be ten times faster than the poles of the controller. The resulting performance of the controller is shown in Figures 5a-c. The success of this control strategy depends on the observability of the system [11].

Control including actuator dynamics In the results so far, we have assumed that any desired change in θ_0 can be made instantly. However, any change in mean pitch angle must take a finite time to complete. A controlled simulation in which the mean pitch angle exhibits first order dynamics with a time constant $\tau_\theta = 1.0$ is shown in Figures 5d-f. The deviations in mean pitch angle are smaller than the cases above, and the system takes slightly longer to converge, but remains stable.

Control with imprecise models The control strategy is generally robust to errors in model and parameter estimates. Control is applied at an advance ratio of $\mu = 0.4$ using a control gain vector estimated for $\mu = 0.3$. The results are shown in Figures 5g-i, they exhibit rapid convergence. Models can be also derived based on data from experiment or simulation, without the benefit of equations of motion. We deliberately misidentify the system as two-dimensional (2-D) and use a least-squares fit to derive an empirical model of the form $\mathbf{x}_{n+1} = A\mathbf{x}_n + \mathbf{h}\eta_n$. The results of successful a control based on this model are shown in Figures 5j-l. Here the transient is long, indicating less effective control.

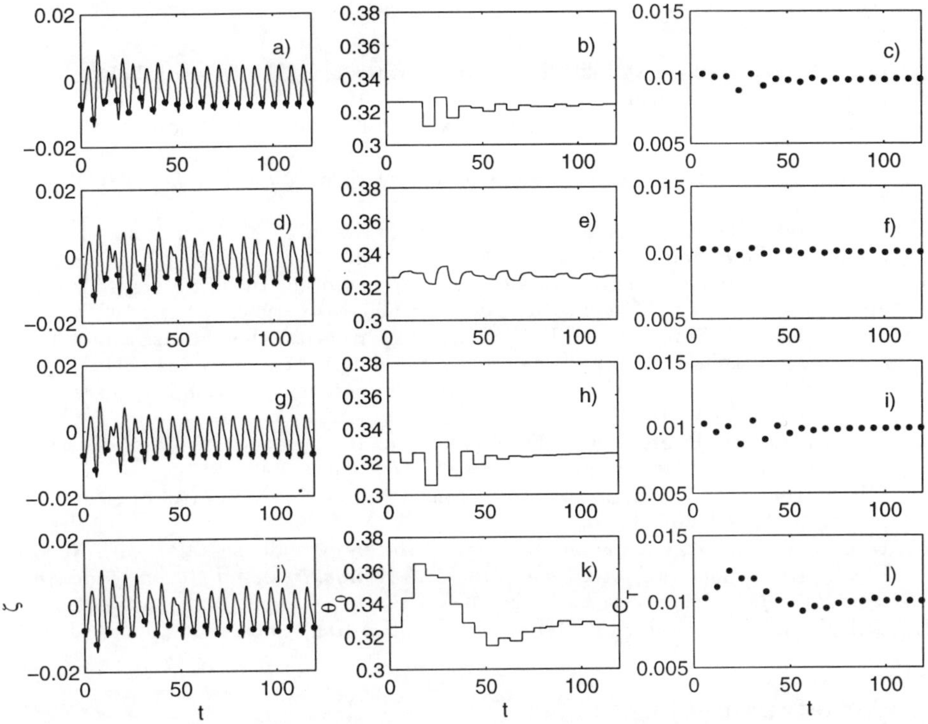

Figure 5. Lag displacement, mean pitch angle, and thrust coefficient for different control cases with advance ratio $\mu = 0.4$. a-c) Observer-based control. d-f) Control of a system including actuator dynamics. g-i) Control using gains estimated for $\mu = 0.3$. j-l) Control based on a 2-D model fitted to noisy data from simulation.

5. Summary

Unstable periodic orbits in simulations of flap-lag dynamics are stabilized by the periodic application of very small perturbations of the blade pitch angle. Transient effects in weakly stable periodic orbits can also be reduced by this method. Detailed knowledge of the equations of motion is not required to implement control, so long as the stability properties of the underlying motion can be estimated. The control law used to stabilize these orbits may also be optimized with respect to different performance or cost functions. The control strategy is applicable even if some states are not measurable, even if control can not be applied instantly, or even if the model is only approximate. The general approach is particularly useful for nonlinear systems with periodic coefficients. As active blade control becomes feasible, this general approach may be practical for improving the performance and simplifying the design of rotorcraft.

References

1. D.A. Peters. Flap-lag stability of helicopter rotor blades in forward flight. *Journal of the American Helicopter Society*, 20(4):2–13, 1974.
2. F.K. Straub and D.J. Merkley. Design of a servo-flap rotor for reduced control loads. In *Proc. American Helicopter Society 50th Annual Forum*, pages 305–314, 1994.
3. E. Ott, C. Grebogi, and J.A. Yorke. Controlling chaos. *Phys Rev Lett*, 64(11):1196–1199, 1990.
4. E.R. Hunt. Stabilizing high-period periodic orbits in a chaotic system: the diode resonator. *Phys Rev Lett*, 67(15):1953–1955, 1991.
5. V. Petrov, B. Peng, and K. Showalter. A map-based algorithm for controlling low-dimensional chaos. *J. Chemical Physics*, 96:7506–7513, 1992.
6. S. Bielawski, D. Derozier, and P. Glorieux. Experimental characterization of unstable periodic orbits by controlling chaos. *Phys Rev A*, 47:R2492–R2495, 1993.
7. W.L. Ditto, S.N. Rauseo, and M.L. Spano. Experimental control of chaos. *Phys Rev Lett*, 65(25):3211–3214, 1991.
8. P. Parmananda, P. Sherard, R.W. Rollins, and H.D. Dewald. Control of chaos in an electrochemical cell. *Physical Review E*, 47(5):3003–3006, 1993.
9. P.V. Bayly and L.N. Virgin. Practical considerations in the control of chaos. *Phys Rev E*, 50(1):604–607, 1994.
10. D.A. Peters and A. Izadpanah. Helicopter trim by periodic shooting with Newton-Raphson iteration. In *Proceedings of the 37th Annual National Forum of the American Helicopter Society*, pages 81–93, May 1981.
11. W.L. Brogan. *Modern Control Systems*. Prentice-Hall, Englewood Cliffs, NJ, 1991.

A. Aerodynamic forces

The dimensionless aerodynamic forces in Equations 1 and 2 are [1]

$$F_\beta = \pm\frac{\gamma}{2}[U_t^2 \sin\theta - U_t U_p(\cos\theta + \frac{c_{d_0}}{a})] \tag{8}$$

$$F_\zeta = \pm\frac{\gamma}{2}[U_p^2(\cos\theta - \frac{c_{d_0}}{2a}) - U_p U_t \sin\theta - U_t^2 \frac{c_{d_0}}{a}] \tag{9}$$

where U_t and U_p are the dimensionless velocities tangent and perpendicular to the plane of rotation of the rotor blades. The force expressions are negative if reversed flow is present, i.e. if $U_t \cos\theta + U_p \sin\theta < 0$. The dimensionless velocities are

$$U_t = (1 + \dot{\zeta}) r \cos\beta + \mu \sin(\psi + \zeta) \tag{10}$$

$$U_p = r\dot{\beta} + \lambda \cos\beta + \mu \sin\beta \cos(\psi + \zeta). \tag{11}$$

The inflow λ is related to the coefficient of thrust by simple momentum theory: $c_T = 2\lambda\sqrt{\mu^2 + \lambda^2}$ (for zero climb rate). The coefficient of thrust is also equal to the integrated vertical force:

$$c_T = \frac{\sigma a}{2\pi\gamma} \int_0^{2\pi} \cos\beta \int_0^1 F_\beta \, dr \, d\psi \tag{12}$$

where the slope of the lift curve $a = 2\pi$, the rotor solidity $\sigma = 0.05$, the Lock number $\gamma = 5$, and the coefficient of drag $c_{d_0} = 0.01$. Frequencies are taken as $\omega_\beta = 0.15$ and $\omega_\zeta = 1.4$. The advance ratio μ is varied.

A MATHEMATICAL MODEL OF PENDELLUFT FLOW IN A SINGLE AIRWAY BIFURCATION

Z.C. Feng
Department of Mechanical Engineering, Massachusetts Institute of Technology, Cambridge, MA

Chisang Poon
Harvard-MIT Division of Health Sciences and Technology, Massachusetts Institute of Technology, Cambridge, MA

Pendelluft is a gas flow phenomenon occurring in parallel respiratory units. When the parallel respiratory units are forded by a periodic pressure, the flow in them becomes asynchronous. This phenomenon plays an important role in mixing and transport of air especially during high-frequency ventilation. Understanding of this phenomenon was based n a mechanical model proposed by Otis et al. (1956). In that model the respiratory units are modeled as electric circuits with lumped parameters. They found that unequal resistances and compliances in the parallel respiratory units could give rise to pendelluft. Ultman et al. (1988) and High, Ultman, and Karl (1991) observed that, for high frequency ventilation, the inertance should be included. They thus expanded the Otis R-C model to the R-I-C model of analogous electric circuits.

The analogous electric circuits are easy to analyze. However, they lose the direct connection between the physiological parameters and the parameters governing the pendelluft flow such as the inertance, the resistance, and the compliance. Furthermore, if nonlinearity in the system is not ignorable, the ease of analyzing the analogous electric circuits is completely lost. There are several sources of nonlinearity. One example is the nonlinear relationship between the pressure (ΔP) and the volume (V)

$$\Delta P = \mathrm{Kel}\ V + K_1 \dot{V} + K_2 \dot{V}^2$$

where Kel, K_1, and K_2 are constant coefficients.

The direct formulation of the mathematical equations based on the mechanical model of Figure 1 has the following advantages. First of all, when nonlinearity is ignored, the derived equations can be solved as easily as the R-I-C circuits. Second, the parameters governing the pendelluft flow can be directly calculated from the physiological parameters such as the length and the cross-sectional area of the airways. Third, when the nonlinearity in the system cannot be ignored, the dynamics of the pendelluft flow can be analyzed using nonlinear dynamics theory. To demonstrate the consequence of the nonlinearity, we show that pendelluft can occur even in perfectly symmetric bifurcations.

F. C. Moon (ed.),
IUTAM Symposium on New Applications of Nonlinear and Chaotic Dynamics in Mechanics, 503–504.

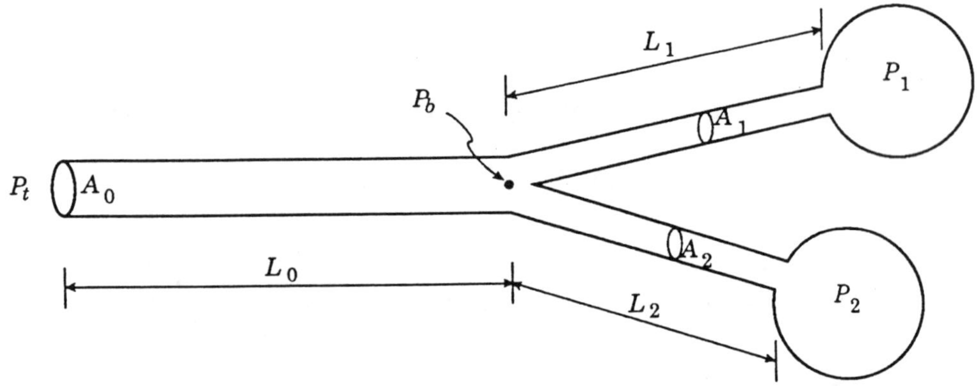

FIGURE 1. A mechanical model for a single airway bifurcation. A_0, A_1, A_2 are the cross-sectional areas of the parent and the two daughter branches and L_0, L_1, L_2 are the corresponding lengths. P_t and P_b are the pressures at the inlet and the bifurcation respectively. P_1 and P_2 are the pressures in the two reservoirs.

References:

High, K.C., Ultman, J.S., & Karl, S.R. (1991) Mechanically induced pendelluft flow in a model airway bifurcation during high frequency oscillation. *J. Biomechanical Engineering* **113**:342-347.

Otis, A.B., McKerrow, C.B., Bartlett, R.A., Mead, J., McIlroy, M.B., Selverstone, N.J., and Radford, E.P. (1056) Mechanical factors in distribution of pulmonary ventilation. *J. Appl. Physiol.* **8**:427-443.

Ultman, J.S., Shaw, R.G., Fabiano, D.C., & Cooke, K.A. (1988) Pendelluft and mixing in a single bifurcation lung model during high-frequency oscillation. *J. Appl. Physiol.* **65**:146-155.

THE PERIODIC TO CHAOTIC WATER WHEEL.

M. E. Kriebel, J. Holsapple, B. Keller, G. Pottlacher* and D. Bridy.
Dept. of Physiology, SUNY Health Science Center, Syracuse, N.Y. 13214 and
Institut fuer Experimentalphysik*, TU-Graz, Austria, A-8010.

1. Introduction

This device was developed as a teaching tool to introduce the concepts of chaos to biology and physiology students. Our purpose was to make a simple mechanical teaching device to generate data in the lecture hall, classroom and laboratory settings in order to demonstrate the characteristics of random, chaotic and periodic behavior. We knew of the "dripping water faucet" developed by Shaw (1984) and we have used it as a model system for transmitter release (Kriebel, et al , 1990). Moon (1992) has developed vibrating devices that project a laser beam onto a wall which dramatically show transitions from periodic to chaotic motions and he describes other chaotic teaching devices appropriate for advanced students. Lorenz (1993) discusses a water wheel which he had proposed earlier which resembles a Ferris Wheel with leaky buckets filled from an overhead stream of water which changes its dynamics as a function of flow rate and generates continuous data. We modified this concept to make a wheel which produces sets of discrete data as sequences of stop angles. One wheel of our device is used as a power wheel and a paddle wheel is used to dissipate energy such that the turning motion is controlled by the single parameter of water depth on the paddle wheel. We developed this water wheel so that students would develop an understanding of chaos theory for an appreciation of a nonlinear dynamic relevant to biological and physiological systems. Even though Sir D'Arcy Thompson wrote **On Growth and Form** over 80 years ago showing that form and structure of organisms resulted from dynamical processes governed by simple rules, there has been little infusion of nonlinear dynamics or chaos theory into the general biological curricula. As a result there are no introductory texts or experimental devices available for our students. Nevertheless, the importance of nonlinear dynamics to biology was reintroduced by May (1976) who used the logistic equation to explain population dynamics. Chaos theory has also been applied to describe heart beat dynamics (Glass and Mackey, 1988) and brain function (Mpitsos, et al, 1988; Freeman, 1991). Concepts such as "sensitivity to initial conditions" and transitions from periodic to aperiodic motions are not taught, but are essential in understanding the complexities and dynamical interactions found in biological and physiological systems.

A cornerstone of physiology is homeostasis as described by Claude Bernard (translation by Hoff, 1974) and Walter Cannon (1932). Homeostasis (better term is homeodynamics) is based on steady state behavior and dynamic processes that regulate the numerous, interactive systems. Biology is currently driven by molecular biological studies which are essentially descriptive. For example, enzymatic actions are described with equilibrium constants which force the idea that cellular processes are also in equilibria and

F. C. Moon (ed.),
IUTAM Symposium on New Applications of Nonlinear and Chaotic Dynamics in Mechanics, 505–512.

not in dynamical change. This notion is exemplified by the 1996 Claude Bernard Lecture of the American Physiological Society presented by Stanley G. Schultz (1996) which has the title *"Homeostasis, Humpty Dumpty, and Integrative Biology"*. Schultz states that *"Cannon's perceptive distinction between steady states, characteristic of living systems, and thermodynamic equilibria, which are anathema to life, is not fully appreciated today; it is this distinction that lies at the heart of the present synthesis of "reductive" and "integrative" approaches"*. Schultz also quotes Schroedinger (1967) from ***What is Life*** *"We must therefore not be discouraged by the difficulty of interpreting life by the ordinary laws of physics. For that is to be expected from the knowledge we have gained of the structure of living matter. We must be prepared to find a new type of physical law prevailing in it"*. Physiology must be rediscovered because molecular biology *"is not quantitative"* (Maddox, 1994) and does not provide a method to study integrative systems and dynamics. Physicists have provided an approach to dealing with problems of complexities including dynamical features inherent in coupled biological and physiological systems. For example, Lorenz (1963) showed that a forced dissipative convection system can be modeled numerically with three coupled non-linear equations.

Our water wheel exhibits either periodic or chaotic sequences of stop angles as resistance to turning changes. The device consists of a drive wheel with four quadrants which is connected to a paddle wheel by an axle that turns in pillow bearings. An overhead nozzle above the axle fills one of the four compartments of the drive wheel and the asymmetric weight causes the drive wheel to turn the paddle wheel. The paddle wheel dissipates the rotational energy causing the wheel to slow and eventually stop. The discontinuous motion of the device permits data to be collected as a series of stop angles. A series of 50 stop angles provides enough data for analysis which is unlike the Lorenz Chaotic Wheel which operates on a continuous basis. The depth of water on the paddle wheel is the major parameter which controls the pattern of sequential turns. High water levels yield stable periodic oscillations which are always attained regardless of starting position for each level of water on the paddle wheel. As the water level is reduced, the rotation angle increases until the device turns past 45 degrees and escapes into a chaotic series of stop angles. At different water levels return maps show the deterministic nature of seemingly irregular sequences of stop angles. Return maps were simply constructed by plotting θ_{n+1} (ordinate) vs θ_n (abscissa). These attractors are similar to those generated by the logistic equation (May, 1976) and drop intervals from a leaky faucet (Shaw, 1984) in that data is discontinuous. Our water wheel illustrates some basic principles of chaos in that the single parameter of paddle wheel resistance determines whether the motion is periodic or chaotic. Several states exist between the simple oscillating pendulum motion and chaotic motion. Chaotic sequences produce characteristic attractors for each water level and there is good short term predictability for three to five turns. However, sets of sequential stop angles starting from the "same" angle deviated after the fourth to seventh turn which demonstrate "sensitivity to initial conditions". Yet, all stop angles fell onto the attractor. We have found that the spectral density of sequences of stop angles was uniform for these cases. The machine becomes a "wheel of fortune" when spun by hand with no water in the paddle wheel bath and the stop angle return maps indicate that the turning motion is probably random. The "Periodic to Chaotic Water Wheels" have proven to be excellent for illustrating basic concepts of chaos theory and how motion of a machine can be dramatically changed from periodic to chaotic behavior with a slight change in one controlling parameter.

2. Methods

2.1. DESCRIPTION OF MACHINE

This device is constructed of plexiglass and consists of a drive wheel and a paddle wheel mounted onto an axle which turns in Delrin^R pillow bearings (Fig. 1A). One pillow bearing is slotted so that axle friction can be adjusted with a screw. Static friction is needed to prevent the device from rocking after a turn and to establish a breakaway turning torque. The bearings are mounted onto the walls of a plexiglass tank which has a center partition to divide the tank into a paddle wheel bath and a spill chamber for the drive wheel (Fig. 1B).

2.2 RECORDING DATA

The drive wheel is divided into four equal compartments such that turning angles need be recorded from only one quadrant. The angle between vertical and any one of the four compartment partitions gives the stop angle. The frequency of turning is slow enough (3-4 turns per minute) so that stop angles can be observed, recorded and predictions made. Stop angles can also be recorded electronically with a serrated disk mounted onto the axle which turns through three photo diodes.

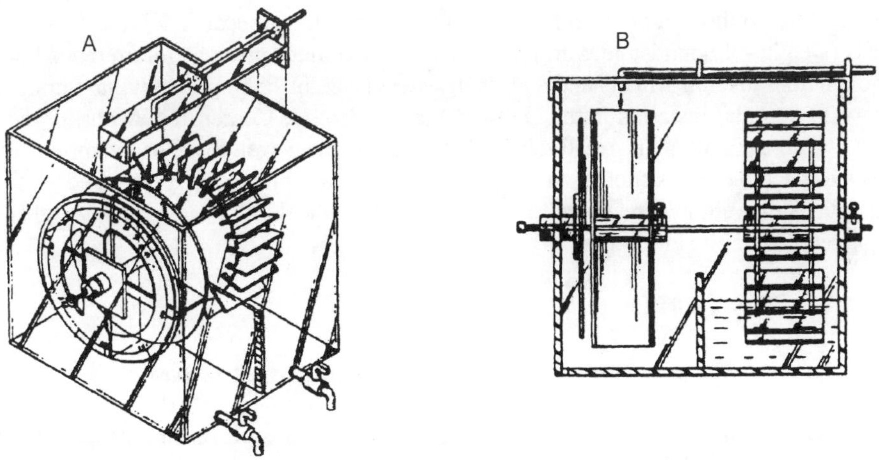

Figure 1. A. Perspective view of **Periodic to Chaos Machine** shows the drive wheel and paddle wheel. Stop angles from 1 to 90 degrees are read from a protractor between vertical and a drive wheel compartment divider in quadrant one.
B. Side view of **Periodic to Chaos Machine** shows paddle wheel bath and raceway at top of wheel. Relative deep water in the paddle wheel bath yields periodic oscillations whereas shallow water produces chaotic series of turns. (After Holsapple and Kriebel, 1990)

2.3. WATER FLOW RATE, TURN AND STOP RATES

Water from an overhead nozzle fills one compartment and the weight asymmetry turns the device. The static resistance of the bearings provides a breakaway turning torque which is similar for all angles. The churning action of the paddle wheel combined with the bearing friction stops the device after each turn. The flow rate from the nozzle is such that the device stops completely before the breakaway torque is reached.

2.4. CONTROL OF TURNING DYNAMICS

The types of angular turning motions are controlled by the depth of water in the paddle wheel bath which in turn controls energy dissipation. Deep water results in periodic oscillations which increase in magnitude as the water level is lowered until a turn rotates past 45 degrees at which depth the stop angle sequences produce a chaotic pattern.

2.5. SIZE OF MACHINE (scaling)

Wheels of 9, 12, 15, 19, 24 and 60 inch diameters yield similar dynamics. The size of the drive wheel hub, the volume of the drive wheel compartments and the surface area and number (24 to 72) of the paddle wheel vanes were approximated from dimensional analysis so that each machine produced either chaotic or periodic turning functions depending on level of water on the paddle wheel. The dynamics are robust because 9 inch student water wheels (weight - 4 pounds; maximum water load - 1 pound) produce return maps like two foot diameter museum wheels and a 60 inch wheel (weight - 250 pounds; maximum water load - 120 pounds) installed in the Hatfield Marine Science Center in Newport, Oregon.

The dynamics are relatively insensitive to wheel weight. For example, two 9" diameter machines constructed with either 1/16" or 1/8" plexiglass performed similarly. Two pounds of weight were added to the rims of a 24" wheel (10% increase in weight) with no noticeable change in motion.

2.6. STALL POSITIONS

At 45 degrees ($\pm$1) the machine stalls with water spilling over two partitions of the drive wheel. At 90 degrees ($\pm$1) the overhead stream of water splits around a drive wheel chamber partition also causing a stall. Stall positions occur every 50 or so turns and a tap on the tank restarts the sequence. For continuous turning suitable for museum exhibits, 8 pie shaped escape compartments were added to the rear of the drive wheel which fill at the stall angles to restart the turning sequence. The escape compartments do not change the dynamics.

2.7. TOLERANCE

The stop angles produced by a given start angle in each of the four chambers of the drive wheel were compared. Except near transition regions well balanced wheels show less than 2% variability in stop angles between chambers and periodic oscillations are essentially the same for each quadrant.

2.8. ANALYSIS OF TURNING

In the general case a data vector consists of a series of stop angles, (θ_n, θ_{n+1}, θ_{n+2}, ...) where θ_n is the stop angle at step (n). In the simplest case a low dimensional attractor was found with a two component vector utilizing a delay of one step so that plots were constructed from pairs of consecutive stop angles (θ_n, θ_{n+1}) yielding a return map. In periodic oscillatory motion the machine steps between two angles producing clouds of points on the return map. In the aperiodic regimes we found linear clouds of points and other patterns. By carefully adjusting the water level a period three state was observed near the transition point between periodic and chaotic behavior.

During periodic oscillations, the device stabilized to two stop angles after a few turns regardless of starting angle. In contrast, during chaotic turning, stop angle sequences starting from the apparently same initial starting angle deviated after 3-7 turns showing the property of "sensitivity to initial conditions".

2.9. USING THE MACHINE AS A "WHEEL OF FORTUNE"

Sequences of stop angles which appear to be random were produced with an empty paddle wheel bath compartment when the wheel was spun by hand. The spin was done with enough force to produce two revolutions while still attempting to make all spins with the same force.

3. Results

3.1 RETURN MAPS FROM 15 INCH WHEEL WATER DRIVEN

Data presented in the plots shown here were taken with an angle detector and transcribed to computer plots. These data and plots are essentially the same as those taken from other sized wheels. The data are slightly inconsistent at the region of transition (as determined by the depth of water on paddle wheel) from clock to chaos as expected due to the sensitivity of the process. On either side of the transition, however, data sets were similar.

At a depth of 15 cm on the paddle wheel the machine oscillated between 32 and 58 degrees illustrating period two behavior (Fig. 2A). At 14 cm the swings expanded an additional 5 degrees but the centers remained about the same as at the depth of 15 cm showing the beginning of a transition state. At 12 cm many swings were at the original 32 and 58 degree locations but an attractor expanded and contracted between the two initial clusters (Fig. 2B). The next data set at a depth of 8 cm shows a limited two line chaotic attractor (Fig. 2C). Decreasing the depth to 6 cm caused the attractors to expand such that essentially all angles were visited (Fig. 2D). A striking attractor was uncovered at a precise depth of 4 cm which shows "multivalued" attractors at approximately 37 and 57 degrees (Fig. 3A). This multivalued aspect probably reflects slight asymmetries in the 4 compartments of the drive wheel. Note, however, that a primary feature is still the two straight lines of the attractor.

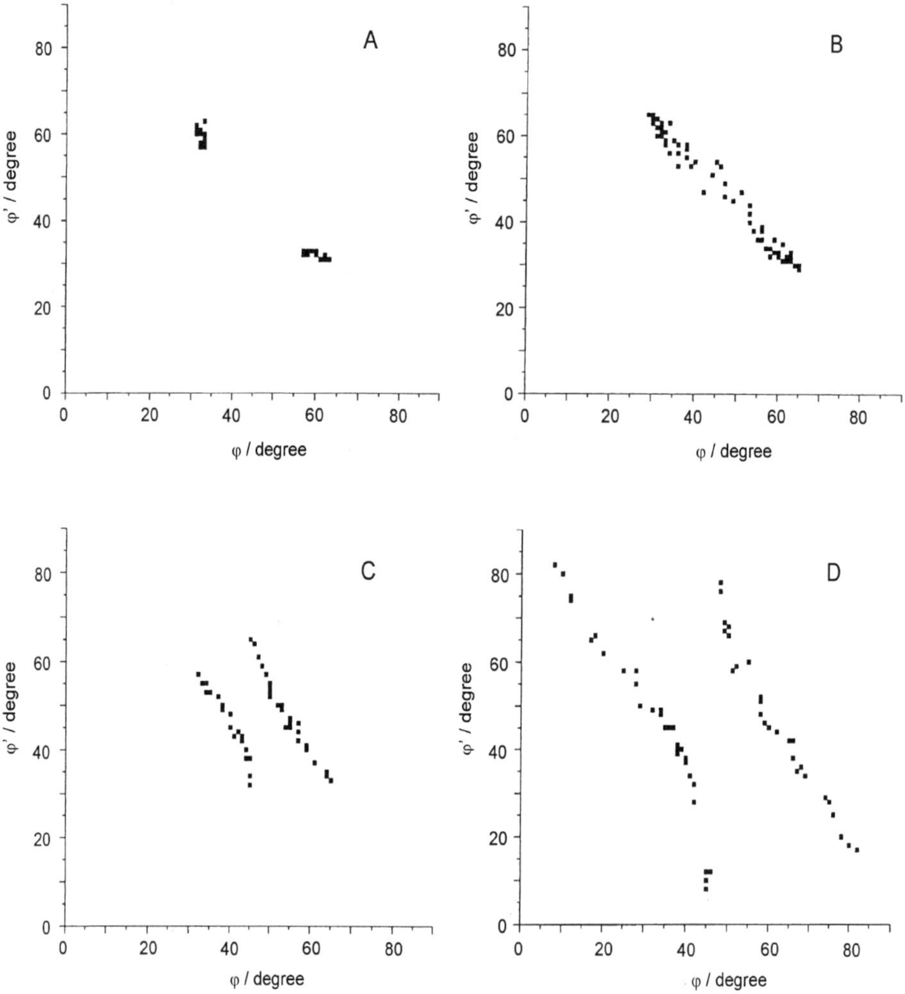

Figure. 2. Return maps of sequential stop angles at different water depths on paddle wheel.

A: 15 cm. Wheel periodically oscillated between 58 and 32 degrees.

B: 12 cm. Wheel periodically oscillated with expanding and contracting stop angles.

C: 8 cm. Wheel turned chaotically such that small angles were not reached which produced a restricted two line attractor.

D: 6 cm. Wheel turned chaotically such that all stop angles were obtained to produce a two line attractor which included all angles.

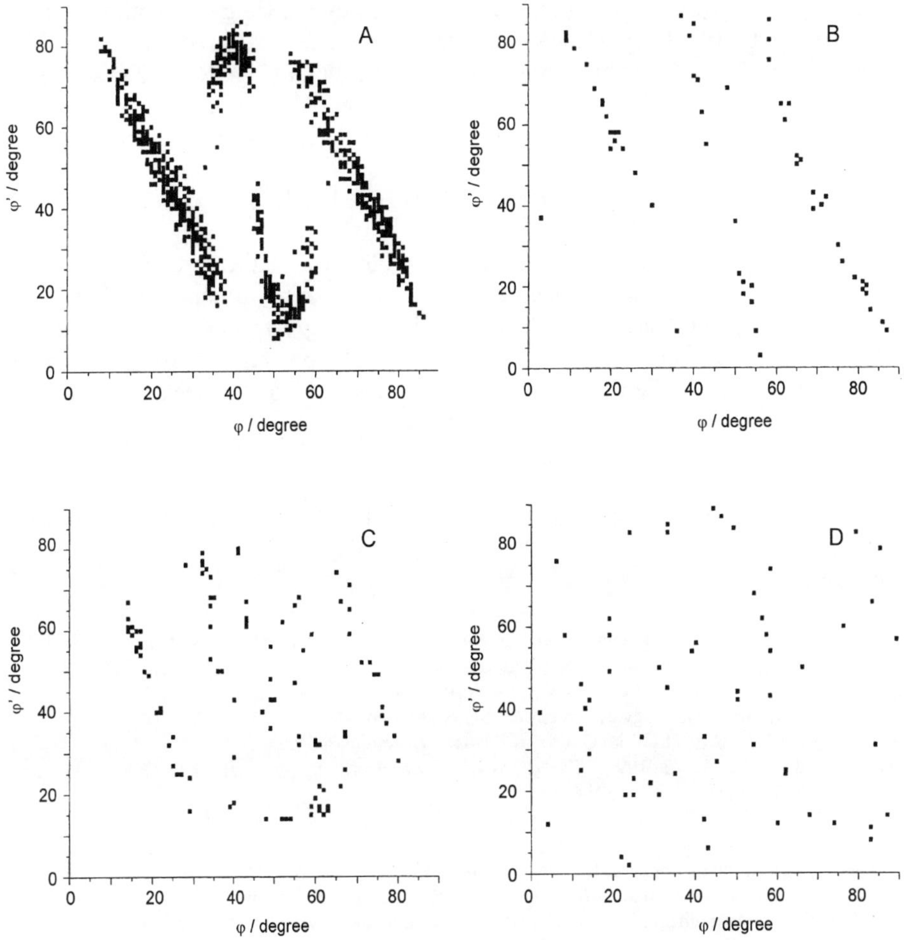

Figure 3: (continuation of Fig. 2). Return maps of sequential stop angles at different water depths on paddle wheel; and, one example of hand spinning (D).

A: 4 cm. Four part chaotic attractor. Edges of the two central, C-shaped parts are multivalued and reflect asymmetries in the 4 compartments of the power wheel.

B: 1 cm. A three line attractor.

C: 0 cm. Water driven. Note that a complicated attractor appears to be developing.

D: 0 cm. Force driven by hand was "constant" and enough to produce at least two turns. No obvious attractor is present.

The center region of the attractor in Figure 3A was lost by reducing the depth to 3 cm and the data produced a two lined attractor similar to Figure 2D. A three line attractor was produced at 1 cm depth (Fig. 3d). For completeness, a plot (Fig. 3C) is shown with no water brake to demonstrate the operation of the wheel with only bearing friction. In the last case there appears to be an attractor developing but the data set is insufficient.

3.2. RETURN MAP OF 15 INCH WHEEL SPUN BY HAND

Wheels were spun by hand with no water brake to observe if either hand spins produced patterns or if there was a preferred stop angle due to a bearing asymmetry. The "same" estimated force was applied for each spin and it was great enough to produce two revolutions. Return maps of 50 to 1000 spins showed no apparent patterns and the points appear to evenly cover the fields produced with either 9 or 15 inch wheels (Fig. 3D).

3.3. MODEL

The wheel has been the subject of a modeling effort and will be the subject of a later paper.

4. References

Bernard, C. Lectures on the Phenomena of Life Common to Animals and Plants, translated by H. E. Hoff, R. Guillemin, and L. Guillemin. Springfield, IL. Thomas, 1974.

Cannon, W.B. 1932. The Wisdom of the Body. New York. Norton.

Freeman, W. J. 1991. The physiology of perception. Sci Amer. Feb.

Glass, L. and Mackey, M.C. 1988. From Clocks to Chaos. Princeton Un. Press.

Holsapple, J. and Kriebel, M. E. 1990. The chaos water wheel primer. Manual. 202 Hillsboro Parkway, SYNTOS, Syracuse, N.Y. 13214.

Kriebel, M. E.; Vautrin, J. and Holsapple, J. 1990. Transmitter release: prepackaging and random mechanism or dynamic and deterministic process. Brain Re. Rev. 15:167-178.

Lorenz, E. 1963. Deterministic nonperiodic flow. J. Atmospheric Sci. 20:130-141.

Lorenz, E. 1993. The Essence of Chaos. University of Washington Press. Seattle.

May, R. M. 1976. Simple mathematical models with very complicated dynamics. Nature 261, 459-467.

Moon, F. C. 1992. Chaotic and Fractal Dynamics. John Wiley & Sons, Inc., New York.

Maddox, J. 1994. Towards more measurement in biology. Nature.368:95.

Mpitsos, G. J.; Creech, H. C.; Cohan, C. S. and Mendelson, M. 1988. Variability and chaos of neurointegrative principles in self-organization of motor patterns. Eds. Kelso, J.A.S.; Mandell, A. J.; Shlesinger, M. F. in Dynamic Patterns in Complex Systems. 162-190 World Scientific, Singapore.

Schrödinger, E. 1967. What is life? Cambridge, UK. Cambridge Univ. Press.

Schultz, S. G. 1996. Homeostasis, Humpty Dumpty, and integrative biology. New Physiol. Sci. 11:238-246.

Shaw, R. 1984. The Dripping Faucet as a Model Chaotic System. Aerial Press. Santa Cruz, CA.

Thompson, D. 1917. On Growth and Form. Ed. by Bonner, J. T. 1961. Cambridge Press.

5. Acknowledgments

We thank B. Fellner and Mike Page for laboratory assistance and graphics.

HELICAL WAVES AND NONLINEAR DYNAMICS OF FLUID/STRUCTURE INTERACTIONS IN A CYLINDER ROW

M. THOTHADRI
Department of Theoretical and Applied Mechanics
Cornell University
Ithaca, NY 14853

AND

F. C. MOON
Mechanical and Aerospace Engineering Department
Cornell University
Ithaca, NY 14853

1. Introduction

An array of cylinders placed in a fluid flowing perpendicular to their long axis has been known to perform large amplitude oscillations when the flow velocity is increased past a critical value. Many researchers have studied this problem and considerable understanding has been gained [1, 2, 3, 4, 5, 6, 7, 8, 11, 12]. [9] provides an extensive review of all the models of fluid elastic instability for cylinder arrays in cross flow. The goal of this study [13, 14] has been to investigate low-dimensional nonlinear models for fluid-structure dynamics of a row of cylinders in cross flow.

The principal results of this experimental-theoretical study are, i) experimental evidence shows that the dynamic instability of the tube row is a sub-critical Hopf bifurcation, ii) the critical flow velocity decreases as the number of flexible cylinders increases, iii) for a range of flow velocities, more than one stable mode (limit cycle) co-exists, iv) the linear model exhibits coupled helical wave solutions in the tube dynamics and v) one of the several nonlinear models of the tube dynamics studied shows a sub-critical Hopf bifurcation with a secondary bifurcation to a torus or quasi-periodic oscillation. The theory of center manifolds and normal forms is used to analyze the bifurcations in the system.

F. C. Moon (ed.),
IUTAM Symposium on New Applications of Nonlinear and Chaotic Dynamics in Mechanics, 513–520.

2. Description of Experiments

The wind tunnel used in this study is a standard blower type, low turbulence air tunnel powered by an AC motor. Figure 1 shows an isometric view of the test section. Details of the experimental set-up and test section can be found in [13]. The cylinder row, consisting of nine cylinders, has seven oscillating cylinders and a fixed cylinder at each end. The pitch to diameter ratio (P/D, where P is the distance between adjacent cylinder axes and D is the diameter of each cylinder) is 1.35. Each oscillating cylinder has two

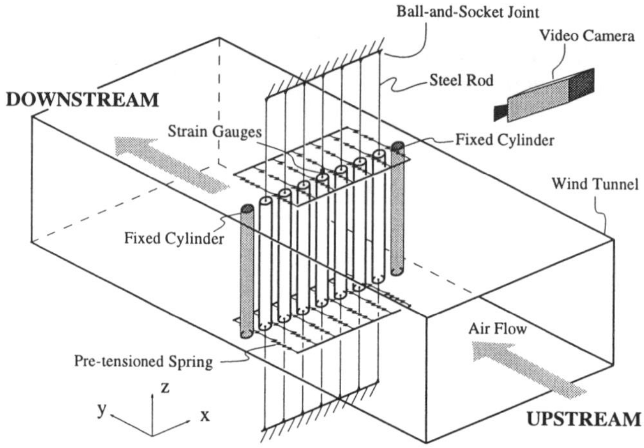

Figure 1. Sketch of the test section. The air flow between the cylinders provides the coupling between the cylinders; no structural coupling exists.

degrees-of-freedom, in-line and transverse to the flow and is supported by a steel rod. The natural frequency of the cylinders in the in-line direction (f_y) is 7.4 Hz and 6.8 Hz in the transverse direction (f_x). The damping ratio in still air was found to be close to 0.013 in both directions.

In our experiments, we quantify *only* the response of the central cylinder. We define *amplitude* as the maximum radial displacement of the central cylinder from the equilibrium position. (Note that *all* of the seven cylinders can oscillate in any direction in the $x - y$ plane). We use $U_r = U/f_yD$ as the non-dimensional form of the gap flow velocity U, and D is the diameter of the cylinders.

Figure 2 shows evidence for a *sub-critical Hopf bifurcation* at the critical flow velocity $U_{cr} \approx 115$. The method used to estimate the amplitude of the unstable limit cycle (see [14] for details) can be roughly justified by the existence of a two dimensional center manifold.

The variation in the critical reduced flow velocity with the number of oscillating cylinders is shown in Figure 3a. As the number of oscillating

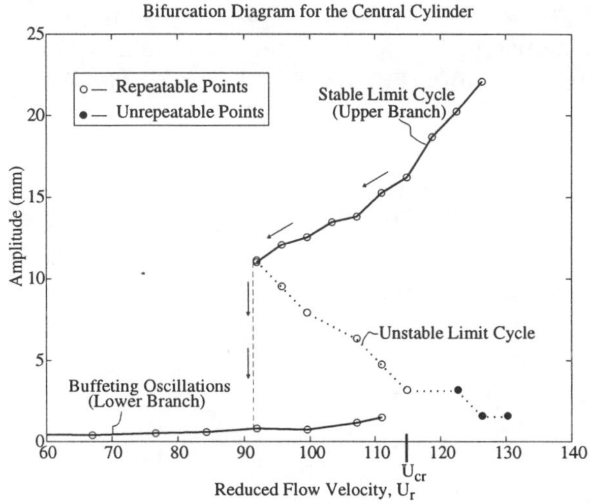

Figure 2. Bifurcation diagram. The amplitude of the oscillations of the *central* cylinder in a row of seven vibrating tubes is shown versus the reduced flow velocity (U_r). A sub-critical Hopf bifurcation occurs as the reduced flow velocity is increased past the critical value of approximately 115. This bifurcation diagram is obtained experimentally.

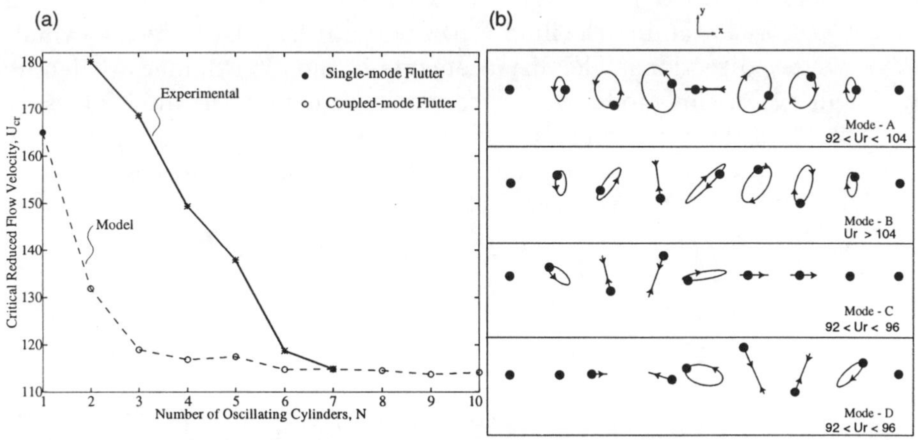

Figure 3. a) The variation of the critical reduced flow velocity (U_{cr}) with the number of oscillating cylinders (N) is shown. For the experimental curve, the value of U_{cr} for a single oscillating cylinder could not be obtained as it is beyond the maximum flow velocity of the experimental set-up. The difference in the U_{cr} between the experiments and the model is large for all except $N = 7$. This is because the cross-coupling fluid forces which were not determined experimentally, were fit for the U_{cr} observed in the $N = 7$ case. b) Schematic illustration of the stable modes of limit cycle oscillations. For some range of the flow velocity, more than one of these stable modes co-exist.

cylinders increases, the critical flow velocity decreases. The value of the critical reduced gap flow velocity for a single oscillating cylinder is greater than 230 (8.5 m/s upstream, the maximum flow velocity of the experimental set-up). This suggests that the coupling between the cylinders drastically reduces the critical flow velocity.

For $U_r \geq 92$, perturbations to the buffeting oscillations result in stable large amplitude modal oscillations of the cylinder row. Figure 3b is a sketch of all the modes observed through a video camera. For a range of reduced flow velocities, *more than one stable limit cycle co-exist.* The linear model of the system presented in Section 3 shows these modes to be similar to helical waves on a string.

3. Theoretical Model and Analytical Results

The equations of motion of nth the cylinder, with respect to a frame fixed at its equilibrium position, is given by

$$\mathrm{M}\begin{Bmatrix} \ddot{x}_n \\ \ddot{y}_n \end{Bmatrix} + \begin{bmatrix} C_x & 0 \\ 0 & C_y \end{bmatrix}\begin{Bmatrix} \dot{x}_n \\ \dot{y}_n \end{Bmatrix} + \begin{bmatrix} K_x & 0 \\ 0 & K_y \end{bmatrix}\begin{Bmatrix} x_n \\ y_n \end{Bmatrix} = \boldsymbol{f}_n, \tag{1}$$

where the terms on the left-hand side are the structural inertia, damping and stiffness terms per unit length of the cylinder and $\boldsymbol{f}_n$ on the right-hand side is a vector containing the fluid forces per unit length on the nth cylinder in the x and y directions. The displacements of the nth cylinder are denoted by x_n and y_n. In this model we ignore motion-independent fluid forces, the negligible fluid inertial terms and assume nearest neighboring coupling.

Using symmetry conditions and linearizing the fluid forces about the cylinder equilibrium positions we get,

$$\begin{aligned} \boldsymbol{f}_n = & -\tfrac{1}{2}\rho \mathrm{DU}\left(\begin{bmatrix} \alpha' & -\sigma' \\ -\tau' & \beta' \end{bmatrix}\begin{Bmatrix} \dot{x}_n - \dot{x}_{n-1} \\ \dot{y}_n - \dot{y}_{n-1} \end{Bmatrix} + \begin{bmatrix} \alpha' & \sigma' \\ \tau' & \beta' \end{bmatrix}\begin{Bmatrix} \dot{x}_n - \dot{x}_{n+1} \\ \dot{y}_n - \dot{y}_{n+1} \end{Bmatrix}\right) \\ & + \rho \mathrm{U}^2\left(\begin{bmatrix} \alpha'' & -\sigma'' \\ -\tau'' & \beta'' \end{bmatrix}\begin{Bmatrix} x_n - x_{n-1} \\ y_n - y_{n-1} \end{Bmatrix} + \begin{bmatrix} \alpha'' & \sigma'' \\ \tau'' & \beta'' \end{bmatrix}\begin{Bmatrix} x_n - x_{n+1} \\ y_n - y_{n+1} \end{Bmatrix}\right). \end{aligned} \tag{2}$$

The single-primed quantities, $(\alpha', \sigma', \beta', \tau')$ are the fluid-damping coefficients and the double-primed quantities, $(\alpha'', \sigma'', \beta'', \tau'')$ are the fluid-stiffness coefficients. The direct coupling coefficients (α'', α', β'', β') were measured by using the ARMA process (Auto Regressive Moving Average) of the system identification. The cross coupling coefficients (τ'', τ', σ'', σ') were determined through heuristic fits based on instability mechanisms (see [13] for details). The translational invariance of the motion-dependent fluid forces is inherent in our model.

We consider a traveling wave solution to Equations 1 and 2 of the for

$$\begin{Bmatrix} x_n \\ y_n \end{Bmatrix} = \begin{Bmatrix} a_1 \\ a_2 \end{Bmatrix} e^{i(qn-\omega t)} \tag{3}$$

where, $q = 2\pi\kappa P$; κ is the wave number (i.e. $\frac{2\pi P}{q}$ is the wave length), ω is the angular frequency and P (pitch) is the periodicity associated with the structure. In the above traveling wave solution, the $e^{-i\omega t}$ characterizes the temporal behavior of the solution for the initial data given by e^{iqn}. The quantity q is the phase difference between adjacent cylinders and hence characterizes the spatial distribution of the cylinders at any instant of time. The dispersion relation for the finite cylinder row system is obtained by

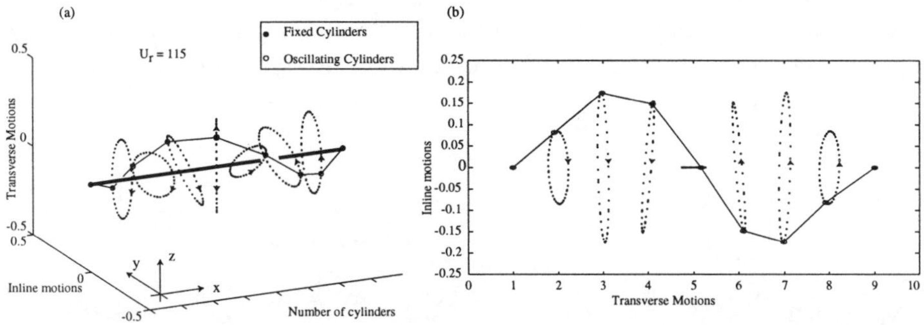

Figure 4. The *least stable* mode is shown for $U_r = 115$. The central cylinder oscillates only in the transverse direction and motions of the cylinders to the left and the right of the central cylinder have a reflection symmetry about the $y-z$ plane of the central cylinder. In part (a) the cylinders to the left rotate in the anti-clockwise direction about the $+x$-axis and those to the right rotate in the clockwise direction. The helical pattern is in the counter-clockwise direction about the $+x$-axis. Part (b) shows the 2-D representation.

substituting equation 3 into the governing equations 1 and applying fixed end boundary conditions. For a particular ω, the wave numbers $\pm q$ result in two waves traveling in opposite directions. The superposition of the two traveling waves results in a standing wave.

As the periodicity of the structure is in x-direction, it is difficult to interpret the deflection of the cylinders in the x-direction. They are easier to interpret if the x-directional motion of each cylinder is represented along the z-axis. This representation renders the depiction of the modes to be similar to that of a string vibrating in the two transverse directions, y and z. For the standing wave, the cylinder row at any instant in time is a *helix* that circles around the $+x$-axis in the counter-clockwise direction. A simple calculation shows that the helical pattern of the standing wave is due to the phase difference between the two constituent traveling waves. This phase difference depends on the coefficients τ'' and σ'' which are in-turn determined by the coupled-mode flutter instability.

Figure 4 shows a helical mode for $U_r = 115$. The central cylinder in this mode oscillates only in the transverse direction and this mode is very similar to mode-A of Figure 3b obtained experimentally.

4. Nonlinear Bifurcation Analysis

Our ultimate goal is to understand what the nature of the nonlinear forces are that lead to a sub-critical dynamic instability. These instabilities are very dangerous in practice. The sub-critical phenomenon however depends on the nonlinear nature of the fluid coupling forces. In the absence of dynamic measurements of the nonlinear part of the fluid forces, we posit two plausible nonlinear models, the second model motivated by the Taylor expansion of the relative velocity. In these models we consider just two tubes; one constrained to vibrate transverse to the flow, and the other constrained to vibrate in the flow direction. We represent the dynamics of this system by the following equations:

Model-1

$$\left\{ \begin{array}{c} \dot{x} \\ \dot{y} \\ \ddot{x} \\ \ddot{y} \end{array} \right\} = \mathbf{A}(U_r) \left\{ \begin{array}{c} x \\ y \\ \dot{x} \\ \dot{y} \end{array} \right\} + (\frac{U_r}{U_{cr}})^2 \left\{ \begin{array}{c} 0 \\ 0 \\ C_{30}\dot{x}^3 + C_{31}\dot{x}^2\dot{y} + C_{32}\dot{x}\dot{y}^2 + C_{33}\underset{\cdot}{o}ty^3 \\ D_{30}\dot{y}^3 + D_{31}\dot{y}^2\dot{x} + D_{32}\dot{y}\dot{x}^2 + D_{33}\underset{\cdot}{o}tx^3 \end{array} \right\} + \frac{U_r}{U_{cr}} \left\{ \begin{array}{c} 0 \\ 0 \\ C_{50}\dot{x}^5 \\ D_{50}\dot{y}^5 \end{array} \right\}, \tag{4}$$

Model-2

$$\left\{ \begin{array}{c} \dot{x} \\ \dot{y} \\ \ddot{x} \\ \ddot{y} \end{array} \right\} = \mathbf{A}(U_r) \left\{ \begin{array}{c} x \\ y \\ \dot{x} \\ \dot{y} \end{array} \right\} + f_1(U_r) \left\{ \begin{array}{c} 0 \\ 0 \\ 1/2C_l\dot{x}^2 + C_d\dot{x}\dot{y} + C_l\dot{y}^2 \\ 1/2C_d\dot{x}^2 - C_l\dot{y}\dot{x} + C_d\dot{y}^2 \end{array} \right\} + f_2(U_r) \left\{ \begin{array}{c} 0 \\ 0 \\ -1/2C_d\dot{x}^3 \\ 1/2C_l\dot{x}^3 \end{array} \right\}$$
$$+ f_3(U_r) \left\{ \begin{array}{c} 0 \\ 0 \\ -1/8C_l\dot{x}^4 - 1/2C_d\dot{x}^3\dot{y} \\ -1/8C_d\dot{x}^4 + 1/2C_l\dot{x}^3\dot{y} \end{array} \right\} + f_4(U_r) \left\{ \begin{array}{c} 0 \\ 0 \\ C_{50}\dot{x}^5 \\ D_{50}\dot{y}^5 \end{array} \right\}, \tag{5}$$

where, $\mathbf{A}(U_r) = \left[\begin{array}{cc} \mathbf{O} & \mathbf{I} \\ -\mathbf{K}(U_r) & -\mathbf{C}(U_r) \end{array} \right]$ is the matrix governing the linearized equations, $\boldsymbol{K}$ is the non-symmetric stiffness matrix, $\boldsymbol{C}$ is the damping matrix and U_r is the parameter in the system. The various elements of the stiffness and damping matrices were obtained from the linear model of section 3. The coefficients C_{ij}, D_{ij}, C_l, C_d and $f_i(U_r)$ in these models are unknowns to be determined. Note that U_{cr} is the critical flow velocity for flutter instability when two of the eigenvalues of $\boldsymbol{A}$ cross the imaginary axis, i.e., zero real parts. At critical speed U_{cr}, the motion of the linearized system approaches the sub-space spanned by the eigenvectors corresponding to those eigenvalues with zero real part. This subspace is invariant under flow generated by the linearized equations. The center manifold theorem (see for e.g. [10, 15]) ensures that for the full nonlinear equations there exists a subspace (center manifold) which is tangential to the subspace spanned by the eigenvectors, at the equilibrium point. The flow close to the equilibrium point approaches the center manifold asymptotically. The symbolic computation for the center manifold was performed using power series expansions. Near-identity transformations of increasing order were used to reduce the flow on the center manifold to a normal (canonical) form.

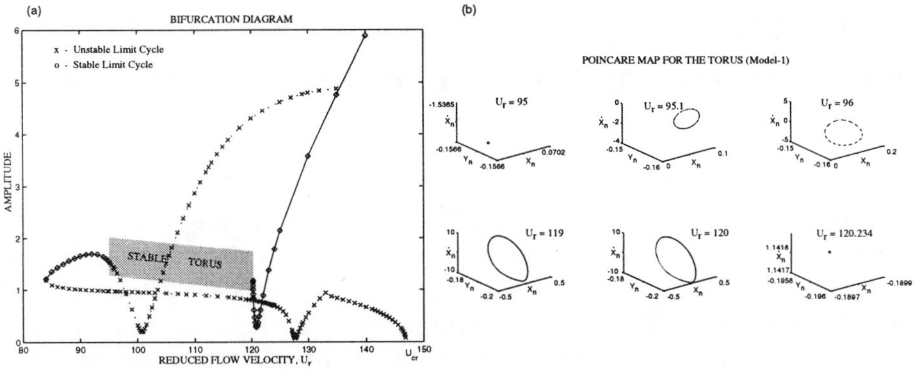

Figure 5. (a) Bifurcation diagram for model-1. A secondary bifurcation at $U_r \approx 95$ results in a torus. (b) The Poincare map of the system (model-1) for $95 \leq U_r \leq 120$ shows the presence of quasi-periodic motion

The relationships between the coefficients that govern the Hopf bifurcation were studied using the normal form of the flow on the center manifold. These relationships were then used to choose the unknown coefficients in the equations 4 and 5. These two systems of equations were studied

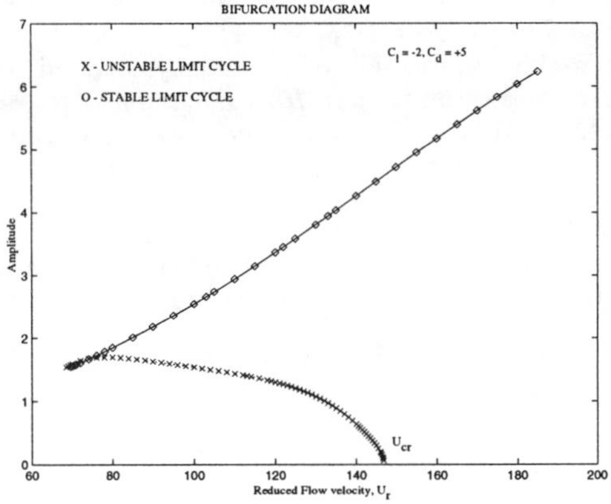

Figure 6. Bifurcation diagram for model-2.

through numerical simulation. Figures 5 and 6 show the bifurcations in these models. The amplitude of the limit cycles in the bifurcation diagram was obtained by using Newton's root finding method on the Poincare map with the Poincare section at $\dot{y} = 0$. The stability of the limit cycles were obtained from the eigenvalues of the Jacobian evaluated at the limit cycle. Figure 5a shows a sub-critical Hopf bifurcation with secondary bifurcation to a torus. The quasi-periodic motion on the torus is shown in the Poincare map of figure 5b.

5. Conclusions

This study has shown that experimental dynamic instabilities in tubes in cross flow can exhibit sub-critical limit cycle oscillations (Hopf bifurcation). These sub-critical Hopf bifurcations are dangerous in practice because large amplitude oscillations can be induced at much lower flow velocities than those predicted by classical linear models for fluid-structure instability. The nature of these sub-critical instabilities depends on the nonlinear fluid and structural forces. We have used an ad-hoc nonlinear model to show the possible complexities in the secondary bifurcations as a function of flow velocity. This study has demonstrated the need to obtain better measurements of dynamic fluid structure forces in order to provide realistic predictive models of nonlinear fluid-structure instabilities.

References

1. R. D. Blevins. *Flow-Induced Vibrations.* Van Nostrand Reinhold Company, 1977.
2. S. S. Chen. Instability mechanisms and stability criteria of a group of circular cylinders subjected to cross-flow; part 1: Theory, part 2: Numerical results and discussion. *ASME Journal of Vibration, Acoustics, Stress and Reliability in Design,* 105:58,253, 1983.
3. S. S. Chen. A general theory for dynamic instability of tube arrays in crossflow. *Journal of Fluids and Structures,* 1:35–53, 1987.
4. H.J. Connors. Fluid-elastic vibration of tube arrays excited by cross flow. In D. D. Reiff, editor, *Flow-Induced Vibration in Heat Exchangers,* pages 42–56. ASME, 1970.
5. J. H. Leaver and D. S. Weaver. A theoretical model for fluid-elastic instability in heat exchangers. *ASME Journal of Pressure Vessel Technology,* 104:147–158, 1982.
6. G. Muntean. *Influence of fluid wake structure on the dynamics of a flexible tube row in cross flow.* PhD thesis, Cornell University, 1995.
7. M. P. Païdoussis and S. J. Price. The mechanisms underlying flow-induced instabilities of cylinder arrays in cross flow. *Journal of Fluid Mechanics,* 187:45–59, 1988.
8. M. P. Païdoussis, S. J. Price, and D. Mavriplis. A semipotential flow theory for the dynamics of cylinder arrays in cross flow. *ASME Journal of Fluid Engineering,* 107:500–506, 1985.
9. S. J. Price. A review of theoretical models for fluidelastic instability of cylinder arrays in cross-flow. *Journal of Fluids and Structures,* 9:463–518, 1995.
10. R. H. Rand and D. Armbruster. *Perturbation Methods, Bifurcation Theory and Computer Algebra.* Springer-Verlag, 1987.
11. B. W. Roberts. Low frequency, aeroelastic vibrations in a cascade of circular cylinders. In *Mechanical Engineering Science, Monograph No. 4.* Institution of Mechanical Engineers, London, September 1966.
12. H. Tanaka and S. Takahara. Unsteady fluid dynamic force on tube bundle and its dynamic effect on vibration. In M. K. Au-Yang, editor, *Flow-Induced Vibration of Power Plant Components,* pages 77–92, New York, 1980. ASME.
13. M. Thothadri. Dynamics of a row of cylinders in cross flow. Master's thesis, Cornell University, 1996.
14. M. Thothadri and F.C. Moon. Helical wave oscillations in a row of cylinders in a cross flow. submitted to *Journal of Fluids and Structures.*
15. S. Wiggins. *Introduction to Applied Nonlinear Dynamical Systems and Chaos.* Springer-Verlag, 1990.

IX. SPATIO-TEMPORAL CHAOS

Experimental evidence for spatial complexity in a periodic elastic structure is presented in the paper by Hikihara and Okamoto. A theoretical paper by Salenger and Vakakis discusses chaos in a periodically supported elastic structure.

A useful model for examining spatio-temporal chaos in coupled map lattices is presented in the paper by Grigoriev.

F. C. Moon (ed.),
IUTAM Symposium on New Applications of Nonlinear and Chaotic Dynamics in Mechanics, 521.

SPATIO-TEMPORAL DYNAMICS OF COUPLED MAGNETO-ELASTIC SYSTEM

Takashi HIKIHARA[1), Yoshinobu OKAMOTO[2)
and Yoshisuke UEDA[1)
1) School of Electrical Engineering, Kyoto University
Yoshida-honmachi, Sakyo, Kyoto, 606-01 JAPAN
2) Department of Electrical Engineering, Kansai University
Yamate-cho 3-3-35, Suita, Osaka, 564 JAPAN

Abstract In this paper the spatio-temporal vibration in a coupled magneto-elastic beam system is discussed experimentally and numerically. The periodical forcing at the end of the beam system induces the several types of vibrations to the magneto-elastic beam through the coupling beams. It is caused by the nonlinearity which depends on the potential well of each magneto-elastic beam and the wave propagation passing through the coupling beams. The temporal vibration of the magneto-elastic beam is explained with relations to the spatial state transition based on the experimental results.

1. Introduction

Recently the dynamics of the highly coupled systems have been widely studied in many fields in order to understand the emergent phenomenon in the complex systems [1]. The coupled map lattice (CML) is considered to be one of the good models that are able to describe the distributed systems by discretizing the space in finite periodical cells [2,3]. These coupled oscillators show the remarkable feature that the populations of coupled pendulum can synchronize to a common frequency [4]. Moreover, the spatio-temporal chaos has also been found in the lattice [3,5]. On the other hand, the periodically coupled lattice structure can also be seen in usual mechanical structures; in airplane, ship, train, and the electrical distributed systems; power line, linear electrical machine, etc. Therefore, the CML is considered to be an substantial model to study the dynamics both in the mechanical structures and in distributed systems.

In our laboratory, one of the authors has studied the dynamics of the magneto-elastic beam system with delayed feedback control [6,7]. Through the studies the time delay input has an ability to make the system stabilize in a new stable orbit which does not exist in the original system caused by the increase of the system dimension. In the distributed systems the vibration induced at an end cell reaches to the next with a time delay, travels through the coupled lattice and reflects at another end. These types of traveling wave in the coupled elastic system have already been reported by Davies [8], Davies and Moon [9]. The wave shows the characteristics of the solitary wave that is usually discussed in the nonlinear distributed system [10].

F. C. Moon (ed.),
IUTAM Symposium on New Applications of Nonlinear and Chaotic Dynamics in Mechanics, 523–531.

On the other hand, the pendulum which is a cell of the coupled lattice shows the variety of vibrations; harmonics, quasi-periodicity and chaos. These temporal vibrations are caused by the wave propagation passing through the coupling beams. The wave which propagates through the system shows the multi-reflections at both ends. Therefore, each magneto-elastic beam is governed by the dynamics excited by a multiple reflected waves with time delay. On this standpoint, the spatio-temporal dynamics of the coupled mageto-elastic beam system (CMES) is experimentally and numerically discussed in this paper.

2. Experimental System and Setup Condition

The CMES is composed of the magneto-elastic beam (truck beam) and the coupling beams. Each beam is the thin ferrite beam that has 0.3 mm thickness. The detail length and width of the beams are shown in Fig. 1. The density of the beam is 7.1 g/cm^3. The natural frequency, the damping coefficient and the spring constant of the truck beam are 22.6 Hz, 0.375, 8.32 N/m and 10.0 Hz, 0.149 and 2.19 N/m for the coupling beam. In the coupled system, eight truck beams are linked by the coupling beam. A truck beam and the coupling beams are joined by plastic screws. Therefore, the system has the continuous elastic structure.

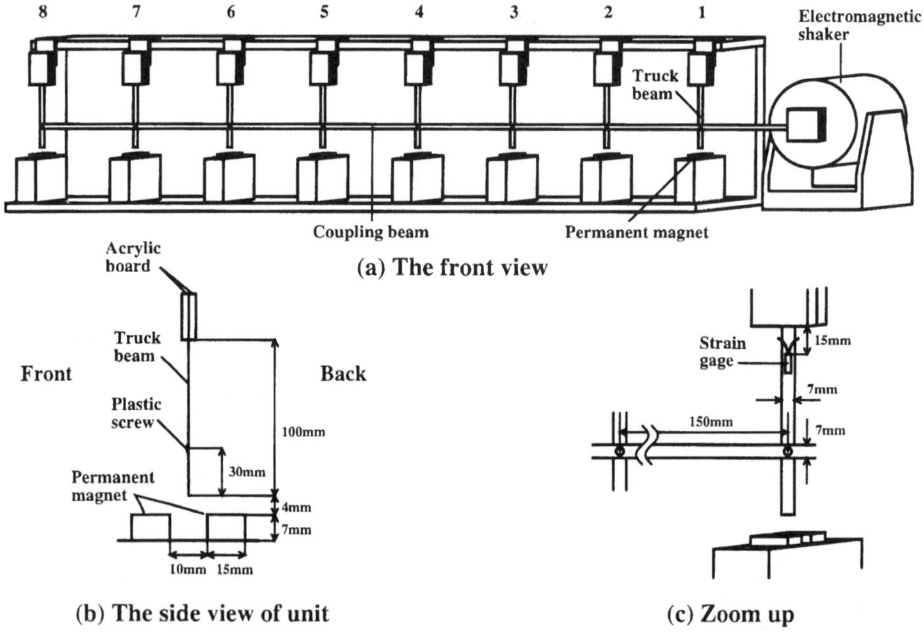

Fig. 1 Coupled magneto-elastic beam system.

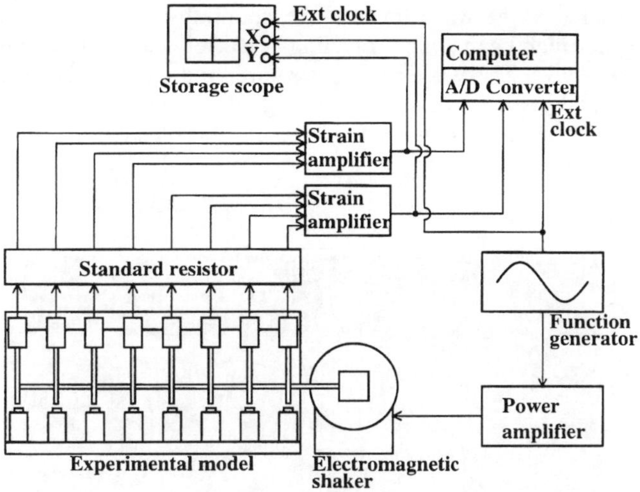

Fig. 2 Experimental system.

The truck beam is sinusoidally driven at the one end by an electromagnetic shaker through the coupling beam. At the bottom of every truck beam two permanent magnets are placed to make the double well-shaped potential. The upper polarities of the magnets are set at the North. Therefore the truck beams have two stable positions over the front magnet and the rear. The single beam system is famous as "Moon's beam," in which the chaotic vibration appears under the sinusoidal driving force [11]. The magnetic flux densities of the magnets are 0.0954 T at the surface. Therefore the system also has the discrete magnetic structure.

The strain gauges are attached on both surfaces of each truck beam to measure the vibration of the beams by the bridge method. The gauges are connected to the bridge circuit with two standard resistors, which are composed of the same strain gauges. The outputs of the bridge circuits are led to the multi-channel strain amplifier, respectively. The eight simultaneous vibration signals are stored by a personal computer at every sampling time (see Fig. 2). Therefore the sampling point is synchronize to the period of the space structure. The Poincaré points can also be obtained by sampling the temporal data at every period of the driving force. As the results, the vibration in the CMES can be measured spatially and temporally at the same time.

3. Experimental Results and Discussion

3.1. IMPACT PROPAGATION

At first we will discuss about the impact propagation. When an impact force is given to one end of the CMES, the vibration propagates to another end with keeping a constant speed. Fig. 3 shows the spatio-temporal contour of the wave propagation. The darkly hatched region implies that the beam is forced to the backward and the white region to the forward. The vibration of the coupling beam between the truck beams is not measured. Therefore the wave form of the coupling beam is interpolated based on the vibration of the truck beam. The traveling wave reflects with keeping the phase at the free end : beam #8. The wave propagates forth and back at the speed around 18.2 m/s over 1.0 second without dissipation. Fig. 4 shows the 3D plot of the wave propagation. The propagation gives truck beams the delayed impulsive excitations repetitively.

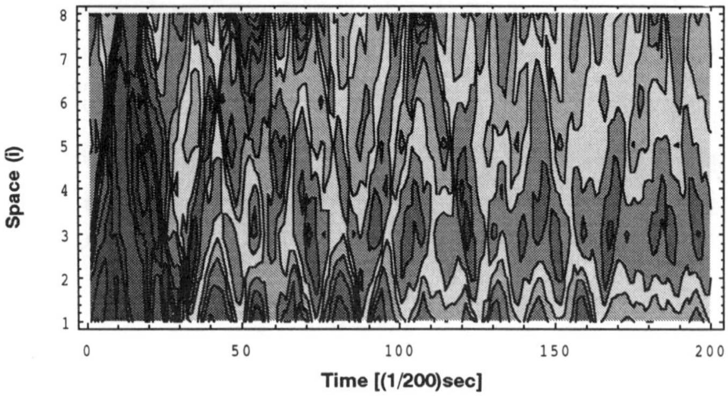

Fig. 3. Impact wave propagation.

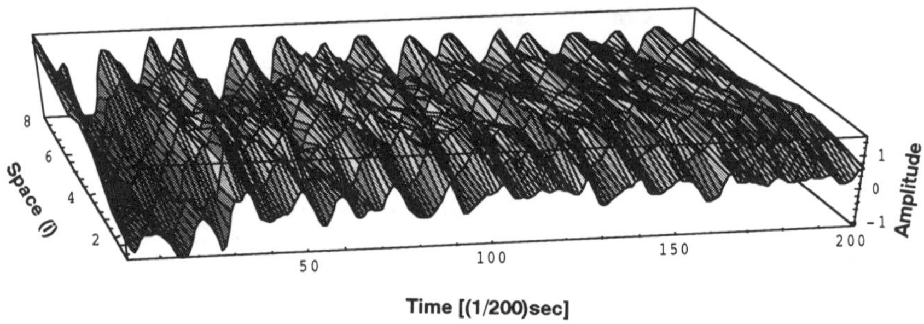

Fig. 4. 3D plot of impact wave propagation.

3.2. QUASI-PERIODIC WAVE PROPAGATION

In the system, there are many natural frequencies caused by the coupling structure. It gives the CMES the possibility of quasi-periodic vibrations. Fig. 5 shows the spatio-temporal contour of the typical experimental results of the quasi-periodic wave propagation. The quasi-periodic wave propagation implies the propagation which gives the quasi-periodic temporal vibration to each truck beam. In these states, the quasi-periodic disturbance superposes on the temporal fundamental wave propagation. Therefore the spatio-temporal pattern of the contour changes gradually.

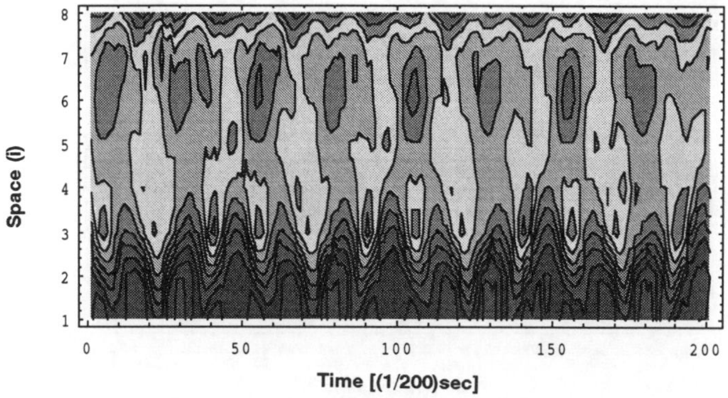

(a) Fundamental quasi-periodic wave (0.9 V, 12.0 Hz)

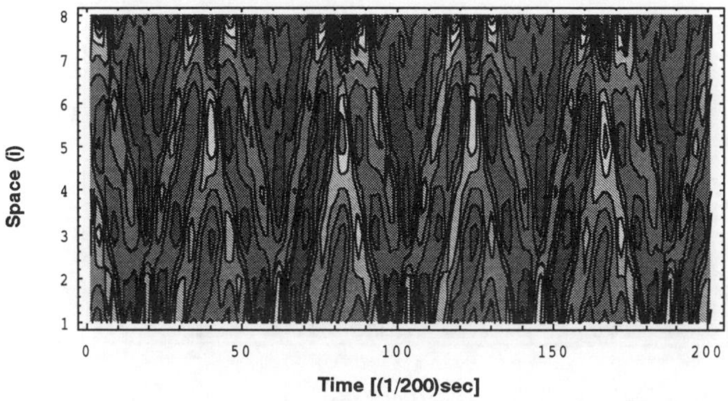

(b) Fourth quasi-periodic wave (2.28 V, 19.0 Hz)

Fig. 5. Experimental quasi-periodic wave propagation.

3.3. CHAOTIC WAVE PROPAGATION

The single truck beam and the permanent magnet structure is well known as " Moon's beam," which can show the temporal chaotic dynamics. In the CMES, the single cell also has the possibility to show the temporal chaotic vibration. Fig. 6 shows the spatio-temporal contour of typical cases of the chaotic wave propagation. In the chaotic wave propagation, the contour shows that the states are completely chaotic temporarily and spatially. The estimation depends on the spectrum analysis, Poincaré map, and the bifurcation diagram as for the time. However, there is no exact way to estimate for the space except the spatial wave form and the delay map (correlation). Because the spatially sampled points are limited in our experimental system. In Figs. 6(b) and (c) the chaotic vibration superposes on the temporal forth periodic wave propagation. In the case the intermittent type chaotic vibration appears. The intermittent chaos is also reported in CML by Crutchfield and Kaneko [2]. However, the mechanism of the intermittency cannot be explained in these experiments.

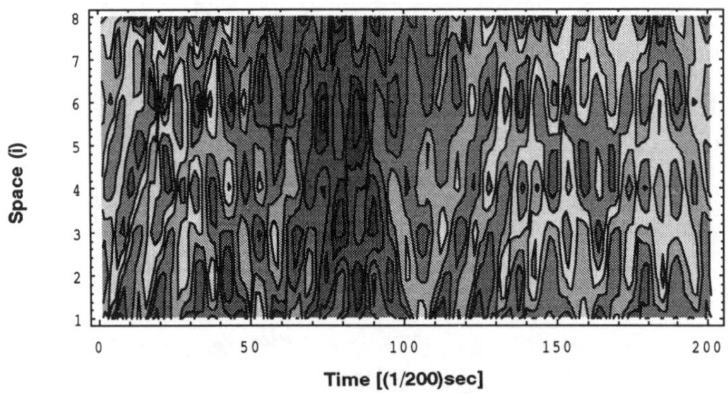

(a) Fundamental chaotic wave propagation (1.8 V, 19.0 Hz)

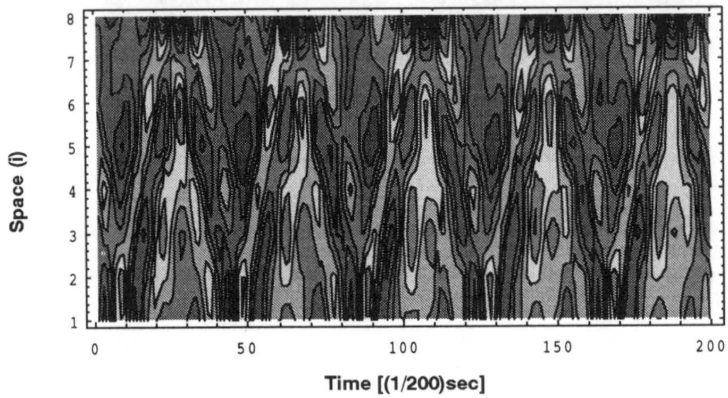

(b) Forth chaotic wave propagation [periodic wave] (1.95 V, 20.0 Hz)

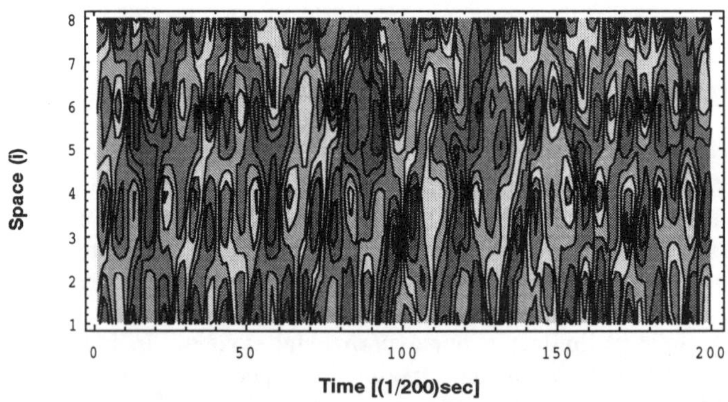

(c) Forth chaotic wave propagation [intermittency] (1.95 V, 20.0 Hz)

Fig. 6 Experimental chaotic wave propagation.

4. Numerical Simulation

The single cell of magneto-elastic beam has already been studied [11,12]. The mathematical model is given by a double well potential Duffing's equation [13]. The CMES has the coupled structure of the beam. Therefore the numerical model is considered to be :

$$
\begin{aligned}
\frac{dx_i}{dt} &= y_i \\
\frac{dy_i}{dt} &= -2\gamma_1 y_i - \alpha_1 x_i - \beta_1 x_i^{\,3} \\
&\quad - 2\gamma_2(y_i - y_{i-1}) - \alpha_2(x_i - x_{i-1}) - \beta_2(x_i - x_i)^3 \\
&\quad - 2\gamma_2(y_i - y_{i+1}) - \alpha_2(x_i - x_{i+1}) - \beta_2(x_i - x_{i+1})^3
\end{aligned}
\qquad (1)
$$

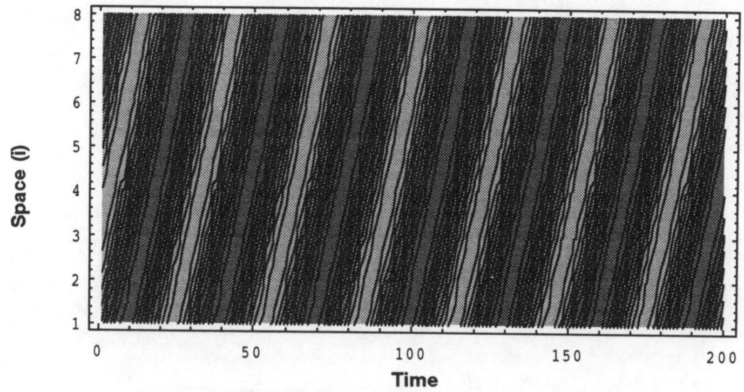

(a) Traveling wave (ω = 2.0 and X_0 = 2.0)

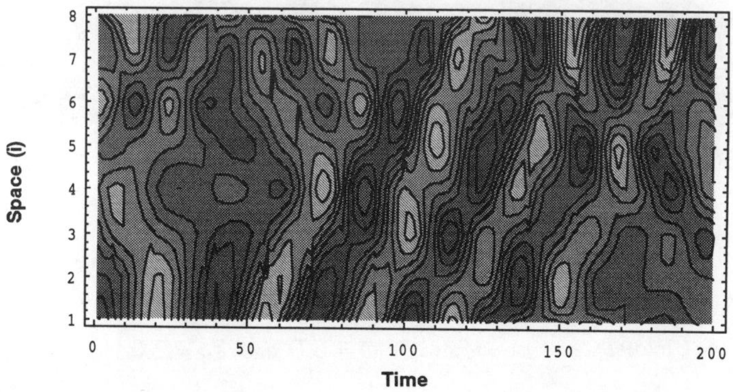

(b) Chaotic wave propagation (ω = 5.0 and X_0 = 2.0)

Fig. 7. Numerical results ($\gamma_1 = 0.0126$, $\gamma_2 = 0.0050$, $\alpha_1 = \alpha_2 = -1$ and $\beta_1 = \beta_2 = 1$)

where $x_0 = X_0 \cos \omega t$, $\dot{x}_0 = X_0 \omega \sin \omega t$, $x_9 = x_8$ and $\dot{x}_9 = \dot{x}_8$. The numerical integration is performed by Runge-Kutta algorithm. Fig. 7 shows the spatio-temporal contour of the numerical results. As shown in the figure, depending on the driving frequency and the amplitude, the spatio-temporal pattern changes drastically. In Fig. 7(a) the wave propagates without reflection. On the other hand, in Fig. 7(b) the temporal and spatial chaotic wave propagation appears. Fig. 8 shows the 3D plot of the wave propagation corresponding to Fig. 7. In Fig. 8(b), the collision of the soliton like wave can be observed.

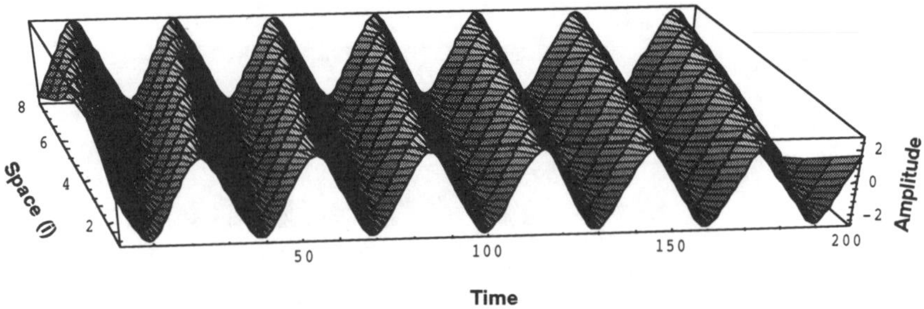

(a) Traveling wave (ω = 2.0 and X_0 = 2.0)

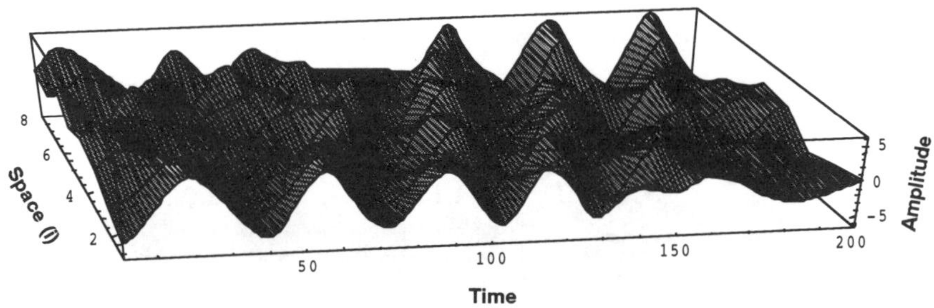

(b) Chaotic wave propagation (ω = 5.0 and X_0 = 2.0)

Fig. 8. 3D plot of simulated wave ($\gamma_1 = 0.0126$, $\gamma_2 = 0.0050$, $\alpha_1 = \alpha_2 = -1$ and $\beta_1 = \beta_2 = 1$)

5. Concluding Remarks

In this paper we have discussed about the temporal and spatial dynamics of a CMES. The behavior of the continuous elastic structure was discussed based on the discretized spatial wave. Through the experiments, the periodical, quasi-periodic and chaotic wave shows the typical feature of the spatio-temporal contour pattern which was observed in CML which is composed of large number of pendulum. The bifurcation of a periodical wave to others has a great dependence on the propagation and the reflection of the waves. As for the CMES, a mathematical model was proposed. The simulated results show the similar wave propagation to the experiment. Especially the soliton like wave propagation can be seen in the chaotic wave propagation. These discussions are limited in the phenomenological considerations. However, we believe they give us some clues to study the occurrence of the bifurcation of the spatio-temporal wave.

Acknowledgment

The authors gratefully acknowledge the fruitful discussions with Prof. F.C.Moon, Cornell University. They would like to thank to Prof. T.Ushio, Osaka University and Prof. H.Nakajima, Kinki University for their wonderful comments and discussions about the infinite dimensional system.

References

[1] P.E.Cladis and P.Palffy-Muhoray Eds., Spatio-temporal patterns in nonequilibrium complex systems, SFI Studies in the Sciences of Complexity (Addison-Wesley, 1995).
[2] K.Kaneko and I.Tsuda, Chaotic scenario of complex system (Asakura, Tokyo, 1996) (in Japanese).
[3] J.P.Crutchfield and K.Kaneko, in: Directions in Chaos (World Scientific, 1987) 272-353.
[4] K.Wiesenfeld, P.Colet and S.H.Strogatz, Synchronization transitions in a disordered Josephson series array, Phys. Rev. Lett., 76, 3 (1996) 404-407.
[5] H.Daido, Strange waves in coupled oscillator arrays: mapping approach, Phys. Rev. Lett., 78, 9 (1997) 1683-1686.
[6] T.Hikihara and T.Kawagoshi, An experimental study on stabilization of unstable periodic motion in magneto-elastic chaos, Phys. Letts. A211 (1996) 29-36.
[7] T.Hikihara, M.Touno and T.Kawagoshi, Stabilization of unstable periodic orbit in magneto-elastic chaos by delayed feedback control, Int. J. of Bifurcation and Chaos (submitted).
[8] M.A.Davies, Spatio-temporal chaos and solitary waves in nonlinear periodic structures, the doctoral dissertation of Cornell University (1993).
[9] M.A.Davies and F.C.Moon, Transition from soliton to chaotic motion during impact of a nonlinear elastic structure, Chaos, Soliton and Fractals, 4(2) (1994) 275-283.
[10] M.Toda, Theory of nonlinear lattice (Springer-Verlag, Berlin-Heidelberg, 1989).
[11] F.C.Moon, Chaotic and fractal dynamics (John Wiley, New York, 1992).
[12] Y.Ueda, H.Nakajima, T.Hikihara and H.B.Stewart, Forced two-well potential Duffing's oscillator, in: Dynamical systems approaches to nonlinear problems in system and circuit, eds. F.M.A.Salam and M.L.Levi (SIAM, Philadelphia, 1988) 128.
[13] F.C.Moon, Experiments on chaotic motions of a forced nonlinear oscillator: strange attractors, ASME, J. Appl. Mech, 47 (1980) 638-644.

SPATIALLY LOCALIZED AND CHAOTIC MOTIONS OF A DISCRETELY SUPPORTED ELASTIC CONTINUUM

G.D. SALENGER and A.F. VAKAKIS
Department of Mechanical & Industrial Engineering
University of Illinois at Urbana - Champaign
Urbana, IL 61801, USA

1. Introduction

In this work we provide analytical computations of discreteness effects in the solitary and standing wave motions, and the chaotic envelope of a forced string supported by an infinite periodic array of springs with a softening cubic nonlinearity. To overcome the difficulty encountered in dealing with the singularities inherent in this system, a nonsmooth transformation, first developed by Pilipchuk [1,2], is applied to the spatial variable. The method, also detailed in [3], was first applied to this problem in [4] for the unforced case, and in [5], to compute solitary waves for a system with hardening cubic and vibro-impact supports and forcing.

This work is a further extension of [5], providing a more general method for computing standing wave solutions in systems with symmetric, conservative, supports. The application of a nonsmooth transformation to the original equation of motion results in a set of smooth nonlinear boundary value problems (NBVPs), which is analyzed using regular perturbation techniques and harmonic balance. As the system is solved at each order, integration constants remain. The equations governing these 'constants' are then solved after imposing the condition that their solutions must remain bounded in space. Finally the analytic solutions are compared to numerical integrations of the original equations of motion.

2. Analysis

2.1. NONSMOOTH TRANSFORMATION OF GOVERNING EQUATION

The system under consideration consists of a linear string of infinite spatial extent, supported by an infinite periodic array of nonlinear elasticities, and forced by a distributed external excitation. The governing equation of motion is given by:

$$\rho\frac{\partial^2 u}{\partial t^2} - T\frac{\partial^2 u}{\partial y^2} + 2f(u)\sum_{k=-\infty}^{\infty}\delta(\frac{y}{\varepsilon} - 1 - 2k) = q(\frac{y}{\varepsilon}, y, t) \quad (1)$$

F. C. Moon (ed.),
IUTAM Symposium on New Applications of Nonlinear and Chaotic Dynamics in Mechanics, 533–542.

where $u \equiv u(y/\varepsilon, y, t)$ denotes the transverse displacement, ρ and T the (uniform) density and tension of the string, $f(u)$ the nonlinear characteristics of the stiffness, and $q \equiv q(y/\varepsilon, y, t)$ the external distributed load. The external force and response are assumed to possess 'fast' and 'slow' spatial scales, defined by the small parameter $0 < \varepsilon << 1$ that governs the distance between adjacent stiffnesses. Moreover, we assume that q is periodic with respect to the 'fast' scale (y/ε), with normalized period $\Pi = 4$.

We seek solutions of (1) that are periodic with respect to the 'fast' spatial variable (y/ε). To this end we employ the following nonsmooth transformations of the 'fast' spatial variable,

$$\frac{y}{\varepsilon} \rightarrow (\tau(\frac{y}{\varepsilon}), e(\frac{y}{\varepsilon}))$$
$$\tau(\frac{y}{\varepsilon}) = (\frac{2}{\pi})\sin^{-1}(\sin(\frac{\pi}{2}\frac{y}{\varepsilon})), \quad e(\frac{y}{\varepsilon}) = \tau'(\frac{y}{\varepsilon}) \tag{2}$$

where prime denotes differentiation with respect to the argument. First, the 'fast' spatial scale in (1) is transformed according to (2); then, the infinite summation of singular terms on the left-hand of (1) is expressed in terms of the nonsmooth variables as follows:

$$2f(u)\sum_{k=-\infty}^{\infty}\delta(\frac{y}{\varepsilon} - 1 - 2k) = -f(u)\,\mathrm{sgn}(\tau)\tau'' \tag{3}$$

Finally, the transverse displacement and the external excitation are expressed in terms of the nonsmooth variables as [4]:

$$u = U(\tau, y, t) + V(\tau, y, t)e, \quad q = Q(\tau, y, t) + P(\tau, y, t)e \tag{4}$$

The dependent variables U and V are functions of the nonsmooth variable $\tau \in [-1,1]$ and of the original variables y and t; in essence, we replace the 'fast' spatial scale with the new set of nonsmooth variables $\{\tau, e\}$.

Following the aforementioned steps and using the chain rule of differentiation, the equation of motion is transformed to the following form,

$$\rho[\frac{\partial^2 U}{\partial t^2} + \frac{\partial^2 V}{\partial t^2}\tau'] - T[\frac{\partial^2 U}{\partial y^2} + \frac{2}{\varepsilon}\frac{\partial^2 V}{\partial y \partial \tau} + \frac{1}{\varepsilon^2}\frac{\partial^2 U}{\partial \tau^2} +$$
$$+ (\frac{\partial^2 V}{\partial y^2} + \frac{2}{\varepsilon}\frac{\partial^2 U}{\partial y \partial \tau} + \frac{1}{\varepsilon^2}\frac{\partial^2 V}{\partial \tau^2})\tau' + (\frac{1}{\varepsilon}\frac{\partial V}{\partial y} + \frac{1}{\varepsilon^2}\frac{\partial U}{\partial \tau})\tau''] = \tag{5}$$
$$= \mathrm{sgn}(\tau)\tau''(R_f + I_f\tau') + Q(\tau, y, t) + P(\tau, y, t)\tau'$$

where $R_f = (1/2)[f(U+V)+f(U-V)]$, $I_f = (1/2)[f(U+V)-f(U-V)]$.

Noting that $\tau'\tau'' = 0$ [4] eliminating the remaining singular terms, and then setting separately terms proportional and not proportional to τ' equal to zero, equation (5) is decomposed into the following set of coupled nonlinear nonhomogeneous boundary value problems (NBVPs) [4]:

$$\frac{\partial^2 U}{\partial \tau^2} = -2\varepsilon \frac{\partial^2 V}{\partial y \partial \tau} + \varepsilon^2 [\frac{\rho}{T}\frac{\partial^2 U}{\partial t^2} - \frac{\partial^2 U}{\partial y^2} - \frac{Q(\tau, y, t)}{T}] \quad (6a)$$

$$\frac{\partial^2 V}{\partial \tau^2} = -2\varepsilon \frac{\partial^2 U}{\partial y \partial \tau} + \varepsilon^2 [\frac{\rho}{T}\frac{\partial^2 V}{\partial t^2} - \frac{\partial^2 V}{\partial y^2} - \frac{P(\tau, y, t)}{T}] \quad (6b)$$

$$V\Big|_{\tau=\pm 1} = 0 \quad (6c)$$

$$-\frac{\partial U}{\partial \tau}\Big|_{\tau=\pm 1} = \pm\frac{\varepsilon^2}{T} R_f = \pm\frac{\varepsilon^2}{2T}[f(U+V)+f(U-V)] \quad (6d)$$

with the new independent variable τ restricted in the range $-1 \le \tau \le 1$. The boundary conditions (6c,d) arise from 'smoothening conditions' that guarantee differentiability of the first and second derivatives with respect to τ of the components U and V. Hence, the problem of computing periodic solutions with respect to the 'fast' spatial scale is reduced to solving the NBVPs (6a-d). This later set of equations does not possess singular terms (in contrast to the original equation (1)), and, more significantly, can be analyzed by *regular* perturbation theory. This is due to the fact that the new set of nonsmooth independent variables $\{\tau, \varepsilon\}$ is bounded in the interval $[-1,1]$, and is periodic with respect to the 'fast' scale (y/ε); this eliminates secular terms normally encountered in standard methods of nonlinear dynamics, where the independent variable is unbounded and non periodic.

The solutions of (6) are expressed in the form:

$$U(\tau, y, t) = \sum_{k=0}^{\infty} \varepsilon^k U_k(\tau, y, t), \quad V(\tau, y, t) = \sum_{k=0}^{\infty} \varepsilon^k V_k(\tau, y, t) \quad (7)$$

where all variables other than $\varepsilon \ll 1$ are assumed to be of $O(1)$. Substituting (7) into (6) and matching coefficients of respective powers of ε, we obtain successive approximations to the solutions.

Terms of U_k and V_k in expansion (2), up to $O(\varepsilon^2)$, were determined in [5] as:

$$U_0(\tau, y, t) = B_0(y, t),\ V_0(\tau, y, t) = 0,\quad U_1(\tau, y, t) = 0,\ V_1(\tau, y, t) = 0$$

$$U_2(\tau, y, t) = -\frac{\tau^2}{2T} f(B_0(y,t)) + H(y)\cos\omega t\,[-\frac{m_0}{2T} +$$

$$+\frac{1}{T}\sum_{j=1}^{\infty}\{m_{2j}\frac{\cos(j\pi\tau)-(-1)^j}{j^2\pi^2} + n_{2j-1}\frac{4\sin[(2j-1)\pi\tau/2]-4(-1)^j}{(2j-1)^2\pi^2}\}] + B_2(y,t) \tag{8}$$

$$V_2(\tau, y, t) = H(y)\cos\omega t\,[\frac{1}{\pi^2 T}\sum_{j=1}^{\infty}\{m_{2j-1}\frac{4\cos[(2j-1)\pi\tau/2]}{(2j-1)^2} + n_{2j}\frac{\sin(j\pi\tau)}{j^2}\}]$$

where the constants of integration, B_0 and B_2, are solutions of the following PDE's:

$$\rho\frac{\partial^2 B_0(y,t)}{\partial t^2} - T\frac{\partial^2 B_0(y,t)}{\partial y^2} + f(B_0(y,t)) = m_0 H(y)\cos\omega t$$

$$\rho\frac{\partial^2 B_2(y,t)}{\partial t^2} - T\frac{\partial^2 B_2(y,t)}{\partial y^2} + f'(B_0(y,t))B_2(y,t) =$$

$$= \frac{1}{6T}[\rho\frac{\partial^2 f(B_0(y,t))}{\partial t^2} - T\frac{\partial^2 f(B_0(y,t))}{\partial y^2} + 3f'(B_0(y,t))f(B_0(y,t))] - \tag{9}$$

$$-\frac{1}{2T}\cos\omega t\,[m_0 + \frac{8}{\pi^2}\sum_{j=1}^{\infty}\frac{(-1)^j n_{2j-1}}{(2j-1)^2}]\{[\rho\omega^2 - f'(B_0(y,t))]H(y) + T\frac{\partial^2 H(y)}{\partial y^2}\}$$

In [5] a system with hardening cubic supports was analyzed, and we were able to compute a solitary wave solution explicitly to $O(\varepsilon^2)$. In this work we find the solitary, standing, and chaotic wave solutions for a system with softening cubic spring supports, $f(u) = \alpha_1 u - \alpha_3 u^3$; $\alpha_1, \alpha_3 \ge 0$.

2.2. SOLITARY WAVE SOLUTION

We first focus on the solitary and standing wave solutions, which are forced with the following external excitation, chosen for the sake of simplicity,

$$q\left(\frac{y}{\varepsilon}, y, t\right) = H(y)\cos\left(\frac{k\pi}{2}\frac{y}{\varepsilon}\right)\cos\omega t \tag{10}$$

time-periodic solutions of (9) are sought using the method of harmonic balance:

$$B_0(y,t) = \sum_p \hat{A}_p(y)\cos(p\omega t),\ B_2(y,t) = \sum_p \hat{C}_p(y)\cos(p\omega t) \tag{11}$$

The leading harmonics in (11) are considered by matching coefficients proportional to $\cos\omega t$. The leading coefficients $\hat{A}_1(y)$ and $\hat{C}_1(y)$ are then governed by the following normalized ordinary differential equations,

$$A_1''(z) - A_1(z) + \mu A_1^{\,3}(z) = 0 \tag{12a}$$

$$C_1''(z) - C_1(z) + 3\mu A_1^{\,2}(z) C_1(z) = F(z) \tag{12b}$$

where the following normalizations apply:

$$\hat{A}_1(y) = A_1(z),\ \hat{C}_1(y) = C_1(z),\ z = \left[-\frac{\sigma}{T}\right]^{1/2} y,\ \mu = \frac{3\alpha_3}{4\sigma},\ \sigma = \rho\omega^2 - \alpha_1,$$

$$F(z) \equiv F[A_1(z), A_1'(z), A_1''(z)] = \frac{1}{\sigma}\{\frac{1}{6T}\alpha_1\sigma A_1''(z)\frac{3}{8T}\alpha_3\sigma A_1''(z) A_1^{\,2}(z) - \frac{1}{12T} * \tag{13}$$

$$*\left[2\alpha_1\sigma - 4\alpha_1^{\,2} - 9\alpha_3\sigma A'^2{}_1(z)\right] A_1(z) - \frac{1}{8T}\alpha_3(\sigma - 11\alpha_1) A_1^{\,3}(z) + \frac{15}{16T}\alpha_3^{\,2} A_1^{\,5}(z)\}$$

and primes denote differentiation with respect to z. In writing the above expressions, the frequency ω satisfies the inequality $\omega^2 < \alpha_1 / \rho$.

The solitary wave solution to (12a) is given by $A_1(z) = \sqrt{2/\mu}\, sech\, z$. To solve for C_1, first note that $\phi_1 = A_1'(z) \Rightarrow \phi_1(z) = sech\, z\ tanh\, z$ is a homogenous solution to (12b). The second homogenous solution may be computed by solving the differential equation satisfied by the Wronskian of the two linearly independent solutions [6]. This was performed in [7] yielding $\phi_2 = (-9\cosh z + \cosh 3z + 12z \sinh z) sech^2 z / 8$.

The particular solution of (9b) is found using variation of parameters and combined with the homogenous solutions to yield a final solution in the following form [6],

$$C_1(z) = [\nu_1 - \int_0^z F(\xi)\phi_2(\xi) d\xi]\phi_1(z) + [\nu_2 + \int_0^z F(\xi)\phi_1(\xi) d\xi]\phi_2(z) \tag{14}$$

where $\nu_1 = 0$ in order to eliminate a time shift in the solution. In order to guarantee boundedness of (14) we set,

$$\nu_2 = -\int_0^\infty F(\xi)\phi_1(\xi) d\xi \tag{15}$$

Using computer algebra we were able to evaluate the integrals in (14) explicitly, and the resulting analytical solutions for $u(y/\varepsilon, y, t = 0)$ and $\partial u/\partial y(y/\varepsilon, y, t = 0)$, up to $O(\varepsilon^2)$ are shown in Figure 1.

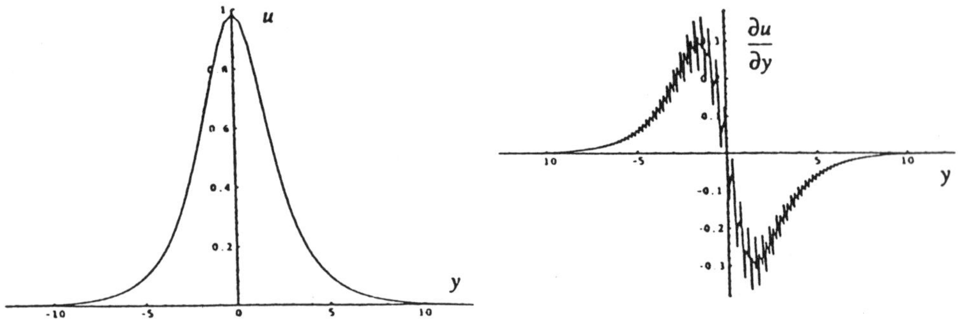

Figure 1. The solitary standing solution $u(y/\varepsilon, y, t=0)$ and $\partial u/\partial y(y/\varepsilon, y, t=0)$ for the system with parameters: $\rho = T = 1$, $\omega = .8$, $\varepsilon = .1$, and external forcing envelope $H(y) = sech\ y$.

2.3. PERIODIC STANDING WAVE SOLUTION

The periodic standing wave solutions of (12) are analyzed similarly to the localized wave solution. Care needs to be taken though in defining the properties of the general solutions $A_1(z)$, $\phi_1(z)$, and $\phi_2(z)$ in order to reach appropriate conditions for v_1 and v_2 in (14) for a bounded solution $C_1(z)$. For initial conditions, $A_1(0) = A_0$ and $A_1'(0) = 0$, $A_1(z)$ is even, periodic, and C^∞ about $z = 0$. Noting that (12a) represents a conservative system, the period Γ of $A_1(z)$ may be calculated by taking advantage of the existence of a first integral of motion. The first homogenous solution $\phi_1(z) = A_1'(z)$ of (12b) is then a continuous, odd, periodic function with period Γ. Finally, the second homogenous solution to (12b) is both even and C^∞ about $z = 0$. Thus it may be concluded that $\phi_1(0) = 0$ and $\phi_2'(0) = 0$. Details of the above are provided in [7].

Once two homogenous solutions of (12b) are determined, the overall solution is given by (14). To guarantee that $C_1(z)$ is bounded we impose the requirement that $C_1(z)$ be a periodic function. This same requirement was imposed in [8] for a similar system solved through elliptic integrals. Noting that the equation (12b) is invariant with respect to a shift in space length Γ, if the following conditions are met:

$$C_1(0) = C_1(\Gamma) \tag{16a}$$

$$C_1'(0) = C_1'(\Gamma) = 0 \tag{16b}$$

then $C_1(z)$ will be periodic with period Γ. The first derivative was set to zero to be consistent with the initial conditions of $A_1(z)$.

It can be shown that (16a) is automatically satisfied regardless of the choice of v_1 and v_2 [7]. To meet condition (16b), differentiate (14) to obtain:

$$C_1'(z) = [\nu_1 - \int_0^z F(\xi)\phi_2(\xi)d\xi]\phi_1'(\xi) + [\nu_2 + \int_0^z F(\xi)\phi_1(\xi)d\xi]\phi_2'(\xi) \tag{17}$$

Imposing (16b) on (17), we compute the constants ν_1 and ν_2 as follows:

$$\nu_1 = 0,\ \nu_2 = \frac{\phi_1'(\Gamma)}{\phi_2'(\Gamma)}\int_0^\Gamma F(\xi)\phi_2(\xi)d\xi \tag{18}$$

3. Numerical Solution

For the verification of the analytical results, a numerical evaluation of the original equation (1) was performed incorporating the discrete nature of the supports. To simplify the numerical problem, the original PDE was reduced to an ODE by assuming solutions with a first harmonic time response:

$$T\hat{u}''(y) + \omega^2\rho\,\hat{u}(y) - 2\hat{f}[\hat{u}(y)]\sum_{k=-\infty}^{\infty}\delta(\frac{y}{\varepsilon} - 1 - 2k) = -\hat{q}(y) \tag{19}$$

where $u(y,t) = \hat{u}(y)\cos\omega t$, $q(y,t) = \hat{q}(y)\cos\omega t$,and $\hat{f}$ represents the first harmonic component of the nonlinear support function f.

All numerical integrations were performed using the *Mathematica* function DSolve. Initial conditions were chosen to match the initial conditions of the analytical solutions up to $O(\varepsilon^2)$. The discrete supports were modeled numerically as a series of finite step functions with a width of l and a height of $\pm l^{-1}$, each separated by a distance of 2ε. For the general scaling used, numerical solutions showed a reasonable convergence for $l < 10^{-3}$, and l was set to $l = 10^{-4}$.

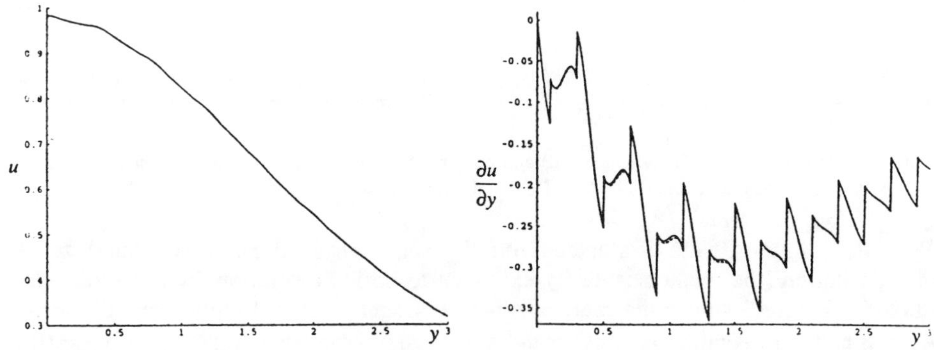

Figure 2. Comparison between analytical (solid) and numerical (dashed) solitary standing solutions.

Figure 2 shows a comparison between the analytical and numerical solutions for the solitary wave solution with the following parameters: $\rho = T = 1$, $\omega = .8$, $\varepsilon = .1$, and external forcing envelope $H(y) = sech\ y$. The two solutions appear to match almost identically for displacement $u(y/\varepsilon, y, t=0)$, and possesses just a few slight discrepancies for $\partial u/\partial y(y/\varepsilon, y, t=0)$. For larger values of y the two solutions begin to diverge with the numerical solution showing signs of periodicity, and the analytic solution remaining true to its solitary wave form. Including higher terms of the analytic solution in the initial condition of the numerical integration should extend the spatial range in which the solutions match.

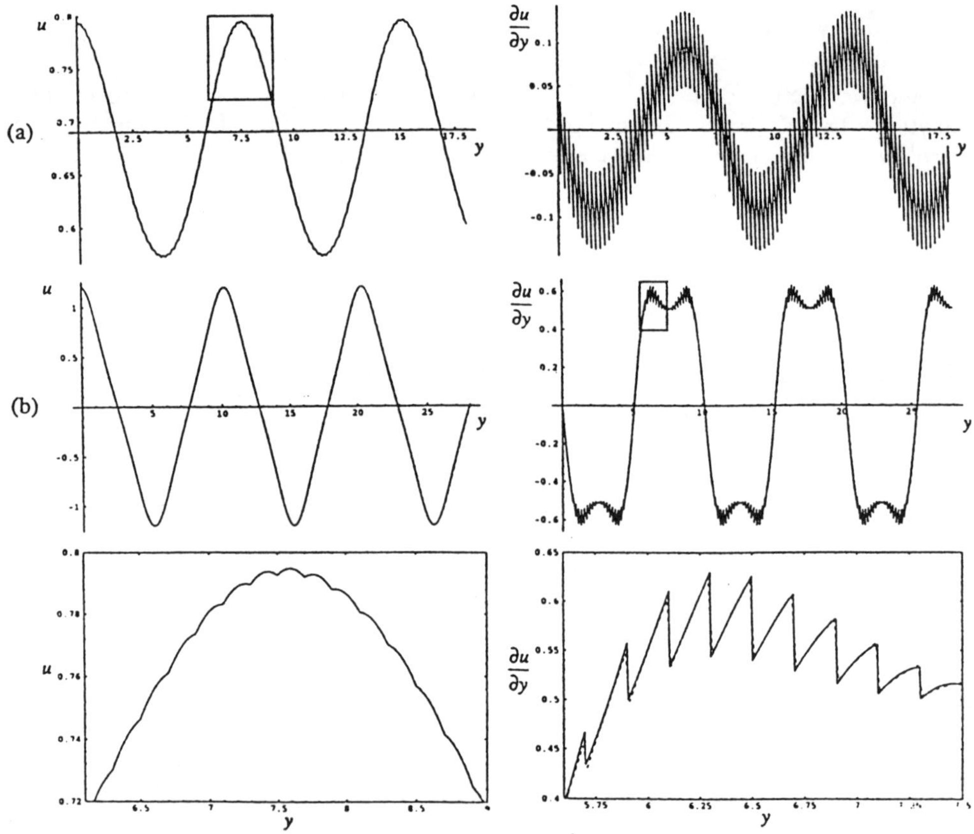

Figure 3. Comparison between analytical (solid) and numerical (dashed) periodic standing wave solutions with first order initial condition $A_0 = .8$ (a) and $A_0 = 1.2$ (b).

Periodic solutions were evaluated with the same physical parameters and zero external excitation (the method is easily extended to periodic solutions with forcing of form (10)). Figure 3 shows the matching between analytical and numerical solutions for $A_0 = .8$ and 1.2 . Again the analytic and numeric solutions match in $u(y/\varepsilon, y, t=0)$ and $\partial u/\partial y(y/\varepsilon, y, t=0)$ for both the 'fast' and 'slow' spatial scales.

4. Chaotic Solution

Applying a forcing of the form,

$$q\left(\frac{y}{\varepsilon}, y, t\right) = F \cos \Omega y \cos \omega t \tag{20}$$

the ODE governing the first order solution becomes,

$$A_1''(z) - A_1(z) + \mu A_1^3(z) = F \cos \Omega y \tag{21}$$

This equation represents well known forced Duffing oscillator, which readily admits chaotic solutions for certain parameters. This equation may be integrated numerically to yield the slow-spatial chaotic envelope for the discrete system. Integrating (12b), while guaranteeing a bounded solution, proved to be a task beyond our capabilities as one of the coefficients of this equation was the aperiodic solution to the first order equation. This was unfortunate as the discreteness effects of the system are computed in the higher order terms. A comparison between the numerical integration of the system before (see Section 3) and after applying NSTT is shown in Figure 4. As with the solitary wave solution, the agreement between the solutions would likely be extended for large y with the matching of higher order terms in the initial conditions of the numerical integration.

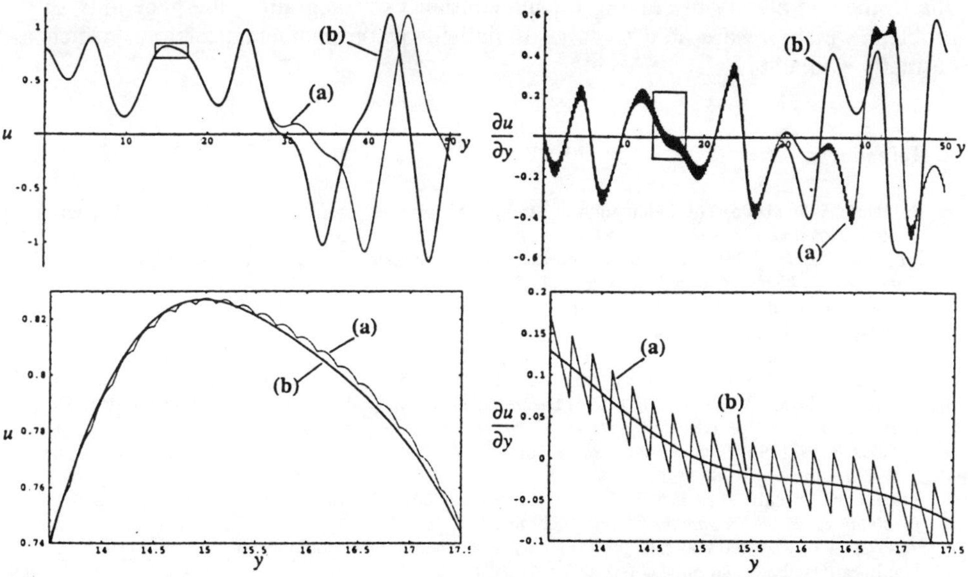

Figure 4. Comparison of chaotic solution to discrete system before (a) and after (b) NSTT is applied. The system parameters are $\rho = T = 1$, $\omega = .8$, $\varepsilon = .1$, $F = .1$, with first order initial condition $A_0 = .8$.

5. Discussion

In this work we analyzed the slow and fast scale spatial distribution of a discretely supported string. The method of nonsmooth transformation was applied to the original equation of motion. This application essentially placed the singularities of the original equation within the nonsmooth variables of a system of smooth NBVP's. These NBVP's are then analyzed through a regular perturbation scheme up to $O(\varepsilon^2)$. This is enabled by the fact that the set of nonsmooth independent variables $\{\tau, e\}$ used in the analysis is bounded in the interval [-1,1], and is periodic with respect to the 'fast' scale (y/ε); thus eliminating secular terms normally encountered in standard methods of nonlinear dynamics [5].

The constants of integration of this analysis are found to be governed by another set of nonlinear PDE's, whose analysis is somewhat simplified by retaining only the first harmonic in a harmonic time expansion. For the general excitation specified in the above analysis, these equations may be solved directly in terms of functionals. Specific properties of these functionals are then used to guarantee that the solution to each integration constant remains bounded for $-\infty < y < \infty$. The examples given involve a softening cubic support; however, the same methods may be applied to a more general type of support, so long as it is symmetric and conservative in nature.

The nearly perfect matching between the numerical and analytical solutions indicate the applicability of the nonsmooth transformations to problems involving periodic discontinuities. Unfortunately, for this system the discontinuities lie at higher order, and due to the complexity of solving for the constants of integration, the generality of the problem which we are able to solve is limited by the mathematical tools which are currently available.

5. References

1. Pilipchuk, V.N. (1985) The Calculation of Strongly Nonlinear Systems Close to Vibrations-Impact Systems, *Prikladneya Matematika i Mekhanika*, Vol. 49(5), 572-578.
2. Pilipchuk, V.N. (1988) A Transformation for Vibrating Systems Based on a Nonsmooth Periodic Pair of Functions, *Doklady AN Ukr. SSR*, Ser. A, 4, 37-40.
3. A.F. Vakakis et al. (1996) *Normal Modes and Localization in Nonlinear Systems*, John Wiley & Sons, Inc., New York..
4. Pilipchuk, V.N. and Vakakis, A.F. (1996) Study of the Oscillations of a Nonlinearly Supported String Using Nonsmooth Temporal Transformation (NSTT), *J. Vib. Acoust.*, (in press).
5. Salenger, G.D. and Vakakis, A.F. (1996) Discreteness Effects in the Forced Dynamics of a String on a Periodic Array of Nonlinear Supports, *Int. J. Nonlinear Mech.*, (in press).
6. Vakakis, A.F. (1993) Exponentially Small Splittings of Manifolds in a Rapidly Forced Duffing System, *J. Sound. Vib.*, Vol. 170 (1), 119-129.
7. Salenger, G.D. and Vakakis, A.F. (1997) Loaclized and Periodic Waves with Discreteness Effects, *Mechanics Research Communications*, (submitted).
8. Vakakis, A.F. and Cetinkaya, C. (1995) Analytic Evaluation of Periodic Responces of a Forced Nonlinear Oscillator, Nonlinea*r Dynamics* 7, 37-51.

OPTIMAL CONTROL OF SPATIOTEMPORAL CHAOS IN COUPLED MAP LATTICES

R. O. GRIGORIEV
Condensed Matter Physics 114-36
California Institute of Technology, Pasadena CA 91125

Abstract. An optimal control approach is used to develop a localized linear control scheme for Coupled Map Lattices. The optimal arrangement of the control sites is shown to depend on the symmetry properties of the system and the controlled state, while the minimal density of controllers depends on the strength of noise present in the system. The method allows stabilization of a variety of steady and periodic states for arbitrary values of system parameters, which is essential for tracking problems.

Systematic application of control theory methods [1] to coupled map lattices (CML) demonstrating chaotic behavior seems very promising. There is a number of reasons for this. On the one hand, due to the explicit spatiotemporal structure, the behavior of CML's is substantially richer than that of the low-dimensional chaotic systems, although many analytic tools developed for the latter can still be successfully applied. On the other hand, CML's are highly symmetric. This property, important in practical applications, is usually missing in the theory of low-dimensional chaos.

The results obtained for the model systems described by a CML can prove highly beneficial for the study of high-dimensional chaotic [2, 3, 4] and, in particular, spatiotemporally chaotic systems [5], with applications to turbulence [6], instabilities in plasma [7], multi-mode lasers [8] and chemical reaction systems [9] and many other practical problems.

Below we will try to uncover the requirements on the control scheme imposed by different properties of the system studied and the state to be stabilized. We will see, that one of the easiest ways to control an unstable reference state is to apply local feedback through controllers placed at certain lattice sites (we will call them pinnings, following Qu and Hu [5]). In particular, we will obtain the limitations on the minimal number and density of controllers and on their possible arrangement.

F. C. Moon (ed.),
IUTAM Symposium on New Applications of Nonlinear and Chaotic Dynamics in Mechanics, 543–550.

Consider the following general coupled map lattice:

$$z_i^{t+1} = f((1-2\epsilon)z_i^t + \epsilon(z_{i-1}^t + z_{i+1}^t)), \tag{1}$$

with $i = 1, 2, \cdots, L$ and periodic boundary conditions $z_{i+L}^t = z_i^t$, or in vector notation:

$$\mathbf{z}^{t+1} = F(\mathbf{z}^t). \tag{2}$$

To be specific, we choose the local map f as a nonlinear function with parameter a,

$$f(z) = az(1-z), \tag{3}$$

but emphasize, that the only result affected by this particular choice is the set of existing periodic trajectories.

Our particular CML has a homogeneous steady state $\hat{z} = 1 - 1/a$, which is unstable for $a > 3.0$ and a variety of unstable time-periodic states embedded in the chaotic attractor. The period-τ trajectory can be found as a steady state solution of the τ-th iterate of the map (1):

$$\hat{\mathbf{z}}^k = F^\tau(\hat{\mathbf{z}}^k), \quad k = 1, 2, \cdots, \tau. \tag{4}$$

It can be easily shown, that due to the translational symmetry of the CML it is impossible to stabilize any unstable trajectory by using the internal parameters of the lattice system (a and ϵ in our case). We can however modify the evolution equation (1) by adding a perturbation $G(\mathbf{u})$, vanishing when no control is applied ($\mathbf{u} = 0$):

$$\mathbf{z}^{t+1} = F(\mathbf{z}^t) + G(\mathbf{u}^t). \tag{5}$$

Linearizing the modified equation around the period-τ periodic orbit $\hat{\mathbf{z}}^1, \hat{\mathbf{z}}^2, \cdots, \hat{\mathbf{z}}^\tau, \hat{\mathbf{z}}^1, \cdots$ we obtain

$$\mathbf{x}^{t+1} = A^t\mathbf{x}^t + B^t\mathbf{u}^t, \tag{6}$$

where we denoted the displacement from the reference trajectory $\mathbf{x}^t = \mathbf{z} - \hat{\mathbf{z}}^t$.

The Jacobian matrix $A_{ij}^k = \partial_j F_i(\hat{\mathbf{z}}^k)$ can be represented as the product $A^k = D^k E$ of the constant coupling matrix

$$E_{ij} = (1-2\epsilon)\delta_{i,j} + \epsilon(\delta_{i,j+1} + \delta_{i,j-1}) \tag{7}$$

and the state-dependent matrix

$$D_{ij}^k = f'((1-2\epsilon)\hat{z}_i^k + \epsilon(\hat{z}_{i-1}^k + \hat{z}_{i+1}^k))\,\delta_{i,j}. \tag{8}$$

The matrix $B^k_{ij} = \partial_j G_i(0)$ defines how the control is applied. If we choose to apply control locally (at the fixed pinning sites i_m), B could be rewritten in the form

$$B_{ij} = \sum_m \delta(j - m)\delta(i - i_m), \tag{9}$$

that depends only on how the pinning sites are placed throughout the lattice. The components u_m of the control vector $\mathbf{u}$ then define the strength of control applied at site i_m, $m = 1, \cdots, M$.

Below (unless otherwise noted) we consider the case of the steady state solution $\hat{\mathbf{z}}$ of (1), which yields constant matrices A and B. In the simplest case of synchronous linear feedback $\mathbf{u}^t = -K\mathbf{x}^t$ equation (6) reduces to

$$\mathbf{x}^{t+1} = (A - BK)\mathbf{x}^t. \tag{10}$$

The state $\mathbf{x} = 0$ can be made stable by a suitable choice of the feedback gain matrix K, if matrices A and B satisfy certain conditions (discussed below). The problem has infinitely many solutions, so in order to make the choice of K unique, we have to impose additional constraints.

Pole placement techniques based on Ackermann's method [10] are inapplicable to the problem of controlling spatially extended systems, because they are numerically unstable [11] and break down rapidly for problems of order greater than 10.

Instead we use the method of the linear-quadratic (LQ) control theory [1], applicable to unstable periodic trajectories as well as steady states. This method is not only numerically stable, but also allows one to *optimize* the control algorithm to increase convergence speed, and at the same time minimize the strength of control. The optimal solution is obtained by minimizing the cost functional

$$V(\mathbf{x}^0) = \sum_{t=0}^{\infty} (\mathbf{x}^{t\dagger} Q \mathbf{x}^t + \mathbf{u}^{t\dagger} R \mathbf{u}^t), \tag{11}$$

where Q and R are the weight matrices, that can be chosen as any positive-definite square matrices.

The minimum of (11) is reached when

$$K = (R + B^\dagger P B)^{-1} B^\dagger P A, \tag{12}$$

where P is the solution to the discrete-time algebraic Ricatti equation

$$P = (Q + A^\dagger P A) - A^\dagger P^\dagger B (R + B^\dagger P B)^{-1} B^\dagger P A, \tag{13}$$

which exists and is unique, if A and B satisfy the *controllability* condition $\text{rank}(C) = L$. The controllability matrix C is defined via

$$C = (B \ AB \ \cdots \ A^{L-1}B). \tag{14}$$

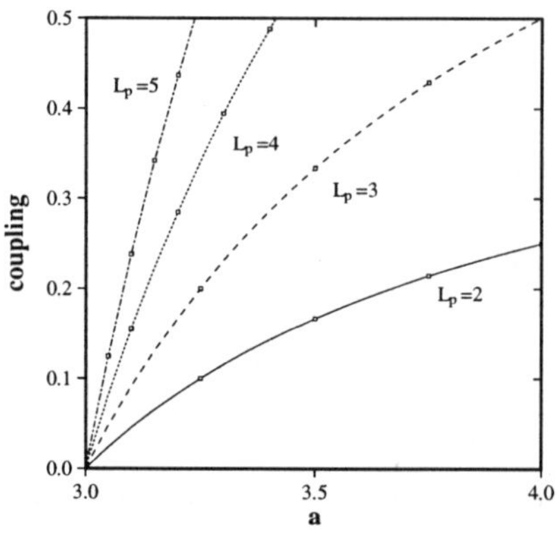

Figure 1. Periodic array of single pinning sites: critical coupling ϵ_{cr} as a function of parameter a. The dots represent the numerical results from figure 2 of Reference [5], with ϵ rescaled by a factor of two to make it compatible with our definition.

For periodic trajectories all the matrices become time-dependent and eq. (13) is replaced by a system of τ iteratively coupled equations.

The controllability condition ensures, that the unstable steady (or periodic) trajectory of the system can be stabilized for the whole range of internal system parameters, where that trajectory exists, using the same control arrangement. For instance, one can easily verify, that the matrix B calculated for a periodic array of single pinning sites $i_m = i_0 + (m-1)L/M$ does not satisfy the controllability condition and therefore spatially homogeneous states are not controllable. Such states could be stabilized however in a restricted area of the parameter space (fig. 1), requiring a prohibitively high density of controllers [12]:

$$L_p = \frac{L}{M} \leq \pi \left[\arccos\left(1 - \frac{a-3}{2\epsilon(a-2)}\right)\right]^{-1}. \tag{15}$$

The numerical results of Hu and Qu [5] support this conclusion completely.

It is possible to extend the limits of the proposed control scheme quite substantially by choosing a different matrix B, i.e. by placing the pinning sites differently. Doing so will enable us to control the system *anywhere* in the parameter space (a, ϵ) at the same time using a *smaller* density of controllers.

First one has to determine the dimensionality of the matrix B, in other words determine the minimal number of parameters required to control the CML (1) of an arbitrary length. It can be shown [13], that the minimal number of parameters required to control a system with degenerate Jacobian is equal to the dimension of the largest irreducible representation contained in the decomposition of the state space representation of the symmetry group of the linearized equation (6).

The coupling matrix (7) has parity symmetry as well as translational symmetry, reduced by the periodic boundary condition, reflecting the symmetries of the original lattice. Since the reference state $\hat{\mathbf{z}}$ in general has a lower symmetry, the symmetry group of (6) is isomorphic to one of the subgroups of C_{Lv} determined by the symmetry of (8).

C_{Lv} has only 1- and 2-dimensional irreducible representations, so the minimal number of control parameters yielding a controllable system for an arbitrary unstable periodic orbit (UPO) in our case is $M = 2$, meaning at least two pinning sites are required. One can easily verify, that the controllability condition is indeed satisfied with an $L \times 2$ matrix

$$B_{ij} = \delta(j-1)\delta(i-i_1) + \delta(j-2)\delta(i-i_2) \tag{16}$$

for a variety of reference states $\hat{\mathbf{z}}$ and for a number of arrangements $\{i_1, i_2\}$, e.g. $i_2 = i_1 + 1$. We choose $i_1 = 1$ and $i_2 = L$ below.

Another complication is noise, limiting our ability to locally control arbitrarily large systems with local interactions, using the minimal number of controllers. Rather simple arguments show [12], that there exist a number of independent theoretical bounds on the size of the controllable system in the presence of noise, such as

$$\begin{aligned} L^{(1)}_{max} &= -\frac{\log(\sigma)}{\lambda_{max}}, \\ L^{(2)}_{max} &= \frac{2\log(\sigma)}{\log(\epsilon) - \lambda_{max}}, \end{aligned} \tag{17}$$

where σ denotes the strength of noise and λ_{max} is the maximal Lyapunov exponent. Similar constraints were obtained by Aranson, Golomb and Sompolinsky for the lattices with asymmetric coupling [14].

Numerical simulations show, that the CML (1) of a moderate size L can indeed be stabilized by the LQ method using the minimal number of controllers (two) in a wide range of parameters a and ϵ (see fig. 3(a) for an example of control of the unstable steady uniform state). The maximal length $L_{max}(\sigma)$ of the system can be obtained by choosing the fixed point as the initial condition and monitoring the evolution of the system in the presence of noise under control (12), calculated by solving eq. (13) with e.g. $Q = I_{L\times L}$ and $R = I_{M\times M}$. This length is quite large for moderate noise level (fig. 2) and agrees with the estimate (17) rather well.

The problem of controlling a large 1-dimensional system with the length $L > L_{max}(\sigma)$, exceeding the maximum allowed for a given noise level, can be easily reduced to the problem of controlling a number of smaller systems with the length $L_p < L_{max}(\sigma)$ [12]. We partition the entire lattice

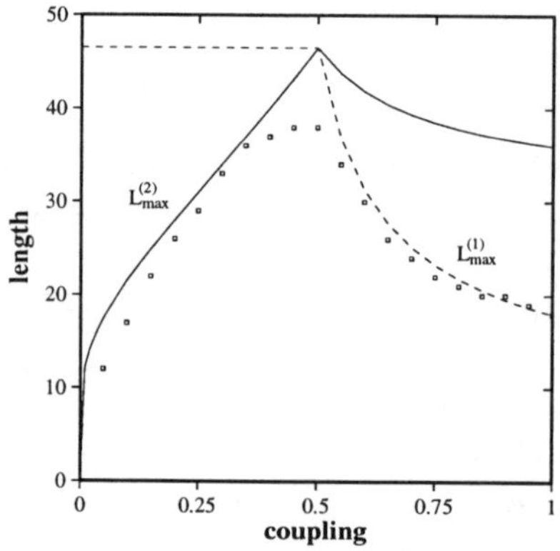

Figure 2. The largest length of the lattice which can be controlled with two pinning sites: theoretical estimates (solid lines) and numerical results (dots) obtained with the uniformly distributed noise of amplitude $\sigma = 10^{-14}$ as functions of coupling ϵ for $a = 4.0$.

$\{z_1^t, \cdots, z_L^t\}$ into $M = L/L_p$ subdomains $\{z^t_{(m-1)L_p+1}, \cdots, z^t_{mL_p}\}$, and control it with an array of pinning sites $i_{m1} = (m-1)L_p + 1$, $i_{m2} = mL_p$, $m = 1, \cdots M$, positioned periodically at the boundaries of subdomains.

The stabilization can be achieved by adjusting the feedback gain

$$\begin{aligned} u^t_{i_{m1}} &= F(z^t_{i_{m2}}, z^t_{i_{m1}}, z^t_{i_{m1}+1}) - F(z^t_{i_{m1}-1}, z^t_{i_{m1}}, z^t_{i_{m1}+1}) \\ &+ \prod_{i=1}^{L_p} \theta(\delta x_i - |x^t_{(m-1)L_p+i}|) \sum_{i=1}^{L_p} K_{1i} x^t_{(m-1)L_p+i}, \\ u^t_{i_{m2}} &= F(z^t_{i_{m2}-1}, z^t_{i_{m2}}, z^t_{i_{m1}}) - F(z^t_{i_{m2}-1}, z^t_{i_{m2}}, z^t_{i_{m2}+1}) \\ &+ \prod_{i=1}^{L_p} \theta(\delta x_i - |x^t_{(m-1)L_p+i}|) \sum_{i=1}^{L_p} K_{2i} x^t_{(m-1)L_p+i} \end{aligned} \qquad (18)$$

to nonlinearly decouple adjacent subdomains (first two terms). Here $\theta(x)$ is a step-function. One can easily see, that in order to calculate the perturbations $u_{i_{m1}}$ and $u_{i_{m2}}$ one needs to know the state of the system only in the immediate vicinity (up to L_p sites away) of the respective controllers. Eqs. (17) give the minimal density of pinning sites.

We demonstrate this approach by stabilizing a number of UPO's of the CML, defined by equations (1,3) with $a = 4.0$ and $\epsilon = 0.33$. In particular, we studied the steady homogeneous state (S1T1), the time-period-2 homogeneous (S1T2), and the time-period-3, space-period-8 (S8T3) orbits. $L = 128$ sites were divided into $M = 16$ subdomains of length $L_p = 8$, each controlled by two pinning sites. The results presented in figs. 3(b) and 4 show the evolution of the system from the initial condition, chosen to be a collection of random numbers in the interval $[0, 1]$, with the nonlinear decoupling control (18).

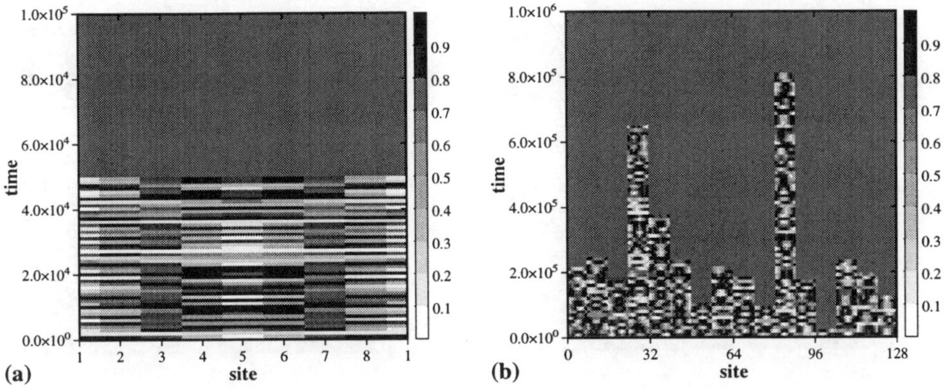

Figure 3. Stabilizing uniform steady state (S1T1): (a) lattice with the length $L = 8$ is controlled by two controllers placed at the sites $i_1 = 1$ and $i_2 = 8$; (b) lattice with the length $L = 128$ is controlled by an array of double pinning sites, placed at the boundaries of subdomains with length $L_p = 8$. The state of the system is plotted at each 10^4-th step.

To summarize, we have shown, that symmetries of the controlled system impose a number of specific constraints on the control scheme, rendering the control more difficult. In particular, it is not possible to control the system perturbing its internal parameters, such as coupling ϵ and local stretching factor a. The dynamics of the system should be modified to effectively break the symmetry, enabling the control of symmetric unstable trajectories.

This can be achieved by placing an array of controllers throughout the lattice. The versatility and efficiency of the control scheme can be improved significantly, if one uses double pinnings instead of single ones. The UPO's become controllable for any values of the internal system parameters and the minimal density of pinning sites is reduced substantially. It is shown, that the maximal distance between pinnings depends on the strength of noise in the system and can be estimated analytically.

Tracking the UPO's in the phase space and/or switching between different orbits is accomplished by readjusting the feedback gain K, without changing the placement of controllers. This stresses the importance of the optimal choice of the control matrix B for systems with symmetries.

And finally, we may suggest, that the dynamics of extended spatiotemporally chaotic systems with local interactions can generically be controlled in such a way, that the information about the state of the system is collected and applied locally. The size of this local region though may depend on the strength of noise as well as the strength of interactions in the system.

The author thanks Profs. H. G. Schuster, M. C. Cross and J. C. Doyle for many fruitful discussions. This work was partially supported by the NSF through grant no. DMR-9013984.

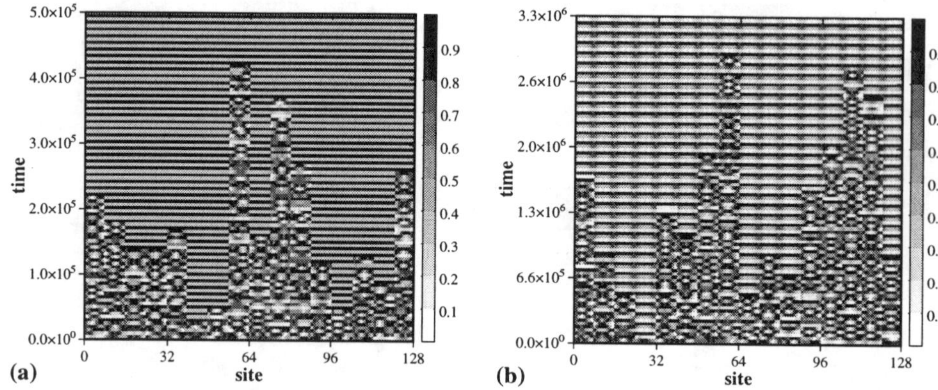

Figure 4. Stabilizing periodic trajectories: lattice with the length $L = 128$ is controlled by an array of double pinning sites, placed at the boundaries of subdomains with length $L_p = 8$. (a) period-2 UPO (S1T2), 2 consecutive states of the system are plotted after skipping each 10^4 steps; (b) period-3 UPO (S8T3), 3 consecutive states of the system are plotted after skipping each 10^5 steps.

References

1. P. Dorato, C. Abdallah and V. Cerrone, (1995) *Linear-Quadratic Control: An Introduction*, Prentice Hall, New Jersey.
2. M. Ding, W. Yang, V. In, W. L. Ditto, M. L. Spano and B. Gluckman (1996) Controlling chaos in high dimensions: Theory and Experiment, *Phys. Rev. E* **53**, 4334-4344.
3. Y. C. Lai and C. Grebogi (1994) Synchronization of spatiotemporal chaotic systems by feedback control, *Phys. Rev. E* **50**, 1894-1899.
4. J. Warncke, M. Bauer and W. Martienssen (1994) Multiparameter control of high-dimensional chaotic systems, *Europhys. Lett.* **25**, 323-328.
5. G. Hu and Z. Qu (1994) Controlling spatiotemporal chaos in coupled map lattice systems, *Phys. Rev. Lett.* **72**, 68-71.
6. C. Lee, J. Kim, D. Bobcock and R. Goodman (1997) Applications of neural networks to turbulence control for drag reduction, *Physics of Fluids* **9**, 1740-1747.
7. A. Pentek, J. B. Kadtke and Z. Toroczkai (1996) Stabilizing chaotic vortex trajectories - An example of high-dimensional control, *Phys. Lett. A* **224**, 85-92.
8. P. Colet, R. Roy and K. Weisenfeld (1994) Controlling hyperchaos in a multimode laser model, *Phys. Rev. E* **50**, 3453-3457.
9. V. Petrov, M. J. Crowley and K. Showalter (1994) Tracking unstable periodic orbits in the Belousov-Zhabotinsky reaction, *Phys. Rev. Lett.* **72**, 2955-2958.
10. E. Barreto and C. Grebogi (1995) Multiparameter control of chaos, *Phys. Rev. E* **52**, 3553-3557.
11. J. Kautsky, N. K. Nichols and P. Van Dooren (1985) Robust pole assignment in linear state feedback, *Intl. J. Control* **41**, 1129-1155.
12. R. O. Grigoriev, M. C. Cross and H. G. Schuster (1997) Pinning control of spatiotemporal chaos, submitted to *Phys. Rev. Lett.*
13. R. O. Grigoriev and M. C. Cross (1997) Controlling physical systems with symmetries, submitted to *Phys. Rev. E.*
14. I. Aranson, D. Golomb and H. Sompolinsky (1992) Spatial coherence and temporal chaos in macroscopic systems with asymmetric couplings, *Phys. Rev. Lett.* **68**, 3495-3498.

APPENDIX I

ADDRESSES OF AUTHORS

Ariaratnam, S.T.
Civil Eng. & Applied Math
Solid Mechanics Division
Faculty of Eng.
University of Waterloo
Waterloo, Ontario
N2L 3G1
CANADA

Bajaj, Anil K.
School of Mechanical Eng.
Purdue University
West Lafayette
IN 47907-1288
USA

Balachandran, B.
Dept. of Mechanical Eng.
Univeristy of Maryland
College Park
MD 20742-3035
USA

Banerjee, Bappaditya
School of Mechanical Eng.
Purdue University
West Lafayette
IN 47907-1288
USA

Bayly, Philip V.
Mechanical Eng.
Washington University
St. Louis
MO 63100
USA

Begley, C.J.
School of Eng.
Duke University
Durham
NC 27708-0300
USA

Belhaq, M.
Faculty of Sciences Aïn Chock
Laboratory of Mechanics
Nonlinear Oscillations & Chaos
BP 5366, Mâarif
Casablanca
MOROCCO

Belyaev, Alexander K.
Institute of Mechanics
& Mechanical Eng.
Johannes Kepler University of
Linz
Altenbergerstr. 69
A-4040, Linz
AUSTRIA

Benedettini, F.
Dipartimento di Ingegneria delle
Strutture
Acque e Terreno
Universita' dell'Aquila
Monteluco Roio
67040 L'Aquila
ITALY

Bishop, S.R.
Centre for Nonlinear Dynamics
& its Applications
University College London
Gower Street
London WC1E 6BT
UK

Boedo, S.
Borg-Warner Automotive
Ithaca
NY 14850
USA

Bridy, D.
Dept. of Physiology
SUNY Health Science Center
Syracuse
NY 13214
USA

F. C. Moon (ed.),
IUTAM Symposium on New Applications of Nonlinear and Chaotic Dynamics in Mechanics, 551–559.

Burns, T.J.
NIST
Gaithersburg
MD 20899
USA

Camp, John
Theoretical & Applied Mech.
Cornell University
Ithaca
NY 14853
USA

Capecchi, D.
Dipartimento di Scienza delle Costruzioni
University of Naples Federico
II. Piazzale V.
Tecchio 80
80125 Napoli
ITALY

Champneys, A.R.
Dept. of Eng. Mathematics
University Walk
University of Bristol
Bristol, BS8 1TR
UK

Chao, Chang-Po
Dept. of Mechanical Eng.
Michigan State University
East Lansing
MI 48824
USA

Chatterjee, Anindya
Theoretical & Applied Mech.
Cornell University
Ithaca
NY 14853
USA

Coleman, Michael
Theoretical & Applied Mech.
Cornell University
Ithaca
NY 14853
USA

Conner, M.D.
School of Eng.
Duke University
Durham
NC 27708-0300
USA

Davies, Huw G.
University of New Brunswick
Dept. of Mechanical Eng.
P.O. Box 4400
Fredericton, N.B.
E3B 5A4 CANADA

Davies, M.A.
NIST
Gaithersburg
MD 20899
USA

Davies, Patricia
School of Mechanical Eng.
Purdue University
West Lafayette
IN 47907-1288
USA

de Kraker, A.
Eindhoven University of Tech.
Dept. of Mechanical Eng.
P.O. Box 513
5600 MB Eindhoven
THE NETHERLANDS

Domokos, Gabor
Applied & Computational Math.
Mechanical & Aerospace Eng.
Princeton Materials Institute
Princeton University
Princeton
NJ 08544
USA

Dowell, E.H.
School of Eng.
Duke University
Durham
NC 27708-0300
USA

Engelbrecht, J.
Institute of Cybernetics
Akadeemia 21
EE-0026 Tallinn
ESTONIA
and
Estonian Academy of Sciences
Kohtu 6
EE0001 Tallinn
ESTONIA

Evans, C.J.
NIST
Gaithersburg
MD 20899
USA

Fedorova, A.N.
Computational Mechanics Group
Inst. of Problems of Mech. Eng.
Russian Academy of Sciences
Bolshoj pr., 61, V.O.,
St. Petersburg, 199178
RUSSIA

Feng, Z.C.
Dept. of Mechanical Eng.
MIT
Cambridge
MA 02139
USA

Fey, R.H.B.
TNO Building & Construction
Research
Center for Mechanical Eng.
P.O. Box 29
2600 AA Delft
THE NETHERLANDS

Garcia, Mariano
Theoretical & Applied Mech.
Cornell University
Ithaca
NY 14853
USA

Glocker, Christoph
Lehrstuhl B für Mechanik
Technische Universität München
D-85747 Garching
GERMANY

Golubitsky, Martin
University of Houston
Dept. of Mathematics
Houston
TX 77004
USA

Gonçalves, Paulo B.
Civil Eng. Dept.
Catholic University - PUC-Rio
22453-900 Rio de Janeiro
RJ - BRAZIL

Grigoriev, Roman O.
Condenced Matter Physics 114-36
California Institute of Technology
Pasadena
CA 91125
USA

Guder, Rabbijah
Technical Univ.-Hamburg-Harburg
Ocean Eng. Section II - Mechanics
D-21071 Hamburg
GERMANY

Hastings, Rachel
Center for Applied Mathematics
Cornell University
Ithaca
NY 14853
USA

Hertzen, R.v.
Helsinki University of Technology
SF02150 Espoo
FINLAND

Hijawi, M.
Wayne State University
Dept. of Mechanical Eng.
Detroit
MI 48202 USA

Hikihara, Takashi
School of Electrical Eng.
Kyoto University
Yoshida-honmachi
Sakyo, Kyoto, 6060-01
JAPAN

Hinrichs, N.
Institute of Mechanics
University of Hannover
Appelstr. 11
D-30167 Hannover
GERMANY

Hogan, S.J.
Dept. of Eng. Mathematics
University of Bristol
UK

Holmes, Philip
Program in Applied &
Computational Mathematics
Mechanical & Aerospace Eng.
Princeton Materials Institute
Princeton University
Princeton
NJ 08544
USA

Holsapple, J.
Dept. of Physiology
SUNY Health Science Center
Syracuse
NY 13214
USA

Homer, M.E.
Dept. of Eng. Mathematics
University of Bristol
UK

Hunt, G.W.
School of Mechanical Eng.
University of Bath
Bath, BA2 7AY
UK

Ibrahim, R.A.
Wayne State University
Dept. of Mechanical Eng.
Detroit
MI 48202
USA

Jensen, Carsten Nordstrom
Es-Consult Ltd
Staktoften 20
DK2950 Vedbaek
DENMARK

Judge, J.
University of Michigan
Ann Arbor
MI
USA

Keller, B.
Dept. of Physiology
SUNY Health Science Center
Syracuse
NY 13214
USA

Knobloch, E.
Dept. of Physics
University of California, Berkeley
Berkeley
CA 94720
USA

Kongas, O.
Institute of Cybernetics
Akadeemia 21
EE-0026 Tallinn
ESTONIA

Kreuzer, Edwin
Technische Universität
Hamburg-Harburg
Arbeitsbereich Meerestechnik II
Besucher-u. Lieferanschrift:
Eißendorfer Straße 42
21073 Hamburg
GERMANY

Kriebel, M.E.
Dept. of Physiology
SUNY Health Science Center
Syracuse
NY 13214
USA

Lakvad, F.
Faculty of Sciences Aïn Chock
Laboratory of Mechanics
Nonlinear Oscillations & Chaos
BP 5366, Mâarif
Casablanca
MOROCCO

Lee, Cheng-Tang
Dept. of Mechanical Eng.
Michigan State University
East Lansing
MI 48824
USA

Li, Y.-Y.
Dept. of Mechanical Eng.
Univeristy of Maryland
College Park
MD 20742-3035
USA

Lord, G.J.
Dept. of Eng. Mathematics
University Walk
University of Bristol
Bristol, BS8 1TR
UK

Macmillen, F.B.J.
Flight Sciences
Bombardier Inc.
Montreal, Québec
CANADA

Matsumoto, T.
Dept. of Electrical, Electronics
& Computer Eng.
Waseda University
Shinjuku, Tokyo 169
JAPAN

McDonald, R.J.
Nonlinear Systems Group
University of Illinois
Urbana
IL 61801 USA

Moehlis, J.
Dept. of Physics
University of California, Berkeley
Berkeley
CA 94720
USA

Moon, F.C.
Mechanical & Aerospace Eng.
Upson Hall
Cornell University
Ithaca
NY 14853
USA

Mureithi, N.W.
Dept. of Mechanical Eng.
McGill University
Montreal, Québec
H3A 2K6
CANADA

Namachchivaya, N. Sri
Nonlinear Systems Group
University of Illinois
Urbana
IL 61801
USA

Nayfeh, Ali H.
Eng. Science & Mechanics
VPI&SU
Blacksburg
VA 24061-0219
USA

Nishi, M.
Electrical, Electronics &
Computer Eng.
Waseda University
Shinjuku, Tokyo, 169
JAPAN

O'Reilly, Oliver M.
Dept. of Mechanical Eng.
University of California, Berkeley
Berkeley
CA 94720-1740
USA

Oestreich, M.
Institute of Mechanics
University of Hannover
Appelstr. 11
D-30167 Hannover
GERMANY

Okamoto, Yoshinobu
Dept. of Electrical Eng.
Kansai University
Yamate-cho 3-3-35
Suita, Osaka, 564
JAPAN

Païdoussis, M.P.
Dept. of Mechanical Eng.
McGill University
Montreal, Québec
H3A 2K6
CANADA

Peterka, F.
Inst. of Thermomechanics AS CR
Doleškova 5,
182 00 Prague 8
CZECH REPUBLIC

Peters, David A.
Mechanical Eng.
Washington University
St. Louis
MO 63100
USA

Pfeiffer, Friedrich
Lehrstuhl B für Mechanik
TU München
Arcisstrasse 21
80290 München
GERMANY

Poon, Chisang
Dept. of Mechanical Eng.
MIT
Cambridge
MA 02139
USA

Popp, K.
Institute of Mechanics
University of Hannover
Appelstr. 11
D-30167 Hannover
GERMANY

Pottlacher, G
Institut fuer Experimentalphysik
TU Graz
Graz A-8010
AUSTRIA

Pratap, R.
India Institute of Science
Bangalore
INDIA

Pratt, Jon R.
Dept. of Eng. Science & Mechanics
VPI&SU
Blacksburg
VA 24061-0219
USA

Price, S.J.
Dept. of Mechanical Eng.
McGill University
Montreal, Québec
H3A 2K6
CANADA

Rand, Richard
Theoretical & Applied Mech.
Kimball Hall
Cornell University
Ithaca
NY 14853
USA

Rega, G.
Dipartimento di Ingegneria
Strutturale e Geotecnica
Universita' di Roma "La Sapienza"
Via
A. Gramsci 53
00197 Roma
ITALY

Ruina, A.
Theoretical & Applied Mech.
Kimball Hall
Cornell University
Ithaca
NY 14853
USA

Salenger, Gary
University of Illinois
at Urbana-Champaign
Urbana
IL 61801
USA

Schiehlen, Werner O.
Institut B of Mechanics
University of Stuttgart
D-70550 Stuttgart
GERMANY

Schmitt, J.M.
Mechanical & Aerospace Eng.
Princeton University
Princeton
NJ 08544
USA

Schmitt, John
Applied & Computational Math.
Mechanical & Aerospace Eng.
Princeton Materials Institute
Princeton University
Princeton
NJ 08544
USA

Shaw, Steven W.
Dept. of Mechanical Eng.
Michigan State University
East Lansing
MI 48824
USA

Soliman, Mohamed S.
Dept. of Eng.
Queen Mary College
University of London
London
UK

Spyrou, K.J.
Centre for Nonlinear Dynamics
& its Applications
University College London
Gower Street
London WC1E 6BT
UK

Steindl, Alois
Vienna University of Technology
A-1040, Vienna
AUSTRIA

Stépán, G.
Dept. of Applied Mechanics
Technical University of Budapest
H-1421 Budapest
HUNGARY

Szemplinska-Stupnicka, W.
Institute of Fundamental
Technological Research
Polish Academy of Sciences
Swietokrzyska 21
00-049 Warsaw
POLAND

Tang, D.M.
School of Eng.
Duke University
Durham
NC 27708-0300
USA

Thompson, J.M.T.
Centre for Nonlinear Dynamics
& its Applications
University College London
Gower Street
London WC1E 6BT
UK

Thothadri, M.
Theoretical & Applied Mech.
Kimball Hall
Cornell University
Ithaca
NY 14853
USA

Todd, M.D.
School of Eng.
Duke University
Durham
NC 27708-0300
USA

Trickey, S.T.
School of Eng.
Duke University
Durham
NC 27708-0300
USA

Troger, Hans
Vienna University of Technology
A-1040, Vienna
AUSTRIA

True, Hans
The Technical University of
Denmark
Dept. of Mathematical Modelling
Building 321
DK-2800 Lyngby
DENMARK

Tyrkiel, E.
Institute of Fundamental
Technological Research
Polish Academy of Sciences
Swietokrzyska 21
00-049 Warsaw
POLAND

Ueda, Y.
School of Electrical Engineering
Kyoto University
Yoshida-honmachi
Sakyo, Kyoto, 606-01
JAPAN

Vakakis, Alexander F.
Dept. of Mechanical &
Industrial Engineering
University of Illinois
at Urbana-Champaign
140 Mech. Eng. Bldg. MC-244
1206 W. Green St.
Urbana
IL 61801
USA

van Campen, D.H.
Eindhoven University of
Technology
Dept. of Mechanical Eng.
P.O. Box 513
5600 MB Eindhoven,
THE NETHERLANDS

van der Heijden, G.H.M.
Centre for Nonlinear Dynamics
& its Applications
University College London
Gower Street
London WC1E 6BT
UK

van Liempt, F.P.H.
TNO Building & Construction
Research
Center for Mechanical Eng.
P.O. Box 29
2600 AA Delft
THE NETHERLANDS

Varadi, Peter C.
Dept. of Mechanical Eng.
University of California, Berkeley
Berkeley
CA 94720-1740
USA

Vestroni, F.
Dipartimento di Ingegneria
Strutturale e Geotecnica
University of Rome La Sapienza
Via Endossiana 18
00184 Roma
ITALY

Vinh, Pham Chi
Hanoi National University
Faculty of Mathematics,
Mechanics & Informatics
90 Nguyen Trai Str.
Dong Da, Hanoi
VIETNAM

Virgin, L.N.
School of Eng.
Duke University
Durham
NC 27708-0300
USA

von Hertzen, R.
Helsinki University of Technology
Otakaari 1
FIN-02150 Espoo
FINLAND

Wiercigroch, Marian
Dept. of Eng.
Kings College
University of Aberdeen
Aberdeen AB9 2UE
UK

Zakrzhevsky, Michael
Institute of Mechanics
Riga Technical University
1 Kalku St.
LV-1658, Riga
LATVIA

Zeitlin, M.G.
Computational Mechanics Group
Institute of Problems of
Mechanical Eng.
Russian Academy of Sciences
Bolshoj pr., 61, V.O.,
St. Petersburg, 199178
RUSSIA

Zemann, Josef
Vienna University of Technology
A-1040, Vienna
AUSTRIA

Zhao, M.-X.
Dept. of Mechanical Eng.
Univeristy of Maryland
College Park
MD 20742-3035
USA

Zounes, Randolph
Center for Applied Mathematics
Cornell University
Ithaca
NY 14853
USA